# WATER RESOURCES AND THE URBAN ENVIRONMENT

PROCEEDINGS OF THE 25TH ANNUAL CONFERENCE
ON WATER RESOURCES PLANNING AND MANAGEMENT

June 7–10, 1998
Chicago, Illinois

SPONSORED BY
Water Resources Planning and Management Division
of the American Society of Civil Engineers

CO-SPONSORED BY
Metropolitan Water Reclamation District of Greater Chicago
ASCE Illinois Section Environmental Engineering and
Water Resources Group
ASCE Illinois Section

EDITED BY
Eric D. Loucks

1801 ALEXANDER BELL DRIVE
RESTON, VIRGINIA 20191–4400

Abstract: This proceedings, *Water Resources and the Urban Environment*, contains extended abstracts presented at the ASCE 25th Conference on Water Resources Planning and Management held in Chicago, Illinois, June 8-10, 1998. while many of the papers address traditional topics such as hydrologic and hydraulic modeling, portection of drinking supplies, and reservoir operation, more recent issues such as groundwater modeling, water resource problems in the developing world, and water quality of urban runoff are also explored. Some of the major topics discussed include 1) urban stormwater management; 2) geographic information systems; 3) hydrologic and hydraulic computer modeling; 4) groundwater management; 5) drinking water quality; and 6) international resources issues.

Library of Congress Cataloging-in-Publication Data

Conference on Water Resources Planning and Management (25th: 1998: Chicago, Ill.)
  Water resources and the urban environment: Proceedings of the 25th Annual Conference on Water Resources Planning and Management/sponsored by the Water Resources Planning and Management Division of the American Society of Civil Engineers; cosponsored by Metropolitan Water Reclamation District of Greater Chicago...[et al.]; edited by Eric D. Loucks.
      p.   cm.
  Includes bibliographical references and index.
  ISBN 0-7844-0343-0
  1. Municipal water supply–Congresses. 2. Urban hydrology–Congresses. 3. Municipal water supply–United States–Congresses. 4. Urban hydrology–United States–Congresses.
I. Loucks, Eric D. II. American Society of Civil Engineers. Water Resources Planning and Management Division. III. Metropolitan Water Reclamation District of Greater Chicago. IV. Title.
  TD201.C584   1998                                                                98-16639
  333.91'15–dc21                                                                        CIP

Any statements expressed in these materials are those of the individual authors and do not necessarily represent the views of ASCE, which takes no responsibility for any statement made herein. No reference made in this publication to any specific method, product, process or service constitutes or implies an endorsement, recommendation, or warranty thereof by ASCE. The materials are for general information only and do not represent a standard of ASCE, nor are they intended as a reference in purchase specifications, contracts, regulations, statutes, or any other legal document. ASCE makes no representation or warranty of any kind, whether express or implied, concerning the accuracy, completeness, suitability, or utility of any information, apparatus, product, or process discussed in this publication, and assumes no liability therefore. This information should not be used without first securing competent advice with respect to its suitability for any general or specific application. Anyone utilizing this information assumes all liability arising from such use, including but not limited to infringement of any patent or patents.
Photocopies. Authorization to photocopy material for internal or personal use under circumstances not falling within the fair use provisions of the Copyright Act is granted by ASCE to libraries and other users registered with the Copyright Clearance Center (CCC) Transactional Reporting Service, provided that the base fee of $8.00 per chapter plus $.50 per page is paid directly to CCC, 222 Rosewood Drive, Danvers, MA 01923. The identification for ASCE Books is 0-7844-0343-0/98/ $8.00 + $.50 per page. Requests for special permission or bulk copying should be addressed to Permissions & Copyright Dept., ASCE.

Copyright © 1998 by the American Society of Civil Engineers,
All Rights Reserved.
Library of Congress Catalog Card No: 98-16639
ISBN 0-7844-0343-0
Manufactured in the United States of America.

PREFACE

*Water Resources and the Urban Environment* presents the proceedings of the 25th annual conference on water resources planning and management held in Chicago June 7 to June 10, 1998. The Water Resources Planning and Management Division of the American Society of Civil Engineers sponsored the conference and the conference chair was Dr. James E. Lindell of Harza Engineering Company. This book contains extended abstracts of many of the papers offered at the conference. These abstracts summarize and, in some cases, expand upon the material presented at the conference.

The major topics and issues addressed in these proceedings exemplify the growth and maturity of the 26-year-old Water Resources Planning and Management Division. The division was founded to provide a professional home for those who recognized that water resources planning is truly multidisciplinary, an indivisible mix of hydrology, hydrodynamics, mathematics, statistics, biochemistry, regional planning, economics and social science. Many papers were devoted to the traditional topics addressed by the division in previous conferences: hydrologic and hydraulic modeling, protection and enhancement of drinking water supplies, decision making strategies and reservoir operation. The division is clearly moving in new directions. Many sessions were devoted to groundwater management and modeling, water resource problems in the developing world and water quality of urban runoff. These are topics that the Division's founders may not have envisioned but they have been embraced by what is clearly the most diverse group in the Society.

This book is divided into ten chapters. Within each chapter, the papers are presented in the order scheduled at the conference. The first chapter, Water Resources Management in Urban Areas, is given first in tribute to papers written to honor the urban theme of the conference. The next four chapters, Hydrologic and Hydraulic Computer Modeling; Drinking Water Systems; Water Policy and Institutions and Planning and Decision Making, are traditional subjects of the annual conference. These are followed by two chapters of papers on the exciting growth areas in the Division. These are Quality Management of Wet Weather Flows and Management and Analysis of Groundwater Systems. The final three chapters represent smaller specialty areas within the division and contain the papers on Sediment and Erosion Control, Remote Sensing and Geographic Information Systems and Rainwater Catchment Systems.

Featured at the conference was the Water Resource Student Capstone Design Competition sponsored by Parsons Brinckerhoff. Students competed for prizes by presenting their design projects during two conference technical sessions. Entrants included teams from Tufts University, University of Nevada-Las Vegas and two teams from Rose-Hulman Institute of Technology.

The photograph on the cover of these proceedings is the winner of the eighth annual Water Resources Planning and Management Division photography contest. Members of all ASCE Student Chapers and Clubs, as well as all Younger Members, were invited to submit original work photography with the theme of the conference. Judging was arranged by the sponsoring Membership Development Committee and the ASCE Headquarters staff.

The winning photograph is of a hydraulic jump at the Oahe Dam outlet taken by Charles J. Baker, a civil engineering student at the South Dakota School of Mines and Technology, who received the first place prize of a full complimentary registration as well as hotel accommodations and travel expenses.

Honorable mention awards were made to Chris Sharek of the University of Central Florida, Shakhan Gamble of North Carolina A&T University, Aisha Niang of Howard University, Cindy Ann Novak of Colorado State University, Wendy Allison of Santa Clara University, and Daisy M. Gee of Sacramento State College.

# Contents

### *Water Resources Management in Urban Areas*

Low-Impact Development Hydrologic Analysis and Design .............................................. 1
    Larry S. Coffman, Mow-Soung Chen, Neil Weinstein, and Michael Clar

*Geomorphic Assessment and Monitoring for the Preservation of a Riverine Resource
    James Gracie

Low-Impact Development Hydrologic Computations Procedure ........................................ 9
    Larry S. Coffman, Mow-Soung Chen, Neil Weinstein, and Michael Clar

Overview of Low-Impact Development for Stormwater Management .............................. 16
    Larry S. Coffman, Michael Clar, and Neil Weinstein

Urban Renewal and Stormwater Management Planning in an Urban Center ................. 22
    K.K. Chin, K.Y. Ng, S.W. Lee, M. Hasni, and A. Choo

Aesthetic Aspects of Urban Stormwater Management System Design ............................ 29
    K.K. Chin, A. Choo, K.Y. Ng, S.W. Lee, and M. Hasni

*Managing Water Supply Source Watersheds: A Taiwan Case Study
    J.T. Kuo, J.Y. Lin, and S.L. Yu

Thailand–U.S. Center for Technology Partnership .......................................................... 36
    James J. Yezzi, Daniel Sullivan, Anthony N. Tafuri

*Improving Coastal Water Quality in the Paracas Marine Preserve, Peru, Through Recoiling in the Fishmeal Industry
    Jan Mueller-Vollmer

Development of an Electronic Knowledge Base To Facilitate Industrial Pollution
Prevention Efforts in Egypt ................................................................................................. 42
    Jim Goggin, Salwa Wahba, Vahid Alavian, and Gilbert Richard

Benefiting the Environmental Bottom Line Through Pollution Prevention in Ecuador,
South America ...................................................................................................................... 49
    Mario Salazar

USAID's Pollution Prevention Program in Latin America ............................................... 55
    Gilbert S. Jackson and Morris S. Israel

*Challenges Facing the U.S.–Canada Border Region
    Isobel Heathcote

*U.S.–Mexico Cooperation in Water Supply and Wastewater Treatment
    David Negrete

*How Texas Views Transboundary Water Supply Problems
    David Eaton

\* denotes manuscripts not available at time of publication

The Role of the Seattle District of the U.S. Army Corps of Engineers in the
International Joint Commission ..................................................................................61
   Mark Killgore

The Storm Over Sturm: A Case Study in Multi-jurisdictional Stormwater
Management, Lake County, Illinois ..............................................................................66
   Fred Royal

*Urban River Restoration: A Community Perspective
   Laurene Von Klan

Retrofitting an Urban Watershed: From Concept to Reality...........................................72
   Patricia S. Werner

*Forgotten Infrastructure: The Natural Stormwater Management System
   Ward Miller

Harborside International Golf and Marina Facility .........................................................78
   Frank L. Kudrna, Jr.

*Local Flooding and Habitat Restoration Program
   Larry Gibbons

*Rehabilitation and Retrofit of Regional Detention Facilities for Enhanced Performance
   Richard Schaefer

*Retrofitting Residential Development for Deficiencies in Surface Water Management
Facilities Within the Existing Constraints: Physical and Regulatory
   Randall Parsons, Don Althauser, and Wendy Kara

Urban Storm Sewage Design Using the Double Detention Pond Concept and a
Modified Rational Formula Approach ..........................................................................86
   Ray-Shyan Wu, Chun Yu, Shu-Liang Liaw, and Ching-Ho Chen

Performance Evaluation of Dry Detention Ponds with Underdrains...............................92
   F.N. Nnadi, K.W. Ashe, and R.C. Sharek

A Comparative Study of Midwestern Detention Basins ..................................................99
   Daniel J. Schuller, Eric A. Casanave, A. Ramachandra Rao

*Implementing Regional Floodplain Management Measures for Alluvial Fans
   Bruce M. Phillips

Sediment Control at Construction Sites: Is it Being Properly Regulated? ....................105
   John A. Jacobson

*Water Resources Management in Stressed Urban Regions: The Case of Thriassion
Plains, Greece
   Christos A. Karavitis, Ageliki Bosdogianni, and Evan C. Vlachos

Drainage Problems in an Urbanized Watershed ............................................................111
   Veronica J.B. Morgan and Michele Good Burton

*Hydrologic and Hydraulic Computer Modeling*

Modeling the Great Lakes Net Basin Supplies Series Using SAMS ................................117
   J.D. Salas, C.H. Chung, W.L. Lane, and D.K. Frevert

The Behavior of Storage Reservoirs in the United States Under Climate Change ........123
Richard M. Vogel, Melissa Lane, and Paul Kirshen

The Use of Indicators To Evaluate Impacts of Global Warming on Water Resources in the United States ............................................................................................................129
Melissa Lane, Paul Kirshen, and Richard M. Vogel

Causes of Increased Net Basin Supply in the Great Lakes Watershed ...........................135
Eric D. Loucks, Orie L. Loucks, and John C. Klink

*Lake O'Hare Facility Planning Study: A Continuous Simulation Approach
Michael Morgan and James Considine

*Application of Hydrologic Simulation Program FORTRAN (HSPF) in DuPage County, Illinois
Tom Price

Applications of the FEQ Unsteady Flow Model in DuPage County and Beyond ...........141
Paula Cooper and H. Sherrie Chang

*Application of the Peak-to-Volume Statistical Analysis in DuPage County, Illinois
Allen Bradley

Operational Modeling System with Dynamic-Wave Routing ..........................................147
A.L. Ishii, T.J. Charlton, T.W. Ortel, and C.C. Vonnahme

*Valley View Flood Control Plan and Analysis
Linda Mele

*Sawmill Creek Floodplain Mapping Using the HSPF/FEQ/PVSTATS Approach
J. William Brown

*Fawell Dam Analysis
Christine Klepp, John Sikora, and Steve Rodgers

Probability of Drought Flows in the Manitoba Hydro System .........................................153
Srinivasan Rangarajan and Harold M. Surminski

Using Digital Topographic Maps for Hydraulic Modeling ..............................................159
Michele Good Burton and Veronica J.B. Morgan

GIS-Based Processing of NEXRAD Rainfall Estimates ..................................................165
C. Bryan Young, Bruce M. McEnroe, and Rebecca J. Quinn

Problems in Calibrating Conceptual Rainfall Runoff Models .........................................171
Paolo Bartolini, Federica Montereggio, and Juan B. Valdés

Two-Dimensional Flood Modeling for a Tidal Canal ......................................................176
P. Michael DePue and Steven R. Halmi

Two-Dimensional Coastal Modeling for Bridge Scour ....................................................182
Steven R. Halmi, P. Michael DePue, R. Wayne Corley, and William H. Hulbert

Hydraulic Analysis Using FESWMS-2DH of the State Route 76 and Hatchie River Crossing in Brownsville, Tennessee ................................................................................188
Edwin W. Watkins and Wayne Seger

Acquisition and Relocation as a Floodplain Management Tool .......................................194
William G. DeGroot and David W. Lloyd

Technical Mapping Advisory Council to Federal Emergency Management Agency:
A Status Report on Its Purpose and Progress ................................................................200
  Mark Riebau

Planning and Design of the New Lenox Community Golf Course ........................205
  Rick K. Suttle and William McCollum

Reduction of Illinois River Flood Stages by Converting a Few Selected Levee
Districts to Managed Storage Areas .................................................................................211
  Abiola A. Akanbi and Krishan P. Singh

Ineffective Overbank Flow in One-Dimensional Modesl..................................................217
  Dirk L. Gowin and Charles Morris

*Kentucky-Barkley Lakes Role in Flood Control on the Lower Mississippi River
  Stanley M. Wisbith and David P. Buelow

Urban Flooding in Coastal Areas: A Case Study of Runoff Management and Storm
Runoff and Tidal Flooding ..............................................................................................223
  Erez Sela

*Analysis and Design for Reata Pass Wash, a Desert Greenbelt Channel, Scottsdale,
Arizona
  Lan Weber

Short-Term Streamflow Forecasting Using Artificial Neural Networks .........................229
  Cameron M. Zealand, Donald H. Burn, and Slobodan P. Simonovic

Investigation of DeLaine's Method..................................................................................235
  Daniel J. Schuller and A. Ramachandra Rao

Using Hydraulic Modeling for System Improvements: State College Borough
(Pennsylvania) Water Authority.......................................................................................241
  Mark V. Glenn and Eric A. Casanave

*Temporal Characteristics of Principal Components and Forecasts of Regional
Droughts
  Thomas T. Burke, Jr.

*Frequency Analysis and Time Distribution of Intense Rainfall Events Over the
Metropolitan Region of Belo Horizonte, Brazil
  Mauro Naghettini and Marcia Guimaraes-Pinheiro

*A Temperature-Based Method To Estimate Actual Lake Evaporation in the Northeast
United States
  Neil M. Fennessey

*Impact of the Spatial Variability of Rainfall and Basin Segmentation on Flood Peaks
  Paolo Barolini and Juan B. Valdés

Non-Linear Parameter Estimation of an Urban Runoff Model Using XP-SWMM32
and PEST .........................................................................................................................247
  Tai Ovbiebo and Anthony W. Kuch

*Drinking Water Systems*

*Systems Analysis Approaches for Multi-Quality Networks
  Uri Shamir

Inverse Chlorine Modeling in Pipe Networks ..................................................................253
    A.S. Al-Omari and M.H. Chaudhry

*Water Distribution Systems Analysis: Quantity–Quality Formulation of Quasi-Steady State
    Charles D.D. Howard and Uri Shamir

Travel Times in Dead-End Mains .......................................................................................260
    Steven G. Buchberger, YeongHo Lee, and Jason T. Carter

Optimal Location of Booster Disinfection Stations for Residual Maintenance ..............266
    Dominic L. Boccelli, Michael E. Tryby, James G. Uber, and
    Lewis A. Rossman

Detecting Accidental Contaminations in Municipal Water Networks: Application ......272
    Avner Kessler and Avi Ostfeld

Using Hydraulic Modeling for Disinfection Effectiveness ...............................................279
    Paul F. Boulos and Imad A. Hannoun

Small Physical Scale Models for the Study of Mixing in Water Distribution Reservoirs ...........................................................................................................................285
    Rolf A. Deininger and Andrew D. Santini

Water Quality Variability in a Dead-End Loop .................................................................291
    Jason T. Carter, YeongHo Lee, Steven G. Buchberger, Lewis A. Rossman,
    and Eugene W. Rice

*Impacts of Random Water Demands on Prediction of Water Quality in Dead-End Watermains
    Lin Wu

Optimal Decoupling of Booster Disinfection Systems in Water Distribution Networks ..............................................................................................................................297
    James G. Uber, Marios M. Polycarpou, and Prathiba Subramaniam

Using Back-Up Sub-Systems for Design and Operation of Reliable Multi-Quality Systems ................................................................................................................................303
    Avi Ostfeld and Uri Shamir

*Water Service Issues in Central Asia
    Ronald M. North

Environmental Systems Rehabilitation in Deschapelles, Haiti .......................................307
    Bruce W. Berdanier

A Community-Managed Water and Sanitation Utility for the Urban Poor:
Cité Soliel, Port-au-Prince, Haiti .......................................................................................313
    Christopher McGahey

*Modeling Water Quality in Water Distribution Systems Under Parameter Uncertainty
    Chengchao Xu and Ian C. Goulter

*Water Loss Evaluation in the Pipe Network of Reggio Calabria (Italy) Through Valve Closures and Parameter Estimations
    Alessandra Bascia and Tullio Tucciarelli

\*Using Distribution System Modeling To Evaluate Treatment Plant Disinfection By-Product Control Strategies
   Joseph C. Reichenberger and Dennis D. LaMoreaux

\*Experimental Verification of Advanced Numerical Modeling Techniques for Leak Detection in Water Distribution Systems
   Angus R. Simpson, John Vitkovsky, and James Liggett

Development of a California Water Use Efficiency Policy Through "Stakeholder" Consensus..........319
   Rick Soehren and Greg Young

\*Urban Water Supply Planning in Developing Countries
   Rameshwar D. Verma

\*Evaluating Water Availability Under the Texas Water Rights System
   Ralph A. Wurbs

\*Seismic Reliability Study of Water Supply Infrastructure: A Disaster Preparedness Approach
   Ira Mark Artz

*Water Policy and Institutions*

Innovative Wet-Weather Flow Management Systems for Newly Urbanized Areas .......325
   James P. Heaney, Leonard Wright, and David Sample

\*EPA BMP Assessment Project
   Ben Urbonas, Eric Strecker, and Jonathan Jones

\*Protocol for Wet Weather Effect Assessment and Technology Performance Evaluation
   Ed Herricks

Wet Weather Flow Designs for the Future..........331
   Robert Pitt, Richard Field, and Chi-Yuan Fan

\*Two-Dimensional Habitat Analysis in the Yellowstone and Missouri Rivers
   Terry Waddle

Effects of Wetlands on Modulating Hydrologic Regimes in Nine Wisconsin Watersheds..........339
   D.L. Hey and J.A. Wickenkamp

\*Creaing In-Lake Wetlands Using Dredge-Filled Geotextile Bag: Coordinating State, Local, and Federal Water Resources and Environmental Regulations When Using Innovation
   Karen C. Kabbes

Water Management in New Hampshire: An Overview and Preliminary Event Duration Analysis of Proposed Instream Flow Rules ..........345
   Neil M. Fennessey

\*Within-State Water Allocation Procedures for Alabama, South Carolina, and Georgia Under the ACT and ACF Agreements
   Kathy Hatcher

The Future of Illinois Water Law ..........351
   Gary R. Clark

Allocation of Water Withdrawals in a River Basin .......................................................... 357
    Jennifer M. Jacobs and Richard M. Vogel

Development of Boundary International Water Quality Standards ............................... 363
    Conrad G. Keyes, Jr.

*Novel Methods of Water Marketing in the U.S.–Mexico Border Region
    David Hurlbut

*Lesson Learned from the International Joint Commission (U.S.–Canada) Water Quality Management Experience
    Walter Lyon

*Private–Non-Profit Transboundary Water and Wastewater Utilities
    David Eaton

Institutional Hurdles in Privatization: The Fairbanks Municipal Utilities System ...... 368
    C. (Kees) W. Corssmit, Carol F. Streiner, and Bill Gordon

*Privatization Options with a Case Study
    Alan Cohen

Managed Competition Sometimes Yields Unexpected Results: Ask San Diego ............. 374
    Ellen R. Bogardus and Kenneth M. Barrett

*Experiences with Client/Server Modeling
    Jeff Wright

*Experiences with Asynchronous Learning Tools
    Jeff Wright

*Virtual Systems Courses Facilitated by Internet Collaboration
    Richard Palmer

Using Readily Available Technologies and Laboratory Experiences To Enhance Course Content ........................................................................................................... 380
    Emmanuel U. Nzewi

Storm Water Permitting in the Milwaukee River Basin ................................................. 386
    James R. D'Antuono

Rural Management for Nonpoint Pollution Control ....................................................... 392
    William C. Hafs

Reducing Nonpoint Source Pollution in Ultra-Urban Areas: Two Case Examples ....... 395
    James A. Bachhuber

*Planning and Decision-Making*

Equity, Cost and Environmental Benefit of Clay Distribution ....................................... 401
    James W. Male and Ellen Hoffman Belk

*The City of San Diego's Competition Program
    Lisa Irvine

*Managed Competition: Process and Results
    George Raftellas

*Cost Savings Through Design-Build Contracts
    Tom Sanders

\*Economic Analysis for FERC Relicensing
John Charbonneau

\*An Integrated Evaluation of the Economic and Natural Impacts of Flaming Gorge Dam Operations
Kirk Lagory

Total Valuation of Grand Canyon Resources ................................................................... 407
Michael P. Welsh and Richard C. Bishop

\*Calibrating Environmental Values
Don Coursey

\*The Colorado River Decision Support System
Steven A. Malers, Ray R. Bennett, William J. Owen, and Douglas B. Greer

D-CORMIX: A Decision Support System for Hydrodynamic Mixing Zone Analysis of Continuous Dredge Disposal Sediment Plumes ........................................................... 413
Robert L. Doneker and Gerhard H. Jirka

Criteria Determination Using a Probabilistic Approach ................................................... 419
James W. Male

Flood Estimation by Combining Gauged and Paleo Data ................................................ 425
Cheng-Shaung Peng, Lucien Duckstein, Donald R. Davis, and Victor R. Baker

Pipe Network Optimization .............................................................................................. 431
Taejin Ahn and G.V. Loganathan

\*Prototype Decision Support Tool for Developing Tropospheric Ozone Control Strategies
Dan Loughlin, S. Ranji Ranjithan, Downey Brill, John Baugh, and Steve Fine

\*Generation of Alternative Strategies for Insightful Decision-Making in Watershed Management
Laura J. Harrell and S. Ranji Ranjithan

\*Post-Storm Hazard Assessment Using Digital Photogrammetry and GIS
Margery F. Overton, Roger R. Grenier, Jr., Elizabeth K. Judge, and John S. Fisher

Potomac River Risk-Based Water Supply Management ................................................... 437
Stuart S. Schwartz

\*Interactively Exploring Efficient Solid Waste Management Alternatives To Meet Environmental Goals
Ken Harrison, Robert D. Dumas, Eric Solano, Morton A. Barlaz, E.D. Brill, and S. Ranji Ranjithan

\*Equity, Cost and Environmental Benefit of Clay Distribution for Landfill Covers
James W. Male and Ellen E. Hoffman

Multimedia Network System for Landfill Siting ............................................................... 443
Jehng-Jung Kao

\*Reservoir Balance Using an Indeterminate Operational Criteria
S. Ali Taghavi and Gene Taylor

\*Decision Support System for Basin Management Planning: A Conceptual Framework
Abdul Q. Khan and M. Najmus Saquib

Protecting Drinking Water Supply Sources: San Francisco's Water Quality
Vulnerability Zones ................................................................................................449
Karen E. Johnson and Edward H. Stewart

\*Basin Planning Management: An Integrated Approach
S. Ali Taghavi, Lyndel W. Melton, and Gene Taylor

Approaches and Levels of Service Analysis for an Area Subject to Interbasin Flow
Transfers ...................................................................................................................453
L. Moris Cabezas

\*Integrated Levels of Service for Watershed Management Planning
Walid M. Hatoum

\*Regulatory Heartburn in the Lowlands
Christian A. Smith

Kazakstan Irrigation and Environmental Regulations......................................459
D.M. Manbeck, M.R. Headrick, and P.L. Ames

Operation and Performance of Reservoir Release Improvements at 16 TVA Dams......465
John M. Higgins and W. Gary Brock

Decision Support System for River Basin Water Management ........................471
Ick Hwan Ko, Darrell G. Fontane, and John W. Labadie

Flood Control Effects of Hwachon Dam in Connection with Peace Dam ......476
Myung Pil Shim, Oh Ig Kwon, and Kyung Tak Kim

\*Risk-Based Reservoir Operation Using Probabilistic Inflow Forecasts
Stu Schwartz, Eric Markstrom, and Bob Steger

*Quality Management of Wet Weather Flows*

\*Effectiveness of a Combined Urban Wetland Detention Basin: First Results
Robert G. Traver

Maintenance of Stormwater Retrofit Projects....................................................482
Gordon England

A Comparison of BMP Requirements: A Case Study .......................................488
Robert A. Sherman and David D. Dee, Jr.

BMP Considerations for a Roadway Project Located Within a Water Quality
Reservoir Watershed................................................................................................494
David D. Dee, Jr., Glenn Bottomley, and John Olenik

Walking a Tightrope Between Developers and Bananas: County Watershed
Management Regulations Versus State and Local Regulations .......................500
Greg Boehm

Flood Management That Revitalizes a Community............................................505
Kevin Shafer and Bob Wolf

*BMPs for the Protection of Underground Potable Water Supplies from Stormwater Pollution: A Case Study in Northern New Jersey
　Erez Sela

Baffle Boxes and Inlet Devices for Stormwater BMPs .................................................... 511
　Gordon England

Implementing Municipal Stormwater Management Program: An Overview of Planning and Administration .................................................................................... 517
　Shih-Long Liao, Richard Field, Daniel Sullivan, and Chi-Yuan Fan

Constructed Wetlands for Stormwater Management: Applications, Design, and Evaluation ........................................................................................................................ 523
　Shih-Long Liao, Richard Field, and Shaw Li Yu

Optimization of Storage and Treatment Systems for CSO Pollution Control ................ 529
　Richard Field and Thomas P. O'Connor

Urban Wet-Weather Flow Toxic Pollutants: Characterization and Enhanced Sedimentation Treatment ................................................................................................ 535
　Chi-Yuan Fan, Richard Field, Thomas P. O'Connor, and Mary K. Stinson

Planned EPA Research in Watershed Modeling ............................................................. 541
　Michael Borst

*CSO Performance Measures: A National Perspective
　Ross Brennan

Performance Measures for CSO Control ........................................................................ 547
　Michael P. Sullivan, Karen T. Gontasz, and Mark P. Hoeke

*Use of Performance Measures and Other Lessons Learned in King County Washington's CSO Control Program
　Laura Wharton

*Use of CSO Performance Measures by Small Communities
　Stephen F. John

Stretching the Life of an Old Sewer System ................................................................... 553
　David J. Anderson and James P. Pistilli

*Sequential Retention/Treatment of Combined Sewer Overflows for Optimizing Control of Variable Wet Weather Events
　Michael D. Waring

*Permit Requirements for CSO Control
　Philip C. Heckler

The Many Faces of Combined Sewer System Performance ........................................... 559
　Kenneth A. Pew

*The Latest Developments in the Testing and Effectiveness of a New Innovative Urban Best Management Quality Practice: Stormceptor
　Vincent H. Berg and Graham J. Bryant

Sheetflow Water Quality Monitoring Device ................................................................. 562
　Stuart M. Stein, Frank R. Graziano, G. Kenneth Young, and Pat Cazenas

Effectiveness of a Combined Urban Wetland Detention Basin: First Results ...............568
   Robert G. Traver

Economic Analysis for Stormwater Quality Management ................................................574
   Orit Wilchfort-Kalman, Jay R. Lund, Dan Lew, and Douglas M. Larson

*The Evaluation and Design of an Alum Stormwater Treatment System To Improve Water Quality in Lake Maggiore, St. Petersburg, Florida
   Jeffrey L. Herr and Harvey H. Harper

Alternative Methods in Stormwater Management .............................................................580
   Isabel C. Escobar, Andrew A. Randall, and Frank E. Marshall, III

*Oxygen Depletion or Hypoxia in the Nearshore Gulf of Mexico off the Louisiana Coast
   Larinda Tervelt

Improved Estimation or Urban Non-Point Pollution .........................................................586
   Stuart S. Schwartz

*Management and Analysis of Groundwater Systems*

A Post-Audit of Optimal Conjunctive Use Policies ............................................................591
   Tracy Nishikawa and Peter Martin

Model Development for Conjunctive Use Planning in Taiwan .........................................597
   Shu-li Yang, Nien-Sheng Hsu, Shiang-Kueen Hsu, and William W-G. Yeh

*Conjunctive Use Modeling: An Edwards Aquifer Case Study
   Daene C. McKinney and David W. Watkins

*Conjunctive Management Model for Yolo Basin, California
   Miguel A. Marino and Hakan Basagaoglu

*Optimal Risk-Based Corrective Action Design
   J. Bryan Smalley

*Negotiating Risk-Based Concentrations for Groundwater: Case Study Strategy Comparisons
   Patricia V. Cline

*Natural Attenuation as a Risk Reduction Mechanism
   Ira May

*A Screening Model for Groundwater Contaminated by LNAPL Spills Based on Human Health Risk
   Gus Williams and James Butler

Application of Groundwater Flow and Solute Transport Models for Groundwater Remedial Design ....................................................................................................................603
   Zafar Adeel, Charles Faurst, Barry Lester, Todd Hagemeyer, and Ron Lantzy

Selection of Remedial Action for a DNAPL Spill .............................................................609
   Lily Sehayek, Terry Vandell, Brent Sleep, and Calvin Chien

*Groundwater Flow Modeling as a Guide in Environmental Decision-Making
   John Glass

Groundwater Modeling for Contaminated Site Closures: Successful Case Studies for Site Assessment, Remediation Design, and Remediation Performance Evaluation .......615
    Chi-Chung Chang

*New Advances in Combining Simulation and Optimization for Solving Groundwater Management Problems
    C. Zheng and P.P. Wang

Operational Definition of NAPL from Soil Analyses ..........................................................621
    Roger H. Page, Michael L. Watkins, Edith Sieber, and
    Mohammad F.N. Mohsen

*Current Trends in Groundwater Management Tools
    B. Prucha and C. Johnson

*Development and Application of a Comprehensive MODFLOW-Based Groundwater Flow and Transport Modeling System
    Sorab Panday, Kiran Khanbham, and Peter Huyakorn

*A Case Study: In Situ Metal Precipitation and Enhanced Reductive Dechlorination of TCE Using Supplemental Nutrient Injections
    Suthan Suthersan, Frank Lenzo, and Stephen Kessel

Practical Approaches for Development of Site-Wide and Local Scale Models at Brookhaven National Laboratory: A Case Study ............................................................627
    Douglas A. Smolensky, Arthur J. Zahradnik, Michael P. Kladias,
    Thomas W. Burke, and William R. Dorsch

* Multi-Objective Optimization of Groundwater Remediation Design Using Pareto Genetic Algorithms
    Alex Mayer and Jeffery Horn

Groundwater Bioremediation Optimization Using Genetic Algorithms.........................633
    Amy B. Chan Hilton and Teresa B. Culver

*Fuzzy Linear/Integer Programming for Hydraulic Gradient Control
    B. Shrestha and C. Shoemaker

*Probabilistic Modeling and Data Worth Analysis for Identification of DNAPL Source Codes
    Stephen Conrad and Brian Borchers

Use of Advanced Hydroenvironmental Modeling and Simulation in Water Resources Management ...........................................................................................................................639
    Robert F. Athow, Jr., and Jeffery P. Holland

*Trickle-Down or Bubble-Up Computing?
    David Dougherty

Continuumization of Spatial Network Data .................................................................645
    John F. Peters and Stacy E. Howington

*Modeling Biodegradation of NAPL Constituents in the Subsurface
    Phil deBlac, Daene McKinney, and Gerald Speitel, Jr.

Visualization for Analysis of Environmental Data ...........................................................651
    Peter H. Townsend

*Making Spatial Network Data a Continuum for the GMS
Patricia V. Cline

*Displaying Uncertainty in Subsurface Data
Gus Williams

*Visualization of Groundwater Flow and Aquifer Salinization Along the Arkansas River Valley, Southeast Colorado
Mark Person

Problems Applying MODFLOW-Style Models to Three-Dimensional Flow ......... 655
Peter O. Sinton

*Modeling of Reactive Barriers
A.J. Rabideau and A. Khandelwal

*A Feasibility Study of Thermal In Situ Bioremediation
Jeremy M. Kosegi and Barb Minsker

*Design of an In Situ PCP Remediation System Using GMS/MS-MVS, Saunders Supply Superfund Site
Klausmeier and J. Yoon

The Effect of Mass Transfer Limitations on the Optimization of Groundwater Remediation Strategies ......... 661
Aysegul Aksoy and Teresa B. Culver

*A Linear Programming Application for Water Resource Management at a Mining Operation
R. Gailey and R. Bartlett

*Coupled Surface Water and Groundwater Management Strategy for an Estuarine Dredge Disposal Area at Craney Island, Norfolk, Virginia
Klausmeier and J. Yoon

*Streamtube-Ensemble Methods for Screening Subsurface Bioremediation Strategies
Timothy Ginn

*The Modeling of Salt Water Intrusion in the Nile Delta Aquifer
Mohamed A. Elganainy

Dynamic Ground Water Management System Based on GIS ......... 667
Werner Erhart-Schippek and Herbert Mascha

*Supply and Demand: Recycled Water Development for Groundwater Protection
Curtis V. Weeks and Jon Sansing

Ground Water Recharge Project, Prescott Valley, Arizona ......... 676
Reed J. Petersen

*Groundwater/Surface Water Interaction Study at Tinker AFB
Kent Duran

Spectral and Bispectral Analysis for Nonlinear Leaky Phreatic Aquifer Systems Subject to Time Variable Inputs ......... 680
Gwo-Fong Lin and Chi-Ming Chen

\*The Effect of Mass Transfer Limitations on the Optimization of Groundwater Remediation Strategies
Aysegul Aksoy and Teresa B. Culver

\*Seepage Beneath Drop Structures Founded on Confined Previous Strata
Mohamed A. Elganainy, F.A. El-Fitiany, M.M. El-Afify, and Z.M. Shuluma

Construction of a New Tollway Below the Groundwater Table with Nitrate Remediation ............................................................................................................686
Scott M. Taylor

\*A Feasibility Study of Thermal In Situ Bioremediation
Jeremy M. Kosegi

Evaluation of the Technical Equivalency of Engineered Phyto-Cover Systems to RCRA Landfill Caps ...................................................................................692
Scott T. Potter, Suthan S. Suthersan, Jeffery A. Smith, and Timothy A. Bent

Comparison of Steady-State and Transient Simulation of the HFBR Tritium Plume at Brookhaven National Laboratory ...............................................................698
Michael P. Kladias, Douglas A. Smolensky, Arthur J. Zahradnik, Jr., and Michael G. Hauptmann

*Sediment and Erosion Control*

Sedimentation Study of Loyalhanna Reservoir Using a Digital Terrain Model............704
Roger T. Kilgore, Daniel C. Lucey, Werner Loehlein, Nick Fusco, and Daniel E. Medina

Reservoir Operations During Sedimentation Removal at Conemaugh River Lake ......710
Walter C. Loehlein

Sediment Transport Capactiy as an Objective of Reservoir Operations .......................716
Robert T. Milhous

\*Evolution of Sediment Production in Urban Watersheds Case Study: The Pampulha Watershed in Belo Horizonte, Brazil
Maria da Gloria Braz de Oliverira

Temporary Erosion Control for Highway Construction in Kansas..................................722
Bruce McEnroe and Brian J. Treff

Development of a Comprehensive Management Plan for Tyler Creek, Elgin, Illinois..................................................................................................................................728
Gary C. Schaefer, T.M. Denning, J.W. Hood, and J.A. Wickenkamp

Erosion-Controlled Radioactive Transport in Agricultural Watersheds.........................734
Reza Khanbilvardi, Vyacheslav M. Shestopalov, and Siamak Esfandiary

*Remote Sensing and Geographic Information Systems*

Engineer's Guide to the Components of GIS and Mapping Using ArcView ..................739
Charles H. Call, Jr.

\*Automated Delineation of Major and Minor Watershed Basins in West Tennessee
Rodney J. Conger and Jerry L. Anderson

**Development of a GIS-Based Watershed Assessment Model: Application to the Kentucky River Basin** ............745
Lindell E. Ormsbee, Lee Colten, and Ted Stumbur

**Stormwater Management of an Urban Watershed** ............751
Wing K. Tam and Blake N. Murrillo

**Watershed Modeling Using Remote Sensing** ............757
Nabil M. Hourani

**\*Satellite Imagery for Delineation of Environmental Contamination**
Bruce W. Berdanier

**\*Remote Sensing Application in Hydrology and Water Resources**
Ghasem Asrar

**\*Remote Sensing Application in Watershed Management**
Yuri Gorokhovich

*Rainwater Catchment Systems*

**Current Status and Best Management Practices for American Catchment Systems** ............763
Dennis J. Lye

**Status of RWCS Development and Progress of the IRCSA** ............769
Yu-Si Fok

**Rainwater Catchment Systems Development Guidelines** ............773
Yu-Si Fok

**Rainwater Catchment Response Surface Ascension** ............779
Richard J. Heggen

**Integrated Dual-Mode Roofwater Collection System for Non-Potable Uses in the NTU Complex** ............785
Adhityan Appan

**Advancement of Rainwater Uses in Taiwan** ............791
K.F. Andrew Lo

*Indexes*

Subject Index ............797

Author Index ............803

Low-Impact Development Hydrologic Analysis and Design

Larry Coffman[1], Mow-Soung Cheng[2], Neil Weinstein[3], Michael Clar[4]

Abstract

    Low-impact Design (LID) uses a detailed hydrologic analysis procedure to analyze the predevelopment and postdevelopment hydrologic functions of the watershed. From this analysis, strategies and Best Management Practices are developed so that the postdevelopment condition emulates the predevelopment hydrologic regime. The design intent is to create a "functional landscape" that incorporates design features that mimic the predevelopment storage and infiltration functions of the watershed. This is done by reducing the change in Curve Number (CN), maintaining the predevelopment Time of Concentration (Tc), and then incorporating distributed retention and, if required, detention BMP's to maintain the runoff volume and peak runoff rate.

Introduction

    Low-impact hydrologic analysis and design is based on procedures and techniques from Urban Hydrology for Small Watersheds (United States Department of Agriculture, 1986). The use of National Resource Conservation Service (NRCS) procedures and techniques was chosen because of "universal" acceptance and to

---

[1] Associate Director, [2] Section Head, Programs and Planning Division, Department of Environmental Resources, Prince George's County, Maryland, [3] Project Manager, Tetra Tech, Inc., Fairfax, Virginia, [4] Executive Vice President, Engineering Technologies Associates, Inc. Ellicott City, Maryland

facilitate the design and analysis procedure. The low-impact analysis and design approach is based on four major elements:

- CN: minimizing the change in postdevelopment CN by reducing impervious areas and preserving more trees and meadows (areas with high storage and infiltration potential) to reduce the storage requirements to maintain the predevelopment runoff volume;
- Tc: maintaining the predevelopment Tc to minimize the increase of the peak runoff after development by lengthening flow paths, flattening grades, and increasing surface roughness;
- Retention: providing retention storage for volume and peak runoff rate control to maintain the pre-development CN, as well as for water quality control; and
- Detention: providing additional detention storage, if required, to maintain the predevelopment peak runoff rate and/or to alleviate existing flooding problems.

Low-impact Development Curve Number

Calculation of the low-impact development CN is based on a detailed evaluation of the predevelopment and postdevelopment land covers so that an accurate representation of the runoff potential can be obtained. This calculation requires the engineer to investigate the following key parameters associated with a low-impact development site; 1) land cover type, 2) percentage and connectivity of impervious areas, 3) hydrologic soils group (HSG), and 4) hydrologic conditions (average moisture or runoff conditions). Figure 1 is an illustration of the comparison between land covers used to calculate a conventional and low impact development CN. Table 1 lists the comparison of percentages of land covers found in the respective analysis. By analyzing the CN in discrete units, a thorough understanding of the effects of change in land cover on the runoff potential of the site is achieved. Low-impact development employs a number of design techniques to reduce the change in CN, or runoff. Therefore, the need for BMPs to maintain the runoff volume and peak runoff rate is reduced. Table 2 is a representative list of low-impact development techniques that can be used to maintain the CN.

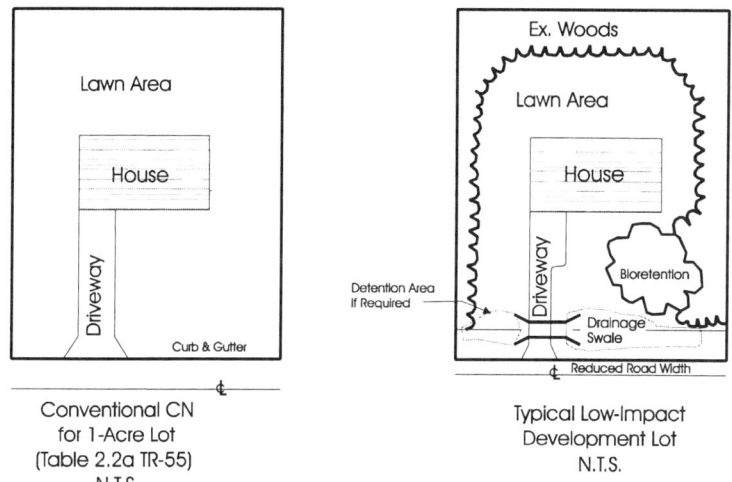

Figure 1. Comparison of Conventional and Low-Impact Development Land Covers for Curve Number Calculations

| Conventional Land Covers (TR-55 Assumptions) | Low Impact Development Land Covers |
|---|---|
| 20% Impervious<br>80% Grass | 15% Impervious<br>25% Woods<br>60% Grass |

Table 1. Comparison of Conventional and Low-Impact Development Land Covers

Maintaining the Predevelopment Time of Concentration (Tc)

In order to maintain the predevelopment peak runoff rate it is critical that the postdevelopment Tc be relatively equal to the predevelopment Tc. If the Tc is equal for the predevelopment and postdevelopment conditions, the storage requirement to maintain the same peak runoff rate is independent of the Tc for retention and detention practices. These concepts are illustrated in Figure 2. Representative techniques to maintain the Tc are listed in Table 3. The travel time (Tt) throughout individual lots or subareas for the postdevelopment condition should be the same as the predevelopment condition, so that the postdevelopment Tc is representative of the predevelopment condition. By incorporating techniques to reduce the impact of development on the CN and by maintaining the Tc, the need

| Suggested Options Affecting Curve Number | Limit use of sidewalks | Reduce road length and width | Reduce driveway length and width | Conserve natural resources areas | Minimize disturbance | Preserve infiltratable soils | Preserve natural depression areas | Use transition zones | Use vegetated swales | Preserve Vegetation |
|---|---|---|---|---|---|---|---|---|---|---|
| Land Cover Type | | | | ✔ | ✔ | | | ✔ | ✔ | ✔ |
| Percent of Imperviousness | ✔ | ✔ | ✔ | | | | | ✔ | | |
| Hydrologic Soils Group | | | | ✔ | | ✔ | | | | |
| Hydrologic Condition | | | | ✔ | ✔ | ✔ | | | | |
| Disconnectivity of Impervious Area | ✔ | ✔ | ✔ | | | | | | | |
| Storage and Infiltration | | | | | | | ✔ | | | ✔ |

Table 2. Low-Impact Development Planning Techniques to Reduce the Curve Number (CN)

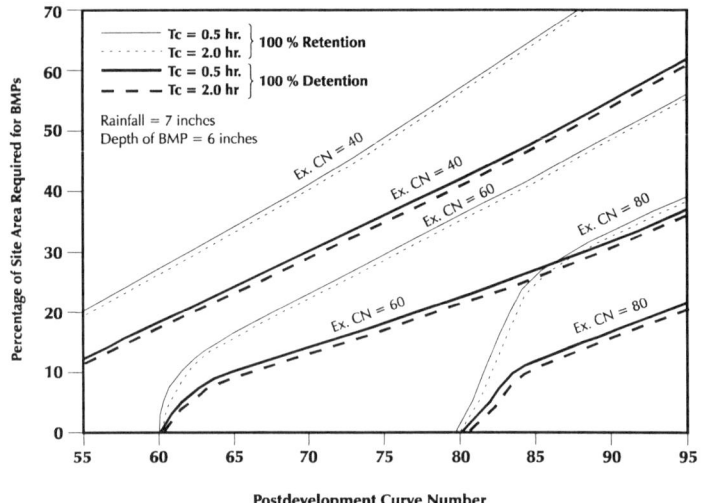

Figure 2. Comparison of Storage Volumes for Various Times of Concentration (Tc)

for stormwater management BMPs to maintain the CN (runoff volume) and peak runoff rate is reduced. Figure 3 is comparison of the runoff hydrographs for; 1) the predevelopment condition (hydrograph 1), 2) the postdevelopment condition using techniques to reduce the change in CN (hydrograph 5), and 3) the postdevelopment condition using techniques to reduce the change in CN and techniques to maintain the Tc (hydrograph 6).

| Low-Impact Development Objective | Low Impact Development Technique | | | | | | | | | |
|---|---|---|---|---|---|---|---|---|---|---|
| | On-lot bioretention | Wider and flatter swales | Maintain sheet flow | Clusters of trees and shrubs in flow path | Provide tree conservation/transition zones | Minimize storm drain pipes | Disconnect impervious areas | Save trees | Preserve existing topography | LID drainage and infiltration zones |
| Minimize disturbance | ✓ | | ✓ | ✓ | ✓ | ✓ | ✓ | ✓ | ✓ | |
| Flatten grades | | ✓ | ✓ | | | ✓ | | | ✓ | ✓ |
| Reduce height of slopes | | | | | | ✓ | | | ✓ | ✓ |
| Increase flow path (divert and redirect) | | ✓ | ✓ | ✓ | | | ✓ | ✓ | ✓ | |

Table 3. Low-Impact Development Techniques to Maintain the Predevelopment Time of Concentration (Tc)

## Maintaining the Predevelopment Curve Number (CN)

After techniques to maintain the postdevelopment Tc and minimize the impact of CN changes have been incorporated into the design, additional reductions in runoff volume must be accomplished through the use of retention BMPs. Unlike conventional BMPs, that utilize end-of-pipe technologies, low-impact development design incorporates small retention BMPs throughout the site in order to mimic the predevelopment storage and infiltration characteristics of the watershed. As the retention storage volume of low-impact development BMPs is increased, there is a corresponding decrease in the peak runoff rate and runoff volume. If a sufficient amount of runoff is retained, the peak runoff rate may be reduced to a level at or below the predevelopment runoff rate. postdevelopment condition for a low-impact development site

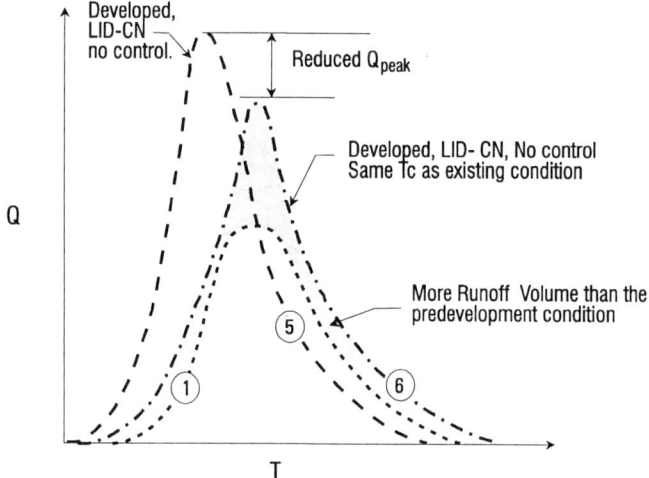

Figure 3. Low-Impact Development Hydrograph That Has a Reduced CN and Maintains the Tc Without Stormwater BMPs

<u>Requirement for Additional Storage to Maintain the Peak Runoff Rate</u>

Even though the postdevelopment Tc and CN are maintained at the predevelopment level, in some cases additional detention storage is needed to maintain the predevelopment peak runoff rated due to the spatial distribution of the retention storage provided (i.e. storage areas not uniformly distributed throughout the site).

The amount of storage that maintains the predevelopment runoff volume might not be sufficient to maintain the predevelopment peak runoff rate. Therefore, additional on-lot detention storage is required. Figure 4 is an illustration of the effect of the combination of retention storage with additional detention storage on the postdevelopment low-impact development hydrograph. Hydrograph 1 is the existing condition, hydrograph 9 is the response of the hydrograph to retention practices alone. The amount of retention storage in hydrograph 9 is not sufficient to control the peak runoff rate. Hydrograph 10 illustrates the effect of providing additional detention storage so that the postdevelopment peak runoff rate is equal to the predevelopment condition.

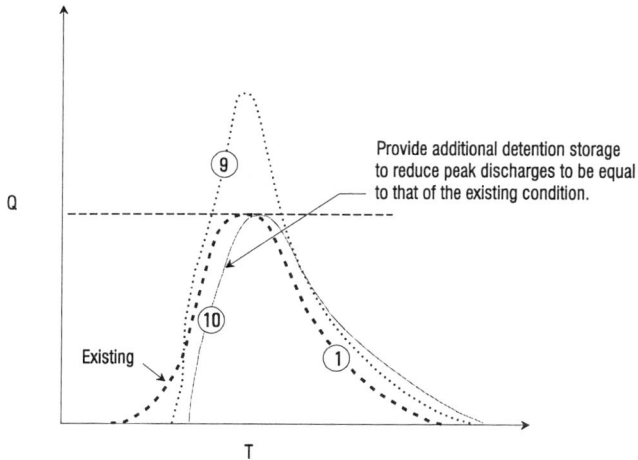

Figure 4. Effect of Additional Detention Storage on Low-Impact Development Retention Practices

Determination of Storage Requirements

Storage requirements to maintain the predevelopment runoff volume and peak runoff rate are determined by the use of design charts found in the Low-Impact Design Manual (Prince George's County, 1997). The charts compare the predevelopment to the post development CN for a specified Type II 24-hour rainfall event in order to determine the percentage of the site required for BMPs for retention control for volume, retention control for peak discharge rate, and detention control for peak discharge rate. Figure 5 is an example of one of the charts that is included in the manual.

Summary

By carefully analyzing a watershed to gain an understanding of the hydrologic functions, strategies and BMP techniques that mimic the predevelopment hydrologic regime can be incorporated into the site plan in order to maintain the runoff volume and peak runoff rate. By maintaining the Tc and utilizing retention storage, in small distributed facilities, impacts of increased volume, duration, and frequency, associated with large conventional end-of-pipe techniques can be avoided.

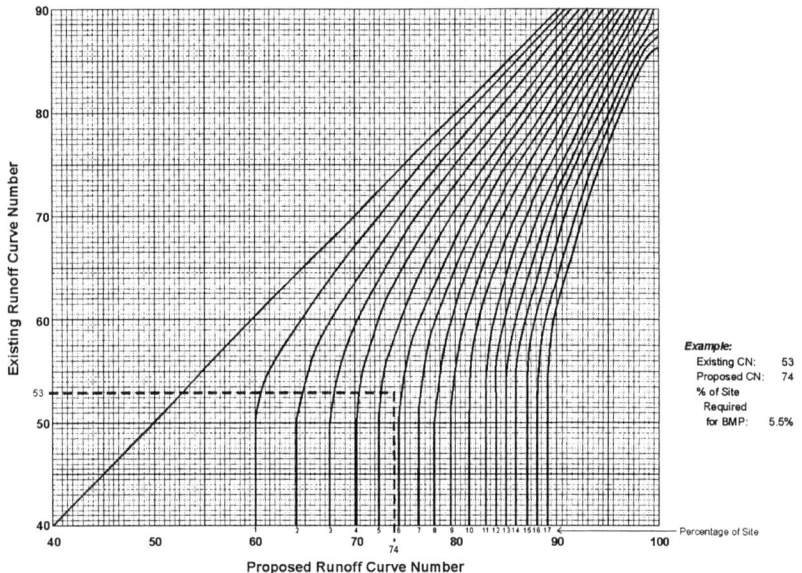

Figure 5. Chart A1 Percentage of Site Required to Maintain Predevelopment Runoff Volume Using Retention Storage for a 2 Inch Rainfall Event- Type II 24-Hour Storm-BMP Depth = 6 Inches

References

Prince George's County, Maryland Department of Environmental Resources, 1997. *Low-Impact Development Design Manual.* Largo, Maryland

United States Department of Agriculture, Soil Conservation Service, 1986. *Urban Hydrology for Small Watersheds, Technical Release 55.* Washington, D.C.

Low-Impact Development Hydrologic Computations Procedures
Larry Coffman[1], Mow-Soung Chen[2], Neil Weinstein[3], Michael Clar[4]

Abstract

Low impact Development (LID) is an alternative approach to stormwater management. LID achieves stormwater control by fundamentally changing conventional site design to create an environmentally functional landscape that mimics natural watershed hydrologic functions (discharge, frequency, recharge and volume). This paper provides a brief overview and introduction the hydrologic computations procedures that are used to design the site to maintain the natural watershed hydrologic functions.

Introduction

The hydrologic analysis of low-impact development is a sequential decision making process that involves a number of sequential steps or procedures. Several iterations may occur within each step until the appropriate approach to reduce stormwater impacts is determined. The procedures for this process are summarized in this paper and are based on the materials presented in the Low-Impact Development Design Manual developed by the Department of Environmental Resources, Prince George's County, Maryland, 1997.

Data Collection

The basic information used to develop the low-impact development site plan and used to determine the hydrologic parameters including the runoff curve number (CN), and the time of concentration (Tc) for the pre- and post development conditions is the same as conventional site plan and stormwater management approaches.

---

[1] Associate Director, [2] Section Head, Programs and Planning Division, Department of Environmental Resources, Prince George's County, Maryland; [3] Project Manager, Tetra Tech, Inc., Fairfax, Virginia; [4] Executive Vice President, Engineering Technologies Associates, Inc., Ellicott City, Maryland.

Runoff Curve Number

The determination of the low-impact CN requires a detailed evaluation of each land cover within the development site. This will allow the designer to take full advantage of the storage and infiltration characteristics of low-impact development site planning to maintain the CN. This approach encourages the conservation of more woodlands and the reduction of impervious area to minimize the needs of the BMP.

In conventional site development, the engineer would refer to Figure 2.2a of TR-55 (SCS, 1986) to select the CN that represents the proposed land use of the overall development (i.e., residential, commercial,) without checking the actual percentages of impervious area, grass area, etc. Because low-impact design emphasizes minimal site disturbance (site fingerprinting), it is possible to retain much of the predevelopment land cover and CN. Therefore, it is appropriate to analyze the site as discrete units to determine the CN. Table 1 lists representative land cover CNs used to calculate the composite "custom" low-impact development CN.

The initial composite CN is calculated using a weighted approach based on individual land covers without considering disconnectivity of the site imperviousness. This is done using Equation 1.

$$CN_c = \frac{CN_1 A_1 + CN_2 A_2 .... + CN_j A_j}{A_1 + A_2 .... + A_j} \qquad \text{Eq. 1}$$

Where: $CN_c$ = composite curve number; $A_j$ = area of each land use cover; and $CN_j$ = curve number for each land cover

Overlay of SCS, hydrologic soil group (HSG) boundaries onto homogeneous land cover areas are used to develop the low-impact development CN. What is unique about the low-impact development custom-made CN technique is the way this overlaid information is analyzed as small discrete units that represent the hydrologic condition, rather than a conventional TR-55 approach that is based on a representative national average. This is appropriate because of the emphasis on minimal disturbance and retaining site areas that have potential for high storage and infiltration. This approach provides an incentive to save more trees and maximize the use of HSG soils for recharge. Careful planning can result in significant reductions in postdevelopment runoff volume and corresponding BMP costs.

When the impervious areas are less than 30 percent of the site, the percentage of the unconnected impervious areas within the watershed influences the calculation of the CN (SCS,1986). Disconnected impervious areas are impervious areas without any direct connection to a drainage system or other impervious surface. For example, roof drains from houses could be directed onto lawn areas where sheet flow occurs, instead of to a swale or driveway. By increasing the ratio of disconnected impervious areas to pervious areas on the site, the CN and resultant runoff volume can be reduced. Equation 2 is used to calculate the CN for sites with less than 30 percent impervious area.

$$CN_c = CN_p + \{ P_{imp}/100 \} \times ( 98 - CN_p ) \times ( 1 - 0.5R ) \qquad \text{Eq. 2}$$

where: R = ratio of unconnected impervious area to total impervious area;
$CN_c$ = composite CN; $CN_p$ = composite pervious CN; and
$P_{imp}$ = percent of impervious site area.

## Time of Concentration

The pre- and postdevelopment calculation of the Tc for low-impact development is exactly the same as that described in the TR-55 (SCS,1986) and other Soil Conservation Service reference manuals.

## Low-Impact Development Stormwater Management Requirements

Once the CN and Tc are determined for the pre- and postdevelopment conditions, the stormwater management storage volume requirements can be calculated. The low-impact development objective is to maintain both the predevelopment volume, peak runoff rate, and frequency. The required storage volume is calculated using three series of design charts such as shown in Figures 1 and 2. The remaining low-impact development hydrologic analysis techniques are based on the premise that the postdevelopment Tc is the same as the predevelopment condition. If the postdevelopment Tc does not equal the predevelopment Tc, additional low-impact development site design techniques must be implemented to maintain the Tc.

Three series of design charts are needed to determine the surface area required to control the increase in runoff volume and peak runoff rate using retention and detention practices. These design charts series include:
- o Chart Series A: Percentage of site area required to maintain the predevelopment runoff volume using retention storage
- o Chart Series B: Percentage of site area required to maintain the predevelopment peak runoff rate using 100% retention
- o Chart Series C: Percentage of site area required to maintain the predevelopment peak runoff rate using 100% detention

These design charts are based on three general conditions; 1) the land uses for the development are relatively homogeneous throughout the site; 2) the stormwater management measures are to be distributed evenly across the development, to the greatest extent possible; and 3) the percent of the site area is based on 1-inch runoff increments. Linear interpolation is used to determine intermediate values. The design charts are also based on a predetermined depth of 6 inches for the management practice, but they can be modified for other depths. However, a 6 inch depth is recommend as the maximum depth for bioretention basins used in low-impact development. Equation 3 is used to determine the percentage of the site required for BMPs for depths other than 6 inches.

% of site area for BMP = % of surface storage area at 6 inch depth x 6 inches     Eq. 3
Alternate depth of storage (inches)

The procedure to determine the BMP requirements are detailed in the Low-Impact Development Design Manual ( PGC, 1997 ) A brief outline is provided in the following sections.

Step1: Determine Storage Volume Required to Maintain Predevelopment Volume or CN Using Retention Storage ( $V_R$ )

Using the pre- and post-development CN values together with the rainfall depth (inches) for the design storm the storage volume required to maintain the pre-development volume or CN can be read directly from chart series A as shown in Figure 1.

Step 2: Determine Storage Volume Required for Water Quality Control

The surface area requirement expressed as a percentage of the site, is the compared to the percentage of the site area required for water quality control. the volume requirement for stormwater management water quality control is based on the requirement to treat the first 1/2 inch of runoff ( approximately 1800 cubic feet per acre) from impervious areas. this volume is translated to a percent of the site by assuming a storage depth of 6 inches. The greater number, or percent, is used as the required storage volume to maintain the CN.

Step 3: Determine Storage Volume Required to Maintain Peak Stormwater Runoff Using 100 % retention ( $V_{R100}$ )

The percentage of the site area or amount of storage required to maintain the predevelopment peak runoff rate is based on Chart Series B: *Percentage of Site Area Required to Maintain Predevelopment Peak Runoff Rate Using 100% Retention.* As in Step 1, the pre- and postdevelopment CN values together with the rainfall depth (inches) for the design storm are used to read the required storage volume directly from Chart series B, as shown in Figure 2.

Step 4: Determine Whether Additional Detention Storage is Required to Maintain the Predevelopment Peak Runoff Rate

Additional detention storage will be required if the percentage of the site required to maintain the runoff volume, Step 1, is less than the percentage of the site required to maintain the predevelopment peak runoff rate using 100% retention, Step 3. The combination of retention and detention practices id defined as a hybrid BMP. The procedure for determining the percentage of the site required for the hybrid approach is described in Step 5.

### Step 5: Determine Storage Required to Maintain Predevelopment Peak Discharge Runoff Rates using 100 % Detention ( $V_{D100}$ )

Chart Series C: *Percentage of Site Area Required to Maintain the Predevelopment Peak Runoff Rate Using 100 % Detention*, is used to determine the amount of site area to maintain the peak runoff rate only. This information is needed to determine the amount of detention storage required for hybrid design, or where site limitations prevents the use of retention storage to maintain the runoff volume. This includes sites that have severely limited soils for infiltration or retention practices. Again as in Step 1, the pre- and postdevelopment Cn values are used together with the design storm precipitation depth (inches) to read the detention storage volume directly from the chart series.

### Step 6: Hybrid Facility Design

When the percentage of the site area for peak control exceeds that for volume control as determined in Step 3, a hybrid approach must be used. For example a dry swale may be used to incorporate additional detention storage. Equation 2.5 is used to determine the ratio of retention to total storage. Equation 2.6 is then used to determine the additional amount of site area, above the percentage of site required for volume control, needed to maintain the predevelopment peak runoff rate.

$$X = 50/(V_{R100} - V_{D100}) \times \{ -V_{D100} + \sqrt{V^2_{D100} + 4 \times (V_{R100} - V_{D100}) \times V_R} \ \} \quad \text{Eq. 4}$$

where: $X$ = Area ratio of retention storage to total storage

$$H = V_R \times (100 / X) \quad \text{Eq. 5}$$

where: $H$ = Hybrid area expressed as percentage of site

### References

Prince George's County, Maryland. Department of Environmental Resources, *Low-Impact Development Design Manual*, Largo, Maryland, 1997

*Urban Hydrology for Small Watersheds*, Technical Release 55, United States Department of Agriculture, Soil Conservation Service, Washington, D.C. 1986

14    WATER RESOURCES AND THE URBAN ENVIRONMENT

Table 1.    Representative LID Curve Numbers

| Land Use / Cover | Curve Number for Hydrologic Soils Groups[1] | | | |
|---|---|---|---|---|
| | A | B | C | D |
| Impervious Area | 98 | 98 | 98 | 98 |
| Grass | 39 | 61 | 74 | 80 |
| Woods (fair condition) | 36 | 60 | 73 | 79 |
| Woods (good condition) | 30 | 55 | 70 | 77 |

[1] Figure 2.2a, TR-55 ( SCS, 1986 )

Figure 1.

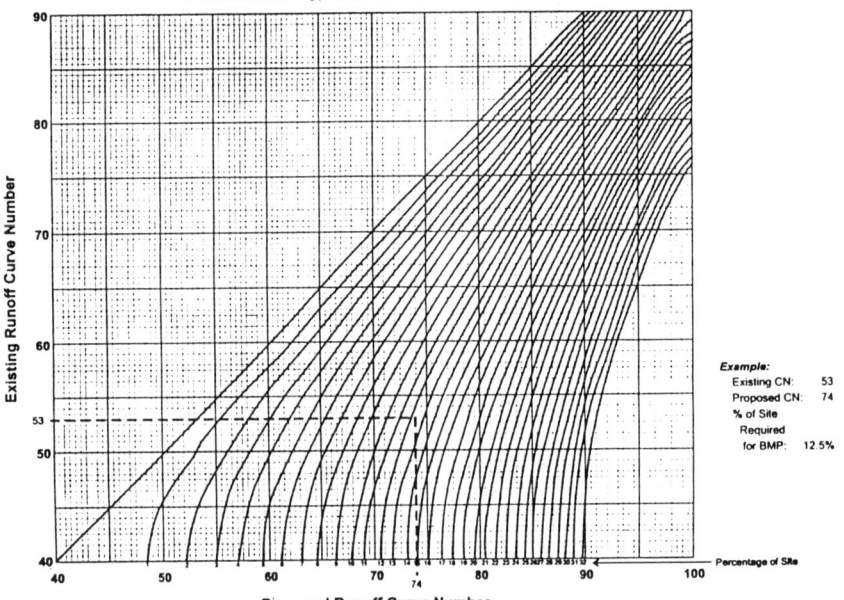

## Figure 2.

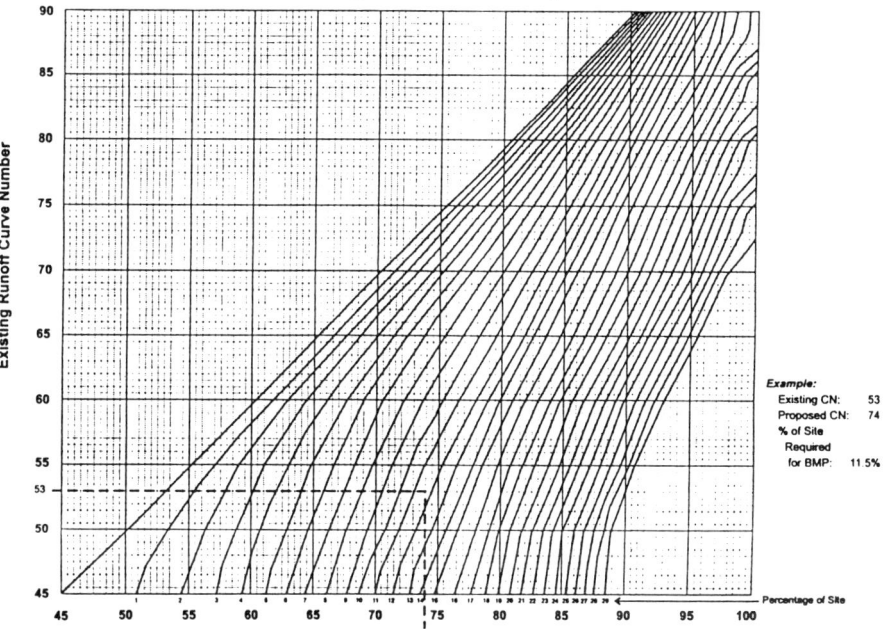

Overview Low Impact Development for Stormwater Management

Larry Coffman [1], Michael Clar [2], Neil Weinstein [3]

Abstract

Low impact development (LID) is an alternative approach to stormwater management. LID achieves stormwater control by fundamentally changing conventional site design to create an environmentally functional landscape that mimics natural watershed hydrologic functions (discharge, frequency, recharge and volume). Runoff is both reduced and controlled at the source. This is accomplished in four ways. First, minimizing impacts to the extent practicable by reducing imperviousness, conserving natural resources / ecosystems, maintaining natural drainage courses, reducing use of pipes and minimizing clearing and grading. Secondly, recreate detention and retention storage dispersed uniformly throughout a site with the use of open swales, flatter slopes, rain gardens (bioretention) and rain barrels. Thirdly, maintain predevelopment time of concentration by strategically routing flows to maintain travel time. Fourthly, implement effective public education programs encourage property owners to use pollution prevention measures and maintain management measures.

Introduction

This paper is only a brief overview of some major aspects of the Low Impact Design Manual developed by Prince George's County, Maryland's Department of Environmental Resources. The manual provides detailed and comprehensive guidance to site planning for use of LID technology as an alternative stormwater approach. The manual explains LID principles, hydrologic analytical methodology, site planning techniques, best management practices, review processes and public outreach programs. The LID design manual builds upon case studies, pilot projects, BMP technology innovations and research produced by the County over the past ten years. The objective of LID is to provide an alternative lower cost effective environmentally sensitive method for stormwater management.

The LID approach combines resource conservation, a hydrologically functional

---

[1] Associate Director, Department of Environmental Resources, Prince George's County Maryland
[2] Executive Vice President, Engineering Technologies Associates IncEllicot City Maryland
[3] Senior Scientist / Engineer, Tetra Tech, Fairfax Virginia

site design with pollution prevention measures to reduce development impacts to better replicate natural watershed hydrology and water quality. Through a variety of on lot site design techniques stormwater is managed in small, cost effective landscape features disbursed throughout a developed site. This source control concept is quite different from conventional end of pipe treatment. Hydrologic functions such as runoff discharges volume, frequency and ground water recharge can be maintained by reducing impervious surfaces, functional grading, open drainage channel sections, disconnection and utilization of runoff, and the use of bioretention or filtration landscape areas.

## Hydrologic Impacts of Conventional Systems and BMP's

Conventional stormwater conveyance systems are designed to collect, convey and discharge runoff as efficiently as possible. The developed landscape is designed and constructed (cleared, graded, piped and paved) with the intent to create a highly efficient drainage system to prevent on lot drainage problems, promote good drainage and quickly convey runoff to a BMP or stream. This efficient drainage system increases runoff volume, decreases ground water recharge and changes the timing, frequency and rate of discharge. Typical use of BMP's as an end of pipe treatment device reinforces and perpetuates the design of the efficient runoff collection and conveyance system.

Conventional stormwater management controls usually maintain peak runoff discharge rate at predevelopment levels for a particular design storm event. However, this approach only controls the rate of runoff allowing significant increases in runoff volume, frequency and duration of runoff from the predevelopment conditions. These hydrologic alterations provide the mechanisms for further degradation of receiving waters.

## Hydrologic Response of Low Impact Development

The objective of the LID site design is to minimize, detain and retain the post development runoff volumes close to the source to emulate predevelopment hydrologic functions. Management of both runoff volume and peak runoff rate is included in the design. This is in contrast to conventional end of pipe treatment.

The LID design approach is to leave as much undisturbed area as practical and optimize infiltration, detention and interception to reduce runoff volume and discharge. LID management (retention / detention controls are integrated throughout the site to compensate for hydrologic alterations of development. The approach of conservation and utilizing areas with high infiltration and low runoff potential in combination with small on lot stormwater retention / detention facility creates a "hydrologically functional landscape". This functional landscape can also enhance a site's aesthetic and habitat value by encouraging greater conservation of trees and use of landscape plant materials.

Wide spread use and uniform dispersion of on lot retention and/or detention to control both runoff discharge volume and rate is key to better replication of predevelopment hydrology. The frequency and duration of runoff are also much closer to the existing condition than can be achieved by typical application of conventional BMP's.

## LID Hydrologic Analysis

The LID site analysis and design approach focus on four major hydrologically based planning principles that are introduced below.

Curve Number (CN) - Minimizing the change in the post development CN by reducing impervious areas and preserving more trees and meadows to reduce runoff storage requirements and maintains the predevelopment runoff volume.

Time of Concentration (Tc) - Maintaining predevelopment Tc reduces peak runoff rates after development by lengthening flow paths and reducing the use of pipe conveyance systems.

Permanent storage areas (Retention) - Retention storage is needed for volume and peak control, as well as water quality control and to maintain the same CN as the predevelopment condition.

Temporary storage areas (Detention) - Detention storage may be needed to maintain the peak runoff rate and/or prevent flooding.

## Minimizing the Change in CN

Calculation of the low-impact development CN is based on a detailed evaluation of the existing and proposed land cover so that an accurate representation of the potential for runoff can be obtained. This calculation requires the engineer or planner to investigate the following key parameters associated with LID; 1) land cover type, 2) percentage of and connectivity of impervious cover, 3) hydrologic soils group (HSG), and 4) hydrologic conditions (average moisture or runoff conditions).

The following LID site planing practices can be utilized to achieve a substantial reduction in the change of the calculated CN; 1) narrower driveways and roads (minimizing impervious areas), 2) maximized tree preservation and/or afforestation, 3) site finger printing (minimal disturbance), open drainage swales, 4) preservation of soils with high infiltration rates to reduce CN, 5) location of BMP's on high-infiltration soils and, 6) construction of impervious features on soils with low infiltration rates. Reducing the change in CN alone will reduce both the post development peak discharge rate and volume.

## Maintaining the Predevelopment Tc

The LID hydrologic evaluation requires that the post development Tc be close to the predevelopment Tc. This is important because LID is based on a homogenous land cover and distributed retention and detention BMP's. The following site planning techniques can be used to maintain the existing Tc; 1) maintain predevelopment flow path length by dispersing and redirecting flows using open swales and natural or vegetated drainage patterns, 2) increasing surface roughness (e.g., preserving woodlands, vegetated swales); 3) detaining flows (e.g., open swales, bioretention, rain gardens), 4) minimizing disturbances (minimizing compaction and changes to existing vegetation), 5) flattening grades in impacted areas, 6) disconnecting impervious areas (e.g., eliminating curb/gutter and redirecting down spouts) and 7) connecting pervious and vegetated areas (e.g., reforestation or afforestation).

Combined use of these techniques, and those to reduced the change in the CN can modify runoff characteristics to effectively shift the post development peak runoff time to that of the predevelopment condition and lower the peak runoff rate.

## Maintaining the Redevelopment Curve Number and Runoff Volume

Once the post development Tc is maintained at the predevelopment conditions and the impact of CN is minimized, any additional reductions in runoff volume must be accomplished through distributed on site stormwater management techniques. The goal is to select the appropriate combination of management techniques that emulate the hydrologic functions of the predevelopment condition to maintain the existing CN and corresponding runoff volume. LID sites use retention practices distributed throughout the site to provide the required volume controls at the source.

Retention storage allows for a reduction in the post development volume and the peak runoff rate. The increased storage and infiltration capacity of retention BMP's allow the predevelopment volume to be maintained. The most appropriate retention BMP's include; 1) bioretention cells (rain gardens), 2) infiltration trenches and 3) rain barrels. Other possible retention BMP's include retention ponds, roof top storage, cisterns and irrigation ponds but, it may be more difficult to distribute these types of controls throughout a development site.

As retention storage volume of LID BMP's is increased there is a corresponding decrease in the peak runoff rate in addition to runoff volume reduction. If a sufficient amount of runoff is stored, the peak runoff rate may be reduced to a level at or below the predevelopment runoff rate. This storage may be all that is necessary to control the peak runoff rate when there is a small change in CN. However, when there is a large change in CN, it may be less practical to achieve flow control using volume control only.

## Potential Requirement for Additional Detention Storage

In some cases where large changes in CN can not be avoided, retention storage practices alone may be either insufficient to maintain the predevelopment runoff volume or peak discharge rates or require too much space to represent a viable solution. In these cases, additional detention storage will be needed to maintain the predevelopment peak runoff rates. Detention practices can easily be integrated into site design features such as swales, roof top storage and shallow parking lot storage.

Determination of Design Storm Event

The hydrologic approach of LID is to retain the same amount of rainfall within the development site as that retained prior to any development (e.g., woods or meadow in good condition) and then release excess runoff as the woods or meadow would have. By doing, so it is possible to mimic, to the greatest extent practical, the predevelopment hydrologic regime to maximize protection to aquatic ecosystems and ground water recharge. This approach allows the determination of a design storm volume that is tailored to the unique soils, vegetation, topographic and recharge characteristics of the watershed.

LID Applicability

LID combines of a wide variety of site design to micro-management techniques to provide stormwater management. The approach can be used for residential, industrial and commercial developments. However, its feasibility and desirability is dependant on several key planning considerations including; 1) need for conservation of the hydrologic regime, 2) reasonable and practical use of the property, 3) conflicts with existing site planning and design regulations, 4) importance of aesthetic and habitat considerations 5) prevention of flooding problems, 6) property owner acceptance an 7) development density must allow space for LID controls.

LID Basic Site Planning Strategies

The goal of LID is to design the site in a way that reproduces hydrologic functions. The first step is to minimize the generation of runoff (reduce the change in the CN). In many respects, this step is very similar to traditional techniques of maximizing natural resource conservation, limited disturbance and reducing impervious areas. The major difference is with LID you must carefully consider how best to make use of the hydrologic soil groups and site topography to help reduce runoff. These considerations would include; 1) maintain natural drainage patterns, topography and depressions, 2) preserve as much existing vegetation as possible in hydrological soil groups A and B, 3) locate BMP's in hydrologic soil groups A and B, 4) where feasible direct impervious areas to soil groups C and D, 5) disconnect impervious surfaces to direct and disburse runoff to soil groups A and B, 6) flatten slopes within cleared areas to facilitate on lot storage and infiltration and 7) revegetate cleared and graded areas.

Public Outreach and Pollution Prevention

Pollution prevention and maintenance of on lot BMP's are two key elements in the overall comprehensive approach. Effective pollution prevention measures can reduce the introduction of pollutants to LID BMP's thereby enhancing their ability to reduce pollutant levels and extend the life of the facilities. Public education is essential to successful pollution prevention and BMP maintenance. Not only will effective public education complement and enhance BMP effectiveness, it can also be used as a marketing tool to attract environmentally conscious buyers, promote citizen stewardship, awareness and help to build a greater sense of community based on common environmental objectives and the unique character of LID designs.

Summary

The effective use of LID site design techniques can significantly reduce the cost of providing stormwater management. Savings are achieved by eliminating the use of stormwater management ponds, reducing pipes, inlet structures, curbs and gutters, less roadway paving, less grading and clearing. Where LID techniques are applicable and depending on the type of development and site constraints, stormwater and site development costs can be reduced by 10 to 25 % compared to conventional approaches. Reduction of the infrastructure also reduces infrastructure maintenance burdens making the LID development more economically sustainable. LID allows for the same or in some cases higher lot yields compared to conventional approaches. LID can achieve greater natural conservation not by regulating protection of more areas but, by using conservation as a stormwater BMP. LID promotes public awareness, education and participation in environmental protection.

References

Prince George's County, Maryland. Department of Environmental Resources. *Low-Impact Development Design Manual.* (PGC, 1997)

Prince George's County, Maryland. Department of Environmental Resources. *Low-Impact Development Guidance Manual.* (PGC, 1997).

Prince George's County, Maryland. Design Manual for use of Bioretention in stormwater Management. (PGC, 1993)

Prince George's County, Maryland. Maryland National Capital Planning Commission. *A Technical Manual for Woodland Conservation with Development.* (Darr, 1990).

Urban renewal and stormwater management planning in an urban center

K K Chin[1], K Y Ng[2], S W Lee[2], M Hasni[2] and A Choo[3]

Introduction

The City of Johor Bahru is situated at the southern tip of Peninsular Malaysia, east of the estuary of Sg Skudai and west of the estuary of Sg Masai. It is within $1°27'-1°33'$ north of the Equator, and $103°40'-103°53'$ east of the Greenwich Meridian. The City of Johor Bahru is now the fourth largest town in Malaysia. As stated in the Johor Bahru Central District Development Master Plan, the City is being established as a service centre for southern Peninsular Malaysia and will act as the 'Mother'City to the proposed larger "Johor Technopolis" encompassing the UTM and the Senai and Skudai Technology Park. The Central District is facing the Johor Strait and Singapore. The land is, low lying and undulating. The existing business district is sandwiched between the Sg Skudai and Sg Tebrau which drain southeast into the Johor Strait. Tributaries to these estuaries cut through the land mass and drain almost perpendicular to them. A few small streams such as Sg Air Molek and Sg Segget which cut through the central district, and estuaries east of Sg Tebrau drain directly into the Johor Strait. The City Centre is undergoing intensive urban development and redevelopment since the early 1980's, transforming the built-up area from the pre-war type of shop houses to high rise commercial buildings. Large scale clearing of land for residential, commercial and industrial development have taken place. The population of the city is projected to reach 760,000 by the year 2010, and is expected to reach one million by the year 2020. Major infrastructure and spatial

---
1. Consulting Engineer, 151 Cavenagh Road, Singapore 229628
2. Respectively Engineer, Director & Managing Director, Sritenaga Jurutera Perunding Sdn Bhd, 27A, 80200 Johor Bahru, Johor, Malaysia.
3. Partner, ESCAPE, 30-20 Jln Molek 1/9, 81100, Johor Bahru, Johor, malaysia.

development are taking place. One major planning consideration is the management of the urban stormwater.

## The Existing Drainage Sysytems

The climate of the Study Area is tropical with temperature generally uniform throughout the year ranging from around $18^0$ to $28^0C$. The mean daily humidity varies from 99.2 % to 60.8% with an overall mean of around 86.7%. The Study Area is influenced by both the Southwest Monsoon and the Northeast Monsoon. Rainfall stations relevant to the Study Area is STOR JPS, JOHOR BAHRU, Station #1437116. A summary of the long term means of monthly rainfall at this Station is given in Table 1. Judging from the available data, high rainfall periods generally coincide with the Southwest Monsoon and the Northeast Monsoon, especially the latter which occurs during the period of November to February of the year.

*Table 1 - 1970-1990 Rainfall Data (Station # 1437116)*

|  | Jan | Feb | Mar | Apr | May | Jun | Jul | Aug | Sep | Oct | Nov | Dec |
|---|---|---|---|---|---|---|---|---|---|---|---|---|
| Mean | 190.0 | 181.5 | 251.2 | 253.1 | 213.3 | 139.0 | 136.3 | 151.5 | 225.3 | 257.1 | 267.3 | 146.4 |
| 1-day max | 159.0 | 108.5 | 136.5 | 93.0 | 118.0 | 96.7 | 167.1 | 101.5 | 83.5 | 104.0 | 118.5 | 96.6 |
| 2-day max | 224.0 | 145.5 | 218.5 | 130.5 | 119.5 | 119.2 | 272.0 | 101.5 | 123.5 | 134.5 | 140.5 | 133.8 |
| 3-day max | 279.0 | 193.5 | 238.5 | 159.5 | 119.5 | 124.0 | 273.0 | 101.5 | 123.5 | 168.5 | 141.5 | 156.5 |

| Max | 167.1 | 272.0 | 279.0 | 453.5 | 505.5 | 515.5 | 598.5 |
|---|---|---|---|---|---|---|---|
| Date | 2/6/79 | 25/7/79 | 9/1/87 | 8/1/87 | 7/1/87 | 1/1/87 | 6/1/87 |

The area between the Skudai and Tebrau estuaries which comprises the greater part of the Johor Bahru municipal area has been divided into 19 drainage basins most of which discharge to these estuaries and some directly to the Johor Strait. These are catchments of relatively small rivers. The Study Area east of Sg Tebrau are 8 other river basins. These rivers drain the hinterland between the Tebrau catchment and the Sg Masai Catchment. Intensive development has taken place at these corridors, and many natural streams have been canalised including Sg Skudai, Sg Tebrau and Sg Masai. Canalisation of these rivers would increase the flood peaks at the lower reaches of these rivers. Feeding to these streams, estuaries and the Johor Strait is a series of roadside concrete drains. These subsidiary drainage systems, each has its own characteristics, are built over the years without much thought for long term development of the adjacent areas. They may not be the main cause of flooding. They do create a more rapid runoff from developed areas increasing the flood peak to receiving trunk drains. Field inspection show some of these drains are in need of repair and upgrading. Currently, local flooding occurs frequently on several streams in the built-up areas of Johor Bahru. The main causes of flooding are a result of increasing runoff from developing areas, restriction of waterways by encroaching development, inadequate bridges and culverts, debris accumulation in the waterways, and silting of streams due to soil erosion from sites cleared for construction. Although improvement has been carried out, much work remain to be done. In recent years, the frequency of flooding of the rivers in the built-up areas has, in fact, increased considerably.

## Urban Stormwater Management Objective and Design Criteria

An integrated approach to planning and design of urban stormwater management systems to reduce flood damages due to development upstream of river basins is the main objective of the study. Information on the existing and projected land use needed for hydrologic and hydraulic study is obtained from the Structure Plan of the MBJB (1994) and the Johor Bahru Central District Development Master Plan (1996).

Standards and procedures used in the Study are in accordance with 'The Planning and Design Procedure No 1 in Urban Drainage Design Standards and Procedures for Peninsular Malaysia, 1975", "Hydrological Procedure No 1, 1973 and 1982" of Jabatan Pengairan dan Saliran's (JPS), and Manual of Department of Irrigation and Drainage (Hydrology, Revised and updated 1988). The design period for the design of (a) river work is proposed to be 20 years at bankfull and 100 year above bankfull, and (b) work for trunk and for secondary drains is respectively to be 5 year and 100 year flood events.

Peak flows of individual trunk drains and waterways are estimated using information available, rainfall and runoff relationship of design storms from stations installed with automatic recorders to monitor individual storm events, and hydrologic and hydraulic simulation study. Several methods may be used to determine the peak rainfall runoff-relationship. The Modified Rational Method has been used in the Hydrological Procedure for flood estimation for urban areas in Peninsular Malaysia by JPS (eg.HP1, HP16). The requirements of the Procedure are well known. Table 2 shows the rainfall-intensity-duration frequency relationships of Johor Bahru.

*Table 2 - Rainfall intensity-duration-frequency data for Johor Bahru (I, mm/h)*

| T-year | Storm Duration (t), h | | | | | | | |
|---|---|---|---|---|---|---|---|---|
| | ¼ | ½ | 1 | 3 | 6 | 12 | 24 | 72 |
| 5 | 130 | 110 | 65 | 30 | 15 | 9.5 | 5 | 2.6 |
| 10 | 160 | 120 | 76 | 32 | 20 | 12 | 8 | 4.2 |
| 20 | 200 | 140 | 95 | 40 | 28 | 17 | 12 | 6.2 |
| 50 | 220 | 160 | 105 | 49 | 34 | 22 | 14 | 7.5 |
| 100 | 250 | 180 | 120 | 73 | 51 | 34 | 21 | 11 |

Using data from HP No.1,the temporal distribution of annual maximum rainstorm for Johor Bahru is given in Table 3.

Hydrologic and hydraulic studies and modelling are used to determine the rainfall-runoff characteristics of river catchments and to provide (a) the hydrographs for routing of the various flood events through the existing and proposed drainage systems during rainstorms, and (b) data for the design and analyses of the existing drainage system and future improvement within the built-up areas.

## Tentative Conceptual Plan

Much work have been carried out on data collection and analysis, engineering field

*Table 3 - Rainfall Design Temporal Pattern*

| Duration, hr | Percentage of Mean Rainfall |
|---|---|
| ½ | 30 minutes duration in 3 periods (%) of 10 minutes 36, 34, 30 |
| 3 | 3 hour duration in 6 periods (%) of 30 minutes 28.6, 26.6, 19.5, 12.6, 9.5,3.2 |
| 6 | 6 hour duration in 6 periods (%) of 60 minutes 34.1, 34.6, 16.4, 7.8, 2.8, 4.3 |
| 12 | 12 hour duration in 6 periods (%) of 120 minutes 57, 18.7, 9.8, 4.4, 6.7, 3.4 |
| 24 | 24 hour duration in 4 periods (%) of 360 minutes 82.4, 3.5, 5.7, 8.4 |
| 72 | 72 hour duration in 6 periods (%) of 720 minutes 28.5, 6.8, 41.1, 6.4, 12.8, 4.4 |

inspection, and the testing and selection of hydrologic and hydraulic computation methods including stormwater simulation models. Route and grid survey of waterways, and detailed survey and assessment of existing storm drains in built-up areas has been initiated. Field survey data are needed for detailed analysis and design of drainage systems.

**Street and Secondary Drains:** Much of the existing drainage systems were built years ago without anticipating the recent and projected growth of the Study Area. In common with many towns in Malaysia as they have grown, the drainage system developed piecemeal to cope with each new development resulting in several distinctly different system often without consideration of downstream flooding. As a result, some parts of the existing drains do not meet with the current standards and code of practice for drainage design. Improvement is needed. The proper design of storm drains paying attention to even minor details such as drainage reserves, inlet and outlet structures, storage junctions and drain gradients can reduce levels of stormwater runoff significantly. This applies to minor street drains and secondary drains. Drain reserves can be made to form parts of the beautification plan and at the same time effectively serve as open storm water storage or infiltration facilities for stormwater runoff volume reduction to cut down flood peak. Stormwater storage facilities are often road side swells or detention ponds. Infiltration facilities include:
(a) permeable pavement for footpath, (b) infiltration trench, (c) infiltration roadside gutter, and (d) infiltration street inlet. Other measures include enhanced direct infiltration of runoff from impervious areas such as roofs, paved streets, or sidewalks. Stormwater runoff can also be temporarily detained by storage on catchment surface such as parking lots, flat rooftops, playgrounds, public gardens, parks, and special flood areas. Properly designed swells, street gutters, catch basins, surface and subsurface detention facilities, and storm drains could reduce the frequency of inconvenience caused by stormwater ponding. The consequences of failure of primary drainage system designed for 1 to 10 year storms are, however, often not significant as long as the secondary and trunk drainage system are functioning properly. Nevertheless, the linkage of the design of

these minor drainage systems to the major ones must be well planned. Some good examples are the river basins at the Central District such as Sg Chat, and Sg Air Molek. These catchments are long and narrow. Both sides of these waterways are on high ground with relatively steep slopes except the coastal strip facing the Johor Strait. During rainstorms, runoff rushing down local drains often spills over to streets causing flash flood at the low lying coastal area. To cut down flood peak, improvement is needed to dissipate the energy to slow down and to provide temporary storage for the runoff. Road side infiltration trenches and suitably located catch basins, channel drops and drop basins are methods to be explored for local drains.

The Structure Plan for the Central District will eventually change the existing pattern of landuse in the near future. The whole coastal area will be transformed to accommodate the Islamic City and the Sultanah Aminah Cultural Urban Core. The proposed inner ring road will start from the coastal strip along Jalan Abdul Samad. Any drainage scheme planned now must cater for these new development. Flash floods from 5-year return storms have so far been of short duration. Construction of Infiltration Trenches or other temporary storage for heavy runoff at open spaces and car parks and widening of the existing drains along some coastal roads will cut down flood peaks and reduce the frequency of flash floods. During dry weather conditions, runoff from these catchments will continue to discharge through the existing system. Tidal levels at Sembawang Shipyard are MLWS 0.7m, MLWN 1.3m, MSL 1.9m, MHWN 2.4m, and MHWS 3.1m chart datum (CD). Records show that only around 45 days of the year tides at the Johor Strait rise 3m above the chart datum (1.1 m RL). As the proposed Islamic City and the Sultanah Aminah Cultural Urban Core will be developed on this site, ground and street levels of all new development is recommended to be raised to above the MHWS or RL 2.6m.

**Trunks and waterways:** The intention of any upgrading work to the existing drainage system is to reduce the peak flow presently carried by the system. Any new system proposed must have the capability of achieving the development targets through well defined stages such that previous capital investments are protected and the performance of the existing installation is not seriously disrupted. Based on investigation carried out thus far, the present level of service of the existing drainage systems is increasingly stressed. A typical example is Sg Air Molek which winds along the west side of Jalan Yahya Awal, through the eastern edge of the Istana Park along Jalan Air Molek, and discharges to the Tebrau Strait. The channel and the numerous culverts and bridges crossing the channel are under capacity to convey the stormwater runoff. Frequent flooding occurs at low lying areas. A large section of the channel is not lined until it reaches the Istana park at the junction of Jalan Air Molek and Jalan Gertak Merah. The 1996 Structure Plan of the Central District targets the Istana Park and the Western Foreshore Precinct for redevelopment. The lower reach of the Sg Air Molek starting from the junction of Jalan Gertak Merah and Jalan Air Molek can be integrated into the park system where recreational ponds in the Park will also serve as stormwater detention facilities to cut down flood peak during high rainfall runoff. The channel upstream can be reshaped to increase its storage capacity, and as much as possible natural lining of the channel is to be retained. Sg Air Molek cuts through a number of streets on its way to the Tebrau Strait. The channel behind each culvert or bridge can be shaped to hold back storm water during high flow especially if the levels of roads are raised. Culverts and bridges can serve as flow control structures. Frequent flooding is not only confined to areas at lower reaches of the basin, but also at areas near Jn Wirawan and Jln Yusuf Taha. All development should take step to ensure, as much as possible, rainfall runoff from the area will not significantly increase to cause flash flood downstream. Infiltration trenches at car parks and along local drains, and storm water detention pond systems will not only cut down flood peaks but also enhance the aesthetic of the new development. Hydrographs of simulation studies show that with some improvement of the waterway and bridges and culverts along its path, the detention time necessary for detention storage to cut down flood peaks should be no more than an hour. Land area requirement and cost for improvement could be lower than the traditional method of widening and concrete lining the channel to accommodate peak discharges.

Drainage improvement has been a continuous process through the building of levees and flood gates, and lining, widening and straightening of waterways. These would temporary relieve flood. For long term solutions, an integrated approach to planning and design of urban stormwater quantity and quality control is essential. The concept, that flood control in urbanised areas should counterweight the effects of urbanisation, is gaining global attention. The benefit of evapo-transpiration and storage of stormwater runoff by vegetation and soil cover lost through urban development must be replaced by technical measures.

As part of the channel improvement program, the design of major drainage systems should take into consideration in total natural streams and valleys, and man made drainage elements, such as channel elements and flood detention facilities. This system has to accommodate runoff from infrequent storms with return periods up to 100 years. Floods with a recurrence intervals of 100 years cannot be controlled without measures such as large detention basins and/or diversions. The Structure Plan for Johor Bahru shows extensive development in the Study Area in years to come. It is expected to have a well-defined city centre and a number of secondary town centres. Two major commercial centres, at Tampoi and Permas Jaya, serve residents of west and east Johor Bahru. A third suburban centre to the north, would be introduced at Kangkar Tebrau. A 'Premier Development Zone' is recommended where low density development in a parkland setting would be planned. The emphasis in this Zone is on development which successfully integrates high quality residential, recreational, industrial and commercial development and achieves architectural and environmental harmony. All round and within the area will be open spaces, trees, plants and water features. The MBJB Revised Structure Plan (1994) states that office space shall be concentrated in the city centre, city downtown and suburban town centres. In addition, low rise, low density, office space, with a low plot coverage will be encouraged in the Premier Development Zone and permitted on extensive sites vacated by government and institutional users. The development of further office space in the form of shop houses will not be permitted except on the upper floors of retail premises in defined commercial centres.

The planners also call for the allocation of adequate land for open space and park land for passive and active recreational activities. Open space can also serve as temporary open stormwater storage facilities in urban areas. The existing swamps and low lying areas serve well as open retention ponds for additional runoff due to development. These detention basins can also serve as park and recreational facilities. Beautification and landscaping forms part of the drainage reserve and flood control plan. This will be carried out along drains and waterways. The 1994 MBJB Structure Plan and the 1996 Structure Plan of the Central District have laid down the criteria for the development of the business district, industry and residences. In tune with the criteria set, the objective of the conceptual plan is to upgrade the waterway environment through landscaping and related work. Beautification of drains and waterways include:

- an analysis of current land use and activities in the proximity of the waterways and drains taking into consideration of the cost and benefit with emphasis on the dual functions of the proposed schemes serving as flood control measures and recreational facilities;
- evaluation of the impact of the conceptual beautification plan on the ecology of the natural systems, especially the water environment;
- assessment of the characteristics of the drain reserves and the modified banks of the waterways as sport and recreational facilities such as pedestrian walks, jogging tracks, promenades, fast food outlets, parking areas etc;
- development of conceptual sketches to illustrate the types of landscaping and related works.

Factors to be considered in the formulation of the plan include (1) the level of minimum acceptable flood risk, (2) town planning and other related guidelines, (3) ease of maintenance, (4) future uses of the waterways, (5) accessibility to the river reserves zone, (6) flora and fauna to be used for the systems, (7) aesthetic, and safety features for recreational purposes.

Conclusion

Flood control measures, when integrated into the overall urban planning process, can help to reduce the flood peaks to close to that of the predevelopment stage. A well planned flood control scheme has the benefit of reducing the frequency of flash floods in urban areas and conserving open space for recreational purposes.

Acknowledgement

The authors appreciate greatly the discussions and help of Mr C C Chan and Mr J T Tan, respectively the Director and a past Director of the Drainage and Irrigation Department of Johor. They also thank the staff of JPS and MBJB for their kind assistance.

Reference

JPS (1975). Urban drain design standards and procedures for Peninsular Malaysia - Procedure No 1. Publication Unit, Ministry of Agriculture and Rural Development, Kuala Lumpur, Malaysia.

Aesthetic aspect of urban stormwater management system design

K K Chin[1], A Choo[2], K Y Ng[3], S W Lee[3], and M Hasni[3]

Introduction

Mersing town is situated at the east coast of Johor. The existing town area of 76,178 hectares is roughly 45 % of the projected study area of about 169,400 hectares. The existing town area is located at the mouth of the Mersing River which is low lying and subject to flooding in the past years. Although the town is well maintained, the drainage system was built some years back and needs major upgrading. Future development will be outward from the existing town at higher ground, upstream of the Mersing River and south of the current built-up area. An integrated drainage system is needed to cope with future developments. A study is being carried out to provide solutions to the existing flooding problems and to cope with problems that may arise due to planned development in the future. The study includes both the present and proposed extension of the local council area covering (1) trunk drains with catchment area greater than 40 hectares and secondary drains serving catchment area between 4 and 40 hectares in the existing built-up areas, (2) trunk drains and secondary drains in those areas proposed for urbanisation, and (3) an adequate drainage system for the remaining areas of the study area. The work carried out is to meet the following objectives:

- To put a comprehensive long-term solution to the drainage problems in the existing built-up areas;
- To provide drainage strategies catering for urban and rural development in the remaining areas up to the year 2015.

Existing conditions

The land within the study area is flat and low lying with a large mangrove, swampy area.

------------------
1. Consulting Engineer, 151 Cavenagh Road #03-161, Singapore 229628
2. Partner, "escape", 30-02 Jln Molek 1/9, 81100 Johor Bahru, Johor, Malaysia
3. Engineer, Director, Managing Director, Sritenaga Jurutera Perunding Sdn Bhd, 80200 Johor Bahru, Johor, Malaysia.

The existing business district is located at the mouth, south of the Sg Mersing. The Sg
Mersing and its tributaries Sg Air Merah and Sg Lengan Baju cut through the Mersing Town with an existing population of 16,398. The business district and the older section of the town are well developed. New housing and commercial development are on-going at both the old and new sections of the town to cater for the growing population and the influx of tourists to the study areas or in transit to off shore island resorts and the surrounding recreational areas.

Mersing Town development is, at present, mainly concentrated on the south-eastern bank of Sg Mersing. Flooding was experienced in the vicinity in the past. The northern bank of Sg Mersing is mainly kampong areas and mangrove swamp. Housing development and some on going construction are observed along the main road, Jalan Endau.

The Sg Mersing Basin and its tributaries with a total drainage area of approximate 232 $km^2$, drains east toward the sea cutting through the built up area of the Town. Major tributaries of Sg Mersing within the study area are Sg Air Merah, Sg Tadahan and Sg Besar at the north bank, and Sg Lengan Baju at the south bank. A network of trunk drains with flood control bunds serves areas west of Jln Endau and north of Sg Mersing. These areas include Kg Bahgia, Kg Air Merah, Kg Sg Muka Dua, Kg Sri Lalang and Kg Sri Lumpur. Another network of trunk drains serves areas south of the Sg Mersing and along both sides of the main road, Jln Jemaluang, till Kg Sri Pantai. At the south eastern end of the study area is the catch basin of a tributary of Sg Jemaluang. A series of primary and secondary drains serves the business district and coastal strips, and these drains discharge stormwaters directly to the mouth of the Sg Mersing or the South China Sea.

The existing drainage system serving the Mersing Town is in need of improvement and expansion to cater for changes that have been taking place. The effect of floods during heavy rainstorms are felt in low lying areas such as Mersing Kechil, along Jln Jemaluang and Kg Sg Muka dua north of the Sg Mersing. Some serious flooding occurred during heavy rain storms at places such as Pengkalan Batu, along Jalan Temanggong, south-west of Bandar Mersing, and at the neighbourhood of Taman Guru.

The flood problems within the Study Areas can be summarised as follows:

- Although still largely rural, Mersing Town has grown to a population of 16,398 and is expected to have a population of 60,505 by the year 2015. Land filling at development sites increases the runoff volume and shorten the time of concentration of rainfall runoff.
- The capacity of the existing drainage systems is inadequate to cope with the rainfall runoff during severe rainstorms.
- Localised flooding is due to a lack of maintenance and/or improper design and construction of storm drains, and a lack of an integrated approach in the planning and layout of the drainage systems. Examples of blockage of channel flow due to the accumulation of debris, improper design of drain junction and bend, and poor maintenance are many. Some tidal gates to the Sg Mersing are not able to cope with heavy runoff from severe storms.

- Inadequate storage capacities of the drainage systems, and outlet capacities of culverts.and flood gates at Sg Lengan Baju catchment, Sg Muka Dua Catchment and Sg Air Merah Catchment.

Although no life loss was reported flooding causes inconveniences, social disruption, depresses the tourist industry and incurs losses in businesses.

Drainage improvement and landscape conceptual plan

The upstream catchments of Sg Mersing and its tributary are essentially undeveloped. Major tributaries of Sg Mersing are Sg Jamari, Sg Mayang and Sg Saran. Much of the catchment of Sg Jamari are primary forests with patches of wet and dry padi. In addition to primary forests, a large track of the catchment of Sg Mayang is swamp forest. Lakes, wet padi and some rubber plantations are among the features at the upper reach of the catchment. Similar features dominate the catchment of Sg Saran and the upstream catchment of Sg Mersing. Most of the Sg Mersing basin is inaccessible by motor vehicles. These rivers are left essentially in their natural states. Although a high percentage of the rainfall will runoff through the study area cutting through the existing Mersing Town, rainfall runoff upstream of the Sg Mersing is not a major flood concern. The runoff drains rapidly to the China Sea. Flood problems are results of poor drainage at the local levels along Sg Langan Batu, Sg Besar and Sg Air Merah. Improvement work has been carried out in these catchments with channel improvement, building of levees along flood prone areas and the improvement of outlet structures.

The Structure Plan of the State of Johor for Mersing focuses on residential development, development related to commercial, servicing and manufacturing industries. The development of recreational facilities catered for the residence and the tourist industry will be a major consideration. To avoid flooding problems an integrated approach to solution of the drainage problems is essential. The intention of any upgrading work to the existing drainage system is to reduce the flow presently carried by the system. Any new system proposed must have the capability of achieving the development targets through well-defined stages such that previous capital investments are protected and the performance of the existing installation is not seriously disrupted. The conceptual plan for drainage improvement is illustrated using the Sg Langan baju basin as an example with emphasis on the aesthetic aspect of urban stormwater management.

Analysis of the Sg Langan Baju basin shows that stormwater would overflow the banks for the stretch of the stream from the discharge point at Sg Mersing up to 400m and 600 m upstream respectively during the 5- and 100-year storms. Concrete lining the channel from the tidal gate to Sg Mersing up to 600 m upstream would increase the channel capacity to cope with peak discharges of up to the 100 year storm. Concrete lining of the channel is, however, not a preferred option at this stage. The land within this stretch of the stream is low and flat with a slope of around 0.001 m/m. Flood water could be contained when stormwater detention facilities are constructed within the drainage reserve
sandwiched between Sg Lengan Baju and Jalan Jamaluang, just upstream of the 8.4m x4.2mx7.5m culvert crossing Jalan Jamaluang, and in the neighbourhood of Taman Sri

Mersing and Taman Baru Mersing. Downstream of this culvert, the current drainage reserve is 2 chains (40 m) along the stream. A 30 m wide in stream-stormwater retention can be constructed along the stretch of the stream till it reaches its discharge point at the tidal gate to Sg Mersing. These in stream detention facilities could greatly reduce flood peaks of major rain storms. It swill also serve as park and recreational areas at dry weather conditions.

The landscape conceptual plan has been formulated to combine the dual functions of the proposed upgrading of the existing drainage reserves and related water retention areas as (1) flood control measures and (2) as recreational facilities. It has been formulated by taking into considerations the existing site conditions and current land uses to minimise adverse impact on the ecology of the natural environment, particularly the complex and sensitive ecosystem of the wetland and river. The site is strategically located, with existing residential area on its northern part and the new development of Bandar Baru Mersing on the southern portion. It is potentially an ideal location for a public park due to the proximity to the residential areas. The proposed park will also provide an identifiable entry gateway to the Mersing Town by creating a sense of arrival to this popular tourist destination. The basic functional requirements of the proposed public park are (1) central activities area, (2) children's playground, (3) pedestrian path, jogging track and par courses, (4) outdoor seating areas/gazebos, and (5) car park area. The retention pond located at the point where Sungai Langan Baju and the trunk drain intersect will be the central activities area and also the focal point of the park. The pond will be permanently filled with water discharging from the river and drain. Water recreational activities such as fishing and remote-controlled model boats can be provided for around the pond area. Natural rock boulders will be placed around the edges and planted with wetland plant species to create a more ornamental character. Gazebos will be located adjacent to the pond as resting and scenic look-out area. A dry creek across the park areas are to be distinctly separated from the pedestrian traffic areas. Pockets of semi-private seating areas are provided as a facility for resting which is scenic and passive.

Selected plants materials shall be hardy and able to survive in the semi-wet condition and withstand the condition of being partially submerged under water during floods. It is intended that the use of shrubs and ground covers be minimised to reduce the on-going maintenance. Proposed tree species shall be of fast growing species that require minimum maintenance. The existing wild Eleochar variegata in the river will be retained and re-established around the edges of the ponds and dry creek bed.

Large canopy trees shalls be strategically located to provide shade from the hot western sun. A formal avenue of oil palms shall be created for framing views to specific point of interests. The existing roadside tree of Eleochar grandeis will be reinforced to partially screen the main vehicular traffic. The proposed list of plant species shall be (1) **Trees:** Eugenia grandis, Saraca thaipingensis, Lagerstroemia speciosa, Melaleuca cajuputi, Dillenia spp., Hibiscus tiliaceus, Pometia pinnata, Samanea saman, Avicennia spp., Rhizophora apiculata; (2) **Palms/Bamboo:** Nypa spp., Elaeis guineensis, Caryota mitis,

Bambusa vulgaris; (3) **Shrubs/Groundcovers:** Ipomoea pes-caprae, Malastoma malabathrieum, Alocasia macrobiza, Colocasia spp., Pandanus odoratissimus. Figure 1 and Figure 2 illustrate the conceptual of the area.

Conclusion

In line with the Government call in 1995 for a "Negara Taman" under the "landscaping the Nation" programme, this study recognises the important of the beautification and the landscaping of drains and waterways to form part of the drainage and flood control plan. This is imminently so in the case of the Mersing Town, a major tourist attraction centre of the East Johor Development Area. Furthermore, the area concerned in this study is targeted for major development under the Structure Plan of Maijlis Daerah Mersing (1995). The landscaping of the drainage reserves and waterways will not only improve the urban drainage system and the aesthetic visual quality of these areas, but it will also promote the image of the Mersing Town as a tourest destination in the State of Johor.

Acknowledgement

The appreciate greatly the Director of the Drainage and Irrigation Departmernt of the State of Johor and his staff for their kind assistance.

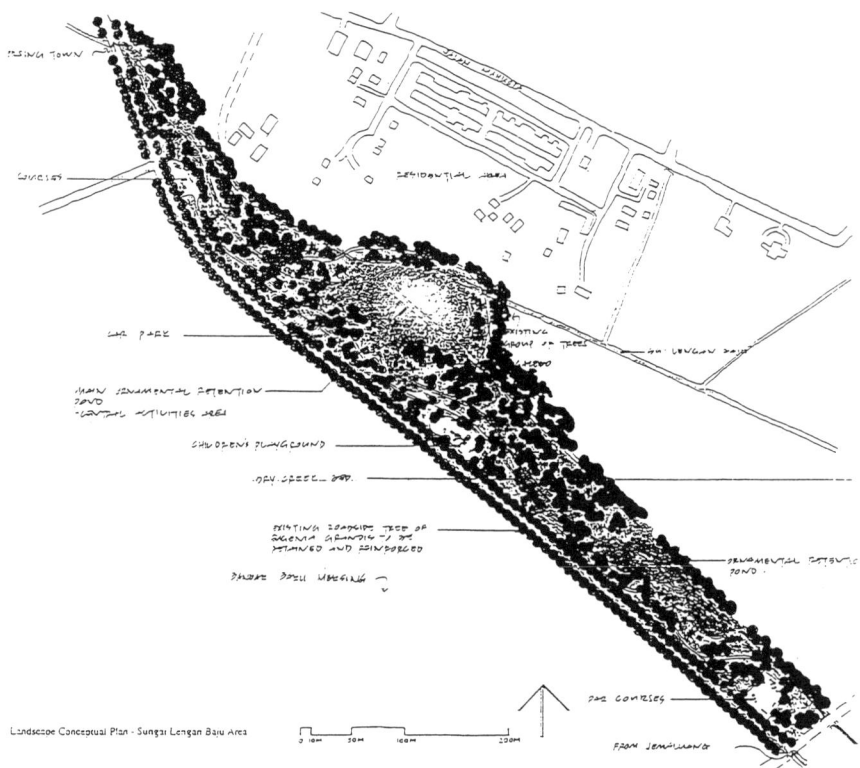

Landscape Conceptual Plan - Sungai Lengan Baju Area

# WATER RESOURCES AND THE URBAN ENVIRONMENT

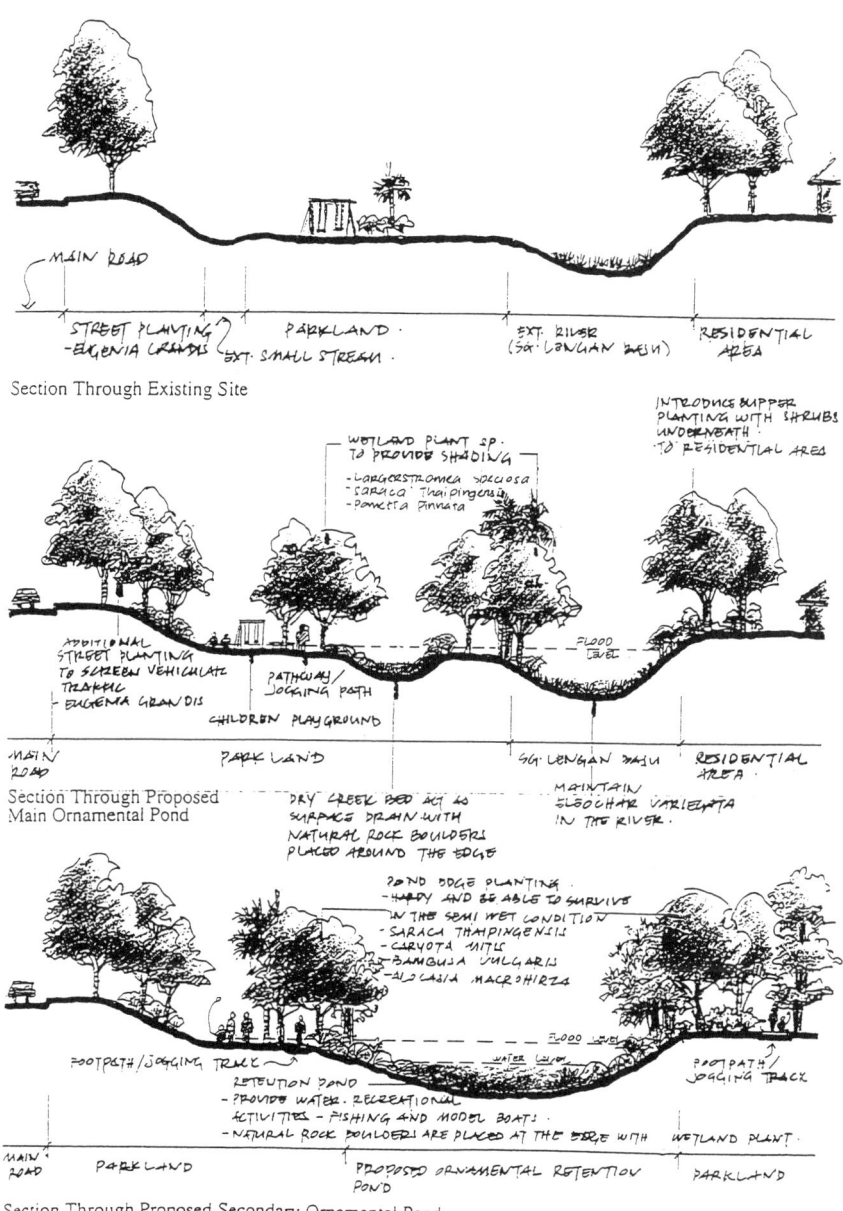

Section Through Existing Site

Section Through Proposed Main Ornamental Pond

Section Through Proposed Secondary Ornamental Pond

Figure – Cross Sections - Sungai Lengan Baju Area

Thailand-U.S. Center for Technology Partnership

James J. Yezzi, Jr.[1], Daniel Sullivan[2], Anthony N. Tafuri[3]

Abstract

A Memorandum of Agreement (MOA, 1997) on Environmental Technology Transfer and Development was signed in May 1997 between the Ministry of Science, Technology and Environment (MOSTE) of the Kingdom of Thailand and the State of New Jersey. The Agreement established a joint Center for Environmental Technology Transfer and Development (CETTAD) for collaboration in the areas of environmental and energy technologies. The CETTAD will provide training; education; technology transfer; joint research, development and demonstration of advanced technologies; and overall engineering support for Thailand's expanding environmental and energy programs. The CETTAD will encourage joint marketing and business opportunities between Thailand, New Jersey and United States industry in the Asian and Indochina regions through their participation in the Center.

Introduction

The EPA Strategic Plan (EPA, 1997) describes five key attributes of an improved approach to deliver consistently better environmental protection at less cost. One attribute is an improved system that reflects strong partnerships; consequently, building and maintaining strong partnerships will continue to be one of the Agency's

---

[1]Physical Scientist, [2]Chief, [3]Team Leader, Urban Watershed Management Branch, Water Supply and Water Resources Division, National Risk Management Research Laboratory, U.S. Environmental Protection Agency, 2890 Woodbridge Avenue, Edison, NJ 08837

highest priorities. For most nations and states, world trade is no longer an option, it is an imperative!

In May 1997, a Memorandum of Agreement on Environmental Technology Transfer and Development was signed between the Ministry of Science, Technology and Environment (MOSTE) of the Kingdom of Thailand and the State of New Jersey. The Agreement established a joint Center for Environmental Technology Transfer and Development (CETTAD) for collaboration in the areas of environmental and energy technologies. The CETTAD will provide training; education; technology transfer; joint research, development and demonstration of advanced technologies; and overall engineering support for Thailand's expanding environmental and energy programs. The CETTAD will also encourage joint marketing and business opportunities between Thailand, New Jersey and U.S. industry in the Asian and Indochina regions through their participation in the Center. The New Jersey Institute of Technology(NJIT) is the US lead organization and will serve to foster the establishment of partnerships.

<u>Framework for Collaboration</u>

CETTAD will function as the mechanism for collaboration between industry, university, and government experts (IUG) representing Thailand and New Jersey together with appropriate representatives from other parts of the United States. This follows a model that was developed by the National Science Foundation(NSF)and implemented earlier by NJIT. The Center's member organizations constitute the International Advisory Board (IAB), which provides guidance and management direction, oversight, identification of expertise, and the forum for the development of partnerships. The specific areas of support and the technologies and techniques to be studied, developed, demonstrated and applied will be reviewed and approved by the IAB.

To support the IAB, Thailand and the U.S. will each establish a national IUG-Advisory Committee for their own routine management and operation of the CETTAD.

Center Organization

The activities of the Center are organized in three divisions: the Engineering Division; the Research and Development (R&D) Technology Transfer Division; and the Laboratory and Field Monitoring Division. The directors of these divisions will be supported and assisted by IAB members or their representatives. As members of project-specific advisory committees, these individuals will have the expertise and desire to develop joint marketing and business opportunities in services, equipment and supplies required for the assessment of problems, the development of alternative strategies, and the implementation of Thai-selected solutions. Participants on the IAB will represent the member organizations sponsoring the Center activities. From the U.S., the IAB is presently represented by NJIT, the State of New Jersey, the U.S. Department of Energy, the U.S. Environmental Protection Agency, the U.S. Army and participating U.S. industry. Representatives from Thailand include the Pollution Control Department(PCD), other governmental agencies, sponsoring Thai industry and participating academic institutions. Other nations, states, and industries are welcome to join the Center.

The IAB voting members will meet twice a year to agree on the Center's research, development, demonstration, testing and evaluation programs and to consider management and policy matters brought to its attention by fellow members or the Center Directors. Each sponsoring organization will appoint an IAB member who can provide their organization's expertise, experience and interest in the Center. In order to gain membership in the Center, industries would be asked to pay an annual fee; along with government funding, these fees would support the activities of the Center.

Potential Areas of Collaboration

Collaboration under the Agreement will address the following activities: exchange of information and data on scientific and technical activities; exchange of scientists, engineers and other specialists to participate in Center activities; organization and participation in
seminars and other meetings; exchange of samples, materials, instruments and components for experiments, testing, demon-

stration and evaluation; execution of joint projects and technology transfer activities, including design, construction and operational activities; and other forms of collaboration as mutually agreed upon.

A potential area of significant collaboration, between the U.S. and Thailand, is the replacement of urban infrastructure to insure the sustainable development of communities well into the future. This requires an assessment of where we are today and where the consequences of our actions will place us tomorrow. The demand for infrastructure worldwide is huge, approximating $3 trillion annually (ASCE, 1996). Many areas of the world are currently experiencing rapid growth, with the associated need for their infrastructure to support that growth. At the same time, developed areas are replacing old and obsolete infrastructures to meet existing and future requirements. Besides the need to address the immediate problems associated with infrastructure, there is a need to make fundamental changes in the way infrastructure is designed, built, utilized, maintained and renewed if we are to achieve economic growth while preserving and enhancing environmental quality.

In 1988, the National Council on Public Works Improvement (NCPWI), a Congressionally mandated special commission, issued a report on the state of infrastructure in the U.S. The Commission looked at eight infrastructure areas: highways, mass transit, aviation, water resources, water supply, wastewater, solid waste, and hazardous waste. Based on the Council's findings, it is estimated that the U.S. would have to invest over $100 billion each year in new and existing public works. Obviously, a tremendous effort must be put forth to ensure that these resources are spent on cost-effective, technologically sound remedies that will meet national environmental goals (NCPWI, 1988).

<u>Initial Areas of Collaboration</u>

An initial series of meetings were conducted with Thailand's Pollution Control Department (PCD) in Bangkok, Thailand, during the period from November 3-14, 1997. The main purpose of these meetings was to address eight priority issues that were identified by the PCD: PCD's lab equipment and management; remedying air pollution problems (volatile organic compounds) in the Industrial Estate of the Mah-ta-put District; eliminating arsenic poisoning in the town of Nakhon Srithammarat; providing emergency response training and developing "train-the-trainers" programs; implementing waste minimization programs for plastics in small industries; controlling fumes and wastewater from electroplating and gold shops; conducting solar energy research; and supporting Thai military research.

Meetings were also held with other Thai governmental agencies, academia, and industries to present the MOA and the role of CETTAD. During the meetings with industry, several additional issues were raised: determining the cost effectiveness of advanced technologies; developing and implementing waste minimization programs; providing education and training for governmental agencies and industry; treating, storing, and reuse of waste oil; characterizing and remediating chemical contamination at active and abandoned hazardous waste sites; standardizing environmental testing data; and managing leaking underground storage tanks.

Intellectual Property Rights

In the case of advanced technologies, intellectual property rights will be respected and protected by all Center participants and their governments according to their respective internal laws and regulations. Internationally recognized standards adopted by the U.S. and Thailand will be followed.

Financial Responsibilities

The administrative, operating and project costs of the CETTAD will be supported by matching funds from Thailand and sources in New Jersey and other parts of the U.S. These funds will be obtained from Thai, State of New Jersey and U.S. government agencies, industry members, foundations, industrial trade associations and non-government agencies. Initial funding for the first year effort under the Agreement will be equally borne by both parties for a total of $800,000 or 29,000,000 Bahts. This financial contribution will be maintained as a minimum each year for at least five years.

## Contact Organizations

Interested organizations are welcome to join (CETTAD) by contacting:

NJIT
University Heights
138 Warren Street
Newark, NJ 07102-1982
Mr. William Librizzi
Librizzi@admin.njit.edu
973-596-5846
Mr. Methi Wecharatana
Wecharatan@admin.njit.edu
973-596-2474

Mr. Sirithan P. Boriboon
Director General of PCD
MOSTE
404 Phahonyothin Center Bldg.
Phahonyothin Road
Phayathai, Bangkok 10400 Thailand
Sirithan.p@pcd.go.th
(662) 619-2316

## Conclusion

If the challenge for our future is how to continue to protect our environment with the least impact on economic development and growth, one answer will be in the development of new technologies. That is why the Agreement and the CETTAD are so important. They will establish partnerships and programs to stimulate international trade between both countries and to create a forum for the exchange of environmental and energy technologies. Collaborative agreements such as these are designed to help the participating countries become cleaner, safer, and economically productive.

## References

ASCE, 1996. "Creating the 21st Century through Innovation," American Society of Civil Engineers, Civil Engineering Research Foundation Report #96-5016.E. 1996.

EPA, 1997. EPA Strategic Plan. U.S. Environmental Protection Agency, Washington, DC 20460. EPA/190-R-97-002. September 1997.

MOA, 1997. Memorandum of Agreement on Environmental Technology Transfer and Development between The Ministry of Science, Technology and Environment, Kingdom of Thailand and the State of New Jersey, United States of America. May 29, 1997.

NCPWI, 1988. "Fragile Foundations: A Report on America's Public Works". National Council on Public Works Improvement, February 1988.

## Development of an Electronic Knowledge Base to Facilitate Industrial Pollution Prevention Efforts in Egypt

## Jim Goggin[1], Salwa Wahba[2], Vahid Alavian[3], and Gilbert Richard[4]

**Abstract**

This paper describes a knowledge base developed through the Environmental Pollution Prevention Project (EP3) in Egypt. The project is a component of the industrial pollution prevention activities funded by USAID in Egypt. The EP3-Egypt Knowledge Base integrates project data, information, and documents with evaluation and display tools into a single software package that is easy to use and distribute. The knowledge base contains documentation of EP3-Egypt pollution prevention activities and products, analysis capabilities, and an extensive pollution prevention training and educational component. The system will facilitate the widespread distribution of the knowledge gained and lessons learned from the EP3-Egypt project to Egyptian industries and other interested parties.

**Introduction**

The Environmental Pollution Prevention Project (EP3), launched in 1993, is a five-year program sponsored by the United States Agency for International Development (USAID) to address urban and industrial pollution and environmental quality in developing countries. The goal of EP3 is to promote the adoption of cleaner production techniques in industry. To achieve this goal and to ensure the sustainability of the program's pollution prevention initiatives, EP3 resources are dedicated to effect changes in industry practices and environmental policy by building capacity, developing partnerships and networks, and encouraging information-sharing.

The EP3-Egypt Project has been involved in a broad range of activities for pollution prevention and environmental awareness in various industrial sectors in

---

[1] EP3-Egypt Project Officer, Egypt Mission, USAID, Cairo, Egypt
[2] EP3-Egypt Project Officer, Egypt Mission, USAID, Cairo, Egypt
[3] Vice President, Rankin International, Inc., 7801 Ember Crest Tr., Knoxville, TN 37938, USA
[4] Principal, Hagler Bailly Services, Inc., 1530 Wilson Blvd., Arlington, VA 22209, USA

Egypt. Over the course of the project, pollution prevention (P2) screenings were conducted for over 600 industries, while detailed Pollution Prevention Diagnostics Analysis (PPDA) reports were generated for some 60 industries. Additionally, case studies, newsletters, training and promotion materials, and workshops were developed and implemented. To ensure that the technical assistance delivered and the knowledge gained in the EP3-Egypt project were documented and made readily available to the industrial community, an electronic knowledge base was developed.

## Objectives:

The purpose of the EP3-Egypt Knowledge Base is to integrate project data, information, and documents with appropriate evaluation and display tools into a single software package that is easy to use and distribute. Specific objectives of the knowledge base are to:

- Provide a mechanism for systematically documenting and retrieving major products and reports resulting from EP3-Egypt activities
- Support training and education on general, technical, and financial aspects of pollution prevention
- Facilitate widespread distribution of knowledge gained and lessons learned from the EP3-Egypt project to Egyptian industries and other interested parties
- Provide a means for interfacing with other databases and the worldwide Web

The EP3-Egypt Knowledge Base will also be available for use by follow-on USAID environmental projects, as well as for further analysis, training, and application of pollution prevention options in Egypt and other countries.

## Users and User Needs

The knowledge base was developed based on an assessment of the primary users of the system and the types of information that would be most helpful to them in addressing pollution prevention issues in Egypt. The primary users of the EP3-Egypt Knowledge Base are expected to be decision-makers and technical staff at:

- Energy and environmental conservation agencies in Egypt
- Industrial plants
- Egyptian Environmental Affairs Agency
- USAID-Egypt Mission and other missions
- Follow-on projects
- Private pollution prevention consultants
- Universities, NGO's and educational institutions

The types of questions these users need to be able to address include:

- What range of options is available for preventing pollution from a specific type of industry?
- What are the cost, effectiveness, and pay back of these options?
- Are there other plants or industries that have used this option and what can be learned from their experience? Are case studies available?
- What is the estimated reduction in pollution that can be expected from a given industry, sector, and/or region with specific pollution prevention options in place?
- What other materials or references are available on pollution prevention options?
- What environmental laws and regulations govern pollution prevention and control?
- What is Law 4 On Environmental Protection and what are its implications?
- What lessons have been learned from the EP3-Egypt project?

**Major Components of the EP3-Egypt Knowledge Base**

To support the needs of the sponsors and users of EP3-Egypt Knowledge Base, the system contains two primary components: a documentation and a training/education component. The **documentation component** includes major products and documents resulting from EP3 activities, such as industry profiles for some 2,600 industries in Egypt, screening and pre-assessment documents, detailed facility assessments outlining pollution prevention options, case studies of successful pollution prevention measures, and newsletters and promotion materials. The documentation component also allows for the analysis of data related to recommended pollution prevention options, as well as the generation of graphics and tables summarizing project results.

The **training/education component** includes general, technical, and financial information on pollution prevention, including general resource and contact information, descriptions of generic options for various industrial sectors, and information on financing and implementation costs. Additionally, the training/education component includes information regarding standards, guidelines, and laws pertaining to environmental protection in Egypt.

**Organization and Major Features of the EP3-Egypt Knowledge Base**

Access to and analysis of the information contained within the EP3-Egypt Knowledge Base are supported by:

- A relational database
- A document repository
- A suite of analytical tools
- Country specific information and data

The types of information and capabilities associated with each of these features are presented in Table 1. Microsoft Access is used as the commercial relational database for the EP3 Knowledge Base. The knowledge base is setup to accept/receive data and information from a variety of sources through a centralized quality control and distribution entity. Standard forms are also available to facilitate the entry of new or additional data. The relational database, in conjunction with the suite of analytical tools, allows for the analysis of data and generation of summary statistics and reports.

Table 1. Major Features and Capabilities of the EP3-Egypt Knowledge Base

| Feature | Capabilities and Types of Information |
|---|---|
| Relational Database | <ul><li>Industrial facility general information, data, contacts, etc.</li><li>Data on resources and utilities usage by industry</li><li>Data on industry pollution (type and numbers)</li><li>Data on industry pollution prevention options (impact, cost/benefit, implementation status)</li><li>Data on industrial pollutants</li></ul> |
| Document Repository | <ul><li>Pollution Prevention Diagnostics (PPDA) Reports</li><li>Follow-up documents on pollution prevention options implemented</li><li>Case studies related to pollution prevention in various sectors</li><li>Manuals, reports, and related materials</li><li>Training, workshop, and course materials</li><li>Pollution prevention promotion materials</li><li>Newsletters</li><li>Photo album</li><li>Pollution prevention training, information sources, web sites, contacts, etc.</li><li>Pollution prevention selected references (general and sector-specific)</li></ul> |
| Analytical Tools | <ul><li>Summary and statistical analysis</li><li>Performance and impact analysis</li><li>Financial and benefit/cost analysis</li><li>Results visualization and graphics</li></ul> |
| Country-Specific Information and Data | <ul><li>Environmental Laws, Regulations, Decrees, and Guidelines</li><li>Applicable Government Policies towards Pollution Prevention</li><li>Industrial planning and development</li><li>Industrial statistics</li></ul> |

Reports and materials within the document repository are in portable document format (.pdf) and are viewed through an Adobe Acrobat file reader. A global document search utility facilitates finding information and data. Larger documents are also indexed for ease of use and viewing.

The version of the Knowledge Base being distributed in 1998 is on CD ROM and includes automatic installation software and user's manual. The EP3-Egypt Knowledge Base has linkage capability to a selected number of pollution prevention related databases currently available in Egypt. The system also allows for direct connection to selected pollution prevention sites around the world via a Web browser. Plans are being made to make the product fully Internet enabled.

Sample Screens from the EP3-Egypt Knowledge Base are given in Figures 1-4.

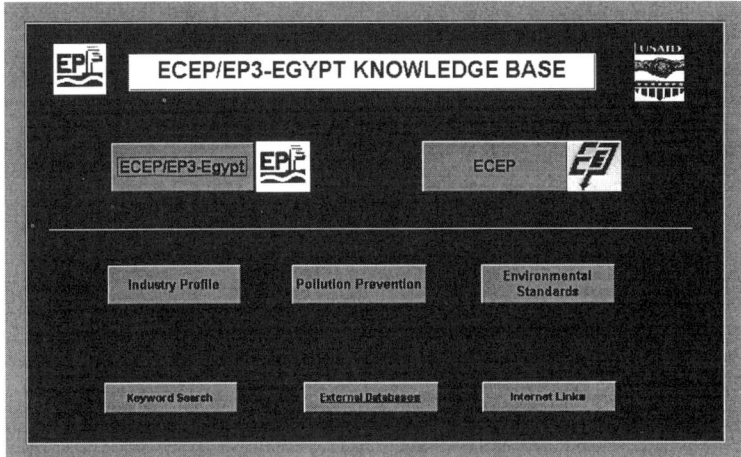

Figure 1. A Knowledge Base Introductory Menu

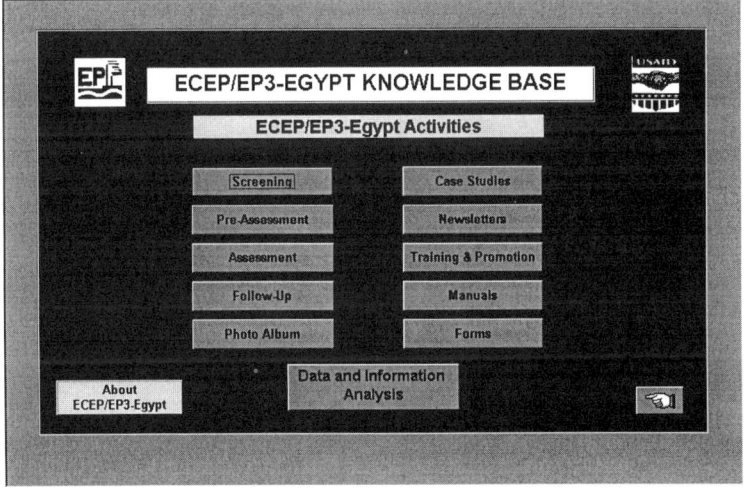

Figure 2. Knowledge Base EP3 Activities Menu

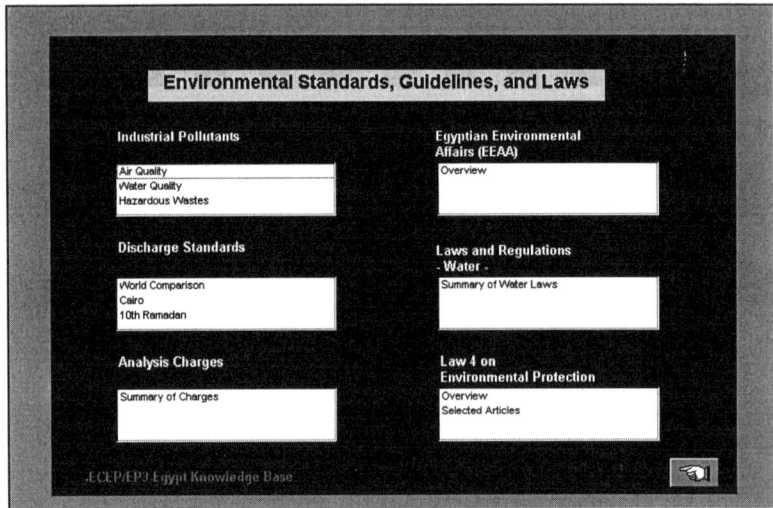

Figure 3. Knowledge Base Menu for Environmental Standards in Egypt

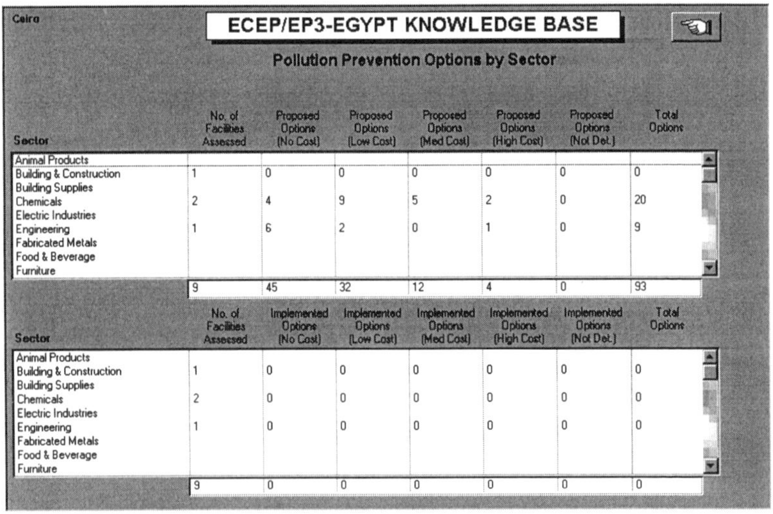

Figure 4. Knowledge Base P2 Options Analysis Sample Output

Benefiting Environmental Bottom Line
Through Pollution Prevention
in Ecuador, South America

Mario Salazar[1]

Abstract

How do you implement environmental protection where there are no regulations to force industry's hand? You show key sectors that prevention pays. The United States Agency for International Development/Global Environment Center has implemented the Environmental Pollution Prevention Project (EP3) world wide since 1993 and it has shown that prevention can be both environmentally and cost effective. The strategy of the project has been to try to influence the developing environmental culture by highlighting the benefits of prevention as the main principle in total environmental management. The scope of the project included collaborations with industry, government, academia, and professional and non-governmental organizations. One of the main principles of the project in Ecuador was to put emphasis on the economic benefits of pollution prevention in a slowly developing environmental regulatory venue. Direct interventions in industrial sites provided evidence that there were plenty of opportunities for savings with no or low investments.

Introduction and Background

In September 1994, EP3 started a pollution prevention program in Ecuador. EP3's counterpart in Ecuador was the "Corporación OIKOS", non-governmental organization (NGO). OIKOS is managed by Dr. Marco Encalada, a well known environmental educator and activist in Latin America.

Ecuador's most important export products are petroleum,

---

[1] Environmental Engineer, Office of Ground Water and Drinking Water (4606), Environmental Protection Agency, 401 M Street, SW, Wash., DC 20460

bananas and shrimp. Other industries that contribute significantly to the economy are textiles, flowers, auto assembly, leather tanning and foodstuffs. Ecuador imports a large variety of products from the US and other countries in the hemisphere. It is dependent on exports for maintaining a viable economy.

Like most developing countries, Ecuador has a slowly developing environmental protection program. While the Ecuadoreans have become more aware of the importance of a healthy environment, economic resources are lacking in the government and private industry for expectation to be fulfilled in this regard. The legislature has created an environmental "framework law", that provides the authority, upon delegation by the central government, for municipalities to implement a comprehensive environmental protection program. Each municipality is then charged with the task of developing local standards and regulations that meet or surpass central government standards. Unfortunately only the more progressive urban centers have requested and have been granted delegation of the environmental mandates. Included in this group of cities are Quito, Cuenca, Ambato and others. One situation that is unusual with respect to the environmental profile in Ecuador is that eco-tourism has been one of the staples of the economy. The Galapagos islands, the Amazonic jungle and the volcano-rich central area have attracted tourist for many years, creating in the population respect for the natural environment. Historical areas like the city of Quito, founded in 1534 and declared a United Nations Human legacy site, also attract many tourists.

While the idea of voluntary compliance is important in regulatory programs, it is imperative in countries lacking resources and expertise in the environmental disciplines. Enforcement budgets are very small and the political and economic power of some of the larger industries are forces that impede the command and control style of implementation of environmental regulations.

<u>Strategy of EP3 in Ecuador</u>

To respond to the conditions in Ecuador, the EP3 developed a three pronged strategy. This strategy included the necessity of working with industry, government and academia. This strategy is similar to the ones implemented by EP3 in other developing countries, and agrees fundamentally with the philosophy of the project. The program included a number of elements that would benefit one or more of the three sectors mentioned above. These elements are:

- Demonstration initiatives at several types of industrial plants;
- Dissemination of successes and techniques to large groups;
- Technical assistance to industrial groups and governments;
- Development of pollution prevention curricula at universities;
- Training in environmental management with emphasis on pollution prevention;
- Recognition of pollution prevention leaders/champions;
- Demonstration tours to the US;
- Cleaner production policy initiative with the government;
- Others as appropriate.

Approach with Industry

Recognizing that industry would have to participate as a partner in any environmental initiative, EP3 set out to develop proof of the effectiveness of pollution prevention. This was accomplished by developing a number of demonstration plants. The method that was used included:
- Analysis of the industrial profile for the country and specific urban centers;
- Selection of the more promising industrial sectors for pollution prevention;
- Screening of a representative number of plants in each sector;
- In plant pre-assessments of a reduced number of plants in each sector to determine potential pollution prevention opportunities;
- Agreement with the plant management for confidentiality and cooperation;
- In depth pollution prevention diagnostic assessment in 2 to 3 plants in each industrial sector;
- Implementation agreement with the plants;
- Implementation support;
- Evaluation of results and publication of case histories.

Analysis of the Ecuador industrial profile indicated that the following industrial sectors offered opportunities for pollution prevention:
- Textiles;
- Tanneries;
- Edible oil;
- Automobile assembly;
- Paper recyclers;
- Metal fabrication and preparation.

During the first two years of the project, over 100 screening visits were made, approximately 50 pre-assessments and 16 individual assessments were performed. Additionally, building on individual successes in the textile and tanning sectors, extensive training and technical assistance efforts were implemented with a large number of plants in these two sectors.

Screenings consisted of one or two hour visits to the targeted plants in an industrial sector. During these visits, the EP3 person gauged the interest of the plant management and potential pollution prevention opportunities. The objectives of the visits were to classify the sector with respect to pollution prevention opportunities and to determine the potential of the plant. Economic impact, duplicability, management interest and pollution prevention opportunities were some of the criteria investigated.

If the screening supported additional consideration of the sector and the individual plant, a pre-assessment visit was planned. This visit was originally done by the local EP3 counterpart assisted by an American expert. The objective was to assure that the plants selected would have the best pollution prevention opportunities and a suitable implementation chance. The pre-assessment visit would last from half a day to a couple of days and included obtaining significant amount of data when warranted. Pre-assessments were done by two engineers from OIKOS after one demonstration visit done in the company of an American expert.

The pollution prevention diagnostic assessment followed the pre-assessment in a very selected number of plants. The assessment would last approximately one week and most of the work would be done at the plant. Normally two US experts would travel to Ecuador to participate in two assessments in plants in the same or similar sectors. Because of the cost, there was a very strong incentive to recruit the best candidates to do the assessment. The assessments were also perceived as an excellent training opportunity, not only for the OIKOS engineers, but for the people at the plants and the participants of a seminar that was usually given once the assessments were done.

EP3 performed 16 in-depth pollution prevention diagnostic assessments in the first two years of operation in Ecuador. These assessments provided the bases for the initiative to access larger audiences. Successes in the individual assessments opened the door for industry sectors to request EP3 training for their technical personnel. The textile and tanners associations requested and were given training in pollution prevention techniques and economic analysis to justify improvements. The initiatives involving the textile and tanning sectors included a large number of plants. Approximately 50 persons representing 23 plants participated in the textiles effort. Three separate training seminars were given over a period of three months to technical personnel from the plants. These seminars encompassed basic information gathering, determining pollution prevention opportunities at the plants and economically justifying the implementation of the opportunities and implementation methods. During the time between seminars, the participants were encouraged to work at their plants on each of the topics covered in the preceding seminar. To assist them a number of consultants were retained. These consultants, with prior training on pollution prevention, visited the plants on a weekly basis to assist the participants in developing data, determining opportunities, justifying them and implementing them.

Seminars were also held after each of the individual diagnostic assessments to disseminate new techniques and other information dissemination activities took place through out the length of the project. Several leadership meetings were also held to recognize several of the plants that had labored to obtain excellent results from implementing pollution prevention opportunities.

<u>Approach with Government</u>

The efforts with government were mostly directed at municipalities. This is because the development and implementation of any environmental effort is the responsibility of municipalities in Ecuador. The strategy consisted in offering government officials the opportunity to participate in all educational activities. They were also consulted while in the process of selecting the target industrial sectors. Local governments were also informed and encouraged to request technical assistance on environmental matters from EP3.

EP3 provided technical assistance to municipalities in 7 different opportunities involving pollution prevention applicable to oil and gas production, water quality laboratories, air pollution sampling, risk assessment, estuary remediation, air pollution at high altitudes and integrated pesticide management. Most of the assistance was provided using EPA experts with no cost to the governments. Goals were accomplished in every case and some were highly successful. All these activities helped in creating a record with key officials in several municipal governments. The recognition of the program helped in establishing a pollution prevention paradigm in the government and helped in accomplishing the goal of making pollution prevention the key concept in environmental management in Ecuador.

Approach with Academia and Others

Introducing the concepts of pollution prevention to young and developing professionals is important in promoting the most effective environmental paradigm. Establishing relevant curricula in universities is also very important. The EP3 provided information and guest lectures in several academic venues. Another important accomplishment was the creation of an elective engineering course on pollution prevention at the Poly-Technical University in Quito. The pollution prevention class was first taught during the second quarter of 1997 and was an unqualified success. A guest lecturer from the US was included in the course for two weeks.

As indicated above, EP3 also worked with several industrial groups associations and other NGOs.

Results

As of this writing, a significant segment of the industrial plants are knowledgeable of pollution prevention and intend to implement opportunities that plant personnel or consultants have identified.

Technical assistance to several municipalities has helped in getting key officials to consider pollution prevention in their environmental protection plans. Excellent attendance to training seminars given to several plants in the textile and tanning sectors was in part the result of local environmental officials supporting these efforts and accepting them as part of the compliance plans for some of the plants. Frequent seminars and new pollution prevention curricula in universities will help to influence professionals and decision-makers into the prevention paradigm. At least six consulting firms in Ecuador are offering pollution prevention services. OIKOS has also joined and is participating in 13 clean technology discussion groups in Latin America.

## USAID's Pollution Prevention Program in Latin America

Gilbert S. Jackson and Morris S. Israel, Assoc. Member[1]

**Abstract**

Pollution prevention (P2) can provide a" win-win" alternative for resolving the seemingly dichotomous conflicts between economic growth and environmental protection in developing countries. The U.S. Agency for International Development (USAID) experience has shown that P2 programs implemented with the support and encouragement of the private and public sectors, including civil society, can improve economic competitiveness and efficiency, while minimizing negative impacts to the environment. This paper documents some of the successes and lessons learned from USAID's recent experiences with creating nascent pollution prevention, clean technology, and waste minimization programs in Latin America and the Caribbean (LAC).

**Introduction**

In most developing countries, economic growth is often pitted against environmental protection. Many argue that private sector participation and structural reforms must first strengthen economies, then address environmental issues. However, USAID's recent experience has shown that private sector participation can and should work closely with the adoption of environmental protection measures, such as pollution prevention, to improve economic competitiveness and efficiency. All too often in the process of incorporating private sector participation, governments, industrial managers,

---

[1]Senior Environmental Officer and Environmental Advisor, respectively, Bureau for Latin America and the Caribbean, U.S. Agency for International Development, Washington, D.C. 20523-5900.

private investors, and financial institutions underestimate the economic advantages of pollution prevention activities.

The economic rationale for pursuing a pollution prevention initiative is well voiced in a *New Partnerships in the Americas: The Spirit of Rio*: "Industries striving to make their practices sustainable agree that reducing waste or pollution at the source is more efficient and less costly than attempting to clean it up later." That is, investments made within production systems offer significant returns on investment because of reduced operating costs and increased productivity, whereas the traditional end-of-pipe (EOP) pollution control investments only add to costs and divert limited capital from productive uses. These are "sunk" costs.

**Status of Pollution Prevention in LAC**
All too often in developing countries pollution control equipment is begrudgingly used in many instances, is viewed by plant managers as less important than production processes, does not function properly after installation resulting in reduced operating efficiencies, and often requires higher maintenance costs than anticipated. In most developing countries with new environmental enforcement programs, independent monitoring of industrial effluents and emissions is so rare that the advantages offered by EOP systems to meet regulations is minimal. It is a classic "lose-lose" situation: capital and operating expenses are wasted, with little or no improvement in pollution control.

Examination of new environmental laws in Latin America reveals little attention to pollution prevention and clean technologies. In fact, increasingly stringent environmental regulations being adopted in may LAC countries will add significant compliance costs to industry when conventional EOP pollution control techniques are required. This is occurring despite the Clinton Administration's policy shift toward pollution prevention, efforts by UNEP and UNIDO's clean production program (including clean technology centers in LAC), and programs of many NGOs to spur interest in clean technologies. Concurrently, government subsidies to industry for water, energy, and sometimes for minerals are being cut back. Market pricing, together with the need to comply with environmental regulations, imply that there are increasingly avoidable costs if clean technologies are used to modernize production processes in existing industries.

While many LAC governments are receiving conflicting advice from donors, it is our experience that the economic benefits of a pollution prevention strategy should be forwarded. The alternative "win-win" strategy promoted by USAID is to use P2 and clean technologies within production systems where they are given higher priority because they are integral to production, and where they increase profitability, competitiveness, and worker safety, as well as reduce or eliminate non-production waste outputs to the environment. Several USAID programs, including the Environmental Initiative for the Americas (EIA), the Hemispheric Free Trade Expansion (HFTE) project, and the Environmental Pollution Prevention Project (EP3) promote awareness of the economic and environmental advantages of P2.

**USAID Pollution Prevention Program in LAC**
At the December 1994 Summit of the Americas, the United States and its Latin American neighbors agreed to form several partnerships to "guarantee sustainable development and conserve our natural environment for future generations." Among these was the Partnership for Pollution Prevention. In response, the U.S. government, through USAID, launched the EIA, the goal of which was to provide a catalyst for USAID Missions to expand their environmental programs to include issues of pollution, urbanization, industrialization, energy, and related topics. The LAC Bureau revised its regional environmental strategy to reflect this shift in program emphasis by including industrial and municipal pollution prevention as one of its three environmental thrusts (the others being sustainable energy and natural resource/biodiversity conservation). Below is a brief description of USAID pollution prevention programs in LAC.

*Environmental Initiative for the Americas (EIA)*
The EIA was a two-year $22.6 million program designed to provide seed money as an incentive for field missions to undertake activities in areas in which the LAC Bureau historically had not invested many resources, such as pollution prevention. The idea was for the initial funds to create an awareness and generate a demand for P2 projects, continued funding for which would be leveraged from the missions themselves and other sources (e.g., governments, multi-lateral banks). Approximately $10.7 million were dedicated to 28 urban-industrial P2 programs. Sample activities include:

**Bolivia**, $0.4 million to address industrial and urban pollution prevention.

**Chile**, $0.25 million to assist small industries develop cleaner production strategies.

**Ecuador**, $0.45 million for industrial pollution prevention audits and outreach seminars.

**Ecuador**, $0.25 million address solid waste problems and best practices in urban environmental management.

**Guatemala**, $0.55 million for improving solid waste management in Central America.

**Haiti**, $0.4 million to address disposal of household waste in impoverished urban community.

**Jamaica**, $1.3 million for self-financing urban pollution mitigation and coastal zone management.

**LAC Bureau**, $0.8 million to promote regional free trade and harmonization of environmental regulation.

**Mexico**, $0.425 million for urban and industrial activities in U.S.-border areas.

**Paraguay**, $0.4 million to assess industrial effluent discharges and for audits and training.

**Peru**, $0.35 million to address industrial pollutants entering the Paracus National Marine Reserve.

Through the EIA, USAID sought to (1) educate managers, government officials, and investors about their strategic options in pollution prevention technologies; (2) perform waste reduction audits to assess exactly what kind of no-cost, low-cost, or medium-cost clean technologies can easily and cost-effectively be used to modernize existing LAC private sector industries; and (3) identify commercial sources of the best available clean technologies and then facilitate purchase, installation, and follow-up maintenance.

The EIA program was a moderate success in generating interest. However, a two-year program is insufficient to generate sustainable solutions. Follow-on activities are essential. Below are some highlights of the program:

- Funding for follow-on activities have been leveraged at a rate 2:1, with 31 funding sources having expressed support.
- 54 entities, including numerous NGOs, were involved in implementation.
- LAC Bureau increased funding for "brown" activities to $8.83 million in 1997 up from $1.0 million in 1995, despite a total environmental budget decrease from $52.0 million in 1995 to $46.34 million in 1997.
- Several Missions modified the focus of their environmental strategy to incorporate P2 activities.

*Hemispheric Free Trade Expansion Project (HFTE)*
The aim of HFTE is to generate and expand environmentally-sustainable free trade in the LAC region by affecting changes to the legal/policy framework; presenting appropriate technical solutions; and "greening" financial investments. HFTE assists local industries in understanding and complying with existing or impending environmental regulations in their country, as well as with

international environmental (ISO 14000) and food safety standards, through improved environmental management practices and processes. One activity under the HFTE project focuses on industrial water use efficiency.

Industrial growth, privatization, and modernization in Latin America has dramatically increased the demand for clean water to support sensitive industrial processes. Latin America has a plentiful water supply. However, inefficient use has had significant impact on many industries. Surface waters are often contaminated, requiring significant treatment prior to use in sensitive industrial processes. As an alternative, many industries obtain process water from aquifers near their facilities. These ground water resources are finite, increasingly costly to pump, and may themselves be contaminated eventually.

Until recently water conservation and water quality management have been low priorities in most industrial facilities. However, there is a growing need for and interest in industrial water conservation being fueled by increasing demand for clean water, increased costs of producing or purchasing water, and an increasing focus on water rights which may restrict future access to clean water supplies.

Water and energy conservation are integral parts of the P2/clean technology approach. The combination of increasing water prices and requirements for wastewater treatment are particularly effective in offering sufficient payback to justify clean technologies. Data suggest that water conservation and protection programs can have a dramatic effect on water use efficiency. In the U.S., the introduction of appropriate waste conservation and recycling technologies in the 1970s reduced water use intensity of some industrial processes by an order of magnitude or more. Recent industrial water conservation efforts in LAC have been successful in reducing water use in some textile plants by up to 25 percent.

USAID will be conducting a series of activities to introduce and strengthen the capability of Andean NGOs to deliver technical assistance to local industries for water conservation projects and to meet the demand for identification and implementation of water reduction alternatives. These activities are designed to introduce process-specific water conservation and use improvement technologies for two water-intensive industries in the Andes region: food processing and tanneries. The approach emphasizes sustainability by empowering local NGOs to act as water conservation promoters and technical assistance providers.

*Environmental Pollution Prevention Project (EP3)*
The EP3 project was designed to introduce the concepts of pollution prevention in LAC, initially targeting eight countries: Mexico, Ecuador, Jamaica, El Salvador, Peru, Bolivia, Chile, and Paraguay. Companies are becoming increasingly aware that if they want access to new international markets, or to maintain good relations with enterprises and countries with which they already conduct business, they will have to comply with international environmental

quality and product quality standards. Reduced costs will increase domestic as well as international competitiveness and can serve as a strong incentive for manufacturers to adopt environmental standards, such as ISO 14000, or food safety standards, which would guarantee product quality and safety and incorporate environmental quality considerations.

LAC food and beverage processors, for example, are upgrading their operations, or in many cases, constructing entirely new plants and facilities to maintain market share. By investing in the best technology available, they are readying themselves to compete not only against foreign products, but in foreign markets. Most cost savings and discharge reducing opportunities in these industries relate to increasing the efficiency of converting agricultural inputs to finished goods, water conservation, and recycling. EP3 experience has identified good opportunities for reuse of waste products and high potential cost savings. Other HFTE activities include the following:

- Provided support for: (a) development of model frameworks for mining policy; (b) clean production workshops and best management practices manual; and (c) analysis of market-based instruments and conduct of regional meeting on greening of private investment in the mining sector.

- Established a program for the provision of trade and business grants through the National Association of State Development Agencies to promote emerging markets in the LAC region for U.S. industrial P2, energy efficiency, and wastewater treatment technologies (17 $10,000 grants awarded; match of $367,234 in cash or in kind; reported sales by 7 companies of $540,032; projected sales in 1998 of $4.9 million)

- Conducted trade and environment workshops in Bolivia and Ecuador reviewing environmental trends in international and regional trade, and the role of clean technologies in improving industrial competitiveness and meeting international standards (ISO 14000).

- Supporting a review of industrial clean technologies internet data bases.

**Conclusions**

If we are serious about helping developing countries improve both their economic and environmental performance to accomplish sustainable growth and to support the hemispheric trend toward democracy and trade liberalization, then we need to develop fast-track technology transfer programs to deliver commercially available clean technologies to LAC countries. The emphasis must be on low- and medium-cost technologies which can be retro-fitted immediately into existing facilities and offer immediate payback. But these technology transfer programs must be accompanied by changes to the legal and policy frameworks as well as to the manner in which the activities are financed.

The Role of the Seattle District of the Army Corps of Engineers in the International Joint Commission

Mark W. Killgore, P.E., M. ASCE[1]

**Abstract**

This paper discusses the participation of the Seattle District of the Army Corps of Engineers in trans-boundary water management with its Canadian counterparts. The District Commander plays a pivotal role as a representative of the U.S. Section of the International Joint Commission. In addition, the District is involved in other water resources operations with trans-boundary implications, such as operations at Ross Dam on the Skagit River under flood conditions.

**Purpose**

The objective of this paper includes summarizing the participation of the Seattle District in the Pacific Northwest on two Boards of Control, including International Osoyoos Lake Board of Control and International Kootenay Lake Board of Control.

**Background**

The International Joint Commission (IJC) was formed by the 1909 Boundary Waters Treaty between the United States and Great Britain on behalf of the Dominion of Canada. As described in an IJC publication (IJC, 1990), the Treaty "provides the principles and mechanisms to help prevent and resolve disputes, primarily those concerning water quantity and water quality along the boundary between Canada and the United States." IJC responsibilities include:

- Issuance of Orders of Approval in response to Applications for the use, obstruction, or diversion of water that flow along, and in certain cases across the boundary, if such uses affect the natural water levels or flows on the other side.
- Investigation of specific issues or monitoring of situations when requested by Governments.
- Provision for Governments to refer matters to the ICJ for binding decision.

---

[1] Seattle District, U.S. Army Corps of Engineers, 4735 E. Marginal Way S., Seattle, WA 98134

The IJC consists of more than 20 Boards of Control of which the Seattle District participates in two. The first is the International Osoyoos Lake Board of Control which was established by IJC Order on September 12, 1946 (IJC, 1997). The second, the International Kootenay Lake Board of Control was established to comply with the IJC Order dated November 11, 1938.

**International Osoyoos Lake**

Osoyoos Lake is located on the Washington-British Columbia border east of the Cascade Mountains in Okanogan County, Washington. The lake has a maximum area of 8.95 square miles at elevation 911 fmsl and is operated over a normal range of 909.0 to 911.5 fmsl. Additional storage between elevation 911.5 and 913 fmsl may be used to provide additional drought storage between April 1 and October 31. Lake levels were set by IJC Orders of Approval dated December 9, 1982 and October 17, 1985. Compliance with the Orders of Approval are measured at USGS Gage "Osoyoos Lake near Oroville".

Lake levels may also rise above elevation 911.5 fmsl when backwater effects from high flows in the Similkameen River are present and/or when high flows occur in the Okanogan River. Zosel Dam controls elevations in Lake Osoyoos. The dam was rebuilt in the 1980s' by the State of Washington and Province of British Columbia. Originally the dam was a timber-crib dam used by Mr. Zosel for his sawmill operations and built in 1927.

The International Osoyoos Board of Control (United States Section) is chaired (acting) by Mr. Gerald Parker, U.S. Geological Survey, retired. The Seattle District is represented as a member by the District Engineer, Colonel James M. Rigsby. The third member of the U.S. Section is Mr. Kris Kauffman, Civil Engineer. Canada has three section members including chair, Dr. Chris Pharo, manager of the Pacific Wildlife Research Center of the Environmental Conservation Branch of Environment Canada. Mr. Pradeep Khare, executive director of the Water Management Division of the Ministry of the Environment, Lands and Parks and Mr. Robin McNeil, head of the Water Resource Inventory Forecast Center of the Ministry of Environment, Lands and Parks.

By 1980, the control structure for the lake, Zosel Dam, had significantly deteriorated and required replacement. The IJC approved its replacement in 1982 and issued a Supplementary Order in 1985 regarding a new location for the dam and ensuring a flow capacity of at least 2,500 cfs through the control structure when Osoyoos Lake is at 913 fmsl. The dam represents a model of international cooperation with both funding and design review provided by the Province of British Columbia and the State of Washington. The new dam was operational in 1987 and a new Board of Control was created to supervise the rule curve for the Lake and determine the need for drought declarations.

The Oroville and Tonasket Irrigation District currently operate the Dam. The owner of the new dam is the Washington State Department of Ecology.

**Kootenay Lake**

The Kootenay River, a Columbia River tributary, originates in the Canadian Rocky Mountains and first crosses the U.S.-Canadian border in northwestern Montana as Lake Koocanusa and then turns north through northern Idaho and returns to British Columbia where it turns into Kootenay Lake. Kootenay Lake is 62 miles long and varies from two to three miles in width. The outlet is located approximately at the halfway point along the western shore. The confluence with the Columbia River is downstream of Nelson, British Columbia.

The origin of International Kootenay Lake Board of Control dates back to an Order of Approval granted by the IJC on November 11, 1938. It gave the West Kootenay Power and Light Company authorization to operate Cora Linn Dam at Granite, British Columbia and store six feet of water in Kootenay Lake. The Order also provides for the excavation of the outlet of the Lake at Grohman Narrows. The dam is located 16 miles upstream from the confluence of the Kootenay River with the Columbia River.

The Board has two primary functions including the operation of Cora Linn Dam at Granite, British Columbia which stores water in Kootenay Lake in accordance with the IJC Order and assure that Duck Lake, a waterfowl nesting area and flood control reservoir, is also operated in accordance with the IJC Order.

The 1938 Order directs that Kootenay Lake be lowered below elevation 1,739.32 fmsl (as measured at Queen's Bay) on April 1 in order to receive the spring runoff. The lake is drawn down to elevation 1,743.32 fmsl (as measured at Nelson) at the end of summer where it is held until August 31. The maximum elevation from September 1 till January 7 is 1,745.32 fmsl as measured at Queen's Bay gauge.

Several IJC Orders (October 12, 1950; April 3, 1956 and March 31, 1970) permitted the construction and improvement of dikes adjacent to the main Kootenay River Channel in order to reclaim flooded lands and develop a waterfowl habitat area near the Idaho-British Columbia Border. Participants included the Creston Reclamation Company, the Duck Lake Diking District and the Creston Valley Wildlife Management Authority.

The Seattle District Engineer chairs the U.S. Section of the Board. Dr. Chris Pharo of Environment Canada chairs the Canadian Section. The two other members include Pradeep Khare, Director Water Management, Water Management Branch, BC Ministry of Environment, Lands and Parks and Derrill Cowing, District Chief, Water Resources Division, U.S. Geological Survey in Boise Idaho. In addition, the board includes a Secretary, Larry Adamache of the Science Division, Environment Canada. The Chief of the Hydrology and Hydraulics Section of the Seattle District, Wayne Wagner serves as Technical Advisor.

Both the Kootenay Board and Osoyoos Board hold an annual public meeting whereby citizens can provide input to the IJC. Each Board of Control writes an annual report to the International Joint Commission. The Seattle District assists in the preparation of the reports.

**Other Duties of the District**

Each year, in conjunction with the Public Meeting and the Annual Meeting of Osoyoos Lake Board of Control, the Seattle District arranges the logistics for meeting places and tours of project facilities. The agenda for the annual meeting is prepared and include such issues as reports from cooperating agencies such as the Washington State Department of Ecology, real-time data management and funding.

**Conclusion**

This past year, the American Society of Civil Engineers (ASCE) sponsored sessions on trans-boundary water resources management at the Ninth Congress International Water Resources Association in Montreal, Quebec. Geoffrey Thornburn of the Canadian Section of IJC presented a thought provoking paper on new directions for the trans-boundary agency and the role of sustainable development. A second paper, US-Canada Cooperation to Protect and Sustain Boundary Waters, by Pete Christich was also presented. The Congress also featured an IJC Commissioner, Dr. Pierre Bourque, at the opening plenary session. Now nearly celebrating its 90th anniversary, the IJC continues to shape trans-boundary water management and influence its development on a global scale. The activities of the Seattle District will continue to respond to changes in priorities and new challenges in managing trans-boundary watersheds.

The IJC has enunciated its vision of the future in The IJC and the 21$^{st}$ Century. Some of the challenges include climate change and its effect on trans-boundary watersheds, loss of habitat and biological diversity, population growth and urbanization and the continuing effects of pollution on both air and water quality. In addition, institutional changes such as downsizing in government and decreasing budgets for environmental monitoring, will make the work of the IJC even more challenging. Five proposals were made by the IJC to the Governments of Canada and the United States and include:

1. The establishment of ecosystem-based international watershed boards;
2. The initiation of broad studies on trans-boundary water quantity and quality, air quality and data acquisition;
3. Review existing IJC orders governing levels and flows to determine if amendments are required based on changed conditions;

4. A Reference on decommissioning of nuclear reactors, interaction of toxic chemicals and radiation the ecosystem and the impacts of use of low-sulfur coal in electric power generation on dispersion of nuclear materials (such coal is higher in thorium);
5. Biennial reports on the state of the trans-boundary environment.

Appendix:   References

Cristich, Pete, *US-Canada Cooperation to Protect and Sustain Boundary Waters*, paper presented at the Ninth World Water Congress, IWRA, Montreal, Quebec, September 1997.

International Joint Commission, Home Page under http://www.ijc.org, 1998.

International Joint Commission, *The International Joint Commission and the Boundary Waters Treaty, Canada~United States*, 1990.

International Joint Commission, *The International Joint Commission and the 21st Century*, 1997.

International Kootenay Lake Board of Control, *Annual Report to the International Joint Commission from the International Kootenay Lake Board of Control for the Calendar Year 1996*, July 11, 1997.

International Osoyoos Lake Board of Control, *Annual Report to the International Joint Commission from the International Osoyoos Lake Board of Control for the Calendar Year 1996*, March 31, 1997.

Killgore, Mark W. and David J. Eaton, *NAFTA Handbook for Water Resources Managers and Engineers*, 1995.

U.S. Army Corps of Engineers, Seattle District, Home page under http://www.nws.usace.army.mil, under other links "International Joint Commission Boards of Control - Kootenay Lake Board of Control, Osoyoos Lake Board of Control"

# The Storm Over Sturm:
# A Case Study in Multijurisdictional Stormwater Management
# Lake County, Illinois

Fred Royal[1]
ASCE Associate Member

## ABSTRACT

A significant number of Lake County, IL residential subdivisions developed since the early 1950's have experienced chronic flood damages due in part to their location in flood hazard areas. Since 1982, there have been 3 federal and 2 state disaster declarations in Lake County due to flooding. The Lake County Stormwater Management Commission (SMC) has performed a multi-phased project on one of these developments designed to minimize future flood damages. The extensive search for an acceptable flood mitigation alternative required multiple hydrologic and hydraulic analysis (H&H), multiple cost/benefit analyses, wetland design, a stream assessment and action plan, property and real estate acquisition, public meetings, permitting and federal and state grant applications. The adopted alternative design includes a combination of measures: modification of detention basins and wetlands, drainage improvements, stream maintenance and acquisition and demolition of flood damaged residential structures. Other factors included in selecting the alternatives were budgetary constraints, political and inter-jurisdictional conflicts of interest, real estate values and homeowner attitudes.

## INTRODUCTION

Over the past 25 years, the residents of Sturm Subdivision in unincorporated Lake County have averaged one evacuation per year due to flooding. Of the 18 homes, 11 are repetitively flooded and the vehicular access is blocked for the remaining 7. The subdivision sits in a natural depressional area with more than 180 acres tributary. Sturm is bordered by fully developed municipalities and hamlets, a state highway and light industrial/commercial parcels. The subdivision has become one of Lake County's more well-known flooding problems with Chicago area media attention after every 2-yr. frequency rain event.

---

[1] Watershed Engineer, Lake County Stormwater Management Commission, 333-B Peterson Rd., Libertyville, Illinois 60048

Solutions to reduce flooding in Sturm were not immediately apparent. The area is not a part of the FEMA FIS study and is not mapped as Regulatory Floodplain. Due to the severity of the flooding, the local elected officials requested that SMC study the problem. It soon became apparent that many factors compounded the flooding problem including: homes with basements and low entry elevations, increased upstream development, poor soil permeability (hydric), poor internal drainage, the lack of an emergency overflow route and resistance from downstream communities to improve drainage conveyance.

For this mitigation project to be implemented, the regulatory requirements and concerns of 9 agencies and/or jurisdictions had to be addressed. The main objective was to achieve on-site flood damage reduction with no adverse downstream or upstream impacts. The major hurdles to achieving this were permitting, grant requirements and agreements with the surrounding municipalities. The development of the localized flood reduction mitigation plan took over four years of study, data collection, engineering design, H&H modeling, permitting, grant applications and negotiated agreements. This resulting plan consists of three (3) phases. The initial phase (Phase I) was designed to improve the internal drainage and create an emergency overflow route. Phase II design will increase the internal detention capacity and conveyance routes. Phase III will include implementation of up and downstream actions recommended as a result of the Phase I study.

Construction and reports, including the installation of storm sewers and swales, modifications to existing storm structures, wetland design and an upstream and downstream assessment and action plan began in May, 1997. Funding sources include a federal Community Development Block Grant (CDBG), SMC grants, local township funding and an unprecedented grant from the FEMA Hazard Mitigation Grant Program (HMGP). This is a voluntary buy-out program for residents who frequently experience flooding. Following a purchase agreement and demolition, the vacant sites will be designed as open space to provide additional detention and stormwater conveyance.

With the completion of Phase I, the level of flood protection was increased from the 2-yr. to the 25-yr. frequency. When the final phase is completed, the remaining residents of this subdivision will no longer experience the levels or frequency of disruption and economic loss experienced prior to this flood mitigation project. The only solution for a higher level of protection for recurring flood damages in the subdivision will be through retrofitting all remaining at-risk structures. Current methods of retrofitting residential structures are: elevation, wet floodproofing, dry floodproofing, floodwalls and levees, relocation and acquisition/demolition. These flood mitigation methods will be pursued to the fullest extent possible for the remaining residents in Sturm Subdivision, as well as in the numerous other subdivisions in Lake County. With a vision of obtainable goals, sound planning and constant communication to create a clear understanding of the issues, a dynamic stormwater management and mitigation project can be successfully implemented.

Figure 1.0
**Sturm Subdivision Flood Mitigation Management Approach**

| Context
Technical/Social/Political/Economic environments |
|---|

| Problem Recognition
Review historical and current circumstances resulting in flooding |
|---|

| Problem Definition
Extent and frequency of flooding, damage assessment,
H&H analysis, upstream/downstream assessment |
|---|

| Generation of Alternative Strategies
Internal drainage, routing, basin expansion, retrofits, HMGP |
|---|

| Analysis/Evaluation/Optimization/Design
Maximum damage reduction by combining alternatives,
permits and agreements, re-design |
|---|

| Implementation
Phasing construction, land purchase, continuous improvement
through HMGP and off-site assessments |
|---|

The framework used in developing the mitigation project considered all of the technical, economic, social and political environmental factors. Next, a progression of activities were incorporated to reach an obtainable solution, based on those factors (Fig. 1.0).

## Context

The technical, social, political and economic environment surrounding Sturm made it possible to proceed with the study and the subsequently-funded project. The chronic flooding at Sturm became a topic of discussion, concern and interest with local elected officials, Chicago area news media, frustrated residents and local government agencies. Sturm was ready for action. Because the subdivision is located in an unincorporated area, the Township Highway Department is responsible for its' public works infrastructure needs, limited to road surface functions and right-of-way drainage services. They are on a community water well system and individual septic systems. The census has categorized the community as low to moderate income, allowing for CDBG eligibility. This grant funding was applied for and awarded as a portion of the total project budget (Fig. 1.1). With the political, budgetary and technical backing in place, the flood mitigation plan was underway.

Figure 1.1

| Total Sturm | Budget |
|---|---|
| Source | Budget |
| SMC FY '97 | 32,500 |
| SMC FY '98 | 40,400 |
| TWSP | 131,194 |
| CDBG | 153,000 |
| DNR/OWR | tbd |
| FEMA | 228,375 |
| TOTAL | 585,469 |

SMC In-kind services not included.

## Problem Recognition

The extent of the overall problem at Sturm was not immediately known. The repetitive flooding in the subdivision had not only caused major disruptions in residents' lives along with economic and property loss, it had also caused septic failure and increasing structural damage to foundation walls. Sturm is located on a 30 acre depressional area consisting of hydric soils. The homes have full basements with window wells and low entry elevations 3-4 feet below the emergency overflow elevation. There exists a 4 acre pond, which routinely overtops causing the subdivision to flood. Other problems resulted from adjacent areas of the watershed. The upstream community, 1-2 acre single family lots and on-line water features, was nearing a fully built-out condition. As is the typical case in Lake County, the development was built on previously agricultural/open space land with a natural meandering stream. In this case, however, there was no drain tile system. The subsequent grading and filling of depressional storage areas and channeling of the stream generally caused an increase in peak flows into the subdivision. The residents here experience yard flooding when the on-line ponds overtop. These conditions made it difficult to easily identify a retrofit opportunity for peak flow reduction. The community downstream from Sturm is similarly "built-out" with larger lots. The stream corridor is severely eroded and several road cross-culverts are in need of replacement. Additionally, several homes in this area have low entry elevations which warrant close attention to proposed BFE calculations. Therefore, to propose an increase of flows out of Sturm into the downstream area was not acceptable without a detailed H&H study.

**Problem Definition**

With the information gathered during the problem recognition phase, a clear and accurate definition of the problem extent was developed. SMC developed computer models (HEC-1 & HEC-2) to provide for the existing stage/storage/discharge conditions throughout the study area. With a calibrated HEC-1 model, the on-site and downstream impacts could be analyzed with a variety of hydraulic conditions. The models were used to generate the mitigation alternatives. Other information was obtained to further define the extent of the problem including: resident questionnaires, further topographic and as-built survey data, a stream assessment and action plan and existing record drawings, detention calculations and permits.

**Alternative Generation and Optimization**

The primary objective was to select the most cost-effective alternative which would decrease flood stage elevations in Sturm while maintaining a 0.1 ft. tolerance to the stage heights downstream. Other hurdles were to meet IDOT, SMC and US Corps of Engineers permitting requirements. All alternatives included an in-kind replacement of the existing internal stormsewer system, structure modifications to existing manholes and junction boxes, and an emergency overflow route (Phase I construction). The remaining alternatives were fully studied and considered in various combinations. The entire tributary area was analyzed down to the outfall point, into the existing pond in the subdivision. The pond was analyzed in a variety of grading and routing scenarios. It is divided into two equal halves by parcel property lines, so the purchase of one or both parcels was also considered in the analysis.

Prior to making a final decision, SMC applied to the Hazard Mitigation Grant Program (HMGP) administered by FEMA and IEMA. Two homes were targeted for acquisition/demolition due to the owners' desires to participate and their optimal location adjacent to the existing easement. Once cleared of structures, the open sites could provide additional detention to further reduce stage heights. SMC was awarded the grant from FEMA for the purchase and demolition of the two homes. With a selected alternative already in place, a site re-design had to be performed to incorporate the two parcels.

The selected alternative for Phase II incorporates a pond-to-wetland conversion plan, outlet reconfigurations, HMGP provided detention and routing and the purchase of one parcel adjacent to the existing pond. Due to the downstream BFE constraints, the level of damage protection was limited. The unimproved conditions had damages beginning with the 2-yr. frequency. The improved conditions have damages beginning with the 25-yr. event (Fig. 1.2). Note the controlling low entry elevation of 809.70.

With every possible alternative examined, the resulting 100-yr. frequency stage height is only reduced by 1.0 ft. This result reinforces the notion that future floodplain management and mitigation plans must include the option of permanently removing structures from these flood hazard areas.

Figure 1.2

| Frequency | Unimproved | Improved |
|---|---|---|
| 2-yr | 809.80 | 807.50 |
| 10-yr | 810.60 | 808.70 |
| 25-yr | 811.00 | 809.60 |
| 50-yr | 811.40 | 810.30 |
| 100-yr | 811.90 | 810.90 |
| | | |
| overflow elev. | 812.00 | 808.00 |
| | | |
| low entry: | 809.70 | |

## Implementation

Phase I construction was successfully completed under budget and on-schedule. The Phase II design is being re-designed due to the HMGP. The anticipated construction date is June, 1998 and will be scheduled to follow demolition by one week. The plan will combine the grading scheme of the pond-to-wetland with the new detention facilities. Phase III recommended actions will be under review during this period. These include stream maintenance and culvert replacement downstream and a wetland expansion opportunity upstream. An HMGP amendment is currently being written by SMC for additional voluntary buy-outs of damaged homes in Sturm and other subdivisions in unincorporated Lake County. To date, there are 18 additional eager participants.

## Retrofitting an Urban Watershed – From Concept to Reality

Patricia S. Werner[1]

**Abstract**

A watershed restoration strategy being implemented in the North Branch of the Chicago River includes projects that retrofit the drainage system to provide multiple benefits including: flood reduction, water quality improvement, natural resource enhancement, and recreation and education opportunities. One of the greatest challenges in urban watershed restoration is generating the support and resources needed to get projects from the conceptual stage, to the drawing board, and ultimately in-the-ground. The North Branch Project is an on-going case study in the "process required" to achieve the in-the-ground "product desired".

**Introduction**

The North Branch of the Chicago River (North Branch) is, as its name implies, an urban river. It is formed from three tributary streams: the 17 mile Skokie River, 24 mile Middle Fork (also known as the West Skokie), and the 14 mile West Fork. From their origins in Lake County, these tributaries flow south into Cook County where they converge to form the mainstem of the North Branch (Figure 1). Outside of the project area, the North Branch of the Chicago continues on a southern route through Chicago, where it is joined by the South Fork of the river. From the city, it flows southwest into the Sanitary and Shipping Canal and the DesPlaines River.

The North Branch Project area covers the northern portion of the watershed. It is 102 square miles or 65,300 acres in size, with north and south boundaries roughly extending from Route 132 (Grand Avenue - Waukegan) in Lake County to Touhy Avenue (Niles) in Cook County. Approximately half of the watershed area, including the headwaters of the river, is located in the rapidly-developing suburban Lake County, the other half is in the more highly developed and densely populated Cook County.

---

[1] Watershed Planner, Lake County Stormwater Management Commission, 333B Peterson Road, Libertyville, IL 60048

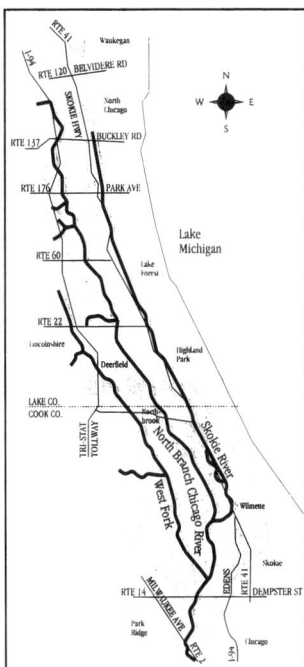

Figure 1: **North Branch Chicago River Watershed**

The North Branch watershed is long and narrow, following the shape of three low glacial moraines that run parallel to the shoreline of Lake Michigan. The sags between the morainal ridges form the shallow valleys through which each of the three tributaries flow southward to Cook County. Prior to development, the landscape of the watershed was composed of a combination of marsh, wet prairie, savanna and woodland. Through the flat valleys, streams with relatively undefined channels flowed slowly southward.

Today, the headwaters and northern reaches of the 3 tributaries are intermittent in flow and interspersed with wetlands. In the developed areas of the headwaters, fingers of the river originate from surface drainage into underground pipes and ditches. A borrow pit excavated for road construction is the beginning of the West Fork. All three forks of the river have been significantly modified for drainage and are presently confined to fairly straight trapezoidal ditches with old spoil piles along the banks. The channels of the river grow larger, are more deeply entrenched and wider, as the tributaries flow southward toward Chicago.

Many cumulative impacts have degraded the North Branch over the past century. The river has a long history of being treated as a sewage ditch rather than a valuable resource. The channel and riparian corridor were drastically altered when the river was ditched for agricultural drainage, and later for urban and suburban development. Now the streambanks are covered with invasive buckthorn and garlic mustard, and the channel is shaped, degraded and eroded by the large volume and velocity of flow it carries from urban stormwater runoff. Flood damage is a common concern among watershed communities[2], and water quality has been degraded by pollutants in the stormwater runoff from parking lots, roofs, streets, lawns and golf courses. Very few areas along the river are accessible and aesthetically attractive to watershed residents.

## North Branch Project
The North Branch Project partnership originated from an ad hoc group coordinated by the non-profit river advocacy organization Friends of Chicago River (Friends). This ad-hoc group co-sponsored a watershed stakeholder workshop in 1991 at the Chicago Botanic Gardens known as *Voices From the Stream*. The workshop was

---

[2] Thirty-five flood damage areas have been identified in the Lake County portion of the watershed.

dedicated to identifying North Branch watershed issues and providing a vision for the future of the watershed. The North Branch Project was born out of the continued interest and commitment demonstrated by watershed partners as they worked on successive restoration projects.

Based on their active involvement in watershed issues, Friends received an Illinois Environmental Protection Agency (IEPA) 319 Grant in 1996 to develop a watershed partnership and strategy for restoring and managing the North Branch watershed. The watershed strategy addresses causes and solutions for non-point source water pollution, flood reduction, and protection and restoration of natural resources. The North Branch Project has four primary components.

1. A watershed assessment, and a management plan for Lake County.
2. A handbook to guide future planning and watershed management in other urban watersheds in the Midwest.
3. An education and public outreach program to educate watershed residents and stimulate stakeholder participation.
4. In-the-ground demonstration projects that provide multiple objectives including water quality improvement.

A partnership has been formed to develop and implement the watershed strategy. Grant-funded co-partners include the Lake County Stormwater Management Commission (SMC) and Northeastern Illinois University (NEIU). Friends, SMC and NEIU staff serve as project coordinators. Other watershed partners include: the Natural Resources Conservation Service, Illinois Department of Natural Resources, IEPA, US Fish & Wildlife Service, the Cook and Lake County Forest Preserve Districts, businesses, environmental groups, watershed municipalities, drainage districts, and county board members.

---

**VIEW OF THE WATERSHED**

**Jurisdictions:**
- 22 municipalities
- 11 townships
- 4 drainage districts

**Population:** (Based on 1990 census data.)
- 385,000 total
- 83% Cook (population density 9.62)
- 17% Lake (population density 2.07)
- 39% population increase forecasted for Lake County by 2020

| Land Use (1990): | Lake | Cook |
|---|---|---|
| residential | 40% | 66% |
| commercial | 6 | 6 |
| industrial | 4 | 3 |
| institutional. | 4 | 6 |
| vacant | 15 | 2 |
| open space | 12 | 12 |
| agricultural | 9 | 1 |
| under development | 5 | .4 |

| Special Resources: | Lake | Cook |
|---|---|---|
| Forest Preserve | 1,582 | 3,696 |
| Wetlands | 4,390 | 1,214 |

- Skokie Lagoons: a series of 7 lagoons (270 acres) created by a dam at Willow Road. An exceptional resource for fishing and recreational activities.
- Chicago Botanic Gardens: 385 acre site with 75 acres of lagoons.
- Plants & Animals:
  23 threatened and endangered species
  16 Illinois Natural Area Inventory sites and Nature Preserves.

An informal watershed steering committee was formed from the ad hoc partnership of stakeholders involved in planning the North Branch grant strategy. This committee has expanded as the project has progressed and more stakeholders have been engaged in project activities. As SMC and NEIU began work on the watershed assessment and planning tasks, an Assessment and Strategy (A&S) work group was formed to provide technical and planning guidance and support. This technical group is made up of federal, state and local government and agency representatives. When project assessment and planning needs declined after the first year of project activity, the A&S work group and steering committee were combined in October of 1997 into a project Planning Committee that continues to meet on a monthly basis.

*Assessing the Watershed Condition - Developing a Management Plan*
Two strategies were used to assess the condition of the North Branch Watershed. The first strategy was to identify and compile relevant information from all existing studies, reports, maps and data. Watershed information was collected from a variety of sources, and maps were produced for purposes of analysis and project reporting. The second was to physically survey the North Branch streams and collect new data on stream condition. The stream survey assessed the condition of the channel, streambanks and riparian corridor; inventoried all of the hydraulic structures by stream reach; and identified areas along the channel in need of remediation. The stream inventory was entered into a database so that the data could be aggregated, manipulated and used in a relational Geographical Information System (GIS).

In addition to the watershed assessment, a watershed management plan is being developed by SMC for the Lake County portion of the North Branch Watershed. The watershed plan will address several key watershed issues identified by the project partners and other stakeholders. They include flooding, poor water quality, erosion, natural resources protection, lack of stream access, poor inter-jurisdictional communication and coordination, and lack of watershed awareness by government, developers, businesses and the general public. To address the broad range of stakeholder concerns and needs, the plan focuses on a multi-objective management strategy (MOM) with project goals for flood reduction, water quality improvement, natural resource protection and enhancement, and opportunities for recreation and education. In addition to including the watershed assessment, the plan will identify opportunities and prioritize recommendations for future multi-objective BMP retrofit projects. The action plan will also match project responsibilities with the appropriate stakeholders and identify sources of funding for multi-jurisdictional projects. A final plan will be completed by summer of 1998 and is scheduled for official adoption by the Lake County Board by the end of the year.

*The Watershed Handbook – A Model for the Future*
Creating the watershed handbook is an exercise in "lessons learned". The handbook is being developed through the Friends. It documents the process and the results of establishing and nurturing the North Branch partnership and developing the North

Branch watershed management plan. The handbook will assess the strengths and weaknesses discovered in the North Branch process and include recommendations that will guide future planning efforts for urban watersheds in the Midwest. The handbook will also serve as a reference guide for the best tools and information available for watershed planning and management.

*Raising Awareness - Public Outreach and Education*
The public outreach and education campaign is designed to both encourage stakeholder participation in the watershed planning process, and to raise awareness and educate a broad range of watershed residents about the river and watershed. To address a varied audience, several public information and education initiatives have been incorporated into the campaign. Slide presentations, village board presentations, and a canoe trip on the river have stimulated general interest and awareness in the project. A two day conference planned for July 1998 at the Chicago Botanic Gardens will be the keystone event for involving a large number of stakeholder groups and watershed residents.

An education campaign for high school students, coordinated by Friends, has been supported by development of a river-based curriculum, teacher training, and field trips for students. An advisory team of interested teachers assists in developing and implementing this aspect of the program. A Chicago River Schools Network uses and promotes the services and materials supplied in the education campaign. The Network is sponsoring the first Chicago River Student Congress in February 1998.

One of the biggest challenges in developing the North Branch plan has been in engaging numerous and varied stakeholders in the planning process. Two strategies have been used to stimulate interest and involvement in the plan. One is to organize events where stakeholders can meet to share concerns and receive information about the watershed, the other is to meet with the stakeholders individually to identify more localized issues and needs. Two large stakeholder meetings were held in 1997 to present the watershed assessment, and to identify and rank watershed issues and opportunities. In addition to these facilitated gatherings, SMC has conducted individual interviews with representatives from each municipality, township, drainage district and all affected county agencies and departments.

Some specialized activities have also been sponsored by the project. Project partners coordinated several field trips for planning committee members and public works and engineering representatives to view applicable BMP projects in neighboring watersheds. Friends and Openlands Project are also sponsoring a one day workshop for watershed golf course superintendents in March 1998.[3]

---

[3] There are 41 golf courses that make up approximately 7% of the land use located in the North Branch watershed. Most of the golf courses are located in the floodplains and along the stream channels of the North Branch tributaries.

*Demonstration Projects – Leading the Way to "In-the-Ground"*
The North Branch Project strategy included implementing four different types of best management practice demonstration projects (BMP) within the first 2 years of the project. The selection criteria include projects appropriate to North Branch conditions that use innovative processes or practices to meet water quality, flood reduction and natural resource enhancement goals while providing opportunities for education. The demonstration projects are visible examples of local leadership in urban watershed retrofitting that serve an educational and motivational purpose in addition to providing in-the-ground benefits. The BMP projects require local sponsors and a 40% non-federal cost-share.

Two demonstration projects have been selected and funded by the North Branch steering committee to-date.
Project 1: A floodplain restoration by Lake Forest Open Lands (LFOL) at Mellody Farms Nature Preserve in Lake Forest. LFOL removed the spoil pile that made up the streambank along the east side of the Middle Fork and graded the area to re-establish the floodplain along 1/2 mile of the river. The restoration included protecting a sedge meadow and putting in native plants.
Project 2: A joint partnership with a local Girl Scout Troop and the Wilmette Golf Course. Three existing ponds on the golf course will be re-graded and converted to more natural wetlands. Native wetland and shoreline buffer plantings will be established. The Girl Scouts will follow-up with a water quality monitoring program for the wetlands and the adjoining Skokie River.
A third project was approved by the committee and outside cost-share funding is being solicited for this project.
Project 3: Co-sponsored by the Union Drainage District on the West Fork of the river in Deerfield, this project will re-meander the low flow of the river within the larger trapezoidal stream channel. Pre-planted coir rolls will be anchored in the stream bottom to create the meanders and stabilize the streambed.

Two additional demonstration projects are being considered for funding by the project committee. One is a new detention pond that includes several wetland pockets and native plantings. The second will restore a section of stream in Chipilly Woods, a Forest Preserve Property in Cook County, by creating a series of pools and riffles.

## Conclusion: From Process to Practice
Watershed stakeholders have expressed concerns about the condition of the North Branch for many years, but because of its large number of jurisdictions, coordination and cooperation opportunities for watershed restoration were limited. Responsibility for poor watershed conditions is diffuse, accountability is difficult to determine. The North Branch project was formed to address the complexity of watershed restoration and coordinate the many stakeholders. The project partnership has grown, and the desired product - projects to improve the condition of the watershed - are being prioritized and completed. Restoration is finally moving from concept to reality.

Harborside International Golf and Marina Facility
Frank L. Kudrna, Jr., F.A.S.C.E[1]

*Abstract*

For decades, the City of Chicago utilized a landfill site adjacent to Lake Calumet for disposal of its municipal refuse. This included general municipal refuse as well as landfill ash. After the City ceased using this as a disposal site, fly dumping occurred on the site and it was considered an extremely serious municipal problem. The Metropolitan Water Reclamation District used the site for a sludge disposal site for a number of years and entered into a settlement agreement with the property owner, the Illinois International Port District, to publicly fund converting the site into a championship golf facility. The Illinois International Port District, owning significant adjacent lands, had the firm of Kudrna & Associates, Ltd. develop a comprehensive plan to construct a golf and marina facility. Dick Nugent Associates, golf course architect, was retained to design the golf course and Kudrna & Associates performed the engineering services and marina design. The facility's first phase, the Teaching Facility, was opened in May of 1994, with the first 18 hole golf course opening in 1995 and the second 18 hole golf course in 1996. Plans are currently being finished for a large Clubhouse/Restaurant/Banquet facility.

The project involved many innovative challenging planning and design issues. For example, over 600,000 cubic yards of clay were needed to close the landfill site and these were obtained by dredging Lake Calumet and creating the marina basin. Additionally, the challenges associated with the growth of golf course specialty grasses within a sludge media were enormously challenging. Construction staging, wetland relocation and fill construction, infrastructure extensions and financing were all major issues in the development of the property.

*Introduction*

From the early 1950's, the 103$^{rd}$ & Doty Street location had been one of the City of Chicago's primary garbage disposal sites. A large incinerator currently inactive was also used to incinerate a portion of the garbage and disposal of ash took place in this site. Fortunately, the underlying lake bed of Lake Calumet contained rich blue clay, providing a natural bottom liner. The site was never permitted for use as a landfill (permits were not required in the periods of initial construction), and the site eventually became the property of the of the Illinois International Port District (Port District) when it was created in 1955. In addition to the City's dumping there had been some historic fly dumping on the site and the site generally was an open, unclosed site which also was unpermitted. The Metropolitan Water Reclamation District (MWRD) (originally the Metropolitan Sanitary District of Greater Chicago) began to use the site for disposal of sludge from its Calumet wastewater plant in the 1980's and working with the Illinois EPA proposed to close the site for recreational purposes. This agreement between the MWRD and the Illinois EPA took place without the property owner's (the Port District) permission. Litigation resulted in a settlement agreement between the MWRD and the Port District where the MWRD would provide $10.4 in funding to close and prepare the site for use as a public recreational facility and the Port District would complete the activity and manage and maintain the facility. This court order, after lengthy negotiations and litigation activities, was entered in 1993.

This was an extremely important settlement to resolve use of the approximately 220 acre landfill site. Because, without a recreational use capable of generating revenues, an enormous long-term expense would have been required of the Port District to maintain and secure the approximately 220 acre site.

## *The golf facility*

In addition to providing a beneficial use with a positive cash flow from the site, the Port District was interested in providing a catalyst for other development in the southeast side. The southeast side of Chicago had an extremely high unemployment rate with the loss of major heavy industry it existed for a quality facility to serve as a catalyst for other development. The Port District board chose to develop a world class golf facility and interviewed the major golf course architects (Palmer, Niklaus, Trevino, etc.) before concluding that Dick Nugent Associates, a local Illinois firm that had previously designed Kemper Lakes, Forest Preserve National and Seven Bridges, among others, was the appropriate firm to be chosen for the golf course routing.

## *Site Description*

The 380 acre site is currently:

> 50% - a 75' deep pile of garbage and sewage sludge. The site has been a primary depository for sludge by the MWRD of Chicago for the past thirty years.

50% - rubble landfill 40' deep. This material was originally rubble from streets and sewers, more recently furnished by independent contractors.

There is no soil on the site.

The site is bordered on the north by a major city incinerator and police automobile impoundment lot; on the west by the interstate highway - Bishop Ford Expressway (formerly the Calumet Expressway); on the south by the Port District facility; and on the east by Lake Calumet - which is directly connected with Lake Michigan.

The general area is primarily heavy industry and associated aging residential.

Golf course design concepts

1. <u>Circulation</u>. Ease and direction of movement through the complex of primary consideration. This is a pay for play facility. The same concepts apply here as any store; once the customer is in the door, he must find it easy and convenient to spend. His golfing experience should be planned every step of the way and he should leave please, happy and satisfied.

The initial arrival has been provided for through use of an off ramp from the Bishop Ford Expressway at 111th Street, extending right into the golf complex where the entrance has been designed to present the all important "sense of arrival". A short distance through a highly landscaped entrance road brings our golfer to the center of the parking area where he can elect to park nearer the clubhouse or the golf academy.

For ease of service, it was determined that four different facilities should each ideally be as close to the center as possible; these being the clubhouse, the golf academy, the maintenance complex and the irrigation pumping plant.

The clubhouse is the hub of all customer related activities and had the highest priority in the placement scheme. The design provided for four nine hole segments of golf to start and finish at or near the clubhouse. The clubhouse as well as being the central staging area for the 36 hole golf complex, also functions as the hospitality area providing food and beverage service. Also, this facility can accommodate golf outings, banquets and other scheduled social activities.

The clubhouse is 30 feet above and overlooking Lake Calumet and is directly above the greens of two finishing golf holes. The elevation differential allows for the design of a two story building with a single story appearing front entrance and a walk-out back. Social activities are provided for on the upper floor and golf on the lower floor. Golf staging areas (carts, bags, etc.) are immediately outside at the lower back level above the golf greens and lake, but not obscuring any views from the upper (social) level of the clubhouse.

Near the clubhouse, located at the opposite end of the clubhouse, is the Golf Academy with its own service buildings. The Academy has been designed as a stand-alone facility complementary to the clubhouse. This Golf Academy has been designed to be the finest facility of its type in the world. It incorporates a large, lighted practice range complete with target greens, four large grass tees backed up by an all weather artificial grass area, a practice putting green, an area for sand practice, a pitching/chipping area with its own green, a full teaching tee with sheltered covered tee areas, a 60 yard practice hole with an actual green and greenside bunkers, plus a separate three hole course dedicated to the Academy where students and beginners can play and develop a minimum proficiency without interrupting play on the regular courses.

Off a frontage road between the golf complex and the Bishop Ford Expressway, there was a parcel of land that had not been used as landfill, was relatively close and accessible to all maintained areas, easily accessed by traffic and utilities, suitably screened and of adequate size where the maintenance facility was located. The maintenance facility has its own entrance and does not conflict with the golf traffic.

Additionally, there is a separate service road to the clubhouse from the frontage road which is between the Academy and the golf course and is suitable screened by grading features. This could also be used as a secondary entrance if desires.

Currently, a main Clubhouse is under construction, approximately 24,000 square foot in size, that will house a major restaurant and banquet facility.

Since it is not feasible to locate the pumping plant and an associated reservoir centrally on the landfill, it was placed on Lake Calumet in a corner of the golf development requiring a run of larger diameter pipe to bring the irrigation water to the center of the distribution system.

2. <u>Constructibility</u>. Construction of the golf facilities has been organized into strategic logical construction phases.

Twenty-four of the 36 holes are on a 75 foot fill of sludge from the MWRD which had been partially capped and was still being actively used.

It was necessary to remove clay from those areas where landfill operations were already completed and capped to achieve the land forms necessary for the golf course. After achieving these forms, the cap was replaced and new clay added from the marina. After the clay cap was completed, an additional two feet of sludge was placed on top of the clay to provide an envelope for irrigation and drainage infrastructure followed by a sand capping to provide an amended growth medium for the turfgrass.

The remaining twelve holes, the three Academy practice holes and the Golf Academy are on the rubble fill portion of the site and require grading of the rubble into land forms, capping with clay and addition of sludge, soil and/or sand to achieve a turfgrass playing surface.

Critical to the golf course construction progress is the integration of the on-going sludge disposal operations, the on-going disposal and grading of rubble interfaced with the marina construction activities and the golf course construction activities.

3. Playability. The design of the golf course seeks to develop a variety of sholmaking possibilities in order to challenge a variety of players and by doing this, develop a strong interest in the course. For example, multiple tees are intended to improve course playability for all skill levels. Multiple tees not only accommodate players of different levels, but also speeds up play. Additionally, wear on tees is reduced which contributes to a more resilient, disease resistant turf and on overall improved tee surface. The courses will play from as long as 7,716 yards from the back tees to be short as 5,146 yards from the forward tees.

Golf as a recreational sport continues to increase in popularity. In no other time in the history of the sport have more women, juniors and seniors started to play golf. This golf course has been designed to accommodate this diverse player profile, expedite play and offer challenge to golfers of all skill levels.

4. Maintainability. Design and maintenance are interrelated. The design concepts include maintenance considerations for each aspect of the design. While construction costs are a one-time expense, maintenance costs are continuous. Low construction costs resulting in high maintenance costs are no bargain. Since this site has no soil and is dependent on the aging of the sludge (a produce known as "new soil") as a growing medium, it presents some special maintenance problems such as:

- Sludge is virtually non-permeable – making root development difficult.
- Water use for irrigation is restricted by law regulating the use of water from the Great Lakes.
- These combine to create a fragile environment subject to heavy use, therefore, control of play must be carefully planned for and closely coordinated with the maintenance program.

Excellent turfgrass and good playing conditions are hallmarks of all great golf courses.

5. <u>Profitability</u>. Bottom-line profitability is directly related to the playability and maintainability of the golf course. It is our intention to develop an attractive golf course that can be maintained at a reasonable expense, will produce increased revenues and offer affordable golf to the broadest player profile.

6. Design of golf course features. The accomplishment of the design development required manipulation of the specific design features particular to golf courses:

<u>Greens</u>. The intent of the design is to establish greens with unique characteristics in size, shape and bunker location, which will provide variation and interest to the players, yet not be unfair.

The green and greenside bunkers reflects the shot value of the approach shot being hit to them; i.e., larger more open greens for holes that require long difficult approach shots and tighter, smaller, more difficult greens for a short approach requiring a more controlled shot.

<u>Tees</u>. Generally, the tees on the Master Plan are shown with pleasant curing shapes. This adds to the eye appeal of the tee surface and allows for easier, more economical maintenance, as well as to provide multiple teeing areas for different classes of players.

<u>Bunkers</u>. A bunker is defined in the <u>Rules of Golf</u> as a hazard consisting of an area of bare ground, often a depression, which is usually covered with sand. Grass covered ground adjacent to or within the bunker is not considered a hazard. The distinction is critical and the edges of the hazard must be clearly defined because of the rules and penalty involved in grounding the club in a hazard.

The primary function of a bunker is to extract a penalty from a miss hit shot so unlucky as to come within its confines. The severity of the penalty is in proportion to the depth of the bunker. Given the lack of trees on this site and the Links design concept and apart from their strategic and penal aspects, bunkers add greatly to the aesthetic qualities of the golf course through the artistic variances in their size and shape.

Mounds. Mounds provide in critical element in the design concepts at Harborside International Golf Links. While not technically a hazard, the strategically placed mounds can cause a ball to take erratic bounces when striking their slopes; or, a ball can come to rest on the side of a mound resulting in an uphill, downhill, or sidehill lie. Mounds also serve as low vertical screening elements and wind breaks.

Plantings. Due to considerations of difficulty in establishing trees on this site because of soil type, avoiding the penetration of methane gas, tree plantings on these courses will be restricted. However, where feasible, trees will be used as vertical elements that must be played over, under or around.

Trees add another element of challenge to the golfer and are also used to help control play by guarding doglegs, forming chutes through which shots must be played, and by forcing players to choose between alternate routes of play or types of shot to play.

Trees and background plantings are used to frame target areas, provide focal points, and aid in gauging distance by providing a sense of proportion and depth perception. Plantings can also be effective in screen out off-site nuisances and visual distractions and give players a sense of seclusion.

Plantings used as screens provide some safety to players and to adjacent properties from errant shots. Screen plantings also act as windbreaks.

7. Aesthetic qualities. A major goal of this project is to heighten the aesthetic quality of the golf course and develop features with a high visual impact. The fact that Harborside International Golf Links is in an aging industrial area provides a positive image for the golf course as a verdant jewel imbedded in the rust belt. The contrast is stunning. The same situation occurs at the Royal St. George in England, site of many British Open Championships.

Additionally, together with the adjacent Historic Pullman Preservation District, this provides the nucleus for urban renewal and neighborhood upgrading.

## Marina

As part of the Closure Plan, the North Turning Basin area had over 600,000 cubic yards of sludge removed in order to close the landfill site. This 600,000 yards was excavated in the dry and provided a 2,000 slip future marina basin, which will join the golf course facility and provide a protected harbor area with access to Lake Michigan (approximately five and a half miles), without any lock structures. The marina phase will include floating dock facilities and full service and support facilities, which include parking, washroom/shower facilities, harbor master and full service marina facilities, as well as commercial support facilities for the marina.

The marina facility will be designed to provide winter storage facilities for the boats as well as winter maintenance and repair activities.

## Summary

This course has inherent problems and opportunities rarely, if ever, encountered in such a diversified format. Several significant lessons were learned from the project.

First, where digested sewage sludge is perfectly suitable for fill material when capped, it has a series of problems when used as surface soil. Specifically, the salt content within sludge is extremely high and where this is not a normal problem for basic cover vegetation, it creates a problem for specialty golf course grasses needed in a championship course. This requires bringing in supplemental topsoil materials and preferably sand in order to provide a suitable material for growth.

Secondly, constructing a golf course on a landfill also requires special consideration for support buildings, i.e., clubhouse, restroom facilities, starter shacks, etc., which are precluded from penetration the clay seal of the site.

Third, implementation of a project with multiple government agency jurisdictions combined with a Closure Plan requires careful project management for a successful project.

The development of the Harborside International golf and marina project provided one of the most challenging undertakings in our careers.

## Urban Storm Sewage Design Using the Double Detention Pond Concept and a Modified Rational Formula Approach

Ray-Shyan Wu,[1] A.M. ASCE, Chun Yu[2], Shu-Liang Liaw[3], Ching-Ho Chen[4]

### Abstract

A modified Rational Formula approach for designing urban storm sewer system, which includes the conventional detention pond and the double detention pond, is proposed in this paper. To illustrate this approach, a typical urban storm sewer system is designed by using the modified Rational Formula with a typical design storm. This approach not only designs the sewer lines but also estimates the hydrographs. Furthermore, these hydrographs may be applied to design the conventional detention pond or the double detention pond. The results show that the double detention pond can save up to 50% of the volume compared to the conventional detention pond.

### Introduction

Since the Rational method was introduced in the United States by Kuichling in 1889, it has become the most widely used method for estimating peak runoff rates in the design of urban storm sewer facilities. Despite a number of criticism, the Rational method would continue to be used in practice, especially for a small watershed (Steel and McGhee 1979).

Due to urban development, stormwater runoff increases and may cause frequent flooding and severe channel erosion downstream. To avoid such adverse effects, detention ponds are applied to control the stormwater in urban storm sewer system. The main purpose of these detention ponds is to reduce the peakflow and thus to assure outflow not exceed the permitted discharge. Clearly, it is necessary to obtain the inflow hydrograph as designing the detention pond. Unfortunately, the Rational formula approach cannot provide such information.

In practice, the required volume of a detention pond needs to be estimated by a trial-and-error procedure. Akan et al. (1987) and Horn (1987) proposed graphical methods that eliminate the trial-and-error procedure to a great extent for a single-outlet detention pond. Then, Akan (1989) developed a semigraphical method that can be used for quick design of a dual-outlet detention pond for

---
[1] Associate Professor and [2] Graduate Student, Department of Civil Engineering, National Central University, Taiwan
[3] Professor, Graduate Institute of Environmental Engineering, National Central University, Taiwan
[4] Lecturer, Department of Civil Engineering, Nanya College of Technology and Commerce, Taiwan

multiple return periods. Furthermore, Akan (1990) developed a desktop method for quick analysis and preliminary design of a detention pond. However, the U.S. Soil Conservation Service dimentionless unit hydrograph was assumed to be the inflow hydrograph. Recently, Wu and Yu (1996) proposed a new type of the detention pond, named the double detention pond, which reduces the volume required and can be directly calculated. The objective of this paper is to develop a design approach of urban storm sewage, which may estimate the size of a conventional detention pond or the double detention pond, by using a modified Rational Formula approach. A case study is presented to address the theorem and the results are discussed.

## Modified Rational Formula Approach

### Linearized Unit Hydrograph

Deriving from the principle of mass conservation, three cases of unit hydrograph are established with respect to the relative conditions of the unit rainfall duration and the time of concentration. In three cases (Chien and Saigal 1974; Design 1984), linear variation in the rising limb and the receding limb of the unit hydrograph for a sufficiently small drainage basin are assumed, as shown in Figure 1. Three unit hydrographs are as followings, in which $t_b$ is the time base of the unit hydrograph, $t_r$ is the unit rainfall duration of the effective rainfall and $t_c$ is the time of concentration for the basin.

**Case 1:** $t_r = t_c$,  **Case 2:** $t_r > t_c$,  **Case 3:** $t_r < t_c$

For Case 1 and Case 2, the peak rate of runoff can be expressed by the Rational formula as

$$Q = CiA / 360 \qquad (1)$$

in which Q is the peak rate of runoff ( in cms ), C is the runoff coefficient, i is the intensity of rainfall ( in mm/hr ) and A is the size of basin ( in hectare ).

For Case 3, the peak rate of runoff is

$$Q = (t_r / t_c) CiA / 360 \qquad (2)$$

The volume of runoff from the rainfall for each case is $Ci\, t_r\, A/360$.

### Time of Concentration

Time of concentration is generally defined as the time required for the runoff from the remotest point of the basin to reach the outlet. For uniform rainfall intensity, this would be the time of equilibrium at which the rate of runoff is equal to the rate rainfall supply. Among various techniques of estimating the time of concentration, the Kinematic wave technique is adopted because of the better results in field verifications (Ragan and Duru 1972). The Kinematic wave formula is

$$t_c = 0.93 L^{0.6} n^{0.6} / (i^{0.4} S^{0.3}) \qquad (3)$$

in which $t_c$ is the time of concentration ( in minute ), L is the length of overland flow ( in meter ), n is the Manning's roughness coefficient, i is the excess rainfall intensity ( in mm/hr ) and S is the overland slope.

### Design Storm

A general type of storm, in which the maximum rainfall intensity occurs at the middle unit rainfall duration, is chosen as the design storm. The design storm is derived from the intensity-duration-frequency (IDF) formula. First, decide the rainfall duration of the storm, T, and the unit rainfall duration, $t_r$. The number of the time interval is $N = T / t_r$. Based on the IDF formula, estimate the rainfall intensity, a, b, c, d,..., with respect to the rainfall time equals to $t_r$, $2t_r$, $3t_r$, $4t_r$, ..., respectively.

**Hydrologic Routing and Design of Urban Storm Sewer System**

In a complex urban storm sewer system consisting of many subbasins and branch lines, the hydrologic routing is required in the design of the sewer system. The purpose of the hydrologic routing is to take the slope and the diameter of sewer line and the travel time of flow between manholes into the consideration of design. For designing, it is convenient to select a tentative grade of the sewer based upon the ground surface and check the velocity of the maximum runoff rate. Usually, the minimum allowable velocity is 0.75 m/s to avoid sediment and the maximum allowable velocity is 3.5 m/s to protect erosion (Steel and Mcghee 1979). The velocity of the actual depth of the flow in the sewer is calculated by the Manning formula. Figure 2 illustrates the hypothetical design procedures.

**Conventional and Double Detention Pond**

The structure of a conventional detention pond and a double detention pond are shown in Figure 3. An upstream inlet provides inflow, and a downstream outlet releases outflow. Usually, the depth of the detention pond, H, may be decided due to topographic circumstance. Then a flood routing scheme for the detention pond, based on solving the continuity equation, is used. The continuity equation is expressed as

$$dS/dt = Qin(t) - Qout(t) \qquad (4)$$

in which, S is water storage in the detention pond, $S = aY^b$, Y is the depth of the water, a and b are constant parameters depending on the shape of the detention pond; Qin(t) is inflow into the detention pond at time t; out(t) is outflow from the detention pond, $Qout(t) = f[Y/D]$, Y is the water depth and D is the height of outlet.

The inflow hydrograph, which is derived from modified Rational Formula approach, Qin(t), is the input variable, and the outflow hydrograph, Qout(t), can be related to the shape and elevation of the outlet. After determining the size of the detention pond and the shape of the outlet, the outflow hydrographs are routed to check whether the performance requirements are met.

A double detention pond (Wu and Yu 1996) is to add an extra detention pond A, within the traditional pond and a set of one-way gates at the bottom of the extra detention pond. Pond A can release water out quickly at the early stage of the flow when the one-way gates are closed due to gravity. When Pond A is full, a part of the inflow overflows from the top of Pond A into Pond B. When inflow is less than outflow, the water pressure in the outside of the gates is larger than that inside. Therefore, the gates open, and water in Pond B pours into Pond A, and subsequently is released from the outlet. It can be shown that by using this flood routing approach, the volume of the double detention pond can be directly calculated.

**Case Study**

A typical urban storm sewer system is selected as a case study, as shown in Figure 4 (Steel and Mcghee 1979). The branch lines were placed to run south on 12th, 13th, 14th and 15th streets, with a main running east on Spruce Avenue. The elementary data are described in Table 1. The IDF formula is given $i = 2590 / (t+17)$ for 2-yr frequency. The runoff coefficient is assumed $C = 0.40$. The times of concentration of the subbasin 1, 5, 9, 13 are 10, 10, 10, 6 minutes, separately. The Manning's roughness of all pipes are n = 0.013. The conventional or double detention pond is set at the end of sewer line 16, and the outflow finally pours into the river downstream.

The design storm must be first accomplished. Each the time of concentration of the subbasin, except the subbasin 1, 5, 9 and 13, may be calculated by Eq.3. Then, according the relationships of the time of concentration and the unit rainfall duration, the subhydrographs of each subbasin can be calculated by Eq.1 or Eq.2. Finally, utilize the hydrological routing and Manning formula to design the size of the sewer, including the velocity and the depth of flow in the sewer. The results are shown in Table 2. The hydrograph at the sewer line 16 is illustrated as shown in Figure 5. Without detention ponds, the maximum peak rate of runoff in the sewer line 16 is 1.517 cms. Assume the permitted discharge downstream is 1.300 cms and the permitted depth of the detention pond is 2.0 meter. A double detention pond with area equals 51 $m^2$ and a squire outlet as 0.51m in width reduces the maximum peak rate to 1.293 cms ( given area of Pond A is 16 $m^2$ and area of Pond B is 35 $m^2$). The results show that the conventional detention pond needs area of 164 $m^2$.

## Summary and Conclusion

The modified Rational Formula approach employs the principle of mass conservation and simple parameters and functional relationships that are very easy to comprehend. In this paper, the kinematic wave time to equilibrium is used as the time of concentration. Moreover, linear variation of the rising limb and the receding limb of the unit hygrograph for a small basin is assumed. The design storm is derived from the IDF formula; and the travel time of the flow in the sewer is calculated by the Manning formula to perform hydrologic routing in the design sewer line. For comparison, the conventional detention pond and the double detention pond are applied to the same case study. The result shows that the double detention pond can save up to 50% of the volume required.

It can be concluded that simplified assumptions are unavoidable and necessary in designing sewer systems for small drainage basins dealing with uncertain urbanizing factors. The modified Rational Formula approach can provide not only a conservative design but hydrographs of each sewer line. This is believed to be a useful tool for the design of a new urban storm sewer system. Moreover, the double detention pond can save a substantial volume compared to the conventional detention pond. Therefore, such a concept should be promoted in engineering practice.

*Acknowledgment. Support for this work was provided, in part, by a grant from the National Science Council of the Republic of China in Taiwan through contract NSC 86-2621-E008-011.*

## Appendix References

Akan, A.O., Al-Muttair, F.F., and Al-Turbak, A.S. (1987), "Design Aid for detention Basins," Design of Hydraulic Structures, Proc., Int. Symp., Fort Collins, Colo., 177-182.

Akan, A.O. (1989), "Detention Pond Sizing for Multiple Return Periods," J. of Hydr. Engrg., ASCE, 115(5), 650-664.

Akan, A.O. (1990), "Single-Outlet Detention-Pond Analysis and Design," J. of Irrigation and Drainage Engrg., ASCE, 116(4), 527-536.

Chien, J. S. and Saigal, K.K.(1974), "Urban Runoff by Linearized Subhydrographic Method," J. of Hydraulic Division, ASCE, vol.100, No.HY8, 1141-1157.

*Design Guidelines for Storm Sewage Detention Basins* ( Draft, in Japanese)(1984), Japan Sewage Works Association.

Horn, D.R.(1987), "Graphic Estimation of Peak Flow Reduction in Reservoirs," J. of Hydr. Engrg., ASCE, 113(11), 1441-1450.

Ragan, R. M. and Duru, J. O.(1972), "Kinematic Wave Nomograph for Times of Concentration," J. of Hydraulic Division, ASCE, vol.98, No.HY10, 1765-1771.

Steel, E.W., and McGhee, T.J.(1979), *Water Supply and Sewerage*, McGraw-Hill Book Co., New York, N.Y., fifth edition, 400-403.

Wu, R.S., and Yu, C.(1996), "A Volume-Saving and Sediment-Concentrating Approach for Designing Detention Pond (in Chinese), "J. of Chinese Soil and Water Conservation, 27(1),22-38.

## Table 1  Fundamental data of the Urban Storm Sewer System

| No. of Line | Subbasin (ha) | Length (m) | Slope | Runoff Coeff. | Time of Concentration |
|---|---|---|---|---|---|
| 1 | 1.32 | 87 | 0.0077 | 0.4 | 10.0 |
| 2 | 1.32 | 87 | 0.0070 | 0.4 | 10.3 |
| 3 | 1.32 | 87 | 0.0074 | 0.4 | 10.1 |
| 4 | 0.54 | 140 | 0.0011 | 0.4 | 23.9 |
| 5 | 1.55 | 87 | 0.0052 | 0.4 | 10.0 |
| 6 | 1.22 | 87 | 0.0147 | 0.4 | 7.3 |
| 7 | 1.22 | 87 | 0.0126 | 0.4 | 7.7 |
| 8 | 0.67 | 140 | 0.0020 | 0.4 | 17.7 |
| 9 | 1.55 | 87 | 0.0084 | 0.4 | 10.0 |
| 10 | 1.22 | 87 | 0.0243 | 0.4 | 7.3 |
| 11 | 1.22 | 87 | 0.0193 | 0.4 | 7.8 |
| 12 | 0.92 | 140 | 0.0009 | 0.4 | 26.0 |
| 13 | 0.43 | 87 | 0.0168 | 0.4 | 6.0 |
| 14 | 0.43 | 87 | 0.0256 | 0.4 | 5.3 |
| 15 | 0.43 | 87 | 0.0248 | 0.4 | 5.3 |
| 16 | 1.05 | 140 | 0.0011 | 0.4 | 18.1 |

## Table 2  Urban Storm Sewer Computation by Modified Rational Formula

| (1) | (2) | (3) | (4) | (5) | (6) | (7) | (8) | (9) | (10) | (11) | (12) | (13) | (14) |
|---|---|---|---|---|---|---|---|---|---|---|---|---|---|
| Line No. | Location | From street | To street | A (ha) | C | Q (cms) | S | D (mm) | Y (m) | $V_{real}$ (m/s) | L (m) | Tv (min) | $Q_d$ (cms) |
| 1 | 15th St. | Locust | Elm | 1.32 | 0.40 | 0.141 | 0.0077 | 380 | 0.28 | 1.59 | 87 | 0.9 | 0.159 |
| 2 | 15th St. | Elm | Beech | 1.32 | 0.40 | 0.277 | 0.0070 | 530 | 0.34 | 1.84 | 87 | 0.8 | 0.369 |
| 3 | 15th St. | Beech | Spruce | 1.32 | 0.40 | 0.412 | 0.0074 | 610 | 0.39 | 2.07 | 87 | 0.7 | 0.552 |
| 4 | Spruce | 15th | 14th | 0.54 | 0.40 | 0.448 | 0.0011 | 840 | 0.63 | 1.02 | 140 | 2.3 | 0.499 |
| 5 | 14th St. | Locust | Elm | 1.55 | 0.40 | 0.165 | 0.0052 | 460 | 0.30 | 1.44 | 87 | 1.0 | 0.218 |
| 6 | 14th St. | Elm | Beech | 1.22 | 0.40 | 0.291 | 0.0147 | 460 | 0.31 | 2.45 | 87 | 0.6 | 0.366 |
| 7 | 14th St. | Beech | Spruce | 1.22 | 0.40 | 0.429 | 0.0126 | 530 | 0.38 | 2.53 | 87 | 0.6 | 0.495 |
| 8 | Spruce | 14th St. | 13th St. | 0.67 | 0.40 | 0.879 | 0.0020 | 990 | 0.70 | 1.52 | 140 | 1.5 | 1.043 |
| 9 | 13th St. | Locust | Elm | 1.55 | 0.40 | 0.165 | 0.0084 | 380 | 0.31 | 1.67 | 87 | 0.9 | 0.166 |
| 10 | 13th St. | Elm | Beech | 1.22 | 0.40 | 0.292 | 0.0243 | 460 | 0.27 | 3.02 | 87 | 0.5 | 0.471 |
| 11 | 13th St. | Beech | Spruce | 1.22 | 0.40 | 0.430 | 0.0193 | 530 | 0.33 | 3.03 | 87 | 0.5 | 0.612 |
| 12 | Spruce | 13th St. | 12th St. | 0.92 | 0.40 | 1.332 | 0.0009 | 1300 | 0.99 | 1.24 | 140 | 1.9 | 1.447 |
| 13 | 12th St. | Locust | Elm | 0.43 | 0.40 | 0.053 | 0.0168 | 305 | 0.16 | 1.84 | 87 | 0.8 | 0.131 |
| 14 | 12th St. | Elm | Beech | 0.43 | 0.40 | 0.104 | 0.0256 | 305 | 0.18 | 2.37 | 87 | 0.6 | 0.162 |
| 15 | 12th St. | Beech | Spruce | 0.43 | 0.40 | 0.156 | 0.0248 | 305 | 0.25 | 2.48 | 87 | 0.6 | 0.159 |
| 16 | Spruce | 12th St. | 11th St. | 1.05 | 0.40 | 1.517 | 0.0011 | 1330 | 1.01 | 1.37 | 140 | 1.7 | 1.600 |

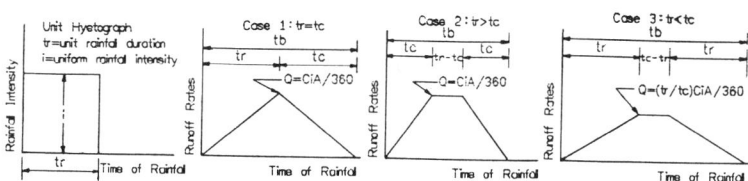

**Fig.1 Schematic Linearized Unit Hydrographs**

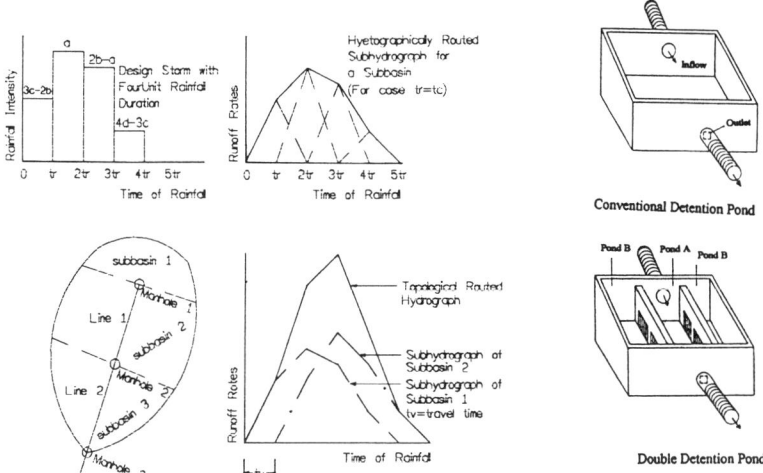

**Fig.2 Hypothetical Design Storm, Urban Storm Sewer System, and Hydrologic Routing**

**Fig.3 Structure of Conventional and Double Detention Pond**

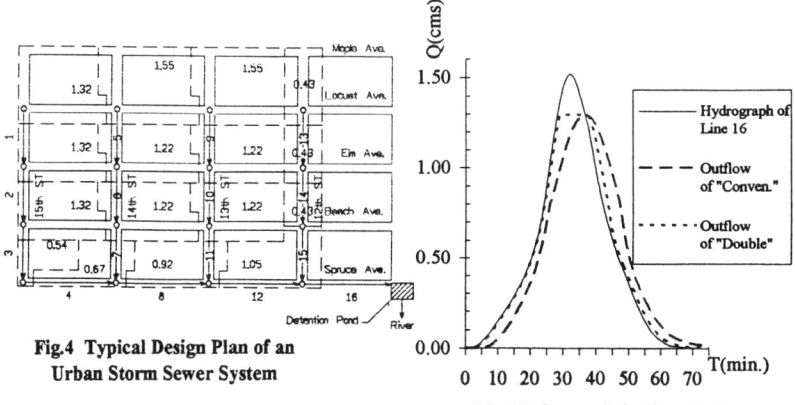

**Fig.4 Typical Design Plan of an Urban Storm Sewer System**

**Fig.5 Inflow and Outflow Hydrograph**

Performance Evaluation of Dry Detention Ponds with Underdrains

F. N. Nnadi[1], K.W. Ashe[2], R. C. Sharek[2]

Abstract

The purpose of this study is to investigate the potential factors that affect the performance of dry detention ponds with underdrains using laboratory, field and forensic approaches. The study evaluates a plan to be used for investigating the performance of the non-functioning dry ponds with underdrains. The results of this study will help in defining the factors that affect the performance of the ponds; therefore, providing the tools required to validate the currently used design criteria as will be seen during permit process. In addition, recommendations for or against, the use of current construction and maintenance methodologies were made based upon the results of this study.

Introduction

A dry detention pond with underdrains is a stormwater basin with the pond bottom above the groundwater table. This basin is designed to detain water no more than 72 hours following a storm event. The Florida Department of Transportation (FDOT) uses this type of system to treat highway runoff. This study focused on the investigation of three non-functioning dry detention ponds in the Central Florida region. Piezometers were installed to monitor local groundwater levels at the ponds. Several soil samples were taken within, and around the basin of all three ponds. Elevations of inlet and outlet structures, underdrains, and pond bottoms were surveyed and compared to those in the design calculations. Pond 1, located at the intersection of State Road 436 and Cassel Creek Boulevard in Casselberry, Florida (Figure 1), was observed to have many of the major drainage problems. The second pond, Pond 2, is located along SR 513 in Indian Harbor Beach, approximately 40 km east of Melbourne, Florida, while Pond 3 is located approximately 500 meters north of Pond 2 along SR 513 (Figure 2).

------------------------------

[1] Assistant Professor, Department of Civil & Environmental Engineering, University of Central Florida, 4000 Central Florida Blvd, Orlando, FL 32826
[2] Graduate Research Assistant

Regional Groundwater Assessment

Prior to field investigation, the groundwater elevation trend in the State of Florida was investigated. Fifteen years of groundwater elevations collected from the U.S. Geological Survey Water-Data Reports for the Central Florida region suggested an increasing trend. Table 1 summarizes the groundwater trends in three of the Central Florida USGS wells.

Table 1 Groundwater Trends in Central Florida USGS wells

| USGS Well I.D. | County | Min (year) | Mean | Max (year) |
|---|---|---|---|---|
| 284147081220201 | Seminole | 12.08 (1989) | 13.42 | 14.89 (1994 & 1995) |
| 281722080543001 | Orange | 16.91 (1981) | 18.98 | 21.18 (1995) |
| 281722080543001 | Osceola | 9.38 (1981) | 9.69 | 9.90 (1995) |

Note: All groundwater elevations are in meters above NGVD

Field Investigation and Results

The field investigation identified several problem areas affecting the drainage performance of the dry detention ponds with underdrains. All three of the non-functioning ponds showed several problem areas that affected their performance. In general, the major problems included the presence of low permeability soils and materials in the pond basin, undersizing due to change in design criteria, elevations that differed from design elevations, and improper maintenance techniques.

*Groundwater Levels*

Based on the general groundwater trend in the region, groundwater levels around each pond were monitored. The results suggested that the groundwater elevations could fluctuate as much as 50 cm within a few months. Throughout the monitoring period, the groundwater levels stayed above the elevations of the underdrains, and were typically above the pond bottom elevations. This affected the ability of the underdrain to drawdown the treatment volume.

*Soil Sampling*

Fine sediments were observed at the bottom of all three ponds. Pond 1 had heavy deposition of soils and litter near the inlet. Across the base of Pond 1, a layer of light sediment containing plants, twigs, and leaves. The bottom of Pond 2 was covered with a 20 to 30 cm deep organic layer. This layer contained a large amount of cattail roots and decaying matter. Pond 3 had a thin layer of low permeability soil covering the pond bottom as well. Table 2 summarizes the permeability rates for the soil samples collected from the bottom of the ponds. These rates were considerably lower than the permeability rate associated with FDOT underdrain filters, which is normally about $10^{-2}$ cm/sec (FDOT Design Standards, 1994). These soils formed a 'cap' at the pond bottoms, which inhibited flow into the underdrain filter material.

Table 2 Pond Basin Permeability Rates

| POND | 1 | 2 | 3 |
|---|---|---|---|
| PERMEABILITY RATE (CM/SEC) | $1.63 \times 10^{-4}$ | $7.47 \times 10^{-5}$ | $4.5 \times 10^{-3}$ |

*Design Criteria*

All three ponds were found to be constructed at elevations not true to the engineer's approved design. A large percentage of the pond bottom and stormwater structures in Pond 1 were not at design elevations. A section of the underdrain invert was observed to be between 5 to 23 cm above the adjacent pond bottom. The pond bottom in both Pond 2 and Pond 3 were also observed to be at or below the seasonal high groundwater table.

*Maintenance*

The St. John's River Water Management District (SJRWMD) guidelines on dry detention system maintenance suggest that nuisance species, sediments, and grass clippings from mowing operations be removed from the pond basin (SJRWMD's Applicant's Handbook, 1996). During the nine-month study period, it was observed that all three ponds were not maintained according to the SJRWMD guidelines. In some cases, the cattails had grown to a height of nearly 3 meters. All three ponds were also observed to contain grass clippings at the pond bottom, which generated most of the organic loading in the ponds.

Summary and Conclusions

Based on the problems identified in the three dry detention ponds investigated, a set of field investigation procedures has been developed as a guide to be used in reviewing non-functioning dry detention pond with underdrains. The guidelines include primary investigation procedures (Phase I) and secondary investigation procedures (Phase II). Phase I is intended to take minimal time and effort, while quickly identifying problem areas and suggesting areas of concentration and the order of evaluation for Phase II (Figure 3). Phase II is divided into four sections, A, B, C, and D, in order of increasing effort, with II-D being the most intensive phase. Figure 3 through 7 presents the guidelines, while Table 3 presents the suggested corrective actions.

Table 3 Suggested Corrective Action

| Corrective Action | Suggested Corrective Action |
|---|---|
| 1 | Clean out or Replace Sediment Trap and/or Skimmer |
| 2 | Clean out or Dredge Inlet |
| 3 | Backwash Underdrains |
| 4 | Clean out or Remove Clogging Material |
| 5 | Scrapping and Removal of Low Permeability Soils |
| 6 | Scrapping and Removal of Low Permeability Soils |
| 7 | Decision Based on Identified Problems, Possibly Redesign |
| 8 | Decision Based, Possibly Replace Filter Media |

References

Saint Johns River Water Management District (1996). Applicant's Handbook: Management and Storage of Surface Waters. Palatka, Florida, August 20.

State of Florida Department of Transportation, (1994). "Roadway and Traffic Design Standards". Tallahassee, Florida.

WATER RESOURCES AND THE URBAN ENVIRONMENT 95

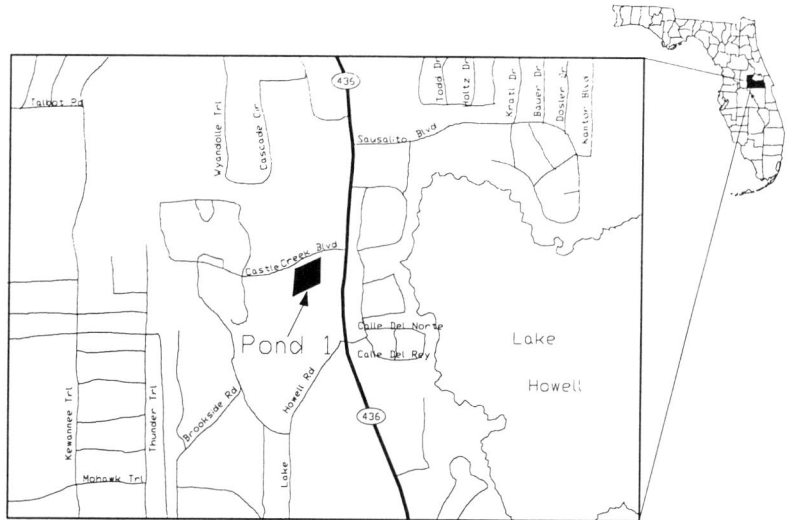

Figure 1  Location Map of Pond 1 in Orange County, Florida

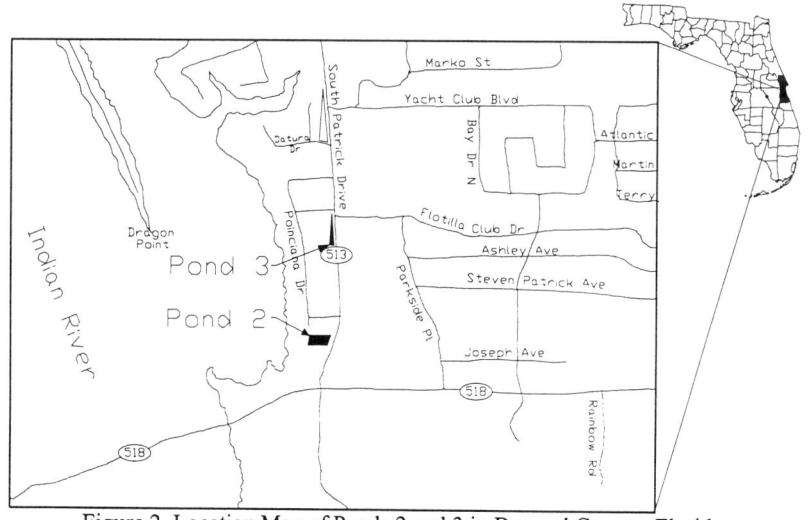

Figure 2  Location Map of Ponds 2 and 3 in Brevard County, Florida

96 WATER RESOURCES AND THE URBAN ENVIRONMENT

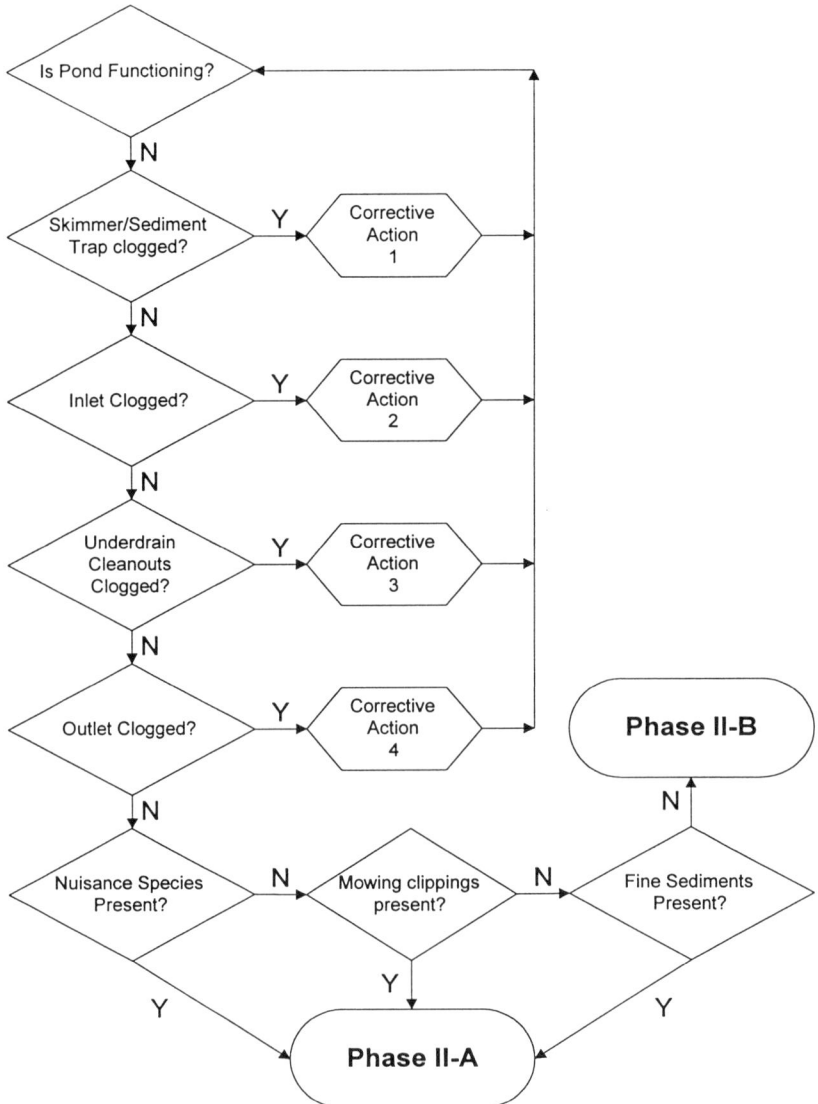

Figure 3 Phase I: Primary Investigation

WATER RESOURCES AND THE URBAN ENVIRONMENT 97

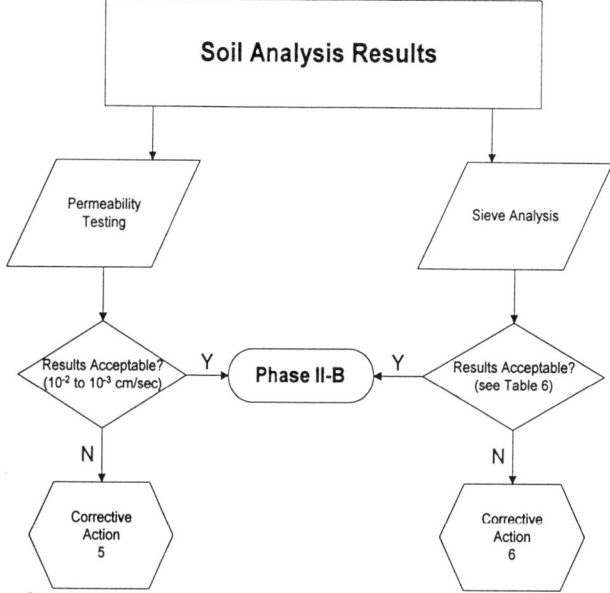

Figure 4  Phase II-A: Secondary Investigation: Soil Analysis

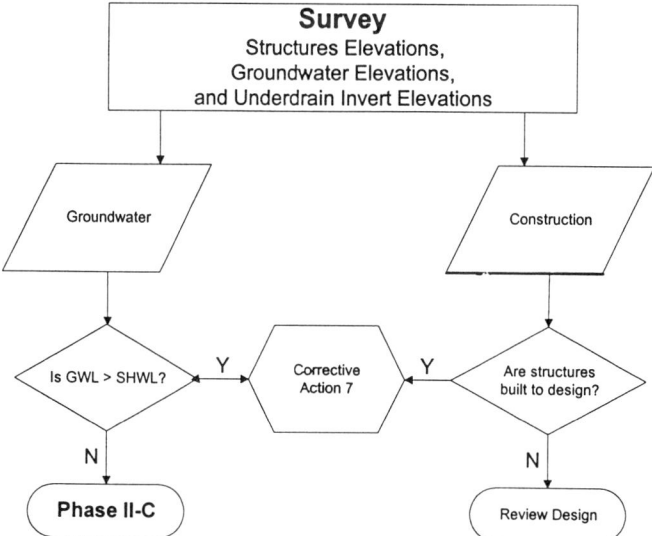

Figure 5  Phase II-B: Secondary Investigation: Survey

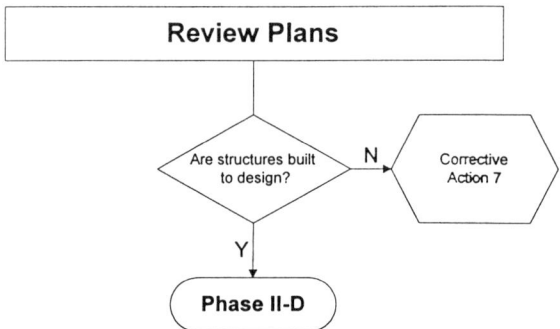

Figure 6  Phase II-C:  Secondary Investigation:  Review Plans

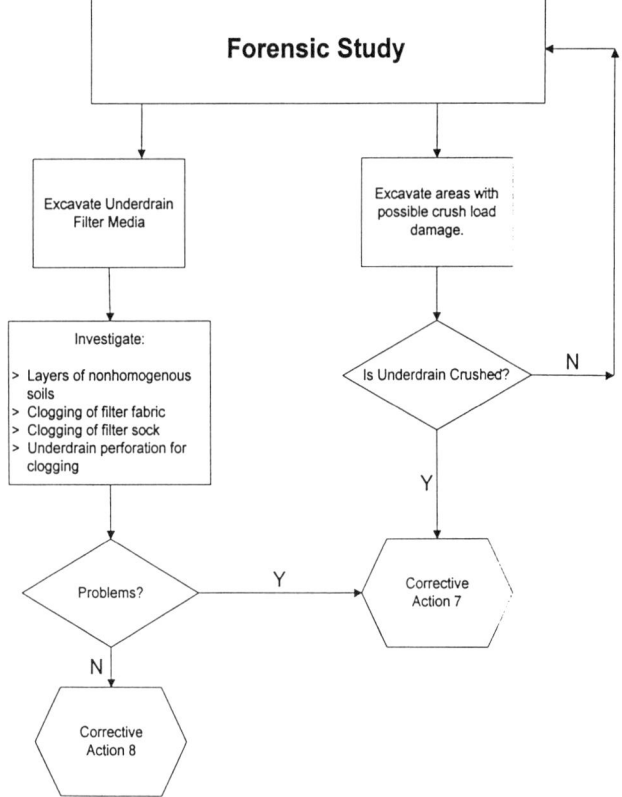

Figure 7  Phase II-D:  Secondary Investigation:  Forensic Study

A Comparative Study of Midwestern Detention Basins

Daniel J. Schuller[1], Student M.ASCE, Eric A. Casanave[2], M.ASCE, and A. Ramachandra Rao[3], M. ASCE

Abstract

The considerable attention given to stormwater management issues in the past twenty years has resulted in countless stormwater management ordinances adopted either by counties, municipalities, or entire regions of a state to limit the amount of excess runoff from developed areas. Developers wishing to construct single or multi-family subdivisions, industrial parks, or other planned communities must work with engineers to create land development plans that meet the stormwater runoff discharge requirements for these ordinances, while at the same time maximizing the number of buildable or saleable lots.

This paper presents the results of a comparative study of four Midwestern stormwater ordinances and their effects on a hypothetical subdivision. The same subdivision was considered in all four cases, and the required detention storage was determined in each case. The study reveals that the size of the detention basin is extremely variable depending on where the subdivision is constructed.

---

[1]Graduate Student, School of Civil Engineering, 1284 Civil Engineering Building, Purdue University, West Lafayette, IN 47907

[2]Civil Engineer, Gwin, Dobson, & Foreman, Inc. Consulting Engineers, 3121 Fairway Drive, Suite B, Altoona, PA 16602

[3]Professor, School of Civil Engineering, Purdue University, 1284 Civil Engineering Building, West Lafayette, IN 47907

## Introduction

Significant attention has been given to stormwater management during the past two decades. It has become common for municipalities, counties, or geographic regions to adopt stormwater ordinances to control the production and release of runoff within the locale. Generally such ordinances require that detention storage be provided by developers to store runoff in excess of the allowable release rate. The developer's goal is to meet the requirements of these ordinances while at the same time maximizing the number of buildable or saleable lots.

In this study, the differences in required detention storage were examined for four counties in the Midwest. These counties - DuPage County, IL, Lake County, IL, Hamilton County, IN, and Tippecanoe County, IN - were selected because, although they are geographically similar, the allowable release rates provided in their respective ordinances differ greatly. In some cases, the release rate is based on the pre-developed hydrology of the site, while in others, it is based solely on a release rate per unit area. Furthermore, different rainfall data are used in each of these counties.

## County Requirements

Hamilton County, Indiana requires staged detention of the 2-, 10-, and 100-year storms. For storms with frequencies less than or equal to 10-years, the allowable release rate from the developed parcel is the rate of flow resulting from a 2-year storm on the pre-developed parcel. For storms with frequencies between 10-years and 100-years, the allowable release rate is the rate of flow from a 10-year storm on the pre-developed parcel. At the time of this comparative study, Hamilton County did not specify the use of particular rainfall data or time distribution of rainfall; these were left up to the designer and were reviewed by the Hamilton County Surveyor's Office on a case-by-case basis. For this study, rainfall depth, duration, and frequency data, as well as the time distribution of rainfall were taken from the study by Purdue et al. (1992).

For sites larger than two hectares (five acres) in Tippecanoe County, Indiana, designers must use the Huff Third Quartile rainfall distribution along with time of concentration and curve numbers calculated using the SCS methodology described in TR-55 (1986). It is required that the SCS computer program TR-20 be used for the detention storage calculations. The Rational Method is acceptable for sites less than two hectares (five acres). The ordinance dictates the rainfall data and Huff Distribution which must be used. These were taken from the County Storm Drainage Manual (Burke, 1983). Several rainfall durations (up to and including the 24-hour storm) are used to determine the largest storage volume required to attenuate the developed runoff from the 100-year storm to a release rate equal to the peak flow from a 10-year storm on the pre-developed site.

DuPage County, Illinois requires that "sufficient storage shall be provided such that the probability of the post development release rate exceeding 0.1 cfs/acre of development shall be less than one percent (1.0%) per year" (DuPage County Stormwater

Ordinance, Section 15-114, p. 43). In other words, the peak release rate from a developed parcel during a 100-year storm shall be less than or equal to 0.1 cfs/acre or 0.00685 m$^3$/(sec*hectare). Rainfall data and time distributions are taken from Illinois State Water Survey Bulletin 70 (Huff and Angel, 1989).

Lake County, Illinois restricts developed release rates to 0.00274 m$^3$/(sec*hectare) (0.04 cfs/acre) and 0.0103 m$^3$/(sec*hectare) (0.15 cfs/acre) for the 2- and 100-year 24-hour storms, respectively. Rainfall depths are taken from isohyetal charts developed by the Lake County Stormwater Management Commission which are based on Illinois State Water Survey Bulletins 70 and 71 (Huff and Angel, 1989 and 1992).

Methodology

In this study, a hypothetical subdivision was investigated in the four counties of interest. The parcel was assumed to be 32 hectares (80 acres) of gently sloping meadow in good condition. The proposed detention basin was located in the southwest corner of the site at the lowest elevation. The site was assumed to consist of soils belonging to Hydrologic Soil Group C. The SCS methodology described in TR-55 (1986) was used to calculate times of concentration and runoff curve numbers for both pre-developed and developed conditions. Pre-developed land use was taken as "Meadow in Good Condition," and resulted in a curve number of 71. Assuming a land use of 0.14 hectare (1/3 acre), single-family residential lots, a post-development curve number of 81 was calculated for the subdivision. The pre-developed time of concentration was 1.63 hours and was heavily dependent on sheet flow time. The same was true for the developed case, although the paved conditions reduced the time of concentration to 1.30 hours. A 2-year, 24 hour rainfall depth ($P_2$ in the TR-55 method) of 70 mm (2.75 inches) was used for all cases. The maximum sheet flow length of 91 m (300 feet) was used in both pre- and post-development cases.

A rectangular detention basin with 4:1 H:V side slopes was used for all counties. The dimensions of the basin were allowed to vary with each county's detention requirements. A maximum water depth of 1.22 m (4 feet), with 0.3 m (1 ft) of freeboard, was used. Each basin was sized such that the water depth was as near as possible to the maximum depth in order to minimize the area of land required for the detention storage facility.

The TR-20 program was used to generate necessary hydrographs and to perform storage routing calculations. Inputs for the program included time of concentration, rainfall depth, storm duration, rainfall distribution (for inflow hydrograph calculation), and basin storage-elevation-discharge relationships (based on assumed basin dimensions and orifice and weir sizes and locations). An iterative approach was utilized, varying basin dimensions and orifice and weir characteristics as necessary to limit the developed outflow rate to the allowable release rate, while minimizing the plan area of the basin. In some cases, it was necessary to perform numerous simulations using various storm durations to determine the maximum storage volume required. This was accomplished in the TR-20 program by repeating the COMPUT/ENDCMP sequence several times in one input set (TR-20 Users Manual, 1982). The iterative approach was continued until the minimum size basin which satisfies all requirements of the local ordinance was determined.

Results

For each county, the allowable and calculated outflows from the detention basin are compared in Table 1. A description of each outlet structure is provided in Table 2. Required storage, basin dimensions, area occupied by the detention basin, and percent of total area occupied by the detention basin are summarized in Table 3.

These tables indicate that, despite its requirement to model various storm durations and use the largest storage volume, the Tippecanoe County drainage ordinance results in the greatest allowable release rate and thus the smallest detention basin size and lowest occupied area. Note that the storage requirement for Tippecanoe County required only the detention of the volume of water necessary to attenuate the 100-year post-development storm.

Interestingly, DuPage County also required only the detention of the 100-year storm; however, the release rate for DuPage County (based only on area and not on hydrologic theory) restricted the outflow much more than did Tippecanoe County. Hence, DuPage County resulted in the largest detention volume of the four counties considered.

Despite its two-stage detention requirements (for both the 10- and 100- year storms), the Hamilton County procedures resulted in the second smallest detention basin and storage volume. Lake County also requires two-stage detention; however, the allowable release rate, like DuPage County, is based only on developed acreage and is not based on pre-developed hydrology as in Hamilton and Tippecanoe Counties. Consequently, using the developed acreage as a basis for the release rate resulted in larger, more conservatively-sized basins than those sized according to pre-developed hydrologic conditions.

Conclusions

Although detention storage is required virtually everywhere in the Midwest, the requirements and the resulting storage basin sizes can differ greatly. In this study, the authors examined the size of the detention storage facility required for the same residential subdivision constructed in four different counties in Indiana and Illinois. In two of the counties, single-stage requirements are enforced, whereas in the other two counties, two-stage requirements are enforced. In the most restrictive case, nearly ten percent of the subdivision area is used for detention storage. In the least conservative case, less than 4 percent of the subdivision is used for detention storage.

Larger basins resulted from ordinances which limited outflows based on area rather than pre-developed conditions. It also happens that the two counties which base outflows simply on developed area are counties which surround Chicago. Due to

Table 1 - Summary of Allowable and Calculated Release Rates

| County | $Q_{allow}$ 2-yr ($m^3$/s) | $Q_{actual}$ 2-yr ($m^3$/s) | $Q_{allow}$ 10-yr ($m^3$/s) | $Q_{actual}$ 10-yr ($m^3$/s) | $Q_{allow}$ 100-yr ($m^3$/s) | $Q_{actual}$ 100-yr ($m^3$/s) |
|---|---|---|---|---|---|---|
| **DuPage - IL** | --- | --- | --- | --- | 0.227 | 0.223 |
| **Lake - IL** | 0.091 | 0.090 | --- | --- | 0.340 | 0.336 |
| **Tippecanoe -IN** | --- | --- | --- | --- | 0.764 | 0.760 |
| **Hamilton - IN** | --- | --- | 0.247 | 0.246 | 0.763 | 0.749 |

Table 2 - Description of Basin Outlet Structures

| County | Outlet Structure |
|---|---|
| **DuPage - IL** | 32 cm orifice at bottom of basin |
| **Lake - IL** | 30.5 cm orifice at bottom of basin, 27.4 cm orifice 36.9 cm above bottom of basin |
| **Tippecanoe - IN** | 61.3 cm orifice at bottom of basin |
| **Hamilton - IN** | 38.4 cm orifice at bottom of basin, 91.1 cm wide weir 80.5 cm above bottom of basin |

Table 3 - Summary of Basin Volumes and Areas

| County | Required Storage ($m^3$) | Basin Length (m) | Basin Width (m) | Area Occupied by Basin (ha) | Percent of Subdivision Occupied by Basin |
|---|---|---|---|---|---|
| **DuPage - IL** | 34370 | 230.1 | 134.1 | 3.09 | 9.7 |
| **Lake - IL** | 22710 | 178.3 | 117.3 | 2.09 | 6.5 |
| **Tippecanoe -IN** | 11890 | 152.4 | 74.7 | 1.14 | 3.6 |
| **Hamilton - IN** | 16630 | 170.7 | 91.4 | 1.56 | 4.9 |

historic flooding, these counties are very conservative with regard to flood control.

It is also interesting to note that the two-stage basins will hold water more frequently than the single-stage basins. This is an important consideration for at least two reasons. First, if a community experiences flooding due to storms with a shorter return period than 100-years, it is imperative to provide storage which is effective for more frequent events. Second, in the case of a wet bottom basin, the basin can have a positive benefit on the water quality of the region by detaining frequent events. By enforcing two-stage detention storage, a community benefits from its detention basins during both routine and extreme storm events.

Due to differences in geography, geology, land use, precipitation, existing development, existing drainage, and vulnerability to flooding, it is not expected that detention storage requirements be identical in all counties in the Midwest. It is, however, a bit startling to find the true variability in the size of the basins resulting from the different requirements.

References

Board of Commissioners of Tippecanoe County. A General Ordinance Establishing Storm Drainage and Sediment Control. Lafayette, IN: Tippecanoe County, 1994.

Burke, C.B. County Storm Drainage Manual. West Lafayette, IN: Purdue University, 1983.

DuPage County Stormwater Management Committee and Environmental Concerns Department: Stormwater Management Division. Countywide Stormwater and Flood Plain Ordinance. Wheaton, IL: DuPage County, 1994.

Hamilton County Surveyor's Office. Letter to All Engineers Working in Hamilton County. 13 June 1994.

Huff, F.A., and J.R. Angel. Frequency Distributions and Hydroclimatic Characteristics of Heavy Rainstorms in Illinois. Bulletin 70. Champaign, IL: Illinois State Water Survey, 1989.

Huff, F.A., and J.R. Angel. Rainfall Frequency Atlas of the Midwest. Bulletin 71. Champaign, IL: Illinois State Water Survey, 1992.

Lake County Stormwater Management Commission. Watershed Development Ordinance. Waukegan, IL: Lake County, 1994.

Purdue, A.M., G.D. Jeong, and A.R. Rao. Statistical Characteristics of Short Time Increment Rainfall. Civil Engineering Report #CE-EHE-92-09. West Lafayette, IN: Purdue University, 1992.

U.S. Soil Conservation Service. Computer Program for Project Formulation - Hydrology. Technical Release No. 20. Washington, D.C.: U.S. Dept. of Agriculture, 1982.

U.S. Soil Conservation Service. Urban Hydrology for Small Watersheds. Revised Technical Release No. 55. Washington, D.C.: U.S. Dept. of Agriculture, 1986.

## Sediment Control At Construction Sites: Is It Being Properly Regulated?

John A. Jacobson[1]

Member, American Society of Civil Engineers

### Abstract

The Clean Water Act of 1987 and the Final EPA Rules of 1990 put in place regulatory measures to control soil erosion and sediment from entering our rivers and lakes. Seven years after implementation of the regulations, soil and sediment control from construction sites appear to be little regulated. This paper begins to investigate this problem and asks why.

### Introduction

Over the winter of ninety-six and ninety-seven a brown liquid flowed down a Ball Mountain stream killing every organism that could not get out of its way. For many months the rains kept coming and man struck over and over again! A contractor, trying to make a deadline, had opened sixty-five acres of highly erodible clays without installing proper soil erosion and sediment control measures. This devastating deed of destruction goes unpunished.

To control this type of devastation, construction of proper soil erosion and sediment control measures during construction projects will take more vigilance by regulators and the public. Contractors and owners need to be fined for their mistakes to help them understand the economic need for proper installation and oversight of construction sites. Simple economic measures put in place will save both fines and construction costs and the environment.

---

[1]Civil Deparrtment Supervisor, URS Greiner, Inc., 3950 Sparks Drive, S.E., Grand Rapids, Michigan 49546

Along the way to supporting these facts, I will request from you, the reader, help in stopping the death and destruction of the organisms caught in this type of a massive event. I wish to enlist your help in asking our regulatory agencies to do their jobs and fine the individuals involved in the carnage. We must document and inform the regulators that we are watching them and they should begin enforcing the regulations that are now in place.

## The Problem

Controlling sediment in storm water from construction sites has always been a problem. "Each year an estimated eighty million tons of sediment washes from construction sites into lakes, rivers and waterways." (Goldman 1.2). Since the Water Quality Act of 1987 shifted the focus from flood control to water quality, controlling pollution caused by dirt picked up by rain water, is one of the most important factor in the protection of the environment. Sediment in storm water is easily controlled with source controls like sediment traps, silt fences, settlement basins, seeding, vegetation and erosion control blankets. However, contractors seldom install proper soil erosion control measures and worse, they never maintain them. Ten years after the Water Quality Act and seven years after EPA's final ruling on storm water discharges (Fifield 54) sediment runoff from construction sites remains a problem and there is little indication that the measures put in place are being enforced.

Controls of erosion of soil on construction sites are the responsibility of the contractor. The contract transfers this task to him. The contractor must conduct the grading operation in a manner as to minimize the threat of pollution. He must install and maintain soil erosion control measures such that sediment, fuels and pollutants are contained to the site and properly handled. Improper installation and maintenance, or the total lack of controls at all, have become the rule at most construction sites.

Several reasons exist for the lack of vigilance by contractors, but three stand out. First, construction pollution and sediment in storm water is hazardous to few, but small animals and fish. Environmental impacts are harder to see. They tend to build slowly and not produce dramatic results for many years, when it may be too late or impractical to correct the problem (Goldman 1.2). With little visible threat, few take the hard line and hold contractors to their responsibility.

The second, and a more significant a reason, is profits. Installation and maintenance of sediment control measures on construction sites takes manpower and time. Manpower and time cost money, especially time. Impacts on the schedule have always driven the way construction projects progress. Sometimes, truth be told, sediment control measures are inadequately or improperly installed. Especially if the sediment control measures hamper construction activities or scheduling of help and equipment (Tilton 32).

Third and most significant is the lack of regulatory oversight. A review of the Michigan Department of Environmental Quality (MDEQ) web site indicated that almost no fines have been levied against contractors or owners for construction site sediment problems. The Surface Water Quality Division (SWQ) of the MDEQ appears to be more interested in Watershed management than site specific controls (Allen). Since 1993, the state logged from twelve to eighteen cases per year. A review of the cases in 1996 and 1997, indicates that only one case resulted in a fine for a soil erosion and sediment control violation and the fine was minor for the blatant violation (Thompson). Without a viable cost implication, contractors choose schedule and profits over environmental impacts.

The effects of not containing sediment and pollution on the construction site is more than the destruction of fish beds and the tainting of drinking water for small animals. Erosion and sedimentation cause both environmental and economic impacts. Sediment from construction sites reduces storage volumes within lakes and streams, leading to the threat of flooding. The economic costs of cleaning reservoirs to lessen the flooding threat is very costly. The environmental impacts include the destruction of stream bottoms and banks, the reduction of spawning grounds and food supplies (Goldman 1.3). Along with the environmental and economic impacts on the society, improper sediment control also places the owners of projects liable for fines as high as twenty five thousand dollars a day, the teeth that was placed in the Water Quality Act of 1987 and the final EPA rules of 1990.

With the Clean Water Act coming up for review and renewal, in this next legislative session, the major future implication of not controlling construction site runoff, is the increased fines and oversight that may be introduces in the proposed regulations. Or at least we can hope for more regulatory oversight. Even if no new regulations are warranted, increased vigilance by regulatory agencies enforcing existing laws could place owners in harms way for the oversights of contractors. All this regulation is not just to satisfy governmental mandates. Not controlling storm water on construction sites leads to destruction of the environment and possibly places the society in harms way of flooding.

## **Call To Action**

Solutions to the soil erosion and sediment problem on construction sites are as varied as the problem. Contractors will argue that soil erosion from construction sites is but a fraction of soil and sediment deposited in the rivers and lakes from agricultural practices and other sources. True as this may be, the sediment runoff from construction sites is as high as twenty times more concentrated (Goldman 1.2) and thus much more damaging to the environment as a single event. The fact is the construction industry and project owners have a conflict of interest and have not stepped up to the plate, as was hoped, and taken action to protect the environment or themselves. It is time for the regulatory agencies to begin enforcing its rules and handing out fines of sufficient size to get someones attention.

Fines levied today are less than the costs of not installing any soil erosion control. We have put a high price on scheduling and profits and a low price on the environment. As indicated in John McCullah's article, "Triumph on the banks of the Sacramento River", the engineer or erosion control specialist can not properly enforce the installation of designed soil erosion control measures without regulatory help. The construction industry only cries conflict of interest and accuses the professional of creating a job for himself when he points out sediment problems or construction concerns at a site (McCullah 60).

Contractors have not excepted the fact that properly installed and timely soil erosion control measures actually saves money on a project. This simple fact is eluding the entire industry and the project owners as well. The main reason for this fact being left out of the cost equation during construction is simple. The environmental and economic costs of the sediment damage have not been transferred back to the project. The cost to the construction project appears the same whether the soil stays on the site or is taken down stream to be handled by the neighbors or society.

Being an independent minded person myself, it is hard to advocate regulatory control. But, the implementation of fines appears to be the only means to transfer the environmental and economic costs of sediment damage back to individual projects. If fines were real and not merely a threat, contractors would realize that soil erosion and sediment control measures can be cost effective. With the treat of real fines, contractors will begin to properly use Best Management Practices (BMPs) and owners will be protected from further liability.

It is time to transfer the environmental and economic costs of construction sediment back to individual projects. It is time to put a price on the mismanagement of erosion control so that contractors and owners will understand, or begin to understand the cost implications. In implementing further and vigilant regulatory action, I am not advocating that the focus of watershed management be changed at the EPA or MDEQ. I am only suggesting that additional regulatory action be taken to get contractor's attention. It is time to study the records that are supposed to be kept on the construction sites that have proper permits and force sites that don't have proper permits to get them (Gilbert 48-53). It is time to find out if the regulations that were implemented in 1990 are sufficient to protect the nations water as was hoped.

## Opposing Views

There are two opposing views to that which are suggested here. They are opposing the positions only because they do not go far enough to address the problem of construction site sediment control. The first is a best management practice (BMP) approach, similar to that which is now being followed, with additional Certified Professional in Erosion and Sediment Control (CPESC) inspectors hired by the owner. This is advocated as the solution in Jerald Fifields article on "Erosion and Sediment Control" (Fifield 55). The second view is a watershed management approach as identified in Edward Wagners article "Watershed Management Addresses Area Water Quality Issues" (Wagner 82). Both of these views have their place in the area of improving water quality and the controlling of sediment within our rivers and lakes. However, these views are not sufficient to get the contractors attention for specific construction sites.

Fifield suggests that the hiring of CPESC professionals will help address the lack of inspection on the site and protect the owner from potential fines. His appeal is to owners as a way to protect themselves against the liability of possible fines. Fines are only a threat if they are actually being imposed. To date, fines have not been used sufficiently to worry owners. Owners like any good businessman, calculate that if there is only a tenth of a percent chance that a fine is possible and my competitor is not hiring CPESCs, why should he? Once more regulatory control is seen, then there may be a place for Mr. Fifields Certified Professional in Erosion and Sediment Control.

Watershed management indeed addresses area water quality issues. Since construction sites are but a fraction of the sediment problem (Goldman 1.2) and other forms of sediment injected into the environment are less identifiable, watershed management is good. Watershed management should be used with regulatory control. Construction sites are short term polluters and very mobile compared to other causes of sediment. Watershed management has a hard time addressing the individual construction site sediment problem. Watershed management and regulatory control (individual fines) should be used together to properly control the sediment within our rivers, lakes and streams.

## Conclusion

The construction of proper soil erosion and sediment control measures during site construction projects will take more vigilance from regulators, design professionals and the general public. Fines must be levied to get the contractors attention and to help the contractor understand the economic advantages for the proper installation of soil erosion and sediment control measures.

The solution to the problem of environmental degradation due to sediment is clear. We all must be more vigilant. Regulators must combine a measure of support and pain. We must show contractors the cost of their mismanagement and help them save money and protect the environment at the same time. Sufficient rules and regulations are in place for the reasonable reduction of water related sediment control problems created by contractors (Fifield). We have only to use them. The regulatory agencies have for the past seven to ten years used voluntary means to try and control the sediment problem. It is time to begin handing out fines of sufficient size to get someones attention.

The creation of the Erosion Control Patrol by the International Erosion Control Association (Trotti 6) allows you, the general public, to get involved in the process of helping to educate the contractor without being in the direct line of retaliation. The vigilance necessary to protect our waterways is all of our responsibility. Join the Erosion Control Patrol and help make a difference.

It's time to stop the devastation and place the cost of soil erosion back where it belongs. The protection of our water and waterways has been put on hold while we were saving the contractors feelings. It's time to get tough. It's time to fine!

## References

Allen, Deb, "Success to Date" Michigans's Nonpoint Source News, Summer 1996, http://www.deq.state.mi.us/swq/newsltr3.htm.

Fifield, Jedrald S., "Erosion and Sediment Control; The Results are not always clear" EROSION CONTROL, Vol. 4, No. 5, pp 54,55.

Gilbert, Anne, "NPDES Stormwater Permit Requirements for Construction in the Southeastern US" EROSION CONTROL, Vol. 3, No. 5, pp 48-53.

Goldman, Steven J., Katarine Jackson, Taras A. Bursyztynsky, Erosion & Sediment Control Handbook, McGraw-Hill, Inc., 1986, pp1.1-1.13.

McCullah, John, "Triumph on the banks of the Sacramento River" EROSION CONTROL, Vol. 4, No. 5, pp 60,63.

Tilton, Joseph Lynn, "Keeping Soil On-Site; Major Metro Construction Challenge" EROSION CONTROL, Vol. 4, No. 7, pp 32-41.

Thompson, Kate, Tom Tohrer, "Penalty Summaries and Settled Cases" Michigan Department of Environmental Quality (MDEQ) Web Page, 4/22/97, http://www.deq.state.mi.us/swq/.

Trotti, John, "Launching the Erosion Control Patrol" EROSION CONTROL, Vol. 4, No. 4, p 6.

Wagner, Edward O., Ronald F. Ott, James W. Dunn, "Watershed Management Addresses Area Water Quality Issues" Pollution Engineering, Vol. 29 Issue 1, pp 82-88

## Drainage Problems in an Urbanized Watershed

by
Veronica J.B. Morgan[1], P.E., M. ASCE
Michele Good Burton[2], A.M. ASCE

### ABSTRACT

In October and December 1994, much of southeast Texas was devastated by widespread flooding. The City of College Station, Texas, by most standards, was spared from this devastation. The city did experience flood damage with much of it occurring within a single watershed. In response to those floods, College Station has undertaken the task of preparing a city-wide drainage master plan. There are seven major watersheds, with more than 25 miles of creeks, all of which contain large floodplain areas, within the city limits of College Station. The most urbanized of these watersheds, Bee Creek, contains approximately 8.7 square miles or approximately 30% of the total area in College Station. The Bee Creek watershed originates on the Texas A&M University campus and is comprised of approximately 55% low, medium and high density residential, 10% commercial development and 35% undeveloped property, some of which is floodplain property. Originally, the creeks had gently sloping banks and most residents considered them an amenity. The majority of development along the main channel of Bee Creek and its tributaries, is low density residential, most of which occurred prior to the passage of the city's drainage ordinance in 1987. As development continued prior to ordinance control, runoff and velocities increased. This decreased bank stabilization and increased floodplain elevations. The challenges and potential solutions in the master plan for Bee Creek lie in the answer to the following question: How do you solve flooding problems within a creek that has few drainage easements, limited available access, homes constructed along both banks and accomplish it cost effectively without looking like a "sea" of concrete? We will discuss these challenges, public involvement, political discussions, recommended solutions, and the current status of the project.

---

[1] Asst. City Engineer, City of College Station, P.O. Box 9960, CS, Texas 77842
[2] Engineering Assistant, City of College Station, P.O. Box 9960, CS, Texas 77842

## Background

In October and December 1994, much of southeast Texas was devastated by widespread flooding. The City of College Station did not experience the more severe flooding that wrought the greater area, but did receive substantial damage to streets, homes and property. Within a 24 hour period on October 16 and 17, College Station received 38 centimeters (14.96 inches) of rainfall, with 26.5 centimeters (10.5 inches) of that occurring within a 6 hour period[1]. Again in December, the city received 15.7 centimeters (6.2 inches) within a 24 hour period[2]. The recurrence interval for the October storm has been calculated at more than a 300 year storm event. In 1994, many cities in Texas received well above their normal annual rainfall amount. Of all the large/medium metropolitan areas, College Station led the state with 145.8 centimeters (57.4 inches) of rainfall for the year, 147% of it's normal annual rainfall[2]. The monthly average rainfall for College Station for the months of October and December is 8.6 centimeters (3.4 inches) and 7.6 centimeters (3.0 inches), respectively[3]. During 1994, the monthly rainfall for October and December well exceeded that amount at 47.8 centimeters (18.8 inches)[1] and 27.2 centimeters (10.7 inches)[2], respectively.

The City of College Station is located within southeast Texas approximately 145 kilometers (90 miles) northwest of Houston and 290 kilometers (180 miles) southwest of Dallas, and is home to Texas A&M University. The city has been rapidly growing since the late 1980's, with a population increase from 40,000 in 1980 to it's present 62,000. The city lies immediately adjacent to and south of it's sister city, the city of Bryan, which has also seen growth in the recent past. Within the combined cities, the current population is approximately 120,000.

The City of College Station consists of seven major watersheds, with more than 40 kilometers (25 miles) of creeks. The most urbanized of these watersheds is Bee Creek and was the one watershed that experienced the most damage to private homes and property during the 1994 storms. Spawned by these storms was increased public awareness for drainage maintenance and improvements necessary to the overall drainage system. Due to past budgetary constraints, the City had very limited funds to allocate to drainage maintenance or capital activity in that area. To compensate for this lack of funding, shortly after the 1994 storms, the City began the process of adopting a drainage utility fee, which was fully implemented in October 1997. The adoption process included discussions with Texas A&M University (TAMU), as drainage from the campus facilities contributes runoff to three of the seven major watersheds in College Station. Rather than being assessed a drainage utility fee based on the amount of impervious cover on campus, the Council agreed to allow TAMU to perform a drainage master plan of the campus, which would determine the current runoff contribution by TAMU and their increase in runoff since the adoption of the 1987 Drainage Ordinance. TAMU agreed to mitigate the increase and detain their runoff on campus to the 1987 runoff amounts. In some cases, they also agreed to mitigate over and above that, decreasing their peak runoff to 80% of the 1987 levels. Also during this time period, the City of College Station began working on a revised Comprehensive Plan (Comp Plan) for the city. It was decided as part of this plan that city staff would address the drainage portion of the plan and it would not be included in the scope of services for the overall Comp Plan. These set forth the planning and ultimate funding mechanisms for drainage capital improvements throughout the city.

The City of College Station adopted it's current Drainage Ordinance in 1987. There were a few floodplain ordinances prior to this adoption, but they were relatively minor in nature. These previous ordinances did not contain the necessary language to qualify the city for inclusion in the National Food Insurance Program (NFIP). The 1987 ordinance allowed that participation in the NFIP and also defined very specific design criteria for drainage systems. Within this ordinance there are design criteria detailing allowable street drainage, acceptable open channel designs as well as easement dedication criteria. The adoption of the 1987 ordinance was in an attempt to standardize the design for drainage systems in College Station and avoid problems in subdivision design that had occurred in the late 70's and early 80's.

## Master Plan Process
*Damage Assessment*
One of the first steps taken as part of the Drainage Master Plan process was to assess all damage that occurred during the 1994 storms. Historical records were evaluated, personal interviews conducted and information regarding street closures, washouts and other emergency activity was obtained from police and emergency operating center records. With this data compiled, a "damage map" was produced locating and categorizing all damage. The damage was grouped into 3 basic categories. *Structural*, which consisted of water damage to homes or other habitable structures and damage to streets which rendered them impassable. *Minor structural/non-structural*, which consisted of erosion problems along creeks which caused bank failures (but did not jeopardize any habitable structures) and damage to fences, retaining walls or flumes. And *nuisance* which consisted of street flooding, yard flooding, and other high water locations that did not cause damage to habitable structures or loss of life, but did cause inconvenience.

*Technical Review Committee/Goals*
With the background information gathered, a technical review committee was established to give staff a sounding board representative of the technical community. This review committee was charged with overseeing the progress of the master plan and assuring it was proceeding in a technically sound manner. In one of the early technical review committee meetings, the goal of the master plan was decided. Given the storm recurrence interval and the type of damage that occurred around the city, the goal was clear. To focus on reducing the structural damage that occurred and if time and funds were available in the future, address the minor and non-structural damage. Occurring concurrently with this process of

assessing damage and focusing efforts on the hydraulics and engineering aspects of the master plan was the adoption of the Comp Plan for the city. Although the Comp Plan did not address drainage in any detailed fashion, recommendations regarding land use, parkland, and open space directly affected drainage facilities. With that in mind, the drainage master plan's focus broadened to include the use of floodplain and drainage corridors for park linkages, open space preservation and multi-use facilities and although not preclude, at least limit, development within these areas.

*Analysis*
Bee Creek was the first of the seven watersheds to be analyzed as part of the Master Plan. Existing HEC-1 runs were reviewed for the basin and found to be somewhat lacking in detail and confusing. The existing HEC-1 runs were produced at various times by several authors, all of which had conflicting results in many key areas. The staff and the technical advisory board agreed that the HEC-1 analysis would be rerun and checked for accuracy. The backwater analysis for the project was completed using the HEC-RAS program in conjunction with the BOSS-RMS program. Cross sections for the model were generated utilizing the BOSS-RMS program and two foot topographic maps of the city produced in 1994. The floodplain location from this analysis was compared to the current Federal Emergency Management Agency (FEMA) floodplain map and the damage map and reviewed again for accuracy.

*Challenges*
With the damage map and floodplain locations identified, it was apparent why the damage occurred. The bulk of the subdivisions within this watershed were developed prior to ordinance control. The homes were constructed within the floodplain, near creeks that have eroded severely over time. In addition, these creeks have little or no access to them and many portions of them were not contained within drainage easements, thereby limiting the amount of maintenance that had occurred within them over time. The access and easement problems were already being addressed for new subdivisions in the city's drainage ordinance, but would not be easy to remedy in these highly developed areas constructed prior to ordinance control. Public support for the project would also be difficult. There would be sections of the creek that would be difficult to reach with any machinery and if any channel improvements would be done, backyards would be decreased in size. But given the current channel bank conditions, with it's bank failures, sloughing and loss of vegetation, something would have to be done within the channel to avoid future erosion that would soon endanger structures.

*Options*
Several different options were analyzed, all with the goal of reducing the floodplain below finished floors of structures in the area. The analysis included culvert improvements downstream and within the area, detention options upstream of the flooded area, and several different types and sizes of channel improvements. Detention, channel improvements, and a combination of the two were successful at

reducing the water surface elevations enough to afford relief. Even with the difficulty of access, the benefit cost analysis revealed that channel improvements would be the most cost effective solution. There was concern that property owners near the creek where drainage easements were not already in place would see any channel improvement to reduce flooding as an intrusion into their personal property, especially if they were not personally damaged during the 1994 floods.

*Public Involvement*
In an attempt to achieve "personal buy-in" by the public into the project, two public meetings were held. The first meeting was a staff directed meeting with the property owners along the creek and the general public to educate them on the damage that had occurred in 1994, the goals of the master plan, the options analyzed, the resulting possible solutions, and finally the recommendations of the plan in the Bee Creek area. There were diverse opinions in the meeting that ranged from "don't work in my backyard" to "please come fix the creek" to "won't detention work upstream or culvert improvements downstream, so you don't have to come in my backyard" to "please don't make this a sea of concrete" to "use concrete so that it is maintenance free". Although the opinions varied, there was consensus that something had to be done to alleviate the problem in the future. Many who feared the intrusion were somewhat appeased with staff's recommendation that a hike/bike path be incorporated into the design of the channel improvements and located at the rear of their property. Such a hike and bike trail would forward the goals of the Comp Plan for the use of floodplains as park linkages, open space and multi-use facilities. Those that were appeased seemed to be those who also supported that goal. Others were not as pleased and viewed that aspect as a further intrusion into their privacy. The second public meeting was held at a city council meeting. The purpose of bringing the issue to the Council was to present the flooding problems, possible solutions, recommendations, and to allow public input with Council present so that a decision could be made involving all levels.

*Political Discussion/Recommendation*
The majority of the discussions regarding the use of floodplain and drainage areas as multi-use facilities, open space and greenbelts occurred as part of the Comp Plan adoption. These public hearings were held prior to those discussing the Bee Creek project. During those lengthy hearings, it was apparent that there was a vocal section of the public that desired the use of floodplains and drainageways for this purpose. The Council recognized this desire and adopted goals within the Plan to forward these ideas. At the same time during the Comp Plan adoption, they assured that floodplain development could still occur but would be limited in several key floodplain areas. Given this, staff's recommendation with the Bee Creek improvements was to proceed with channel improvements in the area with the incorporation of hike/bike paths, a pedestrian bridge to connect the path to the subdivisions and adjacent park, and improve the aesthetics of the channel by minimizing the amount concrete used in the design. After hearing the presentation

and allowing public input, Council directed staff to pursue channel improvements including all of staff's recommendations.

## Status and Conclusion

The project was approved by Council and consultant selection completed in November 1997. The consultant is designing both the Bee Creek channel improvements and the bike path project. As of this writing, both projects are in the preliminary design phase. The project is anticipating dedication of easements wide enough for incorporating hike/bike paths and a low water pedestrian crossing to connect the path system at a city park. The path system will provide linkages for property owners and the public to the arboretum, three city parks, and with the possibility of a future extension to the high school and the city library. Future maintenance access will be taken from these paths with limited street access through the adjacent parks. The preliminary design has also explored using materials other than concrete where necessary for slope stabilization and grassed or vegetative slopes where possible.

The design firm will be conducting two additional public meetings during the early design phases to keep the property owners informed of the project. This is essential to the success of the project due to the absence of adequate access and drainage easements throughout the project area. The property owners will need to be a continued part of the process so they will feel comfortable with the design and understand the need to dedicate easements to the city. Also, considering the divided opinion among these property owners there is a great need for interaction amongst the city, the design firm and the property owners. The project will be the first of its kind in College Station, hopefully with more to come in both retrofit situations and new development standards.

## References

1. Texas Climatic Bulletin, Office of the State Climatologist, College of Geosciences and Maritime Studies, Department of Meteorology, Texas A&M University, Volume 7, Number 10, October 1994

2. Texas Climatic Bulletin, Office of the State Climatologist, College of Geosciences and Maritime Studies, Department of Meteorology, Texas A&M University, Volume 7, Number 12, December 1994

3. The Climate of Bryan/College Station, Texas, Office of the State Climatologist, College of Geosciences and Maritime Studies, Department of Meteorology, Texas A&M University, Revised November 1984

Modeling the Great Lakes Net Basin Supplies Series Using SAMS

J. D. Salas[1], C. H. Chung[1], W. L. Lane[2], and D. K. Frevert[3]

Abstract

SAMS is newly released computer software package that deals with the stochastic analysis, modeling and simulation of hydrologic time series such as annual and seasonal streamflows. It is written in C and Fortran and runs under modern Windows operating systems such as Windows NT and Windows 95. A variety of models and modeling schemes are included in SAMS for both single site and multisite systems such as univariate and multivariate AR, ARMA, PARMA, and disaggregation approaches. An application to the Great Lakes net basin supplies series is performed, in which seasonal disaggregation models using two different modeling-generation schemes are fitted to the historical data and synthetic multivariate realizations are generated and compared.

Introduction

Stochastic analysis, modeling and simulation of hydrologic time series have been widely applied in practice for water resources planning and management studies. Examples include reservoir capacity determination, hydraulic structure reliability evaluation, and irrigation system evaluation under uncertain water deliveries (Salas et al, 1980; Loucks et al, 1981).

Stochastic models are generally needed for synthetic simulation. A number of such models have been suggested in literature (Salas, 1993). Choosing one type of model or another for the data at hand depends on several factors such as, physical and statistical characteristics of the process under consideration, complexity of the system, and overall purpose of the simulation study. Since the statistical characteristics of the data need to be preserved by the model, a standard step in hydrologic simulation is to determine the historical statistics. The next step is to estimate the model parameters and

---

[1]Professor and Graduate Student, respectively, Hydrology and Water Resources Program, Dept. Civil Engineering, Colorado State University, Fort Collins, CO 80523.

[2]Consultant, 1091 Xenophon Street, Golden, Colorado.

[3]Hydraulic Engineer, U.S. Bureau of Reclamation Technical Services Center, P.O. Box 25007, Lakewood, Colorado 80225.

to test whether the model represents reasonably well the process under consideration. Then, based on the fitted model, simulations can be performed.

A variety of mathematical and statistical software have been developed in the last second decades. For example, SAS/ETS, SPSS, ITSM, MINITAB, STATGRAPHICS, STATVIEW, IMSL, and MATLAB, etc., are well known packages for computations of varied degree of sophistication. These packages can be very useful for standard time series analysis of hydrologic processes. However, despite of the availability of such general purpose programs, specialized software for simulation of hydrologic time series such as streamflow have been attractive because hydrologic time series generally have special interest properties such as periodic mean, variance, covariance, and skewness and in many cases long memory. Besides, many of the stochastic models useful in hydrology and water resources have been developed specifically to fit the needs of water resources, for instance temporal and spatial disaggregation models. For example, HEC-4 (U.S Army Corps of Engineers, 1971), LAST (Lane and Frevert, 1990), and SPIGOT (Grygier and Stedinger, 1990) are specifically oriented software for hydrologic time series simulation. The LAST package was developed in 1978 by the Bureau of Reclamation for the purpose of modeling univariate and multivariate hydrologic time series. Even though various additions and modifications have been made to LAST over the past two decades, the package has not kept pace with both advances in time series modeling and in computer technology especially in displaying the output graphically.

SAMS is specially designed for stochastic analysis, modeling and simulation of both annual and seasonal hydrologic time series (Salas et al, 1996). It is written in C and Fortran and runs under PC Windows environment including Windows 95 and Windows NT. It communicates with the user through dialog boxes and shows the results in graphs and tabules. A brief description of the SAMS package is provided below followed by an example of modeling and simulating the Great Lakes net basin supplies series.

Brief Description of SAMS

The main functions of SAMS are: (1) statistical analysis of data, (2) stochastic model fitting including parameter estimation and testing, and (3) synthetic data generation (Figure 1). Both single site and multisite, annual and seasonal data can be analyzed and the results can be displayed in either graphical or tabular forms, or written to output files.

Data analysis consists of time series plots, normality checking,

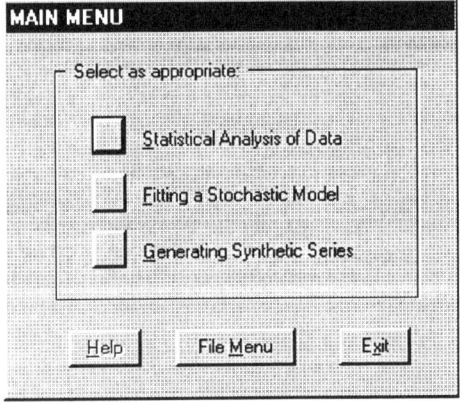

Figure 1. Main menu of SAMS

transformation, and calculation of statistical characteristics. Plotting the data may help detect trends, shifts, outliers, or errors in the data. Probability plots are included for verifying the normality of the data. Data can be normalized by using different transformation techniques such as logarithmic, power, and Box-Cox transformations. SAMS computes a number of statistical characteristics of the data such as mean, standard deviation, skewness, serial correlations (for annual data), season-to-season correlations (for seasonal data), and cross-correlations (for multisite data). They are important in investigating the stochastic characteristics of the data. Figure 2 shows the menu for desired statistical characteristics of the data.

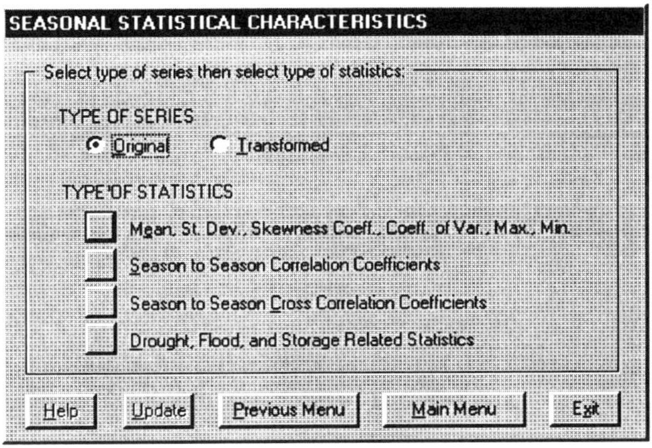

Figure 2. Statistical analysis of seasonal data

Model fitting includes parameter estimation and model testing for alternative univariate and multivariate stochastic models. Currently, SAMS includes the following models: (1) univariate ARMA(p,q), (2) univariate periodic PARMA(p,q), (3) multivariate autoregressive MAR(p), (4) multivariate periodic autoregressive MPAR(p), (5) spatial disaggregation, and (6) temporal disaggregation. Two estimation methods are available, namely the method of

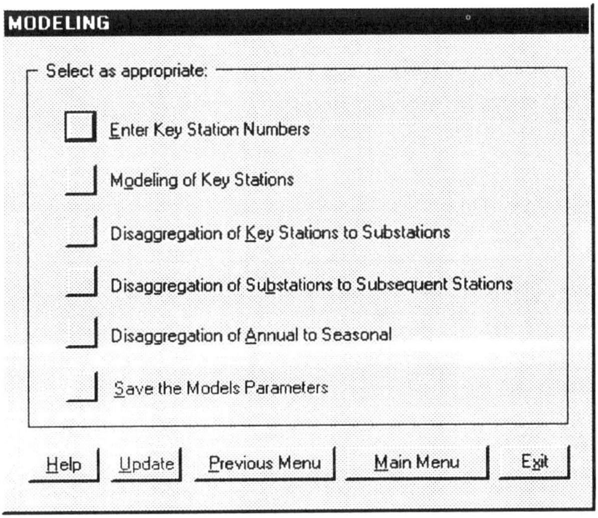

Figure 3. Fitting a seasonal disaggregation model

moments (MOM) and the least squares method (LS). MOM is available for all the models while LS is available only for univariate ARMA(p,q) and PARMA(p,q) models. Regarding annual disaggregation models, MOM is used for parameter estimation based on Valencia-Schaake or Mejia-Rousselle methods while for annual-to-monthly disaggregation Lane's condensed method can be used in addition to the above two. Figure 3 shows the menu for fitting multivariate seasonal disaggregation model.

For stochastic modeling and simulation at several sites in a water resources network system based on disaggregation, two modeling-generation schemes are included which are based on defining a number of key stations, substations, and subsequent stations. Generally the key stations are the farthest downstream stations, substations are the next stations upstream, and subsequent stations are next further upstream stations. Scheme 1 fits a univariate ARMA(p,q) model to the sum of the annual data of all the key stations. Then, that sum is disaggregated into the key stations annual data subsequently, such annual data at key stations are disaggregated into annual values at the substations which in turn are further disaggregated into annual data at the subsequent stations. Scheme 2 fits a multivariate MAR(p) model to the annual data for the key stations and the rest of the disaggregation into substations and subsequent stations is done in a similar manner as in the first scheme. In addition, if monthly data are desired, the annual values at all stations are further disaggregated based on temporal disaggregation approach.

Data generation is undertaken based on the models, approaches, and schemes as mentioned above. The model parameters for data generation can be those which are estimated by SAMS or they can be provided by the user. The statistical characteristics of the generated data are presented in graphical or tabular forms along with the historical statistics of the data. They can be printed and/or written on special output files.

Case Study

To demonstrate the functions of SAMS, an application to the Great Lakes net basin supplies series is performed. The monthly net basin supplies (NBS) records of Lake Superior, Lake Michigan-Huron, Lake St. Clair, Lake Erie, and Lake Ontario, for the period 1900-1989 are used. Lake Superior is the uppermost lake in the system. Its outflow goes into Lake Huron through St. Marys River. Lake Michigan also discharges into Lake Huron, but is considered hydrologically as one water body as Lake Huron because of the close lake levels of the two lakes. Water coming from Lake Michigan-Huron goes into Lake St. Clair through St. Clair River, then through Detroit River into Lake Erie, and finally through Niagara River into Lake Ontario (Quinn, 1992). Seasonal disaggregation models based on Valencia-Schaake and Lane's condensed method were used to fit the multivariate data. Disaggregation Schemes 1 and 2 mentioned above were adopted for modeling and generation. Note that the monthly data were normalized through logarithmic transformation before being fitted by the models. In the spatial disaggregation step, the system was divided into two groups: Group 1 consists of Lake Michigan-Huron as the key station (site N.2) and Lake Superior as the substation (site N.1), and Group 2 consists of Lake Ontario as the key station (site N.5) and Lake St. Clair (site N.3) and Lake Erie (site N.4) as the substations. In the Scheme 1, AR(1), ARMA(1, 1), and ARMA(2, 2) models were used to fit the sum of key stations' annual NBS, and in Scheme 2, the annual NBS of the key stations were fitted by a MAR(1) model. In both cases, the model parameters were estimated by the method of moments (MOM). In the annual-to-monthly disaggregation step, all five sites were considered in one group in order to preserve the monthly cross-correlations between sites.

Based on the fitted models, one hundred samples of synthetic multivariate monthly NBS each 90-years long were generated. The average annual statistics computed from the generated samples including annual mean, standard deviation, skewness coefficient, serial correlation, serial cross-correlation, rescaled range, longest drought, and maximum deficit, were compared with their historical counterparts. The results show that the disaggregation models using Scheme 1 based on ARMA(1, 1) and ARMA(2, 2) models are better in preserving historical serial correlations, cross-correlations, and storage and drought related statistics than the other models. However, non of the models were capable of reproducing well the storage and drought statistics for lake St. Clair. Improving the preservation of such statistics will require either a significantly higher order model or a different modeling scheme such as shifting level models. The analysis of the generated monthly values based on Scheme 1 using either ARMA(1, 1) or ARMA(2, 2) model reproduce reasonably well the monthly statistical characteristics including the means, standard deviations, skewness coefficients, month-to-month correlations, and month-to-month cross-correlations. Figures 4, 5, and 6 show some of the results.

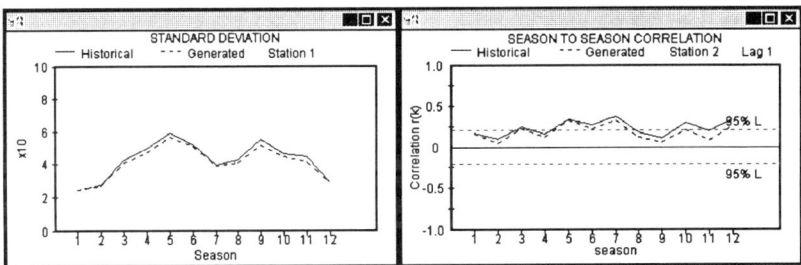

Figure 4. Comparison of monthly standard deviations for Site 1 (Lake Superior) based on Scheme 1 using ARMA(1, 1) model.

Figure 5. Comparison of month-to-month correlations for Site 2 (Lake Michigan-Huron) based on Scheme 1 using ARMA(1, 1) model.

Acknowledgment

This paper was written with support of the U.S Bureau of Reclamation's research program under the advanced Hydrologic Techniques Project and computer support from the CCHE program of excellence in water resources of the Colorado State University Civil Engineering Department. Support from both these sources is greatly appreciated. In addition, partial support from "Stochastic Modeling, Simulation and Forecasting of Hydrometeorological Processes - the Great Lakes System" sponsored by the Great Lakes Environmental Research Laboratory, NOAA is also appreciated.

References

Grygier, J.C., and Stedinger, J.R., 1990.,"SPIGOT, A Synthetic Streamflow Generation Software Package" , technical description, version 2.5, School of Civil and Environmental Engineering, Cornell University, Ithaca, N.Y.

Lane W.L. and Frevert, D.K., 1990. Applied Stochastic Techniques, Personal Computer Version 5.2, User's Manual, Water Resources Sciences, U.S. Bureau of Reclamation, Denver, Colorado.

Loucks, P., Stedinger, J.R. and Haith, D.A, 1981. Water Resources Systems Planning and Analysis, Prentice-Hall, Englewood Cliffs, NJ.

Quinn, F.H., 1992. Special Section on Improving Great Lakes Water Level Statistics, J. Great Lakes Res., 18(1), 199-201.

Salas, J.D., 1993. Analysis and Modeling of Hydrologic Time Series. In Handbook of Hydrology, D.R. Maidment Editor, McGraw Hill Inc., N. York.

Salas, J.D., Delleur, J., Yevjevich, V. and Lane, W., 1980. Applied Modeling of Hydrologic Time Series. Water Resources Publications, Littleton, Colorado.

Salas, J.D., Saada, N.M., Lane, W.L., and Frevert, D.K., 1996. Stochastic Analysis, Modeling, and Simulation (SAMS) Version 96.1 User's Manual, Technical Report No.8, Colorado State University, Fort Collins, Colorado.

U.S Army Corps of Engineers, 1971. "HEC-4 Monthly Streamflow Simulation," Hydrologic Engineering Center, Davis, Calif.

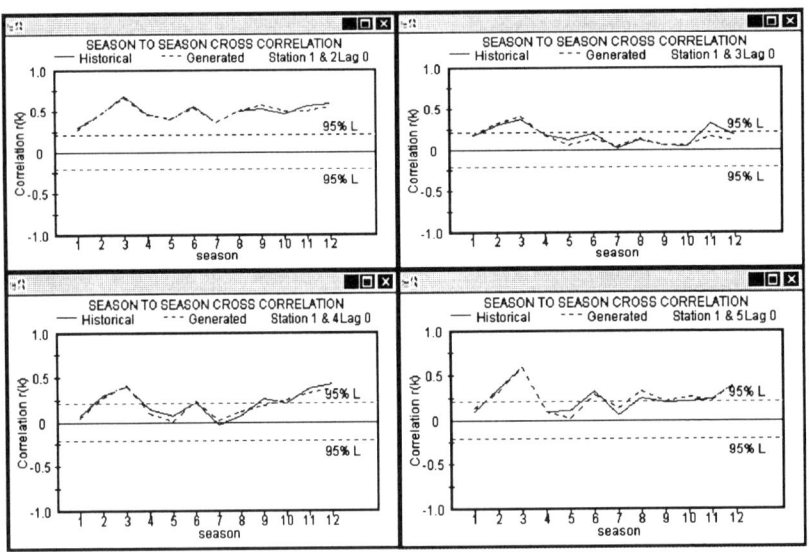

Figure 6. Comparison of lag-0 month-to-month cross-correlations between site 1 (Lake Superior) and the other four sites, respectively, based on Scheme 1 using ARMA(1, 1) model.

## The Behavior of Storage Reservoirs in the United States Under Climate Change

### Richard M. Vogel[1], Melissa Lane[1], and Paul Kirshen[1]

### Abstract

Using a recent inventory of thousands of storage reservoirs across the United States, we explore the behavior of surface water supply systems by region. Using recent advances in our understanding of the behavior of storage reservoirs along with regional hydroclimatologic relationships of streamflow and the reservoir inventory, we are able to summarize the behavior of individual surface water supply systems in terms of their reliability, resilience and vulnerability, by region, under existing and future climate conditions. We also document the hydrologic and water demand conditions that lead to carryover storage requirements.

### Introduction

This study seeks to generalize our understanding of the impact of climate change on reservoir system behavior by exploring relationships among climate, yield, system capacity and system reliability, resilience and vulnerability for reservoir systems across the continental U.S. To generalize such concepts, considerable simplifications are required. The goals of this study are twofold: (1) to compare the performance of surface water supply systems across the U.S. under existing climate conditions and (2) to evaluate the performance of those systems using a climate scenario generated from a general circulation model (GCM).

### Storage-Reliability-Resilience-Yield Relationships

Vogel and Stedinger (1987), Vogel and Bolognese (1995) and others have introduced analytic relationships which approximate the behavior of water supply systems dominated by over-year or carry-over storage requirements. Vogel and Stedinger (1987) summarize relations among reservoir system storage capacity S, mean annual inflow $\mu$, standard deviation of inflows $\sigma$, lag-one serial correlation of the inflows $\rho$, reservoir yield Y, planning horizon N, and N-year no-failure reliability $R_N$, in the form $S/\sigma = f(\mu, \rho, Y, N, R_N)$ for systems fed by AR(1) lognormal inflows. Vogel and Wilson (1996) document that annual streamflows in the United States are well approximated by a lognormal distribution. Vogel et al. (1998a) document that each of the 18 water resource regions of the U.S. shown in Figure 1 are homogeneous in terms of the year-to-year persistence of streamflow.

---

Department of Civil and Environmental Engineering, Tufts University, Medford, MA 02155, Tel: 617-628-5000 ext. 4260, Email: rvogel@tufts.edu

## Indices of Reservoir System Performance

**Within-year Versus Over-year Behavior:** Perhaps the most useful index for classifying the behavior of water supply systems is the index m defined as $m=(1-\alpha)/C_v=(\mu-Y)/\sigma$, where Y is the average annual yield, $\alpha$ ($\alpha=Y/\mu$) is the annual yield as a fraction of the mean annual inflow to the reservoir, $\mu$ and $\sigma$ are the mean and standard deviation of the annual inflows, and $C_v$ is the coefficient of variation of the annual inflows ($C_v=\sigma/\mu$). Ravindiran (1997) documents that as long as the index m is in the range $0<m<1$ and $C_v>0.3$, the system is dominated by year-to-year or carryover storage requirements. Ravinidiran (1997) documents that any reservoir in a region with $C_v$ greater than about 0.8 will be dominated by carryover storage requirements, regardless of the demand on the system!

**Resilience:** Resilience is defined as the probability that the system will recover from a failure once a failure has occurred. That definition is used here, with a failure defined as the inability of the reservoir system to provide its target yield Y in a given year. Using an approach similar to that used by Vogel and Bolognese for AR(1) normal inflows, we derive the resilience r, of a reservoir system fed by AR(1) lognormal inflows.

**Reliability:** Vogel and Bolognese (1995, equation 17) use a two-state Markov model to show that N-year no-failure reliability $R_N$ can be related to the steady-state annual reliability $R_a$ using the relationship $R_N = R_a (1 + r(1 - R_a^{-1}))^{N-1}$, where r is system resilience, and N is the length of the planning period. No-failure reliability $R_N$ is the probability that a reservoir system will provide a constant yield Y, without failure, over an N-year period. Annual reliability $R_a$ is the steady-state probability, in a given year, that the reservoir system will deliver the stated yield.

**Vulnerability:** Reservoir system vulnerability provides a measure of the magnitude of a failure, should it occur. We define vulnerability as the conditional expected extent of a failure. We compute it as simply the conditional mean deficit as a fraction of the yield, D, using a very long simulation experiment. Monte-Carlo experiments were performed to determine an index which could be used to estimate the magnitude of a failure, once it occurs. After many trials, we found the relationship $D=2.22[S/Y]^{1.27}$ where S is storage capacity, Y is system yield and D is the magnitude of the failure as a fraction of the yield of the system. This equation had an adjusted $R^2=0.936$ (in log space), normally distributed residuals, and model coefficients with t-ratio's in excess of 20. One can think of the ratio S/Y as the number of years of water supply yield in storage, when the system is full.

## The Behavior of Storage Reservoirs in the United States

Using a national database of reservoirs, in combination with the indices discussed above and regional hydrologic models of $\mu$ and $\sigma^2$, we determine the distribution of reservoir reliability, resilience, vulnerability, and level of development, by water resource region, under existing and future climate conditions

**Regional Hydrologic Model** - Regional hydroclimatologic regression equations for the continental U.S. developed by Vogel et al. (1998b) for the mean $\mu$ and variance $\sigma^2$ of annual streamflow take the form: $\mu = aA^b P^c T^d$ and $\sigma^2 = eA^f P^g T^h$ where the letters *a-h* are model parameters, $A$ is drainage area, $P$ is mean annual precipitation and $T$ is mean annual temperature. A separate set of regional regression equations were developed for each of the 18 U.S. water resource regions shown in Figure 1. The climate characteristics P and T were estimated for the 1,556 undeveloped watersheds across the U.S. from (2.5 minute) digital grids obtained using the PRISM (Daly et al., 1994) system. Estimates of $\mu$ and $\sigma^2$ and $A$ were obtained from a hydroclimatologic database of streamflows developed by Slack et al. (1993). Each of the 1,556 watershed boundaries were outlined using a digital elevation map and a geographic information system (GIS). The digital climate maps describing the spatial distribution of T and P for the U.S. were then used with a GIS to obtain spatially averaged values of T and P for each of the 1,556 watershed boundaries. The resulting regional regression equations are remarkably precise, with $R^2$ values for the equation for $\mu$ ranging from 80% to 99.7% with an average value of 94% across all 18 regions of the U.S.

The regional hydrologic equations reflect the relationship between hydrology, scale and climate over a wide range of climatic regimes. In each region, the spatial variations in precipitation and temperature associated with the data used to develop the regional equations is analogous to future potential variations in those climate characteristics corresponding to various climate change scenarios. *In other words, it is possible to use these regional regressions to reflect changes in hydrology resulting from future climate conditions, without extrapolation of these relationships.*

**Reservoir Database** - This study employs the national inventory of 75,187 dams developed by the Federal Emergency Management Agency (FEMA, 1995-96) in cooperation with the U.S. Army Corps of Engineers. Excluding dams outside the continental U.S. leaves 74,914 dams. Excluding reservoirs without information regarding their purpose leaves 55,247 dams. Since GIS methods are used to estimate climate and hydrologic inputs to each reservoir, dams without latitude and longitude information and storage volume information were dropped. Data quality assurance procedures were also used to eliminate erroneous data. Since the database is for dams, and it is possible for a single reservoir to contain several dams, duplicate dams were removed, leaving 51,749 dams which met all of the above criteria. Since the purpose of this study is to investigate the behavior of storage reservoirs whose function is to store and release (or regulate) water supply, only reservoirs whose purpose involves flow regulation are included, leaving 23,316 dams. Reservoirs which regulate flow for water supply are interpreted as reservoirs whose purpose involves one or more of the following functions: irrigation, hydroelectricity, navigation, water supply or fire protection. To avoid any extrapolation in the use of the regression equations, dams with drainage areas either larger than the maximum or smaller than the minimum drainage basin used to estimate the regional regression equations were removed leaving the 5,392 dams illustrated in Figure 1.

## The Behavior of Storage Reservoirs in the U.S. Under Existing and Future Climate Conditions

This section applies the reservoir system performance indices to each of the 5,392 storage reservoirs in Figure 1. Our assumption is that each reservoir was designed using the sequent peak algorithm (Rippl mass curve) with a 50 year historical streamflow record. This implies that the no-failure N-year reliability $R_N=0.5$. This mimics the way in which most reservoirs in the U.S. were actually designed. The current climate is assumed to be equal to the historical climate. The future climate corresponds to a doubling of greenhouse gases above current levels, based upon the GISS transient atmosphere-ocean coupled GCM (Russell et al. 1995).

## Results

Figures 2 and 3 use boxplots to illustrate the distribution of standardized inflow m and resilience r, respectively, by region, under current and future climate conditions. Figure 2 illustrates that under both current and future climate conditions, almost all reservoirs considered are dominated by overyear storage behavior in the western regions and most reservoirs in the east are dominated by within-year variations in storage behavior. This was expected because the coefficient of variation is much greater in western regions than in the eastern regions. Figure 3 illustrates the distribution of reservoir system resilience, r, by region, under current and future climate conditions. As expected, reservoirs in eastern regions are much more resilient than reservoirs in western regions. Under future climate conditions, western regions will generally see significant and consistent decreases in system resilience, whereas resilience will tend to increase in some eastern regions. The precipitous drop in resilience for reservoir systems in region 9 results from a significant decrease in system yield and a significant increase in the coefficient of variation of the inflows which will cause reservoir systems in that region to convert from marginally within-year to predominately over-year systems. The increase in resilience in reservoir systems in the east under future climate conditions results because levels of development $\alpha = Y / \mu$ will actually decrease in eastern regions. Under future climate conditions both yield Y and net inflow $\mu$ will decrease everywhere, however, the ratio of yield to net inflow will increase in the west and decrease in the east. This is due to the fact that the yield of reservoirs in the east is more sensitive to drops in net inflow than reservoirs in the east, due to the nonlinearity of the storage-yield relationship.

Figure 4 illustrates the total vulnerability of all reservoirs considered, by region. Here vulnerability is defined using the conditional mean deficit as a fraction of the yield D for each reservoir in the region. Total vulnerability can be interpreted as the magnitude of a failure, in terms of the number of years that region would be without water, assuming no demand management, no imports, and generally no realistic reaction to drought. It can be considered an upper bound on regional vulnerability. Figure 4 documents that the increases in overall vulnerability of each region under future climate conditions is much more dramatic in the western regions than in the eastern regions. Generally reservoirs in eastern regions are subject to either no increase in vulnerability or in some cases slight decreases. Decreases in system vulnerability for some of the eastern regions result again from drops in the level of development $\alpha = Y / \mu$ for reservoirs in those regions.

## References

Daly, C., R.P. Neilson, and D.L. Phillips, A statistical-topographic model for mapping climatological precipitation over mountainous terrain, *Journal of Applied Meteorology*, 33(2), 140-158, 1994.

Federal Emergency Management Agency, National Inventory of Dams, Water Control Infrastructure, Washington, DC, (CD-ROM database), 1995-96.

Ravindiran, R.S., Characteristization of within-year versus over-year reservoir storage behavior, M.S. Thesis, Tufts University, Medford, MA, May 1997.

Slack, J.R., A.M. Lumb, and J.M. Landwehr, Hydroclimatic data network (HCDN): a U.S. Geological Survey streamflow data set for the United States for the study of climate variation, 1874-1988. Water-Resource Investigations Report 93-4076, U.S. Geological Survey, Washington, D.C., 1993.

Vogel, R.M. and J.R. Stedinger, Generalized storage reliability yield relationships, *Journal of Hydrology*, 89, 303-327, 1987.

Vogel, R. M. and R. A. Bolognese., Storage-Reliability-Resilience-Yield Relations for Over-year Water Supply Systems. *Water Resources Research*, 31(3): 645-54, 1995.

Vogel, R.M. and I. Wilson, Probability distribution of annual maximum, mean, and minimum streamflows in the United States, *Journal of Hydrologic Engineering*, 1(2), 69-76, 1996.

Vogel, R.M., Tsai Yushiou and J.F. Limbrunner, The Regional Persistence and Variability of Streamflow in the United States, *Water Resources Research*, (manuscript under review), 1998a.

Vogel, R.M., I. Wilson, and C. Daly, Regional models of annual streamflow for the United States, *Journal of Irrigation and Drainage*, ASCE, (manuscript under preparation), 1998b.

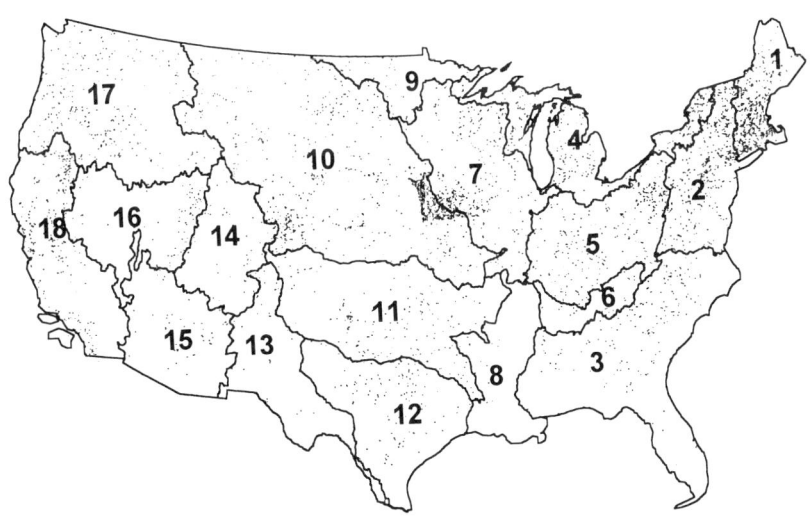

Figure 1 - Location of 5,392 Reservoirs and U.S. Water Resource Regions

128 WATER RESOURCES AND THE URBAN ENVIRONMENT

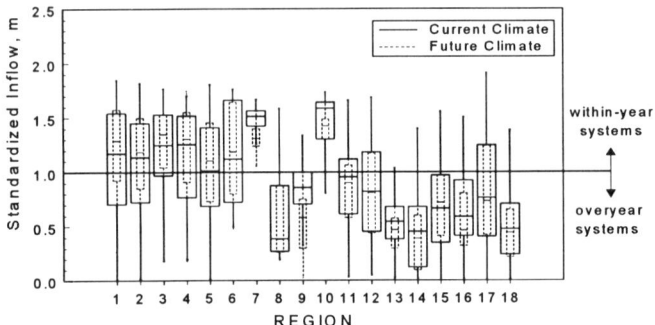

Figure 2 - Box Plots of Standardized Inflow m for Each Region

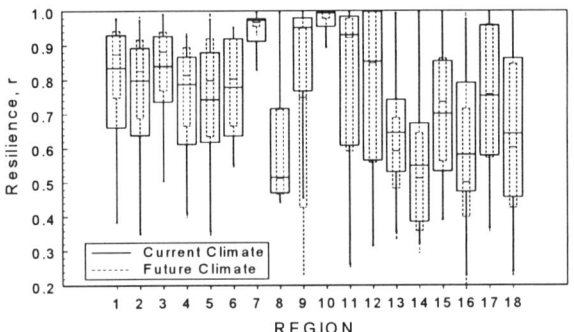

Figure 3 - Box Plots of Reservoir Resilience r for Each Region

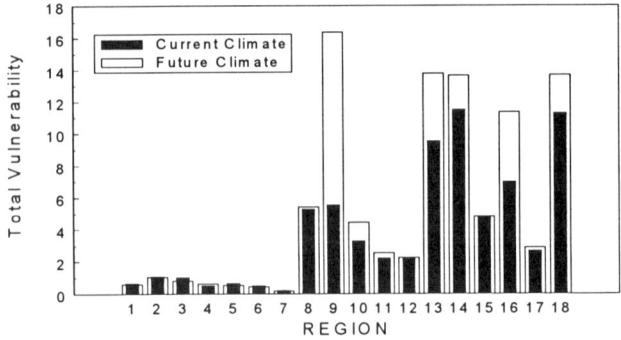

Figure 4 - Total Vulnerability D, of All Reservoirs Considered, By Region

# The Use of Indicators to Evaluate Impacts of Global Warming on Water Resources in the United States

## Melissa Lane[1], Paul Kirshen[1], and Richard M. Vogel[1]

### Abstract

This study explores water resources in the US under current and future climate by aggregating national databases into regional stress indices. Although analysis on the broad level disguises the variability within regions, this investigation is effective for providing a national assessment of water resource stresses. We develop regional water resource indicators for socio-economic and environmental factors. Using these indicators, we conclude that the southwestern regions of the US exhibit the most significant regional stresses and the most sensitivity to climate changes especially when compared to the relatively unstressed northeastern regions.

### Introduction

A changing climate may impact all aspects of water resource planning and management. Many hydrologists have focused their research on evaluation of climate impacts on specific river basins. To our knowledge, no previous studies have quantitatively examined the impact of climate change on water resource stress across the entire United States by studying each river basin.

Even a decade ago, embarking on an analysis of this scale and scope would have required many researchers working with a roomful of computers. Today, a national survey is feasible due to many recent technological advances: the accessibility of government data through the Internet, increased computational power and easy integration of data and software. Specifically, the Geographic Information System (GIS) software, ESRI's ArcView 3.0 was essential for data management, integration and analysis. ArcView GIS software can build a spatial database which allows regional aggregation and the analysis of data layers, a fundamental tool for this type of analysis.

Like Gleick (1990) and Raskin et. al. (1996), this paper explores both hydrologic and economic indicators to evaluate regional stresses because a change in water resource availability could have significant effects on many facets of a region's activities. The goals of this study are twofold: (1) to develop meaningful indicators

---

[1] Department of Civil and Environmental Engineering, Tufts University, Medford, MA 02155, Tel: 617-627-4260, email: rvogel@tufts.edu

and indices of water resource stress and (2) to examine water resource stresses in the United States using both current and hypothetical climate change scenarios. The analysis is conducted for the eighteen Water Resource Regions illustrated in Figure 1 (U.S. Water Resources Council, 1970).

## Indicators

The sixteen environmental and socio-economic indicators are defined and identified below. The data collected to estimate the indicators was gathered in many formats from numerous government agencies.

### ENVIRONMENTAL INDICATORS

**Withdrawal Ratio:** Measure of the intensity of water use in the region. The annual water withdrawals divided by the sum of internal water flow plus water imports, both natural and transfers minus upstream consumptive use.

**Surface Water Stress:** Measure of surface water availability. The annual surface water withdrawals divided by the sum of internal water flow plus water imports, both natural and transfers minus upstream consumptive use.

**Ground Water Stress:** Measure of ground water availability. The annual ground water withdrawals divided by the sum of internal water flow plus water imports, both natural and transfers minus upstream consumptive use.

**Dependence Ratio:** Measure of a region's dependence on upstream flow. The ratio of internal flow to the sum of internal water flow plus water imports, both natural and transfers minus upstream consumptive use.

**Coefficient of Variation:** Measure of the variability in the region's streamflow. The average value of coefficient of variation of annual flows in the region.

**Runoff:** Measure of the average annual streamflow per square mile in the region.

**Biota Stress:** Measure of ecological integrity. The region's number of vulnerable vertebrates divided by the region's total native vertebrate species.

**Water Quality Stress:** Measure of flow weighted biological oxygen demand (BOD) concentration in each region.

### SOCIO-ECONOMIC INDICATORS

**Demand:** Measure of the level of development in the region. The internal consumptive water use in the region plus its exports divided by the sum of internal water flow plus water imports, both natural and transfers minus upstream consumptive use.

**Storage Ratio:** Measure of region's ability to cope with extreme water events. The regional reservoir storage capacity (internal and upstream) divided by the sum of internal water flow plus water imports, both natural and transfers minus upstream consumptive use.

**Import Demand Ratio:** Measure of the significance of interbasin water transfers. The amount of water imported into a region divided by its total internal withdrawals.

**Hydropower:** Measure of dependence on hydropower electricity. The ratio of electricity supplied by hydropower in the region to total basin electricity production.

**Spending Power:** Measure of a region's ability to buy its way out of a environmental crisis.

**Poverty:** Measure of the percentage of a region's population without resources to spare.

**Infant Mortality:** Measure of the overall health of the region.

**Voter Participation:** Measures public feeling of responsibility and empowerment towards local and national issues.

## Climate Change Analyses

The current climate is estimated from U.S. historical temperature and precipitation data using 2.5 arc minute grids (Daly, 1994). Two hypothetical climate change scenarios are developed from average model results from the 1995 Intergovernmental Panel on Climate Change (Houghton et.al., 1996). The scenarios are conservative estimates and designed to test the sensitivity of the indicators to changes in precipitation and temperature individually. In the first scenario, the temperature is altered by increasing each region's average annual temperatures by two degrees Celsius; current precipitation is used. For the second scenario, the temperature remains at a uniform two degrees increase over current levels and a change in precipitation is simulated by calculating a uniform five percent increase from current precipitation across the US.

The third and fourth climate scenarios are derived from the Goddard Institute Space Studies (GISS) general circulation model, GCM (Russell et. al. 1995). The third climate scenario uses the regional precipitation and temperature values from the GISS GCM; while the fourth scenario accounts for both GISS climate changes and economic development. The economic development is a conservative estimate of future growth based on national changes in water sectors from Raskin (1996).

In the GISS and economic development scenario, the future regional values of demand and withdrawals are estimated by calculating the product of the current values and the proportional change in the North American sector. Future BOD is estimated with population growth projections assuming pessimistically that people and pollution are directly correlated regionally and are inversely proportional to changes in streamflow. The ratio of hydropower production to total electric production is estimated by assuming that changes in hydropower production are directly proportional to changes in streamflow.

**Regional Hydrologic Model**

This paper uses models based on 1,556 undeveloped watersheds across the U.S. which relate measurable climate and basin characteristics to regional streamflow statistics (Vogel et.al., 1997). Known climate and basin characteristics are employed to derive regression equations for both the mean annual streamflow and variance of the annual streamflow in each of the eighteen water resource regions. Specific characteristics included in this model are: drainage basin area A, average annual temperature $\mu_T$ and average annual precipitation $\mu_p$. These equations are: $\mu_Q = e^a \bullet A^b \bullet \mu_p^c \bullet \mu_T^d$ and $\sigma_Q^2 = e^i \bullet A^j \bullet \mu_p^k \bullet \mu_T^l$; where $\mu_Q$ = mean annual streamflow and $\sigma_Q$ = standard deviation of the mean annual streamflow. Using these equations, changes in a region's climate can be reflected in the average annual streamflow and that flow's coefficient of variation, assuming that changes in precipitation and temperature reflect all the hydrologic impacts of global climate change.

The regional hydrologic models reflect the relationship between hydrology, scale and climate over a wide range of climatic regimes. In each region, the spatial variations in precipitation and temperature associated with the data used to develop the regional equations is analogous to future potential variations in those climate characteristics corresponding to various climate change scenarios. In other words, it is possible to use these equations to reflect changes in hydrology resulting from future climate conditions, without much extrapolation of these relationships. In order to minimize extrapolation, the USGS 2,111 cataloguing units were used for flow instead of the eighteen water resource regions because those smaller units have drainage areas much closer to the size used to develop the regression equations.

## Comparison of Indicators

We explored three indicator aggregation methods to evaluate the impacts of future climate and development scenarios.

1. **Standardized Indicators** In this method, each regional indicator is standardized from zero to one based on the indicator range in the current climate scenario. Each set of eight indicators is then summed to form the aggregated indices: environmental stress index and socio-economic stress index. These two indices then sum to form a total regional stress index. This method benefits from not having to rely upon somewhat arbitrary warning threshold values. However, it is limited by its standardization range; the indicator range in the current climate does not approximate the true indicator range.

Figure 2 uses the indicator standardization method to illustrate that in the current climate, the stress levels in the southwestern regions of the US are much greater than those in the eastern regions. In comparing the four future climate scenarios to the current climate in Figure 2, the level of stress remains relatively constant throughout the future climate scenarios. Though the eastern regions exhibit a slight increase in stress under each of the climate scenarios, the idea that the water resource stress in the US regions will manifest little change under future climate changes is extraordinary. This suggests that the existing variability in stress levels across regions of the US is much greater than any anticipated changes in water resource stress levels that will occur for individual regions due to climate change.

2. **Threshold Exceedences** For each indicator, a warning threshold exists which distinguishes a non-stressed region from a stressed region. Table 1 displays the warning thresholds, which have been identified by previous work or determined by judgment. A stressed regional indicator is given a score of one; an unstressed regional indicator is a zero. Each set of eight indicator exceedences is summed to form another version of an environmental stress index and a socio-economic stress index. These two indices sum to a total regional stress index.

Table 1 Warning Thresholds for Indicators

| Environmental Integrity | | Socio-Economic Security | |
|---|---|---|---|
| Withdrawal Ratio | greater than 0.2 | Demand | greater than 0.2 |
| Surface Water Stress | greater than 0.2 | Storage Ratio | less than 0.6 |
| Ground Water Stress | greater than 0.2 | Import Demand | greater than 0.1 |
| Dependence Ratio | greater than 0.1 | Hydropower | greater than 0.25 |
| Streamflow Cv | greater than 0.4 | Spending Power | less than 70 |
| Runoff | less than 0.2 | Poverty | greater than 11 |
| Water Quality Stress | greater than 7 mg/L | Infant Mortality | greater than 10 |
| Biota Stress | greater than 0.06 | Public Participation | less than 50 |

Though this method is useful in identifying the indicators that cause the most stress, its effectiveness in detecting impacts of climate change may be limited by the sensitivity of the method to the warning threshold. In addition, because an indicator is either above or below the warning threshold, this method may miss subtle changes in regional stresses. Because the threshold exceedence method yields similar results to those calculated in the indicator standardization method, it is not displayed here.

3. **Percentage Changes** The percent change from the current climate to a future climate scenario is calculated for each indicator. Each set of eight regional indicator percentage changes is then summed to form an environmental integrity index and an socio-economic security index. The two aggregate indices then sum to form a total regional stability index. The assumption is that any change from the current situation merits identification and attention. In calculating the environmental integrity index, the withdrawal ratio is excluded in the sum to avoid double accounting with surface water withdrawals and ground water withdrawals.

Although the results in Figure 2 illustrate relatively marginal changes in water resource stress levels before and after climate change, these results may be misleading. The percentage change method is designed to explore the impact of changes in water resource stress levels in more detail. For example, Figure 3 depicts the results of one such application of the percentage change method to climate scenario 4 which assumes the GISS scenario along with assumptions about economic development. Figure 3 illustrates the changes in the economic security index, the environmental integrity index and the total regional stability index for each region, as compared with current climate conditions. One observes dramatic changes using this method, particularly in the southwestern and mid-northern regions of the US. The large percent changes and resulting high stress levels in these regions result from the relatively large increases in many of the environmental and socio-economic indicators, which make up these indices.

Different weighting systems were explored in combination with this method, however it was found that the weighting system does not alter our conclusions.

An examination of these indicators and their aggregate indices is designed to provide insight into regional sensitivities of water resources to climatic change. These indicators are not prediction tools and caution should be used when viewing them as indicators of the future behavior of water resource regions. Furthermore, they are used most effectively when the indicators are examined in relation to each other, rather than as independent, separate measurements.

## References

Daly, C., Neilson, R.P., and Phillips, D.L., 1994. A statistical-topographic model for mapping climatological precipitation over mountainous terrain. Journal of Applied Meteorology 33 140-158.

Gleick, P.H., 1990. Vulnerability of Water Systems in Climate Change and U.S. Water Resources, P.E. Waggoner, ed., John Wiley & Sons, Inc., New York, NY.

Houghton, J.T., Meira Filho, L.G., Callander, B.A., Harris, N., Kattenberg, A. and Maskell, K. eds., 1996. Climate Change 1995: The Science of Climate Change. for Intergovernmental Panel on Climate Change Cambridge University Press.

Raskin, P., Gleick, P.H., Kirshen; P., Pontius, R.G. Jr. and Strzepek, K., 1996. Water Futures: Assessment of Long-range Patterns and Problems in Comprehensive

Assessment of the Freshwater Resources of the World, Stockholm Environment Institute, Boston, MA.

Russell, G.L., Miller, J.R., Rind, D., 1995. A coupled Atmosphere-Ocean Model for Transient Climate Change Studies. Atmosphere-Ocean 33(4)p 683.

U.S. Water Resources Council, 1970. Water Resources Regions and Subregions for the National Assessment of Water and Related Land Resources Washington D.C.

Vogel, R.M., Wilson, I., Daly, C. and Adams, T., 1997. Regional Models of Annual Streamflow for the Continental United States. Journal of the Irrigation and Drainage Division, ASCE (in preparation).

**Figure 1** US Water Resource Regions

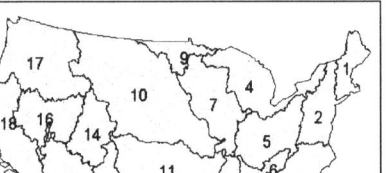

| Region | Region Name | Region | Region Name |
|---|---|---|---|
| 1 | New England | 10 | Missouri |
| 2 | Mid-Atlantic | 11 | Arkansas-White-Red |
| 3 | South Atlantic Gulf | 12 | Texas-Gulf |
| 4 | Great Lakes | 13 | Rio Grande |
| 5 | Ohio | 14 | Upper Colorado |
| 6 | Tennessee | 15 | Lower Colorado |
| 7 | Upper Mississippi | 16 | Great Basin |
| 8 | Lower Mississippi | 17 | Pacific Northwest |
| 9 | Souris-Red-Rainy | 18 | California |

**Figure 2** Standardized Stress Indices in the Current Climate and Future Climate Scenarios

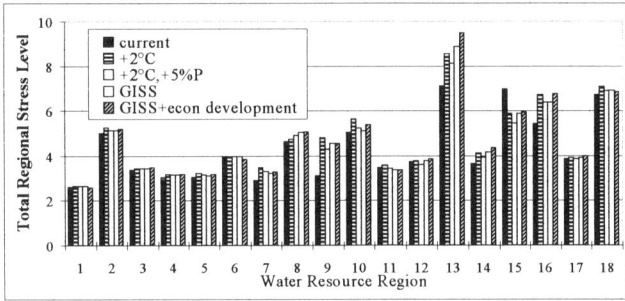

**Figure 3** Changes in Scenario 4 (GISS + Economic Development) vs. Current Scenario

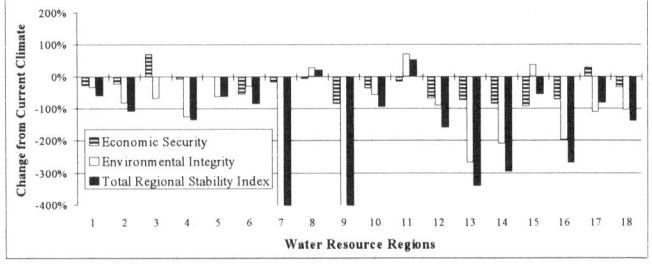

Causes of Increased Net Basin Supply in the Great Lakes Watershed

Eric D. Loucks[1], Orie L. Loucks[2] and John C. Klink[3]

Introduction

The North American Great Lakes have experienced higher than normal water levels over the past thirty years resulting in shoreline damage and economic impacts on coastal interests. The high water levels are an immediate result of above average net basin supplies (NBS) to all of the Great Lakes. This paper considers whether the increased supplies could be due to reduced evaporation and potential evapotranspiration caused by increased levels of atmospheric aerosols.

Net Basin Supply

For a given lake watershed, NBS is the sum of direct overlake precipitation, surface runoff and net groundwater inflow minus direct evaporation from the lake surface. Thus, NBS represents the water the hydrologic cycle makes available to the lake during a specified time period. Through the Lakes' responses in water level and connecting channel flows, the Corps of Engineers has estimated monthly NBS since 1900 (US Army Corps of Engineers, 1995). Great Lakes annual NBS for the period of record, 1900–1995, are summarized in Table 1. NBS can be expressed as a mean annual flow rate or, as a depth averaged over the watershed. Lake Michigan-Huron has the greatest average annual NBS owing to its large watershed while Lake Erie has the least. When expressed as a depth averaged over the watershed, Lake Ontario has the greatest NBS per unit area mainly because precipitation is greatest in the Eastern part of the Great Lakes Basin. However, Lake Superior has the second greatest due to the low levels of evaporation in its watershed. Evaporation is manifest in the NBS in two important ways. There is direct evaporation from the surface of each lake and evapotranspiration reduces land surface runoff the tributary watershed.

---

[1] Water Resources Engineer, Camp Dresser & McKee, Inc., 233 S. Wacker Drive, Chicago, IL, 60606, [2] Professor of Zoology, Miami University, Oxford, OH, [3] Professor of Geography, Miami University, Oxford, OH.

| Parameter | Units | Lake Superior | Lake Mich-Huron | Lake St Clair | Lake Erie | Lake Ontario | Great Lakes Total |
|---|---|---|---|---|---|---|---|
| Mean Net Basin Supply 1900-95 | cms | 2047.9 | 3186.7 | 138.7 | 565.6 | 1028.2 | 6967.2 |
| Standard Deviation | cms | 466.6 | 720.0 | 91.6 | 261.7 | 237.5 | 1396.0 |
| Watershed Area | sq. km | 210,000 | 370,000 | 16,900 | 86,840 | 83,200 | 766,940 |
| Lake Surface Area | sq. km | 82,600 | 117,500 | 1,170 | 25,680 | 19,800 | 246,750 |
| Mean NBS as a Watershed Depth | cm | 30.8 | 27.2 | 25.9 | 20.6 | 39.0 | 28.7 |
| Standard Deviation | cm | 7.0 | 6.1 | 17.1 | 9.5 | 9.0 | 5.7 |

**Table 1—Water Supply and Physical Data Describing the North American Great Lakes**

## Change in Net Basin Supply

Average NBS has increased in each of the Great Lakes watersheds during the period since 1964. Increases in NBS are profound in the three eastern watersheds comprised of the Lakes St. Clair, Erie and Ontario basins. Table 2 compares NBS behavior for the 1900-1964 period with the 1965-1995 period. As shown, Lake Superior NBS has increased 1.8 cm, while average Lake Erie NBS has increased 41 percent or 8.4 cm. Chi-square tests on the historical records before and after 1964 indicate a significance level of 95% for the change in mean NBS for Lake Michigan-Huron and 99% for Lakes Erie and Ontario. The change in mean for Lake Superior is not significant. The combined NBS to the Lake St. Clair, Erie and Ontario watersheds is up an average of 9 cm, or more than 30 percent, since 1964.

| Parameter | Units | Lake Superior | Lake Mich-Huron | Lake Erie | Lake Ontario | Great Lakes Total | Lakes St. Clair Erie and Ontario |
|---|---|---|---|---|---|---|---|
| Mean Net Basin Supply 1900-64 | cm | 30.2 | 26.2 | 17.8 | 36.5 | 27.3 | 26.3 |
| Standard Deviation | cm | 6.7 | 6.2 | 9.0 | 8.0 | 5.4 | 7.4 |
| Mean Net Basin Supply 1965-95 | cm | 32.0 | 29.3 | 26.2 | 44.1 | 31.5 | 35.4 |
| Standard Deviation | cm | 7.4 | 5.3 | 8.0 | 8.9 | 5.3 | 8.0 |
| Change in mean | cm | 1.8 | 3.1 | 8.4 | 7.6 | 4.2 | 9.0 |
| Standard Deviation of Change | cm | 1.6 | 1.2 | 1.8 | 1.9 | 1.2 | 1.7 |
| Percent Increase | | 5.8% | 11.3% | 40.8% | 19.5% | 14.5% | 30.8% |
| Significance | (t) | 1.14 | 2.49 | 4.63 | 4.04 | 3.57 | 5.30 |

**Table 2—Comparison of Mean Annual NBS for 1900–1964 and 1965–1995**

## Increased Cloud Cover and Haze Effects on Evaporation

A series of papers since 1980 (Husar, et al, 1981; Charlson, et al, 1991; Taylor and Penner, 1994) looked at the increases in cloud cover and haze over eastern North America, including the Great Lakes Basin. Figure 1 summarizes the trend in Great Lakes cloudiness since 1900 and shows an increase of about 18 percent from 1930 to 1980. The work of Husar, et al documents the regional increases in haze observed at airports expressed as transparency to the high-energy short-wave radiation. Their data indicate the greatest increases in haze occurred between 1960 and 1975.

The consensus of these authors is that acid gas emissions (sulphur dioxide, $SO_2$ and nitrogen oxides, $NO_x$), subsequently converted to acid aerosol particles (primarily of the salt biammonium sulfate and nitrate), increase cloud cover by serving as cloud seeding agents and along with carbonaceous fine particulates increases the opacity of

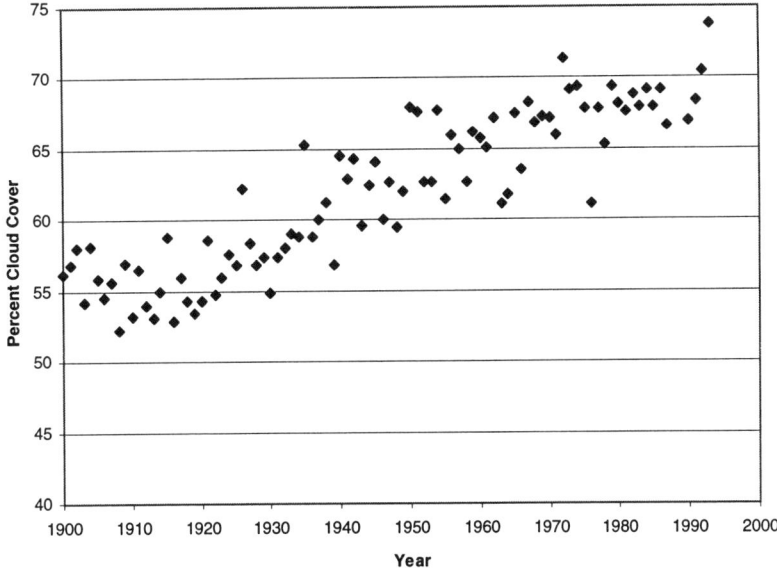

**Figure 1—Average Daytime Cloud Cover at Seven Midwest U.S. Sites 1900–1993**

the lower atmosphere to sunlight. The mechanisms are illustrated in Figure 2. Chemical reactions in the atmosphere lead to the formation, with ammonia, of hydrated sulfate and nitrate aerosols. These particles absorb water and become large enough to scatter incoming radiation and enhance cloud formation under certain conditions.

Both cloud and haze formation, the researchers show, can cause cooling of the climate, thus decreasing the evaporation of water. Our specific goal for this paper then is to estimate how much the increase in cloud cover and aerosols can have affected evaporation and therefore NBS.

Historical and Geographic Patterns of Emissions and Aerosol Washout

A report by the National Research Council/National Academy of Sciences (NRC/NAS, 1988) reviewed the historical progression in emissions of acid gas precursors to acidic aerosols and acid washout in precipitation. Sulphur dioxide emissions experienced peaks in the 1920s, the 1940s and the 1970s. Emissions of nitrogen oxides (NO, $N_2O_2$, $NO_2$) have shown a relatively continuous gradual increase to the present (see Figure 3 in CEQ, 1996). These two acid aerosol precursors added together totaled only 37 million tons in 1960 but reached an unprecedented 51 million tons by 1967. Since then, decreases in $SO_2$ emissions have been slightly greater than the

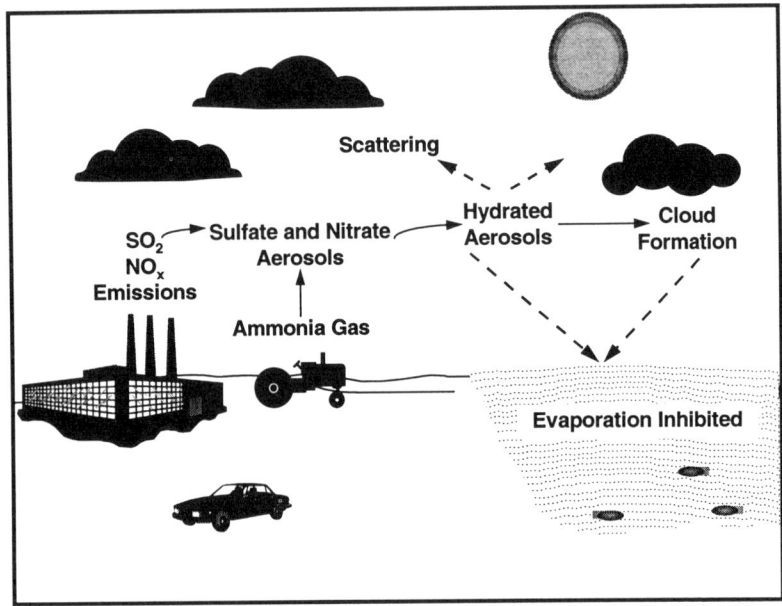

Figure 2—Process by which acid gas emission leads to increased cloud cover and radiation scattering

increases in $NO_x$ emissions resulting in a 1994 total of 43 million tons. During the period 1960 to 1980, U.S. use of nitrogen fertilizers, much of it the second agent in the formation of aerosols, anhydrous ammonia, quadrupled from 2.7 to 11.4 million tons (CEQ, 1996). The fraction of ammonia that escapes from the soil into the air during or following application, while small, becomes an essential component of the aerosol formation process.

During the last 15 years, coordinated measurements of acid washout in precipitation by both the U.S. and Canada, have provided a clear picture of the geographic pattern of aerosols in the atmosphere over the Great Lakes. A composite by Barrie and Hales (1984) is reprinted in Figure 3. A small amount of washout has been recorded over the Lake Superior watershed, while appreciably more is observed over Lake Michigan-Huron and the highest levels of washout are over the Lake St. Clair, Lake Erie and Lake Ontario watersheds. The data in Figure 3 depict sulphate and nitrate in rainfall which we use here as a complementary surrogate for the aerosol-induced cloud and haze patterns cited earlier.

Calculation of the Effect on Net Basin Supply

The decrease in energy input attributed to anthropogenic input emissions by Charlson, et al, (1991), for the whole northern hemisphere is 1.07 W/m². The column

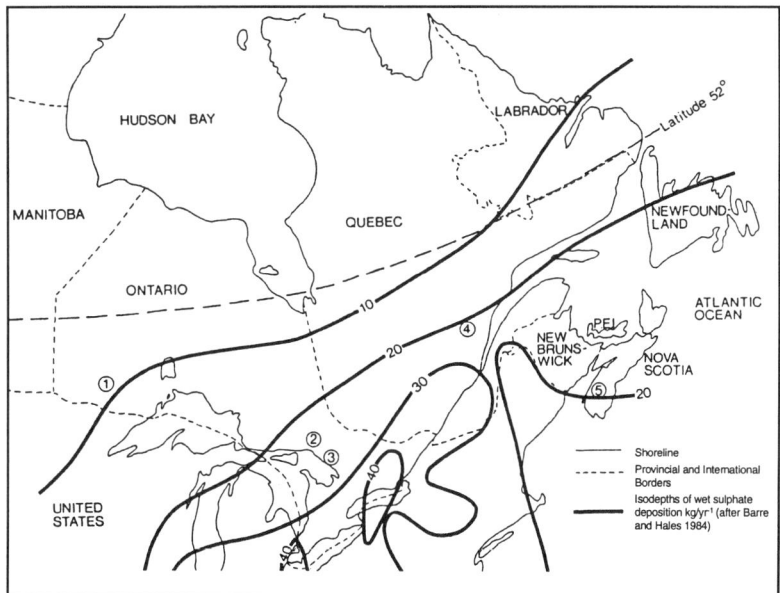

**Figure 3—Distribution of wet sulphate deposition over Eastern North America (Minns and Kelso, 1986)**

burden of sulphate shown in this study for the Great Lakes is 5 to 10 times the hemispheric average, but sulphate alone does not correlate well with the data on cloud and haze; the salts of sulphate and nitrate exhibit a much stronger correlation. Detailed measurements of radiation by Klink (1997) at Oxford Ohio, shows a recent change in warm season (May to September) radiation of minus 10.4 W/m². A corresponding increase in cloudiness has been observed at the Cincinnati (Covington) airport.

It is possible to estimate the impact on evaporation caused by lower incoming radiation using the following calculation. One Watt is equivalent to 0.239 calories per second. It takes 586 calories to evaporate one cubic centimeter of water at 20 C or 597 calories at 0 C so 590 cal/cc is a reasonable heat of vaporization for Great Lakes water. One W/m² is equivalent to $2.39 \times 10^{-3}$ cal/cm²/s or 754 cal/cm²/year. This is enough heat energy to evaporate 1.29 cm of water per year. Thus, a decrease of 10 W/m² in incoming radiation could result in a 13 cm decrease in evaporation. This would be sufficient to explain the 9 cm per year increase in NBS observed in the Eastern Great Lakes for the entire 1965 to 1995 period previously indicated in Table 2.

Our estimate of the change in radiation forcing (in W/m²) depends critically upon good measurements of ground level radiation in the Great Lakes basin during the 1940s and 1950s, but a careful search indicates such data are not available. We can estimate the reference radiative forcing indirectly, however, by noting the recent trend in the relationship of change in radiation to change in cloud cover and working backward for the full historical change in observed cloudiness (Figure 1). We can also check that estimate against modern radiation data for sites with cloud cover comparable to that of the Great Lakes region in the 1940s. We estimate the total change in radiation input to be only 1 or 2 W/m$^2$ in the 1950's to as much as 15 to 20 W/m$^2$ in the Lake Erie basin at the end of the 1970's. This estimate is supported by trends in the ammonia and acid precursor emissions as well as the light extinction coefficients reported by Husar (1981).

Conclusions

This paper presents several compelling points that demonstrate the linkage between emissions of aerosol precursors and increases in Great Lakes Net Basin Supply. Among these conclusions are the following:

- The magnitude of the change in NBS is consistent with reduction in evaporation expected from observed reductions in solar radiation.
- The geographic distribution of airborne sulfates and nitrates is consistent with the geographic distribution of NBS increases.

References

Charlson, R. J., Langner J., Rodhe H., Leovy C. B., and S. G. Warren, Perturbation of the northern hemisphere radiative balance by backscattering from anthropogenic sulfate aerosols, 1991, *Tellus*, 1991, 43AB:152–163.

Husar, R. B., Holloway, J. M., Patterson, D. E., and W. E. Wilson, Spatial and Temporal Pattern of Eastern US Haziness: A Summary, 1981, *Atmospheric Environment* 15(10):1919–1928.

Klink, John C., Evidence of Decreased Global Solar Radiation in Southwestern Ohio, 1997, Association of American Geographers 93rd Annual Meeting.

Minns, C. K. and J. R. M. Kelso, Estimates of existing and potential impact of acidification on the freshwater fishery resources and their use in eastern Canada., *Water Air and Soil Pollution*, 1986, 31:1079–1090.

Taylor, K. E., and J. E. Penner, Response of the climate system to atmospheric aerosols and greenhouse gases, 1994, *Nature*, 369:734–737.

US Army Corps of Engineers Detroit District, Estimated Great Lakes Net Basin Supply 1900 to 1990 and Provisional Net Basin Supply 1991 to 1995.

Applications of the FEQ Unsteady Flow Model in DuPage County and Beyond

Paula Cooper[1] and H. Sherrie Chang[2]

**Abstract**

The inability of traditional steady-state hydraulic models to adequately represent the complexities of current drainage systems is being recognized by municipalities throughout the United States. Use of unsteady flow models to overcome the limitations of steady-state models is increasing. Situations in which use of an unsteady flow model is warranted, and case studies describing applications of the Full Equations (FEQ) unsteady flow model, are presented.

**Introduction to FEQ**

The FEQ model is a one-dimensional unsteady hydraulic model. Inherent in its governing equations is the assumption that flow occurs in one dimension (the downstream direction). Through careful decision-making when developing the model and selecting the locations and orientations of cross-sections, flow in two dimensions can be approximated. The unsteady nature of the FEQ model enables routing of the entire hydrograph through the hydraulic system. In contrast, steady-state models such as HEC-2, WSP2, and WSPRO analyze only a peak flow rate and compute a water surface profile that represents a snapshot in time. Unsteady flow models route the entire flood wave down the valley, simulating the effects of the floodplain filling as flow leaves the channel and is stored and conveyed in overbank areas.

**When to Use FEQ**

There are several situations in which application of an unsteady flow model is necessary to adequately represent the hydraulics of a system. These include:
- Systems having significant storage of floodwaters in the floodplain
- Systems where the relative timing of peak flows in parts of the system affect peak flows in other parts of the system

------------------
[1]Project Engineer, Woodward-Clyde, 1501 Fourth Avenue, Suite 1500, Seattle, WA 98101-1662
[2]Assistant Project Engineer, Woodward-Clyde, 1501 Fourth Avenue, Suite 1500, Seattle, WA 98101-1662

- Systems having reverse flows or strong dynamic backwater effects
- Systems affected by tidal fluctuations

Floodplain Storage

In systems with large amounts of floodplain storage, significant volumes of water are temporarily stored in the floodplain as a flood wave travels down the valley, and the flood wave is attenuated. An unsteady flow model can easily represent this type of hydraulic system. However, steady-state hydraulic models consider only peak flows and represent only one instant in time, so these effects are not simulated. In most steady-state applications, the flow is assumed to be the same along the entire stream reach, or is increased at discrete points along the stream to represent tributary inflows. In unsteady flow models, the peak flow may actually decrease in the downstream direction in areas where floodplain storage is significant.

In the early 1980s, DuPage County in Illinois began a pilot study on Winfield Creek using the FEQ model (Harza Engineering 1983). The County realized that many of its streams and rivers had large volumes of floodplain storage, which were not adequately represented in the typical steady-state hydraulic analyses. In addition, the impacts of changes in floodplain storage, which occurs during floodplain developments involving large amounts of fill, were not realistically represented. While steady-state models can reflect the loss of conveyance caused by floodplain fills, they miss the effects of the loss of temporary floodplain storage. This was the primary concern that caused the County to select the FEQ model for the pilot study.

Timing Issues

An unsteady flow model is needed to represent the effects of the relative timing of flood flows throughout the hydraulic system. Many strategies to reduce flood hazards, such as the construction of storage reservoirs, change the timing of flood peaks reaching the downstream system. In steady-state analyses, the effects of timing changes cannot be represented explicitly in the hydraulic model. The effects can be approximated using a hydrologic model, but successive iterations between the hydrologic and hydraulic model are needed to adequately reflect the effects of timing changes. In unsteady models, these effects are easily represented.

Reverse Flows and Dynamic Backwater Effects

Reverse flow (flow in the upstream direction) does occur in river systems in some situations. When a large flood wave travels past the confluence of a tributary stream, flows in the tributary may reverse direction for a reach of the stream upstream of the confluence. Operation of an adjustable side-flow weir at the Elmhurst Flood Control Reservoir, a large off-line reservoir constructed on Salt Creek by DuPage County, causes reverse flows in Salt Creek. Flow downstream of the weir actually reverses and flows upstream into the reservoir when the weir is lowered.

Dynamic backwater effects also can occur at tributary inflow points. In these situations, the peak flow may not occur at the same time as the peak stage. The peak flow in the tributary may be associated with the flood wave travelling downstream from upstream tributary areas. The peak stage may occur later as the flood wave in the receiving stream passes the tributary confluence.

These types of flow situations can be represented only by an unsteady flow model.

Tidal Fluctuations

Stages and flow rates of many river systems are affected by tidal fluctuations at their confluence, which can affect the hydraulics for a significant distance upstream of the confluence. Only an unsteady flow model can reflect the effects of the changing downstream boundary condition caused by tidal fluctuations.

Other Capabilities of FEQ

In addition to its unsteady nature, the FEQ model has been constructed in a flexible manner, which facilitates the analyses of complex systems characterized by complex hydraulic structures. Looped or branched hydraulic networks are easily represented in FEQ. If carefully constructed, the model can approximate flows that occur in two dimensions. Figures 1 and 2 illustrate a model representation of a complex hydraulic system. The physical system to be modeled is depicted in Figure 1. A road embankment crosses the floodplain of a river; low flows are conveyed through a culvert, and high flows overtop the roadway. A railway embankment runs parallel to the river and its floodplain and has a low spot upstream of the road and a culvert downstream. Figure 2 shows how this situation could be represented in FEQ. Branches represent reaches of the river and its floodplain. Two-dimensional tables represent the hydraulic controls. Level pool reservoirs represent floodplain storage areas.

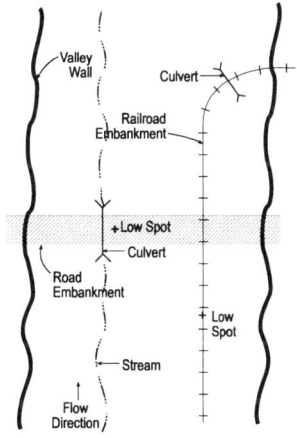

Figure 1. Schematic of Physical System

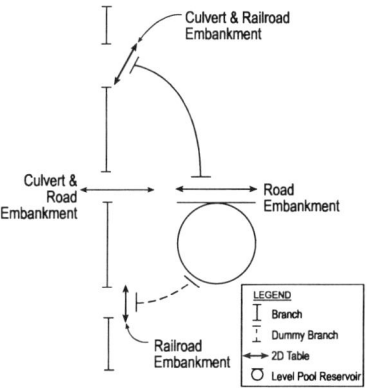

Figure 2. FEQ Model Schematic Representation

In addition, FEQUTL, a utility program which accompanies FEQ, includes routines for modeling the hydraulic behavior of a wide variety of hydraulic structures. Explicit routines exist for bridges, culverts, roadway and levee embankments, weirs, orifices, adjustable gates, expansions and contractions, and pump stations. Other hydraulic structures can be represented by user-defined tables developed through other means.

**FEQ Input and Output**

The required input data for FEQ is similar to that required for other hydraulic models. Cross-section geometry and roughness information is pre-processed with the FEQUTL program to provide tables summarizing the relevant hydraulic properties (area, top width, conveyance, etc.) as functions of depth. Hydraulic structure information is also input to the FEQUTL program to develop rating tables for use by FEQ. Typically, these ratings are in the form of a two-dimensional table relating the headwater and tailwater at the structure to the flow through the structure. The FEQ program includes look-up algorithms to determine the flow for a given headwater/tailwater situation from the two-dimensional tables. This approach increases the computing efficiency of the program, as relatively simple look-up routines are executed at every time step instead of the more complicated routines used to compute the flow based on the structure geometry.

Other inputs to the FEQ model include information describing how the hydraulic components are connected and inflows to the system. Inflows can be specified as hydrographs at specific points or as diffuse lateral inflows distributed along the length of the stream. DuPage County has developed time series files of unit runoff for different combinations of land cover and precipitation gages generated by the Hydrologic Simulation Program-Fortran (HSPF). These files are supplied to FEQ along with the areas of the various land cover/precipitation gage segments tributary to each component of the stream network. From this information, FEQ simulates lateral inflows as occurring diffusely along the stream.

FEQ can generate output in several different formats. The general output file includes information tracking the computations at each time step. Information including flows, stages, velocities and other hydraulic characteristics can be reported at every time step or at a specified interval. This file is commonly used for debugging the model. The 'special output' file format can be used to generate time series files of the flows and stages at selected locations in the system. These files can be used to calibrate the model to a gage record, to perform flow or stage duration analyses, to compute flow volumes, or to develop rating curves. DuPage County has developed a number of other programs that use the special output hydrograph files to determine flood durations for traffic damage estimation and wetland analyses. In addition, the peak-to-volume statistical method (Bradley & Potter 1991) used by the County uses these files for performing flood frequency analysis for their floodplain mapping program.

The 'maximum summary' output file format includes the peak flows, stages, and velocities, and the times that these peaks occur at every point within the system.

This file can be used to map the area flooded during a specific flood event and is especially useful for calibrating the model to observed high-water marks.

Finally, a binary time series file of flows can be written at any location within the stream system. These files can then be used as inflow hydrographs to other FEQ models.

## Case Studies

DuPage County, Illinois

As described throughout this paper, DuPage County has been successfully using the FEQ model to represent its stream and river systems for over 15 years. Almost every major stream system within the County has been modeled as part of a project analysis or watershed plan. Applications within the County include FEQ modeling in support of watershed plans for Salt Creek, many of the tributaries of the East and West Branch of the DuPage River and Sawmill Creek. Projects designed with FEQ include the Elmhurst Quarry, Wood Dale-Itasca and Meacham Grove flood control reservoirs on Salt Creek, the Valley View Flood Control Plan on the East Branch of the DuPage River, and improvements to Fawell Dam on the West Branch of the DuPage River.

The County is using the results of FEQ with an economic program (DEC-1) to compute flood damages. This program is used to assist in evaluating the relative benefits of various flood hazard reduction alternatives.

Additionally, DuPage County uses the peak-to-volume statistical method to perform flood frequency analyses. This method uses FEQ output to produce an estimate of flows and stages throughout the stream network for various return intervals. DuPage County is using this information to map the floodplains within the County.

Pierce County, Washington

The FEQ model was used to represent the complex hydraulics within a system of interconnected depressional areas or 'potholes' in the South Hill area of Pierce County, Washington (Woodward-Clyde 1997). In drier winters, many of the potholes have no outlet other than infiltration into the ground. In wetter years, the potholes overflow and flood surrounding properties. The FEQ model was used in conjunction with the HSPF model to represent the complexities of this system. HSPF was used to generate inflow hydrographs to the potholes and to approximate the effects of groundwater inflows and outflows. FEQ was used to represent the complexities of the hydraulic controls, which were often two-dimensional in nature. (Rating curves used by HSPF are one-dimensional and cannot account for tailwater effects.) In the FEQ model, the potholes were modeled as level-pool reservoirs. The structures were represented as two-dimensional tables. The models were calibrated to observed high-water marks and used to design a pumping station to convey flows out of the system to the Puyallup River.

Snohomish County, Washington

Snohomish County selected the FEQ model for use in developing the Snohomish River Comprehensive Flood Control Management Plan (Snohomish County 1991), due to the complexities of the Snohomish River system. These complexities include branched channel networks, tidal fluctuations at the mouth, and levees which are frequently overtopped, resulting in floodplain flows through agricultural areas behind the levees. The County developed an FEQ model of the river system, using inflow hydrographs at the upstream end and for major tributary inflows. The model was used to develop levee profiles, which results in more equitable flood protection among the County's flood control districts. The plan assumed that levees could not be built high enough to keep all floods out of the agricultural areas. Instead, the levee profiles were adjusted to provide the same level of protection throughout the system. In addition, overtopping sections were designed to provide control of where the levee overtops to prevent levee failures. The resulting plan has been widely accepted throughout the agricultural community and appears to have finally put an end to the 'diking wars' which have been going on for decades.

**Appendix - References**

Bradley, A. Allen and Kenneth W. Potter. 1991. "A New Approach for Flood Frequency Analysis of Simulated Flows." Paper submitted to Water Resources Research, September.

Linsley, Kraeger Associates, Ltd. 1997. Unsteady Flow Solutions - FEQ: A Modeling System for Unsteady Free-Surface Flow in a Network of Channels. Working Draft.

―――. 1997. Unsteady Flow Solutions - FEQUTL: A Function-Table Computation Utility for FEQ. Working Draft.

Snohomish County Public Works Department. 1991. Snohomish River Comprehensive Flood Control Management Plan.

Woodward-Clyde. 1997. South Hill Drainage Improvements Study. Prepared for Pierce County Public Works.

Harza Engineering Company. 1983. Winfield Creek Pilot Study. Prepared for DuPage County, Illinois.

# Operational Modeling System with Dynamic-Wave Routing

A.L. Ishii[1], M. ASCE, T.J. Charlton[2], M. ASCE, T.W. Ortel[3], A.M. ASCE, C.C. Vonnahme[4]

## Abstract

A near real-time streamflow-simulation system utilizing continuous-simulation rainfall-runoff generation with dynamic-wave routing is being developed by the U.S. Geological Survey in cooperation with the Du Page County Department of Environmental Concerns for a 24-kilometer reach of Salt Creek in Du Page County, Illinois. This system is needed in order to more effectively manage the Elmhurst Quarry Flood Control Facility, an off-line stormwater diversion reservoir located along Salt Creek. Near real-time simulation capabilities will enable the testing and evaluation of potential rainfall, diversion, and return-flow scenarios on water-surface elevations along Salt Creek before implementing diversions or return-flows.

The climatological inputs for the continuous-simulation rainfall-runoff model, Hydrologic Simulation Program—FORTRAN (HSPF) are obtained by Internet access and from a network of radio-telemetered precipitation gages reporting to a base-station computer. The unit area runoff time series generated from HSPF are the input for the dynamic-wave routing model, Full Equations (FEQ). The Generation and Analysis of Model Simulation Scenarios (GENSCN) interface is used as a pre- and post-processor for managing input data and displaying and managing simulation results. The GENSCN interface includes a variety of graphical and analytical tools for evaluation and quick visualization of the results of operational scenario simulations and thereby makes it possible to obtain the full benefit of the fully distributed dynamic routing results.

---

[1] Hydrologist, U.S. Geological Survey, 221 N. Broadway Ave., Urbana, IL 61801
[2] Chief Environmental Engineer, Du Page County Dept. of Environmental Concerns, 421 N. County Farm Road, Wheaton, IL 60187
[3] Hydrologist, U.S. Geological Survey, 221 N. Broadway Ave., Urbana, IL 61801
[4] Senior Civil Engineer, Du Page County Dept. of Environmental Concerns, 421 N. County Farm Road, Wheaton, IL 60187

## Introduction

Du Page County, Ill. has designed and built one of the largest non-Federal off-line flood-control reservoirs in the Nation, the Elmhurst Quarry Flood Control Facility along Salt Creek (fig. 1) in northeastern Illinois. The Salt Creek watershed (fig. 1) area is 298 km$^2$ at the U.S. Geological Survey streamflow-gaging station (05531500) located 14.2 kilometers upstream from the mouth. Storage provided in the quarry is about 1,023 hectare-meters. The diversion structure includes a 42.7-meter broad-crested side weir, a 24.4-meter stop-log side weir, and a 2.13-meter by 2.13-meter sluice gate located at an elevation near the bottom of the creek. After the possibility of flooding has passed, the water stored in the quarry is pumped out through an aeration process back into Salt Creek.

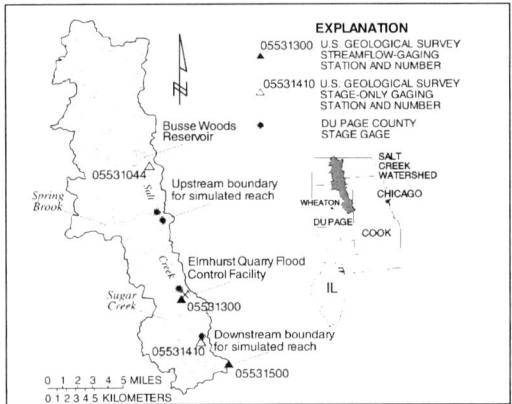

**Figure 1.** Salt Creek watershed and location of gaging stations in Du Page County, Ill.

A near real-time streamflow-simulation system for evaluating diversion alternatives is being developed by the U.S. Geological Survey in cooperation with the Du Page County Department of Environmental Concerns. The rainfall-runoff model, HSPF (Bicknell and others, 1997), was linked to the dynamic-wave routing model, FEQ (Franz and Melching, 1997a and 1997b), to determine and maximize the flood-reduction potential of the Elmhurst Quarry Flood Control Facility. Model simulations indicate that the timing of the flood wave in the lower watershed is highly sensitive to the temporal and spatial distribution of rainfall. Consequently, the ability to generate scenarios in near real time is needed to evaluate and make effective decisions about diversions or return flows. Alternative strategies for operating the sluice gate on the diversion structure can be evaluated and compared prior to taking action. To perform rapid hydrologic and hydraulic simulations needed for operational modeling, automated methods are needed for data retrieval, reformatting, error-checking and data estimation, and storage. Additionally, the outputs from the hydrologic model need to be readily input to the hydraulic model. User friendly graphical and other analytical tools for processing simulation results must be available for quick operational decision-making. These improvements form the basis of the streamflow-simulation system for Salt Creek described in this paper.

## Hydrologic and Hydraulic Models

Continuous rainfall-runoff simulation coupled with one-dimensional, unsteady-flow modeling has been mandated by Du Page County for flood-plain delineation and planning. Design and operational planning have been accomplished using an approach based on the same hydrologic and hydraulic models that are used for flood-plain mapping. The continuous simulation rainfall-runoff model used in Du Page County, HSPF, is being modified by John Kittle, Jr., and others to meet the unique needs of the near real time streamflow-simulation mode (Alan Lumb, U.S. Geological Survey, written commun., 1996). A binary output file of unit runoff time series suitable for direct input as lateral inflows to the one-dimensional, unsteady-flow hydraulic model, FEQ, has been added to HSPF. HSPF also will write the final simulation conditions to a file, so that the model can be stopped and updated without manually listing the initial conditions for the subsequent simulation run.

The FEQ model, developed by Dr. Delbert Franz of Linsley, Kraeger Associates, Ltd., solves the full, dynamic equations of motion for one-dimensional, unsteady flow in open channels and through control structures. In order to make the Salt Creek model operational for rapid simulation for flood-warning purposes, several watershed tributary models that had been routed in detail for mapping and planning purposes required simplification. The effect of routing model detail reductions on the characteristics of the main stem flood wave is under review, and continued testing is planned. Boundary conditions for the dynamic-wave hydraulic model are obtained from telemetered water-surface elevation gages and computed stage-discharge relations.

## GENSCN Graphical User Interface

The application of hydraulic and hydrologic models to near real-time flood simulation requires an interface to streamline input and output, data-processing, linkage between the models, and analysis of model results. The interactive computer program GENeration and analysis of model simulation SCeNarios (GENSCN) has been developed by John Kittle, Jr., and others (Alan Lumb, U.S. Geological Survey, written commun., 1996) in support of several USGS projects throughout the country. A graphical user interface (GUI) is currently under development for use on the personal computer platform. The GUI makes it possible to use the simulation and analysis features of GENSCN in a Microsoft Windows[5]-based environment. Most of the scenario generation and analysis options available in HSPF and ANNIE (Lumb and others, 1990) are implemented in GENSCN. The user can simulate previously developed scenarios, modify scenarios, or create new scenarios from within GENSCN.

---

[5] The use of firm, trade, and brand names in this report is for identification purposes only and does not constitute endorsement by the U.S. Geological Survey.

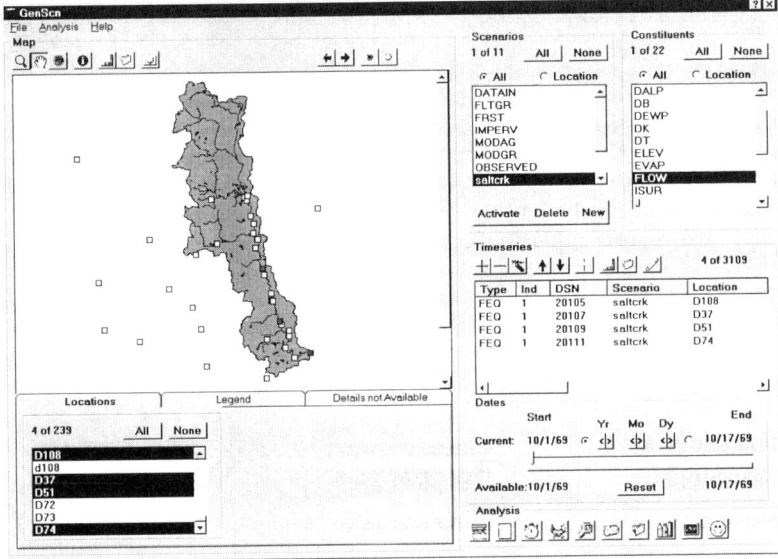

**Figure 2.** GENSCN window showing the selection of the data sets for processing and analytical option buttons (lower right).

Time-series data sets, such as flow, water-surface elevation, precipitation, storage, and volume, are referred to as constituents in the GENSCN interface. The user selects the scenarios (for example, simulated and observed data), the constituent (for example, flow or water-surface elevations), and the location for the desired data. The data sets meeting all criteria are added to the buffer by choosing the add (+) button. The analytical options available for listing, plotting, comparing, and statistical analysis of the selected data are accessed through a drop-down menu or the button bar in the lower right panel (fig. 2). The GENSCN window displays shown in figures 2-3 are intended to illustrate features available in GENSCN and do not indicate citable or quantitative model results.

To utilize the display and analysis capabilities of GENSCN, a compatible output format has been added to FEQ. The special FEQ output file for GENSCN includes water-surface elevation and discharge at each node specified in the FEQ input for each time step. Additional values that can be plotted and listed with time include channel and node hydraulic properties, such as cross sectional top width, area, conveyance, and volume available in level-pool reservoirs. These data can be plotted along one or more river reaches and animated to show the changing values in time. The selection of the standard (time series) plot option is shown in figure 3. For this example, the discharge hydrographs at the model output nodes selected in figure 2 are shown for a particular simulation.

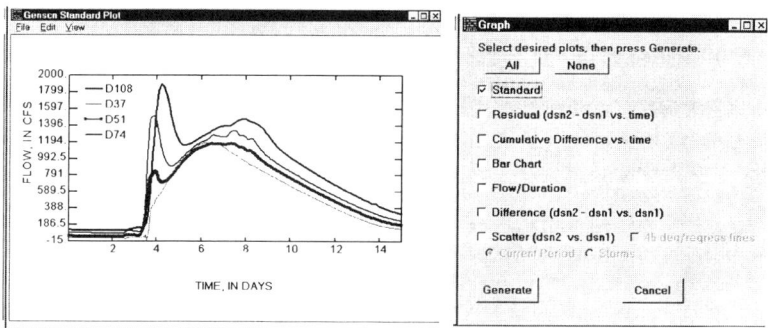

**Figure 3.** Standard plot option for analysis of time-series data.

The example given above is for the uniform rainfall case and does not represent actual streamflow for the simulated period but demonstrates the use of distributed routing for observing the hydraulic characteristics of the watershed. The most downstream hydrographs (at nodes D74 and D108) peak before the flood wave from upstream (D37 and D51) has traveled to the downstream locations for this particular rainfall event, which was of short duration and high intensity. Local inflows from the subwatersheds in the lower watershed are the main source of inflow for the initial peak. Long duration or multiple storms can result in local inflow peaks that coincide with the flood wave traveling from upstream, and, consequently, larger downstream peaks that result at later time.

After the documentation and testing of the GENSCN system for near real-time streamflow simulation are completed, the programs will be released and distributed for application to other watersheds. GENSCN is based on a modular design concept that allows for the efficient linking of other models and addition of new features and enhancements.

## Operational Use

The climatologic and hydrologic data required for the system are retrieved from the Internet and the local base-station computer. The data are checked for errors or missing values and reformatted for input to the data base utilized in HSPF by a pre-processor program. Errors and missing data reports are reviewed, automatic data revisions are either accepted or exchanged for data from other sources or estimates, and forecasted precipitation amounts may be entered to the data base. The GENSCN interface is used to (1) write the data to the data base;

(2) run the hydrologic (HSPF) model using updated initial conditions from the previous simulation; (3) run the hydraulic model under one or more operational scenarios; and (4) review the routed results for discharge, water-surface elevation, and storage at critical locations. Additional forecast precipitation scenarios then may be applied and the process repeated. Future enhancements to the system may include the display and animation of precipitation reported by the radio-telemetered precipitation network and inclusion of National Weather Service Next Generation Radar (NEXRAD) data.

The near real-time streamflow-simulation system for Salt Creek will be operated by the staff of Du Page County. Regular updates of the continuous simulation rainfall-runoff model will be done to maintain the HSPF state variables in preparation for flood events. The fully routed results are not expected to be required on a continuous basis but are to be implemented whenever water-surface elevations along Salt Creek are expected to approach flood stage. The major functions of the system are to (1) simulate Salt Creek main stem elevations based on real-time or forecasted rainfall and snowmelt data, (2) simulate alternative operating strategies for the sluice gates at the Elmhurst Quarry and compare resulting creek elevations, and (3) simulate elevations in the Creek resulting from return flows pumped from the quarry with consideration of precipitation and other climatologic conditions.

## References

Bicknell, B.R., Imhoff, J.C., Kittle, J.L., Jr., Donigian, A.S., Jr., and Johanson, R.C., 1997, Hydrological Simulation Program--FORTRAN, User's Manual for Version 11: U.S. Environmental Protection Agency EPA/600/R-97/080.

Franz, D.D., and Melching, C.S., 1997a, Full equations (FEQ) model for the solution of the full, dynamic equations of motion for one-dimensional unsteady flow in open channels and through control structures: U.S. Geological Survey Water-Resources Investigations Report 96-4240, 258 p.

Franz, D.D., and Melching, C.S., 1997b, Full equations utilities (FEQUTL) model for the approximation of hydraulic characteristics of open channels and control structures during unsteady flow: U.S. Geological Survey Water-Resources Investigations Report 97-4037, 205 p.

Lumb, A.M., Kittle, J.L., Jr., and Flynn, K.M., 1990, User's manual for ANNIE, a computer program for interactive hydrologic analyses and data management: U.S. Geological Survey Water Resources Investigations Report 89-4080, 236 p.

Price, T.H., 1994, Hydrologic calibration of HSPF model for Du Page County—West Branch Du Page River at West Chicago, West Branch Du Page River at Warrenville, East Branch Du Page River at Maple Avenue, Salt Creek at Western Springs; including hydraulic evaluation—Salt Creek at Western Springs, Salt Creek at Rolling Meadows: Northeastern Illinois Planning Commission, 92 p.

## Probability of Drought Flows in the Manitoba Hydro System

Srinivasan Rangarajan[1], P.Eng., M.ASCE, and Harold M. Surminski[2], P.Eng.

Abstract

The reduction in energy generation in the Manitoba Hydro system during extended low flow periods has severe financial implications to the utility. Manitoba Hydro uses the lowest flow period in the 84 year historical record as the critical period for determining energy supply criteria. The probability of recurrence of the critical period is unknown because of the short historical record. This study uses a synthetic flow generation approach to assess this probability of recurrence. A preliminary estimate of the frequency of recurrence using synthetic flow records is about once in 200 years. More work is in progress to use alternate approaches to model basin-wide drought events using synthetic flow generation because anecdotal records as well as tree rings indicate a more frequent occurrence of such events.

Introduction

The availability of hydro-electric energy under drought conditions is critical for predominantly hydro-based utilities such as Manitoba Hydro and Hydro Quebec. Currently, Manitoba Hydro (referred to as Hydro) uses an energy supply criterion based on the extended low flow period recorded in the years 1938-41 (critical period). A second extended drought occurred in the years 1988-91 which was similar in magnitude to that of the critical period. Hydro experiences, on average, a drought of significant magnitude every eight years. The financial loss due to a severe drought can be several hundred million dollars. Therefore, an assessment of the recurrence frequency of the critical

---

[1]Intermediate Water Resources Engineer, TetrES Consultants Inc., 603-386 Broadway, Winnipeg, Manitoba, Canada. R3C 3R6. E-mail: srangarajan@tetres.ca; Phone: (204) 942 2505

[2]Resource Planning and Market Analysis Department, Manitoba Hydro, 820 Taylor Avenue, Winnipeg, Manitoba, Canada. R3C 2P4. E-mail: hmsurminski@hydro.mb.ca; Phone: (204) 474 3170.

period is essential to Hydro in managing financial risks caused by severe droughts.

The critical period event, which occurred once in the last 84 years of historical record, gives an empirical indication that droughts of this magnitude could occur every 84 years. When this event is used for planning studies, the operating strategies are based on a single sequence of inflows that have inherent spatial and temporal variability. Hence, it is difficult to determine whether these strategies are conservative or involve significant financial risks.

Synthetic hydrology is commonly used in the simulation of water resource systems to obtain better estimates of the operation rules, instead of deriving the rules from a single historical flow record (Srinivasan and Vedula, 1990; Grygier and Stedinger, 1990; Vogel and Stedinger, 1986). Staschus and Kelman (1988) used a synthetic hydrology approach to determine the dependable hydro capacity for the Central Valley Project in California.

DeWit (1995) developed a synthetic hydrology approach for the Manitoba Hydro system, and used SPIGOT (Stedinger and Grygier, 1990) as the stochastic flow generation tool. That study concluded that the droughts of magnitude equal to that of the critical period have a recurrence interval of 385 years. In addition to the extended drought event in the years 1988-91, tree ring records (Stockton and Meko, 1983) and archival records of the Hudson Bay company indicate many extended drought periods in the $18^{th}$ and $19^{th}$ centuries. The current study has critically reviewed and has refined the approach of DeWit (1995) to model low flows in the Hydro system.

Manitoba Hydro Energy Generation System

Figure 1 shows the key river basins within the Hydro's generation system. The major rivers are: Red River from the U.S., Assiniboine River and Saskatchewan River from the west, Churchill River from northwestern Manitoba, tributaries on the eastern side of Lake Winnipeg, Winnipeg River from the Canadian Shield, and the local tributary inflows in the Burntwood River and Nelson River in northern Manitoba. These river basins are governed by different catchment and climatological characteristics as evidenced by the low spatial correlations between flows in these basins. Extended low flow events such as the critical period usually result from concurrent occurrence of low flows in many or all of these basins. Hence, the representation of spatial correlations to reproduce basin-wide droughts and the persistence of low flows to reproduce prolonged dry periods was considered as one of the primary tasks in this study.

Another important task is to develop the definition of a drought event pertinent to the Hydro system. Traditionally, droughts are defined by deficiency in flows (Srinivasan and Vedula, 1990; Dracup et al., 1980). This study defines droughts by deficiency in energy generation because system operation and reservoir storage modify the impact of low flow events. The procedure for defining the duration and severity of drought events is described below.

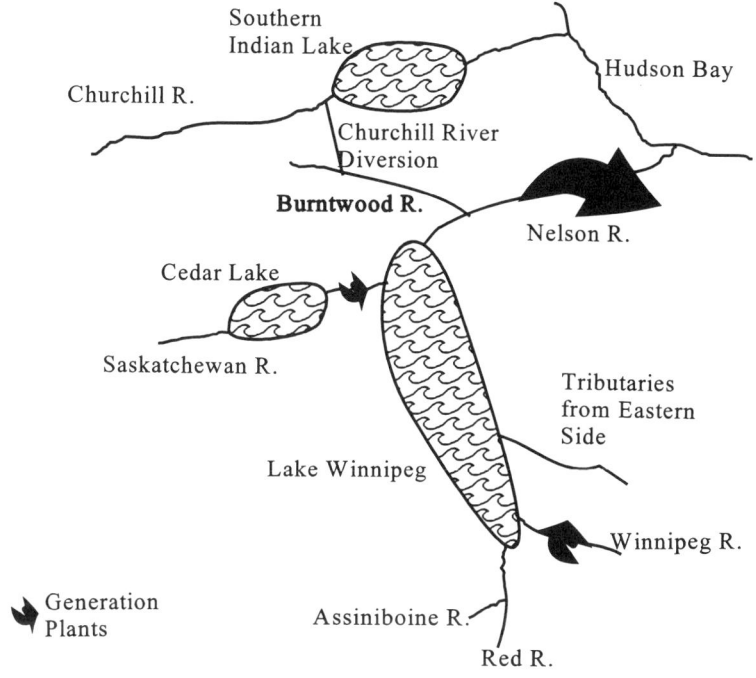

Figure 1: Manitoba Hydro Generation System

Hydro has an installed capability of 5000 MW of hydro-electric generation and 200 MW of coal-fired thermal generation. On average, about 95% of the total energy requirement is supplied from the hydraulic sources and the remaining requirement is met by imports and thermal sources which are heavily utilized in low flow conditions. Referring to Figure 1, Lake Winnipeg is a major reservoir controlled by Manitoba Hydro with a capacity of about six months of average outflow. About 80% of the total energy is generated from the plants downstream of Lake Winnipeg, and about 75% of the required flow is provided by the controlled release of water from the Lake Winnipeg. The second major storage reservoir is Cedar Lake which is controlled by a generating station near Lake Winnipeg. The remainder of the flow comes from the diversion of Churchill River flows and the local tributary flows in the Burntwood River and Nelson River. The flows in Churchill River are controlled downstream of Southern Indian Lake which acts as a third storage reservoir in the system. Energy generation is primarily dependent on the operation rules of Lake

Winnipeg and Southern Indian Lake.

## Definition and Frequency Analysis of Droughts

For this study, a definition of drought was developed to reflect the overall shortage of hydraulic energy in the system in meeting monthly energy demands. As shown in Figure 1, the generation plants are located on the Saskatchewan River, Winnipeg River and the Nelson River. With known inflows from different river basins and the operation rules for Lake Winnipeg and Southern Indian Lake, the energy generation from all hydraulic generation plants in the system can be determined. The difference between the system energy demand and the total simulated hydraulic energy generation, termed as the energy deficit, gives an indication of whether the system will be able to meet its monthly demand using hydraulic resources. The severity of a drought is measured by the accumulation of sustained energy deficits over consecutive months. The end of an extended drought period is reflected by the replenishment of reservoir storage.

Manitoba Hydro uses computer models which simulate reservoir operation and energy generation under various hydraulic conditions (Surminski, 1988). The recently developed SPLASH model uses a chronological sequence of 80 years of monthly inflows from each river basin to determine the energy capability of the system. In order to simplify the simulation of reservoir operation, two heuristic algorithms were developed in this study for the operation of reservoirs at Lake Winnipeg and Southern Indian Lake. These algorithms were calibrated to derive operating rules that were comparable to those derived using SPLASH during historic low flow events.

Using the above definition of drought events, the durations and severities of extended droughts can be computed from the time series of simulated energy deficits. The drought statistics for the historical record can then be analyzed to assess the frequency of recurrence of the critical period by comparing its duration and severity to those of synthetically generated droughts.

## Synthetic Flow Generation

The streamflow simulation software, SPIGOT (Stedinger and Grygier, 1990) was used for generating the monthly synthetic flow records for individual river basins of the Hydro system. In order to generate monthly flows in all the basins, numerous stochastic models which account for spatial correlations between basins and autocorrelations between months are required, resulting in a large number of parameters to be estimated from the limited historical record. Uncertainty in the estimation of parameters has a significant impact on the accuracy of synthetic flow generation schemes. Hence a careful selection of the number of basins and the number of stochastic models is essential in order to accurately model the entire system.

Based on the annual spatial correlations between inflows, six key river basins were chosen for the Hydro system: (1) Winnipeg River; (2) Saskatchewan River; (3) Churchill River at Southern Indian Lake; (4) Sum of local inflows in the Burntwood River and Nelson River; (5) Sum of inflows from the Red River, Assiniboine River and other gauged rivers; and (6) an equivalent basin which accounts for flows from ungauged tributaries that drain into Lake Winnipeg and evaporation from Lake Winnipeg.

One possible approach for synthetic generation is to first model the annual flow in the entire system, and then use appropriate spatial and temporal disaggregation schemes to compute monthly flows in individual basins. However, the flows in the six basins have low spatial correlation which gives an indication that the above aggregation approach is not suitable for the Hydro system. Therefore, the selected approach generates annual flows in each of the six basins in the first step using SPIGOT by preserving the spatial correlations between them. Autocorrelations are used in the second step to disaggregate annual flows into monthly flows in the six basins.

Results and Discussions

A system energy demand corresponding to hydraulic capability under average flow conditions is assumed as the threshold value for computing energy deficits during low flow periods. With known monthly inflows in different river basins, the operation models for Lake Winnipeg and Southern Indian Lake were used to identify the characteristics of significant drought events in the historical record. Table 1 shows the statistics of these events; the potential energy deficit during the critical period is 6569 GWh over a duration of 45 months.

Using the modelling framework discussed in the last section, a synthetic flow record for a period of 1000 years was generated. The operation models were then used to identify the drought events in this record. Table 1 shows the durations and severities of these events. There are five drought events that are at least as severe as the critical period, which gives an indication that the critical period has a recurrence interval of about 200 years (based on the five events in a 1000 year record). Similarly, the 1988-91 drought event has a recurrence interval of about 111 years based on the nine events in the synthetic record that exceed the severity of the 1988-91 event.

The synthetic modelling framework discussed in this paper is one of several modelling configurations that are currently being investigated in order to appropriately model the basin-wide droughts in the Hydro system. This future work will provide more conclusive evidence of the frequency of recurrence of the critical period. A concurrent research and development study of tree rings is being conducted by Manitoba Hydro to obtain additional evidence of this frequency of recurrence.

| Historical Record (84 years) | | | Synthetic Record (1000 years) | | |
|---|---|---|---|---|---|
| Description | Duration (Months) | Severity (GWh) | Rank | Duration (Months) | Severity (GWh) |
| 1938-41 | 45 | 6569 | 1 | 58 | 9799 |
|  |  |  | 2 | 56 | 8589 |
|  |  |  | 3 | 47 | 7873 |
|  |  |  | 4 | 57 | 7390 |
|  |  |  | 5 | 56 | 7029 |
| 1988-91 | 33 | 5199 | 6 | 45 | 6177 |
|  |  |  | 7 | 43 | 5413 |
|  |  |  | 8 | 34 | 5387 |
|  |  |  | 9 | 33 | 5384 |

Table 1: Statistics of Severe Droughts in Historical and Synthetic Records

References

DeWit, W.J. 1995. "Probabilistic Drought Analysis", M.Sc. Thesis, University of Manitoba, Winnipeg, Canada. 139 pages

Dracup, J.A., Lee, K.S., and Paulson, E.G. Jr. 1980. "On the Definition of Droughts", Water Resources Research, 16(2), 297-302.

Grygier, J.C., and Stedinger, J.R. 1991. "SPIGOT: A Synthetic Streamflow Generation Software Package. Technical Description", Ver. 2.6, Cornell University, USA.

Srinivasan, R., and Vedula, S. 1990. "Markov Mixture Models for Drought Lengths", Proc. 3rd National Symposium on Hydrology, CWPRS, India. 157-166

Staschus, K., and Kelman, J. 1988. "Probabilistic Dependable Hydro Capacity: Benefits of Synthetic Hydrology", Proc. *Computerized Decision Support Systems for Water Managers* edited by Labadie et al. 144-157.

Stockton, C.W., and Meko, D.M. 1983. "Drought Recurrence in the Great Plains as Reconstructed from Long-term Tree-Ring Records", J. of Climate and Applied Meteorology, 22(1), 17-29.

Surminski, H.M. 1988. "Computer Simulation of the Long-Term Operation of the Manitoba Hydro System", Proc. *Computerized Decision Support Systems for Water Managers* edited by Labadie et al. 291-300.

Vogel, R.M., and Stedinger, J.R. 1986. "The Value of Stochastic Streamflow Models in Over-Year Reservoir Design Applications", Working paper, Tufts Univ. USA

Using Digital Topo Maps for Hydraulic Modeling

Michele Good Burton[1], A.M. ASCE
Veronica J. B. Morgan[2], P.E., M. ASCE

Abstract

As most hydraulic stream modelers know, generating cross-sectional data for a stream can be time consuming, tedious, and sometimes costly. Boss International has developed two hydraulic modeling systems to make this process and others easier to accomplish. The Boss River Modeling System (RMS) and RiverCAD provide the user with the ability to generate cross-sections from digital topographical data, run the hydraulic model, create floodplain boundaries, and produce maps of the results. In this paper, we will describe the successes and difficulties we encountered with RMS and what improvements were made with the introduction of the RiverCAD system. RMS runs within the AutoCAD environment and utilizes the digital topographical data of AutoCAD drawings. This interaction with AutoCAD provides benefits in both the input and output phase of modeling. In the input phase, it allows the user to generate cross sections by pointing and clicking on contour lines with their mouse. In the output phase, it automatically maps floodplain boundaries by interpolation using the Boss Floodplain Mapping module. RMS uses both U. S. Army Corp of Engineer's (USACE) HEC-2 and HEC-RAS model engines. RMS imports and exports files from HEC-2 and HEC-RAS. This allows the model to be shared with others who do not use RMS by exporting the RMS files to HEC-2 or HEC-RAS files. The disadvantages of using RMS are that it comes at a cost to flexibility in large complex models and that it relies on AutoCAD. This loss of flexibility is a result of the method that RMS uses to identify and relate separate stream reaches. Each individual reach is termed a 'metajob' and operates as an independent unit in the model. This differs from HEC-RAS which allows the user to join separate reaches with junctions. This paper provides an in-depth look at the advantage and disadvantages of using the Boss River Modeling System.

---

[1] Engineering Assistant, City of College Station, P.O. Box 9960, CS,TX 77842
[2] Asst. City Engineer, City of College Station, P.O. Box 9960, CS,TX 77842

## Introduction

Constructing a hydraulic model for a stream can be time consuming and tedious when building the model means cutting cross sections by hand. Digital topographical maps and new software gives engineers a new way to build the model. The new software we will address here is the Boss River Modeling System (RMS)[1]. We have elected to write about the Boss RMS software because we have first hand experience with the program. The City of College Station, Texas purchased RMS early in 1997 and has used it to create the geometry data for its USACE HEC-RAS[2] models. The features of this package are unique in that they focus on the needs of hydraulic engineers. At this time, we are not aware of other software companies with the same type of software. We have come across software that offers some of the same features developed by other software venders. However, it appears from a web search that most software marketed to civil engineers concentrates more on the stormdrain system modeling rather than with modeling of open channels and structures. The scope of this paper, therefore, will be limited to our personal experiences with the RMS package. We have found the program very helpful on many accounts and less helpful than expected on others. We will discuss the advantages and disadvantages of this software package.

## RMS Capabilities

The engineer using RMS has three main advantages over those engineers without a software tool such as this one. One advantage is the use of digital topo maps to develop the geometry of a HEC-2 or HEC-RAS model. Another advantage is that RMS allows the engineer to measure and record reach lengths between cross sections and identify bank stations directly from the topo maps. The program automatically stores the geometric data (cross sections, reach lengths, and bank stations) for the model into the drawing file. All the information about a cross section is linked to the plan view and can be altered at any time. Lastly, RMS will determine the floodplain boundaries and map those boundaries by running a digital terrain model.

## Advantages of RMS

### Cutting Cross Sections

The largest benefit of RMS is the time and energy which can be saved by using RMS to cut cross sections for a model. Of course, the cross sections are only as accurate as the topos themselves. In College Station, the terrain is mostly flat and the streams are meandering. The city topo maps have 2-foot contour intervals and numerous spot elevation to keep the accuracy high. Even with this accuracy, the actual channel cross sections can vary from the topo because of dense vegetation obscuring the ground topography. To avoid inaccuracies due to vegetation in the creek areas, it is easy to make changes to the cross section by adding additional surveyed data.

Before the introduction of software like RMS, engineers surveyed all the necessary cross sections in the field or relied on topo maps and cut cross sections by hand to generate the geometric data for a hydraulic model. Many times USGS quadrangle (quad) maps with their ten-foot contour intervals were used. For day to day use, many cities have replaced their quad maps with maps more accurate topo at 2-foot and 5-foot contour intervals. Having good topographic maps, may save the engineer from surveying an entire area, however, it does not save them from the tedious and slow hand process of cutting cross sections from the maps. Each data point must be located on the map and scaled off relative to the first data point on the cross section. These individual data points must be collected for each cross section and entered into the data file. The tedium of this exercise often leads to an increase in mistakes. Such mistakes range from entering the wrong number to entering the cross section in the reverse order. Of course, a diligent engineer will check for these things, but they can still be overlooked. It is the increased use of more accurate and digital topo maps that make software like RMS practical for the engineering community.

To take a cross section in RMS, the user must first create or use a current metajob. A metajob is what RMS calls an individual reach. All cross sections for a single stream reach will be continued in this metajob. Next the topo map for the cross section location must be brought into the drawing. This can be accomplished with standard AutoCAD functions of external referencing (XREFing) or inserting the topo map into the drawing. Once the topo map is in AutoCAD, the map needs to be configured. This basically consists of telling RMS the contour interval, the layer(s) the contours are stored on, and scale of the topo map in AutoCAD units. The configuration only has to be done once prior to cutting the cross sections. The cross section is drawn from the left to the right while looking downstream. The user must specify the river station number for the cross section and the starting station number for the first data point. To cut the cross section, the user must first select the leftmost point in the cross section by clicking on the first contour line of the cross section. Then move the mouse pointer to the next contour and click again. The location of the cross section is drawn onto the map as you go. After the cross section is cut, it is displayed on one of the view ports. A cross section can be easily changed in the future, either by entering the data points by hand or using the topo map.

*Reach Lengths and Bank Stations*

Another advantage of RMS, is that the user can determine reach lengths along the channel banks and channel bottom. The reach length is measured on the topo map as a line is drawn from one cross section to the next. To measure the reach length from one cross section to another, the topo maps must be XREFed or inserted into the drawing. Then, from the cross section table the user can enter a value or go to the topo map to get the reach length. For instance, to determine the reach length down the center of the creek, the user would start at the creek centerline and click on the cross section. Then continue to point and click following the bottom of the creek until the next cross sections reached. At this time, the user clicks the right hand button to end the measurement. The value automatically appears in the cross section table. The bank stations are easy to select. They can be selected either from the topo map or the cross section within the drawing. The user simply points and clicks at the station location and bank station number is automatically added to the cross section table.

*Floodplain Mapping Feature*
The RMS floodplain mapping module is quite versatile. It has two basic methods to map the floodplain. The first method is the simple water surface method. This method simply locates the provided water surface elevation on each cross section and connects the water surface elevation with straight lines. Thus, creating a general floodplain that can be filled in with a solid color or a hatch pattern. The second method is the advanced water surface method. This method promises what most engineers would be looking for in a floodplain mapping routine. The advanced method uses a digital terrain model (DTM) to calculate the edge of the water surface. This DTM uses the water surface elevations provided at each cross section and the contours on the digital topo map to interpolate the floodplain boundary between the cross sections. The floodplain area can then be shaded or hatched.

### Disadvantages of RMS
*Backwater Analysis*
The Boss RMS package includes HEC-2 and HEC-RAS in separate modules. One of the biggest differences between Boss HEC-RAS and USACE HEC-RAS is that the Boss version uses 'metajobs' and the USACE HEC version uses 'junctions'. A metajob is a separate stream reach that contains the geometric data and flow data for that reach. Each metajob operates independently from other metajobs in the same model. This means that the user must input the boundary conditions for each reach. For instance, let us consider a creek shown in Figure 1 with a tributary. The creek will be called metajob 1 and the tributary will be called metajob 2. To run the hydraulic analysis for this system, first the boundary conditions for metajob 1 (the creek) must be entered as usually required in HEC-RAS. The second step is to run the analysis for metajob 1. Third, find the water surface elevation on metajob 1 that will be the boundary condition for metajob 2 (the tributary). Then, input the boundary condition for metajob 2 and run metajob 2. In contrast, HEC-RAS does not use the concept of 'metajobs', but uses junctions. These junctions provide a link between the water surface profile on the creek (in our example) and the tributary.

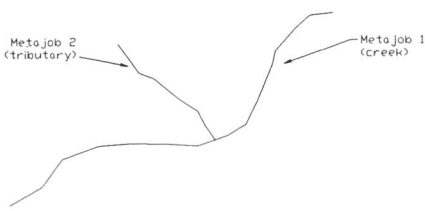

Figure 1. Depiction of Boss metajobs.

*AutoCAD Interface*

RMS uses an AutoCAD interface in order to view digital maps and create accurate drawings. AutoCAD is a very sophisticated program that provides more functions than RMS needs to run. These functions include storing graphical information, modifying this information, and maintaining drawing accuracy. In AutoCAD 12 and 13, some of these functions are memory intensive. The large memory use requires a robust and fast machine to run RMS within AutoCAD. As a result, RMS is dependent on a program that offers more capabilities than it needs and slows down the RMS processing unnecessarily.

**RiverCAD**

RiverCAD was released by Boss last fall, after the city purchased RMS. This program appears to be nearly identical to RMS in its features. It will cut cross sections, determine reach lengths, and floodplain boundaries as well as run the HEC-RAS or HEC-2 analysis. The biggest improvement over RMS is that RiverCAD does not require AutoCAD to run and instead uses its own CAD interface. AutoCAD is a excellent program, but the functions within AutoCAD that are utilized by RMS are minimal and the overhead of running AutoCAD quickly becomes a drawback.

The reliance on AutoCAD limited our experiences with RMS. RMS is currently designed to work with DOS and Windows versions of AutoCAD 12 and 13. A new version of RMS for AutoCAD 14 is promised to be released. The computer we used to run RMS was a Pentium 133 MHz with 16 MB of memory installed with AutoCAD 12 for DOS. We quickly found that the memory was insufficient for our project. Using the topo maps to cut cross sections and reach lengths, frequently required us to have more than one map in use at a time. Our maps divide the city into a grid. As the creeks cross from one map to another, several maps were necessary. Soon we realized that the intensive use of large topo maps under this system was too slow to be practical. As a result, the memory was upgraded to 64 MB and the effect was tremendous. The AutoCAD program, especially the DOS version is extremely memory intensive. Our creek was quite complex and required over 20 metajobs to include all developed reaches. With the increased memory, the combination of AutoCAD and RMS worked reasonably fast.

We have reviewed the demo for RiverCAD and it has an AutoCAD like interface. It provides all the necessary drawing and modify commands such as polyline, circle, trim, and rotate needed to modify drawings as desired. However, if these tools are not sufficient, RiverCad provides the flexibility for the drawings to be saved in AutoCAD or Microstation compatible formats. So, if RiverCAD does not perform all of the CAD functions you are looking for, it is easy to use what ever CAD system you are currently using to make changes. The HEC-2, HEC-RAS, and Floodplain Mapping Modules are very similar to RMS. In fact, we did not find any differences other than the fact that the CAD environment developed for RiverCAD was easy to use, very quick, and would significantly decrease memory requirements.

## Conclusions

As the number of cities with digital topo maps increases, software such as Boss RMS and RiverCAD will be increasingly important. Software tools for engineers are quite common to the industry and will continue to be as new ideas are developed. As for RMS, we have found the software very useful for generating cross sections, reach lengths, and bank stations. With the proper hardware set up, it can be quite fast and an excellent way to build a model from scratch. The use of junctions in HEC-RAS over metajobs in RMS is preferable. The metajobs are a convenient way to store information into each reach length, however, updating the boundary conditions after each model run is inconvenient and unnecessarily time consuming. To avoid this, once the model is built, the data files can be exported to HEC-RAS or HEC-2 and combined. The floodplain mapping module is an excellent concept for hydraulic engineers. We feel that this portion of the software, although we have had limited experience with it promises great things. Overall, we believe there is a real need for this type of software and hope to see more products like these in the future.

## References

[1] Boss International, **BOSS-RMS for AutoCAD: User's Manual**, 1996

[2] U.S. Army Corp of Engineers, **HEC-RAS River Analysis System: Hydraulic Reference Manual**, July 1995

## GIS-Based Processing of NEXRAD Rainfall Estimates

C. Bryan Young[1], S. M. ASCE, Bruce M. McEnroe[2], M. ASCE, and Rebecca J. Quinn[3], M. ASCE

Abstract

The National Weather Service's Stage III precipitation estimates offer some distinct advantages for flood studies. An ARC-INFO GIS application has been developed for processing and analyzing Stage III data. The system provides tools for mapping and contouring rainfall estimates, computing watershed averages of hourly or daily precipitation, and for developing rainfall time series for selected points. Tools are also available for importing, processing, and analyzing other geographic data sets.

Introduction

In studies of historic floods, accurate estimates of rainfall are invaluable. Often, rain-gage data alone cannot provide the spatial or temporal resolution necessary to define the distribution of precipitation over a watershed. The National Weather Service's (NWS) NEXRAD Stage III product offers high-resolution rainfall estimates that can enhance analysis of storm runoff production.

Despite the prevalence of NEXRAD rainfall estimates in every day life, utilization of NEXRAD in water resources engineering has been restricted by several factors. Poor availability of archived data, limited understanding of the radar-rainfall product, and a lack of tools for processing and linking NEXRAD rainfall estimates to rainfall-runoff models have all prohibited wide-spread use of NEXRAD in studies of historic floods.

---

[1] Graduate Student, Dept. of Civil & Environmental Engineering, University of Iowa, Iowa City.
[2] Professor, Dept. of Civil & Environmental Engineering, University of Kansas, Lawrence.
[3] Engineer, George Butler & Associates, Lenexa, Kansas.

Researchers at the University of Kansas have developed an ARC-INFO GIS application for processing and integrating NWS Stage III precipitation estimates with other geographic data. The application provides graphical tools for producing maps of soil characteristics, land-cover types, and slopes as well as for delineating watersheds, mapping and contouring NEXRAD precipitation estimates, and calculating spatial rainfall averages and storm totals. The Kansas Department of Transportation uses this analysis system for investigations of floods on state highways.

NEXRAD Stage III Rainfall Estimates

The NWS River Forecast Centers (RFCs) produce Stage III precipitation estimates using the network of WSR-88D weather radars known as Next Generation Weather Radar (NEXRAD). The Stage III product combines radar-rainfall estimates (produced using a standard Z-R relationship) with rain-gage readings to produce a mosaic of precipitation estimates for each RFC region (Shedd and Fulton 1993). Estimates are produced hourly on the Hydrologic Rainfall Analysis Project (HRAP) grid, which has an approximate grid spacing of four kilometers (Reed and Maidment 1995).

Although the Stage III product is produced operationally by each RFC, not all centers have the capability to archive and make these data available to the general public. Stage III data for southern Kansas are available from the Arkansas-Red Basin River Forecast Center (ABRFC) via the internet. These archives are publicly accessible and date to June 24, 1994. Stage III data files for northern Kansas are obtained by the University of Kansas from the Missouri Basin River Forecast Center by special arrangement, and have been archived there since June 10, 1995. Figure 1 displays the radar umbrellas covering Kansas and the Missouri-Arkansas basin divide.

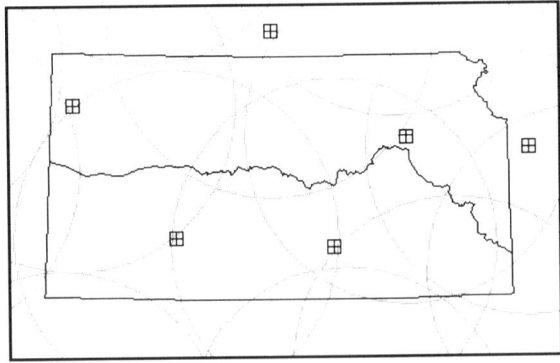

Figure 1: NEXRAD Coverage of Kansas

Before NEXRAD data can be manipulated in a GIS environment, the estimates must be geo-referenced to the HRAP grid. Seann Reed of the University of Texas at Austin has produced coupled FORTRAN and ARC-INFO routines for developing the HRAP grid for a user-specified region (Reed and Maidment 1995). These programs were used to generate an HRAP grid for the entire state of Kansas.

DEMs, DLGs, and Other Geospatial Data Sets

In recent years there has been a proliferation of digital geographic data sets. Important examples include the United States Geological Survey (USGS) digital elevation models (DEMs), USGS digital line graphs (DLGs), the SSURGO soil database produced by the Natural Resources Conservation Service (NRCS), and the Kansas Applied Remote Sensing center's (KARS) land-cover survey of Kansas. These data sources are extremely useful in hydrologic analysis.

Storm Analysis System

The ARC-INFO GIS application developed at the University of Kansas provides a complete, menu-driven interface for processing NEXRAD and other geographic data. The interface is programmed in ARC Macro Language (AML) and FORTRAN, and takes advantage of ARC-INFO's built-in menu system. The application has a modular structure, and menu options may call additional menus or AML macros. The functionality of the application is divided into two major modules: geographic data preparation and rainfall mapping and analysis.

Geographic Data Preparation

The geographic data preparation module provides tools for developing ARC-INFO point coverages and for processing USGS DLGs and DEMs, SSURGO data files, and KARS land-cover files. Figure 2 displays the main data preparation menu. Options for processing DLGs and DEMs are described here.

The 'Process DLG' option generates an additional menu with several options. Tools are available for importing and appending multiple adjacent DLGs, clipping DLGs to a watershed boundary or to an area of interest, and for preprocessing hydrology DLGs for use with the ARC TOPOGRIDTOOL for DEM development (ArcDoc 1995).

The interface provides several options for developing DEMs in the 'Process DEM' sub-menu. DEMs can be imported from USGS 1:250,000 or 1:24,000 scale USGS DEM files, or can be custom-built using TOPOGRIDTOOL. Once a DEM has been developed, several processing tools are available. The user can interactively clip the DEM to reduce size, fill sinks (unnatural depressions due to sampling and other DEM errors), invoke the ARC-INFO GRID FLOWDIRECTION and FLOWACCUMULATION commands, delineate watersheds, clip existing grids to the watershed boundary, and calculate slopes.

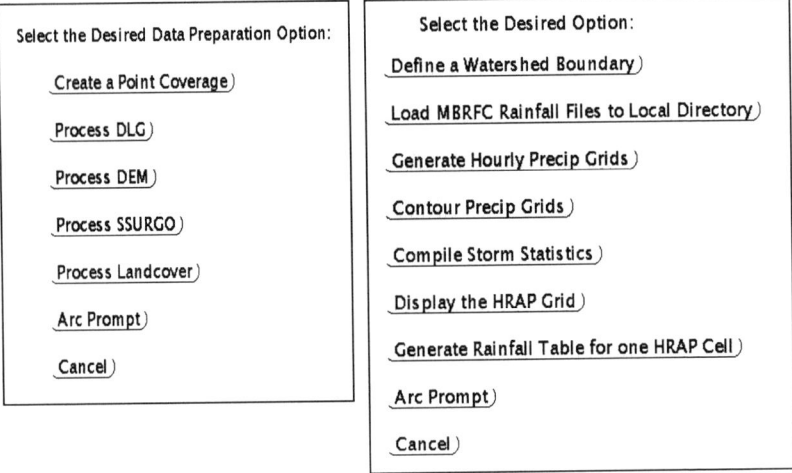

Figure 2: Data Preparation and NEXRAD Processing Menus

NEXRAD Mapping and Analysis

The ARC-INFO rainfall analysis application provides several tools for mapping and analyzing NEXRAD rainfall data. Figure 2 presents the available menu options.

The 'Define a Watershed Boundary' button on the precipitation processing menu allows the user to manually delineate the watershed or area of interest using multiple background coverages. This option is especially useful if the watershed cannot be automatically delineated from the DEM due to the presence of man-made features, inadequate DEM scale, or DEM error, or if the area of interest does not conform to watershed boundaries.

Files containing the hourly rainfall estimates must be loaded to the working directory before the data are processed. Rainfall files for the Arkansas River basin are downloaded from the ABRFC directly, while files for the Missouri River basin are copied to the local directory from the Kansas University archive using the 'Load MBRFC Rainfall Files to Local Directory' function.

The bulk of the rainfall analysis functionality resides in the 'Generate Hourly Precip Grids' tool. This function prompts the user for the dates of the storm, allows the user to interactively select an area that encompasses the study site, opens the relevant Stage III rainfall files (hourly or daily), links the rainfall estimates to the

HRAP grid, and generates a grid of specified cell-size for each rainfall file. Stage III files from the ABRFC are archived in NetCDF format, while those from the MBRFC are unformatted binary files. Archived files are opened using FORTRAN programs which extract the rainfall estimates for the user-selected study area. Extracted rainfall data are referenced to the HRAP grid by an ARC-INFO identification number. A list of these identification numbers and the corresponding rainfall estimates are imported into the ARC-INFO TABLE module, and are then joined to the HRAP grid using the ARC JOINITEM command. Finally, the data are converted from polygon format to ARC-INFO grids using the POLYGRID command. The precipitation grids are then ready for further analysis.

Once NEXRAD rainfall estimates have been converted to grid format, 'Contour Precip Grids' can be used to contour the precipitation field. This option contours the individual and storm total precipitation using the ARC-INFO FILTER and LATTICECONTOUR commands. Figure 3 displays a sample contoured rainfall field for Bluff Creek in Kansas.

In addition to producing graphical summaries of storm rainfall, the user interface provides the 'Compile Storm Statistics' option for calculating hourly spatially averaged rainfall totals for the selected watershed. The averages are saved as a text file, which can then be graphed or used as input to a rainfall-runoff model.

The 'Display the HRAP Grid' option simply allows the user to view the HRAP grid in relation to any additional user-selected coverages.

Often it is useful to compare NEXRAD estimates with rain-gage estimates, or to analyze NEXRAD estimates for a small area. The 'Generate Rainfall Table for One HRAP Cell' option permits the user to select one or more HRAP cells. The tool then generates a time-series of rainfall for that point, which is saved as a text file.

Conclusions

A menu-driven application for processing and analyzing NEXRAD radar-based precipitation estimates and other geographic data sets has been developed in ARC-INFO. This rainfall analysis application provides tools for importing, processing, and analyzing USGS DEMs and DLGs, NRCS SSURGO soils data, KARS land-cover survey for Kansas, and NEXRAD Stage III rainfall estimates for the state of Kansas. The available rainfall analysis tools map and contour hourly or daily precipitation, compute spatially-averaged rainfall over a watershed, and develop precipitation time series for selected points. The Kansas Department of Transportation uses this system for analysis of historic floods on state highways.

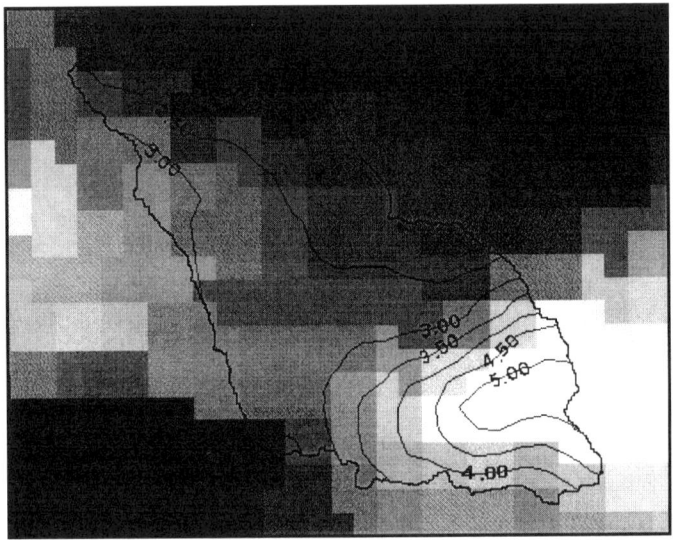

Figure 3: Contoured Storm Totals over Bluff Creek

Acknowledgment

This research was supported by the Kansas Department of Transportation through the Kansas Cooperative Transportation Research Program (KTRAN Project KU-96-7).

References

*ArcDoc Version 7.0.* (1995). Environmental Systems Research Institute, Inc., Redlands, California.

Reed, S. M., and Maidment, D. R. (1995) "A GIS Procedure for Merging NEXRAD Precipitation Data and Digital Elevation Models to Determine Rainfall-Runoff Modeling Parameters." *CRWR Online Report 95-3*, Center for Research in Water Resources, University of Texas at Austin, Texas.

Shedd, R. C., and Fulton, R. A. (1993). "WSR-88D Precipitation Processing and its use in National Weather Service Hydrologic Forecasting." *Engineering Hydrology, Proceedings of the Symposium*, San Francisco, California.

# PROBLEMS IN CALIBRATING CONCEPTUAL RAINFALL RUNOFF MODELS.

Paolo Bartolini[1], Federica Montereggio[2] and Juan B. Valdés[3]

### Abstract

Rainfall-Runoff Models (RRM) allow the use of most of the fast procedures (like the Rational Method) aimed at transferring the probability laws from the rainfalls to the discharges. In fact, the time of concentration of a basin (in order to select a proper rainfall from the IDF curves) as well as the infiltration rate, highly affect the estimation of the T-years discharge; it follows the necessity of calibrating an RRM, for deriving the characteristics of the basin under observation, by using the ones obtained for a similar, gaged, basin. A complete model of the rainfall-runoff transformation (i.e. a model which takes into account the underground portion of the hydrograph, even if the maximum discharge during a flood is mainly due to the surface runoff) is needed. Also it is required to handle the almost unavoidable asynchronism between rainfall and discharge records during the calibration of a model. After a particular RRM is chosen, the proposed solution (i.e., the introduction of an extra unknown, the time-lag between rainfall and discharges, and the contemporaneous fitting of a set of different events) is tested with respect to two small mountain basin in northern Italy.

## 1 Model Selection

Often, during a flood, the subsurface and groundwater components of the hydrograph are a small part of the total runoff (thus making event models the most widely used for simulation purposes). This is not the case for most of the events used in the calibration phase, in which only high frequency events (low return periods) are often at disposal for a comparison with the outputs from the model. It follows the need of a full model, i.e. a model in which the surface water routing is accompanied by infiltration models, able to take into account not only the separation of the surface component of the rain from the infiltrated one, but also to perform the subsequent separation of the last into soil water, interflow and baseflow. The RRM we selected is based on a modified version of the Horton model for infiltration, two linear reservoirs in series-parallel for modeling the interflow and baseflow contributions to channel flow, and the Schaake's version of the Wooding model for surface and channel flow routing. In order to keep the number of parameters as small as possible, this study has been restricted to those basins for which the following assumptions can be made, i.e.:

- the interception, infiltration, interflow and baseflow parameters (7) are the same for all sub-basins;

---

[1]Institute of Hydraulics, University of Genova, Via Montallegro 1, 16145 Genova, Italy; (39)(10)353-2472; 353-2546 FAX; bart@idra.unige.it.

[2]Former student, Institute of Hydraulics, University of Genova, Via Montallegro 1, 16145 Genova, Italy; (39)(10)353-2472; 353-2546 FAX

[3]Department of Civil Engineering and Engineering Mechanics The University of Arizona, Tucson AZ 785721-0072; (520)621-2266; (520)621-2550 FAX; jvaldes@u.arizona.edu.

- the rectangular hillsides have the same Strickler coefficient value for the entire basin, thus giving rise to friction coefficients which are proportional to the square root of their slopes;
- the channels' cross sections may only be triangular (for the first order sub-basins) or rectangular (for the higher orders).

For each type of cross section, the Strickler coefficients are the same all over the basin; moreover, while an unique triangular section is foreseen, the rectangular ones have widths which are dependent on the area of the portion of the basin they drain, by means of a two parameter exponential relationship. It is assumed that the previous assumptions (which lead to a set of 12 parameters) may hold for basin not greater than 50 $km^2$.

## 1.1 Parameter Set.

The parameters of the modified Horton model (Bartolini, 1987) are: the retention volume $V_0$ [$mm$], the dry and saturated infiltration rates $f_0$ and $f_1$ [$mm/h$], the maximum storage in the soil profile $V_{1,m}$ [$mm$] and the corresponding initial value $V_{1,0}$. The interflow and baseflow are modeled by means of two linear reservoirs, in series-parallel; $\alpha$, $\beta$ and $\gamma$ [$1/h$] are the proportionality coefficients between the volumes stored in the reservoirs and the discharges (the interflow $q_1(t)$, the deep percolation $q_2(t)$ and the baseflow $q_3(t)$, respectively). For surface runoff the kinematic wave model was used. For the hillsides, the stage-discharge relationship takes the form:

$$q = \alpha_h y^{5/3} \quad with \quad \alpha_h = k_{s,h}\sqrt{j_h}$$

where $y$ [$m$] is the depth of a sheet of water on the hillsides, and $q$ [$m^2/s$] is the corresponding discharge for unit length. $k_{s,h}$ [$m^{1/3}/s$] is the Strickler coefficient for the hillsides, and $j_h$ is the corresponding slope. The assumption about the nature of the hillsides for the basin under consideration makes the Strickler coefficient the only unknown parameter for the overland flow. For the channels, the stage-discharge relationship takes the form

$$Q = \alpha_c \Omega^{m_c}$$

where $Q$ [$m^3/s$] is the discharge, $\Omega$ [$m^2$] the area of the cross-section, and $\alpha_c$ [$m^{3-2m_c}/s$] a friction coefficient; the exponent $m_c$ depends on the shape of the cross section. It is assumed that the cross-sections of the first-order streams are triangular, with a fixed slope of the banks. It follows that a Strickler coefficient $k_{s,t}$ [$m^{1/3}/s$] may be introduced also for representing the behavior of this kind of streams, without any other geometrical parameter (and with $m_c = 4/3$). The form of the friction coefficient $\alpha_c$ is then

$$\alpha_c = \alpha_t = 0.462\, k_{s,t}\sqrt{j_t}$$

As said before, the higher order streams are assumed to be rectangular, with a width $b$ [$m$] which varies as function of the area $S$ [$km^2$] drained at the mid-point of the last channel of each sub-basin. Since a power law has been assumed to hold between the channel width and the drained area, i.e. a relationship of the form

$$b = cS^d$$

the assumption that the ratio between the depth and the width of the streams is always much less than one enables to write $\alpha_c$ as:

$$\alpha_c = \alpha_r = \frac{\theta_1 \sqrt{j_r}}{S^{\theta_2}} \quad with \quad \theta_1 = \frac{k_{s,r}}{c^{2/3}} \quad and \quad \theta_2 = \frac{2d}{3}$$

and the parameters to be estimated are $\theta_1$ $[Km^{2\theta_2}/(m^{1/3}s)]$ and $\theta_2$ $[--]$. The whole set of parameters is then formed by the value of 12 physical characteristics of the basins:

$$V_0 \quad f_0 \quad f_1 \quad V_{1,m} \quad V_{1,0} \quad \alpha \quad \beta \quad \gamma \quad k_{s,h} \quad k_{s,t} \quad \theta_1 \quad \theta_2$$

An important observation regards the fact that the selected model is based on a transformation of the real basin into a simplified, artificial counterpart, which asks for the definition of artificial, effective physical quantities (which means, for example, that the rectangular hillsides can have unusual values of the Strickler coefficients, with respect to the range of the values commonly foreseen for natural and artificial channels).

## 2 Real World Applications: Calibration Problems and Proposed Solutions.

The proposed model has been used for studying the rainfall-runoff transformation for two small basins in the Ligurian Appennini (northwest Italy). The first basin (Bisagno creek) is 34 $km^2$ wide and is almost entirely included in the sedimentary zone east of Genova, with parent rocks of the limestone type. The second basin (Sansobbia creek) is 32 $km^2$ wide; most of the underlying rocks are of the metamorphic and volcanic type. Each basin has been subdivided into '$V$-shaped' sub-basins, in a totally subjective manner. In order to reduce the computation time, only 4 sub-basins have been considered for the Sansobbia basin, and 6 for the Bisagno. The analysis of the consequences of different number of sub-basins (and of different choices in their composition) will be presented in a forthcoming paper; the problems we met in this work are likely to be the same at whichever spatial (and temporal) scale, at least from a qualitative point of view. As mentioned above, 11 parameters and one initial value must be estimated for each event, for each basin. Ten events have been taken into consideration for each basin; the parameters have been estimated by using the Simple Least Squares (SLS) method, by using both the Powell and the General Reduced Gradient optimization techniques. We are currently using also the SCE method (Yapo et al, 1997). Table 1 shows the result of the calibration for 10 events for the Bisagno basin, by means of the average value ($E$), the standard deviation ($Std$) and the coefficient of variation ($CV$) of the parameters.

Table 1: Parameters for 10 events for the Bisagno creek,

| | Parameters | | | | | | | | | | | |
|---|---|---|---|---|---|---|---|---|---|---|---|---|
| | $V_0$ | $f_0$ | $f_1$ | $V_{1,m}$ | $V_{1,0}$ | $\alpha$ | $\beta$ | $\gamma$ | $k_{s,h}$ | $k_{s,t}$ | $\theta_1$ | $\theta_2$ |
| E | 9.7 | 242.0 | 17.0 | 106.4 | 78.2 | .13794 | .02163 | .00267 | 11.88 | 20.70 | 27.07 | .4970 |
| Std | — | 312.0 | 8.9 | 16.7 | — | .37201 | .03674 | .00556 | 31.84 | 39.80 | 41.31 | .8752 |
| CV | — | 1.29 | .53 | .16 | — | 2.70 | 1.70 | 2.08 | 2.68 | 1.92 | 1.53 | 1.76 |

One parameter depends on the seasons ($V_0$) while the initial value ($V_{1,0}$) is associated to a particular event; their variation from one event to another is not important. As it can be seen from Table 1, the parameters related to infiltration show a better behavior (lower coefficient of variation) with respect to the parameters related to the overland and channel flow. In view of the ultimate goal of this analysis, i.e. the evaluation of the surface contribution to the flood for the basins for which the Rational Method may be acceptable, the spreading of the values of the parameters related to the surface friction is troublesome. The almost unavoidable asynchronism between rainfall and runoff records (and between different rainfall records) has been suspected to be the main cause of the observed problems. The events under consideration have been recorded twenty, thirty years ago, with recording tools (weekly rainfall and discharge records on paper) which did not guaranteed against error in the timing. The question then arises as how to take into account an unknown quantity like the error of the time origin of the records. We decided to consider this time error as an extra parameter to be evaluated; in doing so, shifted hydrographs (observed and calculated) could be compared, in order to look for an otherwise unreachable good agreement. The results of the calibration are shown in the Table 2 for the Bisagno creek.

Table 2: Parameters for 10 events for the Bisagno creek, with shifting of the discharge records.

|  | Parameters | | | | | | | | | | |
|---|---|---|---|---|---|---|---|---|---|---|---|
|  | $V_0$ | $f_0$ | $f_1$ | $V_{1,m}$ | $V_{1,0}$ | $\alpha$ | $\beta$ | $\gamma$ | $k_{s,h}$ | $k_{s,t}$ | $\theta_1$ | $\theta_2$ |
| E | 8.9 | 97.9 | 11.0 | 100.6 | 74.9 | .12549 | .01660 | .00027 | 2.35 | 18.09 | 10.90 | .1444 |
| Std | — | 85.5 | 5.2 | 12.1 | — | .36966 | .03374 | .00030 | 1.57 | 36.68 | 8.74 | .2668 |
| CV | — | .87 | .47 | .12 | — | 2.95 | 2.03 | 1.14 | .67 | 2.03 | .80 | 1.85 |

As it can be seen by comparing Table 1 with Table 2, a better agreement has been obtained, but a certain spreading still persists between the values of those parameters which are not expected to vary from one event to another. This is common in calibration practice, but in this case a further reason for the observed discrepancy is the fact that the time-lag may have had a very important effect on the calibration. For example, if the rainfalls are anticipated with respect to their true time origin, the calibration will tend to produce slower flows, and this speed reduction may be concentrated in the Strickler coefficients for the hillsides or for the channels or for both; the opposite happens if the rainfalls are posticipated. In order to investigate this kind of problem, a number of controlled experiments have been made. The reference (observed) hydrograph was produced by the model, for a given set of parameters; the fitting has been performed for different value of the time-lag $\tau$ [min], for each experiment. The results are shown in Table 3.

Table 3: Parameters obtained by fitting a reference event, with varying time lags.

| Parameters | MSE | | | | | | | | | | | |
|---|---|---|---|---|---|---|---|---|---|---|---|---|
| $V_0$ | $f_0$ | $f_1$ | $V_{1,m}$ | $V_{1,0}$ | $\alpha$ | $\beta$ | $\gamma$ | $k_{s,h}$ | $k_{s,t}$ | $\theta_1$ | $\theta_2$ | $\tau$ |
| 5.5 | 120.3 | 11.5 | 85.0 | 84.1 | .00177 | .00236 | .00020 | 90.64 | 1.82 | 46.28 | .0000 | -40 | 4.297 |
| 9.1 | 120.0 | 11.1 | 89.1 | 80.9 | .00184 | .00203 | .00021 | 22.95 | 2.49 | 13.80 | .0000 | -20 | 1.337 |
| 9.8 | 120.0 | 10.5 | 89.8 | 80.2 | .00163 | .00161 | .00020 | 9.53 | 3.07 | 10.53 | .0000 | -10 | .636 |
| 10.0 | 120.0 | 10.0 | 90.0 | 80.0 | .00150 | .00150 | .00020 | 5.00 | 5.00 | 10.00 | .4000 | 0 | .000 |
| 8.8 | 120.0 | 9.8 | 89.7 | 80.3 | .00145 | .00144 | .00018 | 5.18 | 6.96 | 9.84 | 1.3466 | 10 | .466 |
| 8.7 | 120.0 | 8.9 | 89.6 | 80.4 | .00121 | .00097 | .00010 | 4.94 | 7.46 | 9.62 | 1.6708 | 20 | 1.340 |
| 7.5 | 120.0 | 8.1 | 88.5 | 81.9 | .00099 | .00069 | .00002 | 3.49 | 13.39 | 9.58 | 2.0091 | 40 | 3.227 |

Table 3 shows that sensible differences in the parameters of the model are produced by different time-lags, particularly with reference to the parameters related to the channel flow (the Strickler coefficients). The criterion of looking for the least squares for each event appears not to be able to discriminate between possible values of the time-lag. It seemed necessary to mitigate the effect of the time-lag by constraining the parameters to assume the same value for all the events. Therefore, the last attempt was the contemporaneous fitting of the whole set of events, by keeping 10 parameters with the same value for all the events, and by only allowing three parameters, namely $V_0$, $V_{1,0}$ and the time-lag, to vary from an event to another. The result of the calibration is shown in the Table 4, for the Bisagno creek.

Table 4: Parameters for the Bisagno creek

| Parameters | | | | | | | | | |
|---|---|---|---|---|---|---|---|---|---|
| $f_0$ | $f_1$ | $V_{1,m}$ | $\alpha$ | $\beta$ | $\gamma$ | $k_{s,h}$ | $k_{s,t}$ | $\theta_1$ | $\theta_2$ |
| 20.7 | 5.6 | 102.6 | 0.00126 | 0.00047 | 0.000070 | 0.96 | 11.1 | 8.03 | 0.129 |

The same procedure has been followed for Sansobbia creek. The results are shown in Table 5.

Table 5: Parameters for the Sansobbia creek

| Parameters | | | | | | | | | |
|---|---|---|---|---|---|---|---|---|---|
| $f_0$ | $f_1$ | $V_{1,m}$ | $\alpha$ | $\beta$ | $\gamma$ | $k_{s,h}$ | $k_{s,t}$ | $\theta_1$ | $\theta_2$ |
| 32.0 | 29.8 | 64.6 | 0.00559 | 0.00779 | 0.00071 | 2.44 | 4.4 | 2.36 | 1.231 |

As it can be seen from Tables 4 and 5, reliable values of the parameters have been obtained for the two basin, reflecting the differences in the soil characteristics. The almost impermeable limestones of the Bisagno basin are characterized by a saturated infiltration capacity which is nearly a fifth of the corresponding value for the metamorphic and volcanic rock of the Sansobbia basin. Future researchs will regard the influence of spatial end temporal scales in the values of the parameters, as well as an automatic procedure for selecting the number and position of the *V-shaped* sub-basins, by using Digital Terrain Models.

# 3 Conclusions

Lack in the synchronism between recorded rainfalls and discharges may result in great unreliability for the estimated values of the parameters of a RRM. On the other hand, reliable values of the infiltration parameters, as well as of the surface flow parameters, are needed in order to produce reliable estimates of design discharges in ungaged basins (e.g. by using the Rational Method). In the herein proposed solution the unknown time-lag between rainfalls and discharges is taken into account as an extra parameter to be estimated in the calibration phase, while the last is performed by using an optimization procedure extended to the whole set of events at disposal. Results obtained for two creeks in northern Italy illustrate the benefit obtainable by using the proposed procedure.

# 4 Acknowledgments

This research has been carried out with the financial support of the Provincia of Genova (Italy). Their support is gratefully acknowledged.

# References

1. Bartolini, P.,(1987). "A Possible Definition of the Input Function for a Rainfall-Runoff Model with Lumped Parameters," *Proceedings ASCE Hydraulic Division*, ASCE, New York, N.Y., 299-304.

Two-Dimensional Flood Modeling for a Tidal Canal

P. Michael DePue II[1], M. ASCE and Steven Halmi[2], M. ASCE

## Abstract

Two-dimensional floodplain modeling has become available to the consulting engineer in recent years because of advances in desktop computing power. This type of modeling allows engineers to more accurately simulate floods and to better define the limits of the floodplain. However, some of the 2-D models used in the United States are of limited use in certain real floodplain scenarios because of inherent instabilities and inadequacies in the models. This paper provides a case study of the application of one- and two-dimensional floodplain modeling for a major roadway bridge replacement on the Springfield Canal in Savannah, Georgia.

## Introduction

In the Spring of 1997, the Georgia Department of Transportation contracted with the Newport News office of Post, Buckley, Schuh & Jernigan, Inc. (PBS&J) (then Espey, Huston & Associates, Inc.) to prepare a two-dimensional hydrodynamic model of the Springfield Canal system in Savannah, Georgia. The intent of this model was to analyze the effects of widening and raising Victory Drive and Ogeechee Road on the floodplain limits in an urban tidally-influenced canal system. Victory Drive and Ogeechee Road cross the Springfield Canal in close proximity to one another and Ogeechee Road crosses a tributary of the canal later in the project area.

The Springfield Canal is approximately 6.4 kilometers long and flows northward from the Lynes Parkway in Southwest Savannah. The canal discharges into the Savannah River

---
[1] Engineer, Post, Buckley, Schuh, & Jernigan, Inc., 11838 Rock Landing Drive, Suite 250, Newport News, VA, 23606
[2] Engineer, Post, Buckley, Schuh, & Jernigan, Inc., 11838 Rock Landing Drive, Suite 250, Newport News, VA, 23606

directly beneath the Talmadge Bridge, approximately 22 kilometers upstream of the Atlantic Ocean. The canal has a drainage area of 2118 hectares.

The canal is tidally-influenced. The mean daily tide range at the Talmadge Bridge is approximately 2.2 meters, and the mean tide elevation is 0.25 meters (NGVD29).

The canal is dredged throughout much of its length and is typified by a trapezoidal channel section about 5 meters wide at the bottom and 12 meters wide at the top, with an average bankfull depth of 2.2 meters. The channel bottom elevation varies from -1.5 meters (NGVD29) at the downstream end to 1 meter at the upstream end. The side slopes of the channel are typically 1:1 to 2:1. The floodplain of the canal is typically wide, as much as 2 kilometers in areas, with the limits of effective flow much less due to vegetation and other flow obstructions.

The Springfield Canal system is divided into three distinct reaches. The upper and lower canal are separated by the I-16 roadway embankment, which has a top elevation above the 100-year storm surge water surface elevation. Tributary A flows into the canal just above the I-16 roadway embankment.

The lower Springfield Canal is the area north of the I-16 roadway embankment. The topography in this area is more flat than in the upper portion of the canal and is more tidally-influenced. One kilometer upstream of the outlet of the canal, a flap-type tide gate prevents backflow into the canal. This tide gate is overtopped easily and is being replaced by a new tide gate/stormwater pump station. The new pump station will have a capacity of 9.6 cubic meters per second.

Both the upper and lower Springfield Canal will be improved by dredging and widening in the near future. Several deficient bridges and culverts will be replaced as part of these improvements. The canal will be dredged to a uniform slope to allow the pump station to function more efficiently. Even with these improvements, the canal will continue to flood in the 10-year storm.

The Victory Drive/Ogeechee Road corridor is a major hurricane evacuation route, and maintaining passage in severe storms is a significant concern. The existing centerline of Ogeechee Road has an average elevation of 2.7 meters (NGVD29) in the area of the canal, well below the expected 100-year storm surge flood elevation.

### Development of UNET One-Dimensional Unsteady Flow Model

PBS&J used an existing UNET (U.S. Army Corps of Engineers, 1996) model developed for the design of the new pump station as the starting point for an extended model of the canal system. PBS&J substantially revised the model, added reaches and cross-sections, and substituted a more detailed hydrologic model as the driving force for the simulation. The changes PBS&J made were required to allow much more detailed simulation of the tributary area and the upper canal. PBS&J also wrote a new Visual Basic program to sort and organize the massive amount of data output by the UNET model. This allowed PBS&J to search out the instant in time when the flooding was worst in a given region of the UNET model. This feature was crucial to the preparation of the two-dimensional model.

PBS&J analyzed two versions of the new UNET model for the 100-year storm hydrology. The first model included the canal and all structures as they existed at the time of the study. The second model simulated the effects of all proposed changes to the canal and the downstream structures. The existing condition model was the worst case scenario and produced water surface elevations in the area of the Ogeechee Road/Victory Drive canal crossing of 3.68 meters. The proposed conditions model showed the impact of the pump station and the planned dredging, and resulted in a water surface elevation at the Victory Drive crossing of 3.16 meters. At the Ogeechee Road tributary crossing, the existing and proposed conditions models resulted in water surface elevations of 3.61 meters and 3.31 meters respectively.

The UNET trials showed that the canal would flood in the 100-year storm, with or without the proposed improvements. The UNET analysis was used to provide an estimate of the size of bridge required at the Victory Drive canal crossing and the culvert required at the Ogeechee Road crossing of Tributary A. The least-cost designs for each structure were then modeled in the two-dimensional simulations.

### Development of FESWMS Two-Dimensional Finite Element Model

PBS&J prepared a FESWMS (Federal Highway Administration, 1996) two-dimensional finite element flow model of the upper watershed area. The model domain was designed to allow sufficient distance from the Victory Drive crossing of the canal and the Ogeechee Road crossing of the tributary to the edge of the model, so boundary conditions would not negatively impact the accuracy of the results.

Prior experience with 2-D hydrodynamic models suggested that this simulation would be more likely to converge if the finite element grid was aligned with the general topography of the area and the individual elements were made as similar in shape and size as possible. At the time the modeling was begun, the best way to do this was to digitize the entire topographic model of the area into a surface modeling program and then grid the domain by visual inspection. Spot elevations were computed at each grid intersection and the data was exported to the Surface-Water Modeling System (Brigham Young University, 1997). The points were triangulated, merged, and manipulated to form a 2-D model mesh that was as simple and compact as possible while still representing the key features of the watershed. This mesh contained approximately 13,250 nodes and 5,940 elements. Fine details in the mesh, such as the proposed bridge embankments, were added after initial model testing.

A series of steady-state trials with the mesh was used to optimize the performance and accuracy of the model and to diagnose any convergence problems. Several dozen trials were run with various portions of the mesh and the full mesh. These trials used boundary conditions across the range of water surface elevations and discharges expected in the model to isolate potential problems such as discontinuities. The model performed well in most cases and produced reasonable velocity vectors.

Once the basic model had been tested using steady state trials, the eight major culverts in the study area were added to the model using the procedures described in the FESWMS documentation. The areas around the culvert entrances and exits were refined to encourage the development of realistic flow patterns. Weir elements were added to simulate roadway overtopping.

The model containing the culverts was executed in steady-state mode with the same boundary conditions used earlier. The model ran for several iterations and then stopped. Close examination of the results indicated unrealistic flow patterns at the culverts, including looping through dry elements and mounding of water. Numerous adjustments were made to counter this problem, including: adjustment of the eddy viscosity values and Manning n values, confirmation of all input data, recheck of the weir elements, and staged introduction of the culvert and weir elements into the simulation. Such procedures are recommended in a variety of references and are generally accepted modeling practice.

These efforts did not result in a working steady-state model with culverts. Continued debugging efforts and consultation with other experienced FESWMS modelers did not produce a solution to this problem. Similar problems with the bridge pressure flow routines of FESWMS were also encountered. After significant research, PBS&J concluded that the FESWMS culvert and bridge routines were not suitable for use in the Springfield Canal scenarios and that other methods of solution would be required.

The inability of FESWMS to model culvert and bridge flow in this scenario is of significant concern. The scenario examined in this project is common in coastal areas. This particular model contained several features that caused instability in the culvert routines: the use of multiple barrel culvert crossings in close proximity to one another; the case where the roadway or railway embankment containing the culvert was not significantly overtopped; and the case where multiple barrel culverts had staggered invert elevations or where reverse flow occurred in a single culvert. The culvert flow routines did perform acceptably in cases where the roadway was greatly overtopped, probably because the proportion of flow actually passing through the culvert was very low in comparison to the total flow.

For the Springfield Canal model, it was essential to model all of the culverts as realistically as possible, since the water surface in the canal was controlled by these structures. While equivalent channels could be devised for some crossings, this solution was considered unacceptable, since the purpose of the effort was to prepare a more accurate model.

Ultimately, PBS&J chose to partition the model into submeshes by breaking the full mesh at roadway or railway embankments that were never overtopped. This eliminated all of the problematic culvert features defined above, and allowed the model to run by simulating only cases where the culverts were contained within embankments that were always significantly overtopped by the flood. The culverts connecting the submeshes were analyzed by standard culvert calculation techniques. While this solution necessitates some simplifications, it is the best possible solution under the circumstances.

A dynamic model was impractical to prepare because the connecting culverts had to be computed manually. While the flooding in the canal was clearly a dynamic situation, it would have been exceptionally laborious to model each culvert at each time step and introduce this effect into each submesh. Additionally, the source code for the FESWMS model was not available, so it was not possible to modify the code to allow manual calculation of the culverts.

The results of the UNET models were used to develop a range of worst case conditions for the Ogeechee Road and Victory Drive structures. The times of peak water surface and peak flow rate at each analysis point were determined. Using flow and stage data from the UNET model, point sources and sinks were introduced into the model to provide a steady-state approximation to the worst case dynamic condition. Again, while not optimal, this was the best available solution given the inability of FESWMS to simulate all of the culvert scenarios.

### Conclusion

Ultimately, the results from this approximation proved reliable and reasonable. While they clearly represent a simplification, the results can be substantiated by simple engineering checks and are consistent with one-dimensional dynamic model results. The two-dimensional model did produce different water surface elevations for a given flow rate than the UNET model, but the trend of the results was similar. Floodplain storage was accounted for in a more precise manner and velocities near structures were better estimated. While dynamic effects were considered with a one-dimensional model (UNET), the worst case solutions were analyzed with a two-dimensional model (FESWMS). The replacement structures were designed from these results.

This project demonstrates the need for improvements in the FESWMS two-dimensional modeling software if the program is to be used in practical scenarios. Future models must be able to easily and reliably simulate culverts in tidal canals with variable roadway overtopping.

This analysis shows that the current state of the art in two-dimensional modeling is not up to the level required for straightforward simulation of hurricane effects in urban environments. Reliable, consistent, and inexpensive models that can directly simulate such scenarios are not yet available. While the graphical shells that interface with the two-dimensional models are well-crafted and useful, the simulation engines themselves are not completely capable of performing the day-to-day analysis required by practicing engineers.

### Appendix: References

Brigham Young University - Engineering Computer Graphics Laboratory. <u>Surface Water Modeling System (SMS) Version 5.0.</u> 1997.

Froehlich, David C. <u>Finite Element Surface-Water Modeling System: Two-Dimensional Flow in a Horizontal Plane (FESWMS).</u> 1996.

U.S. Army Corps of Engineers - Hydrologic Engineering Center. <u>UNET One-Dimensional Unsteady Flow Through a Full Network of Open Channels.</u> 1996.

Two-Dimensional Coastal Modeling for Bridge Scour

Steven R. Halmi[1], M. ASCE, P. Michael DePue[2], M. ASCE, R. Wayne Corley[3], and William H. Hulbert[4]

## Abstract

Subject to complex flow patterns and varying tides, urban development in coastal areas can be difficult to design. Two-dimensional finite-element computer models are often the best choice for analysis of a coastal environment. However, practicing engineers must overcome many obstacles when using such inherently unstable models. Running numerous trials to obtain good results can be frustrating. This paper presents a case study of scour analysis for a barrier island bridge near Charleston, South Carolina, where the applicability and limitations of two-dimensional modeling were closely examined.

## Introduction

Scour at highway bridges is an important consideration in the design and continued safety of roadway structures. The effects of scour are particularly important when the bridge is a hurricane evacuation route that must be passable under extreme conditions. Route SC 703 over Breach Inlet, also known as the Thompson Memorial Bridge, connecting the Isle of

---

[1]Engineer, Post, Buckley, Schuh & Jernigan, Inc., 11838 Rock Landing Drive, Suite 250, Newport News, VA, 23606
[2]Engineer, Post, Buckley, Schuh & Jernigan, Inc., 11838 Rock Landing Drive, Suite 250, Newport News, VA, 23606
[3]Assistant Hydraulic Engineer, South Carolina Department of Transportation, P.O. Box 191, Columbia, SC, 29202
[4]Hydraulic Engineer, South Carolina Department of Transportation, P.O. Box 191, Columbia, SC, 29202

Palms and Sullivans Island near Charleston, South
Carolina, is one such structure. During Hurricane Hugo
in 1989, the storm surge inundated both of these
islands and the Thompson Memorial Bridge. Although the
existing bridge survived this immersion, the
superstructure is deteriorating, and a replacement
bridge is being designed. As part of the design
process, the potential for scour must be analyzed in
order to size the substructure of the replacement
bridge.

Breach Inlet (Figure 1) connects a large marshy
estuary and the Atlantic Ocean. Swift currents, widely
variable tides, and moving sandbars characterize the
inlet. The marshy area to the west is connected to the
ocean not only by Breach Inlet, but on two other sides.
To the south, it is connected via the Intracoastal
Waterway to the Cove (at Charleston Harbor), and to the
north, it is connected to Dewees Inlet. Both the Isle
of Palms and Sullivans Island are low, but relatively
stable, barrier islands.

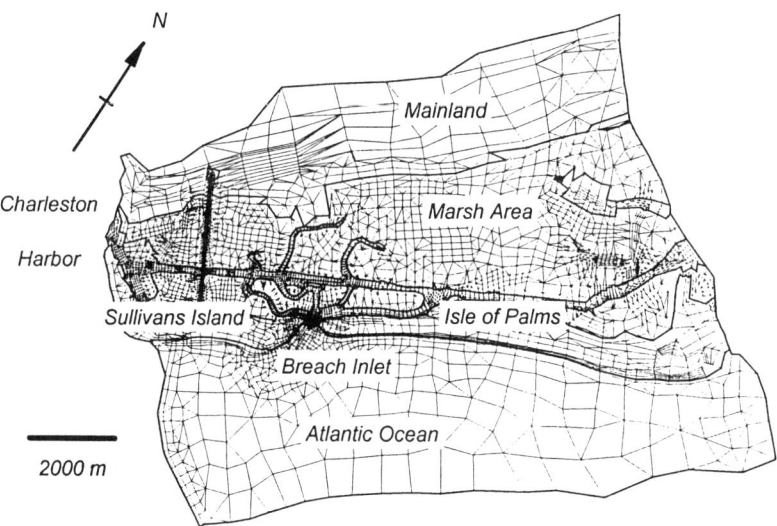

Figure 1. Breach Inlet Location Map and Finite
Element Mesh

Coastal Complexities

There are several physical features which make the
hydraulic study of Breach Inlet much more complicated

than typical riverine analysis. First and foremost, this coastal area is subject to multidirectional tidal flow. The assumption of one-dimensional hydraulics, in which all flow streamlines pass through the bridge in parallel fashion, is not applicable. A two-dimensional hydraulic model is far better suited to the complex eddies, and spreading and converging flow patterns found in and around Breach Inlet.

Careful choice of boundary conditions for the two-dimensional simulation is critical to a successful model. The design event is a hurricane surge phenomenon, which dictates time-varying head (water surface elevation) on all boundaries. A custom surge model was developed to estimate worst-case scour conditions (i.e. highest velocities and heads) in Breach Inlet resulting from a hurricane.

## Model Development

An ideal two-dimensional model would cover an area such that the boundaries are sufficiently distant from the area of interest. With this in mind, a finite element mesh was created that covered approximately 60 square miles centered about Breach Inlet. After an exhaustive search for topographic, bathymetric, and coastal data, elevations were digitized into a database. Once this data was correlated to a common datum, the entire domain was triangulated into a finite element mesh using the Surface-Water Modeling System, SMS (Brigham Young University, 1997). After extensive refinement and manipulation, the result was a mesh containing roughly 25,000 nodes and 10,000 elements (Figure 1).

Another difficulty with two-dimensional modeling is parameter estimation. Both Manning roughness and eddy viscosity are needed as model inputs. Manning roughness, although easily estimated for one-dimensional models, is not as well studied in the literature for two-dimensional flow modeling. Eddy viscosity is a parameter for which there is little guidance available for selecting values. For the Breach Inlet simulation, Manning roughness ranged from 0.015 to 0.250, and eddy viscosity from 2600 to 24,000 Pascal-sec. Both of these parameters can vary throughout the model, and both can be used to calibrate the model if good calibration data is available.

## 2-D Model Features and Limitations

Two computational engines were used to run the simulation: RMA2 (U.S. Army Corps of Engineers, 1995) and FESWMS (Froehlich, 1996). Both of these engines apply the finite element method to solve the differential equations governing two-dimensional, depth-averaged surface water flow. Finite difference models were not considered because they require a very ordered grid of points that does not conform well to natural features.

The ideal solution would be a dynamic (time-varying) record of the entire hurricane surge event, driven by known surge elevation boundary conditions and showing the progression and recession of the storm surge flood over normally dry land. For a simple mesh, this may be possible. But for this project, a much larger mesh (both in area and number of nodes and elements) is required in order to represent the intricate shape of the channels, barrier islands, and bridge layout. A smaller mesh would be an oversimplification.

Solution of such a large mesh is very computationally demanding. The 25,000 node steady-state model ran for nine iterations in six hours on a 133-Mhz Pentium processor, and required 64 megabytes of RAM. Computational demand itself practically prohibits running a dynamic solution. Unlike the steady-state solution, a dynamic solution solves the model for each step in time. The same simulation running dynamically for a 24-hour storm event with 15 minute time step would require 96 time steps, or nearly 600 hours of computer time.

But computer time was not the only problem with running a dynamic simulation. Dynamic runs were attempted for both the full mesh and smaller portions of the mesh. Since hurricane surge water surface elevations change so rapidly, the numerical engine cannot recover from the shock of dramatically varying boundary conditions, and the model diverged. Convergent solutions might be possible by shortening the time step, although more computer time would be required, or by lessening the boundary condition variation, which would not represent the actual situation.

Elemental wetting and drying is another powerful feature of these models allowing the flooding and ebbing of tides and surge onto otherwise dry land.

Although this feature ran successfully for the steady-state model, wetting and drying presented problems for dynamic runs. Especially in this coastal area, where the flat topography causes many elements to become wet or dry simultaneously, sudden gain or loss of large areas of additional storage volume caused the numerical solution to diverge.

The final, accepted steady-state solution was not obtained without problems. First of all, steady-state boundaries must have a constant magnitude, which of course is not realistic. Conservatively, steady-state boundary conditions were chosen that would cause the highest velocities through Breach Inlet. This occurs during the greatest difference between opposing boundary heads. Secondly, the model was found to be not amenable to multiple head boundaries. This tended to overconstrain the problem, and the absence of a flow boundary made it difficult for the model to choose an initial flow direction. Nearly 70 trials were attempted before finally accepting a convergent, steady-state solution.

Solution

The final steady-state solution is shown in Figure 2. Velocity varies widely in the vicinity of the Breach Inlet bridge, and attains a maximum value of 1.6 m/s. Not only could these heads and velocities be used to calculate scour, but design problems such as structural loading, pollutant transport, navigation problems, and beach migration, could also benefit from this solution. In fact, this solution is highly advanced considering the empirical nature of the HEC-18 scour equations (FHWA, 1995), which depend greatly on other parameters such as soil grain size.

Conclusion

Although bridge scour at Breach Inlet is only one case study in two-dimensional coastal modeling, many aspects of this study can be applied to other instances of urban development in coastal areas. Sea walls, jetties, tunnels, underwater utilities, and storm and sewer outfalls are just a few examples of hydraulic structures commonly designed for coastal areas. Water quality issues are also easier to resolve once a comprehensive coastal hydraulic model is established.

Issues such as the selection of model parameters, the location of boundary conditions, the effects of wetting and drying of elements, and the comparison of a dynamic to a steady state solution are common to all coastal models, whatever the purpose. The limitations of two-dimensional finite element models discussed in this paper demonstrate the need for continued improvement to existing numerical methods for solving differential equations, or development of new schemes for modeling the complexities of coastal hydrodynamics.

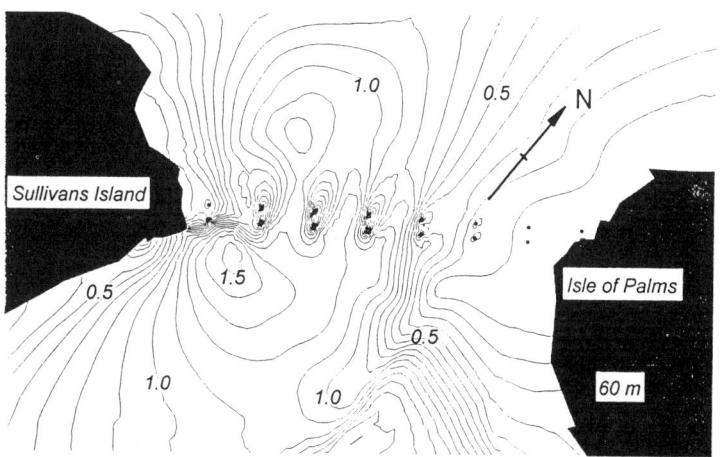

Figure 2. Steady-state Velocity Contours (m/s) in the Vicinity of the Thompson Memorial Bridge

Appendix: References

Brigham Young University - Engineering Computer Graphics Laboratory. Surface-Water Modeling System (SMS), Version 5.0. 1997.

Federal Highway Administration. Evaluating Scour at Bridges, Third Edition. Hydraulic Engineering Circular No. 18. 1995.

Froehlich, David C. Finite Element Surface-Water Modeling System: Two-Dimensional Flow in a Horizontal Plane (FESWMS). 1996.

U.S. Army Corps of Engineers - Waterways Experiment Station Hydraulics Laboratory. RMA2 Version 4.3. 1995.

Hydraulic Analysis Using FESWMS-2DH of the State Route 76 and Hatchie River Crossing in Brownsville, Tennessee

Edwin W. Watkins, P.E.[1], Wayne Seger, P.E.[2]

## Abstract

A hydraulic analysis and scour evaluation of the fourteen bridge crossing over the Hatchie River and Hatchie Bottom floodplain on State Route 76 near Brownsville, Tennessee was performed as part of the State of Tennessee's statewide scour evaluation program. The hydraulics of the crossing are complicated by a five bridge crossing on Interstate 40 approximately 300 m (1,000 feet) downstream of State Route 76, the highly meandering nature of the Hatchie River, and the wide, flat floodplains in the river valley. The original one-dimensional hydraulic model of the crossing using WSPRO proved to be inadequate. Based on the results of the original analysis and the hydraulic complexities presented at this site, the crossing was re-evaluated using the two-dimensional model, FESWMS-2DH. The Surfacewater Modeling System (SMS) was used to develop the element mesh and analyze the results. The element grid consisted of over 10,000 nodes and 2,000 elements. The fourteen bridges on State Route 76 as well as the five bridges on Interstate 40 were included in the model. Predicted flow distributions, scour depths and recommended countermeasures were compared to the original WSPRO model.

## Introduction

The Hatchie River flows from north Mississippi through west Tennessee and discharges to the Mississippi River west of Covington, Tennessee. The Hatchie River is highly meandering and the floodplains in west Tennessee are wide and flat and prone to flooding. The highly publicized failure of the US Route 51 bridge near Covington illustrates the active nature of the river's channel and the potential for scour and channel migration.

---

[1] Water Resources Engineer, Ogden Environmental, 3800 Ezell Rd. Ste. 100, Nashville, TN 37211

[2] Civil Engineering Manager, Bridge Inspection and Repair, Tennessee Department of Transportation, James K. Polk Bldg., Suite 1200, Nashville, TN 37243

As part of the Tennessee Department of Transportation's Bridge Scour Program a study of the fourteen bridge crossing of State Route 76 and the Hatchie River and Hatchie Bottom was undertaken. The results of this study determined the scour potential at each of the fourteen bridges and provided recommended scour countermeasures.

State Route (SR) 76 is a two lane heavily traveled state highway that runs north to south in Haywood County near Brownsville, Tennessee. Near the Hatchie River, SR 76 parallels Interstate (I) 40. The main channel is located at the far north side of the floodplain. Approximately 300 m (1,000 ft.) downstream of the fourteen bridges on SR 76 is a five bridge crossing on I40 (See Figure 1).

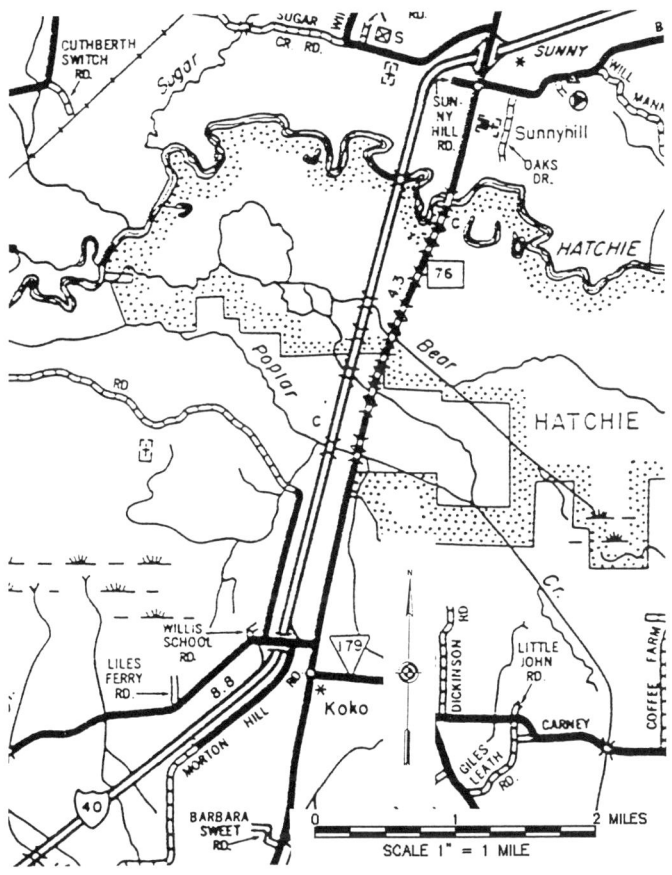

Figure 1 Location Map.

## One-Dimensional Analysis

The drainage area to the fourteen bridge crossing was determined to be 4,812 km$^2$ (1858 sq. mi.). Design flow rates were determined from USGS regional regression equations and compared to gage data upstream and downstream of the crossing. The 100-year (one percent chance) event was determined to be 2,260 m$^3$/s (79750 cfs) and this event was used for the scour evaluation.

The bridge crossing was originally modeled using the Federal Highway Administration water surface profile model ,WSPRO (Sherman, 1990). The modeling procedure was complicated by a ten bridge limitation in WSPRO's multiple opening routine. The crossing was first analyzed combining several of the bridges to reduce the number of openings to ten. Based on the initial flow distribution determined from the ten bridge model, the fourteen bridge crossing was divided into two independent models with seven bridges each. An iterative process was undertaken to balance the predicted water surface elevations between the two models. This process resulted in an agreement of water surface elevations within 0.3 m (1.0 ft.).

The one-dimensional modeling procedure was further complicated by the five I40 bridges downstream and the river course near the bridge. The five bridge opening downstream of the fourteen bridge crossing did not align with the fourteen bridge crossing such that an accurate flow distribution through the I40 bridges could be determined. The main channel also turns sharply downstream of SR 76 and flows parallel to the road for approximately 500 m (1,640 ft.) The distance from the river to the road is less than 60 m (200 ft.) in several places (See Appendix II). Because of the proximity of the river to the road, an accurate exit section for several of the bridges could not be accounted for in the one-dimensional model.

Potential scour for each of the bridges was calculated according to HEC-18 (Richardson, 1993) based on the one-dimensional (Phase II) analysis. The Item 113 codes for each of the bridges was assigned based on these scour predictions and scour countermeasures assigned to the bridges.

## Two-Dimensional Analysis

Based on the results of the Phase II analysis and the hydraulic complexities presented at this site, the crossing was re-evaluated using the FHWA two-dimensional model FESWMS-2DH (Froehlich, 1992). The cost of the two-dimensional analysis was found to be reasonable using the Surface Water Modeling System (SMS) developed by Brigham Young University Engineering Computer Graphics Laboratory and the US Army Corps of Engineers Waterways Experiment Station (ECGL, 1995). SMS provides automated element creation and bookkeeping, as well as a graphics package to display model connectivity and results.

A two-dimensional element grid of the crossing was assembled from survey information and quadrangle map information. The grid covered an area approximately one mile upstream and downstream of SR 76. The grid consists of over 2,000 elements and 10,000 nodes. Grid boundaries were located sufficiently upstream and downstream of SR 76 to minimize the errors that may be introduced in the boundary conditions.

Manning's **n** values were assigned to each element as material properties based on land use and flow restrictions in the element. Five different Manning's **n** values were used. The majority of the elements in the grid were characterized by swampy overbank conditions typical of the Hatchie Bottoms.

Water surface elevations and velocity vectors were determined at each node for the 100-year peak discharge using the flo2dh module in FESWMS-2DH. The boundary condition at the downstream end of the grid (west) was determined based on normal depth of the cross section using the 100-year discharge and the general slope of the Hatchie River and floodplain. The 100-year discharge was used as the upstream boundary condition.

Based on the results from the flo2dh model, flux lines were used to determine flow rates through each bridge opening. The results of the FESWMS-2DH model are compared to the WSPRO results in Table 1.

Table 1. Comparison of WSPRO and FESWMS-2DH Model Results

| Bridge Log Mile | Flow Distribution | | | | Upstream Water Surface Elevation | | | |
|---|---|---|---|---|---|---|---|---|
| | WSPRO | | FESWMS-2DH | | WSPRO | | FESWMS-2DH | |
| | ($m^3$/s) | (cfs) | ($m^3$/s) | (cfs) | (m) | (ft.) | (m) | (ft.) |
| 8.97 | 424 | 14966 | 498 | 17594 | 91.65 | 300.6 | 92.16 | 302.3 |
| 8.83 | 297 | 10478 | 279 | 9849 | 91.65 | 300.6 | 92.20 | 302.4 |
| 8.72 | 141 | 4970 | 113 | 3975 | 91.65 | 300.6 | 92.23 | 302.5 |
| 8.63 | 123 | 4335 | 112 | 3944 | 91.65 | 300.6 | 92.23 | 302.5 |
| 8.46 | 56 | 1963 | 39 | 1387 | 91.65 | 300.6 | 92.26 | 302.6 |
| 8.22 | 201 | 7101 | 252 | 8893 | 91.65 | 300.6 | 92.20 | 302.4 |
| 8.08 | 259 | 9164 | 222 | 7838 | 91.65 | 300.6 | 92.16 | 302.3 |
| 7.95 | 152 | 5370 | 194 | 6838 | 91.52 | 300.2 | 92.16 | 302.3 |
| 7.84 | 162 | 5705 | 113 | 3996 | 91.52 | 300.2 | 92.20 | 302.4 |
| 7.76 | 118 | 4165 | 118 | 4183 | 91.52 | 300.2 | 92.20 | 302.4 |
| 7.61 | 118 | 4173 | 101 | 3567 | 91.52 | 300.2 | 92.20 | 302.4 |
| 7.48 | 62 | 2192 | 64 | 2271 | 91.52 | 300.2 | 92.20 | 302.4 |
| 7.18 | 33 | 1164 | 24 | 856 | 91.52 | 300.2 | 91.89 | 301.4 |
| 7.10 | 99 | 3509 | 129 | 4563 | 91.52 | 300.2 | 91.86 | 301.3 |

Potential scour at each of the fourteen bridges was also calculated using results from the FESWMS-2DH model (Phase III analysis). Scour calculations from the Phase II and Phase III analysis are compared in Table 2.

Table 2. Comparison of Phase II and II Analysis Scour Depths

| Bridge | Phase II (WSPRO) Analysis | | | Phase III (FESWMS-2DH) Analysis | | |
|---|---|---|---|---|---|---|
| Log Mile | Scour Depth* | | Item 113 | Scour Depth* | | Item 113 |
| | (m) | (ft.) | Code | (m) | (ft.) | Code |
| 8.97 | 3.84 | 12.6 | 3 | 6.46 | 21.2 | 3 |
| 8.83 | 3.08 | 10.1 | 3 | 1.74 | 5.7 | 3 |
| 8.72 | 2.83 | 9.3 | 3 | 2.16 | 7.1 | 3 |
| 8.63 | 3.78 | 12.4 | 3 | 1.86 | 6.1 | 3 |
| 8.46 | 3.20 | 10.5 | 3 | 1.86 | 6.1 | 3 |
| 8.22 | 2.23 | 7.3 | 3 | 4.85 | 15.9 | 3 |
| 8.08 | 3.08 | 10.1 | 3 | 3.51 | 11.5 | 3 |
| 7.95 | 3.63 | 11.9 | 3 | 3.54 | 11.6 | 3 |
| 7.84 | 3.54 | 11.6 | 3 | 1.40 | 4.6 | 3 |
| 7.76 | 3.60 | 11.8 | 3 | 3.14 | 10.3 | 3 |
| 7.61 | 3.81 | 12.5 | 3 | 2.38 | 7.8 | 3 |
| 7.48 | 2.96 | 9.7 | 3 | 2.38 | 7.8 | 3 |
| 7.18 | 5.18 | 17 | 3 | 0.76 | 2.5 | 5 |
| 7.10 | 4.60 | 15.1 | 3 | 1.25 | 4.1 | 3 |

*Scour Depths do not include abutment scour.

**Summary**

FESWMS-2DH was used to provide a more detailed analysis of the SR 76 fourteen bridge crossing over the Hatchie River. The Phase III analysis results were used to predict potential scour depths and provide Item 113 coding for each of the bridges. The two-dimensional model predicted higher water surface elevations upstream of SR 76. This increase in water surface elevation is most likely due to a better account of head loss through the combination of the five Interstate and fourteen State Route bridges in the two dimensional model. Because of the position of the bridges, actual flow path lengths were underestimated in the one-dimensional model.

Scour calculations from the Phase III analysis provide different contraction scour predictions from the Phase II analysis due to the change in flow distributions shown in Table 1. Significant reductions in scour at the two bridges at the southern end of the floodplain resulted in a change in the Item 113 code for the Log Mile 7.18 bridge from scour critical (3) to not scour critical (5).

## Appendix 1: References

Sherman, J.O., 1990, "Users Manual for WSPRO," Hydraulic Computer Program HY-7, Federal Highway Administration, US Department of Transportation, Washington D.C.

Froehlich, D.C., 1992, "Finite Element Surface Water Modeling System: Two-Dimensional Flow in a Horizontal Plane, Users Manual," FHWA-RD-92-057, Federal Highway Administration, Washington, D.C.

Richardson E.V. et. al., 1993, Hydraulic Engineering Circular No. 18 (HEC-18), "Evaluating Scour at Bridges," FHWA-IP-90-017, Federal Highway Administration, Washington, D.C.

Engineering Computer Graphics Laboratory, 1995, "Surface Water Modeling System (SMS) Users Manual," Brigham Young University.

## Appendix 2: Velocity Vectors near the Main Channel

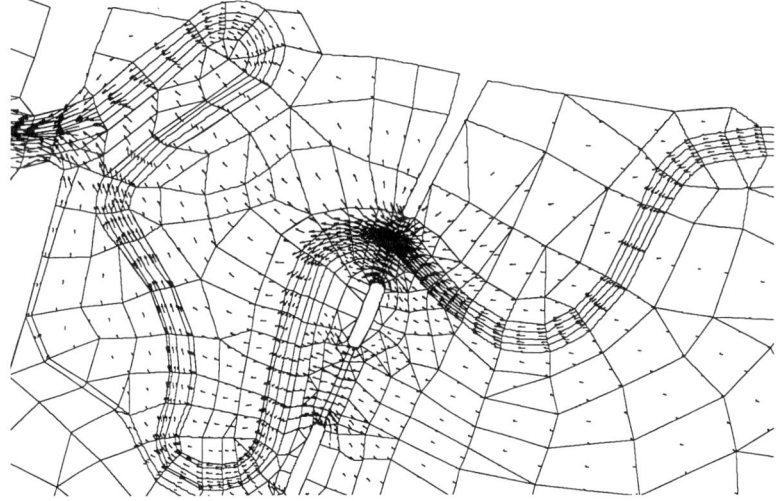

Acquisition and Relocation as a Floodplain Management Tool

William G. DeGroot[1] P.E., F.ASCE and David W. Lloyd[2] P.E.

Abstract

The Urban Drainage and Flood Control District (District) is involved in the planning, design, construction and maintenance of drainage and flood control facilities in the Denver area. One tool available to the District and local governments is the acquisition of flood prone properties and relocation or demolition of flood prone structures. Several successful examples of this concept are presented.

Introduction

The Urban Drainage and Flood Control District (District) is a regional agency established by the Colorado legislature in 1969 to assist Denver area local governments with multi-jurisdictional drainage and flood control problems. The District is involved in the planning, design, construction and maintenance of drainage and flood control facilities; in addition to assisting local governments in a variety of floodplain management activities including floodplain regulation, flood insurance, flood warning and public information.

One floodplain management tool available to the District and local governments is the acquisition and relocation or demolition of flood prone structures, usually before they are flooded. This concept is one that is receiving tremendous support in certain areas of the floodplain management community. Unfortunately, there are a number of problems associated with the implementation of this concept which must be overcome before it can be considered a viable option in most instances. Nonetheless, there have been a number of occasions in which the District

---

[1]Chief, Floodplain Management Program; [2]Chief, Design and Construction Program; Urban Drainage and Flood Control District, Denver, CO.

and its local government partners have been able to employ this concept.

Problems

Among the problems which must be considered with this concept:
- It is difficult to sell to local elected officials the idea of buying and demolishing a perfectly good building for the reason that it might be flooded in the future.
- It is even harder to convince the owner of the building. This brings into play the power of eminent domain, which is not popular with most local officials.
- Voluntary buyout programs can leave a patchwork of structures and vacant lots.
- Maintenance of the vacated properties can be a problem, especially if the above-mentioned patchwork prevails.
- Other, more "traditional" flood control options are usually more cost effective. This is particularly true when others want the acquisition option but the flood control entity is the only one that can bring any dollars to the table.

In spite of these problems the District and its local partners have utilized the acquisition and relocation option in a number of different situations which are detailed below.

Keys to Success

The District has found that the following keys are important to both selling the idea of acquisition to the interested and affected parties; and to actually resulting in a successful project:
- Joint funding involving several partners, each of which contributes funds commensurate to the benefit received for their objectives, can overcome the high cost of acquiring property before it is damaged.
- A master plan which identifies the need for acquisition in certain circumstances establishes an agreed upon approach to solving a problem.
- An after use for the acquired property; which includes land uses consistent with the flood hazard, an owner, and a maintenance funding source.
- A commitment to reducing flood hazards in the community, without which the impetus for acquisition will not be there.

The District has three property tax funding sources which are used in differing ways to acquire property. The Design and Construction Program has a maximum 0.4 mill levy authorization and requires local government sponsors to match District funds. This program acquires property in the most traditional way. Representative projects include the acquisition and evacuation of an 82 unit mobile home park at the confluence of Ralston Creek and Clear Creek and its re-use as a city park, and the acquisition and demolition of 14 single and multi-family flood prone structures in the Boulder Creek floodplain in cooperation with both the City of Boulder and the school district.

The Maintenance Program also has a maximum 0.4 mill levy authorization but no matching fund requirement. This program has established a preservation fund which starts each year at the $600,000 level, which is available for short fuse acquisition of targets of opportunity, such as willing sellers of previously identified desirable acquisitions. Representative projects include the acquisition of several individual houses offered by willing sellers, as well as the prevention of unwise development in the floodplain through acquisition of vacant sites that could otherwise be developed.

The South Platte River Program has a maximum 0.1 mill levy authorization and requires a 25% match from local government sponsors for capital projects and no match for maintenance projects. The program acquired an abandoned filling station site in Denver and turned it into permanent open space. It has also helped private property owners with bank stabilization activities in return for flowage easements which assure permanent open space along the river.

Through these three programs the acquisition and appropriate after-use of flood prone properties has become an important tool in the District's efforts to reduce existing flood hazards while keeping new ones from being created. Examples of successful projects follow.

Valley View Mobile Home Park, City of Arvada

The Valley View Mobile Home Park was located totally within the 100-year floodplain of Ralston Creek at its confluence with Clear Creek. Ralston Creek has a drainage area of 91 mi$^2$, and a 100-year peak discharge of almost 10,000 ft$^3$/sec at its mouth. The mobile home park, which had been located at the lower end of Ralston Creek for many years, had gradually been experiencing more frequent flooding as the upper basin became more fully developed. The existing channel through the mobile home park was not even capable of handling the 2-year event, and the park had been flooded most recently in 1989 and 1991.

In 1995, the City of Arvada embarked on the expansion of a major arterial street in the Ralston Creek floodplain upstream from the mobile home park. The project included construction of a 100-year capacity channel adjacent to the new road. In order to release the discharges from the channel the decision was made to acquire and relocate the mobile home park. In 1995, Arvada, in cooperation with the District Design and Construction Program, entered into a purchase agreement with the owners of the park. The agreement called for the complete removal of the 72 tenant-occupied and 10 vacant mobile homes no later than June 1, 1996. Purchase price of the property was set at $1,575,000. Ownership of the mobile homes and subsequent responsibility to remove them from the site remained with the seller. The now vacant site was flooded again in 1997.

Arvada and the District are now pursuing the development of a city park on the site which will become known as Gold Park, as this is believed to be the site where gold was first discovered in Colorado. This will be accomplished in conjunction with the needed drainage and flood control improvements. Other partners in the development of the park include the Jefferson County Open Space Department and the Colorado Historical Society. The park is expected to be constructed in 1998.

When fully completed, this project will have combined flood control, transportation, recreation, open space and historical site funding into a park celebrating the history of the area; while removing one of the greatest flood damage potential areas in the entire Denver region. The project involved the areas of joint funding, an appropriate after use, an owner and a maintenance funding source.

## Boulder High School, City of Boulder

Boulder Creek, which flows through the heart of the City of Boulder, has a drainage area of approximately 130 mi$^2$ originating at the Continental Divide at 13,500 feet m.s.l. and flowing eastward through about 22 miles of steep narrow mountain canyons before reaching Boulder at about 5,440 feet m.s.l. The 100-year peak at Boulder is approximately 12,000 ft$^3$/sec.

In 1988 the City of Boulder developed a "Comprehensive Drainage Utility Master Plan, Flood Hazard Mitigation Plan" that identified flood prone structures that are of particularly high risk. Since that time the City has embarked on a plan to purchase and remove as many of the high risk structures as budgets and participation from other agencies such as the District will allow.

The Boulder Valley School District's (BVSD) Boulder High School was located in the high hazard area of the 100-year floodplain of Boulder Creek. BVSD wanted to expand the high school while at the same time reducing the flood hazard. In 1990 the City of Boulder and BVSD entered into a cooperative agreement, with funding assistance from the District's Design and Construction Program, for a Joint Use Project (Project). The Project included the purchase and removal of certain flood prone structures upstream from the high school, regrading of school athletic fields between the high school and the creek channel to provide additional conveyance through the Boulder High School property, and construction of a floodproofed addition to the high school.

The Project resulted in the elimination of 44 residential units from the high hazard area; the removal of 224 residential units, 21 non-residential structures, 8 city owned buildings, and the high school from the high hazard area of the Boulder Creek floodplain; and allowed the BVSD to add classroom space and make some much needed changes to their play fields and parking areas. The Project is owned and maintained by the BVSD.

Individual Parcels

As noted above the City of Boulder has identified high risk flood prone structures along Boulder Creek and its tributaries. The city then attempts to buy those structures and remove them. The city does not use eminent domain, but annually sends a notice to structure owners offering to buy the structures if the owners are willing to sell. The District has participated with Boulder in the acquisition of three houses along Boulder Creek in this manner since 1983.

In 1997, Boulder Utilities, Boulder Parks and the District Design and construction Program jointly acquired a residential structure along Goose Creek. The lot will be utilized as a pocket park. This fulfilled the requirements for shared costs, public ownership and maintenance funding.

In 1993 the District and the City of Littleton completed a flood control master plan for Lower Slaughterhouse Gulch, which included the acquisition and removal of four residential structures. Two of the structures have since been acquired from willing sellers utilizing District Design and Construction Program funding matched by Littleton. Littleton owns the properties, which will be utilized as open space along the drainageway.

Along Little Dry Creek in Adams County a high roadway embankment with an undersized culvert creates a large upstream 100-year floodplain. When a flood control master plan was completed it was decided to formalize the detention volume behind the embankment, and to acquire the structures within that flood pool, again only from willing sellers. To date Adams County and the District Design and Construction Program have acquired two structures with matching funds; and the National Flood Insurance Program has acquired one structure which was given to the county.

In 1982 the District utilized its preservation fund to acquire a residential unit on the First Avenue Tributary to Weir Gulch in Denver which had a long history of frequent flooding. The structure was resold to a builder who moved it to a vacant site nearby. The vacant property was then used for drainageway improvements that helped to reduce flooding through the remainder of the neighborhood.

Conclusion

The acquisition and relocation of flood prone structures, and the subsequent appropriate uses of the land consistent with the flood hazard, is a viable remedial flood control option in many instances. The District's experience has been that multiple funding sources, and a sound ownership and maintenance arrangement , are essential ingredients to success. The District will continue to fund similar projects in the future from its three available funding sources.

Technical Mapping Advisory Council
To
Federal Emergency Management Agency
(FEMA)
A Status Report on its
Purpose and Progress

Mark A. Riebau, P.E, Chair[1]
Technical Mapping Advisory Council

**Abstract**

The Technical Mapping Advisory Council was created in 1996 by the Federal Emergency Management Agency (FEMA). FEMA created the Council in accordance with a provision in the National Flood Insurance Reform Act of 1994, P. L. 103-325, Title V, Section 576. Congress directed that the Council be established to evaluate the production, distribution, and use of Flood Insurance Rate Maps (FIRMs) and other mapping products and to make recommendations to the Director of FEMA for improvement of those products. The Council is to address the accuracy, quality, utility and distribution of FIRMs, Letters of Map Amendment (LOMAs), Letters of Map Revision (LOMRs), as well as the standards and guidelines for use in their preparation.

The legislation which prompted the creation of the Technical Mapping Advisory Council authorized the use of a wide array of techniques in order to develop recommendations, including holding hearings, receiving evidence and assistance from Federal, State, or local government agencies, private firms, or individuals. The Council Chairperson is also directed to consult with the Chairperson of the Federal Geographic Data Committee to ensure the Councils recommendations are consistent with national digital data collection and management standards.

Congress was specific in identifying the number and nature of organizations that were to be represented on the Council. Eleven organizations or agencies were spelled out in the legislation. They are:
The Undersecretary of Commerce for Oceans and Atmosphere (or his or her designee); a member of recognized surveying and mapping professional associations or organizations; a member of recognized professional engineering associations or organizations; a member of recognized professional associations or organizations representing flood hazard determination firms; a representative of the U.S. Geological Survey; a representative of state geological survey programs; a representative of state National Flood Insurance Program (NFIP) coordination offices; a representative of a regulated lending institution; a representative of the Federal Home Loan Mortgage Corporation; and a representative of the Federal National Mortgage Association; and, a designee of the Director, Federal Emergency Management Agency.

[1]Manager, Water Resources, Short Elliott Hendrickson Inc., 421 Frenette Drive, Chippewa Falls, WI 54729

The American Society of Civil Engineers (ASCE) was asked by FEMA to nominate a person to represent professional engineering associations and organizations. Other organizations who were asked to nominate candidates to serve on the Council include the American Congress on Surveying and Mapping (ACSM), the National Flood Determination Association (NFDA), the Association of American State Geologists (AASG), the Association of State Floodplain Managers (ASFPM), the Federal Home Loan Mortgage Corporation (Freddie Mac), the Federal National Mortgage Association (Fannie Mae), and NationsBank to represent regulated lending institution. Federal agencies represented on the Council include the Federal Emergency Management Agency (FEMA), National Oceanic and Atmospheric Administration (NOAA), and the U.S. Geological Survey (USGS).

The Technical Mapping Advisory Council met for the first time in May 1996 in Washington, D.C. Since then, the Council has met in September 1996 (Washington, D. C.), January 1997 (Washington, D.C.), May 1997 (Pittsburgh, PA), September 1997 (Washington, D.C.), and December 1997 (Minneapolis, MN). Meetings have alternated with telephone conference calls in July 1996, November 1996, March 1997, July 1997, and November 1997.

At the first meeting, in May 1996, the Council spent considerable time sharing concerns, experiences, and establishing goals to be achieved. The Council is comprised of individuals with a wide diversity of background and knowledge regarding the National Flood Insurance Program, map production, and engineering expertise; yet the goals established demonstrate a clear common purpose:

Goals:

> Increase reliability of data sources and increase accuracy, consistency, security, transportability, and sharing of data.
>
> Ensure that firms reflect flood sources adequately and accurately, and include all current flood data that meet FEMA standards.
>
> Evaluate the need for including multiple hazards on NFIP maps.
>
> Improve the study and review processes.
>
> Identify a standard base map of the United States that can be adapted for a variety of NFIP uses by the addition of overlays depicting area-specific information.
>
> Encourage non-federal entities to improve maps.
>
> Actively promote partnerships with state and local governments to improve FIRMS.
>
> Improve flood-zone determinations by increasing accuracy, adding details and using digitized data.
>
> Make accurate, automated determinations possible.
>
> Improve the distribution systems for NFIP maps, LOMAs, and LOMRs to ensure that current data are made available to local officials, subscribers, and other users.
>
> Improve activities to educate the public and other users of flood maps, and communicate with them regarding the use of the data in the NFIP.
>
> Improve communication among floodplain managers, lenders, investors, FEMA, and its contractors, local communities, states, partners, and other significant stakeholders that have an impact on the national flood effort.

The Council's first Annual Report, published in January 1997, included the following recommendations:

### Recommendations - 1st Annual Report to FEMA, January 1997

1. Retention of Maps and Map Information: an archival system be established for maintaining in perpetuity for historic and legal purposes all FIRMs and supporting technical data.

   The Council acknowledged the cost implications of maintaining records in perpetuity, and FEMA's efforts to archive mapping data electronically. The Council also acknowledged that government regulations mandate retention of records for only three years, but believes three years is an insufficient period of time to maintain FIRMs and supporting data.

2. Distribution Process: letters of map change, including Letters of Map Amendment (LOMA) and Letters of Map Revision (LOMR), be distributed with each map ordered, and that individuals or companies that subscribe to automatic updates receive copies of pertinent letters of map change.

   The Council recognized that an initial outlay of funds would be required to modify the map distribution process, and supported the concept of charging map users a fee or service charge for the added costs of distributing map-amending data or information on how that data could be obtained when maps are distributed. The Council recommended incorporating information into map orders to tell customers whether a map is current (no LOMAs or LOMRs have been issued) or how to obtain LOMAs or LOMRs if any have been issued relating to a map. The Council also recommended FEMA advise customers if they have any knowledge of pending actions to revise a map when an order is placed.

3. Forms: Certification forms required to be submitted with map revision requests be distributed in digital form via the internet or similar electronic means.

   The forms which must be submitted to FEMA to initiate map revisions are available only in hard copy. The Council recommended the forms be made available to the general public in electronic format, and in several formats, or word processing programs. The Council further recommended the forms be made available over the internet through FEMA's homepage and by disc.

4. H.R. 3340: A position should be developed on this proposed legislation, or similar future proposals, when its potential effects have been studied further.

   H.R. 3340 was a bill being considered by Congress in 1996 which would have delegated considerable authority to registered land surveyors and the U.S. Army Corps of Engineers regarding the issuance of Letters of Map Amendment (LOMA) Letters of Map Revision (LOMR). The council recognized that had H.R. 3340 been enacted it would have had an impact on Flood Insurance Rate Maps. The concept embraced by the proposed legislation is an appropriate subject for study, however, the Council was not prepared to take a position in 1996.

5. Scribing: Newer technologies than the scribing method be implemented immediately for the production and dissemination of FIRMs.

   While FEMA has used digital, or electronic, mapping techniques to produce FIRMS, there are still many maps in existence that were produced using scribing techniques. When these maps are updated FEMA uses scribe coat for updates when it appears to be more cost-effective than converting the map to an electronic format. This continued use of scribe coat perpetuates its use. The lack of availability of the materials needed for this process makes investigation into alternatives imperative.

In May 1997, following receipt of the Council's report, James Lee Witt, Director, Federal Emergency Management Agency, asked for a report from his staff with recommendations on what would be necessary to modernize their mapping processes and products. FEMA staff were directed to prepare a report with recommendations and cost estimates. The Technical Mapping Advisory Council was invited to provide suggestions and advice and, in June 1997, several members of the Council met with FEMA staff in Alexandria, VA to review and comment on their draft report.

The report, "Modernizing FEMA's Flood Hazard Mapping Program," spells out a comprehensive plan to improve the accuracy and completeness of the maps. The plan recommends periodic assessment of the mapping needs, addition of updated information, routine maintenance of maps to add new streets, roadways and other features, the production maps for the estimated 3000 communities that currently do not have a FIRM, but need one, and ways to quickly verify the accuracy of, and update maps, following a flood disaster.

The report also recommends the systematic conversion of the complete inventory of existing maps to a digital format and the production of all new FIRMs and revised FIRMs in digital format. The long range goal is to make FIRMs available to all interested users over the Internet. For users who can not read electronic files, hard copy maps would be available through FEMA's map service center by "print-on-demand" services.

Other key features of the plan to modernize FEMA's mapping program is to develop partnerships with state and local government to play a more active role in maintaining and updating Flood Insurance Rate Maps, and to eliminate the technical review of requests for Letters of Map Amendment (LOMA) and Letters of Map Revision (LOMR) that are submitted by licensed land surveyors and registered professional engineers for properties located in floodplain areas where flood elevations have been established and published on a FEMA flood hazard map.
The Technical Mapping Advisory Council received copies of the report, with the recommendations for improving the mapping program, in November 1997. The Council will be evaluating the report and its recommendations throughout 1998 and providing it s own recommendations to FEMA.

The Council also prepared it s second Annual Report to James Lee Witt, Director, Federal Emergency Management Agency. The report, published in January 1998, includes several recommendations for improving FIRMs, identified key issues that will be the subject of further study, and provides a plan for future activities. The recommendations made by the Council in it s second Annual Report include:

## Recommendations - 2nd Annual Report to FEMA, January 1998

1. Improve the flood insurance study process by (a) shortening the study contract process; (b) permitting multi-year contracts to study contractors; (d) ensuring agreement on the base map among the study contractor, Technical Evaluation Contractor (TEC), the State, FEMA, and the community earlier in the process; and (e) providing for intermediate reviews and approvals of mapping elements at key decision points during the project.

2. Base Maps. (a) Base maps need to be improved and existing standards need to be reviewed and updated. In reviewing these standards, FEMA should consult with the Federal Geographic Data Committee (FGDC); (b) FEMA should ensure strict adherence to the standards. Where best available base maps do not adhere to the base map standards, the study, or restudy, must include the development of an accurate base map.

3. Base Mapping Partnerships. FEMA should pursue base mapping partnerships with other public, private, and nonprofit entities, such as the Census Bureau; USGS; state, local, and regional agencies, to achieve cost efficiencies and exchange technical expertise.

4. Digital FIRMs. All new map products resulting from flood insurance studies or restudies, and physical map revisions, should be prepared, produced, and made available digitally.

5. Community Involvement. Community meetings should be held before, during, and after preparation of a new map product, such as a map digitized for the first time or one being converted to a county-wide product. The purpose of such meetings is to enable community and state input to and participation in mapping issues and activities

The American Society of Civil Engineers approved the formation of a special Task Committee of the Water Resources Planning and Management Division in 1997 to facilitate communication and input into the activities of the Technical Mapping Advisory Council. The Technical Mapping Advisory Task Committee members are:

Mark A. Riebau, P. E.     Chair
Short Elliott Hendrickson
Chippewa Falls, Wisconsin

David K. Carlton, P.E.    Vice-chair
KCM, Inc.
Seattle, Washington

William G. DeGroot        Secretary
Urban Denver Flood Control District
Denver, Colorado

Christopher P. Jones, P.E.
Christopher Jones & Associates
Charlottesville, Virginia

Vernon R. Bonner, P.E.
U.S. Army Corps of Engineers
Hydraulic Engineering Center
Davis, California

Wallace A. Wilson, PE
Wison Consulting
Williamston, Michigan

John Ivey, PE
Halff & Associates
Houston, Texas

## Planning and Design of the New Lenox Community Golf Course
## "Creating Success Through Agency Coordination"

Rick K Suttle, William McCollum[1]

**Abstract**

The Park District of New Lenox, Illinois jumped at the opportunity to purchase 230 acres of land located within the center of the community. Its gently rolling terrain and two converging creeks, set against a backdrop of wooded slopes, seemed idyllic for their proposed golf course. Surprises were in store for them. When the challenges began to mount, many thought the Park District had purchased the proverbial "white elephant". These challenges were ultimately overcome through agency coordination efforts, and creative design solutions to create a golf course that is environmentally sensitive and a strong asset to the Community and the Park District.

**Introduction**

The Park District of New Lenox, Illinois was given the first opportunity, much to the dismay of area realtors, to purchase 230 acres of land located within the heart of the Community. While most of the site was agricultural row crop and pasture land, its gently rolling terrain, bordered by two converging streams set against a wooded backdrop seemed a perfect setting for the District's proposed golf course. However, surprises were in store for them. Challenges they were soon to face included:

- Significant archeological resources,
- Relative high quality streams,
- Floodplain over 60% of the site and streams that flood frequently,
- High groundwater,
- Difficult site access
- Several jurisdictional wetlands, and
- Strict budget constraints

---

[1] R.K. Suttle - Director of Landscape Architecture & Environmental Planning, Ruettiger, Tonelli & Assoc., Joliet, IL.    W. McCollum - Director, New Lenox Community Park District, New Lenox, IL.

These challenges were ultimately overcome through close agency coordination and creative design solutions in which several "constraints" were turned into design "assets". Key to the success of the project was the creation of nearly five acres of wetlands around a series of lakes located within a centrally designed flood conveyance and storage corridor (Figure 1).

Figure 1. Site Master Plan

### Challenges and Agency Coordination Requirements

**Water Needs vs Stream Quality**. A key program requirement was the need to develop a dependable water supply as the course would require about 1500 gpm over an eight hour period for irrigation. Due to the limited knowledge of, and potential variances in groundwater, the project team initially proposed to utilize both groundwater and surface water as the water supply sources. A small weir, two feet in height, was proposed to be constructed on the creek to divert water to constructed ponds (Figure 2). The concept to balance impacts between surface and groundwater resources seemed sound. The surface water would help maintain relatively stable pond levels throughout the system, and withdrawal of groundwater would be reduced, which was a concern to the Village since it was the source of community drinking water. An additional benefit would have been a reduction in stream sediment loads as a result of trapping by the weir and ponds. While the concept seemed logical, and proposed withdrawal rates and the weir design were shown to have minimal resource impacts, State Conservation and Regional Fish and Wildlife Agencies would not agree to any development on the stream. After several discussions with the Agencies plans for a weir and diversion were discarded.

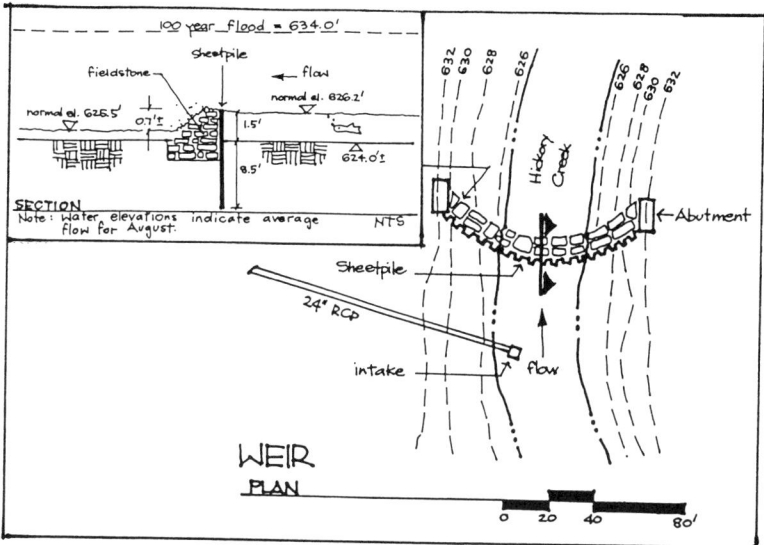

Figure 2. Proposed Weir and Diversion

**Significant Archeological Resources.** The Park District's public agency status and their need to file a Section 404 permit, required that archeological investigations be conducted. Phase II archeological testing identified significant features in select areas of the site. The most notable resource finds were a single Native American burial pit and village site. Consequently, extensive agency consultation was required between the Illinois Historic Preservation Agency, U.S. Army Corps of Engineers, and the Historic Advisory Council in Washington D.C. As a result of the consultation, a mitigation plan and Memorandum of Agreement was prepared to protect 19 acres which were identified as a Nationally Significant Cultural Resource Site. These 19 acres were effectively removed from consideration as part of the golf course development.

**Wetlands.** Wetland field studies identified several acres of forested riparian wetlands adjacent to the stream, and approximately three acres of scrub/shrub vegetation communities found in lower depressional "potholes" in the southwest corner of the site. The scrub/shrub communities were quite degraded due to many years of farming activity surrounding them. Over 80% of the project site was cleared of vegetation and plowed for row-crop production. Soils in the lower lying areas were classified as "Drummer"series, which are hydric soils. Although these soils were hydric, they had been drained and farmed for long periods of time. Close coordination with the Natural Resource Conservation Service helped to determine that based on the level of current and historic disturbance to the site, these hydric soil areas were considered prior converted cropland and not farmed wetlands.

**Floodplains and High Groundwater.** Research into the site's hydrology indicated over

60% of the land was within the 100-year floodplain of Hickory Creek. Golf courses are a permitted use within a floodplain and did not pose a particular problem, except for limiting the location of parking and building locations, which were required to be located outside the limits of the floodplain. An additional constraint imposed on the District, however, was the request from their lending institution that all greens and tees be developed above the 100-year flood elevation. Consequently, detailed hydrologic studies were required to assure that development would provide adequate conveyance and storage of flood waters in compliance with State and County floodplain development ordinances. These ordinances required one and a half acres of storage for every acre of impact. Calculations and plans were submitted to the Illinois State Department of Natural Resources, Office of Water Resources.

Soil borings taken on site, indicated the area was underlain by highly permeable sands and gravels. An ancient streambed was located in the central part of the site as evidenced from old logs that were excavated. Groundwater at this location was less than two feet from the surface. This constraint initially drove up the price of excavation and "scared" off a number of contractors from bidding the project. In the end, however, dewatering was simple, and the gravels made excavation relatively easy.

## The Plan - Turning Constraints Into Opportunities

The project team ultimately developed a plan which resolved many constraints presented during the course of the studies through avoidance or by turning them into opportunities. The need to preserve land for archeological resources and for flood conveyance required that two golf holes be placed on available land across the stream. The golf features were sensitively shaped around existing woodlands and created wetlands which served as regulatory mitigation. Forested wetlands around the stream were avoided and maintained as a buffer. The need for flood storage and conveyance determined the principal design feature of the golf course, however. Seven ponds were designed within a centralized flood conveyance corridor across the course (Figure 3).

The site's high groundwater constraint was turned into an asset. A seven acre pond over sixteen feet deep was created to serve as the course's central attraction and irrigation source. The pond's large volume will minimize drawdown fluctuations from the irrigation operation and reduce potential adverse impacts to created wetland systems. All the ponds within the corridor were interconnected by a combination of underground piping and meandering surface streams. Pond elevations were set so that positive flow would be maintained throughout the system. In order to maintain high water quality within the pond system, storm runoff from parking lots was diverted to grassed depressions. Overflow from these depressions would then flow into grassed "bioswales"before reaching the ponds. The entire system was designed to accommodate rising water levels during a flood event, and then return to normal, typically within 48 hours of the event.

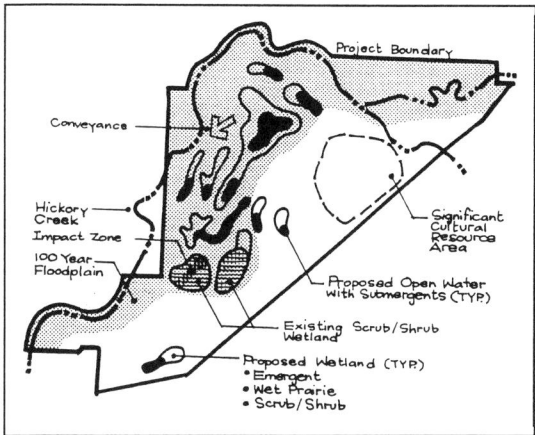

Figure 3  Conceptual Wetland Plan and Flood Conveyance Corridor

## Wetland and Prairie Enhancements

Proposed ponds and the site's high groundwater table provided opportunities to create a variety of wetland communities. Several slopes adjacent to ponds were flattened to create emergent and sedge meadow zones. Hydric soil excavated from on-site was used as a planting medium. Most of the wetlands were designed around the edges of ponds, with the exception of the wetland near the outlet. This emergent wetland was created to fill the narrow pond and serves as a final treatment filter prior to the site's water emptying into the stream. The designed wetland mitigation provides nearly a 3.6:1 wetland mitigation-to-impacted wetland ratio.

An additional natural enhancement to the project was the creation of native grassland sites to serve as buffers between the golf course fairways and the streams and wetlands. A burn management program is planned by the Park District to help maintain these areas.

## A Successful Ending Through Agency Coordination

Agency coordination efforts seemed daunting at first, as the District was squeezed between two agencies' concerns over water usage. On one hand, the State Department of Natural Resources did not like the idea of withdrawing water from nearby Hickory Creek; on the other hand, the Village was concerned over the utilization of groundwater, which is the source of the community's drinking water. This coordination effort took a back seat, however, to the District's efforts required to continue the project in the face of finding significant cultural resources on site. A successful Memorandum Of Agreement was finally achieved due to the coordination efforts of the Army Corps of Engineers working with the Illinois Historic Preservation Agency and the National Historic Advisory Council. In the end, the Park District consulted with 12 different public and quasi-public entities in the pursuit of permits for the project.

The Park District's Sanctuary Golf Course opened in the fall of 1996. In the spring of 1996, the central conveyance corridor was put to the test with a near record rainfall event (nearly 9 inches in a 24-hr period). Even though kayakers paddled past the 16th tee, the corridor worked beautifully and within 48 hours was back within its limits. Wetland plants are filling in the areas designated for vegetation, and the prairie buffers are beginning to emerge. The State's insistence on avoiding the creek turned out to be a good thing for the Golf Course. The groundwater source fills the pond levels consistently, even while pumping for irrigation, and the lower sediment load is easier on the pumps. Although a struggle in the beginning, in the end the New Lenox Golf Course was a success in overcoming constraints and integrating environmentally sensitive wetland and stormwater management systems into a relatively intense land use. The project so far, has been a success ecologically, aesthetically, and economically.

**Improving Coordination**

A primary goal of agency-client-designer coordination is cooperation to achieve a desired end. Of course, the hang-up often is in the perception of what the "desired end" should be. It is important that all parties maintain a "spirit of cooperation", and a willingness to step back and look at projects within a broader future planning context. The New Lenox Golf Course came close to not becoming a reality as a result of the many hurdles the District faced. What is important to reflect upon in the "spirit of cooperaiton" however, is that this site would not have remained undeveloped. If the Park District did not develop it, most of the site would have been purchased by private enterprises and developed for residential housing with far less resource amenities, and much less agency coordination.

## Reduction of Illinois River Flood Stages by Converting A Few Selected Levee Districts to Managed Storage Areas

Abiola A. Akanbi[1] and Krishan P. Singh[2], F. ASCE

**Abstract**

High flood stages in the Peoria-Grafton reach of the Illinois River can be lowered by making limited-sized openings in a few levees so that the areas behind such levees can be used for temporary storage of floodwaters when the flood elevations exceed the elevations of the openings or inflow sections. An unsteady flow model was applied to study the potential reduction in flood elevations due to the conversion of the areas behind selected levees to managed flood storage areas. The analysis also includes the determination of the optimal reduction in flood peaks for various dimensions of the inflow sections and for various combinations of managed storage areas.

**Introduction**

The common response to protect an at-risk levee (overtopped during a 100-year or lower flood) is to raise the levee, but the benefit-cost ratio is about 0.1 (USACOE, 1987). Even if a levee is raised, the additional cost of pumping to keep the water table low in the levee and drainage district (LDD) and the consequent reduction in crop yield make farming in the area behind the levee unprofitable. An alternate approach would be to convert a few selected levees with marginally profitable farmlands to managed flood storage areas so that they provide both flood storage and wetland or conservation functions for the converted LDDs, while also providing greater protection against flooding to agricultural lands served by other levees.

Practically all of the LDDs along the Illinois River are in the lower reach from River Mile (RM) 157.7 at Peoria to the confluence with the Mississippi

---

[1]Professional Scientist, Office of Hydraulics and River Mechanics, Illinois State Water Survey, 2204 Griffith Drive, Champaign, Illinois 61820.
[2]Emeritus Principal Scientist, Hydrology Division, Illinois State Water Survey.

River at Grafton. Extreme flood stages occurring in this reach of the river can be lowered by converting the areas behind a few levees to managed flood storage areas. A suitable, limited section of a selected levee would be lowered to a predetermined elevation so that floodwaters can flow into the area behind the levee for temporary storage (Figure 1). Once the flood elevation in the Illinois River drops below the top of the lowered section, the floodwaters in the storage area will return gradually to the Illinois River.

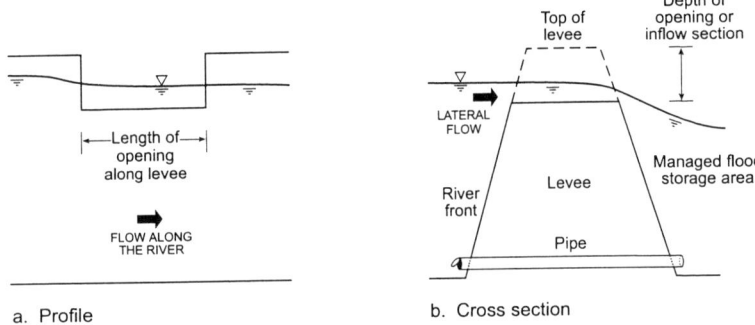

Figure 1. Managed flood storage levee with lateral inflow section

The determination of suitable levees and the optimal reduction in flood peaks that can be attained require an analysis of historical floods to determine flood magnitudes and frequencies at which overtopping occurs at various levees along the lower Illinois River. It also requires an evaluation of the impact and interaction of flow from the tributaries and backwater flow from the Mississippi River.

An unsteady flow model was validated by applying it to simulate historical and design floods along the La Grange (RM 157.7 to 80.1) and Alton (RM 80.1 to 0.0) pools in the lower segment of the Illinois River. The model was also applied to compute water surface elevation (WSE) profiles resulting from conversion of several combinations of LDDs in the La Grange and Alton pools to managed flood storage areas. WSE profiles for various combinations of levees and lateral inflow sections were computed and guidelines were developed for the selection of optimum length (along the levee) and depth of levee inflow sections and the combination of managed storage areas that will provide maximum protection against design floods for other levees. The goal was also to keep the number of converted levees to a minimum.

Details of the analysis and the results are discussed in the following sections.

## Flood Frequency Analysis

The flood frequency analysis involves the development of discharge-frequency relations for Illinois River gaging station records at Marseilles, Kingston Mines, and Meredosia, and the gaging stations on the five major tributaries (Mackinaw, Spoon, Sangamon, La Moine, and Macoupin) to the Illinois River, using both log-Pearson Type III and mixed distributions (Singh, 1996). Stage-frequency relations have also been developed for Illinois River gages at Peoria Lock and Dam (L&D), Kingston Mines, Havana, Beardstown, La Grange L&D, Meredosia, Valley City, Pearl, Florence, Hardin, and Grafton. Singh (1996) and Akanbi and Singh (1997) reported the results of the flood and stage frequency analyses. These results were used to develop the 2-, 10-, 25-, 50-, and 100-year discharge and stage hydrographs required for the boundary conditions in the unsteady flow simulations.

## Unsteady Flow Modeling Approach

The unsteady flow model, UNET (HEC 1993), was validated and then applied to predict design flood stages in the lower Illinois River. A total of 412 cross sections were used to calibrate the model parameters to several historical flood events by matching computed flood stages at the gages on the Illinois River to the observed records for the May 1979 and March 1985 floods. These floods are ranked, respectively, fourth and second at Meredosia, and sixth and third at Kingston Mines. Computed WSEs were compared with the recorded events at eight gaging stations between Peoria L&D and Grafton. Computed WSE profiles for these flood events fit the observed data generally within 0.5 feet for the stations in the La Grange Pool and generally within one foot at the stations in the Alton Pool with the one-reach model (Peoria L&D to Grafton). To improve the fit in Alton Pool, a two-reach model (Peoria L&D to La Grange L&D and La Grange L&D to Grafton) was used. The computed two-reach WSE for the 1979 flood fit the observed WSEs more closely than the one-reach model. The model was further tested by simulating the December 1982, June 1974, April 1973, and July 1993 flood events. Akanbi and Singh (1997) reported the results of these and other simulations.

Analysis of the annual peak flows at Kingston Mines and Meredosia indicates that the 1993 floods at these stations had a recurrence interval of 3 to 5 years. The 1993 high flood stages observed in the Illinois River were not due to a major flood in the river itself. Instead, the high stages were due to the prolonged and unprecedented flooding on the Mississippi River that caused backwater effects upstream on the Illinois River. The upstream extent of this backwater effect was estimated by ranking the annual peak WSE data (1941-1993) for the stations between Meredosia and Kingston Mines. Ranks for the 1993 flood stages at these stations indicated that the backwater effects extended up to Havana (RM 119.6) but not as far as Kingston Mines (RM 145.6).

## Simulation of the Design and Other Flood Profiles

The validated unsteady flow model was used to predict the 10-, 25-, 50-, 100-, and 500-year flood profiles in the La Grange and Alton pools. Other WSE profiles were developed by combining tributary and lateral inflows of various recurrence intervals and adjusting the time lags for the inflow hydrographs. Appropriate time lags for the tributary hydrographs were determined from the timing of the flood peaks for the 1943, 1974, 1979, 1982, and 1985 flood events.

Discharge hydrographs for each tributary station and stage hydrographs at Peoria L&D, La Grange L&D, and Grafton were obtained by normalizing the hydrographs for the top six flood events at each station and plotting them together to determine a representative normalized hydrograph. Each normalized hydrograph was converted to a discharge or stage hydrograph by multiplying the ordinates by the value of the design flood or stage obtained from the flood or stage frequency analysis for the station.

## Managed Storage Reduction of Flood Stages

WSE profiles were also computed to evaluate the potential impact of the conversion of several LDDs in the La Grange Pool and a few in the Alton Pool to managed flood storage areas. Model simulations included prediction of 1) maximum flood stages under no levee failure condition by assuming a vertical wall along the tops of the levees, 2) flood stages along the Illinois River, assuming the levee systems were not constructed, and 3) reduction in flood stages by conversion of some of the levees to managed flood storage areas. The lateral inflow section for a managed storage area was represented by an opening of 250 to 4000 feet along the levee and 2 to 6 feet below the top of the levee. Generally, a 1000-foot length and 4- to 6-foot depth of lowered sections was found to be the most promising in lowering the flood elevations. WSE profiles were computed for various dimensions of the inflow section for all the levees between Peoria and Valley City. Levees below Valley City were not considered because the water surface elevations are governed by the backwaters from the Mississippi River. The results were analyzed and used to select various combinations of levees and dimensions of lateral inflow sections that would produce the greatest reduction in stages at critical sections in the La Grange and Alton pools. An example in Figure 2 shows that practically all the levees are safe against a 100-year flood up to RM 30 when Lacey and McGee Creek LDDs are converted to flood storage areas. The

# WATER RESOURCES AND THE URBAN ENVIRONMENT 215

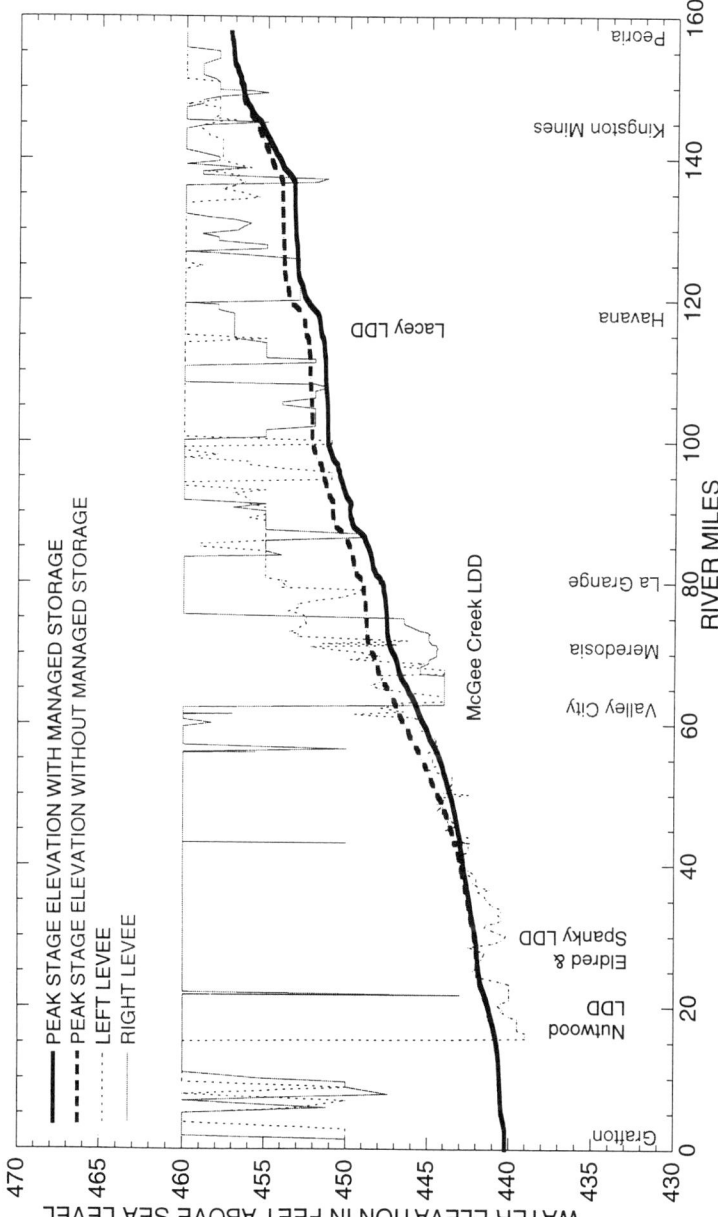

Figure 2: 100-Year Peak Stage Profile without and with Managed Flood Storage at Lacey Levee and Drainage District or LDD (6 ft depth of opening) and McGee Creek LDD (4 ft depth of opening)

reach further downstream of RM 30 is greatly affected by the Mississippi backwaters, and as such the Eldred-Spanky and Nutwood levees in this reach will have to be raised to safeguard them against a 100-year flood.

## Conclusions

This study investigated 1) the potential reduction in flood stages that would result from conversion of a few LDDs along the Illinois River to managed flood storage areas, 2) the selection of optimum dimension of lateral inflow sections along the levees and, (3) the choice of levees that are most suitable for conversion to managed storage areas.

## Acknowledgments

The research, upon which this paper is based, was conducted by the authors as part of their regular duties at the State Water Survey. Partial funding for the project was provided by the Office of Water Resources, Illinois Department of Natural Resources.

## References

Akanbi, A.A. and K.P. Singh, 1997. *Managed Flood Storage Option for Selected Levees along the Lower Illinois River for Enhancing Flood Protection, Agriculture, Wetlands, and Recreation, Second Report: Validation of the UNET Model for the Lower Illinois River*, Contract Report 608, Illinois State Water Survey, Champaign, IL, 110p.

HEC, 1993. *UNET -- One-dimensional Unsteady Flow through a Full Network of Open Channels; User's Manual*, U.S. Army Corps of Engineers, Hydraulic Engineering Center, Davis, CA, 240p.

Singh, K.P. 1996. *Managed Flood Storage Option for Selected Levees along the Lower Illinois River for Enhancing Flood Protection, Agriculture, Wetlands, and Recreation, First Report: Stage and Flood Frequencies and the Mississippi Backwater Effects*, Contract Report 590, Illinois State Water Survey, Champaign, IL, 138p.

U.S. Army Corps of Engineers (USACOE). 1987. *Reconnaissance Study for the La Grange Pool, Illinois River, Illinois*, Department of the Army, Corps of Engineers, Rock Island District.

# INEFFECTIVE OVERBANK FLOW IN ONE-DIMENSIONAL MODELS
by Dirk L. Gowin, PE[1] and Dr. Charles Morris, PE[2]
Full Members of the American Society of Civil Engineers

### Abstract
In a contracting reach at a bridge crossing over a wide, flat floodplain, the stored water at the limits of the floodplain is considered an area of ineffective flow. One-dimensional, standard step, open channel models are widely used and provide reasonable predictions of water surface elevations. Cross sections upstream of a contracting reach are recommended to be placed where the flow lines are parallel and the full cross section is effective within these models. Hydraulic properties near the bridge may be desired, requiring cross sections that are not fully effective. For example, to estimate scour using WSPRO, a cross section one bridge length upstream of the bridge opening is recommended. To accurately predict hydraulic properties within a contracting stream reach using one-dimensional standard step models, determination of areas of ineffective flow is required.

This study compares one-dimensional modeling methods in use to define ineffective flow areas at a cross section one bridge length upstream of a bridge crossing a wide, flat floodplain. In addition, a method is suggested to eliminate ineffective flow areas based upon the premise that a cross section is to be perpendicular to flow at all locations. Flow distribution and energy losses through the bridge (i.e. backwater) are compared with and without consideration of ineffective flow areas in the one-dimensional models. For validation, all results are compared to a two-dimensional model.

### Introduction
In one-dimensional models, recommended placement of a cross section upstream of a constriction is where " . . . the flow lines are approximately parallel and the full cross section is effective."[7] Two variables used in scour computations are flow distribution between the channel and the overbanks as well as anticipated

---

[1] Dirk L. Gowin, PE; Ogden Environmental and Engineering Services, Inc.
9800 W. Kincey Avenue, Suite 190; Huntersville, NC 28078

[2] Dr. Charles Morris, PE; University of Missouri - Rolla
119 Butler-Carlton Civil Engineering Hall; Rolla, MO 65409-0030

backwater. The recommended location of the cross section upstream of the bridge to determine scour in a WSPRO model is one bridge length upstream of the crossing. These stream reaches immediately upstream of a bridge exhibit two-dimensional flow with converging flow lines and areas of stagnation. Therefore modeling techniques are required in one-dimensional modeling to improve output results.

This study investigates commonly used methods and proposes a new method in one-dimensional modeling to eliminate ineffective flow areas at a contracting section one bridge length upstream of a bridge crossing. Since the flow within the channel and water depths impact the prediction of scour, comparisons are made of the velocity within the channel and backwater. These results are validated by comparison with a two-dimensional model.

## Approaches Studied

Several documented methods are available to define ineffective flow regions in a one-dimensional surface water model. Only three of these approaches are considered in this study due to their wide acceptance and are described below.

One-To-One  Because of its wide acceptance and simplicity, "One-to-One" (1:1) contraction methodology is most often applied. From the edges of the contracted section, overbank regions outside of a line projected away from the channel at a 45 degree angle to the flow are eliminated from the model section. Because of its wide usage, One-to-One contraction to determine areas of ineffective flow is studied.

Determination By Velocity  Regions with velocities less than one foot per second are considered storage zones and are determined to be ineffective.[3] This method is less commonly used than 1:1 contraction and is inappropriate for flat overbanks. One dimensional models use conveyance potential, determined by depth of flow and resistance to flow, to predict velocities. Determination of ineffective areas based upon velocities is not used. Since the depths are uniform in the overbank of a wide, flat floodplain, no significant change in velocity or conveyance is indicated at the outer limits of the floodplain compared with the overbank nearer the channel. Therefore, the interface between effective and ineffective areas is not readily apparent by observing velocities.

Matching Channel Velocities To Full-Bank Flow  Channel velocity during out-of-bank flows is considered to be equal to or greater than channel velocity during full-bank flows due to an increase in depth and overall flow.[3] For the design discharge, portions of the overbank are considered ineffective to produce an out-of-bank channel velocity slightly greater than the full-bank channel velocity. Matching out-of-bank channel velocity to full-bank channel velocity is usually considered to provide reasonable results and investigated in this study.

## Methodology

The Federal Highway Administration's (FHWA) one-dimensional flow model, WSPRO, was selected for this study. WSPRO uses a varied flow path length in the overbanks at the constricted section to account for additional losses at a constricted section. Development of the WSPRO bridge routines was based upon a study prepared by the U.S. Geological Survey entitled "Computation of Backwater and Discharge at Width Constrictions of Heavily Vegetated Floodplains."[6] Since this study involved similar situations to the research used in the development of WSPRO, and this model is recommended for scour computations, WSPRO is the model used in this study.

A template section is developed for the floodplain using a trapezoidal channel with side slopes of 2:1. Overbank regions slope slightly by 0.0003 m/m toward the channel (Figure 1). Floodplain widths are set to 4, 6 and 8 times the bridge width, with each overbank being equal at 2, 3 and 4 times the bridge width.

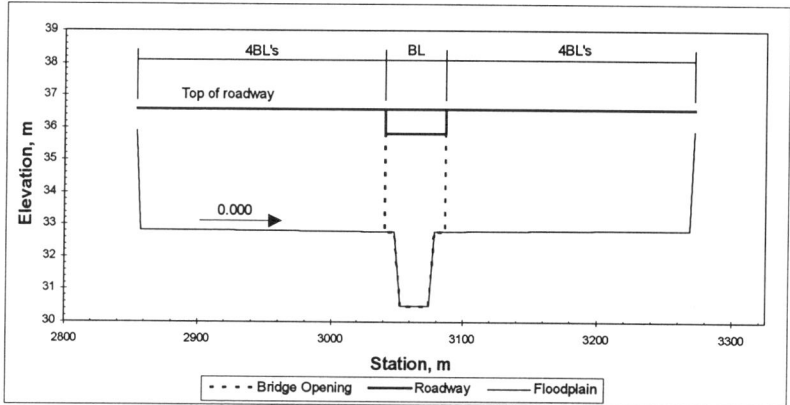

Figure 1. Profile of Floodplain and Bridge Opening.

Manning's n-values for the channel remained fixed at 0.04. Overbank Manning's n-values are set at 0.06, 0.08 and 0.12 for all three geometries. Where flow in the overbank is considered ineffective, rather than change the section geometry, Manning's n-values are raised to 100 to eliminate flow in these areas.

The bridge opening used in this study consists of five 9.14m (30 foot) spans with vertical abutments. Channel geometry through the bridge is consistent with the floodplain sections.

Starting water surface elevations are determined at the downstream section using the slope conveyance method. Normal depth is obtained by using an initial slope equivalent with the valley slope, 0.001 m/m. A flowrate is desired within the model that would inundate the floodplains with one to two meters of water. From trial and error, a flowrate of 226.5 cms (8000 cfs) inundates the floodplains upstream of the bridge with four to five feet of water and is used in all models.

An attempt is made to eliminate regions of ineffective flow based upon the premise that the cross section alignment is to be placed perpendicular to flow direction throughout the entire width of the cross section. Broken or dog-legged sections are recommended to maintain flow patterns perpendicular to the section.[9] A line perpendicular to the flow vectors in the FESWMS-2DH model would be curvilinear in shape. To maintain cross section alignment more nearly perpendicular to the flow direction, a modified cross section is used. This cross section is parallel to the bridge face for the width of the bridge opening and completed with tangential curves using a radius equal to the distance between the cross section and the upstream bridge face, one bridge length (Figure 2). This method is referred to as the Arc Method in this study.

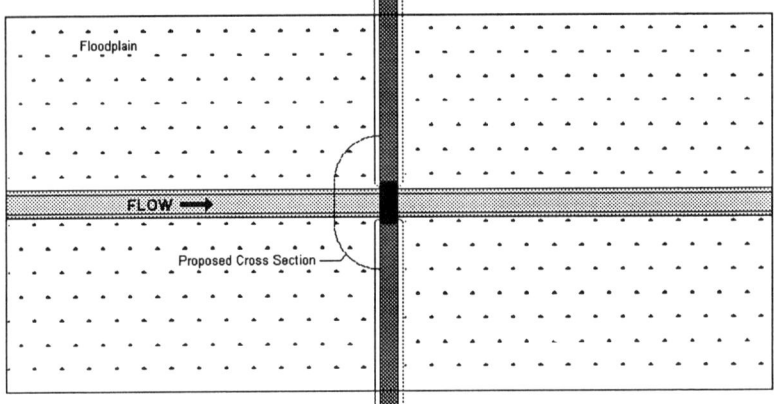

Figure 2. Arc Method

FESWMS-2DH was developed by the FHWA specifically to analyze flow at bridge crossings where complicated hydraulic conditions exist. To validate the results of the one-dimensional modeling techniques, FHWA's two-dimensional model FESWMS-2DH was selected.

## Results

Flow within the channel is used to determine if methods for determination of ineffective flow regions are valid. A comparison of water surface elevations is also made to indicate that each of these methods provides reasonable estimates of water surface.

Channel Flow Comparison  Flow within the channel for the WSPRO models compared with the FESWMS-2DH model at one bridge length upstream of the crossing is presented in Table I. Shaded values are taken from the unconstricted WSPRO model due to channel velocities being equal to or greater than the full-bank channel velocities.

## Table I. Comparison of Flow Within the Channel

| | | FESWMS 2DH | WSPRO | WSPRO 1 to 1 | WSPRO Full Bank | WSPRO Arc |
|---|---|---|---|---|---|---|
| Width of overbank 4 times bridge width | Overbank n: 0.06 Channel n: 0.04 | 125 | 83 | 158 | 120 | 130 |
| | Overbank n: 0.08 Channel n: 0.04 | 131 | 98 | 171 | 123 | 144 |
| | Overbank n: 0.12 Channel n: 0.04 | 136 | 118 | 185 | 118 | 161 |
| Width of overbank 3 times bridge width | Overbank n: 0.06 Channel n: 0.04 | 127 | 97 | 158 | 120 | 130 |
| | Overbank n: 0.08 Channel n: 0.04 | 133 | 112 | 171 | 123 | 144 |
| | Overbank n: 0.12 Channel n: 0.04 | 141 | 132 | 185 | 132 | 161 |
| Width of overbank 2 times bridge width | Overbank n: 0.06 Channel n: 0.04 | 133 | 117 | 158 | 120 | 130 |
| | Overbank n: 0.08 Channel n: 0.04 | 137 | 132 | 171 | 132 | 144 |
| | Overbank n: 0.12 Channel n: 0.04 | 143 | 150 | 185 | 150 | 161 |

<u>Comparison of Water Surface Elevations</u>  Water surface elevations for the WSPRO models compared with the FESWMS-2DH model at one bridge length upstream of the crossing is presented in Table II.

## Table II. Comparison of Water Surface Elevations

| | | FESWMS 2DH | WSPRO | WSPRO 1 to 1 | WSPRO Full Bank | WSPRO Arc |
|---|---|---|---|---|---|---|
| Width of overbank 4 times bridge width | Overbank n: 0.06 Channel n: 0.04 | 34.06 | 34.07 | 33.97 | 34.07 | 34.04 |
| | Overbank n: 0.08 Channel n: 0.04 | 34.09 | 34.10 | 33.98 | 34.10 | 34.05 |
| | Overbank n: 0.12 Channel n: 0.04 | 34.10 | 34.14 | 34.02 | 34.14 | 34.09 |
| Width of overbank 3 times bridge width | Overbank n: 0.06 Channel n: 0.04 | 34.06 | 34.08 | 33.97 | 34.07 | 34.04 |
| | Overbank n: 0.08 Channel n: 0.04 | 34.11 | 34.11 | 33.98 | 34.10 | 34.05 |
| | Overbank n: 0.12 Channel n: 0.04 | 34.19 | 34.17 | 34.02 | 34.17 | 34.09 |
| Width of overbank 2 times bridge width | Overbank n: 0.06 Channel n: 0.04 | 34.08 | 34.10 | 33.97 | 34.08 | 34.04 |
| | Overbank n: 0.08 Channel n: 0.04 | 34.12 | 34.14 | 33.98 | 34.14 | 34.05 |
| | Overbank n: 0.12 Channel n: 0.04 | 34.23 | 34.22 | 34.02 | 34.22 | 34.09 |

## Summary

Water surface elevations between the WSPRO model and the FESWMS-2DH model are essentially in agreement. This indicates that the bridge routines in the WSPRO model are effective in situations involving wide, heavily vegetated floodplains. The greatest discrepancy is indicated with high resistance in a wide overbank, which can be accounted for by the instability within the FESWMS-2DH model for this scenario. Prediction of flow within the channel is poor at this location. Without consideration of ineffective flow areas, WSPRO underestimates flow by an average of 15%. WSPRO severely underestimates the channel flow when the resistance in the overbank is relatively low and the floodplain is wide.

Of the methods used to eliminate ineffective flow regions, 1:1 methodology provided the least accurate results in estimating flow in the channel at a contracting section one bridge length upstream of the roadway. Except when the ratio of the overbank to bridge opening width is large and the ratio of overbank to channel flow resistance is small, WSPRO with no correction provides a better estimate of flow in the channel. One-to-One methodology consistently overestimates channel flow. Water surface elevations are consistently lower using this method, due to the additional constriction of subcritical flow. This departure in water surface elevations when compared to the unconstricted WSPRO model indicates areas of effective flow are being eliminated.

Eliminating regions of the overbank until channel velocities are equal to or greater than velocities during full-bank flow provided the best overall results. When the ratio of the overbank to bridge opening widths is small or the ratio of overbank to channel flow resistance is large, the channel velocities in the unconstricted model are greater than full-bank channel velocities. Therefore, modifications to the cross section to account for ineffective flow regions are not required. This resulted in the greatest error of estimation (shaded values in Tables I and II).

Intuitively, aligning the cross section perpendicular to the flow paths should make the entire cross section effective and improve results. The study results supported this premise. Both channel flow and water surface elevations compare well with the FESWMS-2DH model, except when the overbank resistance is high.

## BIBLIOGRAPHY
1. Chow, V.T., 1959, *Open Channel Flow*, McGraw-Hill, NY.
2. Federal Highway Administration, 1990, *Water Surface Profile Computational Model*, Report No. FHWA-IP-89-027, McLean, VA.
3. Hydrologic Engineering Center, 1988, *Floodway Determination Using Computer Program HEC-2*, US Army Corps of Engineers, Training Document No. 5, Davis, CA.
4. Hydrologic Engineering Center, 1995, *Flow Transitions in Bridge Backwater Analysis*, US Army Corps of Engineers, Davis, CA.
5. Hydrologic Engineering Center, 1990, *HEC-2 Water Surface User's Manual*, US Army Corps of Engineers, Davis, CA.
6. US Geological Survey, 1976, *Computation of Backwater and Discharge at Width Constrictions of Heavily Vegetated Flood Plains*, WRI 76-129, Bay St. Louis, MS.
7. US Geological Survey, 1984, *Techniques of Water-Resources Investigations of the USGS: Computation of Water-Surface Profiles in Open Channels*, Alexandria, VA.

# Urban Flooding in Coastal Areas:
## A Case Study of Runoff Management and Storm Runoff and Tidal Flooding

Erez Sela[1], P.E., M. ASCE

*Abstract*

Traditionally, engineers use floodplain studies performed by the Federal Emergency Management Agency (FEMA) in cooperation with local authorities to establish hydraulic criteria for designing structures in floodplains. However, a case study performed for design of a light rail transit (LRT) system in a coastal watershed in New Jersey indicated that additional detailed hydrologic/hydraulic studies were required to adequately estimate flooding frequency and elevations to enable the development of safe design criteria and efficient drainage system design. In this watershed, which is politically divided among several municipalities, both fluvial and tidal flows cause flooding events.

This paper focuses on a case study performed for the Bellmans Creek watershed extending over three municipalities located near Jersey City, New Jersey. The approach taken in this study accounted for the entire watershed as one integrated unit, regardless of political boundaries, and it also accounted for tidal and fluvial flows. This approach resulted in a more realistic designation of the floodplain and flooding levels and frequencies than available through traditional studies. Whereas traditional studies interpreted Bellmans Creek to be dominated by tides, the case study found that fluvial flooding in Bellmans Creek resulting from storm runoff is dominant on most of its length and that the corresponding design flood level exceeds estimates of previous studies by approximately 4 feet. The case study results were confirmed by field investigations.

This paper reviews a set of symptoms that may indicate the need for additional detailed watershed studies in the design of structures in floodplains. It also reviews planning, management, hydrologic, hydraulic, and field investigation procedures and techniques used in this case study, which provided more

---

[1] Senior Professional Associate, Parsons Brinckerhoff, Inc., 506 Carnegie Center Drive, Princeton, NJ 08540

realistic results than traditional studies. The experience gained and conclusions obtained from the case study are also presented.

*Introduction*

Development of the Hudson-Bergen light rail transit (LRT) system in New Jersey required extending the alignment along an existing Conrail right-of-way located in a low area in the Town of North Bergen and the Borough of Fairview. This low area consisted of wetlands prior to the construction of the Conrail freight tracks and warehouses and currently is an undesignated floodplain. Being at the lowest topographic level in the town—well below nearby residential areas—the alignment is receiving most of the runoff generated in the township. Also, because of its proximity to Bellmans Creek and the Hackensack River, as well as terrain levels very close to sea level, the alignment area is susceptible to flooding from both sources, the runoff generated on the urban lands, and high tide levels resulting from lunar, atmospheric, and oceanic processes and events.

This paper focuses on a case study that was performed for the segment of the Hudson-Bergen LRT alignment located in the Bellmans Creek watershed. The Bellmans Creek floodplain is subject to various hydraulic regimes and flooding sources, and several political boundaries exist within short distances along the alignment and within the small watershed. Additionally, because this location is in a low area previously used for infrequent freight transportation, and more recently for warehouses, the available floodplain designation was not adequate. These factors required preparation of more detailed hydraulic and hydrologic analyses to assess existing drainage conditions as well as the numerous factors that impact area flooding. The primary objectives of these hydraulic and hydrologic studies were to identify the sources of floods, to realistically quantify the flooding parameters, and to design a drainage system that mitigates flooding and protects the LRT's passengers, tracks, and vehicles.

This case study identifies indicators for existing and potential flooding conditions assessed using analytical methods. It reviews field investigations performed to supplement these analytical methods and to check the results of the analyses with actual witnessed flooding occurrences. The case study also identifies a design concept that avoids flooding of the LRT facilities and potential impacts on surrounding properties.

*Indicators for Potential Flooding Conditions*

Existing flood studies of the area indicated that the LRT alignment may be subject to some flooding during a 100-year tidal event in the Hackensack River. A review of the topographic maps of the watershed and a field reconnaissance revealed a potential for frequent and high-rate storm runoff events. The following indicators, when combined, reveal such potential:

- The watershed that contributes runoff to the alignment is relatively large and highly impervious
- The watershed slopes are very steep and many streets extend in the direction of the slope
- The runoff collects on street surfaces and is concentrated on and near one street (91st Street) that extends in the direction of the slope perpendicular to the LRT alignment
- The alignment is located in a low area at the foot of the perpendicular street that receives the concentrated runoff
- This low alignment area appears to be located in the Bellmans Creek floodplain
- Bellmans Creek has a large watershed with steep hills and a hill foot near the LRT alignment that receives high peak stormflows
- Bellmans Creek is subject to tidal flooding and has several bridge crossings that obstruct flows originating from rainstorms

## Study Area Hydrology and Hydraulics

The project alignment is located within Conrail's existing right-of-way between the foot of 84th Street in North Bergen Township in the south and the Susquehanna and Western Railroad crossing in the Borough of Fairview. Approximately 352 acres of upland, mostly impervious area contribute stormflows directly to the east side of the LRT's linear right-of-way that extends in a south-north direction. This right-of-way is currently used for infrequent freight transportation.

Approximately 75 percent of the watershed is occupied by urban residences, approximately 20 percent by commercial and industrial facilities, and less than 5 percent by the exiting rail facilities. The peak overland flows were estimated for 25- and 100-year storms using the hydrograph method developed by the U.S. Soil Conservation Service as outlined in *Urban Hydrology for Small Watersheds, Technical Release (TR) No. 55*. The impervious area on steep hillside slopes with gradients up to 25 percent result in very high runoff rate. The estimated overland peak flows were adjusted to account for the capacities of the existing combined sewer pipes. The adjusted (net) overland peak flows were estimated at approximately 1,200 and 1,640 cfs for the 25- and 100-year storms, respectively. These flows collect on the surface of Tonnelle Avenue at the toe of the steep hillside slope located approximately 1,500 feet from and paralleling the alignment. Most of the flows that collect at the low point of Tonnelle Avenue are concentrated on 91st Street which extends westerly into the alignment area.

Several warehouses are located on the west side of the LRT alignment, partially within the low area and partially on mildly ascending topography to the west. Thus, the right-of-way is located in a low area that turns into a channel whose west ascending side slope deflects the flow north to become a tributary of the tidal Bellmans Creek. Therefore, the study was extended north to the confluence with the Hackensack River.

The Bellmans Creek drainage area at the confluence with the Hackensack River covers 7.7 square miles. Flood Insurance Studies (FIS) of the area indicated that the Hackensack River at that location is dominated by tides. However, the studies performed for Hudson-Bergen LRT project indicated peak flows of 3,625 cfs and 5,235 cfs for the 25- and the 100-year storms, respectively. Backwater analyses using the HEC 2 computer program surveyed cross sections; peak flows indicated that the water surface elevations at the downstream end of the LRT alignment are 11.35 feet and 12.31 feet NGVD for the 25- and 100-year storms,respectively. The case study's water surface elevation estimates for Bellmans Creek indicate that most of Bellmans Creek's reaches, thought to be dominated by the tidal flow regime, are in reality dominated by the riverine flow regime that originates from storm runoff. However, the high spring tide level that commonly occurs at Elevation 5.5—used as the starting water surface elevation for Bellmans Creek—affects the 25- and 100-year flood levels but does not dominate them, nor does the 100-year tide level. The estimated water surface elevations indicated that Conrail's top of rail at 91st Street, at Elevation 7.3, is overtopped by at least 4 feet during a 25-year storm and 5 feet during a 100-year storm as a result of backwater from Bellmans Creek.

### *Existing Hydrologic/Hydraulic Study Data*

Existing hydraulic/hydrologic studies of the area include the FIS performed for the Town of Fairview and for the Hackensack Meadowlands District Commission (HMDC). Even though the Borough of Fairview, the Town of North Bergen and the HMDC have common boundaries and share the Bellmans Creek watershed, the scope of the flood studies performed for each of these political entities was simplified to focus on the perceived dominant flow regime, while no integrated watershed study was performed for Bellmans Creek. Therefore, long reaches of Bellmans Creek have not been studied in detail since their floodplain had not been populated due to frequent inundations that enhanced the wetlands conditions. There has been new commercial development in the last two decades in the Bellmans Creek flood plain. However, even though the flooding concern exist at or near these buildings, no new studies had been performed until now. Hydraulic and hydrologic studies were performed for the Hudson-Bergen LRT project because of the intensive human usage of the corridor via public transportation and potential future development.

The use of existing FISs for design of drainage structures for transportation and other land development projects is common. Projection of flood level estimates, such as those developed in the FIS for the vicinity of the confluence of Bellmans Creek and the Hackensack River to the LRT right-of-way located thousands of feet upstream, is not uncommon. However, the application of this practice to the Hudson-Bergen LRT project was very quickly discarded since it was found to be conceptually incorrect and resulted in a high margin of error. Particularly, the expected intensive human usage of the LRT right-of-way behooved that detailed studies be performed to ensure that adequate drainage and flood protection measures were incorporated into the design.

Previous hydrologic/hydraulic studies provided only a few parameters. The existing FIS provided detailed stage-frequency information for the Hackensack River. It concluded that, in the confluence area, the Hackensack River is dominated by tidal flows. Using the FIS, the 100-year still water tide level was estimated to be at Elevation 9.0 feet NGVD. Other studies performed for the HMDC indicated that the mean high tide level in the area is at Elevation 3.5 and that the spring tide, which commonly occurs in the Hackensack River, is at Elevation 5.5. Therefore, Elevation 5.5 was used as the starting water surface elevation for the confluence with the Hackensack River when performing the backwater computations for Bellmans Creek using the HEC 2 computer program to obtain more realistic flood level estimates.

## Confirmation of the Study Results

As indicated previously, the detailed hydrologic and hydraulic studies performed for the Hudson-Bergen LRT project showed water surface elevations 4 feet and 5 feet above Conrail's top of rail for the 25-year and 100-year storms, respectively. In addition, a field investigation of existing commercial buildings at elevation close to the existing track levels was conducted to confirm the study results . This field investigation provided a definite and full confirmation of the results of the detailed study. The following findings confirmed the study results:

- A vice president of a trucking and storage firm that leases a warehouse located adjacent to the LRT alignment indicated that a flood occurred several years ago in the warehouse and alignment area. An employee who got stranded in the building and crossed the tracks at 91st Street said the water was just above his shoulders. An approximate flood Elevation 12 was estimated for that event.
- The vice president interviewed indicated that 91st Street frequently conveys a large stream during storms.
- An employee of the warehouse showed flood marks on the walls of the warehouse. The floor of the warehouse is raised 4 to 5 feet above grade and the offices are located one story above the floor.
- An employee of the Fairview Department of Public Works showed flood marks on a building located approximately 1,000 feet north of the intersection of the freight tracks and 91st Street, where a survey crew measured the elevation of the highest flood mark. This mark confirmed the information provided by the vice president of the trucking and storage company.

## Summary and Conclusions

The application of an integrated watershed approach for estimating flooding conditions in a detailed hydrologic and hydraulic study of the project site was found to be required after a careful examination of previous studies, topographic maps, and field conditions. This examination revealed indicators for potential significant flooding at the project site. The projection of the results available from the existing FIS to the project site would have resulted in an LRT facility subject to frequent flooding and increased high water levels.

As a result of the hydrologic and hydraulic study, the LRT tracks were designed on a viaduct since a new drainage system that alleviates floods is not practical nor feasible because of the design flood depth of 5 feet, the large peak flow 3,625 cfs, and the backwater from Bellmans Creek. The use of a fill embankment for elevated tracks is not feasible for these reasons. In addition, limited land is available within the right-of-way and an embankment would act as an obstruction and reduce the flow conveyance in the encroached low area, thus exacerbating the potential for flooding.

The detailed hydrologic and hydraulic studies performed for the Hudson-Bergen project drove the designers of the LRT facility to opt for a viaduct structure that protects the passengers, facility, and vehicles from flooding and does not alter the existing drainage conditions.

Short Term Streamflow Forecasting using ANNs

Cameron M. Zealand[1], Donald H. Burn[2], Associate Member, ASCE,
and Slobodan P. Simonovic[3], Member, ASCE.

Abstract
    The research described in this paper investigates the utility of Artificial Neural Networks (ANNs) for short term forecasting of streamflow. The work explores the capabilities of ANNs and compares the performance of this tool to a conventional approach used to forecast streamflow. A number of issues associated with the configuration of the ANN are examined. Some of these issues include the type of input data, the number and size of hidden layer(s), and the composition of the data set. The application of the ANN approach is to a portion of the Winnipeg River system in northwestern Ontario, Canada.

Introduction
    The planning and operation of a water resource system requires forecasts of future events. There is a need for both short term and long term forecasts of streamflow to optimize the system operation or to plan for future expansion. Many water resource systems are large in spatial extent and have a hydrometric data collection network that is sparse resulting in considerable uncertainty in the data that are available. It is thus often difficult to obtain reliable forecasts of future streamflow. As well, the remote location and complex hydraulic relationships of many of the sites contribute to a poor quality streamflow gauging record. There is thus a need for improvements in forecasting techniques.
    The main focus of this paper is on the development of Artificial Neural Network (ANN) models for short term streamflow forecasting, determining which factors have the greatest impact on performance and deriving general methodologies for using these models. Comparisons are made between the

---

[1] Research Associate, Civil and Geological Engineering, University of Manitoba, Winnipeg, MB, Canada, R3T 5V6.
[2] Professor, Civil and Geological Engineering, University of Manitoba.
[3] Professor, Civil and Geological Engineering and Director, Natural Resources Institute, University of Manitoba.

performance of different neural network configurations and a model based on a more traditional forecasting approach.

ANN Characteristics

ANNs differ from traditional forecasting techniques in that they belong to a class of data-driven approaches. Data-driven approaches are suited to complex problems since they have the ability to determine which model inputs are critical. ANNs are relatively insensitive to noisy data as they have the ability to determine the underlying relationship between model inputs and outputs, resulting in good generalization ability. Lorrai and Sechi [1995] verified the possibility of utilizing ANNs to predict rainfall-runoff when only information about the variation of the basic input variables is available. Cheng and Noguchi [1996] obtained better results modeling the rainfall-runoff process with ANNs using previous rainfall, soil moisture deficits, and runoff values as model inputs, when compared with that from a rainfall-runoff model.

ANNs offer valuable characteristics unavailable together elsewhere. First, they infer solutions from data without prior knowledge of the regularities in the data. Second, they learn the similarities among patterns directly from instances or examples of them. Third, ANNs can generalize from previous examples to new ones. Generalization is useful because real-world data is noisy, distorted, and often incomplete. Fourth, ANNs are very good at the abstraction of essential characteristics from inputs containing irrelevant data. Fifth, they are nonlinear and can solve some complex problems more accurately than linear techniques do.

Generally speaking, ANNs map one vector space to another. An input vector is applied to the network and the network produces an output vector. Each vector consists of one or more components, each of which represents the value of some variable (e.g., precipitation, temperature, streamflow, etc.). The architecture of a feed forward ANN can have many layers. The first layer (the input layer) consists of a set of processing elements (PEs) and connects with the input variable(s). The last layer connects to the output variable(s) and is called the output layer. The inputs to neurons in each layer come exclusively from the outputs of neurons in previous layers, and outputs from these neurons pass exclusively to neurons in following layers. Layer(s) of PEs in-between the input and output layers are called hidden layers because they have no direct connection to the outside world. Introducing hidden layers enhances the network's ability to model complex functions.

The PEs in each layer are called nodes. The number of nodes in the input and output layers are dictated by the dimension of input and output vectors presented to the network for training. The number of hidden layers and hidden layer nodes are determined in the network configuration process. Each connection has an associated adjustable parameter called a weight.

Configuring an ANN involves setting the number of input and output nodes to agree with the dimension of input-output vectors in the data set as well as determining the number of hidden layers and the number of hidden nodes per layer. Data analysis helps screen the potential input variables so that only the most telling ones are used to build the training patterns. Preprocessing the data often makes it easier for the network to learn. One important reason for taking care in selecting both the type and number of input/output variables is to keep the network small, so that less time and data are needed to train it.

The main objective of training a network is to produce the desired set of outputs when a set of inputs is fed to the ANN. Training a network is a process during which an ANN passes through a training set (input-output data pairs) repeatedly, changing the values of its weights, according to a predetermined algorithm, to improve its performance. It is important that the training set provide a full and accurate representation of the problem domain; otherwise the network will not meet expectations. Each pass through the training data is called an *epoch*, and the ANN learns through the overall change in weights accumulating over many epochs. This process involves supervised learning in which an output error signal is fed back through the network, altering connection weights, so as to minimize the error between the network output and the target output. During supervised learning, the output predicted by the network is compared with the actual (desired) output, and the mean squared error (MSE) between the two is calculated. The aim of training is to find a set of connection weights that will minimize the MSE.

Application
*Description of Study Area*
The study area consists of a portion of the Winnipeg River system located in the Precambrian Shield region of northwestern Ontario with parts of the watershed in southeastern Manitoba. The Winnipeg River and its main tributary, the English River, have a drainage area of approximately 150,000 square kilometers. The management of the Winnipeg River system is particularly difficult due to interests in control of flooding, water based recreational activities, hydroelectric power generation, agriculture, and water supply. The most upstream lake in the watershed, Namakan Lake, was chosen as the test location in this case study. The Namakan Lake subwatershed has a drainage area of approximately 19,000 square kilometers. There are twenty-nine years of average daily precipitation and temperature values as well as average weekly streamflow values available from 1960-1988 for the Namakan Lake subwatershed. This included data from the period 1965 to 1985 that were used to calibrate the models and the available data before 1965 and after 1985 that were set-aside for model verification.

*Winnipeg Flow Forecasting System (WIFFS)*
The Winnipeg Flow Forecasting System (WIFFS) is a stochastic-deterministic watershed model developed to estimate the quarter-monthly natural inflows into the major lakes in the system. The WIFFS model served as the comparison tool to the ANN method. The WIFFS model contains three basic components: (i) the water input generation model; (ii) the abstraction or loss model; and (iii) the distribution or watershed model. The inputs to the WIFFS model include precipitation and temperature data for the current and the previous seven time periods and an inference of future precipitation and temperature inputs.

The outputs from the WIFFS model are one, two, three or four week(s)-in-advance forecast of inflow to the lake.

*Forecasting Comparison*

Initially, an ANN model was produced to provide a fair comparison between the forecasting accuracy of the WIFFS model and that of the ANN technology. This was done by providing the ANN with the same inputs that are used in the WIFFS model. The ANN model was then trained and tested on the same sets of data that were used to calibrate and verify the WIFFS model. Subsequently, an ANN was built that was not restricted to the input variables used in the WIFFS model. The intent was to build an "optimum" ANN. This "optimum" network involved a smaller number of inputs thus reducing training times. This was done using some of the same inputs as in WIFFS and investigating the use of some new inputs. The ANN model was then trained and tested on the same sets of data used to calibrate and verify the WIFFS model.

In the initial experiment, the number of inputs and outputs were fixed to match those of the WIFFS model. The only part of the network that remained to be configured was the hidden layer. Initial forecasting results indicated that the problem of forecasting streamflow in this river basin could be accomplished with only one hidden layer. The problem then became that of how many nodes would make up this single hidden layer. Cheng and Noguchi [1996] indicate that selecting too many hidden neurons will increase the training time but without significant improvement on training results. Ranjithan et al. [1993] state that the general practice is to determine the number of intermediate units by trial-and-error based on a total error criterion. In this work the trial-and-error process was used.

After training the initial ANN model, two new inputs were added to the data set. The first of these two inputs was the "period of the year" of the forecast. This input provided the network with information on the season in which the forecast was made. The second of the two inputs was the cumulative precipitation from November $1^{st}$ to the time of the current period of the year, up to April $1^{st}$. This input represented a measure of the amount of snowpack that accumulates over the winter and adds to the spring runoff. This input was particularly helpful in accurately forecasting the rising limb of the hydrograph during the spring runoff periods.

With the additional two inputs added to the data set, a sensitivity analysis of all of the input variables was carried out to determine the relative significance of each of the model inputs. The aim of the sensitivity analysis was to delete those inputs that do not have a significant effect on model performance. The sensitivities were used as a guide to decide which inputs to retain and which to delete by applying some degree of judgement.

*Results*

The results of forecasting using the ANN model (with the same inputs used in the WIFFS model) on the test data for forecast lead times from one to four weeks were generally quite good. The ANN model forecasts the magnitude and timing of both the summer peaks and the smaller peaks which occur early and late in the year, quite well. The ANN model forecasts the magnitude of the baseflow very well, but encounters greater difficulty in forecasting the peak flows. Generally, as the forecast lead-time grows from 1 to 4 weeks so does the forecast error. This is expected since these forecasts are beginning to use inferences on the precipitation, temperature and forecasted flow from the previous week(s).

The ANN was able to train to a smaller RMSE than the WIFFS model in all of the four forecast leads. The real test of the ANN was the comparison to the WIFFS model for the testing (verification) periods. The ANN was able to forecast to a smaller root mean squared error (RMSE) than the WIFFS model in all four forecast leads in all eight years of test data. These results are summarized in Table 1, which lists the RMSE for both models and each of the three data sets (one calibration set and two verification sets). The change in the RMSE for the "optimum" ANN model, where the inputs were not constrained to match those used in the WIFFS model, is given in Table 2. The "optimum" ANN was able to train to a smaller RMSE than the initial ANN model for all of the four forecast leads. The optimum ANN was able to forecast to a smaller RMSE than the initial ANN model in all but two of the forecast leads during the 8 years of test data.

**Table 1.** Root Mean Squared Forecast Error

| Period | Forecast Lead (weeks) | RMSE ($m^3/s$) WIFFS | ANN |
|---|---|---|---|
| Calibration | 1 | 36.5 | 35.5 |
| (1965-1985) | 2 | 56.3 | 51.9 |
| | 3 | 69.7 | 68.6 |
| | 4 | 82.3 | 78.4 |
| Verification | 1 | 36.9 | 33.9 |
| (1960-1964) | 2 | 51.3 | 49.4 |
| | 3 | 60.9 | 54.6 |
| | 4 | 68.6 | 58.4 |
| Verification | 1 | 50.4 | 34.1 |
| (1986-1988) | 2 | 78.3 | 52.7 |
| | 3 | 80.1 | 71.7 |
| | 4 | 89.8 | 89.2 |

Conclusions and Recommendations

The ANN model applied to the streamflow forecasting problem achieved encouraging results for the subwatershed under examination. A very close fit was obtained during the training phase and the networks developed consistently outperformed the WIFFS model during the testing phase. The results obtained with ANNs for one through four-week ahead forecasts are better than those reached in the WIFFS model. The initial success of the ANN models developed for the Namakan Lake subwatershed indicates potential for further applications in the Winnipeg River system.

The potential of ANN models for simulating the hydrologic behavior of watersheds has been presented in this paper. The greatest difficulty was determining the appropriate model inputs for such a complex problem. Although ANNs belong to the class of data-driven approaches, it is important to determine the dominant model inputs, as this reduces the size of the network and consequently reduces the training times and increases the generalization ability of the network for a given data set. In the case study considered, sensitivity analyses were used in conjunction with judgement to reduce the number of model inputs that reduced training times and reduced the RMSE for the test data.

**Table 2.** Change in Root Mean Squared Forecast Error for Optimum Model

| Period | Forecast Lead (weeks) | Change in RMSE ($m^3/s$) for Optimum Model |
|---|---|---|
| Calibration | 1 | 1.2 |
| (1965-1985) | 2 | 6.7 |
| | 3 | 7.9 |
| | 4 | 7.5 |
| Verification | 1 | 0.4 |
| (1960-1964) | 2 | -1.8 |
| | 3 | 1.0 |
| | 4 | -1.2 |
| Verification | 1 | 3.5 |
| (1986-1988) | 2 | 5.1 |
| | 3 | 6.3 |
| | 4 | 8.0 |

Acknowledgements

The research reported in this paper was partially supported by a grant from Manitoba Hydro's Research and Development Program as well as an Industrial Post-Graduate Fellowship from the Natural Sciences and Engineering Research Council of Canada (NSERC). This support is gratefully acknowledged.

Appendix - References

Cheng, X. and M. Noguchi (1996). "Rainfall-runoff modelling by neural network approach." Proc. of the Int. Conf. on Water Resour. & Environ. Res. (Vol. II), Oct. 29-31, Kyoto, Japan.

Lorrai, M. and G.M. Sechi (1995). "Neural nets for modeling rainfall-runoff transformations." Wat. Resour. Man., 9, 299-313.

Ranjithan, S., Eheart J.W. and J.H. Garrett, JR. (1993). "Neural Network-based screening for groundwater reclamation under uncertainty", Water Resour. Res. 29(3), 563-574.

Investigation of DeLaine's Method

Daniel J. Schuller, Student Member A.S.C.E.[1]
A. Ramachandra Rao, Member A.S.C.E.[2]

Abstract

An application of DeLaine's (1970) method to data from Indiana watersheds is discussed herein. DeLaine's method provides a procedure to determine unit hydrographs and effective rainfall hyetographs directly from runoff information. An average unit hydrograph was determined for each watershed, and effective rainfall hyetographs were calculated for each storm. The resulting unit hydrographs and effective rainfall ordinates were convolved to compute runoff hydrographs. The computed runoff hydrographs were compared with the measured runoff hydrographs to investigate the utility of the method. Satisfactory results were obtained, thus giving credibility to this methodology for the range of watershed sizes considered in this study.

Introduction

The goal of this study was to derive unit hydrographs for several small watersheds in Indiana. Typically one derives a unit hydrograph by deconvolution, utilizing a runoff hydrograph and an effective rainfall hyetograph (Chow et al. 1988). In this study, a method devised by DeLaine (1970) to compute unit hydrographs by using more than one runoff hydrograph, without recourse to rainfall data, was used. This method also provides effective rainfall hyetographs for each storm. Unit hydrographs were determined for the watersheds, and these were convolved with the calculated effective rainfall hyetographs to calculate runoff hydrographs. To investigate the validity of the method, the calculated runoff hydrographs were compared to the measured runoff hydrographs.

---

[1] Graduate Research Assistant, School of Civil Engineering, Purdue University, West Lafayette, Indiana, 47907-1284

[2] Professor of Civil Engineering, School of Civil Engineering, Purdue University, West Lafayette, Indiana, 47907-1284

## DeLaine's Method

DeLaine's method, which has also been used by Raghavendran and Reddy (1975), is briefly discussed herein. The method relies on the assumption that the rainfall runoff response for a watershed is a linear input-output system. The system is also considered to be time invariant (DeLaine 1970). The watershed as a system can be described by its response, or output, for a unit input. This response is known as the unit hydrograph (for an input of specific duration).

If the input (rainfall), system response (unit hydrograph), and output (runoff) are continuous functions in time $x(t)$, $h(t)$, and $y(t)$, then the output is calculated by the convolution of rainfall and the unit hydrograph:

$$x(t)*h(t)=y(t). \tag{1}$$

Generally, rather than using continuous functions, the rainfall, the unit hydrograph, and the runoff are discretized in the same intervals of time. Rainfall is divided into m parts, the unit hydrograph is divided into n parts, and the hydrograph is divided into (m+n-1) parts. Thus, $x_1, x_2, x_3 \ldots x_m$ are successive ordinates of rainfall (i.e. the rainfall hyetograph), $h_1, h_2, h_3 \ldots h_n$ are successive ordinates of the unit hydrograph, and $y_1, y_2, y_3 \ldots y_{(m+n-1)}$ are successive ordinates of the runoff hydrograph.

The convolution in Equation 1 becomes the following set of simultaneous equations when discrete representation is used.

$$y_1 = x_1 h_1$$
$$y_2 = x_1 h_2 + x_2 h_1$$
$$y_3 = x_1 h_3 + x_2 h_2 + x_3 h_1$$
$$\vdots \tag{2}$$
$$y_{(m+n-3)} = x_{m-2} h_n + x_{m-1} h_{n-1} + x_m h_{n-2}$$
$$y_{(m+n-2)} = x_{m-1} h_n + x_m h_{n-1}$$
$$y_{(m+n-1)} = x_m h_n$$

For DeLaine's method, inputs and outputs are normalized so that the ordinates of x, h, and y add to 1.

$$\sum_1^n h_i = 1 \tag{3} \qquad \sum_1^m x_i = \sum_1^{(m+n-1)} y_i = 1 \tag{4}$$

The runoff hydrograph of a particular storm is the result of convolving the effective rainfall hyetograph of that storm with the unit hydrograph of the basin of interest. The unit hydrograph is assumed to be fixed for a particular basin and; therefore, it is a common component in the calculation of runoff hydrographs for the basin.

Since the data are real, there will be at least one set of real values which satisfy Equations 2, 3, and 4 for each runoff hydrograph which is examined. The number of real and complex sets of solutions which satisfy these non-linear equations is unknown. If one solves Equations 2, 3, and 4 for another runoff hydrograph for the same basin, one will find a second set of solutions. There will be a common set of system response ordinates among the two sets. This is the unit hydrograph. The remaining ordinates from the two sets are the ordinates of the appropriate effective rainfall hyetographs.

DeLaine proposed the following method for performing the above described calculations.

Let
$$x(k)=x_1+x_2k+x_3k^2+ \ldots +x_mk^{m-1} \quad (5)$$
and
$$h(k)=h_1+h_2k+h_3k^2+ \ldots +x_nk^{n-1} \quad (6)$$
thus
$$x(k)h(k)=x_1h_1+(x_1h_2+x_2h_1)k+ \ldots +(x_mh_n)k^{(m+n-2)}$$

combining with Equation 2

$$x(k)h(k)=y_1+y_2k+y_3k^2 \ldots +y_{(m+n-1)}k^{(m+n-2)} \quad (7)$$

Equation 5 is a polynomial whose coefficients are the ordinates of the effective rainfall hyetograph (the effective rainfall polynomial). The polynomial coefficients in Equation 6 are the ordinates of the unit hydrograph (the unit hydrograph polynomial). Equation 7, which is the product of Equations 5 and 6, is a polynomial whose coefficients are the ordinates of the runoff hydrograph. Equation 7 is referred to as the runoff polynomial.

For a given runoff event, a set of roots of the runoff polynomial will consist of the roots of the effective rainfall polynomial and the roots of the unit hydrograph polynomial. For two runoff events on the same basin, one can develop runoff polynomials and calculate their roots. Next, the two sets of roots are compared. Those roots which are common to both sets are the roots of the unit hydrograph polynomial. Those which do not match are the roots of the respective effective rainfall polynomials. From the matching and non-matching roots, the unit hydrograph polynomial and the effective rainfall polynomials are back calculated, and the unit hydrograph ordinates and the effective rainfall hyetograph ordinates for each storm are determined by inspecting the polynomials.

When using DeLaine's method, it is advantageous to use more than two storms to facilitate improved root matching. Otherwise, similar roots due to similar rainfall components in the two storms can be mistaken for matching roots.

Description of Data Used in the Study

Data from several Indiana watersheds were examined in this study. Watersheds from around the state were selected in an attempt to examine the effects of varying topography, soil type, and land cover on the results. All the watersheds are predominately rural. The watersheds cover a range of areas from 7.8 square kilometers to 152.0 square kilometers. A range of areas was chosen in order to examine the effects of watershed size on the applicability of DeLaine's method. For each watershed, hourly streamflow data were obtained from the United States Geological Survey (U.S.G.S.) for the 1994 and 1995 water years.

Results

The complete results of this study are found in the report by Schuller (1997). For the sake of brevity, the results of application of DeLaine's method to data from two watersheds are presented here. In both cases, two-hour runoff data were used, thus reducing the number of roots by one half and making root matching a more reasonable task. Runoff hydrographs for four storm events were examined in each case.

The Galena River watershed is located near LaPorte in northwest Indiana. The watershed area is 38.6 square kilometers. The average unit hydrograph for the basin calculated by DeLaine's method is shown in Figure 1. Effective rainfall hyetographs were determined for the four storms on the Galena River watershed. These were convolved with the unit hydrograph to determine the runoff hydrographs. The calculated and measured runoff hydrographs for the four storms under investigation are compared in Figure 2.

The Galena River 2-hour unit hydrograph resembles a "textbook" unit hydrograph. The calculated and measured runoff hydrographs for Storms 1 and 2 are well matched. Those for Storms 3 and 4 do not match as well, although they are satisfactory.

The Whitewater River watershed is 152 square kilometers and is located near Hagerstown in east-central Indiana. The average unit hydrograph calculated by DeLaine's method is shown in Figure 3. Effective rainfall hyetographs were determined for the four storms on the Whitewater River watershed. These were convolved with the unit hydrograph to compute the runoff hydrographs. The calculated and measured runoff hydrographs for the four storms under investigation are compared in Figure 4.

WATER RESOURCES AND THE URBAN ENVIRONMENT 239

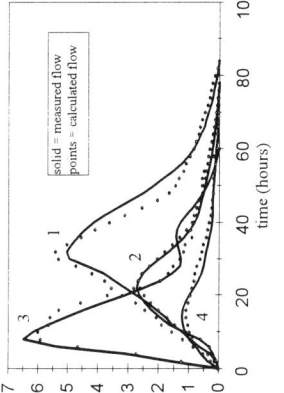

Figure 2 - Comparison of Measured and Calculated Runoff Hydrographs for Four Storms on the Galena River

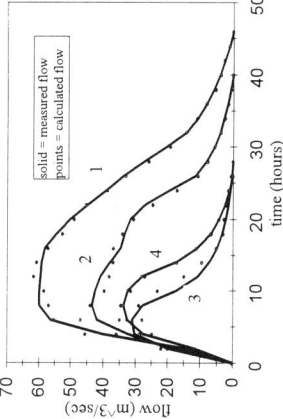

Figure 4 - Comparison of Measured and Calculated Runoff Hydrographs for Four Storms on the Whitewater River

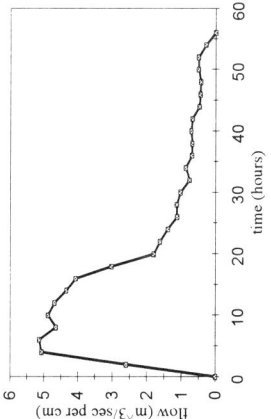

Figure 1 - Galena River Average 2-Hour Unit Hydrograph

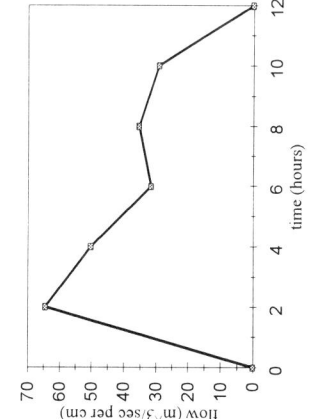

Figure 3 - Whitewater River Average 2-Hour Unit Hydrograph

The 2-hour unit hydrograph for the Whitewater River has a high peak followed by a slight hump on the recession limb. For Storms 1 and 4, convolution of the rainfall hyetographs with the unit hydrograph produces runoff hydrographs very similar to the measured runoff hydrographs. For Storm 2, the peak of the calculated runoff hydrograph has a slightly different shape than, but nearly the same magnitude as, the measured runoff hydrograph. The peak of the calculated runoff hydrograph for Storm 3 has a lower peak which occurs slightly later than that of the measured hydrograph.

Summary and Conclusions

DeLaine's method for the calculation of unit hydrographs and effective rainfall hyetographs from runoff data appears to produce satisfactory results for small Indiana watersheds. In these cases, the generated runoff hydrographs match the measured runoff hydrographs quite well. It is interesting to note that unorthodox unit hydrographs were often obtained, but when they were convolved with the computed effective rainfall, quite "normal" runoff hydrographs resulted. The more defined the original runoff hydrographs (i.e. the more ordinates on the rising and falling limbs), the better the results.

It should be noted that DeLaine's method has drawbacks. Due to errors in the hydrographs, finding matching roots may not be easy. It is an iterative process in which the matching and non-matching roots are determined first. The unit hydrographs and rainfall hyetographs are then calculated. If, upon examining the results, unit hydrographs and hyetographs have negative ordinates, the lists of matching and non-matching roots are revised, and the method is tried again. Although it is a time consuming process, reasonable results can be obtained.

References

Chow, V.T., D.R. Maidment, and L.W. Mays. Applied Hydrology. New York: McGraw-Hill, 1988.
DeLaine, R.J. "Deriving the Unitgraph without Using Rainfall Data." Journal of Hydrology 10 (1970): 379-90.
Raghavendran, R., and P.J. Reddy. "Synthesis of Basin Response with Inadequate Data." Nordic Hydrology 6 (1975): 14-27.
Schuller, D.J., Application of DeLaine's Method to Indiana Watersheds. Civil Engineering Report #CE-EHE-97-6. West Lafayette, IN: Purdue, 1997.

Using Hydraulic Modeling for System Improvements:
State College Borough (PA) Water Authority

Mark V. Glenn[1], P.E., M.ASCE
Eric A. Casanave[2], E.I.T., M.ASCE

## Abstract

The State College, PA area of Central Pennsylvania has experienced significant growth in recent years. Home to Penn State University, the area must provide infrastructure for increased student and residential housing, and rapid retail/commercial construction. The State College Borough Water Authority ("SCBWA") faces the challenge of maintaining a water supply system capable of furnishing potable water and adequate fire protection capabilities to all portions of this expanding service area.

To keep pace with the growth, the SCBWA in previous years installed distribution lines at the extremities of the system, as needed. These connected to smaller, existing lines originally intended to serve outlying areas nearly 40 years ago. Eventually, these "outlying" areas became the primary demand centers of the system. Although this practice opened water supply to new areas, fire protection, tank filling, and system pressure problems began to manifest themselves. As the area grew and demand increased, the SCBWA recognized a need to develop a long term system improvement plan.

To develop such a plan, the SCBWA in 1996 retained the services of its consulting engineer to perform a system-wide hydraulic computer model of the entire system and develop a master plan. This paper presents the results of the modeling effort and discusses the benefits of hydraulic computer modeling as a tool for both short and long term water system management.

---

[1]President - Gwin, Dobson, & Foreman, Inc., Consulting Engineers, 3121 Fairway Drive, Altoona, PA 16602

[2]Civil Engineer - Gwin, Dobson, & Foreman, Inc., Consulting Engineers, 3121 Fairway Drive, Altoona, PA 16602

## Introduction

The system evaluation was intended to develop a plan and program to boost low pressure areas, enhance system redundancy between pressure zones, reinforce the fire protection grid, provide a baseline model for system extensions, plan water system renewals and replacements, and, in general, allow the system to keep pace with the residential and commercial growth. The Cybernet© 2000 computer program was used to model the water system.

The State College system has sufficient water supply capacity for its existing customer base (50,000 connected population serving State College Borough and six (6) surrounding townships). Because of the system's age (75-100 years), however, some high demand areas are served by undersized distribution lines. At one time, these pipes were sufficient but have not kept pace as the system grew. Consequently, the system experiences low available fire flows and difficulties maintaining storage tank levels, which combined to limit growth and development. These factors, combined with the recent addition of several water sources, all necessitated a systematic plan of long term improvements.

The SCBWA distribution/storage system consists of approximately 290 kilometers (180 miles) of transmission and distribution piping and 37,850 cubic meters (10 million gallons) of active distribution storage, spread among eight storage tanks. The distribution network is divided into five pressure zones (Zones 1, 3, 4, 5, and 8), each served by at least one tank.

Five (5) groundwater sources (well fields) serve three of the zones. Four pump stations transmit water between zones. In addition, a 13,300 cubic meter (3.5 MG) Water Treatment Plant, via a high service pump station, supplies treated water to the system. The plant processes water from the Shingletown Reservoir and two of the well fields.

## Distribution System Evaluation

The major study activities undertaken were a computerized hydraulic network analysis, a fire flow analysis and a detailed analysis of tank filling problems in Zone 5.

The initial phase of the study involved refining and updating the Cybernet© AutoCAD map of the distribution system. A macro-level calibration (Ormsbee and Lingireddy, 1997) was performed to establish a working model suitable for large scale modeling purposes. At the outset, most of the modeling focused on solving the tank filling problem in Zone 5.

The next phase involved a detailed calibration effort, the development of a "needed fire flow" map, and an "available" fire flow analysis using Cybernet©. Using the calibrated model, areas of low fire flow were identified, and various improvement alternatives were evaluated. System evaluations for proposed subdivisions were considered throughout the modeling process as well. In some cases, these subdivision evaluations resulted in improvement recommendations.

The Cybernet computer program uses an iterative method to determine pipe flow rates and pressures at pipe junctions (Cybernet© Users Manual, 1992). The calculation method is based on satisfying both the law of conservation of mass (which states that the flow of water into a pipe junction must equal the flow out of the junction, including any water added to or removed from the system at the junction), and the law of conservation of energy (in which the sum of the head losses through a closed loop of pipes must equal zero and the pipe head losses between two constant head sources must equal the difference in head between the two sources).

Cybernet© requires input data for pipes, nodes, loops, valves, and pump and operational characteristics. A node is a junction of two or more pipe segments, the end of a dead end pipe segment, or a water source. Water withdrawals from the system are assumed to occur only at the nodes, not along pipe segments. The data required for each node consists of ground elevations and fixed hydraulic grade line elevations or the water demand. A pipe is the conduit connecting two nodes and is defined by the starting and ending node numbers, length, diameter, Hazen-Williams C factor and a minor loss coefficient.

The Cybernet© network for the existing SCBWA system consists of approximately of 800 nodes, 1,100 pipes and 300 loops. The proposed system contains approximately 1,300 pipes, 860 nodes and 430 loops.

## Calibration

After the network was mapped, it was calibrated using actual fire hydrant flow, and static and residual pressure test data (collected by SCBWA personnel from April to July, 1996), and water consumption data from January 1994 through May 1996. These flow and pressure conditions were matched to system source send-out and storage tank boundary conditions existing at the time the hydrant tests were conducted. Both a static pressure and fire flow calibration were performed.

System demands were developed using the actual SCBWA billing records. For billing purposes, the system is divided into approximately 250 billing zones. Demands for each node were derived by distributing the total usage on record for a particular billing zone among all the junction nodes in that billing zone. Areas of high demand were located also. Accurately locating the high-demand areas greatly aids in the calibration of a computer model (Walski, 1983).

Since system conditions (tank levels, pumps on/off, well fields on/off, etc.) varied as the tests progressed, hydrant tests in each zone were sorted chronologically and the tank level, pump and well field conditions matching each test were determined using chart records provided by the SCBWA. All hydrant nodes having similar operating conditions were grouped together. Individual computer simulations focusing only on these groups were conducted and pipe roughnesses and demands were adjusted until calculated system pressures at most junction nodes (matched to their respective hydrants) agreed with the measured hydrant static pressures within 5 psi.

Following the static pressure calibration, the model was further refined by calculating available fire flows for the same residual pressure as in the hydrant tests. These runs were performed for the same groups of nodes used in the static pressure calibration. Again, pipe roughnesses and demands were adjusted until most calculated fire flows were within 300 gpm of the actual measured flow. The model successfully represented areas of high and low fire flow when matched to the actual test records.

## Modeling Strategy

Each pressure zone was modeled as an independent system. System models for both average daily and peak daily demand were performed using typical tank, pump and well field levels and flows. Ongoing capital construction projects were accounted for using a "provisional" existing system as the base model that assumed the completion of these projects.

Needed Fire Flows (NFFs) for the SCBWA system were calculated based on the Insurance Services Office (ISO) method described in the "Fire Suppression Rating Schedule," (Insurance Services Office, 1980). Where possible, NFFs were obtained directly from the ISO reports (conducted by the ISO in 1996) for State College Borough, and surrounding townships. For areas not covered by these reports, zoning maps were consulted before field viewing these areas.

For the purpose of analyzing the system fire protection availability, the available fire flows were calculated at peak demand conditions. Peak demands were assumed to be twice the average daily demand. Demands representing future build-out in certain areas were used in both the existing and proposed analyses, as recommended by the American Water Works Association (AWWA Manual No. 31, 1989). Future demands were developed in accordance with projections of the Centre Region Planning Commission and from information on projected water demands of subdivisions provided by the SCBWA.

## Recommendations

Water system recommendations were analyzed using the computer model and grouped into three types: system component and undersize main replacements, fire protection grid improvements, and grid reinforcement/redundancy installations. The recommendations were prioritized based on the extent of flow improvement achieved but also some flexibility -- line replacements and installations may be coordinated with future municipal road work and development.

In many areas of the system, undersize main replacements and fire protection grid improvements coincided. Grid reinforcement recommendations were developed to reduce overall system head losses. These were directly related to the age and size of the distribution pipes (in some cases 100 and 152 mm (4" and 6")), and also to the increased demands associated with the system growth. The losses induce low fire flows and system pressures (especially at times of high demand) and may, as in the case of Zone 5, preclude the use of storage tanks at their full capacity.

Redundant connections were modeled and recommended, with appropriate controls, to allow the Authority to efficiently transmit water from any tank or well field to any other tank or load center regardless of pressure zone. Redundancy is important in cases such as well contamination, pump station failure, or general maintenance which require one or more sources to be isolated without seriously jeopardizing water supply for the rest of the system. Redundancy also is important in times of extreme demand loadings where the ability to rely on "backup" sources of water is essential.

The total project cost of the improvements recommended by this study was approximately $11,350,000.

## Conclusions

Computerized water distribution network models allow the development of long-term master plans. Water authorities, regional planning agencies, and municipalities use them a basis for scheduling both regular maintenance and capital construction projects. A well calibrated model enables a user to locate areas of low available fire flow, predict the influence on the system of a future water source, or simply determine whether or not the existing system is capable of supporting a proposed subdivision or large retail establishment.

Many different improvement alternatives can be evaluated, a process which ultimately results in an effective yet cost-efficient (AWWA Manual 32, 1989) set of installation and replacement recommendations. These can be implemented over the long haul and can be factored into future maintenance and capital construction budgets of the authorities. When coordinated with local planning agencies, proposed system components can be strategically located to encourage development and also enable authorities to share the cost of line installations with developers, thereby reducing overall costs to the authority.

## References

1. American Water Works Association (1989). "Distribution System Requirements for Fire Protection," Manual No. 31, 1st. ed., AWWA, Denver, CO.

2. American Water Works Association (1989). "Distribution Network Analysis for Water Utilities," Manual No. 32, 1st ed., AWWA, Denver, CO.

3. Haestad Methods, Inc. (1992). "Cybernet Version 2.0 User's Guide," Haestad Methods, Inc., Waterbury, CN.

4. Insurance Services Office (1980). "Fire Suppression Rating Schedule," Insurance Services Office, Marlton, NJ.

5. Ormsbee, L.E. and S. Lingireddy (1997). "Calibrating Hydraulic Network Models," *Journal of the American Water Works Association*, Vol. 89, No. 2, February, 1997, p.42-50.

6. Walski, T.M. (1983). "Water Distribution Infrastructure Analysis Techniques," Continuing Education Services.

# NON-LINEAR PARAMETER ESTIMATION OF AN URBAN RUNOFF MODEL USING XP-SWMM32 AND PEST

Tai Ovbiebo[1] (M. ASCE) and Anthony W. Kuch[2]
Tai.Ovbiebo@ci.seattle.wa.us   Info@xpsoftware.com

## ABSTRACT

This paper presents a case study of utilizing XP-SWMM32's Automatic Calibration module to calibrate an urban runoff model. In this study a $1.5 million Capital Improvement Project in the Lake City drainage basin is evaluated using the XP-SWMM32 stormwater model. Using the software tools the SWMM model is reliably calibrated in a shorter time than would have been possible using the typical manual, trial and error approach. The software tools presented in this paper allow the modeler to concentrate on initial data preparation and reasonable parameter bounds rather than on tedious model adjustment.

At the core of the Automatic Calibration module is a parameter optimization software program PEST (an acronym for Parameter ESTimation). The calibration software utilizes the models own input and output files and automates the creation of the control files for calibration. The model is calibrated to a multiple storm rainfall event with a duration of two days. Pre-calibration and post calibration plots of computed versus observed hydrographs are presented. The verification by substituting the calibrated parameter set to three other similar events is shown using a comparative plot of all hydrograph peak flows.

## 1.0 INTRODUCTION

The Lake City drainage basin, located in the Northeast area of the City of Seattle receives stormwater runoff from approximately 865 acres. The land use is mostly high density residential with pockets of commercial activity. The surface water drainage system is comprised mostly of ditches and culverts. Stormwater is discharged into Lake Washington on the southeast corner of the basin through a 72-inch diameter concrete storm drain connected to a 90-inch diameter concrete tunnel outfall. In order to reduce local flooding in the area, a Capital Improvement Project was initiated in early 1993. The first phase alleviated flooding by construction of a storm drain along 33rd Avenue North East to divert flow away from the low spot upstream, approximately 600 feet from North East 125th Street intersection. The second phase of the project: 1) opened up the culvert sections of the creek south of North East 125th Street to provide additional capacity, and 2) included construction of a detention pond for storage of peak flows and eventual release into the existing creek system.

XP-SWMM32 was selected as the model for this study for four reasons:
- to analyze system hydrology and hydraulics in one common model,
- to predict flow variations resulting from both past and future storms,
- to process data to aid in the design and construction of chosen alternative solution, and
- to automate most of the calibration effort and the processing of files for PEST.

The first step taken in developing a system representation of the drainage basin for the computer model was to delineate the drainage boundaries based on the topography. The basin was divided into 27 subcatchments of sizes ranging from 7 acres to 56 acres. The ground slopes for most of the basin are relatively flat to moderate slopes (2 to 18 percent) with only about 15 percent of the basin having steeper than 10 percent slopes. Numbered link and node objects represented all major drainage conduits (pipes 12-inches or bigger) and Maintenance Holes. Flow monitor locations were identified at specific nodes over

---
[1] Project Engineer, Seattle Public Utilities, Resource Development Division, Seattle, WA 98104
[2] Software Applications Engineer, XP Software Inc., 5553 West Waters Avenue #302, Tampa, FL 33634

several months during the pre-development and post-development periods. Following development of the system schematic, input parameters representing the physical characteristics of the subbasins, conduits and maintenance holes were compiled and imported into the model from GIS through standard CSV (comma separated variable) files.

## 2.0 MODEL DESCRIPTION

XP-SWMM32 is a link node model encompassing the capabilities of EPA's Stormwater Management Model SWMM, new algorithms and a graphical user interface. The SWMM methodology for the generation of overland flow required the following input data; Manning's roughness, area, width, ground slope, percent imperviousness, depression storage, Horton infiltration parameters and rainfall for each subcatchment (Huber and Dickinson, 1988). 10-minute interval rainfall data for both actual and synthetic storms obtained from the City rain gauge stations and NOAA meteorological data for the Seattle area was used for the model calibration and verification. Calibration of the model for the basin was necessary in order to size and evaluate the performance of both the conveyance and detention systems.

XP-SWMM32 contains an optional interface for automatic calibration. This calibration module expands the software capabilities and allows the user to:

1. Store and plot observations with modeled output.
2. Prepare, edit, and verify external files required in the calibration process.
3. Select and import parameters that will be modified during the calibration.
4. Launch, control and review the model calibration.

At the core of the calibration plug-in is PEST98 (an acronym for Parameter ESTimation) a computer program developed by Watermark Computing. This version of PEST allows parallel processing across a network by utilizing the concept of slave computers to calculate the sensitivity gradients of parameters. The use of multiple computers is advantageous especially with long model run times or a large number of adjustable parameters.

PEST is a nonlinear parameter estimator that uses powerful mathematical techniques. PEST communicates with the model through the model's own input and output files. It implements a particularly robust variant of the Gauss-Marquardt-Levenberg method of parameter estimation (Doherty, 1994). Its innovative optimization algorithm includes a sophisticated derivative calculator that allows successful parameter estimation even for large numerical models such as an entire City watershed model. PEST works by running the model as many times as needed in order to find that parameter or excitation set for which the discrepancies between model-calculated numbers and their field or laboratory counterparts are as small as possible in the weighted least-squares sense. PEST's unique ability to interface with a model through the model's own input and output files adds enormous flexibility to the calibration or interpretation process in a study.

## 3.0 CALIBRATION INTERFACE

The selection of model input parameters to be adjusted during calibration are identified within the XP-SWMM32 interface. Additionally, the model-generated numbers with corresponding field measurements are identified on the model output file. PEST reads these through the use of instruction files and compares these values to observations such as flow gauge records. The optimization process requires repeated adjustment of parameters and preparation of input files for SWMM. PEST accomplishes this by using a template for the SWMM model input. The creation and editing of these files is handled in the Calibration Module interface. The recasting of new inputs to the analysis engine is repeated until the discrepancies between output and field measurements are reduced to a minimum in the weighted least-squares sense. Although most of calibration is automated the user can interrupt the process to evaluate progress, make changes to calibration parameters and continue execution. In calibrating the model, care was taken to resist attempts to estimate too many parameters and avoid increasing correlation between parameters (Kuch, 1997). The minimization of parameters was accomplished by: 1) performing a sensitivity analysis before calibration and selecting only sensitive parameters, 2) omitting physical parameters with low measurement error, 3) examining the SWMM equations and selecting only one of the correlated runoff parameters, and 4) grouping common variables for all subcatchments

## MODEL CALIBRATION ALGORITHM

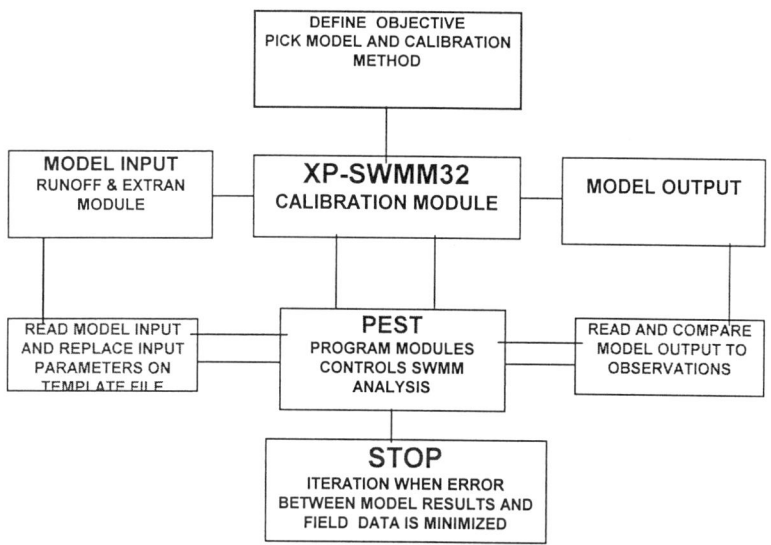

Figure 1. Flow diagram of the calibration process.

5.0 CALIBRATION AND VERIFICATION
To calibrate the model, rainfall data from the City's rain gauge station number 2 located in the watershed was used. Multiple peaked storms were selected from the rainfall TS to correspond with the measured flows from the ISCO brand flow monitoring station number 37. The flow monitor was installed for data collection over a period of 5 years in a maintenance hole represented as node 495 in the model. Four types of storms were used:
- a 25-year SCS synthetic storm was used for initial model set-up;
- the storm of January 19-21, 1993, a 48-hour storm for calibration; and
- the storms of April 16-18, 1992 and October 29-31, 1992 were used for verification.

6.0 CALIBRATION RESULTS
The Calibration involved 96 parameters in 9 groups. The model was adjusted to fit 79 hydrograph points at ½ hour intervals throughout the 46-hour simulation period. All parameters of each group were adjusted equally by being tied to the first, hence when the subcatchment width was adjusted for one subcatchment an equivalent adjustment to the width values of all subcatchments was applied. The following parameters were selected to be optimized: impervious depression storage (WSTOR1), two unique values for pervious depression storage(WSTOR2), percent of the impervious area without depression storage

(PCTZER), maximum infiltration rate Horton (WLMAX), minimum infiltration rate Horton (WLMIN), decay of infiltration rate (DECAY), subcatchment width (WIDTH) and impervious area percentage (IMP). Table 1 shows the starting and final values of each of these parameters.

Table 1: Calibration parameter bounds, starting values and calibrated values.

| Parameter Name | Parameter Bounds | Starting Value | Calibrated Value |
|---|---|---|---|
| WSTOR1 | 0.0 to 1.0 inches | 0.014 inches | 0.06228 inches |
| WSTOR2 | 0.0 to 1.0 inches | 0.3 inches | 0.3 inches |
| WSTOR2 | 0.0 to 1.0 inches | 0.1 inches | 0.1 inches |
| WLMAX | 1.0 to 4.0 inches | 2.0 inches/hr | 2.0 inches/hr |
| WLMIN | 0.05 to 1.0 inches | 0.2 inches/hr | 0.2 inches/hr |
| DECAY | 0.0 to 0.100 $s^{-1}$ | 0.00115 $s^{-1}$ | 0.00115 $s^{-1}$ |
| WIDTH | 80 to 10,000 ft | 374 to 2571 ft | 343.1 to 2358.7 ft |
| IMP | 1% to 99% | 5.6% to 47.5% | 3.96% to 35.35% |
| PCTZER | 3.0% to 50.0% | 25.0% | 3.0% |

For all model runs and throughout the calibration process the model generated no overland flow from pervious areas. Since no overland flow was generated on pervious areas, they were not sensitive parameters (i.e. small changes in these parameters generated no change in output). Only WSTOR1, PCTZER, WIDTH, and IMP were adjusted from their original values as shown in Table 1. For all subcatchments WSTOR1 and PCTZER were adjusted from 0.014 and 25 to 0.0628 and 50 respectively. In the case of PCTZER this was the upper bound of the parameter. The value of WIDTH and IMP differed for each subcatchment but the same adjustment factor was applied to each. At the completion of the calibration WIDTH and IMP decreased by 8.3% and 29.3% respectively from their initial values.

The total sum of weighted squares residuals (error in the fit between the computed hydrograph and corresponding gauged measurements) was 10,921. It was reduced to 2037 or less than 1/5 of the pre-calibration value. A total of 56 model calls were require to calibrate the model and took approximately 1½ hours. Most of the model calls were for generating sensitivity gradients. At each optimization iteration (there were 4) 9 runs were performed for each of the nine adjustable parameter groups. The last iteration required central derivatives or two iterations per parameter group. A total of 10 out of the 56 model runs were used as estimations based on the sensitivity gradients. Figure 2 and 3 show the degree of fit before and after the calibration. Verification followed calibration and the calibrated parameter set was then used with three other storm events. A plot of all of the hydrograph peaks and gauged peaks is shown in Figure 4. Most of the average peak values lie within a 25% error range.

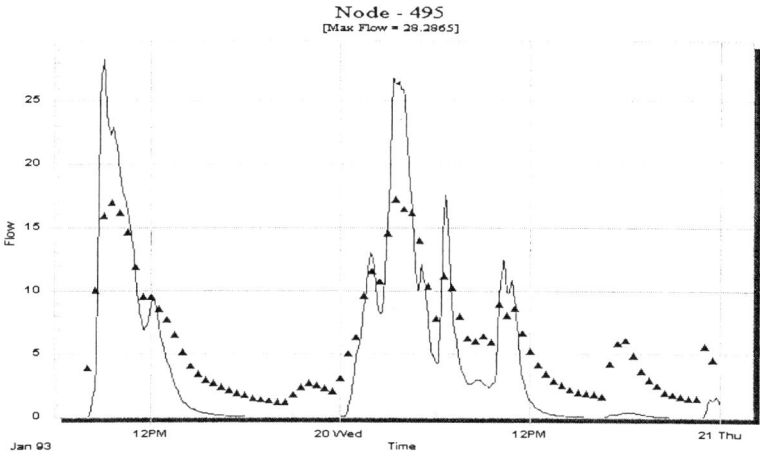

Figure 2: Comparison of observed model flows before calibration Jan 19, 1993.

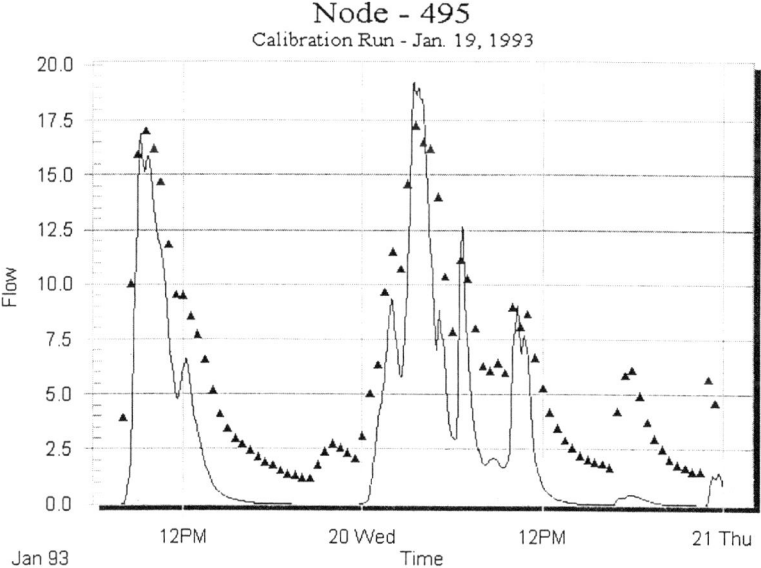

Figure 3: Comparison of observed and calibrated flows.

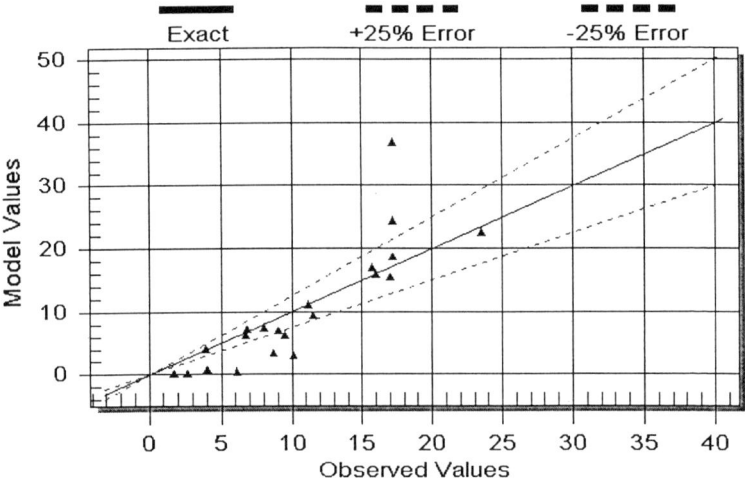

Figure 4: Plot of observed and model hydrograph peak flows for all three storms.

7.0 CONCLUSION
Through the PEST model independence and superior algorithm and the Automatic Calibration Module's user-friendly pre-processors and other features, the task of calibrating the Lake City drainage basin models has been straightforward. Estimating many parameters at once which in the past is almost impossible due to complex dependencies of model outputs on parameters appears to have been made easier (Ovbiebo, 1995.) The PEST methodology represents a quantum leap in the daily problems encountered by modelers from all fields of science and engineering by facilitating the development of innovative and powerful parameterization strategies. Approaching calibration from this point of view recognizes this fact, and implicitly considers various operating conditions.

8.0 LIST OF REFERENCES
Doherty, J. (1994) "Computer Program for Model Independent Parameter Estimation (PEST)", Watermark Numerical Computing, Australia.

Kuch, A.W., (1997) "Sensitivity Analysis and Calibration Decision Support Tools for Continuous Modeling with the Storm Water Management Model" Masters of Science Thesis Presented at the University of Guelph, Guelph, Ontario, Canada.

Huber, W. C. and Dickinson, R. E., (1988) "Storm Water Management Model User's Manual Version 4, EPA/600/3-88/001a (NTIS PB88-236641/AS), Environmental Protection Agency, Athens, GA.

Ovbiebo, T. and She, N., (1995) "Urban Runoff Quality and Quantity Modeling in a Subbasin of the Duwamish River Using XP-SWM" Proceedings of the First International Conference by Water Resources Engineering Division of American Society of Civil Engineers, American Institute of Hydrology, Canadian Society for Civil Engineering.

## Inverse Chlorine modeling in pipe Networks

By Al - Omari, A. S.[1], and Chaudhry, M. H.[2]

**Abstract**

A model that estimates source concentration and chlorine decay coefficient for steady state hydraulic conditions utilizing inverse method is developed. The model considers even - determined, over - determined, and under-determined cases. The model is then applied to an example network from literature. From the computed results, it is concluded that best estimates are obtained for the over-determined case. The under - determined case does not give a good estimate most of the time. Rather, it gives an approximate estimate of the parameter. It is also concluded that the location of the measurement is important in obtaining an accurate estimate of the parameter.

**Introduction**

Water quality, though acceptable as water leaves the treatment plant, deteriorates in the distribution system due to loss of system integrity, chemical and biological transformations and blending of water from different sources (Males, R. M. et al., 1988). In mid 80's, models that simulate water quality in distribution systems were developed. These models predict water quality in the distribution system as a result of mixing of water of different qualities from different sources (Males, M. et al., 1985) and/or simulates water quality under steady and unsteady state hydraulic conditions (Islam, R. M, 1995).

---

[1] Graduate student, Department of Civil & Environmental Engineering, University of South Carolina, Columbia, S. C. 29208.

[2] Chairman and Mr. & Mrs. Irwin B. Khan professor, Department of Civil & Environmental Engineering, University of South Carolina, Columbia, S. C. 29208.

One important measure of water quality in distribution systems is chlorine residual. Water authorities are required to maintain minimum chlorine residuals throughout distribution systems for disinfection. However, maintaining this residual is not an easy task due to the complexity of distribution systems, chlorine decay with time and the formation of chlorine by products like Trihalomethanes (THMs) which are carcinogenic. Several models that simulate chlorine distribution in networks have been developed (Murphy, S. B., 1985; Musa, M., 1991; Islam, R. M., 1995). The model developed by Islam simulates chlorine distribution in networks under steady and unsteady state hydraulic conditions

The model described here utilizes inverse method to estimate chlorine decay coefficient for the desired pipes in the network and to estimate the source/sources concentration to meet the desired chlorine concentration at pre-specified nodes. Even - determined, over - determined and under - determined cases are considered. Since flow simulation is a pre requisite for water quality simulation, the model first simulates the flow under steady state conditions and determines the steady state chlorine distribution in the network. Then, it utilizes the inverse method to estimate chlorine decay coefficient for the desired pipes and the source concentration for the desired sources.

### Steady state hydraulic simulation

The steady state hydraulic simulation is obtained by solving the continuity equations along with the loop equations utilizing the linear theory as described by Wood, D. J. and Charles, C. O. (1972).

### Ultimate chlorine distribution

The ultimate chlorine distribution is obtained by solving the following equations:

1- The Steady state one dimensional advection dominated flow as applied to a pipe

$$Cd_i - Cu_i \, e^{-K_i t_i} = 0 \quad \dots\dots\dots\dots\dots\dots\dots\dots\dots\dots\dots\dots\dots \; 1$$

where $Cd_i$ is chlorine concentration at the downstream end of pipe i, $Cu_i$ is chlorine concentration at the upstream end of pipe i, $K_i$ is chlorine decay coefficient for pipe i, and $t_i$ is hydraulic residence time for pipe i.

2- The upstream concentration of the pipe equals the concentration at the upstream node.

$$Cu_i - CN_j = 0 \quad \dots\dots\dots\dots\dots\dots\dots\dots\dots\dots\dots\dots\dots\dots.2$$

where $CN_j$ is the concentration at node j.

3- The node concentration equals the weighted average of all the downstream concentration of the pipes that meet at that node.

$$CN_j - \Sigma Q_i Cd_i / \Sigma Q_j \quad\quad\quad\quad\quad\quad\quad\quad\quad\quad\quad\quad\quad 3$$

Where, $Q_i$ is the flow of pipe i.

## Determination of the source concentration utilizing inverse method

### Even - determined problem

In an even - determined problem, the number of the pre-specified nodal concentrations equals the number of unknown source concentrations. The governing equations are equations 1, 2 and 3, which govern the forward chlorine distribution. The only difference is that a number of nodal concentrations are known and an equivalent number of source concentrations are unknown. The equations are linear in the unknowns and can be solved by any matrix solver.

### Over - determined problem

In an over - determined problem, the number of the pre-specified nodal concentrations is more than the number of the unknown source concentration. This means that the number of available equations becomes more than the number of unknowns; in other words the problem is over determined. In this case, there is no solution that can satisfy all the equation at the same time. The solution to this problem is obtained by minimizing the difference between the measured and the calculated concentrations in the least square sense.

### Governing equations

The governing equations for this case are equations 1, 2, 3 and the following equation

$$\text{Minimize } E = \Sigma (CN_j^c - CN_j^s)^2 \quad\quad\quad\quad\quad\quad\quad\quad 4$$

Where, $CN_j^c$ is the calculated node concentration and $CN_j^s$ is the pre-specified nodal concentration.

By taking the partial derivative of equation 4 with respect to every unknown source concentration and equating that to zero, M linear equations are obtained where, M is the number of measurements. When combined with the 2NP + NN- M linear equations obtained from 1, 2 and 3, 2NP + NN equations are obtained, where NP is the number of pipes and NN is the number of nodes. The number of unknowns now equals the number of equations, which can be solved by any matrix solver.

### Under - determined problem

In an under - determined problem, the number of the pre-specified nodal concentrations is less than the number of unknown source concentrations. This makes the number of available equations less than the number of unknowns, hereby making the problem under - determined. In this case, more than one solution of the problem is possible. The solution to this problem is obtained by minimizing the length of the solution vector as follows.

$$\text{Minimize } [\Sigma (Co_i)^2]^{1/2} \quad \ldots\ldots\ldots\ldots\ldots\ldots\ldots 5$$

Where, $Co_i$ is the unknown source concentration and L is number of unknown source concentrations. This problem can be viewed as a constrained minimization problem, which can be solved by utilizing Lagrange multiplier as follows.

Minimize $[\Sigma (Co_i)^2]^{1/2}$

Subject to equations 1, 2 and 3.

$$\text{Define } F = [\Sigma(Co_i)^2]^{1/2} + \lambda \, [\Sigma \, (Cd_i - Cu_i \, e^{-Kit_i}) + \Sigma \, (Cu_i - CN_j) + \Sigma \, (CN_j - \Sigma \, Q_i \, Cd_i / \Sigma \, Q_j)] \ldots\ldots 6$$

$Cu_i$, $Cd_i$ and $CN_j$ are expressed in terms of the unknown source concentrations. The number of unknowns becomes NR + 1 where NR is the number of the unknown source concentrations. By differentiating eq. 6 with respect to every unknown source concentration, NR linear equations are obtained. By differentiating equation 6 again with respect to $\lambda$ another linear equation is obtained. The number of equations now equals the number of unknowns, which can be solved by any matrix solver.

### Determination of chlorine decay coefficient by utilizing inverse method

The determination of chlorine decay coefficient can be done by the same manner as the determination of the source concentration exactly except that the equations are nonlinear and have to be solved by an iterative procedure.

### Model application

The model as described here is applied to an example network (Fig. 1) taken from (Islam, R. M., 1995). The network consists of 11 pipes, 6 nodes, a reservoir, a storage tank and a pump. Chlorine concentration at the tank is 0.5 mg/l and at the reservoir is 0.512 mg/l. Other pertinent data are as shown on the figure. The forward chlorine simulation is run first, then the chlorine decay coefficient and the source concentration are determined for even - determined, over - determined and under - determined cases. For the purpose of verifying the model, nodal concentrations as obtained from the froward simulation are considered as measurements. The results obtained are shown in Tables 1, 2 and 3.

Table 1: Nodal concentrations

| Node number | Nodal conc.(mg/l) obtained by author | Nodal conc. (mg/l) obtained by Islam. |
|---|---|---|
| 1 | 0.509 | 0.506 |
| 2 | 0.503 | 0.500 |
| 3 | 0.499 | 0.500 |
| 4 | 0.489 | 0.490 |
| 5 | 0.486 | 0.484 |
| 6 | 0.499 | 0.490 |

Table 2: Source concentration for an even - determined, over - determined and under - determined cases for different measurement nodes.

| | Measurement nodes | Unknown Source concentration | |
|---|---|---|---|
| Even determined case | | 1 | 2 |
| | 2&5 | 0.499 | 0.512 |
| | 5&6 | 0.500 | 0.511 |
| | 1&2 | 0 | 0.512 |
| Over determined case | 1,2,4&6 | 0.500 | 0.512 |
| | 2,4&5 | 0.500 | 0.511 |
| Under determined case | 4 | 0.05 | 0.55 |
| | 5 | 0.55 | 0.44 |

Table 3: Chlorine decay coefficient for an even - determined, over - determined and under - determined cases for different measurement nodes.

| | Measurement nodes | Unknown chlorine decay coeff. |
|---|---|---|
| Even determined case | 4 | $K_5 = 0.11, K_4 = 0.11, K_8 = 0.11$ |
| | 2,3&4 | $K_3 = 0.11, K_4 = 0.10, K_5 = 0.12$ |
| Over determined | 2,3&4 | $K_4 = 0.11, K_5 = 0.13$ |
| | 2,4&5 | $K_4 = 0.09, K_5 = 0.12, K_6 = 0.10$ |
| Under determined | 5 | $K_4 = 0.10, K_5 = 0.13, K_6 = 0.16$ |
| | 4 | $K_4 = 0.06, K_5 = 0.15, K_6 = 0.10$ |

**Conclusions and Recommendations**

A model that estimates source concentration and chlorine decay coefficient is developed utilizing the inverse method. The model considers even - determined, over - determined and under - determined cases. The model is applied to an example network taken from the literature. Results show that the over - determined case gives as good estimates as the even - determined case. The under - determined case does not give good estimates most of the time. However, it gives an approximate estimate of the parameter. Different estimates for the same parameter may be obtained by changing the measurement location. By proper selection of the measurement location, better estimates are obtained, even from the under - determined case.

The model presented here needs to be verified and validated by applying it to larger and actual life networks. The model can be developed to consider the mixed determined problem and to perform error analysis to see how errors in the measurement transform into errors in the estimated parameter. The model can be developed to select the measurement locations to give best parameter estimates.

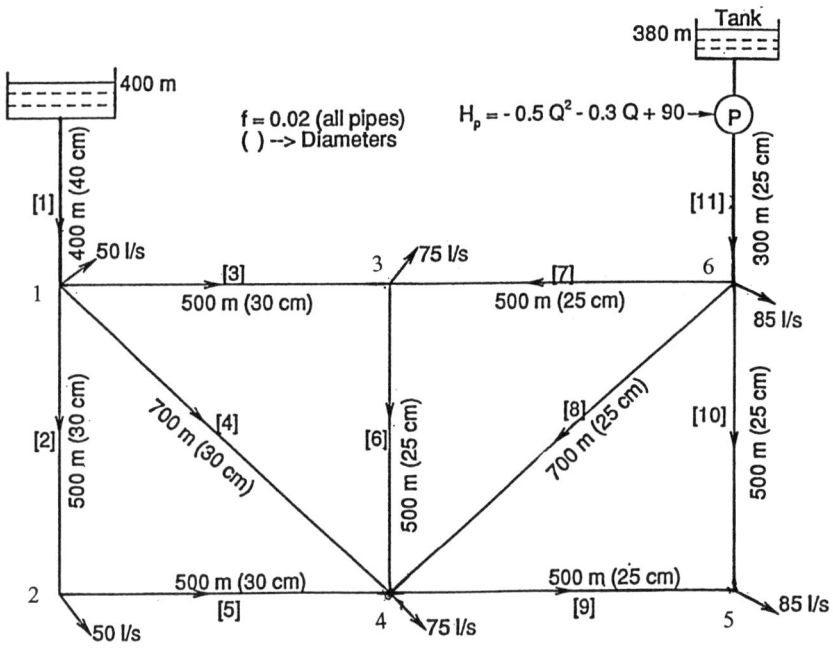

Figure 1: Sample Network (Islam, R. M., 1995)

## REFERENCES

Islam, R. M.(1995), "Modeling of Chlorine Concentration in Unsteady Flows in Pipe Networks", Ph.D. Thesis, Department of Civil and Environmental Engineering, Washington State University, Pullman, WA.

Males, R. M. et al.(1985)."Algorithm for Mixing Problems in Water Systems", *Journal of Hydraulic Engineering*, Vol. 111, No. 2, pp. 206 - 219.

Males, R. M. et al.(1988)."Modeling Water quality in Distribution Systems", *Journal of Water Resources Planning and Management,* Vol. 114, No. 2, pp. 197 - 209.

Musa, M.(1991)."Modeling of Chlorine Concentration in Water Supply Networks", M.S. Thesis, Department of Civil and Environmental Engineering, Washington State University, Pullman, WA.

Murphy, S. B.(1985)."Modeling Chlorine Concentrations in Municipal Water Systems", M.S. Thesis, Department of Civil Engineering, Montana State University, Bozeman, Montana.

Wood, D. J. and Charles, C. O.(1972). " Hydraulic Network Analysis Using Linear Theory", *Journal of Hydraulics Division, Proceedings of the American Society of Civil Engineers*, Vol. 98, No. HY7, pp.1157 - 1170.

# Travel Times in Dead-end Mains

Steven G. Buchberger[1], YeongHo Lee[2], and Jason T. Carter[2]

## Abstract

Residential water use is modeled as a nonhomogeneous Poisson rectangular pulse (NPRP) process. The NPRP premise leads directly to expressions for the probability distribution and moments of number of busy servers, flow rate, and pipe Reynolds number at any point along the dead-end main. Assuming plug flow conditions in the distribution system, the NPRP model also provides the travel time distribution for parcels of water moving to remote points of consumption in the water supply network.

## Introduction

Recent amendments to the Safe Drinking Water Act have led to a proliferation of models for predicting changes in the quality of finished water between the point of treatment and the point of consumption (Clark *et al.* 1993, Rossman *et al.* 1994). Modeling water quality in distribution systems is difficult for many reasons, notable among which are complex dynamic hydraulic conditions. Flow patterns change continuously over time and across space in response to random demands imposed by many consumers dispersed throughout the service area.

The intertwined issues of random flows and uncertain water quality are especially acute in peripheral dead-end regions of the network. Here travel times are long, stagnant conditions are common, chlorine residual may be low and individual consumers measurably influence flow. Taken collectively, these factors make it difficult to apply water quality models to dead-end mains of a distribution system.

---

Associate Professor[1] and Graduate Research Assistant[2], Department of Civil and Environmental Engineering, University of Cincinnati, Cincinnati, Ohio 45221-0071.

This paper highlights some features of a parsimonious stochastic model of residential water use. The model is based on the premise that the intensity, duration, frequency and volume of water demands follow a nonhomogeneous Poisson rectangular pulse (NPRP) process. Focusing on dead-end mains, the NPRP premise leads directly to expressions for the mean, variance, and probability distribution of busy servers, flow rates, and travel times at any point along the dead-end supply line. Only the homogeneous case is presented here. Details on the nonhomogeneous case can be found in Buchberger and Wu (1995) and Wu (1996).

*Water Pulse*

The frequency of residential water use is assumed to follow a Poisson arrival process with a time dependent rate parameter. When a water use occurs, it is approximated as a rectangular pulse of random duration and random intensity as illustrated in Figure 1. It is unlikely that more than one pulse will start at the same instant. Owing to the random duration of each water pulse, however, it is possible that two or more pulses with different starting times will overlap for a limited period. When this occurs, the total water use at the residence is the sum of the individual intensities from the coincident pulses.

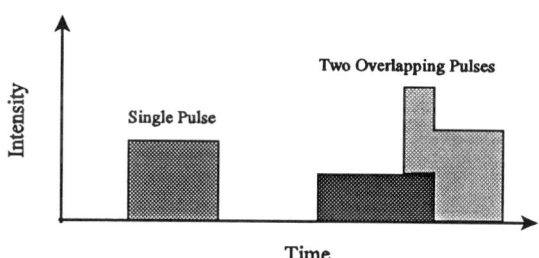

Figure 1. Poisson rectangular pulse process for water demands.

*Dead-end Main*

Dead-end mains are a sequence of links in the distribution network having terminal point(s) that do not rejoin the looped portion of the system as shown in Figure 2. Since water moves through a dead-end main in the downstream direction only, the principle of mass conservation alone is sufficient to estimate flow rates. Dead-end mains are surprisingly common. Utility surveys show dead-end mains usually comprise at least 25 percent of the total infrastructure in a distribution system and tend to service a high percentage of the residential consumer base.

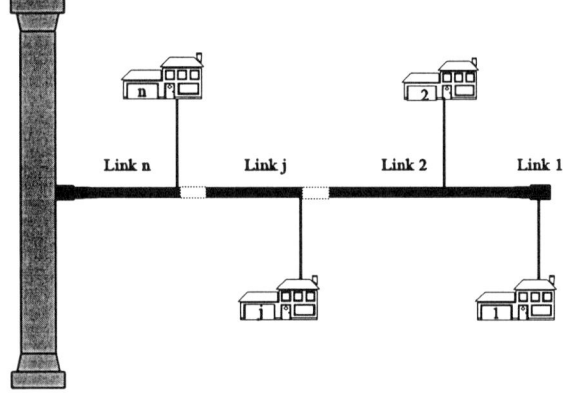

Figure 2. Definition sketch of dead-end main.

## Busy Servers

The approach taken here exploits a rich base of work, intimately connected to queueing theory, that was originally developed to analyze demands on telephone exchanges (Erlang 1917-18). Suppose customers arrive at residence j according to a homogeneous Poisson process with parameter $\lambda_j$ and receive service at rate $\mu_j$. The service rate can be viewed as the inverse of the mean duration of a pulse. Let $K_j$ be the number of busy water servers in residence j under steady state conditions. The total number of busy servers in the block of n residences is

$$K_n^* = \sum_{j=1}^{n} K_j \qquad (1)$$

Note that $K_n^*$ has a Poisson distribution with parameter $\rho_n^* = \Sigma(\lambda_j/\mu_j)$, j=1,...,n,

$$P[K_n^* = k] = \frac{(\rho_n^*)^k \exp(-\rho_n^*)}{k!} \qquad k = 0, 1, 2, \ldots \qquad (2)$$

A typical single family residence has $\rho_1^* \approx 0.05$ (Buchberger and Wells 1996, Schade 1996). From Equation (2), water in a dead-end main is stagnant with probability $P[K_n^* = 0] = \exp(-\rho_n^*)$. Field measurements show excellent agreement between predicted and observed busy servers at 20 single family homes along a dead-end main (Buchberger and Schade 1997).

## Flow Rates

The intensity of water use at a busy server in any residence is assumed to be an independent and identically distributed positive continuous random variable with mean $\alpha$ and variance $\beta^2$. The flow rate through link n (see Figure 2) resulting from water demands occurring downstream at $K_n^*$ busy servers is

$$Q_n^* = \sum_{j=1}^{n} Q_j \tag{3}$$

where $Q_j$ is the flow into residence j. Assuming that the residences are using water independently of each other, the mean and variance of $Q_n^*$ are given by (Buchberger and Wu 1995)

$$E[Q_n^*] = \sum_{j=1}^{n} \rho_j \alpha_j \tag{4a}$$

$$Var[Q_n^*] = \sum_{j=1}^{n} \rho_j (\alpha_j^2 + \beta_j^2) \tag{4b}$$

where $\rho_j = \lambda_j/\mu_j$. Equation (4a) is a generalized version of the expression given by Linaweaver et al (1966) for the expected demand on a water system serving n homes.

## Travel Times

Travel time through a dead-end link depends on three factors, namely, link volume, pulse volume, and customer arrival rate. The arrival rate experienced by link n due to downstream customers demanding water is given by

$$\lambda_n^* = \sum_{j=1}^{n} \lambda_j \tag{5}$$

The total volume of water $V_n^*(t)$ swept through link n during time $\{0,t\}$ is a compound Poisson process

$$V_n^*(t) = \sum_{k=0}^{N_n^*(t)} [X_n^*]_k \qquad t \geq 0 \tag{6}$$

where $N_n^*(t)$ arises from a Poisson distribution with parameter $(\lambda_n^* t)$ and represents the number of water pulses generated by the block of n residences during interval $\{0,t\}$. $X_n^*$ is the volume of a single water pulse and is assumed to be an independent and identically distributed random variable with mean $\phi_n^*$ and variance $(\psi_n^*)^2$.

Assuming plug flow conditions, the cumulative distribution of water travel time through a link can be related to the probability of downstream demands as follows

$$P[T_n \leq t] = P[V_n^*(t) \geq L_n] \qquad (7)$$

where $T_n$ is the travel time through link n and $L_n$ is the total volume of link n. Equation (7) indicates that the travel time through link n is governed by how long it takes cumulative water demands to displace the link volume. The kth moment of travel time follows directly from the distribution function

$$E[T_n^k] = \int_0^\infty k t^{k-1} P[T_n > t] dt \qquad (8)$$

It can be shown that the mean and variance of $T_n$ are given by (Wu 1996)

$$E[T_n] = \frac{L_n}{\lambda_n^* \phi_n^*} \qquad (9a)$$

$$Var[T_n] \approx \frac{L_n}{(\lambda_n^*)^2 \phi_n^*} \left[ 1 + \left( \frac{\psi_n^*}{\phi_n^*} \right)^2 \right] \qquad (9b)$$

Equation (9a) for the mean travel time is exact, irrespective of the probability distribution of pulse volumes. The denominator term $(\lambda \phi)$ is equivalent to the mean flow (see Equation 4a). Hence, Equation (9a) gives $T_n = L_n/Q_n$ which is identical to the well known deterministic result for mean residence time in a closed reactor.

Equation (9b) is exact for exponential pulse volumes. Computer simulations of random intermittent plug flows indicate that Equation (9b) is an excellent approximation for other pulse volume distributions provided $L_n/\phi_n^* > 30$, a condition that is often satisfied in practice.

Using moment generating functions, it can be shown that travel times are asymptotically normal (with mean and variance given by Equation 9a,b) under the assumption of homogeneous arrival rates, plug flow conditions, and exponential pulse volumes (Wu 1996).

## Summary

The premise that residential water use occurs as a nonhomogeneous Poisson rectangular pulse process leads to a reasonable, rigorous and robust framework for solving contemporary problems that arise with water distribution systems. Several results from the NPRP model are outlined. The NPRP approach offers a stochastic complement to deterministic models which are indispensable tools for designing and operating distribution systems to meet water demands and quality standards.

## Acknowledgments

This work was supported by grants from the National Science Foundation (NYI Award: BCS-9257608) and from the American Water Works Association Research Foundation (95-294).

## References

Buchberger, S.G. and L. Wu (1995) "Model for instantaneous residential water demands", **ASCE Journal of Hydraulic Engineering**, 121(3): 232-246.

Buchberger, S.G. and G.J. Wells (1996) "Intensity, duration, and frequency of residential water demands", **ASCE J Wat Res Plan & Mngt**, 122(1): 11-19.

Buchberger, S.G. and T.G. Schade (1997) "Poisson Pulse Queueing Model for Residential Water Demands", 27th IAHR Congress, San Fran, CA, pp. 488-493.

Clark, R.M., et al. (1993) "Modeling contaminant propagation in drinking-water distribution system", **ASCE J Env Eng**, 119(2): 349-364.

Erlang, A.K. (1917-18) "Solutions of some problems in the theory of probabilities of significance in automatic telephone exchanges" **Post Office Elec Engr J**, 10,189-197.

Linaweaver, F.P. Jr., J.C. Geyer and J.B. Wolff (1966) **Residential Water Use**, Johns Hopkins University, Baltimore, Maryland.

Rossman, L.A., R.M. Clark, and W.M. Grayman (1994) "Modeling chlorine residuals in drinking-water distribution systems", **ASCE J Env Eng**, 120(4):803-820.

Schade, T.G. (1996) "Water demand and travel time in a residential dead-end loop", **MS Thesis**, University of Cincinnati, 64 pages.

Wu, L. (1996) "Stochastic model of flow in the periphery of a water distribution system", **PhD Dissertation**, University of Cincinnati, 207 pages.

OPTIMAL LOCATION OF BOOSTER DISINFECTION STATIONS
FOR RESIDUAL MAINTENANCE

Dominic L. Boccelli[1], Michael E. Tryby[2], James G. Uber[3], and Lewis A. Rossman[4]

**Abstract**

Conventional disinfection of drinking water consists of disinfectant addition only within the treatment plant, which must provide adequate residuals within the distribution system. Booster disinfection – the application of additional disinfectant within the distribution system – may reduce the total amount of disinfectant applied to the distribution system while maintaining adequate residuals. A mixed–integer linear programming model was developed to determine the optimal scheduling and location of booster stations, and solved using commercially available software. The optimization model was applied to a sample network to illustrate the potential of booster disinfection to reduce the amount of disinfectant applied to the system while maintaining adequate residuals.

**Introduction**

Conventional disinfection typically consists of disinfectant addition only within the treatment plant. This addition must maintain C*T requirements, provide adequate residuals throughout the distribution system, and satisfy regulations related to disinfectant by–products. Some systems may have problems maintaining residuals at the periphery of the distribution system, which would require a larger dose at the treatment plant possibly leading to taste and odor complaints. Booster disinfection – the application of additional disinfectant within the distribution system – may be able to maintain adequate residuals and avoid taste and odor complaints by producing a more evenly distributed disinfectant concentration. Booster disinfection may also require less disinfectant when compared to conventional disinfection, which may decrease the formation of harmful disinfectant by–products, such as trihalomethanes.

Booster disinfection may achieve these benefits via two different mechanisms: 1) kinetically, from bulk and wall decay, and 2) physically, from the spatio–temporal dis-

---

1. Graduate Research Assistant, Dept. of Civil & Environmental Eng., University of Cincinnati, Cincinnati, OH 45221

2. Graduate Research Assistant, Dept. of Civil & Environmental Eng., University of Cincinnati, Cincinnati, OH 45221

3. Associate Professor, Dept. of Civil & Environmental Eng., University of Cincinnati, Cincinnati, OH 45221

4. Environmental Scientist, Water Supply and Water Resources Division, RREL, U.S. Environmental Protection Agency, 26 W. M. L. King Dr., Cincinnati, OH 45268

tribution of flows within a network. Typical kinetic assumptions consider bulk and wall decay as first order processes with respect to disinfectant concentration, and thus a decrease in average concentration will reduce the disinfectant decay. Additionally, booster disinfection allows the application of disinfectant to smaller flows, as larger flows turn into smaller flows due to external demands and branching.

It would be difficult to determine the optimal locations and dose schedules of booster stations using only distribution system hydraulic and water quality models (for instance, by trial and error). This work presents a framework which optimizes both the scheduling and location of booster stations within a distribution system.

## Problem Formulation

### Scheduling Model

Boccelli et al. (1998) have developed a model to optimize the scheduling of disinfectant mass injections for booster stations. Their model minimizes the amount of disinfectant applied while satisfying upper and lower concentration bounds within the distribution system. The problem is formulated as a linear programming problem by assuming known but time–varying flow rates, first–order disinfectant decay, and linear models of transport in pipes, and mixing in tanks and at nodes. The model formulation is

$$\min_{v} \sum_{i=1}^{n_b} \sum_{k=1}^{n_s} v_i^k \quad (1)$$

subject to,

$$c^{\min} \leq c_j^m(v) = \sum_{i=1}^{n_b} \sum_{k=1}^{n_s} a_{ij}^{km} \cdot v_i^k \leq c^{\max} \quad (2)$$

$$j = 1,\ldots,n_m, \quad m = M,\ldots,M + n_a - 1$$

$$v_i^k \geq 0, \quad i = 1,\ldots,n_b, \quad k = 1,\ldots,n_s \quad (3)$$

where $v_i^k$ is a periodic mass injection at booster station $i$ and time intervals $(an_s + k)\Delta t \leq t \leq (an_s + k + 1)\Delta t$, $a = 0, 1,\ldots,\infty$, $n_s$ is the number of injection periods contained in one scheduling cycle $T_s$ ($T_s = n_s\Delta t$), $n_b$ is the number of booster stations, $c_j^m(v)$ is the concentration at monitoring node $j$ and time $m\Delta t_m$, $\Delta t_m$ is the monitoring time interval, $a_{ij}^{km}$ is the periodic impulse response coefficient which determines the impact $v_i^k$ has on $c_j^m$, $c^{\min}$ and $c^{\max}$ are the minimum and maximum concentration limits within the distribution system, $n_m$ is the number of monitoring nodes, $M$ is the number of time intervals $\Delta t_m$ needed before the initial conditions are negligible, and $n_a$ is the number of time periods contained in one concentration cycle $T_a$ ($T_a = n_a\Delta t$).

The dynamic monitoring node concentrations, resulting from the optimal scheduling solutions, should be appropriate for a long–term planning horizon, which necessitates that $c_j^m$ be periodic over the concentration cycle $T_a$. To ensure concentration periodicity (which includes periodicity of $a_{ij}^{km}$) the hydraulics are assumed periodic over a hydraulic cycle $T_h$, in addition to periodicity of the booster station schedule ($T_s$). Based on these assumptions, $T_a$ is determined by $T_a = \eta T_s = \mu T_h$, where $\eta, \mu$ are integers greater than 0 (Boccelli et al., 1998, Appendix A).

Additionally, a constraint can be written such that the booster stations operate to maintain a specific concentration set point at each booster location $i$:

$$c_i^m(v) = s_i^n \qquad (4)$$
$$i = 1,\ldots,n_b, \quad n = 1,\ldots,n_{p_i}, \quad m \in S_i^n$$

where $c_i^m(v) = \sum_{l=1}^{n_b}\sum_{k=1}^{n_s} a_{li}^{km} \cdot v_l^k$ is the concentration at booster location $i$ and time $m\Delta t_m$, $s_i^n$ is the $n$th concentration set point to be maintained at booster station $i$, $n_{p_i}$ is the number of different set points for booster station $i$, and $S_i^n$ is the set of (usually contiguous) monitoring times that constitute the $n$th concentration set point at booster station $i$.

## Location Model

This work extends the optimal scheduling model to include booster station location. The extension of eqs. 1–3 yields a mixed integer linear programming (MILP) problem (Hillier and Lieberman, 1980):

$$\min_{v,\delta} \sum_{i=1}^{\bar{n}_b}\sum_{k=1}^{n_s} v_i^k \qquad (5)$$

subject to,

$$c^{\min} \le c_j^m(v) = \sum_{i=1}^{\bar{n}_b}\sum_{k=1}^{n_s} a_{ij}^{km} \cdot v_i^k \le c^{\max} \qquad (6)$$
$$j = 1,\ldots,n_m, \quad m = M,\ldots,M + n_a - 1$$

$$\sum_{k=1}^{n_s} v_i^k \le M_v \cdot \delta_i, \quad i = 1,\ldots,n_b \qquad (7)$$

$$\sum_{i=1}^{n_b} \delta_i \le n_b^{\max} \qquad (8)$$

$$\delta_i = \{0,1\}, \quad i = 1,\ldots,n_b \qquad (9)$$

$$v_i^k \ge 0, \quad i = 1,\ldots,\bar{n}_b, \quad k = 1,\ldots,n_s \qquad (10)$$

where $\delta_i$ is a binary variable used to determine whether a new booster station is ($\delta_i = 1$) or is not ($\delta_i = 0$) present, $M_v$ is a "sufficiently" large, positive coefficient ($M_v$ should be as small as possible for computational efficiency, but larger than the maximum sum of doses at any booster station), $n_b$ is now the number of potential, new booster stations, $n_b^{\max}$ is the maximum number of new booster stations allowed ($n_b^{\max} \le n_b$), and $\bar{n}_b$ is the number of potential plus existing booster stations. This work can also include the concentration set point formulation by altering eq. 4 to account for potential booster stations,

$$0 \le c_i^m(v) - s_i^n \le M_s \cdot (1 - \delta_i) \qquad (11)$$
$$i = 1,\ldots,n_b, \quad n = 1,\ldots,n_{p_i}, \quad m \in S_i^n$$

and existing booster stations,

$$c_i^m(v) = s_i^n \qquad (12)$$
$$i = n_b + 1,\ldots,\bar{n}_b, \quad n = 1,\ldots,n_{p_i}, \quad m \in S_i^n$$

where $M_s$ is the maximum feasible concentration allowed at a booster station if $\delta_i = 0$. If, however, $\delta_i = 1$ than eqs. 11 and 12 are equivalent and the concentration at booster station $i$ must be equivalent to $s_i^n$. The concentration set points $s$ are also included as variables of the optimization model.

## Methodology

A code was developed to interface with EPANET (Rossman, 1993), a distribution system hydraulic and water quality model, to set up the MILP problem. The program first determines the hydraulic periodicity $T_h$ which, along with the schedule period $T_s$ (a user defined parameter), allows $T_a$ to be determined. The values of $\alpha_{ij}^{km}$ are then determined by perturbation analysis. Each booster location $i$ and periodic dose interval $k$ is selected in turn and modeled as a periodic dose $\hat{v}$, which is sufficiently large to avoid round–off error. The resulting periodic concentrations $c_j^m$ are used to determine $\alpha_{ij}^{km}$ by $\alpha_{ij}^{km} = c_j^m / \hat{v}$. The solution to the MILP problem can be attempted using widely available software such as lp_solve (Schwab, 1998).

## Application

The model, without the concentration set point, (eqs. 5–10) has been applied to a modified version of the Brushy Plains/Cherry Hill portion of the South Central Connecticut Regional Water Authority network (Rossman et al., 1994). The portion of the network, shown in Figure 1, includes 34 demand nodes, 6 booster station locations, one source node representing a pump station, and one storage tank (total network size ≅ 7 miles of pipe). Consumer nodes are denoted by numerical values, except for nodes 1 and 26 which are the source and tank, respectively. Booster station locations are given by the alphanumeric labels.

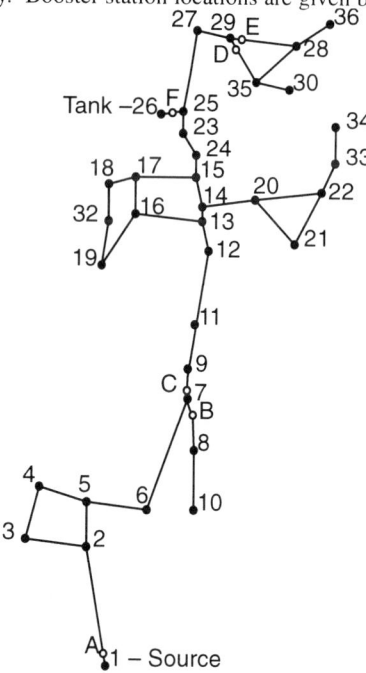

The demands and pump operation were assumed to be periodic on a 24 hour cycle, which coincides with the assumed scheduling period ($T_h = T_s = 24$ hours), thus $\eta = \mu = 1$ and $T_a = 24$ hours. The values of $c^{min}$ and $c^{max}$ were chosen to be 0.2 and 4.0 respectively to reflect current/anticipated SDWA regulations.

Optimal solutions were obtained with one fixed booster location (A) and five potential booster locations (B–F) ($\bar{n}_b = 6$, $n_b = 5$). Different runs were made with the value of $n_b^{max}$ ranging from 0 (simulating conventional treatment) to 5 (all booster stations active). Concentration constraints (eq. 6) were written for all consumer nodes.

## Results

Table 1 contains the booster station locations and their respective mass injection rates for the different values of $n_b^{max}$. For $n_b^{max} = 0$, the total disinfectant mass applied to the system is 2824 g/d. For $n_b^{max} = 1$, booster location F was selected with a resulting total disinfectant mass applied of 1256 g/d, a 55.5% reduction over conventional treatment. Increasing $n_b^{max}$

Fig. 1 Brushy Plains/Cherry Hill portion of the South Central Connecticut Water Authority network.

Table 1. Simulation Results

| $n_b^{max}$ | A (g/d) | B (g/d) | C (g/d) | D (g/d) | E (g/d) | F (g/d) | Total (g/d) | % Reduction |
|---|---|---|---|---|---|---|---|---|
| 0 | 2824 | — | — | — | — | — | 2824 | — |
| 1 | 836.2 | — | — | — | — | 419.4 | 1256 | 55.5 |
| 2 | 753.2 | — | — | — | 1.22 | 399.5 | 1154 | 59.1 |
| 3 | 585.9 | — | 153.9 | — | 1.40 | 393.9 | 1135 | 59.8 |
| 4 | 440.1 | 3.21 | 288.4 | — | 1.48 | 388.5 | 1122 | 60.3 |
| 5 | 440.1 | 3.20 | 292.8 | 0.18 | 0.32 | 383.0 | 1120 | 60.3 |

from 1 to 5 continued to decrease the total amount of disinfectant mass applied, however, the additional booster stations were only able to achieve an additional 5% reduction, beyond $n_b^{max} = 1$, over conventional treatment. The disinfectant additions at booster locations A and F generally decreased as $n_b^{max}$ increased, while the disinfectant addition at the other locations were not as consistant.

## Discussion

While it may be intuitive that the addition of one booster station would reduce the amount of disinfectant applied by the other stations, it may not be as intuitive as to why some booster stations may actually have their injection rates increased by adding a booster station. As an example, increasing $n_b^{max}$ from 3 to 4 produced a significant reduction of applied disinfectant at booster locations A and F, however, there is also a significant increase in the amount of disinfectant applied at booster station C. This occurs since the large injection at A ($n_b^{max} = 3$) is mainly responsible for maintaining adequate residuals at monitoring nodes 8 and 10, and, as a result, provides excess disinfectant which satisfies other monitoring nodes throughout the distribution system. The addition of booster station B ($n_b^{max} = 4$), however, makes it unnecessary for booster location A to primarily satisfy nodes 8 and 10, allowing the injections at A to be decreased and removing the excess disinfectant which originally passed through C, thus requiring an increase in the disinfectant applied by C to make up for the lost disinfectant from A.

The optimal solutions (Table 1) illustrate the importance of location by the addition of booster station F ($n_b^{max} = 1$), which achieved a large reduction in disinfectant mass applied, while the addition of the other four stations did not achieve more than an additional 5% reduction. This also illustrates that an optimal number of booster stations should exist, as the costs associated with each additional booster station are fixed, but benefits achieved by each addition are decreasing.

The importance of location is also seen in the average concentration and variability within the distribution system. Figure 2 shows the average concentration for all monitoring nodes and their variability throughout the 24 hour concentration cycle for three different optimal booster dose scheduling scenarios. Figure 2.a is for conventional treatment ($n_b^{max} = 0$), Figure 2.b is for booster disinfection with booster locations B–E active (an additional run), and Figure 2.c is for booster disinfection at all possible locations. As can be seen, the addition of booster locations B–E reduce the average concentration, but does not appear to decrease the variability. The addition of booster F, however, significantly decreases both the average concentration and the variability.

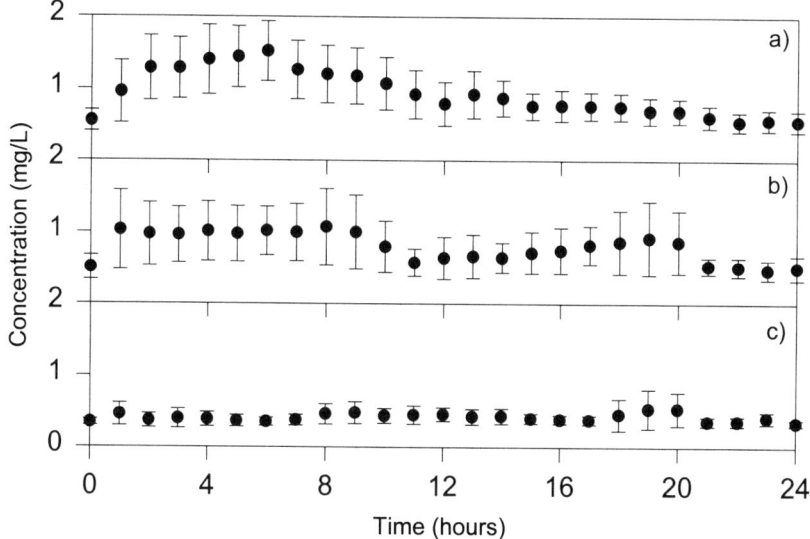

Figure 2. Average concentration and variability for all monitoring nodes over 24 hours with a) booster station A; b) booster stations B-E; and c) booster stations B-F.

## Conclusions

A linear model has been developed for optimal scheduling and location of booster disinfection stations. The linearity of the model is important due to the power and efficiency of available LP/MILP solvers. A simple network example was used to illustrate the methodology and to illustrate significant features of optimal results. Additionally, the example illustrated that booster disinfection, under the stated assumptions, can reduce the amount of disinfectant applied to the system, as well as the average disinfectant concentration and its variability, which may produce other benefits.

## References

1. Boccelli, D. L., Tryby, M. E., Uber, J. G., Rossman, L. A., Zierolf, M. L., and Polycarpou, M. M. (1998) "Optimal Scheduling of Booster Disinfection in Water Distribution Systems." *Journal of Water Resources Planning & Management*, ASCE. 124(2).

2. Hillier, F. S., and Lieberman, G. J. (1980) *Introduction to Operations Research*, Holden–Day, Inc., Oakland, CA, 730–732

3. Rossman, L. A., (1993) *EPANET User Manual*. Risk Reduction Engineering Laboratory, U.S.E.P.A., Cincinnati, Ohio.

4. Rossman, L. A., Clark, R. M., and Grayman, W. M. (1994) "Modeling chlorine residuals in drinking–water distribution systems." *Journal of Environmental Engineering*, 120(4), 803–820.

5. Schwab, H. (1996) "Documentation for lp_solve," Version 2.1. Available from ftp://ftp.es.ele.tue.nl/pub/lp_solve

## Detecting Accidental Contaminations in Municipal Water Networks: Application

By Avner Kessler [1] and Avi Ostfeld [2], Member, ASCE

**Abstract**

This paper is on the application of a recent methodology developed by the authors for the optimal layout of monitoring stations in municipal water networks. The monitoring stations comprise a detection system that is able to discover possible random external pollution intrusions, and which provides the consumers with a prescribed level of service. The methodology is applied on the hypothetical community, Anytown U.S.A. water distribution system.

**Introduction**

A methodology for finding the optimal layout of a set of monitoring stations in municipal water distribution systems that are able to detect possible random external pollution intrusions, was recently developed and demonstrated (Kessler et al., 1997).

The monitoring stations comprise a detection system that provide the consumers with a prescribed level of service, wherein the level of service is measured in terms of the maximum volume of consumed polluted water prior to detection. The methodology involves the establishment of an Auxiliary Network that represents all possible flow directions in the system for a typical demand cycle; the use of the All Shortest Paths algorithm for the identification of domains of pollutions; and a Minimum Covering Set algorithm for the optimal locations of the monitoring stations.

**Application**

To test the capability of the methodology proposed in Kessler et al. (1997) on a more realistic case, the methodology is applied to the hypothetical community, Anytown U.S.A. (Walski et al., 1987).

A schematic representation of the system is shown in Fig.1. The system is made of 34 pipes, 16 nodes, two 56,832 cubic feet elevated storage tanks, one pumping station,

---

[1] Director, Environmental and Water Resources Engineering Ltd., P.O.Box 6349, Haifa 31062, Israel
[2] Senior Engineer, Tahal Consulting Engineers Ltd., 54 Ibn Gvirol St., Tel-Aviv, Israel

and a single well. The tanks are assumed to be cylindrical between their minimum and maximum levels. The water level in the well is maintained at an elevation of 10 ft. The pipes, nodes, tanks, and pumping stations detailed characteristics, as well as the system demand flow pattern, are as in Walski et al. (1987).

The EPANET simulator (Rossman ,1994) was used to establish the Auxiliary Network of the system, with which the layout of the detection system is selected, for a given level of service.

The decision for a detection system is most likely to be governed by a budget limitation from one hand, and a level of service from the other. To choose for the appropriate detection system, the analysis of the tradeoff between the different alternatives of detection systems, versus the level of service provided to the consumers, is essential. Following is such an analysis.

## Tradeoff between different detection systems alternatives and the level of service

Fig. 2 shows the tradeoff between the number of monitoring stations selected and the level of service, for Anytown U.S.A. The establishment of Fig. 2 involved running the model for increasing amounts of maximum consumed polluted water prior to detection (i.e. the level of service), and the examination of the number and location of the monitoring stations chosen.

Analysis of Fig. 2 reveals the following:

1. The number of monitoring stations selected decreases in an exponential manner, as the level of service decreases (i.e. growth of the maximum "allowed" consumed polluted water prior to detection).

2. An even number of monitoring stations (not necessarily unique; explanation follows in 3 below) provide a range of service levels. For example, four monitoring stations cover the level of service between 16530 to 43972 cu-ft (see Fig. 2).

3. The following general property holds for even numbers of monitoring stations that comprise detection systems: as the level of service for which a detection system is selected - decreases, the degree of its invulnerability - increases. The rational for this is that the Domain of Detection of each of the nodes increases as the level of service decreases (i.e. the Pollution Matrix becomes more and more dense as the level of service decreases). As a result of that the Domain of Coverage of each node increases.

The first time the model is shifting to a lower number of monitoring stations as the level of service decreases, is also the first time the model is able to have exactly enough monitoring stations to cover the entire domain, with a lower number of monitoring stations. With this in mind as we continue to reduce the level of service the Domain of Detection of each node increases and thus other sets with the same number of monitoring stations, that give higher invulnerability, can be selected. For example the selection of four monitoring stations that correspond to a level of service of 16530 cu-ft have three nodes for which there is an overlapping between monitoring stations (see Fig. 3), while the selection of four monitoring stations that correspond to a level of service of 43972 cu-ft have eight nodes for which there is an overlapping between monitoring stations, with nodes 80 and 140 covered by three monitoring stations (see

Fig. 4).

The designer is thus faced with the question of choosing among even sets of monitoring stations that differ in their level of service and invulnerability. The answer to this is probably a balance between constraints on putting monitoring stations at different locations in the system, and the reliability of the monitoring equipment.

## Conclusions

A straightforward solution for the detection of pollutants intrusion into a water distribution systems is to locate a monitoring station at each of the system nodes. This, obviously, leads to the detection system of maximum cost, and as such is usually not feasible. The application example presented is for finding the minimum number of monitoring stations that guarantee a prescribed level of service to the consumers in case one of the system nodes becomes a source of pollution.

The combination of extensive hydraulic simulations with graph theory techniques formulates the methodology proposed, which provides a step towards a rational manner for choosing a detection system.

## Acknowledgment

This research project was funded by the Water Research Institute (WRI) at the Technion-Israel Institute of Technology, under research project number 015-053.

## References

Kessler A., Ostfeld A., and Sinai G. (1997). "Detecting Accidental Contaminations in Municipal Water Networks", Proceedings of the Annual WRPMD Conference ASCE, Houston, Texas.

Rossman, L.A. (1994). "EPANET User Manual." Risk Reduction Engineering Laboratory, U.S. Environmental Protection Agency, Cincinnati, Ohio, 107p.

Walski T. M. et al. (1987). "Battle of the Network models: epilogue". Journal of Water Resources Planning and Management Division, ASCE, Vol. 113, No. 2, pp. 191 - 203.

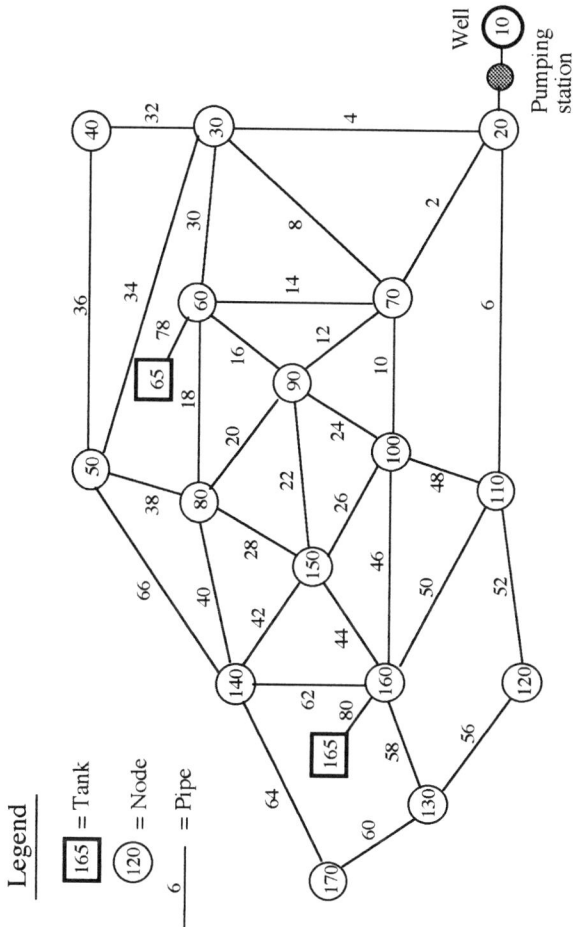

Fig. 1: Anytown U.S.A. water distribution system

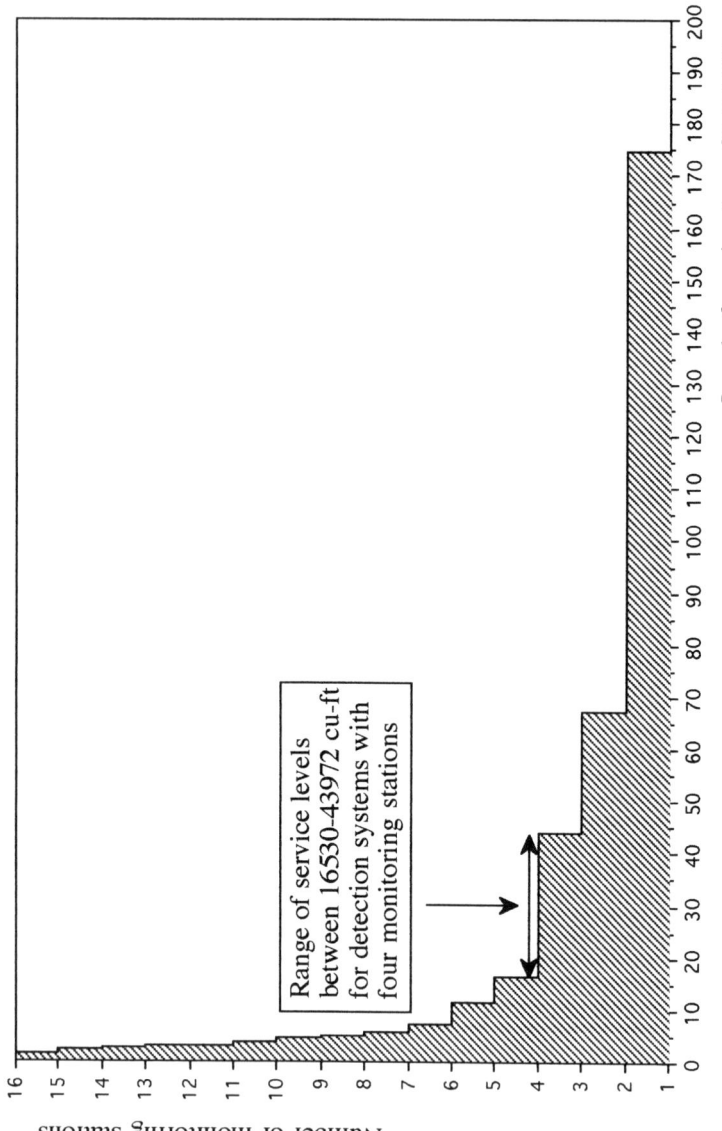

Fig. 2: Tradeoff between the number of monitoring stations and the level of service

WATER RESOURCES AND THE URBAN ENVIRONMENT 277

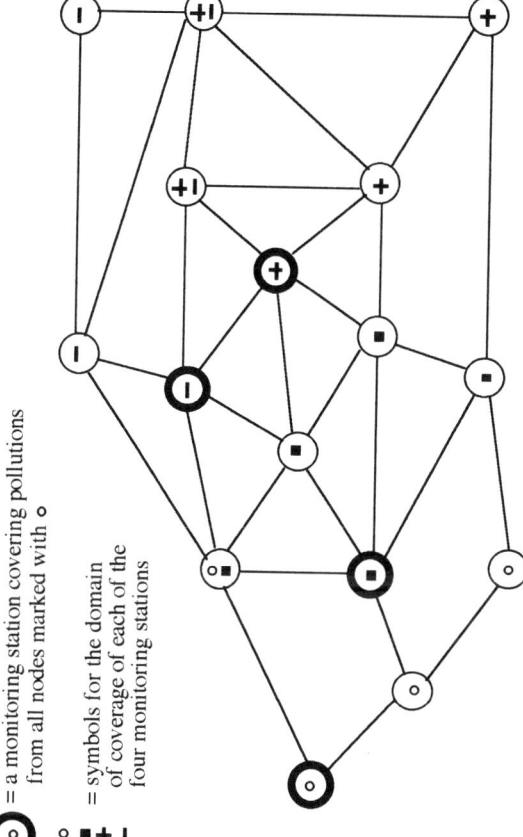

Fig. 3: Optimal detection system for a level of service of 16530 cu-ft

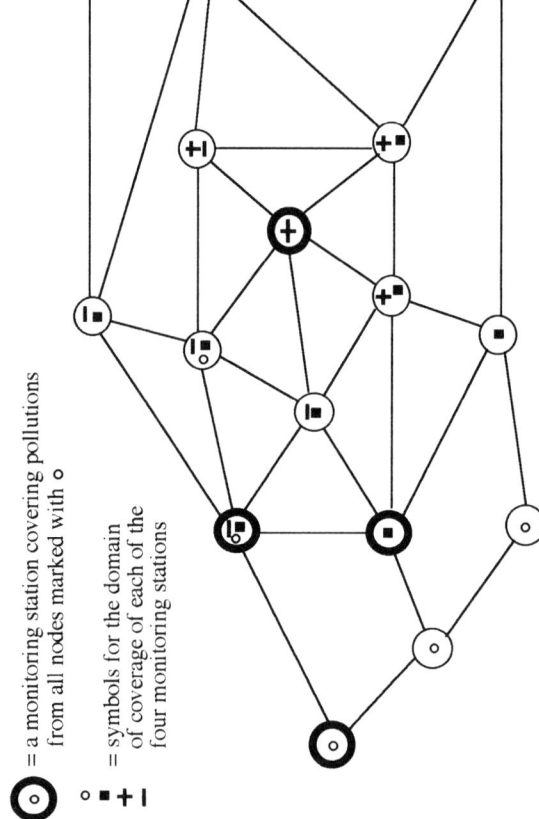

Fig. 4: Optimal detection system for a level of service of 43972 cu-ft

Using Hydraulic Modeling for Disinfection Effectiveness

Paul F. Boulos[1], Member ASCE and Imad A. Hannoun[2]

Abstract

Regulations under the Safe Drinking Water Act and its Amendments (SDWAA) impose some specific and restrictive guidelines for all water treatment plants that treat surface water. Virtually all such plants utilize clearwells for enhanced concentration time values. A continuing challenge in the design of these facilities is to provide for maximum disinfection effectiveness while limiting disinfection by-product formation. These two competing objectives are best realized by optimizing the hydraulic performance and efficiency of water plant clearwells. This paper demonstrates the use of a computer hydraulic model to characterize mixing in water plant clearwells. Comprehensive visualization of flow patterns and mixing regimes and the ease of manipulating the model offer the ability to explore many possible design alternatives for maximum disinfection effectiveness. Corrective modifications could then be tested at low cost. The effective use of the proposed model can lead to better design, operation, and management practices. Such practices will improve the effectiveness of disinfection procedures, increase maintenance of disinfectant residuals, help control bacterial levels in the distribution system, and ensure safe drinking water. It is recommended that the model be used for evaluating the effectiveness of existing designs and the study of novel designs before construction.

Introduction

The Surface Water Treatment Rule (SWTR), as well as the Enhanced SWTR (ESWTR) and the Disinfectant/Disinfection By-Products Rule (D/DPB) impose stringent guidelines for all water treatment plants that treat surface water. Virtually all such plants utilize clearwells to secure disinfection effectiveness and provide adequate operational storage (filter backwash, utilities, and fire fighting). In some cases, these facilities are also used as equalization storage for distribution system demands. They have been traditionally designed based on the concept of plug flow mixing regime, which assumes uniform cross-sectional distribution of incoming

---
[1]Vice President, MWSoft, 300 N. Lake Avenue Suite 1200, Pasadena, CA 91101
[2] Vice President, Flow Science Incorporated, 135 East Hancock Street, Lansdale, PA 19446

water particles and internal movement in parallel paths with constant and uniform velocity (first in - first out). In practice, however; pure plug flow can only be achieved by a combination of baffling arrangement and proper inlet and outlet design and configuration.

The primary goal of the disinfection process in drinking water treatment is the destruction or otherwise inactivation of microbial pathogens. The effectiveness of disinfection is a complex function of several variables including disinfectant dosage, contact time, and mixing characteristics. While increasing the disinfectant dosage will guarantee proper water disinfection, it will also result in inflated operational costs and the potential to induce high levels of DBP formation in excess of the current or proposed standards. Additional storage volumes can lead to smaller doses of disinfectant but with a significant increase in construction costs. In addition to problems of improper mixing and disinfectant distribution (dead space and short circuiting), water plant clearwells can serve as vessels for complex chemical and biological activities exerting undue disinfectant demands and resulting in the deterioration of effluent water quality. Because of the conflicting disinfection requirements and DBP standards, the optimization of disinfection is better achieved by improving the hydraulic efficiency and lowering the required disinfectant dose.

The hydraulic efficiency of a clearwell is characterized by the $T_{10}/T$ ratio. Here $T_{10}$ represents the time in which 10 percent of the water passes through the clearwell and T is the hydraulic detention time. The higher the ratio the more effective the disinfection unit (longer contact time in a smaller volume). It is noted here that for ideal plug flow, $T_{10}/T = 1.0$, indicating that all water particles reside in the facility for an equal time period. Due to velocity gradients in the clearwell, however; the actual $T_{10}/T$ is expected to be substantially below 1.0.

Although maximum flow uniformity throughout the clearwell is desirable, inherent flow velocities are not expected to be uniform. The first source of non-uniformity in the velocity field is due to the presence of the basin sides and bottom, which leads to the formation of boundary layers along the flow boundaries (bottom and side walls). Such boundary-layers result in the water near the boundaries being transported at below average velocities and hence residing in the facility for longer than the average detention time. On the other hand, the water particles away from the solid boundaries moves at above average velocities (and hence shorter residence time). A second source for non-uniformity in velocity (or velocity differentials) is the presence of sharp turns that are necessitated by the placement of baffling walls. After each change in the direction of flow, it is anticipated that the water particles on the outside of a turn will acquire a higher velocity than the water on the inside. This velocity differential also leads to non-uniform detention times. A third source for velocity differentials is the common use of relatively small inlets and outlets. If an inlet consists of a relatively small pipeline, then the velocity of the water particles entering the clearwell will be significantly higher than the average clearwell velocity. Downstream, this velocity excess will be gradually eroded, but nonetheless, it will result in some water particles exiting the clearwell fairly rapidly. Similarly, the use of a relatively small diameter outlet will result in a velocity excess upstream of the outlet.

Maximum flow uniformity throughout the depth of the clearwell can be obtained through clever use of baffling. Ideal baffling design reduces the inlet and outlet flow velocities, distributes the water as uniformly as practical over the cross section of the clearwell, breaks-up large scale eddies, minimizes mixing with the water already in the unit, creates small scale turbulence, and prevents entering water from short-circuiting to the outlet.

Historically, the $T_{10}/T$ ratio has been determined by performing expensive laboratory scale model tests, full scale experiments after the clearwell has been built, or field tracer studies. While model tests can provide valuable results, they have the disadvantages of a relatively high cost and slow turnaround. On the other hand, conducting full scale tests or tracer studies after a clearwell has been built can only provide data for the as-built basin. If the measured $T_{10}/T$ ratio for the as-built basin falls below the target value, a substantial cost may be incurred in modifying the clearwell design, or alternately, the maximum clearwell flow rate will be limited to below design levels.

Computer hydraulic models represent the most cost-effective and viable means of assessing flow distribution patterns and internal mixing conditions inhibited in water plant clearwells. In this method of analysis, The $T_{10}/T$ ratio is calculated using a three-dimensional hydrodynamic computer model. The model can be effectively used to reproduce mixing conditions within the unit, to identify the cause of insufficient mixing, and to determine the effectiveness of corrective modifications to improve contact time with minimal structural changes and cost. This applies to both new and existing facilities. The effects of baffles, porous walls, perforated outlet diffusers, and various inlet and outlet configurations on residence time distribution and hydraulic efficiency can also be evaluated. The versatility of the model is demonstrated by application to an actual water treatment plant clearwell.

Computer Hydraulic Model

**Methodology** - The analysis of the hydraulic efficiency of water plant clearwells is derived from hydrodynamics principles. In this method of analysis, the three-dimensional Reynolds equations that describe turbulent fluid flow are discretized over a Cartesian grid and solved using the finite difference method. The equations of momentum and continuity are solved using the finite-volume approach with a Poisson-like equation for pressure. Furthermore, a two-equation k-ε turbulence closure model is used to evaluate the turbulent eddy viscosity. All equations are solved in a time-dependent fashion and the solution proceeds until a steady-state equilibrium is obtained. Details of the method can be found in Hannoun and Boulos (1997). From the calculated velocity field, the path and residence times of the water particles throughout the discretized flow domain are determined. The particle residence time distribution is then computed, and the $T_{10}/T$ ratio is evaluated.

**Computer Implementation** - The solution methodology is embodied in a computer program written in FORTRAN 77 (Hannoun and Boulos 1997). Typically, a computational mesh is established by subdividing the domain into discrete computational volumetric cells (grid points) of various sizes. The smaller the mesh size, the more accurate is the model in simulating the unit hydrodynamic behavior. The structure and resolution of this mesh are critical factors in achieving a good solution; the mesh must be dense enough to resolve the geometric details. However, the number of geometric cells also determines the number of equations that need to be solved and therefore dictates the computational run time. As a result, there is generally a compromise between the number of volumetric cells required and those that can be solved in a realistic amount of time. The optimum mesh density is the one that best compromises modeling accuracy and execution time. The computer model automatically creates any user-specified mesh distribution size which can be manually edited at any time for finer adjustments. Finer adjustments are generally required in regions of sharp velocity gradients and high turbulence levels such as in corners, near walls and baffles, and inlets/outlets. Input to the model consists of a geometrical description of the clearwell, a reaction rate expression for the selected disinfectant, a set of initial and boundary conditions (e.g., inlet/outlet velocities, inlet concentrations, and temperatures), and a set of user-selected variables (e.g., velocity, concentration, temperature, and pressure). The model output produces spatial and temporal solutions for the selected variables in graphical format.

Application

The proposed computational fluid dynamics model has been successfully applied to improve the hydraulic efficiency of a number of rectangular and circular clearwells throughout the United States. The versatility of the model is demonstrated herein by application to the design of a new rectangular clearwell in Little Rock, Arkansas. A schematic of the clearwell is shown in Figure 1. The proposed rectangular clearwell has a volumetric capacity of 18,940 m$^3$, a length of 76.2 m, an overall width of 54.9 m, and an average water depth of 4.5 m. The proposed design features internal baffles that effectively force the flow through a series of 180 degree turns. The resulting width of each individual channel (due to use of baffles) ranges between 10.4 and 15.5 m, and the effective length of the water path is approximately 300 m. This corresponds to an effective length-to-width ratio of 300:13, or equivalently, 23:1. The peak design inflow rate is 2.19 m$^3$/sec, thus providing an average residence time of 2.4 hours at peak flow conditions. The inlet consists of two 1,524 mm diameter pipelines and the effluent is withdrawn from a rectangular sump with an area of 10.45 m$^2$. The particle paths for the originally proposed configuration are shown in Figure 1. As a result of the baffle placement, the flow establishes a serpentine path through the clearwell. After each sharp turn, the flow generally separates from the wall, thus creating "dead water" zones. This design featured a

$T_{10}/T = 0.45$, well below the target value of 0.65. This relatively low value is attributed to the presence of several moderate size recirculation zones behind the various flow turns.

In an effort to increase the $T_{10}/T$ ratio, cross baffles were implemented. The first baffle was placed a short distance downstream of the inflow, whereas the remaining three baffles were placed downstream of each of the 180 degree turns. These cross baffles (shown in Figure 2) are placed perpendicular to the direction of flow and feature variable porosities. The porosities were chosen such that the velocities through the baffles were maximized, while keeping the total head loss in the clearwell below a prescribed value. Figure 2 depicts the computed particle paths for this configuration. In general, the flow is similar to that in the rectangular clearwell without cross baffles. However, the size of the various recirculation zones is greatly reduced - a direct result of flow equalization across the width of each channel resulting from the porous baffles. The computed $T_{10}/T$ ratio for this configuration is 0.68, which represents a 51 percent improvement. This case study clearly demonstrates that a simple modification to the original design would result in significant improvements in clearwell detention characteristics and hydraulic efficiency.

Conclusions

Water plant clearwells are a major and integral component of water distribution systems that treat surface water. Proper design of these facilities entails many decisions based on cost, constructability, environmental factors, structural integrity, and disinfection effectiveness. Securing disinfection credit under the SDWAA necessitates optimizing the hydraulic efficiency and performance of these units. Poor hydrodynamic characteristics can either result in adverse effluent water quality or require the costly incorporation of additional volumes to meet regulations.

In this analysis, a three-dimensional computational fluid dynamics model was used to characterize residence time distribution and hydrodynamics of mixing within clearwells. The model can be effectively used with limited measurement data to assess modifications and evaluate improvements to detention time characteristics of existing or new clearwells. Models with dissimilar geometric, baffling methods, and inlet/outlet configurations and characteristics, and under differing operating and seasonal conditions can be inexpensively evaluated in a relatively short time. Such capabilities will greatly assist in advancing and improving water quality management practices and will ensure that cost-effective solutions can be identified.

References

Hannoun, I.A., and Boulos, P.F. (1997). "Optimizing distribution storage water quality: a hydrodynamic approach." *J. Applied Mathematical Modelling*, 21(8), 495-502.

284  WATER RESOURCES AND THE URBAN ENVIRONMENT

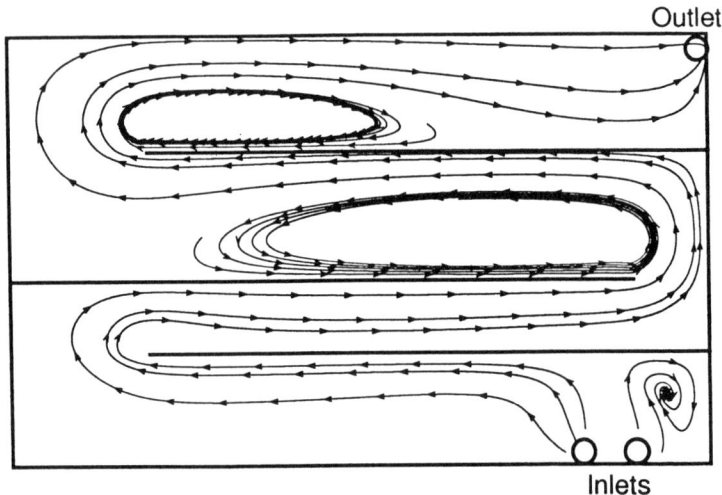

Figure 1: Particle Paths for the Originally-Proposed Configuration

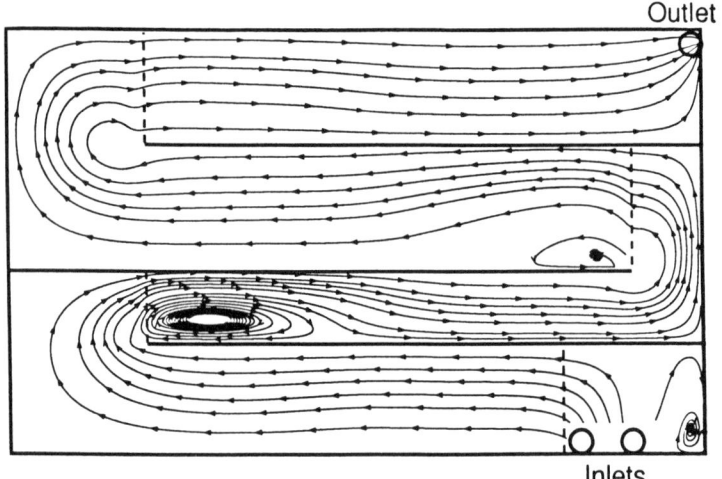

Figure 2: Particle Paths for Chosen Configuration

## Small Physical Scale Models for the Study of Mixing in Water Distribution Reservoirs

Rolf A. Deininger[1], and Andrew D. Santini[2]

**Abstract**

Distribution System Reservoirs need to be designed and operated to encourage complete mixing during the fill cycle. Small scale physical models are ideal and economical to study alternative designs of new reservoirs, and the retrofitting of existing reservoirs. The basic principles of these studies are discussed, and it is shown how the mixing in circular and rectangular reservoirs can be improved.

**Introduction**

In water distribution systems there are basically two types of reservoirs, the ones at the treatment plant, and the ones in the distribution system. Reservoirs at the treatment plant are usually designed to maximize the detention time to satisfy the disinfection requirements. The product of the concentration of the disinfectant and the detention time (CT) are prescribed by EPA regulations and depend mostly on the disinfectant chosen. In this situation the desired flow through the reservoir should be as close as possible to the plug flow.

In contrast, a distribution system reservoir should be designed for complete mixing because complete mixing will guarantee the highest disinfectant residual when the water leaves the reservoir and enters the distribution system. The time it takes for a reservoir to mix is primarily a function of the flow into the reservoir. Studies on circular reservoirs by Rossman (1) show that the mixing time can be reasonably predicted by a formula developed by Okita (2). The time it takes to complete mixing is as follows:

------------------

[1]Professor, School of Public Health, The University of Michigan, Ann Arbor, MI, 48109
[2]Student, Civil and Environmental Engineering, College of Engineering, The University of Michigan, Ann Arbor, MI, 48109

$$t = 4.6 \, H^5 \, D^{1.5} \, M^{-5}$$

where  t =   mixing time in seconds
       H =   height of the water in reservoir in meters
       D =   diameter of reservoir in meters
       M =   momentum of inflow in $m^4 \, sec^{-2}$
             (product of flow $m^3$/sec and velocity m/sec)

If the mixing time is less than the normal fill time, complete mixing will be achieved.

The above formula can be used also for rectangular reservoirs using an equivalent diameter.

## Model Laws

A physical scale model is a reduced-size representation of the prototype system of interest. Hydraulic models are usually designed to look like their prototypes, but the important factor is that they operate in a manner similar to the prototype. Various similitude criteria must be followed so that the results of the model mimic the real world system.

The laws governing models are based on a consideration of the forces of gravity, inertia, viscosity and surface tension. Dimensionless groups used in hydraulic experimentation are mainly the Froude, Reynolds, and Weber number. The Froude number represents the ratio of the inertial force to the gravity force; the Reynolds number represents the ratio of the inertial force to the viscous force; the Weber number is the ratio of the inertial to surface tension force.

The Froude number is: $F = \dfrac{V^2}{Lg}$

The Reynolds number is: $R = \dfrac{VL\rho}{\mu}$

The Weber number is: $W = \dfrac{V^2 L \rho}{\sigma}$

Where:   V = characteristic velocity of system
         L = length characteristic, diameter or depth
         g = acceleration of gravity
         $\rho$ = density of fluid
         $\mu$ = viscosity of fluid
         $\sigma$ = surface tension of fluid

Similitude exists between two geometrically similar systems when the Reynolds, Froude and Weber numbers are the same for the model and the prototype. In using physical scale models to predict the behavior of a prototype system it is seldom possible to achieve simultaneous equality of the various force ratios. And therefore the scaling laws are then based on the predominant force, and strict dynamic similarity is not attained. Reynolds modelling is used in studies of flows in pipes, and in flows of submerged jets. The Froude number is the governing factor in flows with a free surface since gravitational forces are predominant. Hydraulic structures such as spillways, weirs and water reservoirs are modelled according to the Froude law.

The modeling of the flow in a drinking water reservoir, has been based primarily on Froude's law. The fundamental relationships are the following:

$$\begin{aligned}
\text{Length Scale} &= L \\
\text{Velocity Scale} &= L^{0.5} \\
\text{Area Scale} &= L^2 \\
\text{Flow Scale} &= L^{2.5} \\
\text{Time Scale} &= L^{0.5} \\
\text{Volume Scale} &= L^3
\end{aligned}$$

Once a length scale is chosen, all other scales are fixed. For example, if a model is built with a length scale of 1:100, then the width, length and height of the model will each be 1/100th of the dimension of the actual structure. In this case, the time scale then is $(1/100)^{0.5}$ or 1/10 of the real-time value. Thus, in the operation of the model, 6 minutes would correspond to 60 minutes in the real world.

## Model Construction

The general aim of the construction of the scale models was to make them as inexpensive as possible. For this purpose all materials used were purchased at local hardware stores, and the lowest cost materials were selected. The floors and walls of the models were dictated to some degree by the wood sizes of beams and sheets, and thus a 4' x 8' plywood sheet, and a 2" x 4" beam became the major determinants of the models. Marine plywood is available at most lumber yards, but only on special order, and is roughly 3 times as expensive as regular plywood. And thus the floor of the reservoir was constructed by attaching a fiberglass or tile board to the plywood sheet using construction grade adhesives. Walls were constructed in a similar manner. Figure 1 shows a small square reservoir used in our studies. The dimensions of this model are 1.2m by 1.2m by 0.2m.

Figure 1. A small physical scale model

The entire plumbing of the inlets and outlets was done with garden hoses, plastic hoses, and plastic plumbing material which can be obtained in any hardware store. Again, the basic emphasis was to construct inexpensive models.

Since wood responds to temperature changes, sometimes cracks developed which had to be resealed. Generally speaking, water should not be left in the models when not used in an experiment, and the model should be drained promptly after every experiment.

Dyes can be added to the inflowing water in two forms, either as continuous feed or as an instantaneous addition. We have used both of these approaches, one through the use of a feed tank in which the water was colored with a dye, or in the form of addition of the dye with a syringe. Since the water feed lines were garden hoses or other plastic hoses, the injection of few cc of dye is very easy. The setting up of an experiment takes time, and we have found that injection of a dye allows two experiments to take place in one setting by first injecting a say, red dye, and after suitable time to follow up with a second, say blue dye. This allows an observation on how the first dye and water mixes, and then how the second dye mixes.

## Problem Reservoirs

Problem reservoirs, as described in the following, are reservoirs where the present configuration is not conducive to complete mixing. Typical examples are shown in the 4 figures following, and solutions to improve the mixing are shown. The structural modifications to encourage mixing are minimal, and very easily attainable.

Figure 2(a) shows in sketch form a problem reservoir since the tangential jet will cause the entire body of water to spin in circular motion. While the outer layers are mixed, the inner part of the reservoir mixes very slowly. There are two possible solutions. Figure 2(b) shows that a rotation of the inlet pipe to point it towards the center will cause good mixing. Figure 2(c) shows that a central inlet with the jet being directed straight up in the middle of the reservoir leads to very predictable and complete mixing in the shortest time.

Figure 3(a) shows an essentially rectangular reservoir with a diagonal baffle wall which is staggered based on the location of the roof support pillars. An improvement in the mixing can be achieved by removing the short sections of the baffle wall (see Figure 3(b)), and, of course, the best solution is shown in Figure 3(c).

Figure 4(a) shows a square reservoir with a central baffle wall. Better mixing of the reservoir can be achieved by splitting the flow into two jets as shown in Figure 4(b). Figure 4(c) shows mixing can be improved by removing almost entirely the baffle wall.

Figure 5(a) shows a square reservoir with 3 baffle walls. Removing the baffle walls and directing the inlet jet toward the middle of the reservoir will lead to the best mixing.

## Conclusions

The aim of achieving a complete mixing during the fill cycle can be accomplished with minimal structural modification. The use of small scale physical models is a very low cost method to get an estimate in the determination of mixing times.

## References

1. Okita, N. And Oyama, Y., Japanese Chemical Engineering, 1963, 1.

2. Rossman, L and Vidal, R, "Experimental Studies of Mixing in Water Storage Tanks", Proceedings, CSCE-ASCE Environmental Engineering Conference, July 23-25, Edmonton, Alberta, Canada 1997.

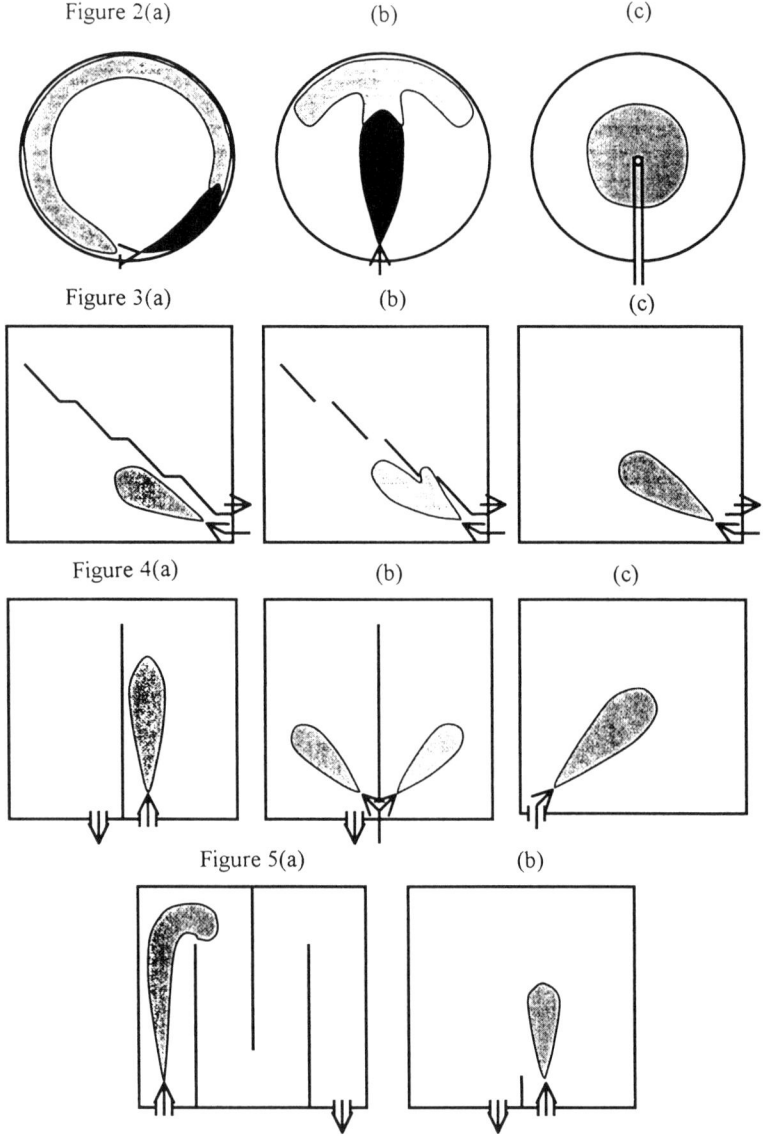

Figure 2(a) (b) (c)

Figure 3(a) (b) (c)

Figure 4(a) (b) (c)

Figure 5(a) (b)

## Water Quality Variability in a Deadend Loop

Jason T. Carter[1], YeongHo Lee[1], Steven G. Buchberger[1],
Lewis A. Rossman[2], and Eugene W. Rice[2]

### Abstract

Continuous water quality measurements were taken at the entrance and apex of a residential deadend loop. Parameters include pH, temperature, turbidity, oxidation-reduction potential, dissolved oxygen, conductivity, nitrate, and free chlorine. In addition weekly grab samples were analyzed for heterotrophic plate count, total organic carbon, and assimilable organic carbon. Unique hydraulics in deadends "trap" segments of water in the pipe line resulting in a large variability in water quality. Intensive monitoring may be required to adequately characterize regions such as these in drinking water distribution systems.

### Introduction

Aspects of the Safe Drinking Water Act (SDWA) require that utilities meet water quality standards at the point of use in addition to the point of treatment. Areas in distribution systems that experience high residence times, such as deadend mains, allow the water to increase in age before being consumed by a user. During the period between treatment and consumption, the concentration of disinfectant residual may decrease significantly allowing the regrowth of potentially harmful bacteria (Donlan and Pipes, 1988).

Within the Cincinnati area, about 23 percent of the drinking water distribution system consists of deadend mains (Schade, 1996). Deadends are sections of the distribution system which have only one connection to the looped portion of the water supply network. These water lines serve predominantly residential water users. Owing to relatively low localized usage, water in deadend mains can be stagnant for long periods of time (Buchberger and Wu, 1995).

---
[1]Department of Civil and Environmental Engineering, University of Cincinnati
P.O. Box 210071, Cincinnati, OH 45221-0071
[2]United States Environmental Protection Agency, Cincinnati, OH 45268

Most deadend mains are single stems characterized by uni-directional flow. Occasionally, a deadend main closes back on itself to create a deadend loop. Like any looped portion of the distribution system, deadend loops permit bi-directional flow in response to local demands. When bi-directional flow occurs in a deadend loop, it is possible for some parcels of water to wander back and forth in the loop and become trapped between users. These trapped parcels of water can attain a great age before being removed from the loop and consumed. A study of flows through deadend loops may offer some insight into the relationships between residential demands, travel time, and water quality in municipal distribution systems.

## Study Site

The site for this study is part of the drinking water distribution system for the City of Milford, Ohio--a community of 6,000 people just east of Cincinnati. Milford draws its water from an unconfined sand and gravel aquifer bounded on the north and west by the Little Miami River, the primary source of recharge. Three wells in the aquifer provide raw water to Milford's softening plant. The treatment regiment consists of initial volatile organic compound removal through the use of an air stripper, lime softening, sedimentation, rapid sand filters, and retention in a 55,000 gallon contact chamber where chlorine gas and fluoride are injected. The treated water is pumped to two reservoirs, which provide storage for the city. The first, Wallace Street Tank, is a 640,000 gallon standpipe, which services the city proper, including the region used in this study. The second, an elevated tank, serviced the rest of the distribution system.

The deadend loop consists of approximately 1900 feet of 6 inch cast iron main and connects to the main distribution system by a one way 6 inch stem (Figure 1). Schade's 1996 study found that the homes at the entrance of the loop extracted water with a shorter travel time than those homes located near the apex of the loop. The water extracted at the home nearest the loop entrance experienced a mean travel time of 8.2 hours after entering the loop. The house designated as the "DLA" experienced the highest travel times with a mean of about 30.9 hours.

To characterize water quality along the loop, monitoring equipment was placed at a booster pump station (BPS) near the loop entrance and in a home near the apex of the deadend loop (DLA). The continuous monitoring included the following parameters: pH, nitrate ($NO_3$), temperature, turbidity, conductivity, oxidation reduction potential (ORP), free residual chlorine, and dissolved oxygen. Less frequent measurements of heterotrophic plate counts (HPC), assimilable organic carbon (AOC), and total organic carbon (TOC) were also taken. In addition, approximate travel times and flow regime have been modeled and recorded.

## Water Quality Monitoring Design

The first monitoring station was established at the booster pump station approximately 50 feet north of the deadend loop connection. The placement was

Flosearch transmitter, a Rustrak Ranger II datalogger, a 2400 baud per second (bps) modem, and a dedicated phone line for data transmission. Figure 1 shows the schematics of the water demand monitoring equipment.

The water meter had a nutating disk in the base of the register head. The meter could monitor a continuous flow rate with a minimum resolution of 0.26 gallons per minute (gpm). Each revolution of the nutating disk was converted into a pulse signal and was sent to the data logger by the Flosearch transmitter. Every second, the data logger counted and saved the number of signals received from the transmitter. On a daily basis, a control computer at the University of Cincinnati accessed each data logger via modem and downloaded all water demand data

Downloaded data were analyzed to determine flows in the deadend loop using Hardy-Cross network analysis. Results from the Hardy-Cross calculations were used to estimate travel time of water in the deadend loop. The loop was discretized into 0.1 gallon cells, and water entering the loop was tracked until it was removed from the pipe by a consumer. Results from hydraulic modeling reveal that the predicted travel time of the water to residences along the loop ranged from 6 to 100 hours.

## Results of Water Quality Monitoring

Data collection efforts generated over 30,000 observations for each water quality parameter. Time series profiles of the water temperature and the free residual chlorine have been provided to illustrate the dynamic character of distribution systems (Figures 2 and 3). Measurements of pH and conductivity resulted in fairly constant readings throughout the year, 9.00 and 440 $\mu$S/cm, respectively. The discontinuities in the plots are a result of calibration windows, system failure, or censored data.

Typical distribution system monitoring efforts rely on infrequent grab samples. A grab sample taken at different points through out the course of the year may be misleading even for relatively stable parameters such as pH or conductivity. Currently many large utilities are considering the use of continuous water quality monitoring for this reason.

The annual profile of water temperature leads to a very interesting observation. The two monitoring sites reacted quite differently to the changes in air and ground temperature through the course of the year. In the winter months, water temperatures at the DLA were lower than water temperatures at the BPS. This may be due to the water main being closer to the ground surface at the DLA than at the BPS. As the summer approached, the condition reversed. This behavior suggests that DLA is more sensitive than the BPS to air temperature owing to relatively long residence times of water in the deadend loop. It is also worth noting that the variance in temperatures also increased during the summer. What effect might these changes in the variability of a particular water quality parameter have on monitoring strategy?

The level of residual free chlorine is an important indicator of water quality in a distribution system. Of all the parameters characterized by a grab sample, it may also be the most misleading measurement. As is apparent in the Figure 3, two

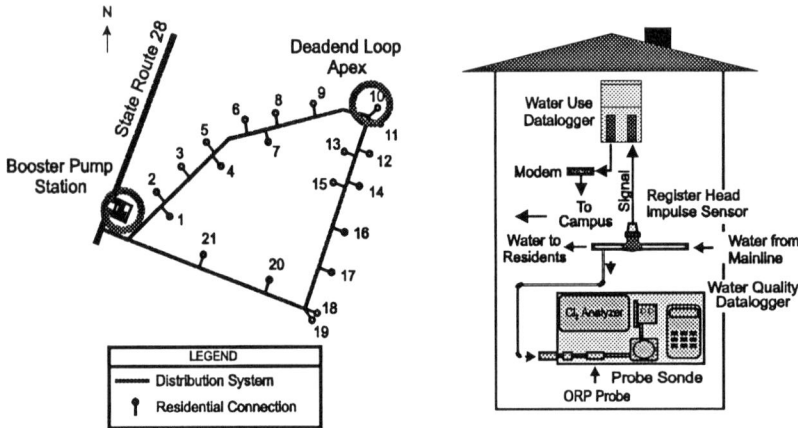

Figure 1: Site and Water Monitoring Schematic

dictated by accessibility to the 8 inch main line running parallel to the highway. A tap was set into the main and a slow bleed was drawn and sent through the water quality monitoring unit. A home near the apex of the loop was chosen as the second location and a 1/2 inch copper line was installed in the house to draw the bleed necessary for the water quality monitoring probes.

The two water quality monitoring units were constructed in an identical manner (Figure 1). A flow control valve (FCV) was used first to regulate the flowrate through the unit. The FCV water passed through a 1/2 inch Teflon tube connected to a PVC tee housing the Sensorex ORP probe. The water then traveled to a flow-through cell via 1/2 inch Teflon tubing and polyethylene pressure fittings. A Solomat 803 PS water quality probe sonde was screwed into the cell. The sonde had electrodes, which measured: pH, $NO_3$, temperature, dissolved oxygen, conductivity, and turbidity. The water then passed through an acrylic constant head flow-through cell where an ATI polaragraphic membrane probe monitored the free chlorine residual.

Measurements from the monitoring unit were taken continuously as the bleed traveled through it. During this process each probe was queued by the datalogger at which point the reading was recorded. A recording cycle through the probes occurred every ten minutes. Grab samples for HPC were taken weekly and samples for AOC and TOC measurements were taken bi-weekly to characterize bacterial growth at each location.

## Water Demand Monitoring Design

Water demand was recorded at all 21 houses along the deadend loop. Each house was equipped with a 5/8-inch Model T-10 Neptune water meter, a Neptune

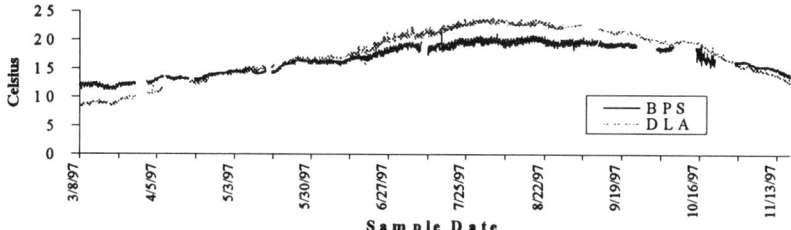

Figure 2: Comparison of Water Temperature at the BPS and DLA

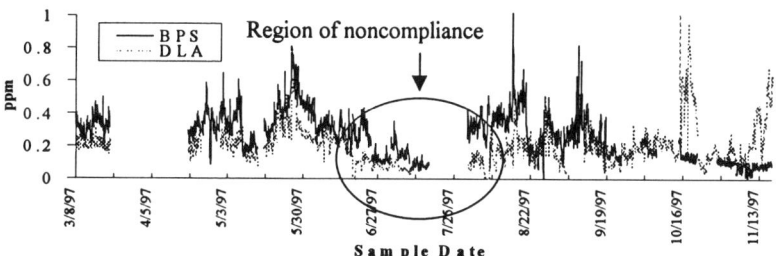

Figure 3: Comparison of Free Chlorine Residual at the BPS and DLA

samples taken at different times would provide very different levels of chlorine residual. The majority of the time, the BPS held a higher residual than did the DLA. During the summer months the residual was not maintained adequately and fell beneath 0.2 ppm standard. However, these excursions went undetected due to the current grab sampling monitoring technique. Inadequately disinfected water entering the distribution system represents a significant health hazard to the public through microbial regrowth in the pipe network.

A total of 162, 119, and 78 measurements of HPCs, TOC and AOC, respectively, were made of bulk water samples taken from the distribution system. Despite the decreasing trend in the chlorine residual, the bacterial numbers detected in the bulk fluid decreased by a factor of 10 between the BPS and DLA when averaged over the course of the year. Overall consumption of AOC was also very small. With the advent of spring and summer the number of bacteria increased as well as the consumption of AOC (Figures 4 and 5) at both locations in the loop. During these warmer months water trapped in the loop by the demand patterns may allow the bacterial numbers to increase in small sections of the pipe network. A continuous monitoring scheme would have allowed a quick response to the decrease in the disinfectant residual or an increase in the bacteria numbers present in the water line. This would enable the operators to implement improved management techniques when dealing with deadend regions.

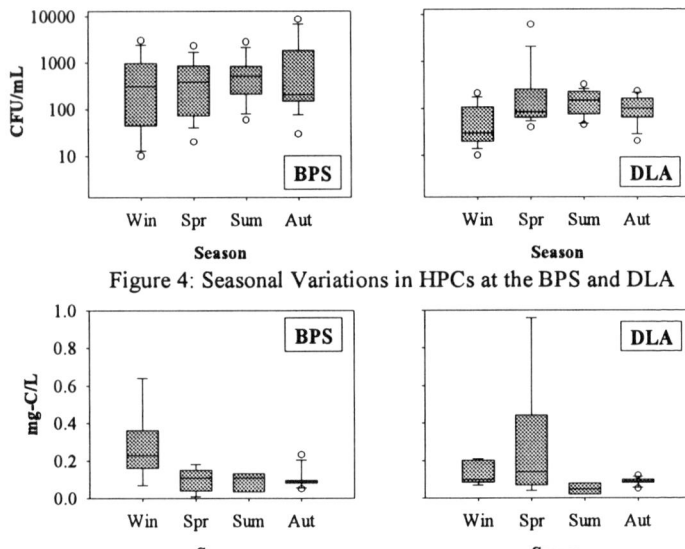

Figure 4: Seasonal Variations in HPCs at the BPS and DLA

Figure 5: Seasonal Variations in AOC at the BPS and DLA

## Summary

Water quality within a distribution system may not be adequately characterized with intermittent grab samples. Failures to maintain disinfectant residuals may go undetected under this standard method of sampling. Deadend loops allow water to experience widely varying residence times, which allow almost complete disappearance of the disinfectant. Without a disinfectant, residual bacterial can regrow in these regions of the distribution system and become a problem. Continuous monitoring in the pipe networks will provide a way to understand the time dependent processes occurring in each water distribution system. Techniques such as flushing or changes in disinfectant residual maintenance can be optimized to provide a constant supply of high quality drinking water to the public.

## Acknowledgments

Support was provided by the National Science Foundation (NYI Award BCS-9257608) and by the American Water Works Association Research Foundation.

1. Buchberger, S. G. and L. Wu. Model for Instantaneous Residential Water Demands, *Journal of Hydraulic Engineering*, 121:3:232-246 (1995).
2. Donlan, R. M. and W. O. Pipes. Selected Drinking Water Characteristics and Attached Microbial Population Density, *AWWA Journal of Research and Technology*, 80:70 (1988).
3. Schade, T. G. Water Demands and Travel Times in a Residential Dead-End Loop, *MS Thesis*, University of Cincinnati, 64 pages (1996).

## OPTIMAL DECOUPLING OF BOOSTER DISINFECTION SYSTEMS IN WATER DISTRIBUTION NETWORKS

James G. Uber[1], Marios M. Polycarpou[2], and Prathiba Subramaniam[3]

**Abstract**

Booster disinfection involves the dosing of a disinfectant, such as chlorine, at points located within a distribution network (i.e. after treatment and before consumption). This paper formulates three related optimization models for locating booster disinfection stations in water distribution networks, using as inspiration the class of maximum covering location models. All three models seek to minimize the number of booster stations required, and to maximize the independence of booster station influences on the spatial and temporal distribution of disinfectant residual. By decoupling the influence of each booster, it will be easier to develop reliable feedback control algorithms based on residual concentration measurements, since then there exists a unique relationship between residual concentration at a monitoring point and dosing at a booster station. While in general it may not be possible to achieve such a high level of booster decoupling, two of the models attempt to maximize the degree of decoupling, as much as allowed.

**Introduction**

Drinking water delivered to consumers should contain a small disinfectant residual to reduce the risk of human exposure to pathogens. This residual should not be excessively large anywhere in the distribution system (at consumption points), because excess amounts of the most common disinfectant — chlorine — may not appreciably reduce the health risk of pathogen exposure (compared to small residuals in the 0.2 mg/l range), while at the same time may increase unnecessarily the health risk associated with disinfection by–product exposure. As a result, distribution system disinfectant residual man-

---

1. Associate Professor, Dept. of Civil & Environmental Eng., University of Cincinnati, Cincinnati, OH 45221

2. Assistant Professor, Dept. of Electrical & Computer Eng. & Computer Sci., University of Cincinnati, Cincinnati, OH 45221

3. Graduate Research Assistant, Dept. of Civil & Environmental Eng., University of Cincinnati, Cincinnati, OH 45221

agement is receiving increasing attention, and methods are being developed to minimize the disinfectant required to maintain small residuals at each point of consumption.

Any desire to accurately manage disinfectant residuals in the distribution system would be practically impossible if using only traditional control options. These options are limited to adjustment of disinfectant residuals at points of entry into the distribution system (i.e., at treatment plant clear wells or at groundwater wells). In order to maintain small residuals at locations far from the source, or associated with large disinfectant decay, one would anticipate higher than needed residuals in other areas of the distribution system. One approach that should allow more precise spatial control of disinfectant residuals is booster disinfection, where disinfectant is added to the drinking water at points located in the distribution system [Tryby et al 1997, Boccelli et al. 1998]. By allowing disinfectant control at locations throughout the distribution network, one might more precisely control the spatial and temporal residual distribution.

The work presented here is motivated by feedback control of disinfection residuals through the use of booster disinfection. Specifically, we envision a scenario whereby a number of booster stations are located in a distribution network, each with its associated controller. These controllers would rely on feedback about the state of the network, as provided by, e.g., flow rate, pressure, and (most important) residual concentration measurements. Each controller maintains target residual concentrations at selected monitoring locations in the presence of varying consumer demands and water quality parameters.

In the general case, the residual at any monitoring location will include significant contributions from a number of booster stations. Indeed, because of the looped nature of water distribution networks, the presence of storage reservoirs, and variations in water consumption rates, the disinfectant residual at any monitoring point can be a complicated superposition of historic doses occurring at various booster locations [Boccelli et al. 1998]. From a controls perspective, the coupling of residual with disinfectant doses at multiple booster locations would imply an interdependence between the dose controllers. This interdependence might be modeled explicitly as part of the controller, or be handled as an external disturbance; either way the interdependence of booster station controllers would add complexity to the controller design problem. Thus as a practical issue one might wish that such interdependence is minimized. Below we formulate an optimal booster location model that attempts to do just that.

## Maximum Decoupling Design Model

### Basic Model

We seek to compute the minimum number of booster stations, and their locations in the network, such that the residual concentration at each monitoring location is (within limits, to be discussed below) affected by dosing at one, and only one, booster location. In this case, we could then approach the design of booster dose controllers as if they were independent, each regulating the residual concentration at a subset of monitoring locations. This maximum decoupling design (MDD) problem is formulated as a maximal covering location model [Church and Revelle, 1974], which has also motivated methods to locate wells in groundwater monitoring networks [Meyer and Brill, 1988].

Consider a number of discrete potential booster station locations ($n_b$), and discrete monitoring locations ($n_m$) where disinfectant residual is to be controlled (in application, the potential booster and monitoring locations will correspond to nodes in a network model). For design purposes, the network flow rates and disinfectant concentrations are assumed to be time dependent and periodic, and we sample the disinfectant concentrations over a set of discrete time intervals, $k = 1, 2, \ldots, K$, corresponding to one period of the dynamics (Boccelli et al. [1998] give a rationale for this, as well as conditions under which concentrations are periodic). Further, define a set of potential booster locations, $N_j^k$, such that booster location $i$ is included in $N_j^k$ if and only if disinfectant dosing at $i$ elicits a "significant response" at monitoring location $j$ and time interval $k$. Thus $N_j^k$ is the set of all potential booster locations that are assumed to have an effect on residual at monitoring location $j$ and time $k$. We discuss the definition of $N_j^k$ below, as it is central to the formulation of the MDD model as a maximal covering location model.

Given the above definitions, we seek to minimize the number of booster stations (a surrogate objective for minimizing the cost of booster station installation and operation):

$$\min_x \sum_{i=1}^{n_b} x_i \tag{1}$$

where $x_i$ is a binary variable equal to 1 if a booster station is to be constructed at location $i$, and zero otherwise. In minimizing the number of booster stations, we require that each monitoring location and time include a significant response from at least one selected booster location, else it will be impossible to control residual concentration at that monitoring point. Mathematically, we have:

$$\sum_{i \in N_j^k} x_i \geq 1, \quad j = 1, \ldots, n_m, \quad k = 1, \ldots, K \tag{2}$$

Solution of (1) subject to (2) will yield the minimum number of booster locations that provide spatial and temporal "coverage" for the network, in that the residual concentration at each monitoring location and time is influenced by at least one selected location. The effect of each booster location will not necessarily be "decoupled" in any sense as described above. To ensure that each monitoring location is "served" by the dose at one and only one booster location, we define a new set of booster locations, $N_j$, that contains all booster locations eliciting a significant dose response at monitoring location $j$, at *any* time:

$$N_j = N_j^1 \cup N_j^2 \cup \cdots \cup N_j^K \tag{3}$$

We allow only one booster location to be selected out of each set $N_j$:

$$\sum_{i \in N_j} x_i = 1, \quad j = 1, \ldots, n_m \tag{4}$$

The basic MDD model is defined by minimizing eq. (1), subject to eqs. (2) and (4), with the added constraint that $x$ is binary:

$$x_i = \{0, 1\}, \quad i = 1, \ldots, n_b. \tag{5}$$

Note that (4) implies that (2) will be binding, but we leave (2) as an inequality to facilitate modifications to the basic MDD model presented below. These modifications address the concern that the above model may not possess any feasible solutions, because of the severity of requirement (4). Yet even if it is impossible to identify the response at each monitoring location with a single booster station, it may be meaningful to minimize the number of contributing locations as much as possible.

*Multiobjective MDD Model — Average Contributing Booster Locations*

We define a multiple objective MDD problem that minimizes a weighted sum of the number of booster stations (cost) and the average number of booster stations contributing to each monitoring point residual (decoupling):

$$\min_{x,y} \quad w \sum_{i=1}^{n_b} x_i + (1-w)(\frac{1}{n_m} \sum_{j=1}^{n_m} y_j) \qquad (6)$$

where $y_j$ is the number of booster stations contributing to residual concentration at monitoring location, $j$, and $0 \le w \le 1$ is a constant weight that regulates the emphasis on cost versus decoupling. Eq. (6) is minimized subject to (2) and (5), plus constraints similar to (4) that define the $y_j$:

$$\sum_{i \in N_j} x_i \le y_j, \quad j = 1, \ldots, n_m \qquad (7)$$

Note that (7) will be binding at the optimum of (6), and thus $y$ will be binary if $x$ is binary.

*Multiobjective MDD Model — Maximum Contributing Booster Locations*

The final model considered is similar to the above multiobjective formulation, but instead of considering the average number of contributing booster stations we consider minimizing the maximum number over all monitoring locations:

$$\min_{x,\hat{y}} \quad w \sum_{i=1}^{n_b} x_i + (1-w)\hat{y} \qquad (8)$$

where $\hat{y}$ is the maximum number of contributing booster stations, over all monitoring locations. Eq. (8) is minimized subject to (2) and (5), plus constraints that define $\hat{y}$:

$$\sum_{i \in N_j} x_i \le \hat{y}, \quad j = 1, \ldots, n_m \qquad (9)$$

Eq. (9) will be binding for some subset of the monitoring locations $j$. These locations are associated with the maximum number of contributing booster locations, or the minimum degree of decoupling.

*Dose Response and Significance Level*

In the above development we introduced the notion of whether, or not, dosing at a booster location elicits a significant response at a monitoring location and time. This notion of significant response is addressed here, along with definition of the sets $N_j^k$ (and thus also the sets $N_j$).

Define the response coefficients, $\alpha_{ij}^k$ ($[M/L^3]/[M/L^3]$), as the disinfectant concentration observed at monitoring location $j$ and time $k$, due to maintaining a unit set–point concentration at booster location $i$ for all time $m < k$ [Boccelli et al. 1998, Zierolf et al. 1998]. Thus the concentration at location $j$ and time $k$ due to an historic set–point concentration $v_i$ at booster location $i$ is simply $c_j^k = v_i \alpha_{ij}^k$. This development assumes that disinfectant decay in the distribution system behaves as a first–order reaction (a common assumption), in which case linear superposition implies linearity of dose response and summation of individual dose effects.

One measure of whether, or not, a dose at booster location $i$ has a significant effect on monitoring location $j$ and time $k$ is simply whether, or not, $c_j^k = v_i a_{ij}^k > \hat{c}$, where $\hat{c}$ (M/L$^3$) is an assumed dose response significance level. For example, we might assume that an individual booster station contribution less than 0.05 mg/l is "insignificant" ($\hat{c}$ = 0.05 mg/l). In the absence of *a priori* information to suggest otherwise, we assume the same set point concentration, $\hat{v}$, for all potential booster locations, so that $c_j^k = \hat{v} a_{ij}^k$. Our definition of significant response is then $\hat{v} a_{ij}^k > \hat{c}$, or more simply $a_{ij}^k > \eta = \hat{c}/\hat{v}$, where $\eta$ is a dose response significance level interpreted as the monitoring concentration resulting from a unit set–point concentration. Thus the notion of dose response significance is reduced to whether or not the response coefficients exceed the threshold $\eta$.

Using the above definition, we define the significance sets $N_j^k(\eta)$ as the set of all booster locations having a dose response at monitoring location $j$, and time $k$, that exceeds the significance threshold $\eta$:

$$N_j^k = \{i \mid a_{ij}^k > \eta\} \tag{10}$$

The significance sets $N_j$ are then defined by eq. (3).

It is clear from eq. (10) that the MDD model solutions depend on the somewhat arbitrary threshold parameter $\eta$. If one of these solutions is implemented, and if a dose controller specifies an actual set point different than unity, then the degree of coupling (at threshold level $\eta$) among the booster stations may be different from that computed by the MDD model. Two cases are possible: (1) If the actual set point is greater than unity, then two or more booster station controllers that were decoupled in the MDD model solution may in reality be coupled together; (2) If the actual set point is less than unity, then two or more booster station controllers that were coupled may in reality be decoupled. Given that actual doses are not known *a priori*, there seems no way to avoid interactions between the optimal dose control and booster location problems — short of solving them simultaneously. One can, however, check the degree of booster station response coupling, for any particular MDD locations, by simulating the actual doses that are anticipated given those locations.

### Solution of the Model

The MDD model formulations are in the class of integer linear programming (ILP) or mixed integer linear programming (MILP) optimization models. Solution efficiency can be problem dependent, but similar models (in other application contexts) have been solved using linear programming plus heuristics, linear programming plus branch and bound implicit enumeration, and simulated annealing [Church and Revelle 1974, Meyer and Brill 1988, Meyer et al. 1994].

### Acknowledgements

The authors appreciate ongoing assistance from Dominic Boccelli and Michael Tryby, of the University of Cincinnati, related to implementation and application of the

MDD model, and many discussions with Tim O'Connell and Amy Petticrew, also at the University of Cincinnati, related to control issues in water distribution systems.

**References**

1. Boccelli, D., Tryby, M., Uber, J., Rossman, L., Zierolf, M., and M. Polycarpou, "Optimal Scheduling of Booster Disinfection in Water Distribution Systems," *J. of Water Res. Plan. Manag.*, A.S.C.E., 124(2), Mar/Apr, 1998.
2. Church, R., and C. Revelle, "The Maximal Covering Location Problem," *Pap. Reg. Sci. Assoc.*, 32, 101–118, 1974.
3. Meyer, P., and E. D. Brill, Jr., "A Method for Locating Wells in a Groundwater Monitoring Network Under Conditions of Uncertainty," *Water Resour. Res.*, 24(8), August, 1988.
4. Meyer, P., Valocchi, A., and J. W. Eheart, "Monitoring Network Design to Provide Initial Detection of Groundwater Contamination," *Water Resources Research*, 30(9), 1994.
5. Tryby, M., Boccelli, D., Uber, J., and L. Rossman, "Optimal Scheduling of Booster Disinfection," *Proceedings, AWWA Annual Conference and Exposition*, 1997.
6. Zierolf, M., Polycarpou, M., and J. Uber, "Development and Auto–Calibration of an Input–Output Model of Chlorine Transport in Drinking Water Distribution Systems," *I.E.E.E. Trans. Control Sys. Tech.*, 6(2), March, 1998.

## Using Back-Up Sub-Systems for Design and Operation of Reliable Multi-Quality Systems

By Avi Ostfeld [1], Member, and Uri Shamir [2], Fellow, ASCE

**Abstract**

A methodology for integrating Back-Up Sub-Systems into the design or operation of reliable multi-quality water distribution systems is formulated. Back-Up Sub-Systems are sub-systems of the entire network which survive when a failure occurs and whose performance is considered explicitly in the design or operation processes. This paper focuses on the stages involved in using Back-Up Sub-Systems, and on its properties.

**Introduction**

Reliability of a water distribution system is a measure of how well it performs under emergency conditions, such as failure of its components. For single quality systems the performance criteria are meeting consumer demands with adequate pressure. In multi-quality systems an additional criterion is that quality of the water supplied must be within specified limits.

The quantity, pressure, and quality standards which must be met when a failure occurs, and until it is corrected, may be somewhat reduced from those which must hold under normal operating conditions, since the failure condition is only temporary (the criteria under failure may be the same as for normal times, or reduced, if desired). When failure occurs, it is the "residual system", reduced from the full system by the failed element(s), which must provide the emergency level of service.

This is the basis for our approach to the design of reliable multi-quality systems: we seek a system which delivers the normal level of service (quantities, pressures, qualities) when it is fully operating, and any single-link failure has a residual "Back-Up Sub-System" that delivers the specified emergency level of service - all at minimum total cost.

The field of reliability-based design for distribution systems is still not fully developed, although recent years have seen considerable advances. We are still seeking measures of reliability which are meaningful to the consumers and yet can be used practically and

---
[1] Senior Engineer, Tahal Consulting Engineers Ltd., 54 Ibn Gvirol St., Tel-Aviv, Israel
[2] Professor, Department of Civil Engineering, Technion - I.I.T, Haifa 32000, Israel

effectively in the process of optimal design and operation. This paper is based on a new concept: performance under failure is incorporated directly into the optimal design model, rather than being the result of some post-optimality analysis.

## Methodology

When a water distribution system component fails (e.g. pump, pipe) there are two results: (1) isolation of the failed component by valve closure to allow its repair or replacement, and (2) redistribution of the flows in the remaining system.

The ability of a system to meet its consumers demands during failure depends on its inherent redundancy, and on the hydraulic capacity of its functional remaining components.

The use of Back-Up Sub-Systems (referred to as Backups) is for designing or operating a water distribution system in a way which will a priori guarantee an adequate level of service once a component, or portions of the system are out of service.

The first to introduce the idea of using Backups were Kessler et al. (1990) and Ormsbee and Kessler (1990) who presented a model for explicitly incorporating reliability in the optimal design of water distribution networks fed by a single source. We followed and generalized these ideas for single and multi-quality water distribution systems (Ostfeld and Shamir, 1996).

The following stages constitute the methodology for inclusion of Backups in the optimal design or operation of single or multi-quality water distribution systems:

(1) Formulation of a model for optimal design or operation of the system considered, minimizing costs of construction and/or operation and maintenance.

(2) Identification of Backups which will maintain a prescribed level of service when a failure occurs.
   A Backup is a sub-set of links and nodes of the full system; two Backups can be defined such that if any one link in the full network fails, then one of these two Backups "survives", i.e. remains intact.

(3) The hydraulic laws and consumer demands are formulated separately for each of the Backups and loading conditions considered, together with the level of service required for each.

(4) The models for the Backups are added to the model of the complete system, and an optimization model is solved.

This formulation creates an expanded design or operation model that minimize total cost, subject to constraints on residual system performance. Explicit inclusion of the required redundancy of the system is addressed by inclusion of the Backups as part of the constraints. Still, one remains with the problem of how to select Backups in a rational manner.

A set of Backups is good if it has two properties:

(1) It gives a proper meaning to reliability, i.e. the Backups are able to supply the required residual service to the consumers.
(2) The optimal solution of the design or operation model with a given set of Backups is competitive to other sets that guarantee a similar level of residual system performance.

The selection of a good set of Backups is analogous to the selection of a good system layout, for which there is no quantitative model, and is primarily based on engineering judgement and experience.

One way to select the Backups, given a water distribution network layout, is by searching for two spanning trees whose distance is maximum, where the distance between two trees is defined as the number of arcs contained in one tree but not in the other (Kishi and Kajitani ,1969). This procedure defines two Backups with the following properties:

1. The set of nodes of each of the Backups equals the set of the system nodes.
2. The union of the sets of arcs of the two Backups equals the set of the system arcs.
3. Each of the Backups is a connected graph.
4. The number of common arcs of the two Backups is minimal.

The tradeoff between cost and reliability can be evaluated by changing the selection of the Backups for given consumers demands, or by reducing the consumers requirements for a given set of Backups.

Using a transportation simplified model (i.e. a model that does not take explicitly the hydraulics into consideration) for selecting the Backups, based on topological considerations (e.g. maximum distance between spanning trees), the consumers demands, and the cost of designing or operating the system -- can be another way for choosing Backups.

## Conclusions

Much remains to be done in the domain of distribution system reliability, for single- and multi-quality systems. There is a very large number of reliability measures which are relevant to the various consumers. But if we wish to include reliability considerations directly into the design process, this requires selecting specific criteria and developing models which incorporate them into the constraints and/or objective function. The approach presented in this paper, applied to multi-quality systems, provides a way of doing this.

## Acknowledgment

This work was partially funded by the Water Research Institute (WRI) at the Technion-Israel Institute of Technology, under research project number 015-056.

## References

Kessler A., Ormsbee L. E., and Shamir U. (1990). "A methodology for least cost design of invulnerable water distribution networks". Civil Engineering Systems, Vol. 1, pp. 20 - 28.

Kishi G. and Kajitani Y. (1969). "Maximally Distant Trees and Principal Partition of a Linear Graph". IEEE Transactions on Circuit Theory, Vol. CT - 16, No. 3, pp. 323 - 330.

Ormsbee L. E. and Kessler A. (1990). "Optimal Upgrading of Hydraulic - Network Reliability". Journal of Water Resources Planning and Management Division, ASCE, Vol. 116, No. 6, pp. 784 - 802.

Ostfeld A. and Shamir U. (1996). "Design of Optimal Reliable Multiquality Water Supply Systems", Journal of Water Resources Planning and Management Division, ASCE, Vol. 122, No. 5, pp. 322 - 333.

# Environmental Systems Rehab in Deschapelles, Haiti

Bruce W. Berdanier, Ph.D., P.E.[1]

## Abstract

Hospital Albert Schweitzer (HAS) was established in Deschapelles, Haiti in 1954 by Dr. and Mrs. William Larimer Mellon. HAS was conceived, planned and built as a state-of-the art hospital for a tropical region. The environmental support systems of the hospital included: water collection, treatment, storage, and distribution; sanitary sewerage collection system; wastewater treatment; and solid waste incineration. The work planned for the subject project is for rehabilitating and replacing the HAS environmental support systems. Although these systems were well conceived and built in 1954, their typical useful design life is approximately 30 years. The staff has kept the systems operational with maintenance and design changes through the years, but major rehabilitation and replacement is now required for continued support of HAS operations and to protect the health of the staff and of the residents of the region.

## Introduction

Hospital Albert Schweitzer (HAS) was established in Deschapelles, Haiti in 1954 by Dr. and Mrs. William Larimer Mellon. The 120 bed hospital provides medicine, pediatric services, and surgery to an area of 620 square miles and approximately 220,000 people in the Artibonite Valley region of Haiti. In addition to the main hospital complex, HAS also staffs and directs several outlying dispensaries and conducts health training through it community health program. Education in areas of agriculture, tree planting, and well drilling and latrine construction assistance are also supplied to the service area through the HAS community development program. HAS is supported today by the Grant Foundation in Sarasota, Florida.

HAS is located in the village of Deschapelles approximately 90 miles northeasterly of Port au Prince. The terrain is mountainous with severe erosional characteristics due to continuing drought conditions and almost total lack of any trees. Less than 2% of Haiti's 10,700 square miles is estimated to be covered by forests today. Agricultural development programs, such as the HAS tree planting operation, have planted millions of trees over the past ten years many of which have been harvested for the production of charcoal. Haiti is the poorest country in Americas with an average per capita income of perhaps $300 US.

---
[1]Assistant Professor of Civil and Environmental Engineering, South Dakota School of Mines and Technology, 501 East St. Joseph Street, Rapid City, SD 57701-3995.

The hospital was conceived, planned, and built as a state-of-the-art hospital for a tropical region. The environmental support systems of the hospital included: water distribution and treatment, sanitary sewerage collection system, wastewater treatment, and solid waste incineration. The hospital building structure is basically in good condition for the age of the facility due to good initial planning and construction methods. A major reorganization of the entrance and outpatient ward was just recently completed to better serve the users by separating administrative service areas from medical service areas.

The work described in this presentation is for rehabilitating and replacing the environmental support systems. Although, these systems were well conceived and built in 1954, their typical useful design life is approximately thirty years. The staff has kept the systems operational through maintenance, and small design changes through the years, but major rehabilitation and replacement is now necessary for continued support of HAS operations and to protect the health of the staff and of the residents of the region. The cost of the work required to complete necessary upgrades is projected to be approximately $360,000 which is well beyond the annual operation funds which are available for system maintenance (Berdanier and Beck, 1997).

In September of 1995, Martin Hettich, a European educated electro-technical engineer working in the Engineering and Maintenance division of the hospital wrote a "white paper" to the Board of Directors of HAS (Hettich, 1995). Hettich's treatise was a fairly detailed effort to quantify the different waste streams produced by HAS operations each year. An informational matrix was constructed delineating the HAS department producing the waste, level of toxicity of the waste, and amount produced each year.

HAS conducted an environmental feasibility study (Berdanier, March 1996) to develop an overall conceptual plan of the work that needed to be accomplished. The 1996 study evaluated all of the systems and prepared preliminary cost estimates for materials based on experience of construction in the United States.

Wastewater Treatment Plant

The original wastewater treatment plant was a primary digester followed by dosing pumps and sand filters. At some point over the past 40 years the sand beds were replaced with lagoons. The lagoons are a better concept for this region for wastewater treatment and for simplicity of operation. However, the current system is almost completely nonfunctional at the present time. The primary digester is bypassed (due to valve, pump, and piping failure) into the first lagoon. The walls and bottom of both lagoons have lost all integrity due to age and weathering conditions.

The recommendation for the wastewater treatment plant rehabilitation is to rebuild the two lagoons and provide piping for them to be operated in series or parallel mode. The primary digester should be reused if possible after cleaning. All necessary piping and control valves between the digester and lagoons will have to be replaced. A simple submerged rock filter could be provided at the lagoon outlets to provide additional treatment. Pumping should be eliminated if sufficient head is available for gravity operation. Two small sludge drying beds could be built to dry the sludge which is periodically removed from the primary digester. Again, it would be best if this operation can be accomplished without pumping.

Sanitary Sewerage Collection System

The sanitary sewerage system serving the southern wing of the building is currently non-functional. This portion of the system which serves the laboratories, pediatrics, surgery, and clinics is currently bypassed into a nearby surface water canal. A rehabilitation of the system in the area of the garage was accomplished in 1996 to alleviate backup of the sewerage system in the rest of the hospital.

Concrete manholes need to be established at strategic points throughout the hospital sewerage system to allow maintenance personnel reasonable access to the system to clean and maintain it with the equipment that is available to them. Additionally, the system serving the southern wing of the hospital, and the outlet portion of the system from the hospital to the wastewater treatment plant need to be replaced and relocated. Once manholes are established on the system within the hospital allowing proper inspection of the system, it may be determined that additional line replacement will be required.

Water Tower

The elevated water tower is a 50,000 gallon storage tank 125 feet in height which was built in 1954. The last complete painting of the tank exterior was in 1983. An inspection in 1988 recommended that the tank be repainted in 2 - 3 years. There is considerable surface rust and pin hole leaks developing on the outside of the tank indicating the need for complete cleaning and some welding of the tank bowl.

The tank should be drained to accomplish the interior work. The tank will be cleaned with a detergent and wire brush hand tool or sand blasted to remove rust. Holes will be welded. The interior of the tank will be painted with a two coat epoxy paint system which will last for approximately 15 years. The tank should then be chlorinated and put back on line. The exterior of the tank can be completed while the tank is on line using moisture cured urethane paint system. The outside of the tank will also require cleaning or sandblasting to remove rust and existing paint system. Due to the specialized nature of the work, the elevation of the tank above the ground, and the specifications for the paint systems, a specialized company will be required to do this work (possibly a foreign company).

Water Distribution Lines

The water distribution lines come to the HAS area from two different raw water sources. The main supply source for the hospital is called the La Forge spring while the source for the village and the housing area of the HAS campus is the Ca Charles spring. Additionally, there are several deep wells that are tied into the system at various locations. The untreated water delivery lines from La Forge were constructed in 1954. However some of the campus distribution lines and the Ca Charles line predate Dr. Mellon's arrival and were constructed in the 1940's when this area was a banana plantation.

The hospital and citizens of the village participate in joint source water protection projects through the community development division at HAS to protect the raw water quality. Spring boxes collect the surface water from the gorges above the village and the hospital and feed into the gravity water lines to the village and the hospital. After all available data was reviewed, the water system for this region was found to be extremely complicated due to the interconnections of the different systems and lack of any substantial records of the piping networks. The water storage reservoirs for the village and the hospital are the only clearly separate components of the systems. A further more complete field study will need to be accomplished on this portion of the project before any major corrections can be made in the system design or operations.

The system has many leaks both in the untreated water lines delivering water from La Forge and Ca Charles and in the treated water distribution lines on the HAS campus. Additionally, cross connections have unintentionally been created in the system over the years due to lack of records and attempts to maintain pressure in the system. Major renovation work was completed at the La Forge spring box in 1996 to keep flood waters from contaminating the spring box.

Considerable rehabilitation of the system can be completed by replacing obviously broken lines in both systems and installing new control valves and boxes at proper places in the system. The piping needs to be properly selected as ductile cast iron waterline pipe for use in this area. In some areas in the untreated water supply system sections of the line need to be relocated, replaced , and properly buried due to problems caused by erosion in the gorge route from the spring box to the hospital. Considerable information for analysis of the system could be gathered by mapping the system while designing the repair locations.

## Incineration

The original HAS incinerator which is no longer functional was a refractory lined, two chamber, external fuel injected system. The system suffered a major explosive breakdown approximately a decade ago and has been nonfunctional since that time. The system was replaced with two open grate masonry buildings. One of these suffered a burn out some years ago due to overheating resulting in only one building still being operational. The heat generated from self combustion of all of the HAS waste is insufficient for safe destruction of the medical wastes generated in the hospital. The current site is too close to the hospital and the open canal causing severe risk to the staff, patients, and residents of the area due to both air and water contamination.

Use of the current burning building should be eliminated, and two new steel incinerators should be installed at a further distance from the hospital. Final location will be recommended in the design phase. The incinerators need to be very functional in design and simplistic in operation. The design should be a one chamber, steel fabricated, manual waste feed, with an external fuel oil feed that does not rely on a sophisticated injection system. The stacks should be supported on their own foundations and should be approximately 10 meters high. Additionally waste segregation should be conducted as a management policy to compost as much of the waste as possible.

## Additional Considerations

The feasibility study of 1996 also recommended water treatment plant rehabilitation. It is recognized that a new filtration plant would provide considerable protection against protozoa in the drinking water supply which could be introduced by animal fecal contamination. However, the rehabilitations recommended above in the water distribution and sanitary collection systems will go a long way toward helping this situations. Additionally, the current manual addition of chlorine is probably the simplest method to provide the most protection against bacterial contamination in the system. Disinfection potential of the chlorine could be improved through management of the chemical addition. Also, the safety of the system can be improved through elimination of cross connections between treated and untreated water systems. Filtration should still be considered at some time in the future.

Waste oil contamination of the soil and existing landfill areas were also discussed in the 1996 study. Oils from the generators and automobile maintenance operations have been generally disposed of through infiltration in the soils near the hospital. These areas are also important and still need to be addressed. Waste oil contaminated soil can be cleaned up during the sanitary sewerage construction as can the landfill areas. Future contamination of soil with waste oil should be prevented by selling all waste oil to in-country factories for reuse.

## Conclusions

The overriding concerns for the final project design were simplicity of installation, operation, and maintenance. The systems need to be very durable in their ongoing operation. The remoteness of this region and lack of a stable economy make it very difficult to operate and maintain systems that require any type of fairly sophisticated control mechanisms. Replacement parts can be very difficult to obtain in any reasonable period of time for a reasonable cost. This market instability for product acquisition leads to continual on-site modifications of mechanical systems by the maintenance staff whenever repairs are required.

## References

Berdanier, B.W., 1996, " Report of HAS Site Visit for Environmental Issues", Prepared for Board of Directors, Grant Foundation.

Berdanier, B.W., and Beck, U.J, 1997, "Proposal for Rehabilitation of Environmental Systems at Hospital Albert Schweitzer, Deschapelles, Haiti", Prepared for the Board of Directors, Grant Foundation and the International Relief Association, Swiss Government.

Hettich, M., 1995, "Report to the HAS Board of Directors", Deschapelles, Haiti.

# A Community-managed Water and Sanitation Utility for the Urban Poor: Cité Soleil, Port-au-Prince, Haiti

## Christopher McGahey, Ph.D.[1]

## Abstract

The U.S. Agency for International Development's (USAID) Environmental Health Project, worked with a non-governmental organization, the Centres pour le Développment et la Santé (CDS), to plan and implement a water and sanitation district in Cité Soleil, an informal settlement of 200,000 inhabitants in Port-au-Prince, Haiti. The USAID activity enhanced a United Nations-funded project to construct an independent water supply system for Cité Soleil.

The USAID project supported CDS in establishing and operating an autonomous organization, or district, capable of managing the water supply system and providing environmental sanitation services with revenues from the sale of water. Design of the activity began in October 1995 and inauguration of the district, under Haitian supervision, began in April 1997.

After six months of operation, the district was successfully covering its operating costs with revenue generated for the retail sale of water. At that time, the district had not yet generated sufficient monthly income to completely implement its environmental sanitation mandate, but it was scaling up its services from a targeted focus to broader community coverage. Community ownership of the operation and maintenance of the system was strong and clearly visible, and no significant illegal tapping of the system was evident which could not be addressed by the community.

## Background

In recent years, lack of fertile land, political instability, repression, and embargoes have resulted in the rapid migration of the Haitian population from rural to urban areas. For example, the population of the capital, Port-au-Prince, has increased from an estimated 1 million to 1.6 million since 1990. Yet this city is completely without a sanitary sewer system. This rapid urbanization has had significant negative impacts on the urban environmental conditions in the city. Migrants to the city often live in crowded urban slums with appalling environmental conditions. The environmental sanitation and health conditions in the slums of Port-au-Prince are particularly bad.

---

[1] Associate, Water Resources and Environmental Sanitation Specialist, ARD Inc., 110 Main Street, Fourth Floor, Burlington, Vermont 05401 USA

In these areas, there is no safe piped water supply and no adequate disposal of solid or human waste. Diarrheal disease incidence is very high, and diarrhea remains the main cause of morbidity and mortality among children under five. In response to the environmental situation in one urban slum, Cité Soleil, USAID/Haiti and the USAID Regional Housing and Urban Development office for the Caribbean (RHUDO/CAR) designed a project to support a UN-funded effort constructing a water supply system for Cité Soleil. The United Nations Development Program (UNDP) provided funds and contracting for the construction.

Purpose of the Project

UNDP contracted private construction firms through a local NGO, the Centres pour le Développement et la Santé (CDS). CDS also signed an agreement with CAMEP, the Port-au-Prince metropolitan water authority, to manage the system for three years. At that time, USAID saw an opportunity to support CDS in establishing a community-managed, autonomous district to manage the new water supply system and at the same time provide environmental sanitation services financed through revenues from the sale of water to the population of Cité Soleil. An implementation plan, developed with assistance from USAID's Environmental Health Project (EHP), detailed the requirements for setting up a district to manage the water supply system and provide environmental sanitation services.

Description of Cité Soleil

Since 1973, Cité Soleil has undergone continuous population growth. The present day population of 200,000 lives in an area of approximately two square kilometers of habitable land transected by two major open drainage canals. Cité Soleil suffers from large doses of nearly every environmental threat imaginable. It is located at the western end of the principal runway of the Haitian National Airport and consequently suffers regular noise pollution. Its open areas are barren of vegetation and combine with tremendous accumulations of solid waste to generate significant dust on a daily basis. The drainage canals which cross the area are largely fed from outside the community and carry pollutants from the manufacturing zones, roadways, petroleum storage facilities, and electrical generation plants which abut the eastern and southern borders of Cité Soleil. Groundwater can be found from 1 to 5 feet below the surface and is fed by the surrounding ocean and by infiltration of the community's waste and polluted drains. The general lack of sanitary facilities, only 30 to 40% of the households have access to a latrine, combines with ever-present mounds of solid waste to generate a large and health-threatening population of flies and other pests. The entire area is then placed under enormous environmental pressure because the residents of Cite Soleil are directly exposed in their homes on a regular basis to the overflowing and flooding of the intensely polluted drainage canals which are obstructed by vase amounts of solid waste.

Cité Soleil receives no services to collect solid waste. The responsible authority is unable to provide this service, and private contractors are reluctant to enter the area due to its hostile reputation and the fact that there is ample waste for them to pick up in other better paying communities. Cité Soleil is confronted with a variety of problems resulting from the generation of large amounts of solid waste including proliferation of pests and clogging of drainage canals.

Prior to the initiation of the district, household water was supplied primarily by private vendors and brought into Cite Soleil by truck. Other sources include (1) an organization which provides free water brought in by trucks and (2) a deteriorated distribution network which provides intermittent contaminated water. The UNDP-financed project was designed to provide water through 76 communal water fountains distributed throughout the community. The fountains are constructed of concrete block and are each equipped with four faucets, a water meter, and security features. It is estimated that the system can deliver 437,000 gallons of water daily. The supply and storage components of the system can provide sufficient water to service the entire population with 20 to 25 liters of water per person per day, however, distribution capacity is limited by the number of communal fountains.

Creation of an Autonomous Water and Sanitation District

The administrative and environmental goals of the district are to:
- Efficiently manage the water supply system;
- Provide solid waste collection services to all residents of Cite Soleil, using revenues generated from selling water; and
- Improve excreta disposal services available to residents of Cite Soleil.

An effort to establish a community-managed, autonomous district for any very poor and very large informal community is ambitious and optimistic. The social and political environment in Cité Soleil is volatile and could prove at any point to be an obstacle to success. Also, if the community does not support its effort, the district will not be successful and environmental conditions within Cité Soleil will continue to be among the most severe in this hemisphere. The challenges are daunting. Nevertheless, the willingness of the population to support the district and the amount of money people were paying for water before initiation of the district was so large that the potential for financial and institutional sustainability is very real.

Establishment of the water and sanitation district in Cité Soleil was based upon four key principles:
- The district will have autonomy in key aspects of its operation, including policies and objectives, investment decisions, budgets, tariffs, hiring and firing of personnel, salaries, and investment decisions;
- The district will be run with substantial involvement by the community. The district will be structured so that the community plays a major role in key decisions such as setting tariffs, investments, and selection of water vendors;
- The private and nongovernmental sectors will be involved where possible. Community organizations will be responsible for management of the fountains and solid waste services, and private sector contractors will be used for certain maintenance tasks to keep permanent staff levels and associated costs low; and

- The district will be entirely self-financing and thus will be able to control all factors important to its operation. It is expected that the fees for water will generate sufficient revenues to pay the operating costs of both the water system and the sanitation services.

## Organizational Structure of the Water and Sanitation District

The water and sanitation district in Cité Soleil is organized into four departments: (1) operation and maintenance of the water supply system, (2) water sales, (3) solid waste, and (4) administration. A previously established division of the community into seven zones was adapted to the district to maximize existing organizational structures. Community organizations in each of the zones are united by the district through zone committees which are the daily link between district administration and community members. Solid waste collection is managed by the district entering into contracts with local community organizations. Water is sold by entering into concession agreements with these community organizations.

The size of the district's staff has been maintained as small as possible to minimize operating costs. The central administration consists of the following staff positions: (1) general manager, (2) operations supervisor, (3) solid waste supervisor, (4) community organization specialist, (5) accountant, (6) cashiers, (7) secretaries, (8) plumbers, (9) seven zone committee coordinators, and (10) seven zone committee clerks. Fountain operators are not employees of the district. Instead, they are chosen by the fountain committees and paid by them. Solid waste collection crews are not employees of the district either.

The district oversees solid waste collection services by working with the community to mobilize salaried collection teams and ensure rapid, efficient payment of private contractors' for the removal of solid waste. The strategy for community involvement in solid waste management is similar to the one for water supply management. The strategy is to create at least one team in each of the seven zones in Cité Soleil. These 20-person teams are managed by the zone committee with day-to-day management by the zone coordinator. The district has budgeted a set amount for the salaries of the team members, and the zone committee is responsible for paying the team on a regular basis. There is no household or user fee for solid waste collection. Instead, collection and disposal services are paid for from water revenues. An agreement between the district and zone committees has been formalized in a written contract which addresses roles and responsibilities, payment, and monitoring and enforcement.

## Community Participation and Behavior Change

This social component was crucial to the development and initiation of the water and sanitation district in Cité Soleil. This component is divided into two distinct but related areas. The first is participation of the community in management of the district. The second area is behavior change. For the purposes of this paper, emphasis will be placed on the role of the first in the establishment and operation of the district.

The objectives of the community participation component are to:
- Create a sense of ownership of the water supply system by the community so residents have a real stake in its success;
- Place significant responsibility for managing water and sanitation services on the community; and
- Protect the system against illegal connections and other abuses.

Initial plans call for construction of 76 public fountains in Cite Soleil. The strategy is to involve residents directly in management of the district in three ways:
- Each fountain is operated under a neighborhood fountain committee;
- Seven zonal committees, composed of representatives from the fountain committees, act as liaisons between the district and the fountain committees; and
- Management of solid waste collection.

Fountain committee members are elected by local residents to manage each fountain. Each committee supervises the work of one or more fountain operators, insures cleanliness around the fountain, resolves problems as they arise, and promotes participation in education and mobilization meetings. Each five-member committee elects a president, secretary, and treasurer; selects one fountain operator; and names a delegate to the zonal committee. The agreement between the district and each fountain committee is formalized in a written contract. The contract contains provisions concerning payment to the fountain committee, roles and responsibilities, and monitoring and enforcement.

Zone committees then serve as intermediaries between the fountain committees and the district. Zone committees are composed of one delegate from each fountain committee in the zone. Seven zone coordinators are full-time, paid employees of the district. Their responsibilities are to insure efficient, helpful communications between the district and the zonal and fountain committees; provide sound financial and administrative management of the zone; help fountain committees resolve problems as needed; and promote community participation in education activities.

Finance and Rates

Before inauguration of the district, most residents of Cité Soleil obtained water from local vendors who store water trucked in from nearby private wells. Total daily purchases were estimated to be at least 200,000 gallons based on turnover times of water vendors' supplies. Under these original conditions, a family of 6 paid about US$3.80 per month for water based upon typical usage and water price. At the prevailing price of water, US$84,200 of water sales were estimated to be generated in Cité Soleil monthly, or slightly more than US$1,000,000 per year. This very conservative estimate is quite remarkable for such an impoverished population.

Significant profits can be made in the water vending business in Cité Soleil. Water purchased at private wells cost truckers $1.61 for one 3000 gallon truckload.

Typically, vendors in Cité Soleil would buy that same truckload of water for $20. In two days, the vendor normally sold that water for nearly $60. The total marketing margin is, therefore, about 35 times the price at the well. This left ample pricing margin for the district.

The existing commercial water prices and the volume purchased provided hard evidence of the residents' capacity to pay for water. The preliminary cost estimates assembled as part of EHP's work demonstrated that a well-managed local water and sanitation district clearly can provide Cité Soleil with high quality water while also financing solid waste removal. These activities are helping to resolve environmental and health problems in Cité Soleil and provide the country with a prototype of how to sustain such activities in similar areas.

Conclusion

The establishment of a community-based, autonomous, financially self-sufficient water and sanitation district in Cité Soleil is very ambitious. There is little question that the district has the potential to be sustainable. Before its initiation, people were paying for water an amount that could finance the operation of the water supply system and provide solid waste services as well. What will determine the success of the project are the institutional and community components.

In September of 1997, 90 percent of the proposed fountains had been constructed and were operating, and the district was selling over 100,000 gallons of water per day with approximately 19% unaccounted-for-water. In addition, the community had shown strong ownership of the system as demonstrated by their initiative and ability to operate and maintain the fountains and to penalize and disconnect households which had made illegal connections to system distribution lines. In summary, the district did an excellent job in getting established. Significant accomplishments were visible in both institutional and technical areas including community mobilization, training of community members, job creation, water delivery, financial accounting, and operation and maintenance. And it should be noted, the district is off to an excellent start financially. In July and August 1997, enough water was sold to pay for its core administrative and water supply operation and maintenance costs. Careful management of expenses and USAID support resulted in a net balance of over US$16,000 to enable the district to begin funding solid waste activities in the community.

**Acknowledgements**
The author wishes to acknowledge the contributions made to this paper by Fred Rosensweig of USAID' Environmental Health Project. In addition, he wishes to acknowledge the funding and assistance provided to this work by the U.S. Agency for International Development (USAID), USAID's Environmental Health Project, Mr. Henri Supplice, Ing. Michel Genois, and the Haitian staff of CDS. Further information regarding this work can be obtained in A Plan for CDS to Establish A Water and Sanitation District in Cité Soleil, Haiti, USAID Environmental Health Project Activity Report No. 21, Environmental Health Project, 1611 North Kent Street, Suite 300, Arlington, VA 22209, USA (phone 703-247-8730, fax 703-243-9004, internet ehp@access.digex.com).

# Development of a California Water Use Efficiency Policy through "Stakeholder" Consensus

Rick Soehren[1] and Greg Young, P.E. [2]

## Abstract

Increased efficiency in the use of limited water resources for a diverse set of beneficial uses is part of the necessary water management solution in a state where conflict among these uses continues. Working through a process with extensive public and stakeholder input, the CALFED Bay-Delta Program has drafted a policy framework to improve water use efficiency throughout California.

## Introduction

The CALFED Bay-Delta Program is a joint state-federal effort developing a long-term comprehensive plan to restore the ecological health and improve water management for beneficial uses of the San Francisco Bay/Sacramento-San Joaquin Delta. The Program is addressing several categories of Bay-Delta problems including ecosystem quality, water quality, and water supply reliability. Efficient use of developed water supplies can contribute to solution of problems in all of these categories. The purposes of the water use efficiency component of the Program are to improve the reliability of water supplies by reducing the mismatch between supply and beneficial uses, and improve water quality and enhance ecosystem health through improvements in local water use management.

Many local water agencies in California already have strong water use efficiency programs. The greatest current challenge in water use efficiency is finding ways to encourage more water users and water suppliers to implement the proven cost-effective efficiency measures that are being used successfully by their peers throughout the state.

## Component Definition

Efficiency has several definitions. One is a traditional view of water use efficiency defined in terms of physical efficiency: the ratio of water consumed to water applied. Efficiency can also be defined in economic terms: deriving the greatest economic output from a given input such as a unit of water. For the purpose of developing and implementing a

---

[1] Program Manager for Water Use Efficiency, CALFED Bay-Delta Program, 1416 Ninth Street, Suite 1155, Sacramento, California 95814

[2] Agricultural Engineer, Water Resources, CH2M HILL, Inc., 2485 Natomas Park Dr., Suite 600, Sacramento, California, 95833

water use efficiency component, CALFED has defined efficiency somewhat differently: **efficient water use is characterized by the implementation of local water use management actions that increase the achievement of CALFED objectives.** This definition includes physical efficiency but is not limited to this narrow definition.

Increases in physical efficiency and increases in the achievement of CALFED objectives through improved water management will be direct results of the component. Increasing economic efficiency -- which might result in a reallocation of water -- is not a specific objective of the CALFED Program and the Program will not take direct action to increase economic efficiency.

Collectively, the role of CALFED agencies will be twofold. First, they will offer support and incentives such as programs to provide planning, technical, and financing assistance. Second, the CALFED agencies will play an important role in providing assurances that cost-effective efficiency measures will be implemented.

Component Development

The broad topic of water use efficiency was divided into three elements to facilitate discussion and development of CALFED Program approaches: urban water use efficiency, agricultural water use efficiency, and water recycling. The first two elements correspond to traditional water use sectors of urban and agriculture. Water recycling is treated separately for the sake of expediency, because urban water recycling has traditionally been approached separately from urban water conservation, and is often the responsibility of different agencies.

The water use efficiency component was developed with extensive input from the public and from "stakeholders" - interest groups with a stake in the solution of problems in the Bay-Delta region. With extensive stakeholder input, implementation objectives were identified to guide development of appropriate policy. These objectives are intended to reflect and protect the various stakeholder interests regarding local water use management and efficiency. General objectives include:

- Ensure a strong water use efficiency component in the Bay-Delta solution - During the CALFED scoping period and at numerous public meetings, the general public as well as stakeholders said local water use management and efficiency improvements should play an integral role in the Bay-Delta solution.

- Emphasize incentive based actions over regulatory actions - The CALFED Program's approach to water use efficiency emphasizes incentives to encourage efficient use. Principal incentives include planning, technical, and financing assistance to local water agencies. Additional incentives include access to potential benefits of the Bay-Delta Program such as increased water supplies and increased ability to convey transferred water. Regulatory actions provide necessary assurances of efficient use as well as mitigation for third party impacts that may result from incentive-based approaches.

- Preserve local flexibility - During the CALFED Bay-Delta Program's scoping period and at numerous public meetings, stakeholders stressed the desire to maintain the flexibility of implementing water use management and efficiency improvements at the local level. The CALFED Program's approach to local water use management and efficiency provides necessary assurances of improved efficiency while maintaining the flexibility to tailor implementation to local conditions.

- Remove disincentives and barriers to efficient water use - Water agencies and water users may be discouraged from implementing conservation measures as a result of various disincentives. Examples of disincentives include poorly planned water wholesaler drought water allocation plans, negative impacts to agency operation budgets resulting from reduced water sales, and inability to pass some conservation costs along to customers (as occurs with some investor owned utilities). Removal of disincentives can allow agencies and their customers to implement conservation measures that otherwise could not be justified. However, removal of barriers must support the original purposes of the institutions associated with the measure.

- Offer greater help in the planning and financing of local water use management and efficiency improvements - To implement efficient water management practices, some water users need information about proposed measures and may also need the ability to finance implementation of such measures. Greater levels of technical, planning, and financing assistance are essential to improve local water use management and efficiency. This assistance will help agencies use integrated resource planning methods and common approaches to cost-effectiveness determinations, will help agencies recognize the value of conservation, and will allow them to make more informed decisions regarding implementation of such measures.

Assurance Mechanisms

The CALFED Bay-Delta Program solution alternatives include a variety of programs, policies, and actions to provide assurance that appropriate water management planning is carried out by local agencies and that cost-effective efficiency measures are implemented. Some specific assurance mechanisms and assurance needs are described in the sections that follow. In addition, CALFED and the CALFED agencies will implement three general policies to provide assurance of efficient use. Demonstration that appropriate water management planning is being carried out and that cost-effective efficiency measures are being implemented will be necessary prerequisites for a local water supplier to be eligible to:

- receive any "new" water made available by a Bay-Delta solution,

- participate in a water transfer that requires approval by any CALFED agency or use of facilities operated by any CALFED agency, and
- receive water through the Department of Water Resources Drought Water Bank.

A unique feature of the water picture in California is the existence of stakeholder forums that have developed consensus on appropriate conservation measures for local water suppliers. Urban water suppliers and public interest groups have defined conservation measures called Best Management Practices and agreed on appropriate levels of implementation. Agricultural irrigation districts and environmental groups have agreed on Efficient Water Management Practices for districts to consider, and analytical methods to select measures for implementation. These stakeholder forums would be given important powers to certify or endorse the water use efficiency efforts of local water suppliers as a prerequisite for suppliers to be eligible to receive benefits of the CALFED program

Urban Water Use Efficiency Policy Actions

The urban areas of California currently use over seven million acre-feet of water each year. The majority of this demand is met by diverting water from the Bay-Delta system. As populations continue to grow, the demand will also grow. The CALFED Bay-Delta Program will help the urban sector meet its future water needs and improve supply reliability through a number of programs, one of which is to facilitate implementation of cost-effective water use efficiency measures.

The urban approach recognizes a responsibility to carry out local water management planning and establishes a process for recognition of adequate planning efforts and recommends a balanced process for recognition of conservation implementation.

1. Conservation Implementation, Reporting, and Certification
   Rely on a stakeholder forum to provide a uniform, verifiable, locally-directed process for urban Best Management Practices (BMP) implementation and reporting.

2. Certification of Water Management Planning
   Help urban suppliers prepare, adopt, and implement useful water management plans and comply with the requirements of the Urban Water Management Planning Act (California Water Code 10610 et. seq.).

3. Technical and Planning Assistance
   Ensure that lack of technical and planning expertise does not impede implementation of cost-effective measures by providing easily accessible assistance for planning and implementing local water management programs.

4. Funding Assistance
   Ensure that lack of financing ability does not impede implementation of cost-

effective measures. Provide easily accessible funding for planning and implementing water management programs.

5. Assurances for Urban Water Management and Conservation
Provide specific assurance that urban water suppliers will carry out good water management planning and implement cost-effective conservation programs.

Agricultural Water Use Efficiency Policy Actions

Agriculture is an important part of California's economy. This $24-billion-a-year industry produces about 11 percent of the total U.S. agricultural value and 40% of the nation's produce on 9.1 million irrigated acres. The CALFED Bay-Delta Program, by solving interrelated problems of the Bay-Delta system, will help to preserve the viability of agriculture in California. The Program's approach to agricultural water use efficiency will be to encourage cost-effective water use efficiency measures and to achieve other CALFED Program objectives in ways that are compatible with agriculture.

The agricultural approach recognizes a clear standard for agricultural water management planning and a balanced process for recognition of planning and implementation.

1. Water Management Planning and Implementation
Rely on a stakeholder forum to provide a uniform, verifiable, locally directed process for agricultural water management planning and provide a balanced process for review and endorsement of water management plans.

2. Technical and Planning Assistance
Ensure that lack of technical and planning expertise does not impede implementation of cost-effective measures by providing accessible assistance for planning and implementing local water management and efficiency improvements.

3. Funding Assistance
Ensure that lack of financing ability does not impede implementation of cost-effective measures. Provide easily accessible funding for planning and implementing local water use management and efficiency improvements.

4. Management Improvements to Achieve Multiple Benefits
Help to meet CALFED objectives, including those related to ecosystem health and water quality, by encouraging districts to identify opportunities for improvement when preparing water management plans, and giving incentives for implementation.

5. Assurances for Agricultural Water Use Efficiency
Provide specific assurance that agricultural water supplies are used at highly efficient levels.

## Water Recycling Policy Actions

The use of recycled water as a water supply source is projected to grow. In 1996 the California Department of Water Resources conducted a Survey of Water Recycling Potential to help identify and quantify recycling plans. The survey identified actual recycling of nearly 350,000 acre-feet annually in 1996, and projected recycling of 1.48 million acre-feet annually by 2020. It should be noted that these projected reuse totals, although large, represent less than half of the total recyclable waste stream that could be available for reuse.

1. Water Recycling Planning and Implementation
   Provide a uniform, verifiable, locally directed process for recycled water market identification and integrated water and wastewater project planning for water recycling.

2. Water Recycling Technical and Planning Assistance
   Ensure that lack of technical and planning expertise does not impede implementation of cost-effective water recycling projects by providing easily accessible assistance for planning and implementing local water recycling market evaluations, integrated water and wastewater project planning, and financial evaluations leading to accessing special water recycling funding opportunities.

3. Funding Assistance
   Ensure that lack of financing ability does not impeded implementation of cost-effective measures. Provide easily accessible funding for planning and implementing local water recycling projects.

4. Identify and encourage regional water recycling opportunities that maximize reuse at minimum cost
   Provide opportunities for local water and sanitary agencies to join together to plan regional projects to their mutual benefit.

5. Assurances for Water Recycling
   Provide specific assurance that urban water and sanitary agencies will carry out good water recycling analysis and planning and implement cost-effective recycling programs.

## Conclusion

The water use efficiency program proposed by CALFED relies heavily on stakeholder forums to define appropriate water use efficiency measures and levels of implementation. State and federal agencies would offer assistance programs to local water suppliers and condition eligibility for additional water supplies based on implementation of cost-effective efficiency measures. Greater efficiency of use is expected to result in greater water supply reliability as well as improved ecosystem health and improved water quality.

# INNOVATIVE WET-WEATHER FLOW MANAGEMENT SYSTEMS FOR NEWLY URBANIZING AREAS

By James P. Heaney[1], Member, ASCE, Leonard Wright[2], Associate Member, ASCE, and David Sample[2,] Associate Member, ASCE

## ABSTRACT

Highlights of the initial phase of a study of innovative urban stormwater management systems for the 21st century are presented. Urban stormwater management is viewed as a subsystem of the urban water infrastructure and watershed management systems. Results of an extensive literature review and preliminary evaluations of alternative future scenarios are presented.

## INTRODUCTION

The purpose of this paper is to present an overview of the results of the initial phase of projects sponsored by the U.S. Environmental Protection Agency to propose innovative urban stormwater systems for new urban developments in the 21st century (Heaney et al. 1998). The initial phase of this effort was to conduct a thorough literature review of contemporary and projected urban stormwater management practices. Based on this review and discussions with leading experts, a framework for evaluating the effectiveness of innovative stormwater management systems for the 21st century is presented. Selected results are presented in this paper. First, a discussion of historical and future urban land use is presented. Next, urban stormwater management is viewed as a subsystem of urban water management. Water budgets are used to identify the relative importance of each component of the urban water budget. The pros and cons of alternative collection systems have been debated for the past century and this debate is expected to continue into the 21st century. BMP effectiveness is evaluated. Onsite control of

---
[1] Professor, Dept. of Civil, Environmental, and Architectural Engineering, and Faculty Associate, Center for Advanced Decision Support for Water and Environmental Systems, Campus Box 421, U. of Colorado, Boulder, Colorado 80309-0421.
[2] Graduate Student, Dept. of Civil, Environmental, and Architectural Engineering, U. of Colorado, Boulder, Colorado 80309-0421.

urban stormwater is discussed and a monthly water budget is used to provide a preliminary estimate of how effective urban runoff from roofs and driveways would be in reducing irrigation needs. Real-time control (RTC) will be an integral part of future stormwater systems. A brief overview of RTC is provided. Lastly, a summary and conclusions are presented.

## URBAN LAND USE AND WATER INFRASTRUCTURE

Prior to World War Two (WW 2), U.S. cities were developed around the concept of mixed neighborhoods as part of villages, towns, and cities. Suburbia began to dominate urban America beginning in the late 1940s. Urban land use is expected to revert to pre-WW 2 models with higher density, mixed neighborhood developments and this type of land use will probably predominate in the latter part of the 21st century. The impact of telecommunications on future land use may reverse current trends of people migrating to cities and may result in clusters of smaller settlements, or it may induce widespread sprawl of people over wide areas, as physical proximity becomes less important.

With regard to urban water systems, Grottker and Otterpohl (1996) list the following general principles for providing sustainable development:

1) Minimize the distance of water and wastewater transportation.
2) Use stormwater from roofs preferably for water supply instead of infiltrating or discharging it.
3) Do not mix the human food cycle with the water cycle. Do not mix waste waters of different origin.
4) Decentralize urban water systems and do not allow human activities with water if a local integration into the water cycle is not possible.
5) Increase the responsibility of the human being for his impacts on local water and wastewater systems.

An additional principle is that is essential to redevelop existing urban areas rather than continuing to expand to greenfields on the periphery of existing urban areas.

Stephenson (1996) compares the water budgets of an undeveloped catchment with an urbanized catchment in Johannesburg, South Africa. The results show the expected increase in direct runoff and the need to import water for water supply. Mitchell et al. (1996) describe a water budget approach to integrated water management in Australia. The results of test applications in two developments indicate 41 to 49% reductions in the demand for imported water, 49 to 56% reductions in stormwater runoff, and 8 to 11% reductions in wastewater discharges.

Urban water use, wastewater, and urban stormwater are interdependent. Virtually all of the indoor water use is discharged to separate or combined sewers. The total quantity of wastewater is strongly influenced by infiltration and inflow (I/I) which often increase as a result of wet-weather conditions. Outdoor urban use for irrigation of plants makes pervious areas wetter and reduces the potential soil moisture storage available during wet weather periods. However, properly managed, a significant portion of urban stormwater can be directed onto pervious areas to reduce irrigation needs. Indoor residential water use has been shown to be quite constant with an average of 60 gpcd (Harping 1997). Conservation-oriented hardware changes such as switching to low-flush toilets should reduce indoor usage to 35-40 gpcd. Toilet flushing represents the only black water source of wastewater. It constitutes about 30 % of indoor water use. In more arid and warmer parts of the United States, lawn watering is the largest single use on an annual average and is the dominant component of peak daily and hourly use during the summer months. Unlike indoor water use, outdoor water use is quite variable. Sewer I/I are major issues in urban water management. I/I is 20 to 60% of the average annual flow in sanitary sewer systems and it is responsible for most of the peak flows in these systems.

## COLLECTION SYSTEMS

An interesting development regarding combined sewer systems (CSSs) is that, due to contaminated stormwater runoff from urban areas that requires treatment, combined systems are now at least being considered for new urban areas in some parts of Europe. Combined systems may in fact discharge less pollutant load to a receiving water than separate systems where stormwater is discharged untreated and sanitary wastewater is treated. In Skokie, Illinois and southern Germany, CSSs are being designed with state-of-the-art BMPs to reduce the volume of stormwater entering the system. With reduced stormwater input, the number and volume of overflows are reduced over a traditional "old-fashioned" CSO, thus only discharging CSO during large, infrequent events, when the receiving water is most likely to be at high flow conditions.

Separate sanitary sewers serve a large portion of the sewered population in the United States. These sewers are of smaller diameter than combined or storm sewers, and serve residential, commercial, and industrial areas. While sanitary systems are not specifically designed to carry stormwater per se, stormwater and groundwater do enter these systems. This is a common and complicated problem for sewer owners. So common in fact, that the design of sanitary sewers must include capacity for I/I, which may actually exceed pure sanitary flowrates. The trend towards lower housing densities results in more sewer pipe per capita, thereby exacerbating the I/I problem. The capacities of many collection systems are being exceeded well before the end of their design life, resulting in bypasses, overflows, surcharging and reduced treatment efficiency. Without I/I, peak wastewater flows in sanitary sewers have a very low peaking factor since this use is relatively constant. Current design practice calls for using ratios of 2 to 10 times the dry-weather flow. A review of ten case studies (US EPA 1990) indicates that peak waste flows ranged from 3.5 to 20 times the average DWF. System surcharges would typically occur as the ratio reached 1:4 or 1:5 (US EPA 1990). For the sewer owner of the 21st century, it is imperative that measures be taken to ensure that the construction and design contractors have a vested interest in the acceptable long-term performance of the collection system.

## STORMWATER STORAGE-TREATMENT-REUSE SYSTEMS

Because of the dynamic nature of stormwater flows and water quality, most control systems are a hybrid of temporary storage and high-rate treatment. For a given level of stormwater control, the engineer can accomplish this objective using various combinations of storage and treatment. Heaney and Wright (1997) provide a current summary of these methods. High-rate operation of WWTPs during and following wet-weather events is an important option to evaluate as part of the overall stormwater management program for combined and separate systems that are affected by I/I (Field and O'Connor 1997). There is much interest in local management of stormwater from smaller, more frequent events. The primary on-site option is to encourage infiltration of this stormwater from roofs, driveways, parking lots, and streets. This infiltrated water increases the moisture in the unsaturated zone and raises the groundwater table which can provide benefits in terms of increasing base flows in streams and providing storm water to help meet ET needs of the local vegetation. By examining the water balance of one residential parcel in differing climatic zones, the efficacy of the option of on-site reuse of stormwater was evaluated. Most eastern and west coast cities were able to satisfy all of their ET needs. The Rocky Mountains and semi-arid southwest were able to achieve over 90% and the desert southwest was able to achieve 24%.

## MONITORING AND REAL-TIME CONTROL SYSTEMS

Monitoring is the fundamental component of a sophisticated urban water management system. The most advanced simulation models cannot replace having actual field data. Fortunately, it has become increasingly cost-effective to do direct monitoring and control. Management systems of the 21st century can be expected to be based on sophisticated direct monitoring of the system with associated real-time control. Direct measurement of the performance of stormwater systems is an essential element in improved design and operation. Given direct measurements of the status of the system, control options can be classified as 1) manual control, 2) local automatic control, 3) supervisory systemwide control, and 4) integrated systemwide automatic control (WEF 1997).
RTC systems will be an integral component of future stormwater management systems of the 21st century. Data centered approaches using sophisticated data measurement, transmission, and analysis systems will be used in conjunction with RTC to truly optimize the performance of real systems.

## SUMMARY AND CONCLUSIONS

The purpose of this paper is to provide a preliminary look at innovative 21st century urban stormwater management systems for new developments. Urban stormwater management needs to be viewed within the context of overall urban water management. The major unknown for dry-weather flow is the amount of infiltration and inflow (I/I) which is related to wet-weather conditions.Future sewer systems should be designed to much more effectively prevent I/I by reducing I/I and promoting higher density development that reduces the length of sewers required per capita. Urban water reuse is an attractive possibility for 21st century systems. On-site BMPs should be incorporated in future urban developments. These on-site controls will permit complete capture of the smaller, more frequent, storms that constitute the majority of the annual volume of rainwater falling on cities. Cities in the east and midwest generate enough stormwater runoff from their roofs to meet their lawn watering needs. Only cities in the arid southwestern U.S. do not receive enough stormwater to meet a significant part of their lawn watering needs. On-site devices are needed to effectively store this water since the plants do not need the rainwater until subsequent dry periods. With contemporary technology it is possible to monitor and control urban stormwater systems. Monitoring and RTC should become cornerstones of 21st century technology.

## APPENDIX. REFERENCES

Field, R. and O'Connor, T.P. (1997) "Control strategy for storm generated sanitary sewer overflows". *Jour. of Environmental Engineering*, Vol. 132, No. 1, p. 41-46.

Grotter, M., and Otterpohl, R. (1996) "Integrated Urban Water Concept". *Proc. 7th Int. Conf. on Urban Storm Drainage*, Hannover, Germany, p. 1801-1806.

Harpring, J. S. (1997) *Nature of Indoor Residential Water Use*. MS Thesis, Dept. of Civil, Environmental, and Arch.Engg. U. of Colorado, Boulder, Colorado.

Heaney, J. and Wright, L. (1997) "On integrating continuous simulation and statistical methods for evaluating urban stormwater systems." Chap. 3 in James, W. (Ed.) *Advances in Modeling the Management of Stormwater Impacts*. Vol. 5. Computational Hydraulics International, Guelph, Ontario, Canada.

Heaney, J.P., Wright, L., Sample, D., Urbonas, B., Mack, B.W., Schmidt, M.F., Solberg, M., Jones, Clary, J., and T. Brown. 1998. *Development of Methodologies for the Design of Integrated Wet-Weather Flow Collection/Control/Treatment Systems for Newly Urbanized Areas*. Volume 1: Draft Technical Report to US Environmental Protection Agency, Edison, NJ.

Lemmen, G., de Bijil, D., and Maessen, M. (1996) "A New Development in the City of Dordrecht: Sewerage and Drainage Masterplan of Buitenstad." In

Sieker, F. and H.R. Verworn (Ed.) *Proc. 7th Int. Conf. on Urban Storm Drainage*, Hannover, Germany, p. 1235-1240.

Mitchell, V.G., Mein, R.G., and McMahon, T.A. (1996) "Evaluating the Resource Potential of Stormwater and Wastewater: an Australian Perspective." *Proc. 7th Int. Conf. on Urban Storm Drainage*, Hannover, Germany, p. 1293-1298.

Schilling, W. (1996) "Potential and Limitation of Real Time Control". *Proc. 7th Int. Conf. on Urban Storm* Drainage, Hannover, Germany, p. 803-808.

Stephenson, D. (196) "Evaluation of Effects of Urbanization on Storm Runoff." *Proc. 7th Int. Conf. On Urban Storm Drainage*, Hannover, Germany, p. 31-36.

US EPA (1990) *Rainfall Induced Infiltration into Sewer Systems*. Report to Congress, EPA/430/09-90/005, Washington, D.C.

WEF (1997) *Automated Process Control Strategies*. Alexandria, Virginia.

Wet Weather Flow Designs for the Future

Robert Pitt[1] M. ASCE, Richard Field[2] M. ASCE, and Chi-Yuan Fan[2] M. ASCE

Introduction

Throughout history, many strategies have been implemented to control wet weather flows (WWF). Major incentives have included flood control, water quality improvement, aesthetic improvement, waste removal, and others. An urbanizing region must choose various management, control, and treatment alternatives specific to its circumstances in order to protect the receiving waters (both surface water and groundwater). A guidance manual for wet weather flow systems in newly urbanized areas is being developed as part of a cooperative agreement between the Urban Watershed Management Branch of the U.S. Environmental Protection Agency and the University of Alabama at Birmingham (Pitt, *et al.* 1997). This manual will examine the history of wet weather flow management, and will present some recommended strategies for newly developing areas. A related effort is also being conducted by the ASCE and the University of Colorado.

Existing Problems with Sanitary and Storm Drainage Systems

The continued use of combined sewer systems is common in many parts of the world, and the U.S. has many existing combined systems still in use. In addition, separate sewer overflows (SSOs) are also common in many urban areas that only have separate systems. Overflows of raw sewage during wet weather is therefore unfortunately common in many areas of the U.S. In addition, there is renewed interest in the use of combined sewer systems in the U.S. under specific conditions, where their use (in conjunction with improved treatment facilities) may result in reduced, and more cost-effective, WWF discharges. Heaney, *et al.* (1998) for example, found that combined systems may discharge a smaller pollutant load to a receiving water than separate systems in cases where the stormwater is discharged untreated and where the sanitary wastewater is well treated.

The debate on the use of combined sewers has been long. In the late 19th century, Hering (1881) visited Europe and made recommendations to the U.S. National Board of Health concerning the use of combined sewers. He recommended that combined sewers be used in extensive and closely built-up districts (generally large or rapidly growing cities), while using separate sanitary systems for areas where rainwater did not need to be removed in underground drainage conveyance systems. His recommendations were largely ignored. Combined sewers were extensively used in many of the older U.S. cities because of perceived cost savings.

---

[1] Dept. of Civil and Environ. Engineering, University of Alabama at Birmingham.
[2] Wet-Weather Flow Management Program, USEPA, Edison, NJ

Of course, the existing combined sewer systems in the U.S. are now mostly located in the most dense portions of central cities, along with some of the older residential areas. However, current separate sewer systems actually may operate as combined systems due to excessive infiltration of sewage into stormwater systems, or by direct, illegal, connections of sewage into stormwater systems.

Current interest in illicit or inappropriate connections to storm drainage systems is an outgrowth of investigations into the larger problem of determining the role urban stormwater runoff plays as a contributor to receiving water quality problems. Besides direct runoff from rains, urban runoff also includes waters from many other sources which find their way into storm drainage systems.

Sanitary sewage finds its way into separate storm sewers in a number of ways. Direct cross-connections may tie sanitary lines directly to storm drains (relatively rare), or seepage from leaking joints and cracked pipes in the sanitary collection system can infiltrate storm sewers (much more common). Surface malfunctions and insufficiently treated wastewater from septic tanks may contribute pollutants to separate storm sewers directly or by way of contaminated groundwater infiltration. Seepage of sewage or septic tank effluent (septage) into underground portions of buildings may be pumped into separate storm sewers by sump pumps.

There are situations in which the sanitary system is so connected to the stormwater system that good intentions, vigilance, and reasonable remedial actions will not be sufficient to solve the problems. In an extreme case, it may be that while it was thought that a community had a separate sanitary sewer system and a separate storm drainage system, in reality the storm drainage system is acting as a combined sewer system. When recognized for what it really is, the alternatives for the future become clearer: undertake the considerable investment and commitment to rebuild the system as a truly separate system, or recognize the system as a combined sewer system, and operate it as such, without the disillusionment that it is a problem-plagued storm drainage system which can be rehabilitated.

It may be more cost-effective and result in the least pollutant discharges to operate separate drainage systems that are badly in need of repair as actual combined sewer systems, compared to costly and ineffective repairs to the separate systems. However, proposed construction of new combined sewer systems would be very controversial in the U.S. and it would be very difficult to overcome resistance to their construction. The main areas of resistance relate to the massive efforts expended in the last several decades in reducing the number and severity of combined sewer overflows (CSOs), usually under court order. In addition, current interest and massive correction efforts to control separate sewer overflows (SSOs) in many cities would also result in a great deal of resistance from engineers, municipalities, regulatory agencies and environmental groups to the construction of new combined sewer systems. The political resistance to the construction of new combined sewer systems in the U.S. is therefore considered almost insurmountable.

As pointed out by Hering in 1881, combined sewer systems may be suitable in dense urban areas, where the sanitary sewage flow is relatively high per area. Of course, any use of a combined sewer must be accompanied with provisions to reduce any untreated overflows to almost zero. In reality, the current level of untreated sanitary sewage discharges in urban areas from badly functioning separate systems is likely much higher than anyone acknowledges or considers when conducting wet weather flow management projects. The major concern with combined sewer systems is the overflow discharges of dangerous levels of pathogenic microorganisms, and nuisance conditions associated with floatable debris and noxious sediment accumulations. Discharges of potentially dangerous medical wastes and drug paraphernalia is also of great public concern. However, it may be possible to construct a new combined sewer system that would operate with fewer annual untreated discharges of sewage than many current separate systems, plus provide treatment of stormwater.

One option may be termed a shared sewer system as the two flows (stormwater and sanitary wastewater) are not co-mingled at the same time in the single drainage system, but are kept separate as much as possible. This option, commonly used in England in the later part of the last century, and recently re-introduced by Pruel (1996) would require an adequately sized storage tank that could hold household wastewater for specific periods of time (depending on rain durations, conveyance capabilities, and treatment rate available). Reyburn (1989) shows just such a system in an old drawing of sanitary fittings and drains from a $19^{th}$ century catalogue from Thos. Crapper & Co., Ltd., Sanitary Engineers, Chelsea, England.

Preul (1996) calculated the needed on-site storage volumes for this "shared sewer" concept. His "combined sewer prevention system" (CSPS) was investigated for locations in Cincinnati, Ohio, and in Toronto, Ontario. He found that storage tanks capable of detaining household sanitary wastewater on-site for 6 hours in Cincinnati would prevent about 90% of the CSO occurrences. The Toronto location would only require on-site detention capabilities of 3 hours for similar benefits. A household storage volume of 55 L would provide 6 hours of storage and 90% control of CSO, while a 220 L storage capacity per household would virtually eliminate all CSOs in Cincinnati. Required household storage capacities in Toronto would be even less, with 30L storage tanks providing almost complete control.

Another option is basically a separate sanitary sewerage system that is constructed to be very tight. The sanitary sewerage system may best be a vacuum or small diameter pressurized system. The stormwater would be conveyed separately, emphasizing on-site reuse and infiltration, through either open channels if compatible with the land use, or through a separate drainage system. Critical source area controls would be utilized, along with end-of-pipe treatment, as appropriate. With a tight conveyance system, no extra stormwater could enter the sanitary sewerage, greatly lessening the threat of overflows during wet weather.

Stormwater Drainage Design Objectives

There are four major functional aspects of a drainage system, each reflecting distinct portions of the long-term rainfall record. Figure 1 is an example of observed rainfall and runoff observed at Milwaukee, WI, (Bannerman, *et al.* 1983) as monitored during the Nationwide Urban Runoff Program (EPA 1983). This observed distribution is interesting because of two unusually large rains that occurred during the monitoring program. This figure shows the accumulative rain count and the

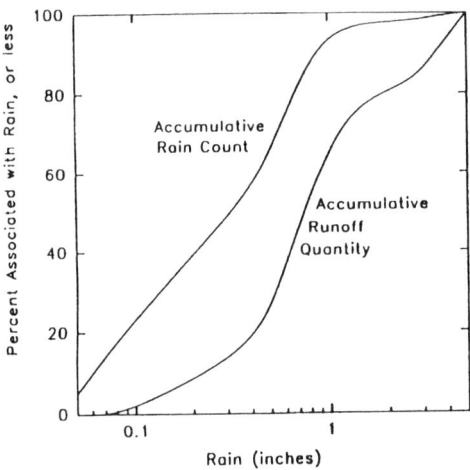

Figure 1 Milwaukee rainfall and runoff probability distributions

associated accumulative runoff volume for a medium density residential area. This figure shows that the median rain, by count, was about 0.3 inches, while the rain associated with the median runoff quantity is about 0.75 inches. Therefore, more than half of the runoff from this common medium density residential area was associated with rain events that were smaller that 0.75 inches. The very large storms (about 3 and 5 inches in depth) distort this figure because, on average, the Milwaukee area can only expect one 3.5 inch storm about every five years. If these large rains did not occur, such as for most years, then the significance of the small rains would be even greater. Similar plots of the accumulative loadings of different pollutants that were also monitored during the Milwaukee NURP monitoring effort are very similar to the runoff plot, indicating that runoff volume is a very most import factor affecting pollutant discharges.

These rainfall/runoff distribution plots for Milwaukee can be divided into four categories, separated by rain depth:

• <0.5 inch. These rains account for most of the runoff events, but little of the runoff volume, and are therefore easiest to control. They produce much less pollutant mass discharges and probably have less receiving water effects than other rains. However, the runoff pollutant concentrations likely exceed regulatory standards for several categories of critical pollutants, especially bacteria and some total recoverable heavy metals. They may also cause large numbers of overflow events in uncontrolled combined sewers. These rains are very common, occurring once or twice a week (accounting for about 60% of the total rainfall events and about 45% of the total runoff events that occurred), but they only account for about 20% of the annual runoff and pollutant discharges. In most situations, runoff from these small rains should be totally captured and either re-used for on-site beneficial uses or infiltrated in upland areas.

- 0.5 to 1.5 inches. These rains account for the majority of the runoff volume (about 50% of the annual volume for this Milwaukee example) and produce moderate to high flows. They account for about 35% of the annual rain events, and about 20% of the annual runoff events. These rains occur on the average about every two weeks during the spring to fall seasons and subject the receiving waters to frequent high pollutant loads and moderate to high flows. The small rains in this category should also be used on site for beneficial uses, or infiltrated to replenish the lost groundwater infiltration associated with urbanization. The runoff from the larger rains in this category should be treated to prevent pollutant discharges from entering the receiving waters.
- 1.5 to 3 inches. These rains likely produce the most damaging flows, from a habitat destruction standpoint, and occur every several months (at least once or twice a year). These recurring high flows, which were historically associated with much less frequent rains, establish the energy gradient of the stream and cause unstable streambanks. Only about 2 percent of the rains are in this category and they are responsible for about 10 percent of the annual runoff and pollutant discharges. Typical storm drainage design events fall in the upper portion of this category. Extensive pollution controls designed for these events would be very costly. The infiltration and other treatment controls used to handle the smaller storms in the above categories would have some benefit in reducing pollutant discharges and flows during these larger, rare storms.
- >3 inches. The smallest rains in this category are included in design storms used for drainage systems in Milwaukee. These rains occur only rarely (once every several years, to once every several decades, or less frequently) and produce extremely large flows. The monitoring period during the Milwaukee NURP program was unusual in that two of these events occurred. Less than 2 percent of the rains were in this category (typically <<1% would be), and they produced about 15% of the annual runoff quantity and pollutant discharges during the monitoring period. During a "normal" period, these rains would only produce a very small fraction of the annual average discharges. However, when they do occur, great property and receiving water damage results. The receiving water damage (mostly associated with habitat destruction, sediment scouring, and the flushing of organisms great distances downstream and out of the system) can conceivably naturally recover to before-storm conditions within a few years. These storms, while very destructive, are sufficiently rare that the resulting environmental problems do not justify the massive controls that would be necessary for their reduction. The problem during these events is massive property damage and possible loss of life. These rains typically greatly exceed the capacities of the storm drainage systems, causing extensive flooding. It is critical that these excessive flows be conveyed in "secondary" drainage systems. These secondary systems would normally be graded large depressions between buildings that would direct the water away from the buildings and critical transportation routes and to possible infrequent/temporary detention areas (such as large playing fields or parking lots).

The above specific values are given for Milwaukee, WI, selected because of the occurrence of two very rare rains during an actual monitoring period. Obviously, the critical values defining the design storm regions would be highly dependent on local rain and development conditions. Computer modeling results (using SLAMM and long term rain records) for 24 urban locations throughout the U.S.are shown on Table 1. These results indicate how these rainfall and runoff probability distributions can be used for more effective storm drainage design in the future. In all cases, better integration of stormwater quality and drainage design objectives will require the use of long-term continuous simulations of alternative drainage designs in conjunction with upland and end-of-pipe stormwater quality controls. The complexity of most receiving water quality problems prevents a simple analysis using a small set of "design storms." The use of simple design storms, which was a major breakthrough in effective drainage design more than 100 years ago, is not adequate when receiving water quality issues must also be addressed.

Conclusions
There are many questions that remain concerning the "best" wet weather flow drainage and treatment systems that should be used in newly developing areas. Of course, there is no one "best" answer for all areas and conditions. A wide variety of options exist and an engineer must select from these depending on numerous site specific situations. In most cases, conventional separate sanitary wastewater and stormwater drainage systems would seem most appropriate. However, these systems have shown to be of reduced value in many cases.
The following list indicates some possible wastewater collection scenarios for several different conditions for the future:
• low and very low density residential developments (<2 acre lot sizes). Sanitary wastewater should be treated on site using septic tanks and advanced on-site treatment options. Domestic water conservation to reduce sanitary wastewater flows should be an important component of these systems. Most stormwater should be infiltrated on site by directing runoff from paved and roof areas to small bio-retention areas. Disturbed soil areas should use compost-amended soils and should otherwise be constructed to minimize soil compaction. Roads should have grass swale drainage to accommodate moderate to large storms.
• medium density developments (¼ to 2 acre lot sizes). Separate sanitary wastewater and stormwater drainage systems should be used. Sanitary wastewater collection systems must be constructed and maintained to eliminate I/I. Most stormwater should be infiltrated on site by directing runoff from paved and roof areas to small bio-retention areas. Paved areas should be minimized and the use of porous pavements and paver blocks should be used for walkways, driveways, overflow parking areas, etc. Disturbed soil areas should use compost-amended soils and should otherwise be constructed to minimize soil compaction. Grass swale drainages should be encouraged to accommodate moderate to large storms for the excess runoff in residential areas, depending on slope, soil types, and other features affecting swale stability. Commercial and industrial areas should also use grass swales and bioretention areas, depending on groundwater contamination potential and available space. Wet detention ponds should be used for controlling runoff from commercial and industrial areas. Special controls should be used at critical source areas that have excessive pollution generating potential.

• high density developments. Combined sewer systems could be effectively used in these areas. On-site infiltration of the least contaminated stormwater (such as from roofs and landscaped areas) is needed to minimize wet weather flows. On-site storage of sanitary wastewaters during wet weather (using Preul's CSPS), plus extensive use of in-line and off-line storage, and the use of effective high-rate treatment systems would minimize the damage associated with any CSOs. The treatment of the wet weather flows at the wastewater treatment facility would likely result in less pollutant discharges in these areas than if conventional separate wastewater collection systems were used.

References

Bannerman, R., K. Baun, M. Bohn, P.E. Hughes, and D.A. Graczyk. *Evaluation of Urban Nonpoint Source Pollution Management in Milwaukee County, Wisconsin*, Vol. I. Grant No. P005432-01-5, PB 84-114164. US Environmental Protection Agency, Water Planning Division, November 1983.

EPA (U.S. Environmental Protection Agency). *Results of the nationwide urban runoff program, volume I - Final report.* Water Planning Division, Washington, D.C. 1983.

Field, R. T. O'Connor, and R. Pitt. "Optimization of CSO storage and treatment systems." Proceedings: *A Global Perspective for Reducing CSOs: Balancing Technologies, Costs, and Water Quality.* July 10 to 13, 1994, Louisville, Kentucky. Water Environment Federation. Fairfax, VA. 1994.

Heaney, J.P., L. Wright, D. Sample, R. Pitt, R. Field, and C-Y Fan. "Innovative wet-weather flow collection/control/treatment systems for newly urbanizing areas in the 21st century." Proceedings of: *Sustaining Urban Water Resources in the 21st Century.* Edited by A.C. Rowney, P. Stahre, and L.A. Roesner. Malmo, Sweden. Sept. 7 – 12, 1997. An Engineering Foundation/ASCE Conference. To be published by ASCE in 1998.

Hering, R. "Sewerage systems." *Transactions of the American Society of Civil Engineers*, 10, 361-386. 1881.

Pitt, R., M. Lilburn, S.Nix, S.R. Durrans, and S. Burian. *Guidance Manual for Integrated Wet Weather Flow (WWF) Collection and Treatment Systems for Newly Urbanized Areas (New WWF Systems).* First Year Project Report. Urban Watershed Management Branch. Wet Weather Flow Program. U.S. Environmental Protection Agency. 1997.

Preul, H.C. "Combined sewage prevention system (CSCP) for domestic wastewater control." Presented at the 7th *International Conference on Urban Storm Drainage*, Hannover, Germany, 1996. pp. 193 – 198. 1996.

Reyburn, W. *Flushed with Pride, the Story of Thomas Crapper.* Pavilion Books, Ltd. London. 1989.

WATER RESOURCES AND THE URBAN ENVIRONMENT

Table 1. Rainfall and Runoff Distribution Characteristics for Different Locations from Throughout the U.S.

| | Median rain depth, by count (in) | % runoff < med. rain | Rain depth for med. runoff (in) | Lower break-point rain (in) | % rain events < lower break-point | % runoff volume < lower break-point | Upper break-point rain (in) | % rain events < upper break-point | % runoff volume < upper break-point | % runoff volume between break-points | % rain events between break-points |
|---|---|---|---|---|---|---|---|---|---|---|---|
| Boise, ID | 0.07 | 3 – 5 | 0.30 – 0.35 | 0.10 | 52 | 9 - 11 | 0.91 | 99 | 89 - 93 | 80 - 82 | 47 |
| Seattle, WA | 0.12 | 4 – 6 | 0.62 – 0.80 | 0.18 | 60 | 8 - 11 | 3.4 | 99 | 92 - 96 | 84 - 85 | 39 |
| Los Angeles, CA | 0.18 | 3 – 5 | 1.2 – 1.5 | 0.29 | 64 | 7 - 10 | 3.5 | 99 | 92 - 98 | 85 - 88 | 35 |
| Reno, NV | 0.07 | 3 – 5 | 0.35 – 0.41 | 0.10 | 61 | 8 - 10 | 1.7 | 99 | 93 - 95 | 85 | 38 |
| Phoenix, AZ | 0.10 | 4 – 6 | 0.55 – 0.68 | 0.19 | 64 | 9 - 12 | 2.3 | 99 | 94 - 98 | 85 - 87 | 35 |
| Billings, MT | 0.06 | 2 – 4 | 0.55 – 0.60 | 0.12 | 64 | 8 - 10 | 1.6 | 99 | 89 - 93 | 81 - 83 | 35 |
| Denver, CO | 0.08 | 2 – 4 | 0.50 – 0.60 | 0.19 | 71 | 13 - 17 | 1.8 | 99 | 91 - 95 | 78 | 28 |
| Rapid City, SD | 0.06 | 2 – 4 | 0.50 – 0.55 | 0.15 | 69 | 10 - 13 | 1.9 | 99 | 92 - 96 | 82 - 83 | 30 |
| Wichita, KS | 0.13 | 2 – 5 | 1.1 – 1.4 | 0.31 | 65 | 10 - 13 | 3.0 | 99 | 88 - 93 | 78 - 80 | 34 |
| Austin, TX | 0.14 | 2 – 3 | 1.4 – 1.8 | 0.50 | 72 | 8 - 12 | 6.0 | 99 | 88 - 94 | 80 - 82 | 27 |
| Minneapolis, MN | 0.11 | 3 – 5 | 0.73 – 1.0 | 0.22 | 65 | 9 - 13 | 2.8 | 99 | 94 - 96 | 83 - 85 | 34 |
| Madison, WI | 0.12 | 3 – 5 | 0.78 – 0.98 | 0.23 | 65 | 9 - 13 | 3.5 | 99 | 97 - 99 | 86 - 88 | 34 |
| Milwaukee, WI | 0.12 | 2 – 4 | 0.9 – 1.1 | 0.25 | 65 | 9 - 12 | 2.5 | 99 | 89 - 95 | 80 - 83 | 34 |
| St. Louis, MO | 0.14 | 4 – 6 | 1.0 – 1.2 | 0.31 | 65 | 10 - 13 | 2.8 | 99 | 90 - 95 | 80 - 82 | 34 |
| Detroit, MI | 0.20 | 7 – 11 | 0.72 – 0.81 | 0.20 | 50 | 7 - 11 | 2.4 | 99 | 92 - 95 | 85 - 84 | 49 |
| Buffalo, NY | 0.11 | 2 – 4 | 0.61 – 0.72 | 0.12 | 64 | 8 - 12 | 2.1 | 99 | 88 - 93 | 80 - 81 | 35 |
| Columbus, OH | 0.12 | 3 – 5 | 0.80 – 1.0 | 0.22 | 63 | 8 - 12 | 2.2 | 99 | 85 - 91 | 77 - 79 | 36 |
| Portland, ME | 0.15 | 2 – 4 | 1.1 – 1.5 | 0.30 | 64 | 8 - 12 | 4.5 | 99 | 90 - 96 | 82 - 84 | 35 |
| Newark, NJ | 0.28 | 6 – 12 | 1.2 – 1.5 | 0.33 | 54 | 8 - 12 | 3.3 | 99 | 89 - 94 | 81 - 82 | 45 |
| New Orleans, LA | 0.25 | 3 – 5 | 1.7 – 2.2 | 0.45 | 62 | 7 - 11 | 4.0 | 99 | 88 - 93 | 81 - 82 | 37 |
| Atlanta, GA | 0.22 | 3 – 5 | 1.2 – 1.7 | 0.32 | 58 | 5 - 9 | 4.0 | 99 | 91 - 95 | 86 | 41 |
| Birmingham, AL | 0.20 | 3 – 5 | 1.2 – 1.5 | 0.40 | 64 | 8 - 13 | 5.0 | 99 | 90 - 96 | 82 - 83 | 35 |
| Raleigh, NC | 0.18 | 4 – 6 | 1.0 – 1.2 | 0.26 | 60 | 7 - 11 | 2.5 | 99 | 87 - 93 | 80 - 82 | 39 |
| Miami, FL | 0.13 | 3 – 5 | 1.2 – 1.6 | 0.30 | 67 | 9 - 13 | 4.0 | 99 | 87 - 93 | 78 - 80 | 32 |

# Effects of Wetlands on
# Modulating Hydrologic Regimes in Nine Wisconsin Watersheds

D. L. Hey[1] and J. A. Wickenkamp[2]

## Abstract

While few people question that the hydrology of the Great Lakes basin has been affected by agricultural and urban development, the effects of wetland losses have not been clearly quantified. This study was designed to test the hydrologic importance of wetlands to the Great Lakes. Nine watersheds in southeastern Wisconsin, all tributaries to Lake Michigan, were analyzed and the presence of wetlands was related to the hydrologic characteristics of these basins. A number of hydrologic and statistical measures were selected to assess the effects. Findings also show that the hydrologic regime is most sensitive to the presence or loss of wetlands in the range of 1 to 10 percent of the watershed.

## Introduction

Wetlands are important to the ecological structure surrounding the Great Lakes and their tributary streams. The wholesale loss of wetlands, over the past 300 years, contributes to the disastrous lake fluctuations, degraded water quality, and declining wildlife populations. To many, the environmental damage resulting from the loss of wetlands in the Great Lakes basin is obvious, but to others it is not so clear.

To test the hypothesis that wetlands are important in defining the hydrologic characteristics of a watershed, nine watersheds in southeastern Wisconsin, all tributaries to Lake Michigan, were analyzed. The presence of wetlands was related to the hydrologic characteristics of these basins. These characteristics in turn relate to flooding and water quality. A number of hydrologic and statistical measures were selected to assess the effects.

This analysis also attempted to answer the question, "How many wetlands are enough to improve the water quality or reduce flooding in a watershed?" This question has been previously addressed by others (Novitski, 1982; Johnston et al., 1990;

---

[1] Senior Vice President, The Wetlands Initiative, 53 W. Jackson Blvd., Chicago, IL 60604.
[2] Water Resources Engineer, Hey & Associates, 627 N. Second St., Libertyville, IL 60048

Hey and Philippi, 1994). The results of these analyses, although of insufficient scope to be universally conclusive, indicate that the hydrologic regime is most sensitive to the presence of wetlands in the range of 1 to 10 percent of the watershed. The argument that wetlands along first order streams are as important as wetlands along higher order streams appears to be true as well.

## Methods

Nine watersheds were selected along the southeast Wisconsin shore (Table 1). These basins were geographically grouped to avoid climatological differences; they all fall within the glaciated portion of Wisconsin. They have similar watersheds, except their land use varies as required by the analyses. Wetlands are present in each basin, ranging from 2 to 19 percent of the surface area (Table 1). The degree of urbanization, as represented by imperviousness, varies from 1 to 16 percent. The drainage system of each basin has been altered by humans. In the upper watershed areas, most of the defined channels are man-made.

Table 1. Watershed parameters and characteristics.

| Watershed | Watershed Area (acres) | Percent Hydric Soil | Percent Wetlands | Percent Impervious ness | Mean Flow (cfs) | Annual Yield (in/yr) | Flow >0.1 Percent of Time (cfs) | Flow > 85 Percent of Time (cfs) | Excursions Per Year Above Flow >50 Percent of Time |
|---|---|---|---|---|---|---|---|---|---|
| N. Milwaukee River | 388,480 | 22.3 | 17.0 | 3.1 | 473.8 | 10.7 | 3944 | 134.2 | 8.6 |
| Manitowoc River | 336,640 | 22.5 | 18.8 | 3.8 | 374.3 | 9.9 | 4195 | 36.6 | 4.8 |
| Sheboygan River | 267,500 | 22.8 | 14.8 | 4.0 | 313.8 | 9.8 | 3900 | 58.4 | 8.4 |
| Root River | 120,806 | 21.0 | 5.2 | 6.7 | 176.0 | 12.6 | 2339 | 17.2 | 11.2 |
| Kewaunee River | 81,280 | 13.9 | 7.8 | 1.4 | 102.4 | 10.5 | 3020 | 15.4 | 7.7 |
| Menomonee River | 78,720 | 17.6 | 7.7 | 16.4 | 117.2 | 12.6 | 2156 | 18.8 | 24.2 |
| East Twin River | 70,400 | 13.1 | 15.6 | 2.6 | 87.6 | 11.0 | 1440 | 13.6 | 7.6 |
| Pike River | 24,640 | 26.3 | 2.4 | 8.3 | 32.1 | 11.3 | 690 | 1.2 | 15.1 |
| Oak Creek | 16,000 | 19.1 | 2.8 | 12.2 | 26.9 | 14.8 | 512 | 3.2 | 20.0 |

The watersheds are independent drainage units (i.e., one is not tributary to another). U.S. Geological Survey (USGS) stage recorders at the outlets have been in operation for at least twelve years. The period of record covers water-years 1981 through 1993. Discharges are computed from measured water levels (stages) and well-defined stage-discharge relationships. The record for each watershed is nearly complete. In cases where there were missing data, they were estimated by interpolation.

Hydric soils and wetland data for each watershed were derived from the Natural Resources Conservation Service's STATSGO soil associations, and the Wisconsin Department of Natural Resource's (WDNR) wetland inventory. Both data sets were obtained from GIS databases.

Four hydrologic measures were used to evaluate the influence that wetlands exert on the water resources of a watershed:
- Yield (inches);
- Unit base flow, as defined by the flow value exceeded 85 percent of the time divided by basin area (cubic feet per second per acre);
- Peak unit flow, as defined by the flow value exceeded 0.1 percent of the time

divided by the area of the watershed (cubic feet per second per acre); and
- The frequency of excursions above the flow value exceeded 50 percent of the time (number of excursions).

These measurements were determined for each watershed and then correlated to the percent of wetlands, hydric soils, and imperviousness in the watershed.

Yield reflects the productivity of a watershed. It measures the amount of water produced by the climatic and hydrologic conditions of the watershed. As an engineer named Charles Ellet, Jr. (1852) observed, in a report to Congress about flooding on the Mississippi River, yield increases when wetlands are drained and stream channels made more hydraulically efficient. Increases in yield also result from increases in impervious surface.

The base flow characteristic also reflects the effects of wetland losses. Groundwater recharge and discharge to the surface drainage system are affected. The hypothesis being, in this case, that the loss of wetlands reduces infiltration, thereby reducing groundwater storage and base flow.

The effects of wetlands on peak flow events are represented by the flow value exceeded 0.1 percent of the time. This exceedence level approximates the mean annual flood. This statistic reflects the propensity of the watershed to produce extreme events, high and low. It should be noted that the flow value is the result of direct observation and not derived from statistical computations such as would be required to determine the 100-year event.

A relatively large number of excursions above the flow value exceeded 50 percent of the time is indicative of unstable hydrologic conditions, as would result from a loss of storage. In the absence of storage, such as provided by wetlands, small amounts of precipitation and snowmelt produce significant runoff, which immediately enters the drainage system, causing a rise in discharge. The number of excursions above the 50 percent flow value is used to indicate hydrologic instability.

As wetlands are drained, and storage, infiltration, and evapotranspiration reduced, the high flows become higher and the low flows lower. The previous two statistics measure the extremes. The flow range measures the differences. It is defined by the difference between the flow value exceeded 85 percent of the time and the flow value exceeded 5 percent of the time. As such, it characterizes simultaneously the influence that wetlands have on the extremes.

Daily discharges computed by the USGS were used. These values were adjusted by subtracting the sanitary flow entering each drainage system. These monthly flow values—evenly distributed over the days of the month—were provided by the WDNR. The average daily values were then subtracted from the computed discharge values. In cases where the average sanitary flow exceeded the computed discharge, the resulting negative flow value was replaced by a zero. This occurred in only a few cases.

The land use data also were provided by the WDNR. Wetland areas and imperviousness were provided in a GIS file from which the relevant land cover statistics were derived (Table 1).

## Results

### Yield

The relationship between annual yield and percent of wetlands is best represented by a nonlinear function of the general form $y=ax^b$. Given the regression coefficients, $a=14.8$ and $b=-0.124$, the relationship explains 52 percent of the variation (i.e., $r^2=0.52$). The general form also reasonably represents the relationship between yield and percent imperviousness. In this case, the coefficient of determination equals 0.53. On the other hand, virtually no relationship exists between the presence of hydric soils and yield, the coefficient of determination being less than 0.01.

These results are supported by the observed physical relationships. The hydric soils, which remain in the watershed, largely have been drained. Agricultural tiles and outlet ditches remove the water very quickly from these areas; therefore, the influence of hydric soils on yield has been greatly diminished. In regard to imperviousness, these surfaces cut off infiltration and reduce evapotranspiration, thereby increasing surface runoff. The associated storm drains and channelized stream reaches offer little storage, moving the water rapidly downstream. Wetlands, in turn, reduce yield by detaining water longer and providing the opportunity for greater infiltration and evapotranspiration.

Further, the regression lines relating percent wetlands and imperviousness to yield are properly sloped. The slope of the line relating wetlands and yield is negative and that for imperviousness is positive. As the percent of wetlands increases, increasing the opportunity for evapotranspiration, yield is reduced. As imperviousness increases, evapotranspiration is reduced and yield is increased.

### Base Flow

The base flow characteristic is represented by the unity flow values (cfs/acre), which is exceeded 85 percent of the time. The least-squares estimate of the relationship between percent of wetlands and the unit base flow, $y=7.67 \times 10^{-5} x^{0.368}$, is positively sloped, indicating that as the area of wetlands increases, base flow increases. On the other hand, a similar relationship between imperviousness and yield, $Y=2.03 \times 10^{-4} x^{-0.126}$, is negatively sloped. As impervious surface increases, infiltration and groundwater recharge are reduced, leading to a reduction in base flow. As before, little or no relationship exists between hydric soils and base flow.

### Peak Flow

The influence of wetlands on extremely high flows is characterized by the flow value exceeded 0.1 percent of the time. The exponential form, again, best represents the relationship, $y=0.463 x^{-0.415}$. This function explains 54 percent of the variation. The slope of this relationship is negative: As the percent wetlands increase, extreme flows are reduced.

It is interesting to note that there is little or no relationship between the extreme flow statistic and imperviousness. The coefficient of determination for both the exponential and linear functions is less than 0.2. Consequently, less than 20 percent of variation is explained by imperviousness. The low coefficient of determination can be explained by the following: Extreme events are usually related to saturated soil conditions in the watershed (i.e., antecedent moisture conditions are high); consequently, infiltration is greatly reduced, and the soil's surface then acts like the impervious surface, both converting a high percentage of precipitation or snowmelt to runoff.

### Excursion Frequency

The relationship between percent wetlands and the frequency of excursions above the flow value exceeded 50 percent of the time ($Q_{50}$) is of the same form as that for the other measures. The exponential function, $Y=29.2 \times ^{-0.486}$, best represents the hydrologic behavior of the nine watersheds. It explains 52 percent of the variation. As with yield, the relationship between imperviousness and excursions above $Q_{50}$ is positively sloped, while that for excursions and percent wetlands is negatively sloped. Again, hydric soils explain less than 0.01 percent of the variation.

### Conclusions

The percentage of wetlands needed in a watershed depends on a variety of issues. First, the specific need for the wetlands must be identified (e.g., flood storage or water quality treatment). In-stream flow needs, recreation, and aesthetic quality can be converted to required acres of wetlands. Such needs can lead to very specific site locations and design criteria.

In answering the question, "How many wetlands are needed?", there is, however, another perspective offered by the law of diminishing returns (Samuelson, 1967). The law refers to the declining amount of output (e.g., the reduction of peak flows or increase in base flow) that results from equal amounts of input (e.g., increase in the percent of wetlands in a watershed). Based on this law, wetlands should continue to be added to a watershed until the desired output is no longer sufficiently increased. For example, as equal units of wetlands are added, the extreme flow value should decrease (Table 2). But, after wetlands comprise 5 or 6 percent of the watershed, the extreme value decreases at a declining rate.

All of the exponential functions derived for the nine watersheds are relatively more steeply sloped when wetlands are present in lower percentages than in higher. The slope begins to flatten very rapidly above 10 percent (Table 2). This is true for all measures studied: yield, peak flow, base flow, excursion frequencies, and flow range. Increasing wetland percentages above 10 percent may be undesirable since less change in the desired characteristic results with each percentage increase.

Having selected the percent wetlands that a watershed should contain, the next question is where these wetlands should be constructed. This depends on where the benefits are most needed. As shown by Ogawa (1983), the closer to the need that wetlands are reconstructed, the greater the benefit. The distribution of wetlands is of considerable importance, given, for example, the variable precipitated patterns and hydrologic conditions that lead to flooding. Better to have the wetlands distributed over a broad range so as to minimize the adverse effects throughout the watershed than having them concentrated in

Table 2. Predicted results from the exponential function $Y=ax^b$ for percent wetlands.

| Hydrologic Characteristic | | Percent Wetlands | | | | | | | | |
|---|---|---|---|---|---|---|---|---|---|---|
| | x= | 2 | 4 | 6 | 8 | 10 | 12 | 14 | 16 | 18 |
| Yield | Y= | 1.36E+01 | 1.24E+01 | 1.18E+01 | 1.14E+01 | 1.11E+01 | 1.08E+01 | 1.06E+01 | 1.05E+01 | 1.03E+01 |
| | dy/dx= | -8.43E-01 | -3.87E-01 | -2.45E-01 | -1.77E-01 | -1.38E-01 | -1.12E-01 | -9.45E-02 | -8.13E-02 | -7.12E-02 |
| | %change= | | -8.3 | -4.9 | -3.5 | -2.7 | -2.2 | -1.9 | -1.6 | -1.5 |
| | cum chg= | | -8.3 | -12.8 | -15.8 | -18.1 | -20.0 | -21.5 | -22.8 | -23.9 |
| Baseflow/acre | y= | 9.90E-05 | 1.28E-04 | 1.48E-04 | 1.65E-04 | 1.79E-04 | 1.91E-04 | 2.03E-04 | 2.13E-04 | 2.22E-04 |
| | dy/dx= | 1.38E-05 | 8.89E-06 | 6.88E-06 | 5.74E-06 | 4.98E-06 | 4.44E-06 | 4.03E-06 | 3.70E-06 | 3.44E-06 |
| | %change= | | 29.1 | 16.1 | 11.2 | 8.6 | 6.9 | 5.8 | 5.0 | 4.4 |
| | cum chg= | | 29.1 | 45.2 | 56.3 | 64.9 | 71.8 | 77.7 | 82.7 | 87.1 |
| Peak flow/acre | y= | 3.67E-02 | 2.75E-02 | 2.33E-02 | 2.07E-02 | 1.88E-02 | 1.75E-02 | 1.64E-02 | 1.55E-02 | 1.48E-02 |
| | dy/dx= | -7.62E-03 | -2.86E-03 | -1.61E-03 | -1.07E-03 | -7.81E-04 | -6.04E-04 | -4.85E-04 | -4.02E-04 | -3.40E-04 |
| | %change= | | -25.0 | -15.5 | -11.3 | -8.8 | -7.2 | -6.2 | -5.4 | -4.8 |
| | cum chg= | | -25.0 | -36.6 | -43.7 | -48.7 | -52.5 | -55.4 | -57.8 | -59.8 |
| Excursion frequency | y= | 20.9 | 14.9 | 12.2 | 10.6 | 9.5 | 8.7 | 8.1 | 7.6 | 7.2 |
| | dy/dx= | -5.1 | -1.8 | -1.0 | -0.6 | -0.5 | -0.4 | -0.3 | -0.2 | -0.2 |
| | %change= | | -28.6 | -17.9 | -13.0 | -10.3 | -8.5 | -7.2 | -6.3 | -5.6 |
| | cum chg= | | -28.6 | -46.4 | -59.5 | -69.7 | -78.2 | -85.4 | -91.7 | -97.3 |

one subwatershed, leaving the others without benefit.

The foregoing analysis, in terms of the nine watersheds studied, clearly indicates the importance of wetlands in modulating hydrologic regimes. With only limited wetlands, peak flows are more extreme and low flows, lower. The frequency of excursions above the 50 percent flow value are far greater with fewer wetlands than with more wetlands. This implies that there are less stable hydrologic conditions in watersheds with fewer wetlands. This analysis indicates that an increase in wetlands will change the hydrologic characteristics of a watershed, reduce the risk of flooding by providing storage, improve water quality by increasing detention time, and provide for in-stream flow needs by stabilizing the flow regime.

## References

Ellet, Jr., C. 1852. Overflows of the delta of the Mississippi. Washington, D. C.: The War Dept.

Hey, D. L., and N. S. Philippi. 1994. Reinventing a flood control strategy. Chicago: The Wetlands Initiative.

Interagency Floodplain Management Review Committee. 1994. Sharing the challenge: Floodplain management into the 21st century. Washington, D. C.: Administration Floodplain Management Task Force.

Johnston, C. A., N. E. Detenbeck, and G. J. Niemi. 1990. The cumulative effect of wetlands on stream water quality and quantity. *Biogeochemistry* 10.

Moor, I. D., and C. L. Larson. 1980. Hydrologic impacts of draining small depressional watersheds. *J of the Irrigation and Drainage Division* Dec: 106 (IR4).

Novitzki, R. P. 1982. Hydrology of Wisconsin wetlands. University of Wisconsin-Extension: 40.

Ogawa, H., and J. W. Male. 1983. The flood mitigation potential of inland wetlands. Water Resources Research Center, University of Massachusetts, 138.

Samuelson, P. A. 1976. Economics. Seventh Ed. New York: McGraw-Hill Book Co.

# Water Management in New Hampshire: An Overview and Preliminary Event Duration Analysis of Proposed Instream Flow Rules

Neil M. Fennessey[1], Member, ASCE

Abstract

The New Hampshire Department of Environmental Services (NHDES) has developed draft rules to establish how instream flows should be managed in the state. If approved by the State Legislature, the Instream Flow Rules (IFR) will apply to all consumptive withdrawals located within 250 feet of designated rivers or river segments. The IFR stipulate that the regulated community must reduce, even cease withdrawing water depending on how real time flows compare with season specific trigger streamflow rates. The focus of this paper is to explore the need to examine the historic frequency and the future likelihood that a consumptive user would have been/would be required to either reduce or cease their withdrawal.

Introduction

The NHDES has devised draft rules for the State Legislature which establish how the State's streams and rivers should be managed in real time as required by the New Hampshire Rivers Management and Protection Act, RSA 483. The IFR will apply to all consumptive water users with withdrawal points located within 250 feet of 11 designated rivers or river segments representing 110 watersheds. Affected rivers or segments include: the Ashuelot; the Connecticut; the Contoocook; the Lamprey; the upper and lower Merrimack; the North Branch; the Pemigewasset; the Piscataquog; the Saco; and the Swift. This work has been discussed by Fennessey (1997).

Instream Flow Rule Development

The Instream Flow Rules (IFR) were developed by the Rivers and Management Advisory Committee established by RSA 483, and a special statewide task force known as the Instream Flow Working Group. The Instream Flow Working Group represents the interests of: Public Water Supply; Fisheries and Wildlife; Business and Industry; Hydropower; Historic and Archeological, Conservation; Environment; Recreation; Ski Industry and Agriculture.

[1] Assistant Professor, Dept. of Civil and Environmental Engineering, University of Massachusetts, North Dartmouth, MA. 02747. Email: nfennessey@umassd.edu

### The Regulated Community

The IFR will apply to all consumptive withdrawals, including public water suppliers, which are located within 250 feet of a designated river or river segment. A withdrawal is deemed non-consumptive if the discharge following use is returned to the withdrawal location in the same quantity and quality. Consumptive users will be required to reduce or cease withdrawing depending upon the real time daily streamflow rate estimated by gauging at the lowest point in the designated watershed. Regulatory flow rates upstream of this point will be calculated using historic gage data if available or by way of regional regression analysis as necessary.

### Instream Flow Rules and Trigger Flows Defined

The state has been subdivided into two geographical regions which depend upon the time of typical spring snow melt and runoff. The year is subdivided into four flow seasons. Within each season, four IFR trigger streamflows have been established which are referred to as Phase I, II, III and Phase IV flow rates. These flows respectively correspond to season specific $Q_{50}$, $Q_{70}$, $Q_{80}$ or $Q_{90}$ where for example, $Q_{50}$ is the estimated median daily flow of that season and $Q_{90}$ is that rate which is equaled or exceeded with a probability of 0.9 (or 90% of the time).

The IFR rules become effective when measured flows equal or fall below Phase II, Phase III or Phase IV streamflow rates for 7 consecutive days. After 7 days, the NHDES Commissioner will order the following: with sub-Phase II flow conditions, the total of all consumptive uses in a basin must be reduced to not more than 5% of the seasonal $Q_{70}$; under sub-Phase III flow conditions, the total consumptive use in a basin must be reduced to not more than 2% of the seasonal $Q_{80}$; and given sub-Phase IV flow conditions, all consumptive withdrawals must cease.

### An Event Frequency and Duration Analysis

By definition, flows historically fell below $Q_{70}$ 30 percent of the time (during the period-of-record employed in the analysis) and that flows historically fell below $Q_{90}$ 10 percent of the time. However, the proposed IFR flow rates say nothing as to how long any sub-threshold event might last, either historically or in the future. As such, one serious shortcoming of the draft IFR, as major water resources management policy instrument, is that the regulated community has no idea as to how often or for how long they will be actively regulated by the Commissioner. What's missing in the IFR discussion to date is a comprehensive crossing property analysis (see Salas, 1993) of the historic stream gauge records. This information should be fundamental to the rule making process, and as yet, no such analysis has been performed.

Consider that the regulated community might wish to pose the following sorts of questions: (a) how many times did the average daily streamflow historically equal or fall below $Q_{50}$, $Q_{70}$, $Q_{80}$ and $Q_{90}$ for 7 or more consecutive days in some given season, in a designated watershed? In other words, how would they have been regulated in the past? (b) what is the likelihood of regulated withdrawers having to reduce or cease withdrawals for *"n"* or more consecutive days during any given season?

To begin to answer these questions, estimates of summer and winter season $Q_{70}$ and $Q_{90}$ were made by sampling the IFR season defined daily flow records for a

USGS gauge on the Merrimack River, at Lowell, MA (4635 mi$^2$), near the New Hampshire-Massachusetts border and a gauge on the Saco River near Conway, N.H. (385 mi$^2$). Flows observed on the Saco might be representative of the behavior of streamflow in smaller watersheds located in the IFR northern region while flows observed at the Lowell gauge of the Merrimack would be representative of the behavior of a large designated river basin located in the IFR southern region

Figure 1 illustrates how daily flows rose above and fell below historic estimates of $Q_{70}$ and $Q_{90}$ at these gauge locations on the Merrimack and Saco Rivers during the 5 summers of 1960-1964 inclusive. How many times and for how long would a withdrawer located in either watershed have been required to cease withdrawing ?

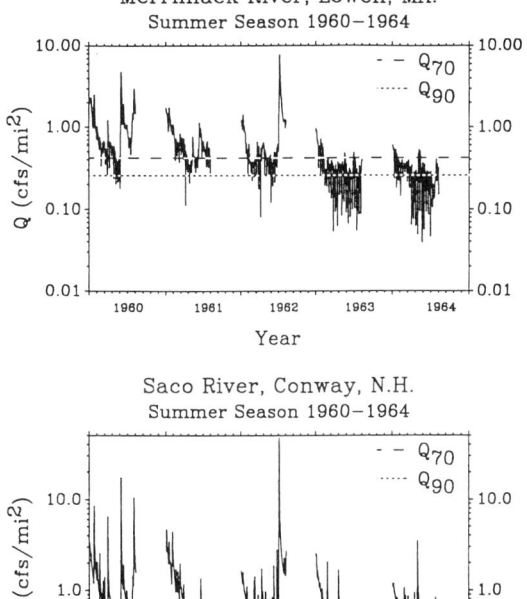

**Figure 1.** *Summer season flows in the Merrimack and Saco Rivers.*

Figure 2 is an event duration frequency histogram for summer season $Q_{90}$ at these two gauge locations. For example, flows equaled or fell below $Q_{90}$ (1190 cfs) during 318 separate one day long events and twice for 26 consecutive days at Lowell on the Merrimack River during the summer over the period-of-record used. By comparison, there were only 19 one day long events when Saco River flows were less than or equal to $Q_{90}$ (148 cfs) and one 37 consecutive day event during this period-of-record.

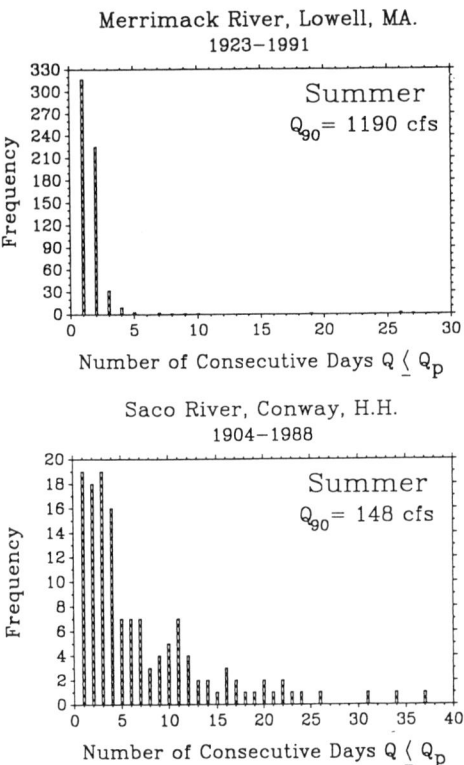

**Figure 2.** *Frequency and duration of sub $Q_{90}$ flows in Merrimack and Saco Rivers*

One can view this issue with an eye towards the future by analyzing the problem in a probabilistic framework. In this case, the regulated community could then know the probability, or likelihood that daily streamflows will equal or fall below $Q_p$ (p=50, 70, 80 or 90) for an event duration of "n" days or more in a designated watershed.

Recall that the regulated community will have to respond to a specific NHDES Commissioner order only if flows equal or fall below Phase II, III or IV triggers for 7 or more days. As shown on Figure 3, a horizontal line drawn on each graph corresponds to 7 days on the vertical axis. A vertical line is drawn from the point of intersection between this reference line and the plotted data downward to the "P" horizontal axis. One can easily read the estimated probability of an event in the watershed controlled by that gauge lasting 7 or more days: the probability or likelihood of being required to reduce or cease withdrawing water.

Figure 3 indicates that during the IFR summer season, a Lower Merrimack River withdrawer has a 1.5% chance of being ordered to cease withdrawals whereas a withdrawer located in the Saco River watershed has a 37% chance of a similar order when flows fall below Phase IV $Q_{90}$.

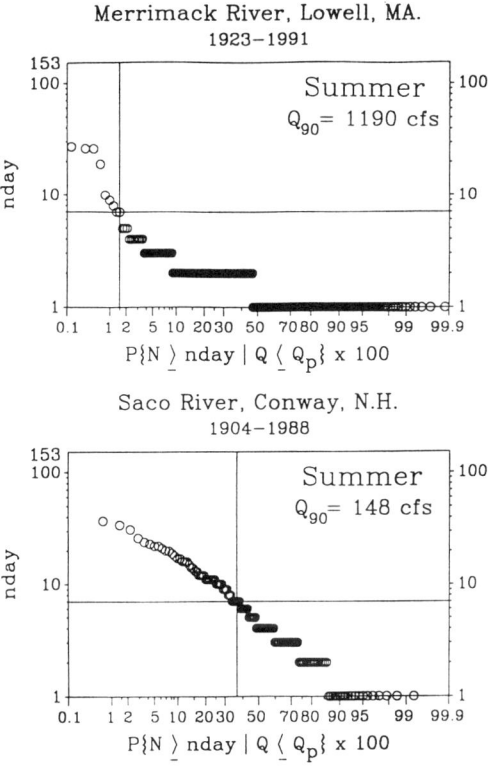

**Figure 3.** *Probability of sub $Q_{90}$ summer flows in the Merrimack and Saco Rivers*

A similar analysis of winter season flows, shown in Figure 4, suggest that a lower Merrimack River withdrawer has a 10% chance of being ordered to cease withdrawing whereas those in the Saco have a nearly 50% chance of being so ordered. An analysis of Phase II $Q_{70}$ flows during both seasons at these two locations suggests that withdrawers in the Lower Merrimack have a 12% chance of having to reduce withdrawals versus a 43% chance in the Saco during the summer. During the winter season, there's a 20% chance that Lower Merrimack River withdrawers will be required to reduce their withdrawal compared with those in the Saco watershed who have a 55% chance of being required to reduce their withdrawals.

Discussion

On the face of this very limited analysis, it would seem that members of the regulated community located in the Lower Merrimack River watershed have a clear regulatory advantage over those located in smaller tributaries. Withdrawers within the Lower Merrimack are at least one third less likely to be subject to any regulation, whether reduction or cessation, when compared with those in upland watersheds.

It is likely that a full analysis of New Hampshire streamgauge records would show that the frequency and duration of IFR sub-trigger flow events vary widely across the state. It is very important that a crossing property analysis as discussed herein be performed at all of New Hampshire streamgauge sites, for each season's Phase I, Phase II, Phase III and Phase IV flows. Until that study is complete, the Instream Flow Rule makers and the State Legislature can not fully debate the central issue as to whether the IFR are equitable to the regulated community statewide.

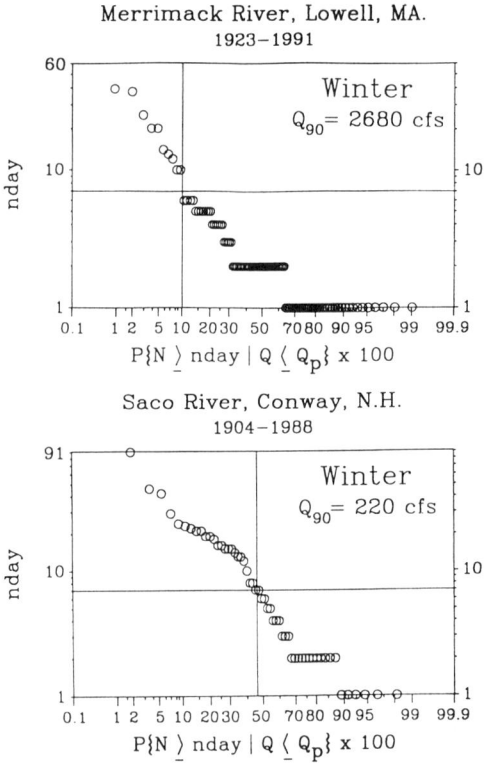

Figure 4. *Probability of sub $Q_{90}$ winter flows in the Merrimack and Saco Rivers*

References

Fennessey, N.M. (1997). "An event duration analysis of New Hampshire's proposed instream flow rules", J. New England Water Works Assoc. Vol. CXI (2):107-126.

Salas, J.D. (1993). "Analysis and modeling of hydrologic time series", Chapter 19 in the *Handbook of Hydrology*, edited by D. Maidment, McGraw-Hill, New York.

The Future of Illinois Water Law

Gary R. Clark,[1] M. ASCE

Abstract

During the 1990's, numerous citizens based committees concluded that the State's water use law was inadequate and recommended that legislative changes be developed. Due to these concerns, the Department of Natural Resources undertook studies to identify the inadequacies of the State's law and develop options for change. Through the use of focus groups, the Department's consultant identified specific problems and conflicts in the areas of surface water law, groundwater law and state and local water management institutions. Consideration for change in these areas are now underway under the direction of the Natural Resources Coordinating Council. Potential legislative initiatives are now being developed specifically in the area of instream flow protection which is considered as one of the major areas of conflict and inadequacy in Illinois water use law.

Background

In May of 1992, Governor Edgar appointed a Water Resources and Land Use Priorities Task Force comprised of 25 citizens with wide ranging expertise in agriculture, conservation, recreation, water resources, business, and land use. The purpose of the task force was to bring together diverse interests to air their differences, find common ground, reach agreement on certain principles, and make recommendations to the Governor on ways to address the growing conflicts over water and land use.

The report of the Task Force's Water Work Group (Water Resources and Land Use Priorities Task force, 1993) stated that "water rights in Illinois are poorly defined" and that "the lack of an adequate, single water resources act is the root cause of most of the water use conflicts during recent droughts." The Task Force further stated that "outside of regulated surface water, there is no policy to manage surface water usage and only

---

[1] Chief of Policy and Research, Office of Water Resources, Illinois Department of Natural Resources, 524 South 2nd Street, Springfield, Illinois 62701-1787

limited regulation and policy for groundwater." The Task Force therefore concluded that "the historical water law in Illinois is inadequate to meet present and future needs" and "conflicts about water usage will continue to grow, and their resolution will be determined more easily with a new, comprehensive state water use act." The Task Force recommended the following:

*Prepare a comprehensive water resources act to replace the inadequate collection of statutes and court decisions scattered throughout Illinois law.*

As a first step in a longer-term strategy to rewrite state water law, it was recommended that expertise outside of state government should be retained to undertake a comprehensive evaluation of current Illinois water use laws to identify inadequacies and conflicts. It was determined that Illinois needs to review its system of water rights and water resources management, define its strengths and weaknesses, compare its system with the evolving systems being implemented in other states, and develop recommended legislative and administrative responses to address inadequacies in the law and water resources management structure.

Comprehensive Study

As part of the State's Conservation 2000 initiative, the Illinois Department of Natural Resources (IDNR) was charged with implementing the recommendations of the Task Force regarding the assessment of the State's water law. IDNR contracted a two-part investigation, the purpose of which was to evaluate needed reforms in water management in Illinois and to determine the need to move the State toward a comprehensive water resources act. The initial effort, completed in 1995, consisted of a survey of water law in the eastern states, many of which have riparian-based legal systems similar to that of Illinois (Foran, 1995). The later effort, conducted by the consulting firm of Planning and Management Consultants (Beck et al., 1996) assessed the current system of water law in Illinois. This second study specifically conducted: (1) a review of Illinois system of water rights and water resources management, (2) an identification of its strengths and weaknesses, and (3) comparison of its system with the evolving systems being implemented in other states.

The assessment of Illinois water law focused on Illinois state law. Federal laws were discussed only to the point required to differentiate state and federal roles and responsibilities in water management. The assessment of Illinois water law also focussed solely on water quantity issues, even though it was recognized that many water quality-related laws, regulations, and programs can have significant impacts on the availability, control, and distribution of the state's water resources.

Methodology

The method used by the consultant involved the identification of inadequacies in state law based on: (1) the technical analysis of water management issues and conflicts and

(2) the legal analysis of the state's statutory law and case law. This method combined the reviews of state water management literature and state law with the expressed concerns of stakeholder focus groups to conduct an issue-oriented legal analysis of state water law. The legal analysis also drew upon other technical resources such as the Survey of Eastern water Law and the Regulated Model Water Code (American Society of Civil Engineers, 1997). The purpose of the legal analysis was to generate optional legal responses for further consideration by state governmental leaders or other entities to improve the effectiveness of state water law in addressing the identified water management issues and conflicts.

Focus Groups

The consultant's literature review was followed by a telephone contact of water management stakeholders. The stakeholders were then invited to attend focus groups meetings designed to: (1) identify and explore issues and conflicts in water management and (2) to develop and consider alternative solutions to these management problems. The groups also gave an opportunity to explore new issues and conflicts that were not documented in the literature. The focus group participants represented a wide variety of water management interests, including agriculture, environment, recreation, municipalities, utilities, drainage districts, and special purpose water authorities. The initial list of stakeholders was supplied by the State Water Plan Task Force and was expanded during the telephone survey based on referrals from the initial list of stakeholders. Over 50 stakeholder were invited to four separate meetings. Three of the focus groups meetings did not include state officials in order to allow stakeholders to openly discuss water management issues and conflicts without he presence of state agency representatives.

Identified Water Management Issues

Following the completion and analysis of the focus group meetings, the consultant identified key water management conflicts and issues based on the results of the literature review, focus groups meetings and legal survey. Figure 1 was developed by the consultant to illustrate the three major issue areas identified as surface water, groundwater and institutional. These issue areas are summarized below.

Surface Water

In the issue area of surface water management, the two key issues identified were instream flow and public water supply development. The instrean flow protection issue concerns the need to preserve a base stream flow that is protected from existing and future water withdrawals. This base stream flow would be set to protect downstream users, commercial and recreational navigation and aquatic ecosystems. A concern quite opposite to the instream flow issue addresses the problem that it is becoming increasingly more difficult to develop new water supplies such as reservoirs. A multitude of state, local and federal permits in addition to opposition of environmental groups, rural landowners and neighboring communities cause numerous delays, costs increases and a high level of uncertainty to the process of developing the needed expansion of public water supply systems.

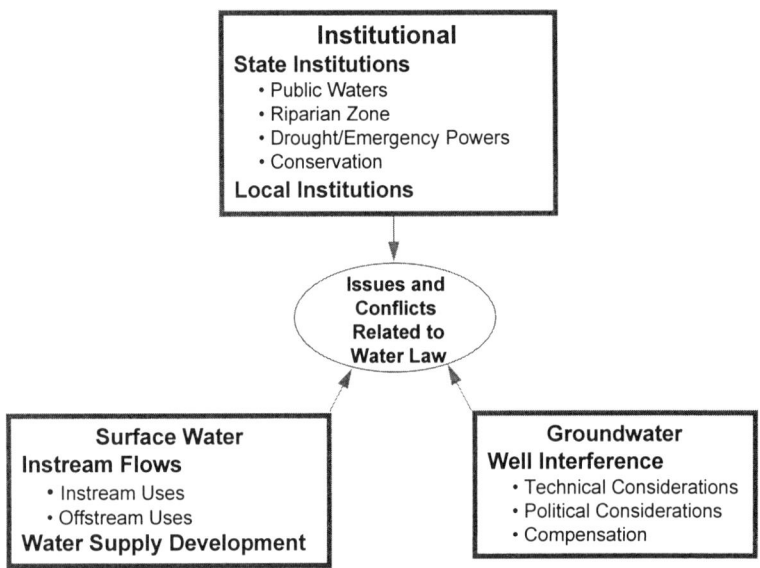

Figure 1. Management Issues Related to Illinois Water Use Law as Identified by Consultant

Groundwater

One of the fundamental groundwater water issue in Illinois concern well interference due to inadequate well spacing or more commonly, inadequate well development concerning well depth and screen location. The other two aspects to groundwater issues concern the rural/urban conflicts over the location of well fields and the issue of compensatory assessments for damages. During the drought of 1988 and 1989, numerous conflicts arose when communities expressed a desire to expand their well fields further out into rural areas. Rural landowners strongly objected based on their concern that they might lose the future rights to the use and the related economic benefits of the groundwater resource located under their land. These same landowners were also concerned that they and their neighbors may not be adequately compensated for any impacts due to the interference caused by nearby high capacity wells.

## Institutional

Focus groups expressed a concern regarding the appropriate future roles of state agencies to manage the state's surface and groundwater resources. Some concerns regarding whether the state through its agencies, or whether local governmental entities empowered through state law, would more adequately fund and manage the local water resource. Concerns were also expressed regarding the apparent limited definition used by the state to define the "public waters" over which the state exercises broad control and management. There was also concern expressed over the state's ability to protect riparian habitat as well as the state powers to respond in a timely fashion to conflicts which may occur during drought emergencies. Some stakeholders also felt the state should provide more leadership in the area of water conservation.

## Process for Change

The process to implement change in Illinois water law began with a day long conference to discuss the issues and solutions to the water use conflicts which were documented through the consultant's investigation. The Directors of the State's natural resources agencies requested that the State Water Plan Task Force review the proceedings of the conference along with the consultant's studies and select the key water quantity issues that warrant further consideration through legislative and administrative initiatives. The Water Plan Task Force was also directed to develop, along with the list of initiatives, a selection of optional responses for each initiative. The resource agency directors through the Natural Resources Coordinating Council approved the issues selected by the State Water Plan Task Force in December of 1997. The issues approved for further implementation are as follows:

- Emergency Powers and Drought Management
- Instream Flow Protection
- Codification of Water Quantity Laws
- Future Needs for Public Water Supplies
- Recreational Stream Access

The Natural Resources Coordinating Council assigned a separate lead state agency with the responsibility for implementing each one of the five water quantity issues and to report on the progress towards implementation at each meeting of the Council.

The key test for whether significant change to Illinois water use law is possible will be the willingness of key interest groups and the Illinois General Assembly to work towards the passage of legislation designed for the protection and management of instream flows within Illinois.

## References

American Society of Civil Engineers. 1997. The Regulated Riparian Model Water Code, New York, N.Y.

Beck. R.E., Harrington, K.W., Hardy, W.P., and Feather, T.D., 1996. Assessment of Illinois Water Quantity Law. Planning and Management Consultants, Carbondale, IL.

Foran, P.G., J.A. Beecher, and L.J. Wilson, 1995. Survey of Eastern Water Law. Lueders, Robertson, and Konzen, Granite City, IL and School of Public and Environmental affairs, Indiana University.

Water Resources and Land Use Priorities Task Force, 1993. Report to the Governor of Illinois. Springfield, IL.

## Allocation of Water Withdrawals in a River Basin

Jennifer M. Jacobs[1], Associate Member, ASCE,
and Richard M. Vogel[2], Member, ASCE

Abstract

    A general approach is suggested for allocating and permitting water withdrawals in a river basin. A mathematical programming methodology facilitates optimal streamflow allocation while maintaining desired levels of instream flow. The approach uses a graphical tool, the flow duration curve, to illustrate the quantity and frequency of joint streamflow withdrawals in a river basin. The methodology is unique because while it uses mathematical programming methods, it is implemented using a spreadsheet optimization tool, Microsoft Excel Solver™, and the solution is illustrated in a graphical form so that non-technical individuals can easily understand the methodology results.

Introduction

    The purpose of this paper is to present a general method for allocating consumptive water uses which reflects the goals (objectives) established by the permit system and the natural limitations (constraints) of unregulated streamflow availability, yet is easily implemented using common software and readily summarized in graphical terms. A case study for a hypothetical basin using a popular optimization tool, Microsoft Excel Solver™, documents the implementation of the proposed methodology.

    One of the most widely used optimization techniques in the field of water resource management is linear programming (LP) which includes methodologies that can introduce statistical constraints that account for random variables such as streamflow availability. A simple hydrologic tool, the flow duration curve (FDC), describes the variable nature of daily streamflow. The FDC has interesting graphical properties and features. The FDC specifies the relationship between streamflow and reliability. Reliability or the exceedance probability is defined as the probability that

---

[1] Asst. Prof., Dept. of Civil Engrg., Univ. of Florida, Gainesville, FL 32611.
[2] Assoc. Prof., Dept. of Civil and Envir. Engrg., Tufts Univ., Medford, MA 02155.

the average daily flow will be greater than or equal to the corresponding streamflow. The area underneath the FDC represents the average daily streamflow (Vogel and Fennessey 1994).

This paper uses FDCs to quantify the streamflow available for allocation, as did Male and Mueller (1992) and Fennessey (1998). The following application of FDCs to allocating withdrawals in an unregulated basin is in spirit similar to the methodology introduced by Alaouze (1989, 1991), but differs from that of Alaouze in three respects: (1) the withdrawals are distributed throughout the basin rather than withdrawn from a single point; (2) other constraints are introduced, such as instream flow requirements; and (3) other objectives exist, such as minimizing basin consumptive use and prioritization according to use category.

Streamflow Allocation Models

The problem is finding the allocation that maximizes the objective function, subject to restrictions based on streamflow availability, individual withdrawal requirements, instream flow requirements, and reliability limitations. The nonlinear streamflow availability constraints due to nonlinear FDCs are transformed to linear constraints using a piecewise linearization technique. The general chance-constrained model for maximizing water allocation for productive use is given as

$$\text{Max } Z = \sum_{i=1}^{N} w_i q_i \tag{1}$$

subject to

$$q_i \leq a_i, \quad \forall i = 1,2,\ldots,N \tag{2}$$
$$q_i \geq p_i, \quad \forall i = 1,2,\ldots,N \tag{3}$$
$$r_i \geq r_{min,i} \quad \forall i = 1,2,\ldots,N \tag{4}$$
$$\Pr[q_{i,TOT} \leq Q_{r_i}] = r_i \quad \forall i = 1,2,\ldots,N \tag{5}$$
$$q_i \geq 0 \quad \forall i = 1,2,\ldots,N \tag{6}$$
$$r_i \geq 0 \quad \forall i = 1,2,\ldots,N \tag{7}$$

here $w_i$ is the weight for site i, $q_i$ the permitted withdrawal quantity for site i, N the number of withdrawals, $a_i$ the withdrawal amount requested for site i, $p_i$ the existing permitted withdrawal at site i, $r_i$ the streamflow reliability for withdrawal i, $r_{i,min}$ the minimum acceptable streamflow reliability for withdrawal i, $q_{i,TOT}$ the total streamflow allocated to site i including instream flow, consumptive upstream use and withdrawal at the site and $Q_{ri}$ the streamflow available with reliability $r_i$ at site i.

Constraint set (2) limits the quantity allocated to each user to that requested by the user. Constraint set (3) protects existing permit quantities. Constraint set (4) establishes a minimum value for the reliability of each permitted withdrawal. Constraint sets (6) and (7) ensure the decision variables $q_i$ and $r_i$ are nonnegative.

The remaining streamflow constraint set (5) uses a chance-constraint to establish the probability that the amount of streamflow allocated will not exceed the amount of streamflow available. The FDC represented by the function, $Q_{ri}$, may be made piecewise linear with mixed integer linearization (Loucks et. al., 1981). Linearization is achieved by defining K segments, each having slope, $s_{i,k}$, horizontal length, $r_{i,k}$, beginning at reliability $R_{i,k}$. The reliability, $r_i$, is then defined as the sum of the K values of $r_{i,k}$

where
$$r_i = \sum_{k=1}^{K} [R_{i,k} z_{i,k} + r_{i,k}] \qquad \forall\, i = 1,2,...,N \qquad (8a)$$

$$\sum_{k=1}^{K} z_{i,k} = 1 \qquad \forall\, i = 1,2,...,N \qquad (8b)$$

and $z_{i,k}$ is an integer. An additional constraint must be added to limit the length of each $r_{i,k}$ to its maximum length

$$r_{i,k} < (R_{i,k+1} - R_{i,k}) z_{i,k} \qquad \forall\, i = 1,2,...,N,\, \forall\, k = 1,2,...,K \qquad (8c)$$

The result of the formulation is that at most one of the $r_{i,k}$ and one of the $z_{i,k}$ are greater than zero so that if $z_{i,k}=1$, then $r_i = R_{i,k} + r_{i,k}$. The linearized form of constraint (5) is then

$$q_{i,TOT} \le \sum_{k=1}^{K} [s_{i,k} r_{i,k} + z_{i,k} Q_{i,k}] \qquad \forall\, i = 1,2,...,N \qquad (9)$$

where each linearized FDC has K segments with slope $s_{i,k}$ and $Q_{i,k}$ is the streamflow available with reliability $R_{i,k}$. The final optimization model is solved for objective function (1) constrained by (2-4, 6-9).

The methodology was implemented by means of a computer tool that is easy to employ and readily accessible to the water permitting community. The Microsoft Excel Solver™ tool can optimize this model using mathematical programming.

Model Implementation and Results

To illustrate the model, the methodology was applied to a hypothetical unregulated river basin. The river basin is composed of two separate streams, S1 and S2, that converge downstream to form a single stream S3. There are three possible withdrawal locations. Site 1 on S1, site 2 on S2, and site 3 on S3. Table 1 shows the reliabilities and corresponding streamflow for the linearized FDCs as well as the instream flow requirement. The FDCs for site 1 and site 2 are the same. The proposed methodology was solved for 2 different allocation request scenarios. The results can be represented graphically. For any withdrawal location, the allocated streamflow and corresponding reliabilities are represented by category (e.g., instream flow, $q_s$, upstream allocation, $q_u$, that is, streamflow not available for use due to present and/or future consumption by an upstream user, and point of withdrawal streamflow allocation, $q_w$) using a FDC.

Table 1. Linearized flow duration curves for potential withdrawal sites.

| Site | Instream Flow Requirement | Streamflow $Q_r$ for Reliability, r | | | | | |
|---|---|---|---|---|---|---|---|
| | | 0.1 | 0.5 | 0.6 | 0.7 | 0.8 | 0.95 |
| 1 | 0.5 | 80.10 | 3.36 | 1.80 | 0.92 | 0.42 | 0.06 |
| 2 | 0.5 | 80.10 | 3.36 | 1.80 | 0.92 | 0.42 | 0.06 |
| 3 | 1.0 | 200.25 | 8.40 | 4.50 | 2.30 | 1.05 | 0.15 |

Scenario 1 considers withdrawal requests (a = 2) from site 1 and site 2. For scenario 1, sites 1 and 2 have the same reliability requirements (r = 0.6), and consumptive loss coefficients (c = 0.75). The optimal allocation for this scenario gives the identical allocation to site 1 as to site 2. Site 1's FDC and allocation for this scenario is shown in Figure 1. Instream flow was allocated first and given the highest priority. The remaining streamflow available with reliability greater than or equal to 0.60 is 1.3. As this amount is less than the 2.0 that was requested, the entire 1.3 was allocated to the withdrawal request.

Scenario 2 has withdrawal requests from all three sites. For this scenario, the withdrawal requests from sites 1 and 2 are the same as the requests in scenario 1. Site 3 has a 0.6 reliability requirement, and a 0.75 consumptive loss coefficient. For

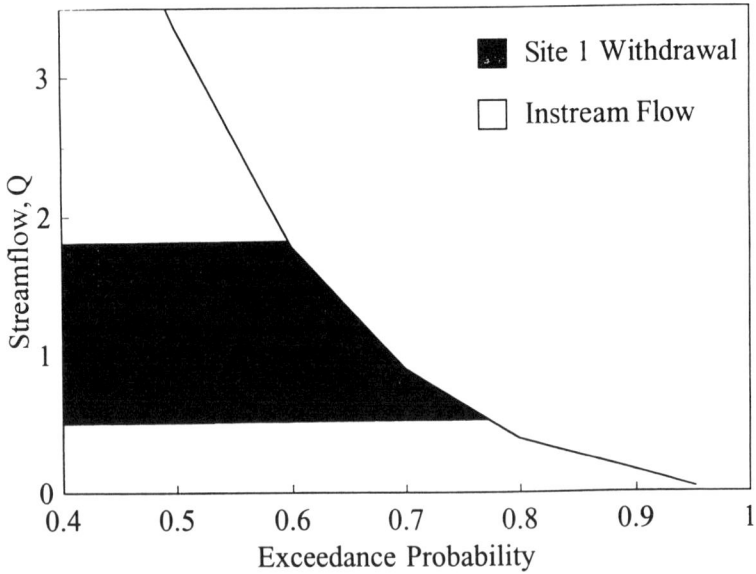

Figure 1. Withdrawal allocations for scenario 1 as illustrated by site 1's FDC.

scenario 2, the streamflow allocated to sites 1 and 2 was identical to scenario 1 (see Figure 1). The streamflow available for allocation at site 3 was q = 1.5. Figure 2 shows how the water used by sites 1 and 2 reduced the streamflow available at site 3 in this scenario. Site 3's FDC displays the allocations for the instream flow requirement, the upstream consumptive withdrawals, and the site 3 withdrawal.

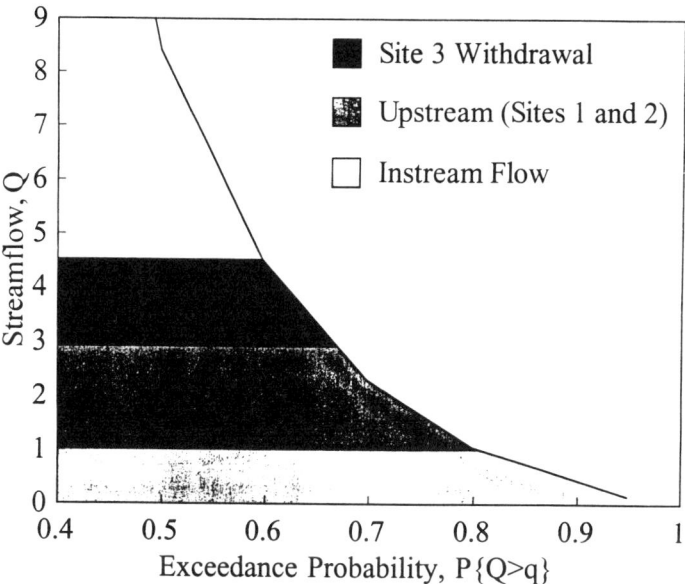

Figure 2. Withdrawal allocations for scenario 2 as illustrated by site 3's FDC.

Conclusions

An objective methodology was developed using flow duration curves and a chance constrained mathematical program to determine the allocation of streamflow to competing multiple users in a river basin. The methodology provides the optimal allocation expressed as withdrawal rates and corresponding reliabilities. A flow duration curve for each site is used to illustrate the site's withdrawal and any upstream withdrawals. Allocation problems are solved using an interactive optimization algorithm, Microsoft Excel Solver™. This methodology is flexible and may be modified to include specific goals associated with permit programs that were not considered here.

Appendix: References

Alaouze, C.M. (1989). "Reservoir Releases to Uses with Different Reliability Requirements." *Water Resour. Bull.*, 25(6), 1163-1168.

Alaouze, C.M. (1991). "Transferable Water Entitlements Which Satisfy Heterogeneous Risk Preferences." *Australian J. of Agric. Economics*, 35(2), 197-208.

Fennessey, N. (1998). "Water Allocation in Massachusetts: A Case Study." *J. Water Resour. Plng. Mgmt.*, ASCE, (in review).

Loucks, D.P., Stedinger, J.R., and Haith, D.A. (1981). *Water Resource Systems Planning and Analysis*. Prentice-Hall, Englewood Cliffs, N.J.

Male, J.W., and Mueller, F.A. (1992). "Model for Prescribing Ground-water Use Permits." *J. Water Resour. Plng. Mgmt.*, ASCE, 110(4), 543-561.

Vogel, R.M., and Fennessey, N. (1994). "Flow Duration Curves I: New Interpretation and Confidence Intervals." *J. Water Resour. Plng. Mgmt.*, ASCE, 120(4), 485-504.

Development of Boundary International Water Quality Standards

by

Conrad G. Keyes, Jr., ScD, PE/PS, F.ASCE
Engineer Advisor, Rio Grande Compact Commission
State of Texas, P.O. Box 1917, El Paso, TX 79950-1917

Abstract

The presentation will mention ASCE environmental systems and water policy statements used to help develop certain standards by the Society. The recent intensive water quality investigations for the US and Mexico and the environmental and water resources standard review process to be used by the new Environmental and Water Resources Institute will be discussed. The recommended method for the development of future ASCE Border International Water Quality (BIWQ) standards is provided.

Introduction

Any committee of ASCE may issue position papers in its name on public policy issues. On issues of national concern, ASCE, through action of its Board of Direction, has adopted policy statements, position papers, or resolutions on many subjects for which the civil engineering profession has expressed concern. This cumulative effort is intended to maintain a substantial reference for any public pronouncements or comments on public issues. As new policies are adopted, their availability is announced to the membership (ASCE Official Register 1998, pg. 95). The appropriate ASCE Board of Direction policy committee reviews most policy documents every three years.

In April 1980, the Board of Direction approved ASCE Rules for Standards Committees to govern the writing and maintenance of standards developed by the Society. A consensus standards process managed by the TAC's Codes and Standards Activities Committee develops all such standards. The consensus process includes balloting by the balanced standards committee made up of Society members and non-members, balloting by the membership of ASCE as a whole, and balloting by the public. All standards are updated or reaffirmed by the same process at intervals not exceeding five years (ASCE Official Register 1998, pg. 103).

The following sections of this paper will describe how certain standards are used in some of the proposed water related standards of the Society, the proposed process for future environmental and water resources standards of the Society, and the recommended method for the development of future ASCE Border International Water Quality (BIWQ) standards.

## ASCE Environmental Systems and Water Policy Statements

The annual ASCE document on Policies & Priorities is designed to familiarize the member or non-member of the Society with the scope of ASCE policy activity and to provide a referenced abstract of the full policy document text that is available upon request from the web page at http://www.asce.org or from the ASCE Washington Office. The policy abstracts comprise the ASCE public policy agenda for the civil engineering profession.

The environment systems policy statements (PSs) and position paper (PP) to be reviewed by the National Environmental Systems Policy Committee of the ASCE Board of Direction during 1998 includes: PS 286-Waste Assimilative Capacity of Ocean and Coastal Waters, PS 332-Water Reuse, PS 338-Public Awareness of the Environment, PS 345-Storage Tanks Risk Management, PS 360-Global Climate Change, PS 378-National Wetlands Policy, PS 395-Control of Combined Sewer Discharge, PS 396-Infrastructure Impacts of Population Growth, PP 420-Clean Water Act Reauthorization, PS 437-Risk Management, and PS 438-Endangered Species Act Reauthorization.

The water policy statements and position paper being reviewed by the National Water Policy Committee during this same period includes: PS 275-Atmospheric Water Resources Management, PS 330-Coastal Data Collection Program, PS 360-Global Climate Change, PS 378-National Wetlands Policy, PP 420-Clean Water Act Reauthorization, PS 421-Flood Plain Management, and PS 438-Endangered Species Act Reauthorization.

Each of these set of policy statements and the position paper was scheduled to be reviewed during the ASCE Policy Week of March 3-6, 1998 and forwarded to the ASCE Board of Direction in April 1998 by the Board Committee on Policy Review. Since many of the policy documents are revised every three years, all ASCE Standards will be subject to revision in accordance with the five-year review of standards.

The Regulated Riparian Model Water Code of the Water Regulatory Standards Committee, of the Water & Environment Standards Council of the ASCE Codes and Standards Activities Committee, will go through the initial standards committee balloting process during 1998. This water code references the Clean Water Act and many more of the US authorization acts affecting water allocations and water quality within watersheds throughout the US and its boundaries. Some of the policy statements and position papers of ASCE; particularly PS Ground Water Management, PS 275-Atmospheric Water Resources Management, PS 308-Surface Water Data Collection, PS 312-Federal/State/LocalCooperation on Water Resource Projects, PS 337-Water Conservation, PS 348-Emergency Management Planning by Water Providers, PS 361-Implementation of Safe Drinking Water Regulations, PS 407-Desalination, PP 420-Clean Water Act Reauthorization, PS 421-Floodplain Management, and PS 422-Watershed Management; will definitely change aspects of future ASCE Codes and Standards (ASCE 1998 Policies & Priorities).

Recent Intensive Water Quality Investigations by US and Mexico

In response to the need for comprehensive information, the US and Mexico agreed to an intensive water quality investigation of the Rio Grande/Rio Bravo from El Paso/Ciudad Juarez to Brownsville/Matamoros. The Mexican and WE Sections of the International Boundary and Water Commission (IBWC) conducted coordination between the two countries. The IBWC developed IBWC Minute Number 289, dated November 13, 1992, which approved the study design and addressed binational cooperation for the water quality investigation.

The main objective of the study was to screen the system for the occurrence and impact of toxic chemicals. The initial sampling and analysis was conducted during 1992-93 and involved sampling at 19 mainstream sites and 26 tributaries along the reach of the river which forms the international boundary between the two countries. All available information was used to identify sites and chemicals of potential concern, to facilitate water quality management decisions during 1994-2000 and future monitoring efforts during 1995-2000.

In the initial study, potential problems were more prevalent in tributaries, which was not surprising since some of them transport wastewater in relatively undiluted form. According to US results, samples from 14 of the 26 tributaries produced significant adverse effects in at least one phase of the toxicity tests. Results from Mexico's analyses indicated eight potentially toxic chemicals that exceeded their water quality standards. Low-level human health criteria were exceeded in water and/or edible fish tissue at 22 of the 45 sites.

Although each country's water quality standards were not the same in 1992-93, the same chemicals identified by the US that exceeded screening levels were in agreement with the results obtained by Mexico. Grouping by high, medium and low priority helped to identify sites of potential toxic chemical impacts on humans and fish.

## Environmental and Water Resources Standard Review Process by EWRI

The draft business plan of the Environmental and Water Resources Institute (EWRI) has proposed "project teams" which will be the working units of the Institute. They will be formed for specific purpose and for a limited duration, which could be associated with the length of time that an environmental or water related standard would be in existence. The standard committee of the Institute would adhere to the ASCE Rules for Standards Committees as published in the ASCE Official Register and approved by ANSI.

The standards committee using the consensus process provided in the Rules would develop each draft standard. This process requires a committee ballot on each draft of the standard until the committee accepts the document. Then the document goes through the ASCE Public Ballot process until the standards committee settles all negative votes. The final ASCE document then must go through the ANSI public ballot process for it to become an ANSI standard. None of these processes are to be changed under the EWRI organization under ASCE. The only procedures that are proposed for change deal with the planning and budgeting process to start, continue, and terminate standards committees. Some version of the present Water and Environmental Standards Council for administration of the environmental and water resources standards committees of the EWRI should be in existence after the "wet divisions" of ASCE move under the umbrella of the Institute.

The consensus document that is prepared by members and nonmembers of the Society according to the Society's Rules for Standards Committees shall be either mandatory or nonmandatory. A mandatory standard shall be enforceable by the appropriate public body and shall be worded so that a person auditing its use or application can point out where it has been followed or where it has not been followed, or the extent to which it has or has not been followed. Standards intended for adoption in codes or other regulatory documents shall be written in mandatory language. A non-mandatory standard is written in permissive language (i.e., should, may) and may not be suitable for adoption in codes or other regulatory documents. The title of the standard shall clearly indicate the non-mandatory nature of the standard through the use of such terms as "guide, guideline, or consensus guideline." Nonetheless, a non-mandatory standard is developed in accordance with the Rules (ASCE Codes and Standards Form and Style Manual, 1998, in progress).

## ASCE BIWQ method for 1998

In 1993, several Mexican engineers and members of many "wet divisions" of the Society suggested that the Special Standards Division (now Water & Environment Standards Council) of ASCE should develop standards in water quality management, pollution prevention, waste engineering or other topics such as border international water quality standards.

After formulation of the ASCE Border International Water Quality (BIWQ) standardization activity in fiscal year 1995, the Standards Committee was developed by the author. At least 12 members or nonmembers of the Society were approved for membership on the BIWQ Standards Committee. Using a small amount of funds for the initial conference call of the committee in October 1996, a brief working meeting of the Texas members of the committee in February 1997, and a brief funded meeting of the committee in Houston in April 1997, created a method for the establishment of the committees first draft document. Another working meeting of four members of the committee in late September 1997 provided the draft outline for this initial document of the committee.

Assuming that a water code for a transboundary would progress from an agreement to development of guidelines and then to the code for protection of the water source, it is first appropriate to develop an annotated outline based on the table of contents of the ASCE Model Water Sharing Agreement – Comprehensive that has been prepared by the Shared Use of Transboundary Water Resources (SUTWR) task committee of the Water Regulatory Standards Committee. This new version would emphasize international water quality regulations. The main articles of this first draft would cover: 1) declaration of policies and purpose, 2) general obligations and definitions, 3) administration, 4) water quality management, 5) financing, 6) dispute resolution, 7) effectuation, and 8) references.

The working group schedules the first draft of the proposed Border International Water Quality Agreement for completion in late March 1997. The document was to be mailed to the full standards committee during April 1997. The agenda for the June 1998 meeting of the BIWQ Standards Committee in Chicago includes discussion of items in the draft document. After revision of the document in committee, it is possible that the draft can be sent out for the first committee balloting process. Most likely, the earliest ASCE Public Ballot on this draft BIWQ Agreement could occur in fiscal year 1999. It is possible that the document could be published under the direction of the EWRI in 2000.

Summary

The future of ASCE policies and standards on boundary international water quality standards rests with the ASCE and its proposed Environmental & Water Resources Institute. The National Environmental Systems and/or Water Policy Committee under the Board of Direction will renew most national ASCE policies every three years. However, it is possible the EWRI could create a policy statement or position paper for use by its Board of Governors before approval of the policy by the ASCE Board of Direction. The codes and standards of the EWRI shall be in accordance with the ASCE Rules on Codes and Standards and the Procedures established by the ASCE Codes and Standards Activities Committee.

Institutional Hurdles in Privatization: The Fairbanks Municipal Utilities System

C. (Kees) W. Corssmit, Ph.D.[1], Carol F. Streiner[1],
and
Bill Gordon[1]

Abstract

The City of Fairbanks, Alaska, offered its Fairbanks Municipal Utilities System (FMUS) for sale to private bidders in 1996. FMUS operates a water, wastewater, electric, steam heat, and telephone utility in the different parts of the greater-Fairbanks area. A consortium of private entities was selected by the City to negotiate the sale of the utilities. Three different entities would assume ownership of the water and wastewater, electric and steam heat, and the telephone utility. The privatization, as negotiated by the parties, was ratified by a comfortable majority of the city's electorate during the Fall of 1996. This paper discusses the subsequent institutional hurdles the new owner of the water and wastewater utilities, Fairbanks Sewer and Water, Inc. (FSW), had to overcome to finalize the proposed transaction. Significant efforts had to be mustered by FSW with the State of Alaska Public Utilities Commission (APUC) and the United States Environmental Protection Agency (EPA) to secure the final authorization to proceed with the change in ownership status.

Introduction

Ten years under study and two years under public scrutiny in a bitterly contested process, it came down to a flurry of details to complete a series of transactions. On October 7, 1997, the last day allowed under a self-imposed contract deadline, the City of Fairbanks transferred its utilities to three buyers. Among the assets privatized was its Regional Wastewater Treatment Plant, perhaps the first ever EPA-approved complete privatization of a publicly funded wastewater plant in North America.

---

[1]Integrated Utilities Group, Inc., 5200 DTC Parkway, Suite 530, Greenwood Village, CO 80111

[1]Utility Services of Alaska, Inc. P.O. Box 80370, Fairbanks, AK 99708-0370

As laid out in *FMUS Privatization* (Corssmit and Brunsdon), appearing in *Aesthetics in the Constructed Environment*, Proceedings of the 24$^{th}$ Annual Water Resources Planning and Management Conference, Water Resources Planning and Management Div./ASCE, April 6-9, 1997, motivation behind the complete privatization of FMUS utilities was threefold:

1. The value of the telephone utility was deteriorating due to competitive pressure in telecommunications.
2. The noncompetitiveness of the electric utility's production cost would result in great losses if service area protection was removed.
3. The City was unwilling to significantly raise water and wastewater rates to cover the cost of service and pay for badly needed capital improvements.

The City realized that their best option to capitalize on the fleeting opportunity to obtain a significant endowment for a trust fund was to completely privatize the utilities. An all-or-nothing purchase package was offered to interested private sector parties. That way, the City would not be left in the unenviable position to sell the only utility with real value but keep its problem utilities. FSW, consisting of a group of local investors with a strong interest in solving the regional water and wastewater utility problems, was the entity to take over those systems. The wastewater system, constructed with various funds including grants from the Federal government, is in need of significant repairs and upgrades. This paper describes the institutional hurdles FSW overcame to finalize the transaction by the deadline to advance the sale of all FMUS utilities.

City of Fairbanks Wastewater Treatment Facility

Because the City of Fairbanks is located at a very northern latitude and for other engineering reasons, the Fairbanks wastewater treatment facility was designed as an entirely enclosed treatment plant. Being enclosed, the corrosive atmosphere within the plant caused serious deterioration in the electrical components. This deterioration resulted in incomplete grounding of the components, requiring immediate and expensive attention. In addition, more than 100,000 feet of the wastewater collection system was constructed using defective Techite pipe, now the subject of a lengthy lawsuit by the City. Much of the wastewater collection system's inflow and infiltration (I&I) (as high as 30 percent of total flow) was attributed to the Techite pipe.

The City's unwillingness to increase water and wastewater rates, the unwillingness of voters to approve revenue bonds, and the unlikelihood of additional and future grant funding from Federal or State governments left only FSW to salvage the water and wastewater utilities.

## Institutional Hurdles

Selling a publicly owned water and wastewater system, whose operation is perceived by some to be a duty of government to protect the health and safety of the community, was not without opposition. FSW encountered opposition both from the public and the regulating agencies in its attempt to purchase the utilities. Three areas discussed below are:

- Local opposition
- Environmental Protection Agency
- Alaska Public Utilities Commission

### Local Opposition

Under the City of Fairbanks charter, any sale of the utilities requires ratification by a majority of the City's voters. Opposition to this sale developed prior to the scheduled election. Opponents claimed that a sale would stifle economic development and raise future rates through the loss of federal and state grants and tax exemptions. Given the highly unusual nature of privatizing wastewater utilities, some people believe that only public ownership of these services is acceptable because of the importance of achieving public health standards.

Opponents challenged the sale in an expensive campaign stressing four main points:

- buyers were not paying enough and would reap a windfall profit
- water and wastewater rates would skyrocket
- future economic development would be stifled
- public health would be compromised

On October 7, 1996, however, the voters approved the sale of all five utilities by a 54 to 46 percent margin.

### Environmental Protection Agency

Because certain wastewater assets were funded by Federal government grants, the privatization of the wastewater utility must be approved by the Environmental Protection Agency. The EPA, among other roles and in conjunction with state environmental agencies, regulates issues such as the treatment and discharge requirements of wastewater treatment plants. The question of how to handle "old" EPA grants to recipients who would now like to sell federally funded assets has not been dealt with in the past.

The City of Fairbanks turned to Presidential Executive Order 12803 to guide its wastewater utility privatization plan. President George Bush approved this executive order in 1992, which directed agencies to "assist and aid local and State government in their efforts to privatize" state or locally owned infrastructure assets. The EPA, however, had not yet established guidelines for selling publicly funded facilities and became inactive.

The City attempted to encourage the EPA to make a decision based on Executive Order 12803 in an application to the EPA in December 1996. No action was taken by the EPA, despite repeated appeals by the City.

Due to a contractual termination date of October 7, 1987, the City and FSW enlisted the support of the Alaska Congressional Delegation to encourage EPA officials to consider the City's application seriously and quickly. Substantial legal efforts were needed to discuss and negotiate transaction rules with the EPA, which of course led to a great deal of money and time. The following points supported the City's proposal:

1. No proceeds are due the Federal Government under EO 12803 because the full grant amount associated with the infrastructure assets less the accumulated depreciation is zero.
2. A market mechanism, a legally enforceable agreement, and regulatory mechanisms will insure that the infrastructure assets will continue to be used for the originally authorized purpose as long as needed.
3. Alaska Public Utilities Commission's oversight will assure the user charges will be consistent with current Federal conditions that protect users and the public.

Following the guidelines established in Executive Order 12803, the EPA eventually approved the transfer of the federally funded wastewater treatment plant, adding a requirement that FSW lease the federally funded portion until 1999 when, according to EPA, federal grants would be fully depreciated, at which point a purchase option could be exercised.

Alaska Public Utilities Commission

The Alaska Public Utilities Commission (APUC) is the regulatory agency responsible for monitoring rates or tariffs. Although FMUS utilities were not regulated by the APUC, all the utilities would be fully economically regulated after privatization. APUC hearings regarding the privatization of the water and wastewater utilities were scheduled for May 1997. APUC staff hired experts to critique the applications, who apparently did not like the idea that the private sector would end up with "public" assets and "enrich" themselves. Their opposition was puzzling because the rates were capped at lower levels than cost-of-service rates recommended in a comprehensive study performed by Integrated Utilities Group for FMUS prior to the privatization hearings, and service would be improved. Five weeks of hearings were necessary to cover all the political, constitutional, and operational issues.

Financial models created by APUC experts maintained that the application was not in the public's interest. FSW, however, brought several experts to the stand to demonstrate that the staff's experts' models were flawed and in reality, the public would benefit from privatization. The APUC finally issued an order on September 24, 1997 approving the application for the privatization of the water and wastewater utilities. This order provided the final approval necessary to complete the sale of all three utilities as the other two applicants' approvals were simultaneously granted. In its order, the APUC granted the requested inception rates for water and wastewater, but failed to explicitly approve the commensurate rate base underlying FSW's negotiated agreement with the City. The APUC Commissioners decided to postpone the rate base issue until the applicant files a full rate case, expected in about three years.

Thus, the hearings took much longer than expected at a much higher cost than planned and the level of the rate base established for the water and wastewater utilities was not finalized. The closing of the sale contracts negotiated by the City and FSW was conditioned upon the establishment of a guaranteed rate base. Failure of the Commission to decide the rate base issue gave FSW an option to withdraw–a decision which would jeopardize the entire sale.

A few days before the closing deadline, FSW was faced with a critical decision–either accept the APUC's decision and be open for business in a few days or terminate the sale process. The latter choice would have caused the City much financial distress and possible insolvency. After lengthy meetings, all parties including FSW agreed to close the transaction. This finalized the first known complete privatization of publicly held utilities.

Conclusion

The finalization of the privatization process, of not only the wastewater utility but also the other FMUS utilities, was clearly in the public interest and of paramount importance to the City of Fairbanks. Considering the economic realities of the utility systems, the new competitive pressures, and the admittedly inefficient nature of the City's operation, the City's best alternative was privatizing its utilities. Its citizens and two government regulatory agencies agreed.

The purchaser of the water and wastewater utilities, FSW, learned that the privatization process is not easy. In its effort to overcome the institutional hurdles described in this paper, four important lessons were learned.

1. Significant efforts in time and expense are associated with obtaining EPA and APUC approvals, especially for something as rare as wastewater utility privatization.
2. Ingredients needed to finalize a privatization transaction are: heavy duty legal input (several law firms were involved), a detailed understanding of local utility conditions, needs, and operations, and substantial expert witness analysis and testimony.
3. The actual process to finalize the transaction took about six months longer than originally planned.
4. Transaction costs were perhaps ten times higher than originally estimated when including all labor, expenses, communication, and hearing time needed.

Managed Competition Sometimes Yields Unexpected Results: Ask San Diego

Ellen R. Bogardus[1] and Kenneth M. Barrett[2]

Introduction

The water and wastewater industry across the country is experiencing significant changes in its operating and management environments. Managers, faced with reduced financing support, increased water quality regulations, higher customer expectations, and more critical oversight, must make their operations more efficient, competitive, and customer responsive. Competition has become the themesong of governments seeking to cut costs and increase service quality.

As local governments evaluate privatization opportunities, many jurisdictions have given public employees the opportunity to compete for contracts with private companies through the managed competition process. While public employees have typically enjoyed a protected monopoly, allowing employees to bid has reduced political opposition to competitive contracting. And yet, a debate continues that it is almost impossible to create a level playing field in public-private competitions. Public bidders, for example, pay almost no taxes, do not have to earn a profit, have cheaper cost of capital, and do not have to bear the costs associated with the financial, liability and compliance risks commonly borne by the private operator. On the positive side, no matter what the outcome, the competitive challenge forces local government to decide how to be the best caretaker of its utility's assets, how to be the best steward of its ratepayers' money, and how to most cost-effectively deliver the service required.

This paper reviews the San Diego Competition Program for the water and wastewater utilities. In June 1994, the San Diego City Council established the Competition Program to ensure not only that the City was competitive, but also that it provided the highest quality service at the optimum cost to residents, businesses and visitors.

---

[1] Vice President, Environmental Services, Infrastructure Management Group, Inc., 4733 Bethesda Avenue, Suite 600, Bethesda, MD 20814
[2] National Director, Competitive Services for Water and Wastewater, HDR Engineering, Inc., 128 S. Tryon St., Suite 1400, Charlotte, NC 28202-5001(mgd).

The San Diego Water Department went through the managed competition process in 1996-97, while the San Diego Metropolitan Wastewater Department, in 1997, rejected bidding against private companies in favor of a new process called "bid-to-goal". While there was considerable initial employee enthusiasm for both programs, implementation has been slowed by a number of institutional and political issues. This has resulted in employee disillusionment, and a concern that without the threat of competition staff and management will revert to the more comfortable former ways of doing business.

## The City of San Diego Water Department

There are three water treatment plants under the jurisdiction of the Water Department's Water Operations Division - Alvarado, Miramar, and Otay. These three facilities provide the City with a daily average of 198 million gallons of water a day. The combined rated capacity of these three plants is 300 mgd. The three facilities were reviewed under the Competition Program in 1995-96, but the projected savings were not considered to be "good enough". This led to the announcement by the City Manager that a Request for Proposal (RFP) would be issued in March, 1997, and that the City employees would be able to compete against the private sector in this managed competition process.

In the spirit of the City's Competition Program, a cross section of employees were assembled six months prior to the issuance of the RFP to form the Bidding Team. The Team included management and administrative representatives; supervisors and staff level operations and maintenance professionals from each of the three water plants; representatives from the American Federation of State, County and Municipal Employees (AFSCME) Local 127 and the Municipal Employees Association (MEA); and consultants from HDR Engineering, Inc. The Bidding Team's acronym was $SDH_2OT$ (San Diego Water Operations Team).

*Three Plants versus One*

During the competition assessment, it was assumed that contract O&M proposals for all three water treatment plants would be solicited at once. While a three-at-once process would take less time, the City decided that bidding all plants simultaneously would limit the benefit of using the competitive process. Specifically, the incentive to continue to improve performance would disappear once the process was finished. San Diego, therefore, decided to issue an RFP for the 40-MGD Otay Plant first, with Alvarado and Miramar to follow within a year or so. While the Otay plant was the smallest in the system, it could become one of the largest water treatment plants in the United States to undergo managed competition. The primary reasons for selecting Otay included:1) relatively new condition; 2) minimum potential for major repairs, thus minimizing unforeseen contractual expenses; 3) large capacities and capital improvement upgrades and expansions at Alvarado and Miramar presented circumstances different from other water systems currently utilizing contract operations; 4) City's risk would be reduced with the contract O&M process beginning with the least strategically critical facility; 5) a total system competition could result in the displacement of City employees with few options for reassignment.

## The HOTCO Proposal

The proposal written by the SDH$_2$OT Team envisioned that a separate accounting entity would be established within the Water Department assuming the contract was awarded to the City employees. This entity was to be known as SDH$_2$OT Contract Operations (HOTCO). The HOTCO proposal was based on converting a traditional municipal operation to "public contract operations" whereby public employees would be bound to a performance contract. HOTCO would be held to the same economic and performance expectations as the private bidders. The implementation of the HOTCO workplan projected savings of approximately $5,200,000 over the nine-year term of the agreement.

Historic labor needs were proposed to be altered by the use of automation, including a distributed control system using programmable logic controllers (PLC), and a graphics-based supervisory control and data acquisition (SCADA) system using WonderWare man-machine interface software. The team proposed to shift from reactive to reliability based maintenance strategies. Shifts were to be restructured and the Water Operations Supervisor was to be included in the shift rotation. Non-core activities including janitorial and grounds maintenance services as well as specialized maintenance functions were proposed to be outsourced to either non-city sources or a Central Maintenance Pool.

HOTCO proposed to use load management systems to take advantage of off-peak electrical rates. Although power costs were not part of the requested bid, HOTCO was committed to minimizing energy costs as part of its broad goal of demonstrating efficiency. HOTCO, also, proposed to make use of a revised chemical management strategy while still producing high quality drinking water.

Performance goals were established for the maintenance staff that included targets for preventative, predictive and reactive maintenance. The goals were to be tied to a gainsharing pool to reward activity that reduced overall repair and replacement costs.

## *Proposal was Never Submitted: City Council Cancels Procurement*

The first harbinger of things to come was the delay of the March 31, 1997 issuance of the RFP for the Operations, Maintenance, and Management of Otay. The RFP was issued May 19, 1997 with a due date of August 8, 1997. Based on a shortlist developed from the Statement of Qualifications, three firms(OMI, PSG, United Water/Cal America ) and the City received the RFP.

The HOTCO Team worked long, hard hours on the proposal and was prepared to submit copy to the printer on August 1, when the City Competition Team announced on July 31 that the proposal deadline was being extended until August 22, 1997. The delay was caused by AFSCME Local 127's lawsuit, filed in early July, challenging the validity of the City's competition program and specifically the RFP for Otay. AFSCME claimed that the process for preparing the City's bid was poorly defined and did not appear to accommodate the "meet and confer" legal requirements. AFSME's key issue was that the proposal predetermined terms and conditions of employment.

The proposal deadline was delayed for a second time until October 22 in hopes that the City Attorney's office could find a way to settle the lawsuit with AFSCME. Two weeks prior to October 22, 1997, all proposers were notified that the San Diego City Council had voted to cancel the Otay procurement. AFSCME felt they had "won". Staff at Otay and in the Water Department were left somewhat speechless, but agreed the Otay proposal should be implemented at the plant as soon as possible.

## The Metropolitan Wastewater Department

The Metropolitan Wastewater Department Operating and Maintenance Division (MWWD) was selected to undergo a two year competitive assessment in July 1996 as part of the City of San Diego's Competition Program. In early 1997, MWWD launched an effort to show it could meet the City's efficiency demands without having to compete against the private sector. The managed competition process was rejected in favor of a new pilot process called "bid-to-goal". With this concept, MWWD agreed to target a specific dollar amount of savings and to incorporate more private-sector techniques for both operation and maintenance. Management and labor partnered in submitting a proposal to the City Manager in May,1997 which reflected a goal of $77 million in savings for the City over a six-year period.

### The Wastewater System

MWWD administers, operates and maintains a wastewater system for a service area of 450 square miles which includes the City of San Diego and surrounding jurisdictions, municipalities and areas in the region. MWWD has approximately 1,800,000 customers and there are 15 participating agencies. MWWD is an enterprise-funded department of the City. It receives no general tax funds, but is funded entirely through customer charges and user fees for services provided.

The wastewater system has a plant processing capacity of approximately 240 million gallons per day (mgd). Wastewater flows through 2,500 miles of transmission pipelines and 81 pumping stations ( 6 major ones ), ultimately reaching two major interceptor sewer pipelines that roughly parallel the coastline under the City streets. The interceptor sewers deliver wastewater to Pumping Station No. 2 (located next to the City's airport at Lindberg Field) where it is pumped through two force mains to the Point Loma Wastewater Treatment Plant on the west side of Point Loma. Anaerobically digested biosolids resulting from this advanced primary plant are piped approximately 8 miles to the Fiesta Island Sludge Dewatering Facility. In Spring, 1998 the residuals will begin to be processed at the new Metropolitan Biosolids Center(MBC), roughly 14 miles from the Point Loma WWTP, and MWWD's use of Fiesta Island for biosolids processing will end.

## *Getting to "Bid-to-Goal"*

MWWD faced a dilemma. Management and labor were concerned about the direction that the competition would take if they followed the guidelines of the City's Competition Program. This uncertainty was reinforced by the Water Department's experience in developing a competitive plan and subsequently being directed to bid against private companies. If MWWD employees did their best to develop a competitive plan, that plan, when released, could serve as a blueprint for private companies to follow in developing counter proposals. However, if MWWD hedged its risk by holding back strategy for fear of disclosure to potential private competitors, MWWD might be directed to bid in a managed competition process.

From the beginning MWWD's objective was to deliver to ratepayers the kind of savings they could expect from a private operation. In late fall, 1996 MWWD management and labor agreed to work together and submit a proposal to the City Manager that would reflect their joint interests, as well as enhance and expedite the potential savings to the City and ratepayers. The resultant bid-to-goal concept involved the establishment of the following three criteria that City forces must meet: achievement of a goal; compliance with service requirements; and compliance with a schedule

In the event that any of the three criteria are not met, the City has the option of moving to competitive bidding through the managed competition process. The key question before anyone could commit to the bid-to-goal process was, "What should the goal be?" MWWD used a three-step process which included extensive benchmarking to compare MWWD with other public and private utilities across the country; determining the likely cost of O&M services and labor requirements for a comparably operated private facility; and identification of a competitive target level. An effort was made to adjust this competitive target from the floor of a competitive range to a point that reflects a private company's optimal O&M cost, plus their costs of a modest profit, corporate overhead, administration, insurance, risk, and transition costs. Based on an evaluation of recent private sector contract operation bids, a private company generally includes 20 to 30 percent markup of base O&M costs in the development of a private contract operations bid. Using a conservative 15 – 25 percent overhead assumption for combined private sector profit and overhead, the following annual costs were assumed for what a competitive bid would be in a Managed Competition Process:
- Base O&M for Targeted Facilities: $33,600,000
- Overhead and Profit(17%): $5,745,200
- Non-Operating Section Costs: $9,576,800
- **Competitive Cost: $48,922,000.** This, therefore, would be the estimated low bid for the portion of the O&M Division currently under competition, under ideal circumstances in a competitive marketplace.

The current 1997 MWWD financing plan for FY 98-2003 was the baseline for savings computations. Baseline costs were adjusted to remove the cost of facilities not included in the competition program. The 1997 Baseline budget was

$73,414,417. Through a series of savings strategies, MWWD staff developed a competitive cost proposal for FY 98 – 2003 that represents the employee offering to the City Manager. Bottom line is an 18.3% savings or $77 million savings goal over the six-year period.

## *The MOU and the Bid*

The MWWD "Bid-to-Goal" Memorandum of Understanding (MOU) is the primary governing reference document and it sets forth the basic policy ground rules. This MOU was developed in partnership with management and employees represented by AFSME, MEA and the O&M Division's Management Team. The City Council will hopefully approve it in March, 1998.

MWWD will then present the MWWD Bid to the City Manager for approval. This Bid was developed by the Bid Development Sub-Committee, then reviewed and approved by the MWWD O&M Competition Coordinating Committee/Task Force. Detailed projections concerning influent and effluent characteristics form the basis of the Bid. The Bid defines the levels of service for each operation or support section of the O&M team to each of the other operating or support sections or to MWWD as a whole. The Bid also details performance specifications for operations; regulatory reporting requirements; periodic report requirements including an annual auditor's report on both financial and O&M performance; and a thorough list, with examples, of institutional and cost risks and who retains responsibility for each risk.

## *Year One*

The original intent of bid-to-goal was to launch the program with the start of FY 1998. Employees were motivated to start. Heavy maintenance was to be centralized; chemical feed optimization was planned; inventory management was to be centralized; increased reliance on automation was anticipated; staffing was to be reduced through attrition; supervisors were to be accountable to budgets they had developed; a gainsharing program was to be initiated; and savings were to be tracked. However, it took the City close to eight months to get final agreement on the MOU and City Council approval. Year One of the Bid, therefore, did not start with the energy that had been anticipated and results will be compromised.

## **Lessons Learned**

- Delays in process, either proposal delays or policy maker approval delays, diminishes the momentum to make changes as well as decreases employee trust.
- Participation by front-line/operations staff in cost development allows for cost control to be delegated to the lowest levels and budget contingencies to be managed at the highest levels.
- Detailed risk allocation facilitates more precise budgeting.
- Bid-to-goal affords the opportunity for timely transition to staffing at optimum levels since a two year transition phase was part of the strategy to reach the savings goal.

# Using Readily Available Technologies and Laboratory Experiences to Enhance Course Content

Emmanuel U. Nzewi[1], A.M., ASCE

**ABSTRACT**
This paper describes the use of readily available and easy-to-implement technologies, as well as, laboratory experiences and field trips to enhance course content. The desired result is to improve teaching effectiveness thereby increasing learning capacity and the technical proficiency of students. In this paper, technologies like Geographic Information Systems (GIS), computer-enhanced overhead presentation, the use of the Internet, and laboratory exercises including simple instructional multimedia modules are described. Their use in the introductory junior- and senior-level water-related courses taught by the author are discussed to show how these activities can be used to enhance course content. These activities were designed to facilitate the development of a water resources curriculum which produces technically more competent students and empowers them to solve engineering problems using innovative techniques and appropriate emerging technologies.

**INTRODUCTION**
The desire of every teacher is to get the message across effectively --- and to make each lecture or laboratory session exciting and stimulating. This is referred to as effective teaching. Effective teaching requires quite a bit of preparation time and other resources; and a good knowledge of what helps to get the message across, effectively. This paper describes on-going efforts at the author's University to improve water-related courses in the Civil Engineering program.

In 1993, an evaluation of the author's program revealed several areas of concern. In order to addresses these weaknesses, initiatives geared toward improving a number of introductory civil engineering courses were proposed. In general, these initiatives addressed course content, delivery systems and the use of easy-to-implement technologies including the introduction of state-of-the-art engineering design and analysis tools for in-class presentations and laboratory exercises. Specifically, the following were recommended for implementation: (1) the reorganization of the curriculum to provide a more effective sequencing of courses (2) the introduction of new courses to improve technical proficiency; (3) a new laboratory course in

---

[1] Associate Professor, Dept. of Civil Engineering, North Carolina A&T State University, Greensboro, NC 27411.

Hydrology; (4) the development of an introductory civil engineering Hydraulics course (with a laboratory course as co-requisite) to replace a Fluid Mechanics course which was taught as a service course by the Mechanical Engineering department.

The need for a laboratory course in Hydrology has been noted by many sources in the literature (MacDonald [1992a], MacDonald [1992b], Groves and Moody [1992], Mack [1992], NRC [1991]). For example, in a book by the National Research Council (NRC) entitled, *"Opportunities in the Hydrologic Sciences,"* it is stated that, "the lack of field and Laboratory experience at the undergraduate level, is a situation that has reached crisis proportions" (page 287, NRC [1991]). In addition, data reported by Groves and Moody (1992) show that out of 73 undergraduate courses in introductory Hydrology, only six (6) include a weekly laboratory period. Further, only two (2) courses listed any field or experimental work. Groves and Moody (1992) state that, *"This finding is particularly disheartening because it is during field work that students become acquainted with the difficulties and uncertainties surrounding basic data collection, and can fully recognize the limitations of model approximations of actual behavior of hydrologic systems. Further, field experience gives life and excitement to the facts, equations, computer models, and theories learned in the classroom, sharpens observational skills, and may help stimulate the pursuit of a career in water resources."* They indicate that new strategies for teaching some engineering courses should be explored. This line of thinking is not new but one that was advocated by Jon Amos Comenius (AD 1592-1671) in his writings. Comenius (Joyner, 1997) is recognized as the father of modern education. Comenius believed that learning is facilitated by using all of the senses to interact with our environment. According to Comenius, a curriculum should employ seeing, touching, handling, smelling, instead of just hearing. A good example of this notion is embodied in the well known saying: "a picture is worth a thousand words." In this vein then, a delivery and learning system which involves the use of more of the senses would be superior to another that involves fewer of the senses, all things being equal.

## OBJECTIVES

The main objective of this paper is to describe delivery techniques and laboratory experiences that were introduced to improve the introductory Hydrology and Hydraulics courses in the author's program. The activities described here were particularly designed to produce sustained student interest in the water-related courses in the program. The other objectives of the program are listed below and include:
(1) to generate more interest in the higher-level (senior and graduate) water-related courses in Civil Engineering,
(2) to increase learning and therefore the technical literacy of students, and
(3) to stimulate interest in students so that they would consider pursuing post-graduate degrees in water resources engineering or related areas.

## IMPLEMENTATION

The equipment and other physical resources required to implement this program are presented in this section along with a brief description of each program activity. Some of the resources were provided under the National Science Foundation's Instrumentation and Laboratory Improvement program.

### Equipment

Several computer (including software) and non-computer (including laboratory hardware) equipment were purchased to achieve the goals outlined in the previous section. Some of the major and strategic acquisitions have been described in Table 1 below.

### Table 1 - Description of Major and Strategic Equipment

| Description of Equipment | Number | Function |
|---|---|---|
| Rainfall Hydrographs Unit | 1 | To provide for the simulation of rainfall and runoff relationships. |
| Microcomputers | 3 | Power Macintosh computers and a portable computer for field work/classroom demonstrations. These computers have been instrumental for in-class presentations and provides students with workstations. |
| Digitizers | 2 | To be attached to desktop/portable computers for digitizing maps and watersheds. This is also used for in-class presentations |
| GIS Software (MapInfo) | 3 | For performing GIS analysis necessary for Hydrologic of watersheds. This software is both powerful and inexpensive. A manual and a tutorial (Nzewi and Peele, 1996) was developed to facilitate learning the use of the software. |
| Flow probe | 2 | Measurement of flow rates in open channels (especially in the field) and for the calibration or other metering devices. |
| Groundwater Flow Demonstration Unit | 1 | Visualizing ground water flow characteristics and computing ground water flow parameters. This system is an inexpensive demonstration system. |
| Allegiant SuperCard | 1 | Multimedia authoring software |
| Armfield Hydraulics Bench, Deadweight Tester, Pipe Friction and Bernoulli Equation Setups | 4 | To enhance the experience of students in the Hydraulics laboratory course and to provide more hands-on experiments. |

### Laboratory Experiences

As stated above, Groves and Moody (1992) and MacDonald (1992) have argued very strongly for including field exercises in water resources courses. MacDonald (1991a) indicates that because much computer literacy is necessary in engineering, field work enables the student to understand the limitations of hydrologic and hydraulic models proposed as well as the difficulties and complications involved in data collection. To this end, the program has included the following activities or strategies:
    (1)   Field trips -- To visits hydraulic structures, talk to dam operators, etc.
    (2)   Field experiments ---specifically, streamflow measurements (late Spring)
    (3)   Enhancement of existing Fluid mechanics laboratory.
    (4)   Development of a hydraulics course laboratory manual (Nzewi, 1997).

The field measurements needed to study the rainfall-runoff process are very difficult and time consuming to complete, thus an instrument for simulating the rainfall/runoff process was purchased (See Table 1). The acquisition of this equipment was deemed necessary because a good understanding of the rainfall-runoff process is a very important topic in engineering Hydrology.

A groundwater flow and contaminant transport demonstration model was acquired to facilitate the visualization and ground water flow characteristics --- including the dispersion of pollutants and aquifer characteristics. This remarkable instrument was developed at the University of Wisconsin, Stevens-Point. It has been very effective for use in describing difficult concepts in groundwater hydrology. The setup is inexpensive, portable and easy to maintain.

### Emerging Technologies

To incorporate the use of emerging technologies in the hydrology course, the use of a desktop Geographic Information System (vendor --- MapInfo) was provided to facilitate computer-aided watershed analysis. This is not a new idea but one that was introduced to the author in a presentation by Snow (1992). Snow outlined the benefits of using a such a system for hydrologic analysis. He indicated that the use of digitizers for spatial data acquisition not only reduces analysis time but also greatly increased student participation and understanding in the fields of area, volume, and spatial overlays (soil type, land use, vegetative cover, etc.) as applied to water resources. In a final project required to be completed in the Hydrology course, students are required to use this system in the hydrologic analysis involved. This experience, not only exposes them to a new and rapidly evolving technology, but it also empowers them to explore a wider array of possibilities --- allowing for a more robust solution of the engineering problem.

To encourage and introduce self-paced learning, multimedia software modules are under development (by customizing already published modules or generating new ones). The Apple Macintosh platform provides a rich environment for the integration of multiple media sources in the development of self-paced learning or instructional modules. For example, Hotchkiss (1992) presented the development of a self-paced, interactive, computer-based multimedia teaching module developed with Allegiant (formerly Aldus SuperCard) SuperCard. Photographic slides, videotape, and computer-recorded audio were integrated in its development. The idea here is not to re-invent such modules, where they exist, but to acquire and modify appropriate modules for use in the computer laboratory. A dedicated water resources laboratory has also been developed to house these computational and instructional resources.

Overhead projection systems attached to portable computers have been used for in-class presentations and to provide computer-based tutorials. These presentations have been very well-received by the students. In one particular course, such hands-on presentations have greatly encouraged computer use in problem solving. Short cuts and simple, but powerful, computer-based techniques were effectively demonstrated; thus showing the students how empowering effective computer use could be.

### The Internet

A webpage has been established for water-related courses. This webpage is maintained on an Apple Power Macintosh server. The server is under the full control of the author and thus is easily accessible for updates and general webpage maintenance. The webpage is designed to disseminate course information, and other resources including data and handouts (in Portable Document Format). The webpage will obviously be utilized more heavily as Internet technologies mature and as access to the Internet becomes common place and the Internet becomes more reliable.

**Network-Based Computational Resources**

Over several years, the author has gathered certain instructional programs and provided access to them through the College's Engineering Computer network. Only one command is needed to access these modules. These design and instructional tools are available to all students. It has been a very efficient way to teach and to introduce several design and analysis concepts to students. These modules (see Table 2, below) require minimal Unix shell programming and the use of compilers that are widely available on Unix computer systems.

Table 2 - Design and Analysis Master Menu

```
==================================================
::::::::::   DESIGN   Master   Menu    ::::::::
::::::          W   A   T   E   R          ::::::
:::::::: Water Resources Design Modules ::::::::
==================================================
              Version 2.00 04-08-94
    By  Dr. Emmanuel U. Nzewi, Civil Engineering
        North Carolina A&T State University
        Greensboro, NC 27411
==================================================

    Select the Design or Analysis Module You Need:

    [1]   Sanitary Sewer Design
    [2]   Storm Sewer Design
    [3]   SCS Unit Hydrograph Method for Runoff Hydrographs
    [4]   SCS TR-55 Graphical Method for Peak Runoff Analysis
    [5]   INTERACTIVE Unit Hydrograph Convolution Program -- CONVOL.I
    [6]   BATCH Unit Hydrograph Convolution Program -- CONVOL.B
    [7]   S-Curve Unit Hydrograph Program -- SCURVE
```

### ACKNOWLEDGMENTS

The author gratefully acknowledges the support of the National Science Foundation for the funding of part of this project under the Instrumentation and Laboratory Improvement program, grant number DUE 9351736.

### REFERENCES

1. Groves, J. R. and D. W. Moody, "A Survey of Hydrology Course Content in North American Universities," Water Resources Bulletin, AWRA, Vol. 28, No. 3, pp. 615-621, June 1992.
2. Hotchkiss, Rollin H., "Computer-Aided Instruction in Water Resources: A Multimedia example," presented at 28th Annual AWRA Conference and Symposium, Reno, Nevada, Nov. 1-5, 1992.
3. Joyner, Rick, "Three Witnesses," MorningStar Publications, Charlotte, NC, 1997.

4. MacDonald, L. H., "Water Resources Education -- A Need to Get Wet and Dirty?," presented at 28th Annual AWRA Conference and Symposium, Reno, Nevada, Nov. 1-5, 1992b.
5. MacDonald, L. H., "Hydrologic Education: The Need for a Field Component," presented at Water Resources and Environment: Education, Training and Research Conference and Workshops, Colorado State University, Fort Collins, July 13-17, 1992.
6. Mack, M. J., "Curriculum Issues in Applied Hydrology and Environment," presented at Water Resources and Environment: Education, Training and Research Conference and Workshops, Colorado State University, Fort Collins, July 13-17, 1992.
7. NRC, "Opportunities in the Hydrologic Sciences," Water Science and Technology Board, National Research Council, National Academy Press, Washington, D. C., 1991, 348 pp.
8. Nzewi, E. U. and L. E. Peele, "A Manual of Digitizing Instructions Using MapInfo on the Macintosh," Civil Engineering Department, North Carolina A&T State University, March, 1996.
9. Nzewi, Emmanuel U., "Hydraulics Laboratory Manual," Civil Engineering Department, North Carolina A&T State University, Third Edition, August, 1997.
10. Snow, P. E., "Use of Digitizers for Hydrologic Analysis," presented at 28th Annual AWRA Conference and Symposium, Reno, Nevada, Nov. 1-5, 1992b.

Storm Water Permitting in the Milwaukee River Basin

James R. D'Antuono[1]

Abstract

The Wisconsin Department of Natural Resources has undertaken an aggressive approach to storm water permitting in the Milwaukee River Basin. Over the last year, 26 local units of government, including cities, villages, a new baseball stadium and a state fair grounds have begun this process. Previous storm water management efforts laid the foundation for the cooperative approach under which this new permitting program was initiated. This unique approach to comprehensive storm water permitting serves as a national model under which the challenge of cleaning up urban storm water runoff can be addressed.

Introduction

Fifty years ago society faced the challenge of reversing decades of using many of the nation's rivers and streams as sewers. A century of industrial growth, a World War and a post war society hungry to make up for lost time aggravated the already dreadful condition of our surface waters. State and federal government agencies passed laws to begin stricter regulation of "point sources" of pollution.

Cleaning up urban runoff may prove to be an even more difficult environmental challenge to solve. Society has asked the federal government and in turn state governments to address the problem. The Clean Water Act amendments of 1987 identified the need to regulate the quality of urban storm water runoff. It meant that the nation's largest cities and most populous counties would begin regulating runoff generated in their

---

[1]Milwaukee River Basin Sub-Team Leader, Wisconsin Department of Natural Resources, P. O. Box 12436, Milwaukee, WI 53212

jurisdiction. Further, construction sites larger than 5 acres would also need to have management practices to reduce the amount of sediment leaving a site. Finally, a large number of industries would also have to meet minimum performance standards to help ensure that pollutants in runoff generated on these sites could be minimized.

This paper is a discussion of the progress made in the Milwaukee area in dealing with the issues of permitting storm water runoff from many of the area's local units of government. The role of the Wisconsin Department of Natural Resources and local government are discussed. The methods for undertaking a comprehensive approach to regulating urban storm water runoff on a regional basis is described.

Methodology

The life blood of the Milwaukee Basin pulses through its 500 miles of streams, 100 lakes, and 60,000 acres of wetlands before flowing into the waters of Lake Michigan in the City of Milwaukee. Here the Milwaukee, Menomonee, and Kinnickinnic Rivers join together forming the Milwaukee River estuary. The Milwaukee metropolitan area contains the largest concentration of urban land uses and population in both the Lake Michigan and Lake Superior watersheds (storm water in the greater Chicago metropolitan area drains away from Lake Michigan).

Continental glaciers shaped the distinctive topography of the 900 square mile Milwaukee River Basin. This area gives birth to the Milwaukee River as it starts its 100 mile journey to Lake Michigan. Along the way it touches the lives of many of the nearly 1 million residents who call the area home. Four major watersheds embrace portions of 7 counties, 30 townships, 12 cities, and 24 villages. Rural land uses (farmland, woodlands, and wetlands) occupy more than 70% of the area. The remaining 30% of the Basin contains urban land use types including residential, commercial, and industrial.

Surface water quality ranges from excellent in many near pristine headwater trout streams to poor in the most developed areas in the lower Basin reaches. Overall, since the 1960's and 1970's when pollution abatement programs began, water quality has dramatically improved.

No where has the connection between surface water quality and human health been felt more keenly than Milwaukee. During an April 1993 outbreak of Cryptosporidiosiis (a disease with severe, flu-like symptoms), almost 400,000 Basin residents were affected and about 100 people died. In response, the Milwaukee Water Works voluntarily adopted new water quality standards that are more stringent than those required by state or federal law. By 1998, the waterworks will complete three major improvements at a cost of $89 million to virtually eliminate the possibility of a reoccurrence of such an outbreak.

The Milwaukee area has long recognized the importance of clean water. The City of Milwaukee was the first in Wisconsin to construct and install public sewer and water utilities. The nation's first activated sludge sewage treatment plant was constructed in 1925 at the mouth of the Milwaukee River. The Milwaukee Metropolitan Sewerage District continues these efforts, spending nearly $3 billion to upgrade sewage treatment plant facilities and increase capacity with a 17 mile long interceptor tunnel to collect urban runoff and sewage.

Urban and rural storm water runoff is the leading source of pollution to the Basin's streams, lakes, wetlands and the Milwaukee River Estuary. Since 1985, a voluntary approach for addressing urban and rural nonpoint source pollution has enjoyed significant success. These efforts are part of state sponsored program that is administered locally. Termed the Milwaukee Basin Priority Watersheds Program, rural landowners and municipalities have worked together to curb runoff pollution. To date, more than 500 rural landowners and 26 local governments have signed on with the program to do their part as guardians of the Basin's river, lakes and streams. The local and state investment under this voluntary program totals an estimated $40 million.

Water quality management activities in the Basin have been recognized nationally. They were highlighted at the prestigious Watershed '96, Moving Ahead Together Conference in Baltimore, Maryland. The Milwaukee River Basin was one of only four projects singled out for special recognition at this meeting. A video conference, broadcast internationally, was used to share accomplishments in the Basin with a live audience.

The City of Milwaukee initiated work to comply with the federal PHASE 1 storm water permitting requirements in the late 1980's. Over the next several years, the two part application process was completed and the Wisconsin Department of Natural Resources (DNR) issued a Wisconsin Pollutant Discharge Elimination System permit to the City in October 1994. This was the first municipal storm water permit issued to a Phase 1 community in the midwest (U.S. EPA Region 5). This permit covers the discharge of storm water from nearly 200 major outfalls to area streams and Lake Michigan. The permit requires a significant amount of wet weather monitoring, storm water management planning, aggressive street sweeping and catch basin cleaning, enforcement of construction site erosion control and storm water management ordinances, and the conduct of a comprehensive public information and education program.

Also in the fall of 1994, the DNR completed a two year effort to develop and approve rules governing the conduct of the state's storm water permit program. Under Chapter 216 of the Natural Resources Code (N 216), the DNR was given authority to regulate the discharge of storm water from municipalities, construction sites, and industries. Elements of the code provided two mechanisms for communities to request the DNR to include other local units of government in a municipal permit program. These were the identification of neighboring communities as part of a urban storm water planning area or a more formal process by which the DNR could be petitioned to regulate a communities' storm water discharges.

Milwaukee's permit contained a section in which the City identified their municipal storm water planning area. In general, the City identified 29 communities which either had physical connections to Milwaukee's storm sewer system and/or which contributed drainage to streams into which the cities storm sewer system drains. Correspondence from Milwaukee also requested the DNR to consider the relationship of the Cities' efforts to improve the quality of area rivers to the contributions of pollutants by surrounding communities.

Results

The DNR recognized that a comprehensive urban storm water permitting program was needed for the Milwaukee River Basin. Water quality in area streams and the near shore area of Lake Michigan did not meet state standards. Aggressive voluntary efforts to control sources of rural and urban nonpoint pollution sources did not achieve the desired results.

The advent of NR216 provided an impetus to try and address the problem of urban runoff pollution on a broad scale. The springboard was the City of Milwaukee request that the DNR consider additional communities for permitting in the cities' urban watershed planning area. A strategy was developed which laid out a systematic approach to begin the process. The main goals were to: 1) Foster a cooperative atmosphere by providing timely information and applying decision making criteria fairly; 2) Provide opportunities for feedback and discussion and 3) Ensure that the goals of the permitting program were understood by the affected communities.

Implementation steps were developed to try and meet the goals. An important element of the implementation plan was the work that the DNR staff had undertaken with many of the local units of government over the past five years. The above referenced Milwaukee River Priority Watersheds Program had been a catalyst for many of the communities to become involved in urban storm water management efforts. DNR staff provided technical assistance and in many cases grant funds. State dollars were matched by local municipalities to complete comprehensive storm water management plans, design structural best management practices, initiate non-structural best management practice programs, and to construct a variety of structural practices. These included wet detention ponds, wetland treatment systems, and underground treatment devises.

DNR staff developed an implementation program which consists of the following:

1. Provide information simultaneously about the permitting process and schedule.
2. Develop preliminary criteria to determine which communities would be required to apply for a municipal storm water permit.
3. Hold a series of informational meetings to discuss the process and seek input on the decision making criteria.
4. Determine which communities should be permitted and share the results.
5. Provide opportunities for feedback.
6. Develop written training materials about the application process.
7. Hold training sessions for elected officials, municipal employees and DNR staff.
8. Provide technical and where possible, financial assistance to permitted municipalities.

To begin the permitting process, the DNR worked with the 29 affected local units of government to develop evaluation criteria to determine who should be permitted. The criteria included: 1) extent of physical interconnections between municipal separate storm sewers or ditches; 2) location of the discharge of storm water to a designated municipality; 3) the amount of critical acres (land uses generating the most pollutants in urban runoff); 4) the type of recommendations needed for new urban development; and 5) population density. Point values were assigned to each criteria and the relationship of the individual municipal contributions was evaluated.

This evaluation process resulted in the identification of 21 of the original 29 communities in the original Milwaukee urban storm water management planning area. Subsequently, 5 additional communities not in the original list provided by the City of Milwaukee were added by the DNR.

Currently, these 26 local units of government have begun a two and one half year process during which a pre-application will be submitted to the DNR for review and approval. Subsequently, the DNR will assist the local units of government to complete a full storm water permit application. The communities have two years to submit this information. Individual permits will be issued for a 5 year period. An annual fee of $5,000 will be paid to the DNR by each permitted community.

Discussion and Conclusion

A fundamental element of the storm water permitting approach is the customer service the DNR has provided and will continue to provide to help make this program a success. The response to this new program has been encouraging. None of the 26 communities participating in the permit program has filed any legal challenges. A devastating June 1997 flood in portions of the Basin has raised the awareness of the general public and elected officials to the importance of storm water management. The permit program will be an important catalyst to changing the philosophy of the means for managing both the quality and quantity of storm water runoff for existing and new development in this part of the state. We anticipate that this approach will have application in other areas of the state.

DNR staff anticipates that the first permit applications will be submitted this spring. Individual WPDES permits will be issued before the end of the year. Communities continuing their strong working relationship with the DNR will be well positioned to help write their own permit. The proposed management program included as part of the application will be used extensively by DNR staff to prepare the permits. Communities which have not undertaken any storm water initiatives or which have not established a working relationship with the DNR on this issue will face more challenges in complying with the application process and fulfilling the permit requirements.

Rural Management for Nonpoint Pollution Control

William C. Hafs

The most significant stressors to the Green Bay ecosystem are nutrients and sediment. Twenty-seven dump truck loads of sediment are delivered to the Bay daily (one hundred fifty thousand tons per year). The cost of dredging the Green Bay Harbor has averaged one and six tenths million dollars per year for the last ten years. Two hundred thousand cubic yards per year is (ten-year average) dredged for shipping channel maintenance and placed in a contaminated disposal facility. The cost is Eight dollars and eight cents per cubic yard of dredge sediment which is twenty-eight percent solids. The actual cost of sediment removed (dry weight) is Twenty-eight dollars and eighty-five cents per cubic yard.

Agriculture is the number one source of sediment. Ninety-one percent of the sediment delivered to Green Bay is the result of agriculture runoff and poor farming practices. Beyond harbor and shipping problems, sedimentation causes the water to become turbid and transports other nutrients (phosphorus) and pesticides to the Bay. Extensive inventories and studies of this area from 1986-1996 identified the upland sites contributing most of the problem.

The Fox-Wolf Drainage Basin is sixty-four thousand square miles and is the second largest tributary contributor of sediment to Lake Michigan. Recent studies have shown that two subwatersheds near the mouth of the Fox, East River and the Duck/Apple/Ashwaubenon while only seven percent of the land area of the Fox-Wolf Basin, contribute sixty-five percent of the sediment and fifty-five percent of the phosphorus loads to the ecosystem.

To combat agricultural runoff, Brown County entered into cooperative agreements with the State of Wisconsin (DNR and DATCP) on programs aimed at prevention and clean-up of nonpoint pollution. These programs included: NR243 (1984) administrative rules relating to animal waste management, whereby landowners

---
County Conservationist, Brown County Land Conservation Department, 1150 Bellevue Street, Green Bay, WI 54302

who pollute the waters of the state may be required to take corrective action. The Wisconsin Farmland Preservation Program (1986) where landowners receive tax credits for keeping land in agriculture vs. developing the land. Continuous eligibility requires conservation standards related to erosion control on their land. Brown County Animal Waste Storage Facility Ordinance (1986) regulating design and construction of animal waste storage facilities. County Manure Management Water Pollution Control Plan (1986) and Erosion Control Plan (1988). Five active priority watershed nonpoint abatement projects are currently underway. East River ('89), Red River ('95), Branch River ('96), Duck Creek ('97), Apple/Ashwaubenon ('97). A county Streambank Protection Ordinance ('91) was adopted to protect riparian areas.

Most efforts to date in combating nonpoint pollution on rural landscape have focused on animal waste problems. Barnyard practices and animal waste storage facilities are the most popular, most implemented and most expensive. In 1996 for example, $4.65 million dollars was spent on nonpoint Best Management Practices in the eleven county area of Northeast Wisconsin. Sixty-eight percent of this cost-share funding to landowners was spent on animal waste storage facilities ($3.16 million dollars), seventeen percent was spent on barnyard practices, and only fourteen percent was left for engineering practices that deal with upland sediment control. In Brown County alone, one hundred fifty-four animal waste storage facilities were installed in the last ten years. While animal waste storage and barnyard has received a majority of the attention and effort from an engineering standpoint, recent watershed studies indicate animal waste is only thirty percent of water quality problem, recently an evolution of the program has begun. Management practices dealing with soil erosion and sedimentation have increased in popularity. Traditional conservation practices that address sedimentation were analyzed, concluding that they were not being effective for various reasons:

<u>Weaknesses With Conventional Erosion Control Practices</u>
(1) Poor incentive packages (cost-share) offered to landowners.
(2) The cost of implementation of practices is not worth the economic return.
(3) Major changes in farming operations needed to implement practice.
(4) Inconsistent government programs, some requiring a T value or Tolerable Soil Loss vs. sediment delivery in other programs.
(5) Practice may require additional expensive equipment to be purchased.
(6) Conservation practices may not fit topography.
(7) Staff time and knowledge required to sell and implement inadequate for comprehensive application.
(8) Not enough research data to show practice effectiveness on various soil types.
(9) Limited correlation between T values or Tolerable Soil Loss to water quality.

A need for alternative approaches that address sedimentation has resulted in new management practices being promoted. Land Conservation Departments have traditionally employed engineers and technical staff to design concrete walls, steel placement, sieve analysis on soils, concrete inspection, concrete slump tests, hydrology, soils investigations, etc. Management practices such as green manure cover crop, conservation tillage, nutrient management address large acreages of land that are delivering soil and nutrients to streams. New staff that work on these land management practices are Agronomists, Planning Specialists, and Information and Education personnel.

A new conservation practice that reduces sediment delivery is the Riparian Vegetated Buffer Strip (Buffers). A strip of grass as narrow as thirty feet placed between row crops and a stream can reduce sediment delivery as much as ninety percent. Only a few research papers on the water quality benefits of buffers were published prior to the early 1970's. There are now over four hundred such research papers and the rate of publication is thirty to thirty-five papers per year. Brown County Land Conservation Department aggressively promoted a County Shoreline Protection Ordinance which requires buffers on all one thousand two hundred miles of intermittent and perennial streams. The ordinance was adopted in October, 1991. The ordinance requires a minimum of thirty-five feet free of row crops, planted to grass, to be maintained between farm fields and all mapped (USG) intermittent/perennial streams. Over the last four years, two hundred miles of buffers have been installed with only two staff years of time. Over seven thousand tons of sediment and fourteen thousand pounds of phosphorus were reduced annually. Landowners are paid seventy percent of installation costs such as seeding and shaping, and Five hundred dollars per acre (one time payment) incentive. Because we have an ordinance, the buffer is perpetual and attached to the deed.

It costs $28.85 per cubic yard to remove dredge soil from the Green Bay Harbor. Using buffers, the cost can be reduced to $2.00 per cubic yard. Adequate incentives ($500/acre) have prevented landowners being opposed to the practice, as the two hundred miles of buffer installed in the last four years have been installed voluntarily without use of regulation.

Brown County has one thousand two hundred miles of streams. When we began, seven hundred miles needed buffers. Five hundred miles of buffers that remain (thirty-five feet on each side) equates to four thousand two hundred forty-two acres (at $500 per acre) or $2.12 million dollars and will reduce sediment delivery fifty to seventy percent. For the same dollars, we can build one hundred six barnyards (At Twenty thousand dollars each) or sixty animal waste storage facilities (at Thirty-five thousand dollars each), which do not address the sedimentation problem.

Reducing Nonpoint Source Pollution in Ultra-Urban Areas: Two Case Examples

James A. Bachhuber[1]

Abstract

Many municipalities are challenged with requirements to reduce nonpoint source pollution from intensively developed areas with little or no space for conventional best management practices (BMP's). These areas often include commercial and/or industrial land use; and these land uses are typically identified as significant sources of urban nonpoint pollution.

Within the 16.2 ha (40 acre) festival grounds (Henry Maier Park) in downtown Milwaukee Wisconsin, the creation of a small landscaped island provided an opportunity for managing nonpoint source pollution. A flow splitter diverted the first flush stormwater from a 1.2 ha (3.0 acre) area to a constructed landscaped island. The landscaped area was graded as a depression and lined with a layer of peat over a layer of fine graded sand. Under lying the sand perforated collection pipes intercepted the filtered stormwater and returned the water to the storm sewer system.

At St. Mary's Hospital in Green Bay, WI a pressurized, two stage filter system was designed to reduce pollution from a 2.6 ha (6.5 ac) parking lot. The system proved more economical, treated a greater amount of runoff, and also removed a greater percentage of annual pollution than gravity driven sand filtration systems.

Introduction

Federal and State Stormwater Permit Programs have been in effect for several years; and with the advent of the EPA's Phase 2 program; more communities of mid-size populations (50,000 - 100,000) will be required to comply with the permit program. The State of Wisconsin was delegated authority to implement the federal program. With the adoption of Administrative Code NR 216 in 1994, Wisconsin's urban stormwater permit program included selected communities with populations greater than 50,000. Nonpoint

---

[1]Senior Water Resources Project Manager, Rust Environment & Infrastructure, 1020 North Broadway, Milwaukee, WI 53202

source pollution control in the suburban and newly developing areas can generally be achieved through conventional BMPs (vegetated swales, wet detention basins, infiltration basins, etc.). These BMPs require relatively large land commitments, and can be accommodated in the less intensively developed areas.

Previous monitoring efforts (Bannerman et al, 1993, Bannerman, et al, 1983) consistently show that the highest pollutant loadings from urban land uses are generated from commercial, industrial, and "downtown" areas of large and medium size cities. These are most often also the areas with the highest land values, and least open space available for conventional types of BMP's. These areas have been identified as "ultra-urban" lands (Bell, et al 1995). They represent the most critical lands in terms of pollutant loadings, and also the most challenging lands for implementation of practical BMPs. Various forms of sub-surface media filters have been employed in ultra-urban areas, most notably in the Chesapeake Bay area of the United States.

Rust Environment and Infrastructure recently completed the engineering and design of two different BMPs to meet the nonpoint source pollution control needs in ultra-urban settings in Wisconsin.

Henry Maier Park - Milwaukee, WI

Over 800,000 people attend the eleven day Summerfest celebration located on the 16.2 ha (40 acre) Henry Maier Park adjacent to Lake Michigan in Milwaukee, WI. Many more attend one or more of the nine ethnic and cultural festivals each year summer at this same site. The park is about 95% imperious surfaces (asphalt, rooftops, and concrete). Activities at the festivals include amusement rides, music stages, street shows, and restaurants.

Because of the extremely dense human use of the park; many factors were considered in developing acceptable BMPs. The factors included: public safety, aesthetics, and minimal space requirements. In addition the site is located on fill material, adjacent to Lake Michigan, and has groundwater within 1.5 - 2.0 m (5 - 6 ft.) of the surface.

The festivals and grounds are managed by a private enterprise - Milwaukee World Festivals Inc. (MWF). As part of a renovation project, MWF intended to remove the asphalt from a 560 m$^2$ (6,000 ft$^2$) area to create a landscaped island. Rust worked with the MWF and their architect (Eppstein-Uhen) to create a landscaped area that would double as a stormwater filtration area.

Components of the System

The proposed landscaped site was near a storm sewer that collected runoff from a 1.2 ha (3.0 ac) area of the park. The stormwater treatment system consisted of the following components:

- Flow Splitter: A manhole with a weir structure was designed to be cut into the storm sewer near the landscaped feature. The weir height was set to divert the first flush flow to the filtration area. The "first flush" was defined as the first 1.25 cm (0.5 in) of runoff from the drainage area. Runoff greater that the first flush flowed over the weir and out the existing storm sewer to Lake Michigan. Weir height was critical to minimize potential surface ponding within the drainage area (from the increase in the hydraulic grade line). The weir height also fixed the elevation requirements of the filtration site.

- Low Flow Pipe: A 40.5 cm (16 inch) diameter pipe was designed to carry the low flow runoff from the flow splitter to the filtration basin.

- Filtration Basin: Design calculations determined that a 560 $m^2$ (6,000 $ft^2$) filtration area required 0.46 m (1.5 ft) of storage height (above the top of the filtration media) to hold the runoff volume from the 1.25 cm (0.5 in) of runoff. The filtration media consists of 7.7 cm (3 in.) of peat over 15 cm (6 in.) of fine graded sand (average particle size of 0.2 mm). Beneath the fine sand layer are perforated PVC 15 cm (6 in.) diameter collection pipes embedded in a 30 cm (12 in.) sand layer. Beneath the entire filtration area is an impermeable synthetic membrane to separate the surrounding fill material and groundwater from the filtration basin. The infiltration basin will be vegetated with grass and other flowering species.

- Return Flow Pipe: The perforated collection pipes feed into a 20 cm (8 in.) diameter return pipe. The return pipe enters the flow splitter structure, downstream of the weir. This treated runoff is then discharged to Lake Michigan via the existing storm sewer.

Design Challenges

The site presented many constraints that were addressed for the BMP to be feasible. The constraints included:

- Groundwater within 122 - 150 cm (48 - 60 in.) of the surface: This factor determined the maximum depth of the collection pipes at the base of the filtration practice. This depth, in turn, determined the depth of filtration media, and the maximum storage elevation for the first flush.

- Elevation of Storm System Inlets: Inlet rim elevations upstream of the flow splitter, generally were within 1 m (3 ft) of the storm sewer's crest. If the hydraulic grade line exceeded the rim elevation, temporary ponding would occur during storm event. The length of time water would pond, and the location of the ponding, was critical to public safety and convenience of the park users.

- Fluctuating Lake Levels: Lake Michigan levels (above normal elevations) controlled the water levels within the storm sewer system. The lake can fluctuate by 0.5 m (1.5 ft) within a year, and greater fluctuations can occur over a period of years. The lake level can affect the permanent water elevations within the filtration site.

- Practice Maintenance: It is expected that the media will need to be replaced on a periodic basis. The frequency of replacing the media depends on the pollutant characteristics of the drainage area, and tolerance of the MWF for the length of time water remains ponded within the filtration basin. Initial calculations show that the water will remain ponded (above the peat surface) for a maximum of four hours. It is expected that as pollutants are trapped, the residence time will increase.

Construction of this BMP is expected in the spring or fall of 1998. Preliminary construction cost estimates range from $55,000 to $70,000.

## St. Mary's Hospital, Green Bay, WI

The St. Mary's Medical Center campus covers approximately 9.7 ha (24 ac). Almost 80 percent of the campus is taken up by buildings and parking lots (impervious surfaces). The St. Mary's Hospital has a long history of being "environmentally pro-active". Even though the facility is not permitted under current stormwater regulations, the administration wanted to treat the pollution from it's parking lot runoff. Faced with restrictions on land available for conventional stormwater management facilities, for the control of urban nonpoint source pollution, the St. Mary's Medical Center chose to build an innovative pressurized filtration system. According to the manufacturer of the filtration components of the system, this is the first application of the technology to stormwater management.

The designed system treats runoff from a 2.6 ha (6.5 ac) drainage area consisting of employee/patient parking lots, and building rooftops. The system is completely underground except for a 28 $m^2$ (300 $ft^2$) housing structure for the filtration system and pump. The system is pressurized by a 0.056 cms (2 cfs) pump. The system treats 100% of the flow from the 1.25 cm (0.5 in) of runoff (defined as the first flush). The first flush from larger events are treated; and the remaining portions of the runoff are bypassed to the existing storm sewer system. The system treats about 70% of the average total annual runoff from the site. Particles down to 5 micron size are removed.

Components of the System

Flow Splitter: A manhole which receives runoff from the drainage area via a 46 cm (18 in) diameter pipe, has two outlet pipes: a low flow, 38 cm (15 in) diameter pipe, and a high flow by pass, 46 cm (18 in) diameter pipe. The high flow pipe is set 0.54 m (1.8 ft) higher than the low flow pipe.

Holding Tank: The low flow pipe discharges to an 252 m$^3$ (9,000 ft$^3$) subsurface concrete holding tank. The volume of the tank is sized to completely hold the runoff from first flush (taking into account the pumping rate of 0.056 cms [2 cfs]). The stormwater in the holding tank is withdrawn by a self-priming solids handling pump.

Mechanical Housing Unit: This structure is built partially below grade, and is approximately 20 m$^2$ (300 ft$^2$). The unit houses the pump, a two phase filtration system, a backwash tank, and the system's control panel. The house has electrical power and the treated stormwater is discharged back to the storm sewer system.

The Filtration Process

The stormwater treatment process consists of two filtration systems, under pressure from the pump. The first filtration process consists of a cluster of commercial disc filters manufactured by Arkal Filtration Systems (U.S. distribution by Zeta Flow Inc. of Florida). The disc filters have been used for many industrial and agricultural applications. The redundant system allows for simultaneous filtration with one cluster, while another cluster is back washing. This means that the system can maintain a constant filtration flow rate of 0.056 cms (2 cfs). The disc filters can be modified to filter various size particles in the stormwater down to a minimum of 50 micron size. The backwash water is temporarily stored in a backwash tank and subsequently discharged to a sanitary sewer. The filtered stormwater is sent to a second filtration stage.

The second filtration stage consists of a series of five enclosed sand filter tanks also manufactured by Arkal Filtration Systems (U.S. distribution by Zeta Flow Inc. of Florida). The 1.2 meter (48 in) diameter tanks are also a redundant system so that one or more tanks can filter the stormwater, while other tanks are back washed. The tanks are enclosed to maintain a pressurized flow system. This feature allows a continuous filtration process at a rate of 0.056 cms (2 cfs). The backwash water is temporarily stored in a backwash tank and subsequently discharged to a sanitary sewer. The filtered stormwater is discharged to the storm sewer system. The total backwash volume from both filter systems is about 5.6 m$^3$ (200 ft$^3$), or about 1.5% of the total volume from a two year, 30 minute event.

Advantages/Disadvantages

This management practice is an active, processing system for treating stormwater, and it may not be practical in all "ultra-urban" situations for stormwater quality treatment. The potential disadvantages of this BMP include:

- The mechanical systems require routine maintenance.
- A power source is necessary.
- A sanitary hookup is necessary.
- Electrical and sanitary fees are estimated at $4000/year

For the application at the St. Mary's Hospital site, this stormwater treatment approach provided many advantages over other sub-surface, gravity driven, media filters including:

- The system is flexible to handle various size drainage systems. The most cost effective treatment can be designed by balancing the size of the holding tank, the pumping rate, and the number of disc filter clusters and media tanks. For example; a smaller holding tank can be built through the use of a higher pumping rate, and more filtration equipment.
- The rapid filtration rate allows for a greater percentage of the annual runoff to be treated compared to gravity driven filter treatment systems
- Filter discs can be adjusted to remove variable particle size pollutants depending on the pollutant characteristics of the runoff.

This system is scheduled to be constructed in the summer of 1998. The estimated construction cost is $200,000.

Conclusions

Runoff from ultra-urban areas is a critical source of nonpoint pollution, and is also one of the most difficult settings to implement treatment systems. Generally, the costs per acre treated are high when compared to conventional BMPs. The two examples described in this paper emphasizes the importance of incorporating the specific site's opportunities and restrictions into the design process.

References

Bannerman, R.T., K. Baun, M. Bohn, P.E. Hughes, and D.A. Graczyk, 1983. Evaluation of Urban Nonpoint Source Pollution Management in Milwaukee County Wisconsin, Vol. I. EPA, Water Planning Division, PB 84-114164.

Bannerman, R.T., D.W. Owens, R.B. Dodds, and N.J. Hornewer, 1993. Sources of Pollutants in Wisconsin Stormwater. Wat. Sci. Tech. 28(3-5):241-259

Bell, Warren, L. Stokes, L.J. Gavan, and T.N. Nguyen, 1995. Assessment of the Pollutant Removal Efficiencies of Delaware Sand Filter BMPs. Department of Transportation and Environmental Services, City of Alexandria, VA.

# Equity, Cost and Environmental Benefit of Clay Distribution

James W. Male[1] and Ellen Hoffman Belk[2]

## Abstract

This paper considers how agencies can address decisions regarding the equitable distribution of resources, and how the question of equity can be addressed in concert with other goals. The general approach is applied to the excavation and removal of clay from the Central Artery and Tunnel Project (CA/TP) in Boston, and distribution to solid waste landfills in Massachusetts for use as capping material. Cost, equity and environmental benefit objectives are considered.

## Background

**Central Artery/Tunnel Project.** The Massachusetts Highway Department's (MHD) Central Artery/Tunnel Project (CA/TP) will replace Boston's elevated Central Artery with an underground expressway. Excavation as part of the Project will yield approximately 2.9 million cubic yards (mcy) of Boston blue clay and glacial till. The clay has qualities which make it well-suited for landfill capping (CDM 1991).

**Unlined Landfills.** The Massachusetts Department of Environmental Protection (DEP) is currently monitoring approximately 210 municipal landfills, representing a total area of about 3,100 acres, which will need to be capped to prevent leaching of contaminants into nearby drinking water sources (DEP 1995). Of these municipalities, 82 (with approximately 1,500 acres of uncapped area) have had stockpiling and capping plans approved by DEP. Excavation from the CA/TP will generate enough clay to cap approximately 975 acres with a thickness of 18 inches (MA DPW 1990).

---

[1]Edwin and Sharon Sweo Professor of Engineering, University of Portland, Portland, OR 97203, formerly, Professor, Department of Civil and Environmental Engineering, University of Massachusetts Amherst.

[2]Remedial Project Manager, U.S. Environmental Protection Agency, Region 2, New York, NY 10007.

**Clay Distribution Goals.** Each organization involved with the CA/TP clay distribution program has slightly different goals. MHD wants to provide for reliable, timely and cost effective disposal of clay. DEP is interested in minimizing the negative impact of unlined landfills on public health, welfare, and the environment. For political reasons, both DEP and MHD want to provide assistance to a wide geographical distribution throughout the state.

**Purpose of Study.** Agency personnel sought assistance in formulating policy issues, including: (1) how to address the goals of the agencies, (2) how to quantify criteria to address those goals, and (3) the relative impact associated decisions would have on the distribution of the clay.

## Quantification of Objectives

A series of discussions with agency personnel identified the overall goals and assisted in how to more specifically define and quantify objectives

**Equity.** Equity refers to the fairness, impartiality, or equality in the distribution of a resource. Quantifying equity requires agreement regarding the following issues: the general principles of equity to be investigated and the more specific equity parameters and equity measures to be implemented in the model (Male and Franz 1986). An equity parameter represents the quantity or value that one participant might realize with a particular distribution alternative. It is a quantifiable entity which becomes the basis for comparison with respect to equity. The measure of equity quantifies the definition of an equitable distribution. An equity measure makes use of a selected equity parameter in specifying how differences in equity will be determined. The summed deviations measure (Brill et al. 1976, Rossman and Graham 1979), was used in conjunction with two equity parameters: the amount of clay received and the fraction of clay received by each landfill. The amount of clay is measured in cubic yards, while the fraction is the amount of clay received divided by the total amount of clay needed by a particular landfill. Analyses and results for only the first case are presented here.

**Environmental Benefit.** Personnel from MHD and DEP agreed that environmental benefit could not be represented easily by a continuous measure. Therefore, based on DEP criteria, landfills were divided into two groups: those posing a Significant Threat to the Environment (STE), and those posing a Potential Threat to the Environment (PTE). 19 landfills were determined to pose a Significant Threat to the Environment. The greatest benefit to the environment results when all of the clay needed is allocated to each STE landfill. A less beneficial result is more likely when clay is distributed without regard to the environmental benefit objective.

**Cost.** Cost estimates for clay distribution consisted of the: administrative costs, stockpile site preparation costs, round trip travel time, stockpiling and maintenance costs, and contingency costs. (Atkinson 1995, Hoffman 1996, Nessen 1995).

## Model Development

The problem was modeled as a two objective formulation (Equity and Cost) and solved twice, once for maximum environmental benefit and once not emphasizing environmental benefit. The general formulation using the equity measure of volume of clay and not emphasizing environmental benefit is shown below:

$$\text{Minimize: } Z_1(e_i) \quad (1)$$

$$\text{Subject to: } x_i - \bar{x} = e_i \quad \forall i = 1\ldots82 \quad (2)$$

$$x_i \leq v_i \quad \forall i = 1\ldots82 \quad (3)$$

$$c_1 x_1 + c_2 x_2 + \ldots + c_{82} x_{82} \leq C \quad (4)$$

$$x_1 + x_2 + \ldots + x_{82} = V \quad (5)$$

$$x_1, x_2, \ldots, x_{82} \geq 0$$

where $Z_1$ represents the Equity objective and the five constraint equations incorporate the problem's restrictions and the Cost objective. The variable $x_i$ is the volume of clay allocated to the $i^{th}$ landfill, $\bar{x}$ is the mean volume of clay delivered, $e_i$ is a measure of equity for the clay delivered to the $i^{th}$ landfill, and $v_i$ refers to the maximum volume of clay the $i^{th}$ landfill can receive. C is the total cost of distributing all of the clay and $c_i$ describes the cost for delivering clay to the $i^{th}$ landfill. V is the total volume of excavated clay. A similar formulation can be developed where $x_i$ is the fraction of clay allocated.

Constraint set 2 defines equity for each model. Constant set 3 limits allocation to any landfill to be less than the approved amount. Cost is incorporated as an ε-constraint in constraint 4 (C is equivalent to ε and different values can be used to obtain different solutions). Constraint equation 5 requires that all excavated clay be delivered.

When environmental benefit is maximized constraint set 3 is changed to:

$$x_i = v_i \quad i \in \text{STE} \quad (6)$$

$$x_i \leq v_i \quad i \in \text{PTE} \quad (7)$$

The ε-constraint method (Cohon and Marks 1973) was used to solve the problem using a range of cost values for the right-hand side of the cost objective, C. They were varied from a minimum value that was just beyond an infeasible region (where no solution was possible because there was not enough money to distribute all of the clay excavated, even to the least costly landfills), to a maximum value associated with a non-binding solution (where the solution did change with increased cost).

On a two-dimensional plot with axes of Cost, C, and Equity, $Z_1$, two tradeoff curves were developed, one not emphasizing Environmental Benefit (Version A) and one emphasizing environmental benefit (Version B).

## Results

**Trade-Offs Among Objectives.** The results of the application are shown in the accompanying figure. Note that both axes improve away from the origin. The upper curve represents results of the formulation which treats all landfills equally with respect to Environmental Benefit (Version A). The lower curve shows the results for Version B (the 19 landfills classified as an STE are provided the entire amount of clay needed for capping and the remainder of the clay is allocated to PTE landfills).

The mean volume of clay distributed to landfills is approximately 35,000 cy. All landfills which receive more or less than this amount contribute to the total inequity for a given solution. Using this equity parameter, it is not possible to achieve 100% Equity for either Version A or B since a landfill must not receive more clay than it requires for capping and all of the excavated clay must be delivered. As a result, many smaller landfills receive the full amount of clay needed for capping, but since this amount is less than the mean volume of clay delivered, these differences contribute to the total inequity for this model.

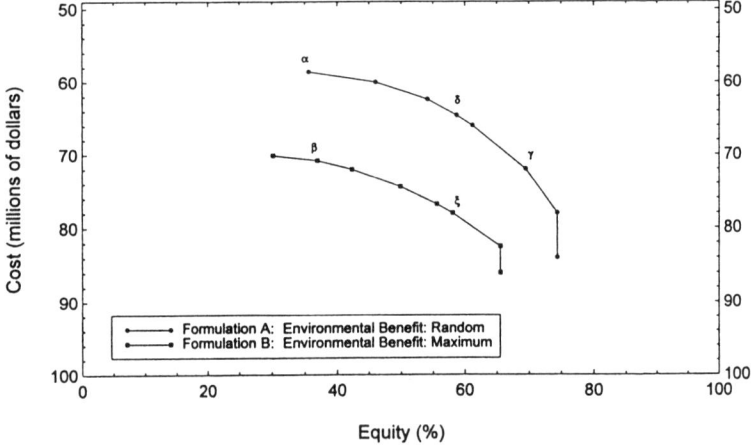

**Tradeoff Curves for Cost and Equity for Two Environmental Benefit Objectives.**

**Representative Solutions.** The representative solutions labeled α, β, γ, δ, and ξ in the tradeoff figure are further illustrated below showing the allocation of clay volume among the 82 landfills as a function of their distance from the CA/TP site.

## Conclusions

Based on the research performed, the following conclusions are drawn.

(1) Quantification of the goals set forth by DEP and MHD was possible and effectively represented desires of the two agencies.

(2) Multiobjective techniques proved to be useful in analyzing the problem of clay distribution, particularly in its ability to portray tradeoffs among objectives.

(3) Exploration of two definitions of the equity parameter allowed study of the impact those definitions have on distribution of clay. These results were useful in assisting the agencies in policy decisions regarding equitable distributions.

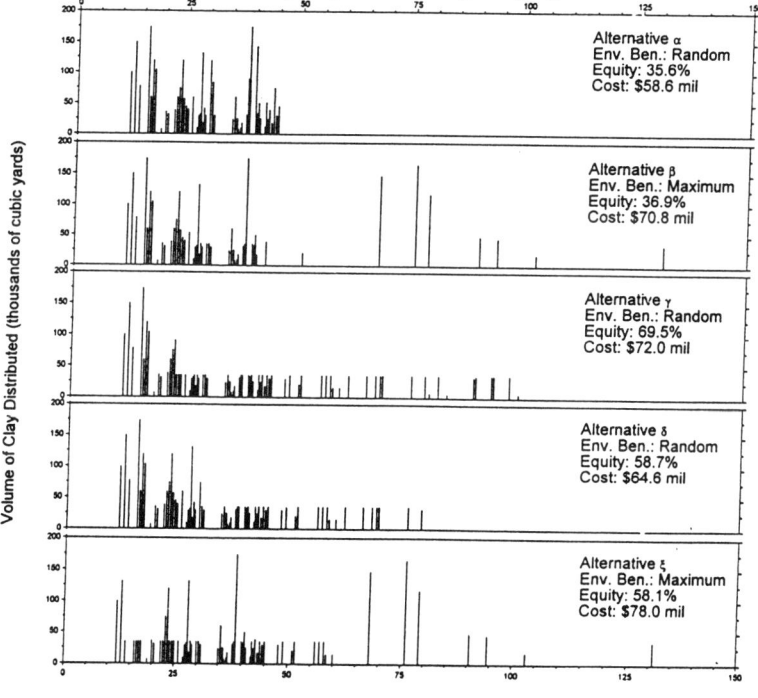

**Volume of Clay Distributed vs. Distance from CA/TP Site.**

(4) When the amount of clay was used as the equity parameter and when environmental benefit was allowed to be random, optimal clay distribution solutions ranged in equity from 35.6% to 74.4% with a corresponding range in cost from 58.6 to 78.0 million dollars. Optimal solutions with maximum environmental benefit had a slightly lower range of equity (30.1% to 65.5%) and associated costs which were significantly higher (70.1 to 82.5 million dollars).

(5) When the fraction of clay was used as the equity parameter, optimal clay distribution alternatives with random environmental benefit ranged in equity from 53.7% to 100% and in cost from 58.6 to 81.9 million dollars. Optimal solutions for the same model with maximum environmental benefit had a slightly lower range of equity (52.0 - 88.0%) and associated costs which were significantly higher (70.1 to 88.5 million dollars).

(6) For both formulations, results for the random environmental benefit formulations (the A-versions) distribute to STE landfills approximately one-half of the clay they require. Solutions to the maximum environmental benefit formulations (the B-versions) provide nearly twice as much clay to those landfills.

## References

Atkinson, D.S. (1995). "Comparative Analysis of Capping Alternatives." Presented at SWANA Conference, Boxboro, MA, January 19, 1995.

Brill, E.D., Jr., Liebman, J.C., and C.S. ReVelle (1976). "Equity Measures for Exploring Water Quality Management Alternatives." *Water Resources Research*, Vol. 12, No. 5, pp. 845-851.

Camp Dresser & McKee, Inc. (CDM) (1991). "Environmental Consequences of Utilizing Boston Blue Clay in Landfill Closures", January 25, 1991.

Cohen, J.L. and D.H. Marks (1973). "Multiobjective Screening Models and Water Resource Investment." *Water Resources Res.*, Vol. 9, No. 4, pp. 826-831.

Department of Environmental Protection (1995). Municipal Landfill Database, February 23, 1995.

Hoffman, E.M. (1996) "Use of Systems Analysis in Modeling Distribution of Clay for Capping Municipal Landfills." M.S. Project Report, Department of Civil and Environmental Engineering, University of Massachusetts Amherst.

Male, J.W. and S.L. Franz (1986). "Allocating Funds For Repair Of Leaky Water Distribution Systems." *TIMS Studies in the Mngt Sci.*, Vol. 22, pp. 183-203.

Massachusetts Department of Public Works (1990). Final Supplemental Environmental Impact Statement, Part II, Vol. 1, Chap. 4.

Nessen, B. (1995). "Clay Distribution TAC Proposal (For Discussion Purposes)." Paper prepared for a Meeting of the CA/T Clay Distribution Technical Advisory Committee, January 11, 1995.

Rossman, L.A. and P.A. Graham (1979). "Distributing Regional Services Costs." *Journal of the Urban Planning and Development Div.*, Vol. 105, No. UP1.

Total Valuation of Grand Canyon Resources

Michael P. Welsh[1] and Richard C. Bishop[2]

**Abstract:** Glen Canyon Dam is located on the Colorado River, just upstream from the Grand Canyon National Park. The Bureau of Reclamation's Glen Canyon Environmental Studies (GCES) program has documented significant impacts of the operation of the hydroelectric facility at Glen Canyon Dam on beaches along the river, vegetation, Native American cultural sites and endangered fish species. Consequently, changes in the operation of Glen Canyon Dam are likely to benefit downstream resources. GCES researchers determined that dam operations benefitting downstream environmental resources will reduce the economic value of the power produced at Glen Canyon Dam.

The presenters will discuss a valuation study to determine the benefits associated with ecosystem preservation in the Grand Canyon. The study shows that enhancing the river-related environment of Grand Canyon has substantial benefits, both to Americans generally and to individuals who will face higher electricity bills if dam operations are changed. The magnitude of these benefits suggest that gains from enhancing the river-related environment are larger than the reductions in the value of hydroelectricity produced at Glen Canyon Dam.

**Valuation of Environmental Resources**

Many regulatory decisions of government affect environmental resources. Using benefit-cost analysis to support such decision making is challenging because most environmental amenities are not traded in markets. Where goods and services are traded in markets, resulting market prices provide hard evidence about economic values. However, new or modified regulations may affect air and water quality, fish and wildlife populations and habitats, and other environmental resources for which market prices are not available to serve a basis for valuation of benefits and costs.

---

[1] Michael P. Welsh is a Senior Associate at the Hagler Bailly Consulting, Inc., 455 Science Drive Madison, WI 53711.

[2] Richard C. Bishop is a professor in the Department of Agricultural and Applied Economics at the University of Wisconsin-Madison, Madison, WI 53706.

Where can the analyst—and ultimately the decision maker—turn when market evidence is not available?

Depending on the circumstances, several techniques might be useful. For example, if recreational resources are affected, travel and related expenditures may be used to estimate benefits and costs through application of the "travel cost method." Where the environmental effects of regulations influence property values, "hedonic price" analyses can yield insights into environmental values. Where regulations affect health, changes in expenditures for health care and income gained or lost due to changes in days worked constitute part of the values that can be counted in a benefit-cost analysis, although additional values may be associated with pain and suffering.

None of these methods work well when the primary effect of the regulations is to non-use values. Non-use values are values that members of the public hold for environmental resources for reasons other than personal use. People may value resources they do not personally plan to use because they wish to see others, either present or future generations, have the opportunity to use them; because they are sympathetic to the plight of animals; because they simply enjoy knowing that those resources will continue to exist; or for yet other reasons.

In cases where the regulations affect non-use values, the contingent valuation (CV) method is often used. CV uses survey methods to value environmental benefits and costs. In personal interviews, telephone interviews, or mail surveys, respondents are asked questions designed to elicit their economic values. CV has been applied in literally hundreds of cases, both in North America and abroad.

Dissecting a CV survey would reveal three parts of its anatomy: the "scenario," the valuation question, and other survey questions. The scenario tells potential respondents what they are being asked to value and under what circumstances. When the topic is a proposed regulatory change, the scenario provides respondents with a description of the proposal and how they will be affected by it. Verbal descriptions of the proposal and its potential effects, possibly aided by photographs, maps, diagrams, and other visual aids, are central to any CV scenario. Respondents usually also like to know something about the circumstances under which their values will be expressed. Thus, another major role of the scenario is to convey information to respondents about the so-called "context of valuation". For example, valuation exercises are often framed as referenda. Respondents are asked how they would vote on a proposal if passage of the proposal resulted in the changes described in the scenario at some specified cost to them. Or, as another example, suppose that the proposed regulatory actions affect amenities at a recreational site. Then, the context of valuation might involve asking the respondents to decide whether or not they would visit a site with improved or degraded amenities if their costs for such a visit were higher or lower.

The scenario serves as the foundation for the valuation question itself. For example, if the context of valuation involves a referendum, then the question asks how respondents would vote if given the opportunity to do so. Various question formats can be used. Assuming that the question is posed as a referendum, respondents could be asked to vote yes or no given that the regulatory change will cost them a specific amount specified in the valuation question, much as they vote for or against a proposal in a real referendum. Or, another commonly used format would ask respondents simply to write in a blank the maximum increase in their taxes they would accept and still vote positively. So-called payment cards are sometimes used, where a range of amounts appears in the survey and respondents are asked to mark the maximum amount they would be willing to pay for positive environmental changes or to avoid environmental changes that would adversely affect them.

CV surveys invariably include questions in addition to the valuation question. Such questions may provide additional information to decision makers by providing data on respondents' characteristics and their preferences and opinions about the proposed change in regulations. They may also provide data to be used in evaluating the validity of the study in ways described below.

Designing a successful CV survey involves several steps. To begin with, the proposed changes in regulations and their potential environmental effects need to be fully understood by the investigator. This normally involves collaboration between economists, decision makers, and environmental scientists. As the scenario begins to take shape, initial contacts with the sorts of people who will eventually complete the survey are often made. This may take place in focus groups, where a moderator engages a group of eight to ten people in discussions that help to investigate what sorts of language eventual respondents use, what attributes of the environment are important to them, and how they react to possible information that might eventually appear in the scenario. Once the survey instrument is in semifinal form, individual interviews and debriefings may supplement focus groups. Next, surveys are often pretested on small samples. Pilot surveys on larger, more representative samples may be used to finalize the instrument.

Administration of a CV survey to be used in actual public decision making typically involves choosing a representative sample from the relevant group of citizens. Once the sample is chosen, survey administration procedures are implemented that are designed to obtain a response rate sufficient to support extrapolation of the sample results to the population. Analysis and reporting of results complete the process.

## An Application

The Glen Canyon Dam is located on the Colorado River just upstream from the Grand Canyon. It stands more than 600 feet above the original river bed and backs water up for 200 miles from northern Arizona into southern Utah to form Lake Powell. It is part of a system of Colorado River dams and reservoirs designed to store water and provide electric power and outdoor recreation. The hydroelectric power plant at Glen Canyon Dam has the capacity to generate more than 1,300 megawatts of electricity, the equivalent of a modest nuclear plant.

As Glen Canyon Dam was completed in the early 1960s and went into regular operations, it changed the downstream environment of the Colorado River. Floods that once raged through the Grand Canyon in spring as snows melted in the distant Rocky Mountains were eliminated or much moderated. Huge loads of sediment that once passed through the Grand Canyon were trapped in Lake Powell. As the reservoir filled, water released from the dam came from deeper and deeper within Lake Powell and hence was much colder, particularly in the summer months, than it had been previously.

Operations of the power plant itself also affected the riverine and riparian environments. Hydroelectric power is particularly valuable because output can be more inexpensively and quickly changed in response to changes in electricity demand than is the case at power plants using fossil fuels and nuclear energy. Thus, there are strong economic incentives to use a facility like the Glen Canyon Power Plant to generate power "on peak," i.e., when daily power demand is at higher levels. However, on-peak hydroelectric generation at Glen Canyon Dam caused dramatic fluctuations in downstream water levels as power generation increased and decreased to meet power demands over the course of the typical day.

Concerns about possible adverse environmental impacts of dam operations on Grand Canyon resources led, in the early 1980s, to a large, decade-long interdisciplinary series of research projects known as the Glen Canyon Environmental Studies. Eventually, an Environmental Impact Statement (EIS) on Glen Canyon Dam operations was prepared.

The EIS identified several adverse impacts of previous operations of Glen Canyon Dam:

- Native fishes were being adversely affected by the dam. Of the eight native fish species of this part of the Colorado River, only three remain relatively common and two, the humpback chub and razorback sucker, are officially listed as endangered. A third species, the flannelmouth sucker, is a candidate for listing as endangered.

- Riverbed sand and hundreds of acres of sandbars or "beaches" along the river were affected by the dam operations. Sediments now stored in Lake Powell are no longer available to replenish sandbars. Under patterns of dam operations prior to preparation of the EIS, many sandbars were eroding. Daily fluctuations in water levels resulting from on-peak power generation were exacerbating this problem.

- Loss of sandbars was deemed an environmental concern because they support vegetation and backwaters that provide habitat for wildlife and fish. Erosion of sandbars was reducing these habitats.

- Sandbars and other riparian areas contained archeological sites and Native American traditional cultural properties, several of which were being lost to erosion.

- The dam created conditions that allowed introduction of nonnative trout species and establishment of a sport fishery. However, daily fluctuations in water releases reduced food supplies and spawning success. Under historical dam operating procedures, the fishery remained heavily dependent on stocking of fish from hatcheries.

- Adverse effects on the enjoyment and safety of white-water boating in the Grand Canyon as well as fishing downstream from the dam were identified.

The Glen Canyon Dam Environmental Impact Statement (GCDEIS) proposed a number of dam operation alternatives that would ameliorate the environmental impacts of dam operation. Each alternative involved restrictions in the magnitude of daily fluctuations in water releases from the dam. The non-use values of three of these alternatives were estimated using the contingent valuation method. The study was carried out for two distinct populations: individuals living in areas served by power produced at Glen Canyon Dam (the marketing area) and for the nation as a whole (the national sample). The results are summarized in Tables 1 and 2.

**Table 1**
**Summary of Population Weighted Average Willingness-to-Pay, Per Household, Per Year**

| Water Release Alternative | National | Marketing Area |
| --- | --- | --- |
| Moderate fluctuating flow | $13.56 | $22.06 |
| Low fluctuating flow | $20.15 | $21.45 |
| Seasonally adjusted steady flow | $20.55 | $28.87 |

**Table 2**
**Summary of Population Weighted Average Willingness-to-Pay, Aggregated Across Households, Per Year (Millions of Dollars)**

| Water Release Alternative | National | Marketing Area |
|---|---|---|
| Moderate fluctuating flow | $2,286.4 | $62.2 |
| Low fluctuating flow | $3,375.2 | $60.5 |
| Seasonally adjusted steady flow | $3,442.2 | $81.4 |

Changes made in operations to benefit the downstream environment and the quality of recreation will also reduce the value of power produced at the dam. This conflict over dam operations can be partially evaluated by measuring the relative economic value placed on electric power, and preservation of river-related resources downstream from Glen Canyon Dam. Values for electric power were estimated using a power system model that predicted the cost of supplying power over a large area of the West depending on constraints on Glen Canyon Dam operations. Like the non-use values, the power values were measured relative to baseline conditions, which were defined as operations under the criteria that applied prior to the beginning of the EIS process. Negative numbers in the power value column indicate losses relative to baseline, while the positive numbers in the non-use value column indicate gains. The range presented for the power values represents a range of assumptions that were used in the modeling of the power system. The comparisons between the lost power values and gains in non-use values are shown in Table 3.

**Table 3**
**Annual Power Values and Market Area Non-Use Values (Millions of Dollars)**

| | Power Values | Marketing Area Non-Use Values | National Non-Use Values |
|---|---|---|---|
| Moderate fluctuating flow | -$36.7 to -$54.0 | +$62.2 | +$2,286.4 |
| Low fluctuating flow | -$15.1 to -$44.2 | +$60.5 | +$3,375.2 |
| Seasonally adjusted steady flow | -$88.3 to -$123.5 | +$81.4 | +$3,442.2 |

## D-CORMIX: A Decision Support System for Hydrodynamic Mixing Zone Analysis of Continuous Dredge Disposal Sediment Plumes

Robert L. Doneker[1], Member ASCE, and Gerhard H. Jirka[2], Fellow ASCE

### Abstract

D-CORMIX is a near- and far-field mixing zone model for negatively buoyant discharges with suspended sediment loads. It contains a classification scheme that determines the important flow regimes in a given situation, and applies the correct jet-integral or length-scale model appropriate to each regime. Included is the ability to predict surface attached jets (both Coanda and wake attachment), fully mixed jets into shallow water, surface fall down (i.e. the plunge point), internal trapping, density current behavior on a flat bottom, inclined bottom, or trapped density level. It simulates submerged, surface, and above surface discharge configurations. Sedimentation is modeled after bottom contact using Stokes settling, with five particle size class sizes (large chunky solids, sand, coarse silt, fine silt, and clay). The ambient water body may have one or two zones (near-shore and off-shore), where different uniform ambient velocity, bottom slope, and Darcy friction factor may be specified. Ambient density stratification is specified with up to 3 stable layers. Output includes plume shape, trajectory, total solids and tracer concentration, dilution, particle size concentration, and sediment accretion rates. Initial validation suggests overall agreement with field and laboratory data. Recent developments include post-processor linkages for assessment of deposited sediment quality.

### Methodology Description

D-CORMIX is a near- and far-field mixing zone model for negatively buoyant discharges with suspended sediment loads resulting from continuous dredge disposal operations (Doneker and Jirka, 1997). The model uses the same expert-system user-interface as CORMIX (Jirka, et al, 1996), and the classification scheme and computational modules are similar to those used in CORMIX1 (Doneker and Jirka, 1990) and CORMIX3 (Jones et al, 1997) with a number of modifications and

---

[1] Assistant Professor, Department of Environmental Science and Engineering, Oregon Graduate Institute, PO Box 91000, Portland, OR 97291-1000.

[2] Director, Institute for Hydromechanics, University of Karlsruhe, Karlsruhe, D-76131, Germany.

additions. In general, D-CORMIX uses a classification scheme that has been inverted from the CORMIX flow classifications. Whereas CORMIX1 is primarily concerned with buoyant discharges near the bottom, the model assumes a negatively buoyant discharge near the water surface. D-CORMIX allows for sedimentation to occur in flows contacting the bottom. Simulation modules are modified to represent negatively buoyant density currents with sedimentation on a sloping bottom in a crossflow.

The strength of the model is its classification scheme that determines the flow regimes that are important in a given situation, and applies the correct jet-integral or length-scale model appropriate to each regime. Central to the methodology is the ability to predict dynamic boundary interaction of jets and plumes. Boundary interactions include near-field attachments, fully mixed jets in shallow water, surface fall down, internal trapping, and three-dimensional density current trajectory on either a flat bottom, inclined bottom, or trapped density level. A summary of ambient and discharge assumptions appears in Figures 1 and 2.

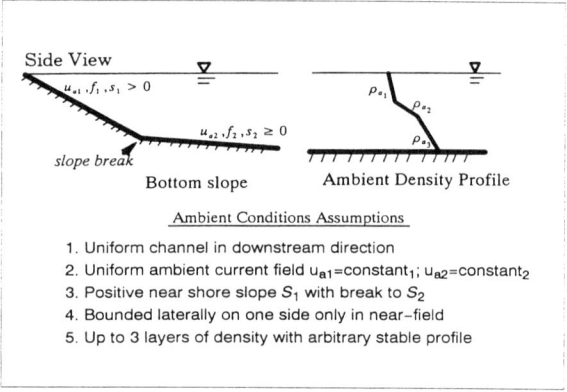

Figure 1: D-CORMIX ambient conditions assumptions.

Several different discharge configurations are considered, including submerged, surface, and above surface discharges. Above surface discharges can either be sprayed i) over a large area, ii) in a single jet, or iii) onto a deflector plate. Below surface jet discharges are restricted to the upper half of the water column. Shoreline discharges are issued from a pipe or canal at the surface of the ambient water body. Particle settling is accounted for after bottom contact, and is modeled using Stokes settling with five particle class sizes (large chunky solids, sand, coarse silt, fine silt, and clay). Hindered Stokes settling and a conservative or non-conservative ($1^{st}$ order decay or growth) tracer pollutant may be specified. Furthermore, the model can be run without any sediment discharge.

Figure 1 shows the ambient water body may have either one or two zones (nearshore and offshore), where different ambient velocity $u_a$, bottom slope S and Darcy

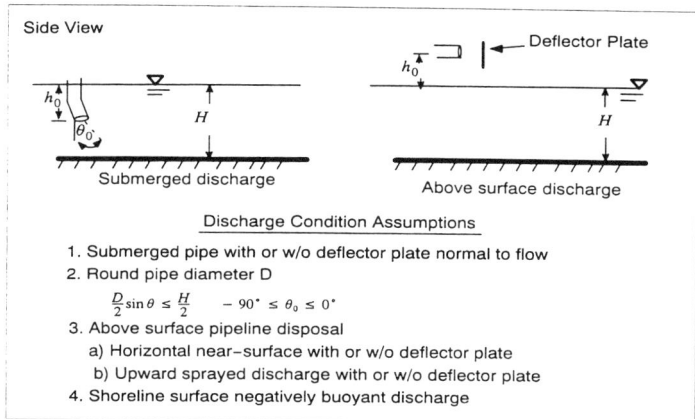

Figure 2: D-CORMIX discharge assumptions

friction factors $f$ may be entered. The offshore slope $S_2$ may be flat or inclined, while the near-slope $S_1$ must have some inclination. Ambient stratification is described by a surface density value, and up to three density values (other than the surface) which are known at different submergence levels below the water surface.

## Comparison with Laboratory Data

An exhaustive search of literature provides a limited set of data for model validation (Alavian 1986, Christodoulou 1994, Hauenstein, 1983, Luthi, 1981). Most experiments use saline solutions, which permit validation of general density current trajectory and dilution, but prohibit analysis of sedimentation effects. Several studies use low velocity discharges (low Reynolds numbers) which do not represent typical dredge discharge scenarios. Low velocity discharges produce laminar flows dominated by viscous forces. No experiments include an ambient crossflow, and confined laboratory tanks often prevent fully steady-state conditions.

Overall, model predictions give both very good and weak agreement with experimental results. However, this data furnishes the basis for an initial validation of the fundamental physical processes modeled by D-CORMIX. In particular, these data verify two important fundamental physical processes simulated: 1) density currents with sedimentation, and 2) the behavior of a density current along a sloping bottom.

Figure 3 shows effective dilution vs. distance along plume centerline for an experiment by Hauenstein (1983). Temperature decay is directly analogous to plume dilution. Although plume width and depth predictions are under-predicted for this case, model predictions of effective dilution show strong agreement with data.

Plume sediment flux values calculated from measured sediment deposition data from Luthi (1981) is compared to model predictions in Figure 4. Although plume predictions of width and depth are in good agreement with experimental results, only modest agreement is indicated for plume suspended sediment predictions.

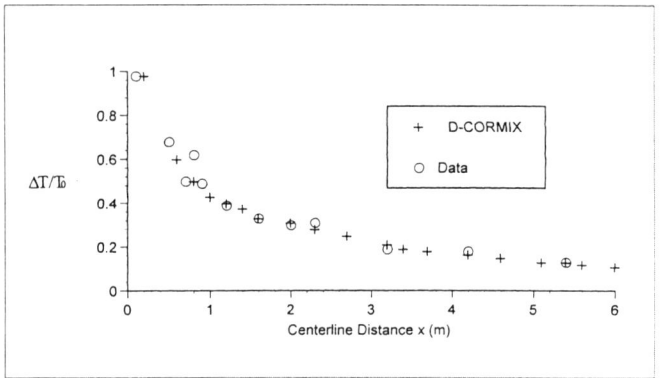

Figure 3: Temperature deficit decay (dilution) along centerline for Hauenstein (1983) experiment U03 (Fr = 4.7, $S_0$ = 2.53°, $u_a$ = 0).

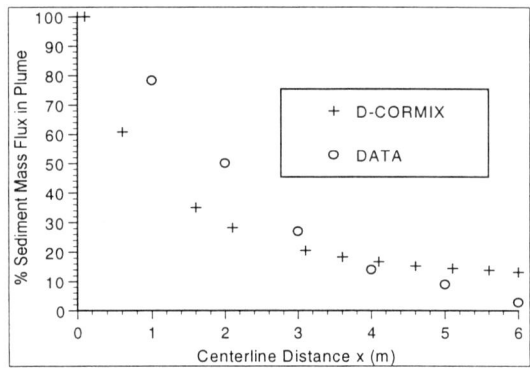

Figure 4: Luthi (1981) calculated plume sediment mass flux determined from deposited sediment vs. D-CORMIX predictions (Fr = 0.8, $S_0$ = 5.2°, $u_a$ = 0). Lateral tank boundary interaction prevents steady-state plume conditions after $x \cong 4$ m).

## Comparison with Field Data

In addition to the laboratory data, an exhaustive search was conducted to obtain field data from continuous dredge discharges. Presently, no reasonably complete set of field data is available for full methodology validation. Most field studies were not intended for hydrodynamic model validation, so critical parameters are often missing. In addition, it is difficult to assess natural discharge, tidal, and spatial variations in parameters that can occur with field studies in coastal areas.

The most complete field data set available is the report on the pipeline dredge *Louisiana* in Mobile Bay, Alabama (Clarke and Miller-Way, 1992). The weakness of the data set for validation is poor discharge source and ambient site characterization. The discharge port orientation, location, and local bathymetry are not reported. The vertical sampling is also limited, and at times did not track the sediment plume.

For D-CORMIX simulations, the reported discharge of 2.039-$m^3$/s containing 130 g/L of suspended sediment discharging downward through an angled pipe section 30° from horizontal 1-m below surface in 3-4 m of water was assumed to be oriented in the direction of ambient flow. The study did not report on discharge particle size distribution, however data (Nichols, et al 1979) indicates it can be characterized as 10% coarse silt, 20% fine silt, and 70% clay.

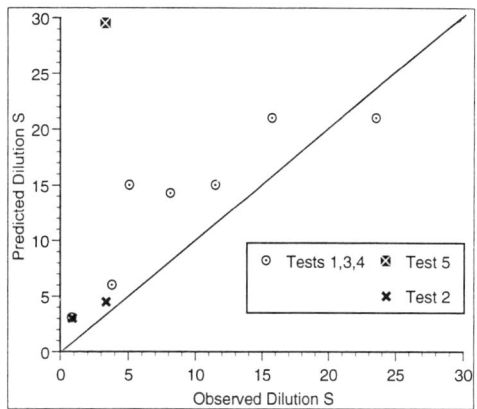

Figure 5: Alabama average plume suspended sediment dilutions and D-CORMIX predictions

The data clearly indicate formation of bottom density current formation through reported suspended sediment values. Figure 5 shows that overall, D-CORMIX appears to give reasonable, if slight under-prediction, of suspended sediment concentration in plumes when compared to the data.

**Conclusions/Recommendations**

D-CORMIX appears to capture the essential physical processes of dredge sediment plume behavior. Compared to limited laboratory data, the model presents both very good and weak agreement with data. Most laboratory data is difficult to evaluate because of limited physical scales and poor approximation of dynamic parameters typical of dredging operations. Overall the model appears to give good predictions of sediment plume behavior and dilution in field environments, slightly under-predicting suspended sediment concentrations, especially near the discharge source.

Researchers have reported hindered settling in plumes with concentrations greater than 10-g/L (Henry et al, 1978). Since the overall results of the laboratory and field data suggest that D-CORMIX may under-predict suspended sediment concentrations, especially in locations near the discharge source, simulation of hindered settling should be evaluated.

The complete validation of D-CORMIX requires additional laboratory and field experiments. A better understanding of the physical processes in sedimentation plumes is essential. Laboratory experiments are preferable to field data collection because they allow for control of the physical processes. Additional information about D-CORMIX can be found at http://ese.ogi.edu/ese_docs/doneker.

## Acknowledgements

This research is supported through grants from the Office of Water, Office of Science and Technology, U.S. Environmental Protection Agency.

## References

Akar, P. J. and G. H. Jirka (1991), "CORMIX2: An Expert System for Hydrodynamic Mixing Zone Analysis of Conventional and Toxic Discharges", USEPA, EPA/600/3-91/073.

Alavian, V. (1986), "Behavior of Density Currents On An Incline", J. of Hyd. Eng., ASCE, Vol. 112, No. 1. Jan., pp. 27-42.

Christodoulou, G. and F. Tzachou (1994) "Experiments on 3-D Turbulent Density Currents", 4th Annual. International. Symposium. on Stratified. Flows, V. 3., Jun. 29-Jul. 2, Grenoble, France.

Clarke, D. and T Miller-Way, (1992), "An Envir. Assess. of the Effects of Open Water Disp. of Main. Dredge Mat. on Benthic Res. in Mobile Bay, Al.", USCOE, WES, Vicks. MS, Misc. Pap. D-92-1.

Doneker, R.L. and G.H. Jirka (1997), "D-CORMIX Continuous Dredge Disposal Mixing Zone Water Quality Model Laboratory and Field Data Validation Study", Tech. Rep. DMX-1-97, Oregon Graduate Institute, PO Box 91000, Portland, OR 97291-1000.

Doneker, R. L. and G. H. Jirka (1990), "Expert System for Mixing Zone Analysis of Conventional and Toxic Single Port Discharges: CORMIX1", USEPA, EPA/600/3-90/012.

Jones, G.R., Nash, J.D., and Gerhard H. Jirka, (1996), "Cormix3: An Expert System For Mixing Zone Analysis And Prediction Of Buoyant Surface Discharges,". To be pub. by USEPA, Office of Water, 1997. Available at: http://ese.ogi.edu/doneker/cmxpg1.html

Hauenstein, W. and T. Dracos (1983) "Investigation of Plunging Density Currents Generated by Inflows in Lakes", Jour. of Hydraulic Res., Vol. 22., No. 3, pp. 157-179.

Luthi, Stefan, (1981) "Experiments on Non-Channelized Turbidity Currents and Their Deposits," Marine Geology, Vol. 40 M59-68.

Nichols, M. M., and G. S. Thompson, (1978) "A Field Study of Fluid Mud Dredged Material: Its Physical Nature and Disposal", USCOE, WES, Vicks. MS, D-78-40.

Thevenot, M. M., Prickett, T. L., and N. C. Kraus (1992), "Tylers Beach, Virginia, Dredged Material Plume Monitoring Project 27 September to 4 October 1991", USCOE, WES, Vicks. MS, DRP-92-7.

# Criteria Determination Using a Probabilistic Approach

James W. Male[1]

## Abstract

A procedure is presented that utilizes a probabilistic approach (Monte Carlo analysis) to assist in the determination of criteria for environmental contaminants. The risk-based approach allows incorporation of distributions for input characteristics, eliminating the need to determine average or conservative (protective) values for these parameters. The result is a distribution of contaminant criterion values from which decision makers can select an appropriate value based on a desired level of reliability. The approach is illustrated with the potential for risk to human health from the ingestion of seafood taken from the waters overlying contaminated sediment. Results show that the approach provides an attractive means to assist regulators in determining criterion values.

## Background

Regulatory agencies are faced with the problem of determining criteria for a variety of pollutants, many of which are toxic. There is considerable uncertainty among the scientific community and regulators about the potency of these contaminants. In addition, there is also uncertainty and variability in other parameters that determine the extent of human exposure to toxic chemicals. Yet for regulatory purposes, a single value is often needed upon which to base decisions.

**Application.** The State of Washington has taken action to prevent future contamination of sediments and to clean up sites where pollution is unacceptably high. The state's Department of Ecology recently adopted the Sediment Management Standards addressing sediment contamination and potential adverse impacts on the environment (aquatic and other organisms) and on human beings.

---

[1] Edwin and Sharon Sweo Professor of Engineering, University of Portland, Portland OR 97203, formerly Professor, Department of Civil and Environmental Engineering, University of Massachusetts.

**Intent of Paper.** This paper illustrates the use of a probabilistic approach (Monte Carlo analysis) in determining specific criterion values. The procedure incorporates the variability and uncertainty associated with various input parameters, allowing decision-makers the opportunity to place a reliability on the result.

### Risk Assessment

Risk is one measure of the threat to humans from exposure to toxic chemicals. It involves the extent of the chemical's toxicity (potency) and the degree of exposure.

**Exposure.** Exposure is based on hypothesized pathways that a chemical can take from a source of pollution to a human being. It is usually determined on a daily basis averaged over an extended period of time and normalized by human body weight (mg/Kg/day). This paper is based on exposure to chemicals in sediment via seafood consumption. Determination of exposure via seafood consumption includes: (1) the sediment concentration, SC, (2) seafood ingestion rate in grams of fish tissue per day, IR, (3) exposure duration - the length of time over which a person fishes in years, ED, (4) the body weight of a human in Kg, BW, (5) the averaging time - the time over which the chemical may act on the body in years, AT, and (6) the concentration in fish tissue, which is based on the biota sediment accumulation factor, BSAF and the fish lipid content, FL.

**Toxicity.** For carcinogens, EPA has determined Cancer Potency Factors (CPF) that represent their potency. Units are the inverse of mg of chemical per Kg of body weight per day of exposure $((mg/Kg/day)^{-1}$

**Determination of Hazard.** The hazard posed by a carcinogen is usually expressed as risk; more specifically, the incremental increase in the risk of an adverse outcome resulting from exposure to a carcinogen. For a carcinogen, risk is calculated by multiplying the potency (CPF) by the estimated exposure:

$$R = (CPF) x (Exposure) = \frac{(CPF)(SC)(BSAF)(FL)(IR)(ED)}{(BW)(AT)} \quad (1)$$

A commonly used acceptable risk level is one in one million ($10^{-6}$). EPA uses a range of acceptable risk from $10^{-6}$ to $10^{-4}$ for Superfund site evaluation

### Regulatory Criteria

Regulators must often determine acceptable concentrations for carcinogens, frequently based on a risk of one-in-a-million.

**Calculation of Sediment Criteria.** To determine an acceptable sediment contamination, Equation 1 is rearranged to allow calculation of SC:

$$SC = \frac{(R)(BW)(AT)}{(CPF)(BSAF)(FL)(IR)(ED)} \tag{2}$$

The result of this equation is a level of chemical concentration in sediment above which the risk of adverse effect would be greater than the predetermined acceptable value of R.

**Probabilistic Approach.** In the determination of risk, the usual approach is to use estimates of extreme values for one or two input parameters and estimates of the means or medians for the others. Many ingestion studies use high estimates for IR and CPF. The combination of the means and extreme values in the calculation of R provides a result that is conservative and therefore protective. A probabilistic approach, such as Monte Carlo analysis, uses distributions for most or all of the input parameters. Use of a probabilistic approach allows the variability and uncertainty associated with some or all of the input variables to be included in the calculation. The result is a distribution of SC representing the uncertainty and variability associated with determination of the SC criterion, based on the uncertainty/variability of the input parameters. From this distribution decision makers could select a criterion value for SC based on their overall desired reliability.

**Interpretation.** For a carcinogen, the results are most easily stated in the following way; With x percent reliability, a sediment concentration of SC will result in a risk, R, on average, of an adverse effect to an individual of the targeted population. Or alternatively, with x percent confidence, the risk to the target population of an adverse effect for a concentration of SC is less than R.

**Application.**

An illustrative example is presented for Puget Sound, assuming the only pathway is via seafood consumption.

**Input Distributions.** Distributions for all of the input parameters were determined based on available data and reasonable assumptions. They are summarized below and details can be found in Male (1994). EPA's Exposure Factors Handbook (1989) provides guidance on use of distributions in Monte Carlo analyses.

**Analysis.** Monte Carlo analyses, with Latin Hypercube sampling, were performed using a commercial software package, @RISK (Palisade 1990). From the results, sediment concentration values were determined corresponding to values exceeded 1, 2, 5, 10, 20, and 50 percent of the time. In all, 50 chemicals were evaluated, however, illustrative results are shown for one chemical, hexachlorobenzene. Results of three alternatives were compared: (A) Distributions were used for all inputs with the exception of BSAF and R, (B) Distributions were used for all inputs with the exception of BSAF, R and CPF, and (C) All inputs were point values. Alternative C was included to provide a comparison to the conventional case where all input parameters are determined and one value results for the sediment concentration. Table 1 summarizes the distributions and point values used for all input parameters.

**Table 1. Summary of Distributions and Point Values Used for Example Calculation (hexachlorobenzene)**

| Parameter | Input Characteristics Distribution | Point Value | Units |
|---|---|---|---|
| R | Point: $10^{-6}$ | $10^{-6}$ | - |
| BW | Empirical | 70 | Kg |
| AT | Empirical | 75 | years |
| CPF | Equation 6 | 1.6 (IRIS) | $(mg/Kg/day)^{-1}$ |
| BSAF | Point: 2.0 | 2.0 | g. of sed.(o.c.) per g. fish lipid |
| FL | Triangular (0.003, 0.01, 0.04) | 0.04 | - |
| IR | Empirical | 26.1 | grams/day |
| ED | Uniform (5,75) | 30 | years |

**Results.** Statistics of the resulting distributions for SC are shown in Table 2 for Alternatives A and B. Alternative C resulted in an SC value of approximately 0.05 ug per gram of sediment (o.c. normalized)

**Table 2. Statistical Results for Carcinogenic Action of Hexachlorobenzene**

| | Alternatives A | B |
|---|---|---|
| Mean | 5.09044 | 0.44263 |
| Standard Deviation | 21.0680 | 1.43182 |
| Skewness | 10.7943 | 8.92931 |
| Percentile Values | | |
| 1 | 0.00209 | 0.00157 |
| 2 | 0.00489 | 0.00203 |
| 5 | 0.01740 | 0.00440 |
| 10 | 0.03556 | 0.01000 |
| 20 | 0.09304 | 0.01632 |
| 50 | 0.45803 | 0.08066 |

Note: Alternatives correspond to different inputs:
A - $R=10^{-6}$, BSAF = 2.0, all others are distributions,
B - $R= 10^{-6}$, BSAF = 2.0, CPF = 1.6 (IRIS value), and all others are distributions.

**Discussion.** Interpreting the results for Alternative A; If decision makers were to select a desired reliability of 95 percent, the sediment concentration should not exceed the five percentile value, approximately 0.017 ug per gram of sediment (o.c. normalized

Several interesting aspects can be gleaned from the results shown in Table 2. For both alternatives the median values are much less that the means. This fact is also represented by the positive values for skewness, meaning the distribution has a very long tail for high concentration values. In addition, values in the low-percentile range (five and below) are less than one percent of the mean values. There is a striking difference among the distributions for the two alternatives. Mean and percentile values for Alternative B are much smaller than Alternative A. These results are not unexpected. Alternative B had two inputs (R and CPF) that were point values corresponding to values in the conservative tails of their respective distributions while Alternative A had one (R).

It is interesting to compare Alternatives A and C. Alternative C used point values for both R and CPF are at the conservative ends of the tails of the distributions used in Alternative A. The sediment concentration resulting from the Alternative C calculation (0.05 ug/g) is the same as a value corresponding to one between the 95 and 98 percent reliable values from Alternative A. These results imply that the conventional approach provides a fairly reliable result.

**Summary**

The approach eliminates the need to specify point values of input parameters and allows the incorporation of distributions representing the best knowledge of the parameters for the application at hand. The approach also eliminates the need to specify one or more conservative input parameter values to allow for a protective result. Rather, decision makers can choose from the resulting distribution of contaminant concentrations associated with a desired level of reliability. The procedure has several advantages for use in the realm of regulatory criteria determination; it removes the need to answer some of the policy questions (e.g., specification of a point value for body weight), it allows for a more visible display of the input and output characteristics, it provides a better link between the final criterion value and its reliability, and the computational tools are readily available.

**The Larger Context.** The Washington State Department of Ecology is exploring a two-tier approach to determining sediment criteria for protection of human health. Tier 1 is a "generic" approach utilizing criteria developed for most situations. As such the criteria should be conservative, or err on the side of being protective of human health. Tier 2 would be more site specific and allow the particular situation to influence the criteria, as long as an established procedure was followed. For use in Tier 1, decision makers would need to be sure that the reliability level selected from the resulting distribution of SC assures that conservative level of protection. The important consideration is that the approach is not meant to determine criteria for all situations. It is meant to be a procedure that is capable of assisting decision makers to select criterion values based on a depiction of the likely distribution of sediment criteria.

## Acknowledgments

The author would like to acknowledge the contributions of the following people from the Washington State Department of Ecology: Keith Phillips, Rachel Friedman-Thomas and Laura Weiss; from the Washington State Department of Health: Joan Hardy, Denise LaFlamme, David McBride, and Glen Patrick.

## References

Male, J.W. (1994) "Development of Human Health Sediment Criteria Using a Distributional Analysis," report to Washington State Department of Ecology, 18 January 1994.

Palisade Corporation. (1990) "@RISK - Risk Analysis and Simulation Add-In for Microsoft Excel: User's Guide," 1990.

U.S. EPA (1989) "Exposure Factors Handbook," EPA/600/8-89/043, July 1989.

# Flood Estimation by Combining Gauged and Paleo Data

C. Peng[1], L. Duckstein[2], D. R. Davis[3], V. R. Baker[4]

Abstract

In this paper, the Salt River basin possessing paleoflood study has been selected for flood estimations. By plotting the flood data on log-normal probability paper, it is found that the trend of high floods, whose peaks are greater than 2,400 m$^3$/s, deviates from the straight line fitted to the gauged flood data and the trend of these high floods is similar to that of the paleofloods. Thus, it is reasonable to assume that these high floods belong to different probability distributions than that of the rest. The distributions for the high values of gauged flood data and the paleoflood data are assessed by linear regression using their plotting positions on probability paper. Then, a Bayesian statistical approach is applied to combine the gauged flood data and paleoflood data. The floods estimated by the combined distribution are much lower than those obtained by the conventional methods. For example, the 100 year flood estimated by the combined distribution is about 60% of that obtained by a Pearson type III distribution.

Introduction

Flood records are usually short so the high flood predictions require extrapolation beyond the observed records. Such extrapolations are very much affected by model and parameter uncertainties. Some researchers have applied a Bayesian statistical or decision theoretic approach to account for model and parameter uncertainties in flood frequency analysis (Bernier 1967; Davis et al. 1972; Wood & Rodriguez-Iturbe 1975; Vicens et al. 1975). Another way to

---

[1] Department of Hydrology and Water Resources, University of Arizona, Tucson, AZ 85721, USA
[2] Department of Syatems and Industrial Engineering, University of Arizona, Tucson, AZ 85721, USA
[3] Department of Hydrology and Water Resources, University of Arizona, Tucson, AZ 85721, USA
[4] Department of Hydrology and Water Resources, University of Arizona, Tucson, AZ 85721, USA

improve flood estimation is to combine historical flood data and paleoflood information whenever it is available ( Hosking & Wallis, 1986; Stedinger & Cohn 1986). In the past decade, paleoflood studies have been carried out in the lower Colorado River basin in the United States (Enzel et al. 1996). The paleoflood data provide valuable information about major floods that occurred before gages were established. Accordingly, the research in this paper seeks how to employ the available paleoflood information for improving flood estimations. The Salt River basin in central Arizona has been selected as a case study. First, the conventional flood frequency analysis is used to estimate floods. Then, a Bayesian point of view is used to combine the gauged flood data and paleoflood information, leading to the selection of a different probability distribution for high floods. Finally, the flood estimates obtained by conventional flood frequency analysis are compared to those obtained by combining gauged flood data and paleoflood data.

The Study Area and Available Data

The selected flow station locates on the Salt River near the Roosevelt reservoir with a drainage area 11,150 km$^2$ (Station No. 09498500), which is selected to match the study area in the paleflood studies (Partridge and Baker, 1987). Table 1 shows the sample statistics of the annual peak flows, from 1925 to 1995.

Table 1    Sample statistics of annual peak flow, 1925-1995

|  | Annual Peak Flow (m$^3$/s) | log$_{10}$(Annual Peak Flow) |
|---|---|---|
| Mean | 771.79 | 2.6194 |
| Standard Deviation | 916.26 | 0.4935 |
| Skewness | 1.7860 | 0.1491 |

Tables 2 shows the available paleoflood data in this area (Partridge and Baker, 1987). The estimated age and magnitude of the paleofloods were stated with upper and lower bounds.

Table 2    Available Paleoflood data for the Salt River Basin[*]

| Estimated Flood Age | Calculated Discharge (m$^3$/s) | Return Period (years) |
|---|---|---|
| 1980 A. D. | 2300~3700 | 30 |
| 1950~1958 A. D. | 2500~4500 | 200 |
| 1542 A. D. | 2900~ ** | 300 |
| 1200~1400 A.D. | 3000~5400 | 600 |
| 1000~2000 years B. P. | 3300~6500 | 1000~2000 |

* Prepared from Table II, III, & IV in Partridge & Baker, 1987.
** The Absolute Maximum value for this flood was not assessed.

Conventional Flood Frequency Analysis

To check if a probability distribution fits the flood data, the first step is to plot the data on specially designed probability paper. A linear trend of the data on a given probability paper indicates that the data may be appropriately described by the distribution which corresponds to the probability paper. To plot the flood data, plotting positions are assigned to them. Several plotting positions have been proposed. Cunnance (1978) showed that the Blom plotting position is best for the normal distribution and the Gringorten for extreme value type I.

The log-normal distribution and the Blom plotting positions are adopted to model the annual peak flood in this area because of their performance. The flood estimations for return periods of 50, 100, and 200 years by a log-normal distribution are listed in Table 3. For a comparison, also the flood estimations by a log-Pearson type III distribution. Note that the log-normal distribution is a log-Pearson type III distribution with zero skew.

Table 3 Flood Estimation of selected returned period

| Return Period | T = 50 years | T = 100 years | T = 200 years |
|---|---|---|---|
| Log-Normal | 4,294 cms | 5,853 cms | 7,772 cms |
| Log-Pearson III | 4,698 cms | 6,627 cms | 9,115 cms |

Combination of Gauged and Paleo Flood Data

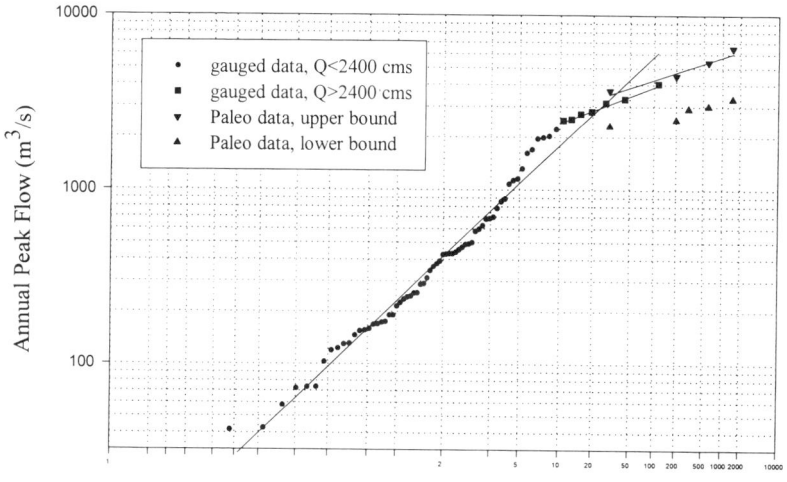

Figure 1 Annual Peak Flow and Paleo Flow with Return Period

In figures 1, the paleo and gauged data are plotted on the same probability paper. It can be seen that the trend of plotting positions for high values of gauged floods, say Q > 2,400 m$^3$/s, deviates from the straight line fitted to the whole population of the gauged data. This trend coincides to that of the paleoflood data. Thus, using a Bayesian viewpoint to embed this information into the likely existence of an upper limit to flood value (Enzel et al. 1996), it is reasonable to assume that these high floods belong to another distribution than that of the floods whose peaks are less than 2,400 m$^3$/s. Empirically, we can get a distribution for high floods by linearly regressing the plotting positions of these high flood data on the probability paper. Because the small size of these samples, the parameter uncertainty of the distribution is high. Bayesian statistical procedures (Bernier 1967; Davis et al. 1972; Wood & Rodriguez-Iturbe 1975) can be used to reduce the parameter uncertainty of the distribution for high flood estimates by combining the gauged flood data and the paleoflood data. The detailed calculations to get a combined high flood distribution are as follows:

1. Assess the likelihood function of the parameter from gauged flood data above the threshold of 2,400 m$^3$/s
   Let y = log(Q), $f(y) \sim N(\mu_y, \sigma_y)$, where $\mu_y$ is unknown but $\sigma_y$ is assumed to be known, for simplicity sake. There are seven points whose magnitudes are greater than 2,400 m$^3$/s. The linear regression for these floods is
   $$y = \alpha + \beta z = 3.1232 + 0.2028\, z \; ; \; r^2 = 0.99008$$
   where z is a standard normal variable, $z = \Phi^{-1}(1-p) = \Phi^{-1}(1-\tfrac{1}{T})$
   The intercept of the line is the mean of the distribution and the slope of the line is the standard deviation, that is,
   $$f(y | y > \log(2400)) = N(3.1232, 0.2028^2)$$
   The unbiased estimate of the variance of errors is given by
   $$s^2 = \frac{1}{7-2}\sum_{i=1}^{7}(y_i - \alpha - \beta z_i)^2 = 6.7056 \times 10^{-5}$$
   The mean of parameter $\mu_y$ may be taken as $\bar{y} = 3.1232$ and the variance of $\mu_y$ is determined by
   $$\sigma^2_{\mu_y} = \frac{s^2 \sum_{i=1}^{7} z_i^2}{7\sum_{i=1}^{7}(z_i - \bar{z})^2} = 0.0159^2$$
   Therefore, the likelihood function of the parameter is
   $$f(\mu_y | y > \log(2400)) = N(3.1232, 0.0159^2)$$

2. Assess the prior distribution from paleoflood data
   For a conservative estimate, only the upper limits of the paleodata are considered to produce the prior distribution of the parameter. With these four points, the linear regression of the paleoflood data is
   $$y = \alpha + \beta z = 3.2382 + 0.1719\, z \; ; \; r^2 = 0.95067$$
   The unbiased estimate of the variance of errors is given by
   $$s^2 = \frac{1}{4-2}\sum_{i=1}^{4}(y_i - \alpha - \beta z_i)^2 = 8.1603 \times 10^{-4}$$
   The mean of parameter $\mu_y$ may be taken as $\bar{y} = 3.2382$ and the variance of $\mu_y$ is determined by
   $$\sigma^2_{\mu_y} = \frac{s^2 \sum_{i=1}^{4} z_i^2}{4\sum_{i=1}^{4}(z_i - \bar{z})^2} = 0.0744^2$$

Therefore, the prior distribution of the parameter is
$\pi_0(\mu_y | paleodata) = N(3.2382, 0.0744^2)$

3. Calculate the posterior distribution of the parameter using Bayes' theorem:
$$\mu = \frac{0.0159^2}{0.0159^2 + 0.0744^2} \times 3.2382 + \frac{0.0744^2}{0.0159^2 + 0.0744^2} \times 3.1232 = 3.1282$$
$$\sigma^2 = \frac{0.0159^2 \times 0.0744^2}{0.0159^2 + 0.0744^2} = 0.0155^2$$
$\pi(\mu_y | y > \log(2400)) = f(\mu_y | y > \log(2400)) \pi_0(\mu_y | paleodata)$
$= N(3.1282, 0.0155^2)$

The mean is very close to that of the likelihood function, the variance is smaller as expected.

4. Calculate the Bayesian, or marginal flood distribution
$$f_Y(y | y > \log(2400)) = \int f(y|\mu_y) \pi(\mu_y | y > \log(2400)) d\mu_y$$
$$= \int N(\mu_y, 0.2028^2) N(3.1282, 0.0155^2) d\mu_y$$
$$= N(3.1282, 0.2028^2 + 0.0155^2)$$
$$= N(3.1282, 0.2034^2)$$

The mean is the same as that of the posterior, the variance reflects the uncertainty of the sample mean as expressed by the prior.

5. Estimate flood magnitude using the Bayesian flood distribution

The Bayesian distribution can now be used to estimate high floods for specific return periods. Table 4 lists the flood estimations for return periods 50, 100, and 200 years.

Table 4 Flood Estimation by Combined Gauged and Paleo Data

| Return Period | T = 50 years | T = 100 years | T = 200 years |
|---|---|---|---|
| Flood ($m^3/s$) | 3,515 | 3,994 | 4,489 |

Discussion and Conclusions

Floods come from precipitation. It is believed that there exists a limiting value of precipitation called the Probable Maximum Precipitation in a region; one would thus expected that there is also an upper boundary to extreme floods. This implies that the one-distribution flood estimation model may not applicable to high flood estimations.

Paleoflood studies give a solid support to the hypothesis of an upper boundary to flood magnitudes. The gauged flood data from the Salt River in central Arizona also show that the high floods have a different property from that of the rest. The trends in these high flood magnitudes coincide to those of the available paleoflood data. Therefore, we have proposed on using two distributions to fit the flood data.

The fitted distribution for high values of gauged flood data can be combined with the fitted distribution for paleoflood data using a Bayesian statistical approach to reduce the uncertainty of the parameters of the distribution and increase the accuracy of flood estimation.

As a result of using two different distributions, there is a substantial decrease in the estimates of large floods with respected to the values obtained using a unique distribution. For example, the estimated 100 year flood by the combined method is about 60% of that determined by a log-Pearson III distribution. For the 200 year flood, 50%.

The proposed distribution for high floods still does not provide an upper limit to the magnitudes of extreme floods but these extreme floods are far beyond our concern. For return periods ranging from 50 to 1000 years, the proposed distribution is supported by both the high values of gauged flood data and of paleoflood data for the study area.

The design return period of floods for major water control structures is usually 50 to 100 years. Given that the marginal cost of a water control structure is increasing with its size, the reduction in design flood can considerably reduce the cost of building a water control structure. It is believed that the experience from the paleoflood studies in the lower Colorado River basin can be extended to other sites.

References

Bernier, J., Les methodes bayesiennes en hydrologie statistique, Proceedings of the International Hydrology Symposium, pp. 459-470, Colorado State University, Fort Collins, 1967.

Cunnane, C., Unbiased plotting positions—a review. Journal of Hydrology, Vol. 37, pp 205-222, 1978.

Davis, R. D., C. C. Kisiel, and L. Duckstein, Bayesian Decision theory applied to design in hydrology, Water Resources Research, Vol. 8, No. 1, pp. 33-40, 1972.

Enzel Y., L. L. Ely, P. K. House, and V. R. Baker, "Magnitude and frequency of Holocene paleofloods in the southwest United States: A review and discuss of implications", Geological Society Special Publication No.115, pp. 121-137, 1996

Hosking J. R. M., and J. R. Wallis, Paleoflood hydrology and flood frequency analysis, Water Resources Research, Vol. 22, No. 4, pp. 543-550, 1986a.

NERC, Flood studies report. National Environment Research Council, London, United Kingdom, 1975.

Partridge, J. and V. R. Baker, Paleoflood Hydrology of the Salt River, Arizona. Earth Surface Process and Landforms, Vol. 12, pp. 109-125, 1987.

Stedinger J. R., and T. A. Cohn, Flood frequency analysis with historical and Paleoflood information, Water Resources Research, Vol.22, No.5, pp. 785-793, 1986.

USWRC, Guidelines for Determining Flood Frequency. Bulletin 17, U. S. Water resources Council, Washington, DC, 1976.

Vincens, G. J., Rodriguez-Iturbe, I. and Schaake, J. C. Jr., A Bayesian framework for the use of regional information in hydrology, Water Resources Research, Vol. 11, No. 3, pp. 405-414, 1975.

Wood, E. F., and Rodriguez-Iturbe , I., A Bayesian approach to analyzing uncertainty among flood frequency models, Water Resources Research, Vol. 11, No. 6, pp. 1043-1056, 1975.

## Pipe Network Optimization
Taejin Ahn[1] and G. V. Loganathan[2]

Abstract

The water distribution system problem consists of finding a minimum cost system design subject to hydraulic and operational constraints. In this paper a paddy irrigation system and a municipal (Hanoi) water distribution system are considered. Because the problems are nonconvex, a global search scheme called Stochastic Probing method is employed to search among the various local minima. The paddy irrigation system is solved for multiple loadings. The optimal solution for the Hanoi network has significantly smaller cost than the ones previously reported.

Introduction

It has been well established that the pipe network design problem has multiple local minima. Comprehensive reviews are given in Loganathan et al(1995) and Savic and Walters(1997). In this paper two types of distribution systems, (i) an irrigation distribution network and (ii) a municipal water distribution system are considered. Mathematical formulation for the paddy irrigation pipe network is similar to that of the municipal water distribution system in that the objective function of the model is to minimize the total pipe costs and pumping costs while satisfying the continuity and minimum head requirements at each node. The differences between them include the way water demand is assessed at a node as well as the pressure head for fire demand, which is not considered in an irrigation system.

---

[1] Assistant Professor, Department of Civil Engineering, Ansung National University #456-749, 67, Seokjeong-Ri, Ansung-EUP, Ansung-KUN, Kyunggi-DO, Korea.
[2] Associate Professor, Department of Civil Engineering, Virginia polytechnic Institute and State University, VA 24061.

## Model Formulation

Consider a pipe network comprised of $N$ nodes. Let $S$ be the set of fixed head nodes. Let $\{N\text{-}S\}$ be the set of junction nodes and $\mathbf{L}$ be the set of links. $Q_{(i,j)}$ is the steady state flow rate through link $(i,j) \in \mathbf{L}$. Let $L_{(i,j)}$ denote the length of link $(i,j) \in \mathbf{L}$ and $D_{(i,j)}$ be its diameter which must be selected from a standard set of discrete diameters $D = \{d_1, d_2, \cdots, d_M\}$. In the present formulation it is assumed that each link $(i,j)$ is made up of $M$ segments of unknown lengths $x_{(i,j)m}$ (decision variable) but of known diameter $d_m$ for $m = 1, 2, \ldots, M$. Let $C_{(i,j)m}$ be the cost per unit length of a pipe of diameter $d_m$. Let $r_k$ be the path from a fixed head node (source) to demand node, $k$. Let $\mathbf{P}$ be the set of paths connecting fixed head nodes and basic loops. Let there be $P_\ell$ loops and $b_p$ denote the head difference between the fixed head nodes for path $p$ connecting them; and it is zero corresponding to loops. Let $H_s$ be the fixed head and $H_k^{\min}$ be the minimum head at node $k \in \{N\text{-}S\}$. Let $q_i$ be the supply at node $i$ which is positive; if it is demand, it is negative. Let $J_{(i,j)m}$ be the hydraulic gradient for segment $m$ (partial length of a link with diameter $d_m$) which is given by

$$J_{(i,j)m} = K[Q_{(i,j)}/C]^{1.85} d_m^{-4.87} \tag{1}$$

in which: $K=8.515(10^5)$ for $Q_{(i,j)}$ in cfs and $d_m$ in inches; $C$ = Hazen-Williams Coefficient. The pipe network problem may be stated as follows: Problem P1:

$$\text{Minimize } f(x) = \sum_{(i,j)} \sum_{m=1}^{M} C_{(i,j)m} x_{(i,j)m} \tag{2}$$

Subject to:

$$\sum_j Q_{(i,j)} - \sum_j Q_{(j,i)} = q_i \quad \text{for } i \in \{N\text{-}S\} \tag{3}$$

$$H_s - H_k^{\min} - \sum_{(i,j) \in r_k} \pm \sum_m J_{(i,j)m} x_{(i,j)m} \geq 0 \quad \text{for } s \in S \text{ and } k \in \{N\text{-}S\} \tag{4}$$

$$\sum_{(i,j) \in p} \pm \sum_m J_{(i,j)m} x_{(i,j)m} = b_p \quad \text{for } p \in P \tag{5}$$

$$\sum_m x_{(i,j)m} = L_{(i,j)} \qquad \text{for } (i,j) \in L \qquad (6)$$

$$x_{(i,j)} \geq 0 \qquad (7)$$

In Problem (P1) pipe cost objective function (2) is minimized; constraint (3) represents steady state flow continuity; constraint (4) is the minimum head restriction; constraint (5) represents the sum of head losses in a path which is zero for loops; constraint (6) dictates that sum of segment lengths must equal link length; constraint (7) is the non-negativity on segment lengths. The decision variables are: $Q_{(i,j)}$, $J_{(i,j)m}$, and $x_{(i,j)m}$. Problem (P1) is nonlinear, nonconvex programming problem which has several local minima. The following two stage strategy Problem (P2) is suggested for the solution of Problem (P1).

Problem (P2): $\operatorname*{Min}_{Q_{(i,j)}} \left[ \operatorname*{Min}_{x \in X} f(x) \right]$ (8)

in which $Q_{(i,j)}$ are the perturbed flows of an underlying near optimal spanning tree of the looped layout satisfying constraint (3), $X$ is the feasible region made up of constraints (4)-(7), and $f(x)$ is the objective function of Problem (P1) for a fixed link flows. It is observed that the inner Problem (P3) of Problem (P2) given by

Problem (P3): $\operatorname*{Min}_{x \in X} f(x)$, for fixed flows (9)

is a linear program which can be solved efficiently by using commercially available codes. This phase represents the local minimizer.

The stochastic probing method is used to search the location of the global loop flow vector $\varepsilon^*$. The method begins with the construction of a probing probability distribution, with the density function $P \sim N(\varepsilon, \sigma)$, where $\varepsilon$ is the location parameter (loop flow vector) and $\sigma$ is a scaling parameter, respectively. The costs of $f(\varepsilon)$ are evaluated at a few loop flows, $\varepsilon$ sampled from the density function. The updating location of loop flow $\varepsilon$ and scale $\sigma$ are based on Gibbs-like distribution and the entropy of the current distribution, respectively. If there is an improvement in the objective value, the updated location is selected as the mode from the sampled points. If there is no improvement in the objective value, the sampling reinitiated.

Analysis of Example Networks

The first application is the optimal design of a paddy irrigation pipe network. The study area Haenam estuary basin is located in

southwestern part of the Korean peninsula. The Haenam agricultural development project has recently been completed by the Rural Development Corporation (RDC) in Korea. This project includes construction of one sea dike with sluice gates, two pumping stations, irrigation pipelines and canals, and drainage channels. In addition to the construction of these irrigation facilities, tidal land reclamation is part of the project. The salt concentration of water in a reservoir is the same as that of sea water when the reservoir is formed by final closing of sea dike. The original sea water within the reservoir will be diluted by the inflowing fresh river water. Thus the proper design and operation of drainage system is crucial problem to succeed in the tidal land reclamation project because the process of desalting the reclaimed tidal land and the reservoir is mainly dependent on the drainage capacity of the project area. Directly diverting water from the reservoir to the sea through pumps or culverts is also adopted to shorten the desalinization period of the reservoir.

In order to develop an irrigation water source and tidal land it is first required to construct a sea dike with sluice gates across the mouth of the bay. The reservoir is then separated from the sea. The tidal land submerged at high tide is exposed, which will be reclaimed as arable land for paddies. The newly reclaimed paddy fields are irrigated from the reservoir through an irrigation conveyance system. The conveyance system consists of a pipe nework and a system of pumps. There are 221 nodes in the pipe network. The minimum pressure at each node is 0.6 $N/m^2$. Link length, minimum head, and demands are availbale. As given in the Haenam project report (RDC, 1988), Hazen Williams friction coefficient is 140 for all links and exponents for discharge and diameter are 1.852 and - 4.87 respectively. Polyethylene (PE) pipes are adopted in the study area and the pipes are buried over 1.5 m deep. If the diameter of a pipe exceeds 700 mm, coated steel pipes for preventing from corrosion are used.

In the present study Problem (P1) is solved for two loading conditions. In addition to the source pumps, two booster pumps are considered in the system. The advantages of booster pumps are: (a) to avoid designing source pumping station for abnormally high operating head; (b) to reduce maximum hydraulic heads over large service area; and (c) to reduce energy costs. Saving energy is important for pumping water distribution networks. The selection of undersized pumps would lead to the violation of minimum pressures within the distribution system.

The selection of oversized pumps would lead to unnecessary capital and operational costs, and excessive pressures at nodes. Thus an optimal pump selection is necessary to provide adequate pressure in the distribution system. Problem (P1) yields optimal pumping heads; operating both source pump and booster pumps results in smaller costs than operating only source pumps.

The second example network is the planned water distribution network (Fujiwara and Khang, 1990; Sonak and Bhave, 1993) in Hanoi,

Vietnam, which is solved using the proposed procedure. The network has 1 source node, 31 demand nodes, 34 links, and three loops. As given in the previous studies, the following data are used: Hazen-Williams; C = 130 for all links; conversion factor K = 162.5 for flows in cubic meters per hour and diameters in inches; exponents for discharge and diameter are 1.85 and -4.87 respectively. Commercially available diameters are 12, 16, 20, 24, 30, and 40 inches and the cost per unit length of pipes is given by $.1d^{1.5}$ in which $d$ is the diameter in inches. The minimum required flow in a link is 5 cubic meters per hour.

The decision variables of the Hanoi optimization model based on Problem (P1) are the unknown segment lengths of known six different candidate diameters. The algorithm TREESEARCH ( Loganathan et al., 1990) is applied to obtain the optimal tree layout and the optimal tree link flows. The global tree network has a cost of $ 5,812,889. The procedure Stochastic Probing is implemented to search the feasible region in the outer problem of (P1) beginning with the perturbed optimal tree link flows as the initial flows. The optimal three loop flows result in a cost of $6,032,548. For the local minimizer of (P1), the loop flows are further refined by a gradient search to obtain a cost of $6,031,807. However, the latter solution violates the minimum required flow, 5 $m^3 / hour$. The present approach yields a cost of $6,032,548 which is an improvement over $6,319,000 of Fujiwara and Khang, and $6,045,500 of Sonak and Bhave.

Conclusions

The results of the well established test problems for municipal water distribution system (Hanoi network) attest to the efficiency of the method in finding better local optima. In the irrigation system design, the advantage of simultaneous multiple loadings is established. The traditional analyses have employed peak loading only which lasts only for

a small period. In this study it is shown that if the average loading lasts over long time, it will dictate the design over the peak loading. In the paddy irrigation network two loading conditions are considered: one loading for the peak water demand, the other loading for the average demand. It is also observed that providing booster pumps may prove advantageous both in term of reducing head on the source pump as well as in selecting pipe sizes. The method accommodates various pipe fittings in terms of their minor losses, pumps, and elevated tanks; rehabilitation of pipes is accounted for by using the appropriate roughness ( C factor) for cleaning and lining of pipes and putting new pipes; multiple loadings are included by appropriate addition of the constraints corresponding to each loading pattern while retaining the pipe length variables to be the same in all loadings. The pipe length variables remain the same because the same pipes are utilized under all loadings. In conclusion, the proposed method is comprehensive and efficient in providing an optimal network design.

References

Fujiwara, O., and Khang, D. B., "A Two-Phase Decomposition Method for Optimal Design of Looped Water Distribution Networks", *Water Resources Research*, Vol. 26, No. 4, Apr., 1990, pp 539-549.

Laud, P. W., Berliner, L. M., and Goel, K. G., "A Stochastic Probing Algorithm for Global Optimization", *J. of Global Optimization*, No. 2, 1992, pp. 209-224.

Loganathan, G. V., Sherali, H. D., and Shah, M. P., "A Two-Phase Network Design Heuristic for the Minimum Cost Water Distribution Systems under a Reliability Constraints", *Engineering Optimization*, Vol. 15, 1990, pp. 311-336.

Loganathan, G. V., Greene, J.J and Ahn, T., "A Design Heuristic for Globally Minimum Cost Water Distribution Systems," *Journal of Water Resources Planning and Management, ASCE*, 121(2), pp. 182-192, 1995.

Rural Development Corporation (RDC) of Korea, *Feasibility studies on Haenam Agricultural Development Project*, Seoul, Korea, 1988.

Savic, D.A., and Walters, G.A., "Genetic Algorithms for Least-Cost Design of Water Distribution Networks", *Journal of Water Resources Planning and Management, ASCE*, 123(2), pp. 67-77, 1997.

Sonak, V. V. and Bhave, P. R., "Global Optimum Tree Solution for Single-Source Looped Water Distribution Networks Subjected to a Single Loading Pattern", *Water Resource Research*, Vol. 29, No. 7, 1993, pp. 2437-2443.

## Potomac River Risk-based Water Supply Management
Stuart S. Schwartz Ph.D.[1]

### ABSTRACT

Current capability in hydrologic forecasting provides real-time risk-based information that can substantially improve the utilization of available resources. Emerging technologies in hydrologic forecasting and real-time reservoir operation supports higher utilization and multiple uses through the non-structural management of hydrologic risk in real-time. This presentation describes nonstructural risk-based management of the Potomac River that supports substantially increased production rates, while satisfying regional commitments of reliable refill and drought availability.

The use of real-time probabilistic hydrologic forecasts in conjunction with mathematical optimization is shown to provide robust, dynamic reservoir operating rules. By explicitly incorporating reliability goals into real-time reservoir operations, hydrologic forecasting supports dynamic reservoir operation that conserves valuable storage as drought risks grow, and more aggressively utilizes the full operating potential of existing facilities when hydrologic drought risks are low. By actively managing risk, as well as reservoir storage, the utilization of existing water resources can be dependably expanded while meeting system-wide reliability goals through non-structural means.

Examples of real-time reservoir operation based on real-time hydrologic risks demonstrate the potential for dramatically increasing the operational benefits that can be reliably achieved in existing water resource systems.

---

[1]

Director, Section for Cooperative Water Supply Operations, Interstate Commission on the Potomac river Basin. **Current Address**: Hydrologic Research Center, 12780 High Bluff Drive, Suite 250, San Diego, CA 92130-2069

## Introduction

Municipal water supply for the Washington D.C. Metropolitan area is provided by three major water suppliers: The Washington Aqueduct Division of the U.S. Army Corps of Engineers, the Washington Suburban Sanitary Commission, and the Fairfax County Water Authority. During times of adequate natural flow, these suppliers operate independently, drawing upon the Potomac, Patuxent, and Occoquan rivers as sources for raw water. The Fairfax County Water Authority's Occoquan Reservoir is an integral part of the regional water supply system for the Washington D.C. Metropolitan area. Under non-drought conditions the Occoquan Reservoir and its associated treatment facilities are operated in conjunction with the Authority's Potomac River treatment plant to economically meet demands in the FCWA system. During drought conditions withdrawals from the Potomac River, as well as the joint utilization of storage reservoirs on the Patuxent and Occoquan rivers are coordinated in order to maximize regional water supply reliability.

As part of the cooperative agreements guiding regional operations, the major suppliers have committed to operate all reservoir storage with a 95% probability of refilling to 90% of capacity by June 1. In order to achieve this reliable refill goal, operating recommendations on withdrawal rates have been developed specifying maximum withdrawal rates based on reservoir storage throughout the year. While these refill recommendations achieve the reliable refill goal established in the cooperative regional drought plan, the recommendations also impose extremely conservative operating limits, under non-drought conditions. In particular, these operating recommendations place severe limitations on the opportunity to utilize the abundant supply available from the Occoquan during most years.

While drought preparedness and risk management is an essential constraint on utilization of the Occoquan, the current operating recommendations fail to utilize any hydrologic forecast skill or anticipation of operational responses to changing hydrologic conditions. Reexamination of current operational practices to achieve reliability-based operational goals, emphasized the opportunities that exist to utilize currently available hydrologic forecasting techniques and real-time operating rules in order to achieve effective risk-based operation through non-structural means.

### Real-Time Risk-Based Operating Rules

Currently, operating recommendations for Occoquan Reservoir are determined and implemented as an open loop feedback operating strategy. The extremely conservative nature of these rules results from both the lack of any hydrologic forecast skill, and the lack of anticipation of future operating decisions. Current operating recommendations for the Occoquan Reservoir are derived through solution of the problem:

**P1**:

$$\text{Max } R$$

subject to:

$$S_{t+1} = S_t + Q_t - R \qquad t = t_0, t_0+1, \ldots \tau \qquad (1)$$

$$\Pr(S_\tau \geq 0.9\, S_{cap}) \geq 0.95 \qquad (2)$$

Problem P1 essentially identifies on day t, the single constant value of Occoquan release, R that will result in a 95% chance that the storage in the Occoquan will exceed 90% of capacity, $S_{cap}$, at the end of the operating horizon, designated as time $\tau$. Here and throughout, the end of the operating horizon corresponds to June 1. Equation 1 is the continuity equation (ignoring losses) that simply relates reservoir storage at any time, t, to inflow $Q_t$ and the release. Additional constraints are also enforced to assure that water supply demands will be reliably satisfied during the low flow season. The conservative nature of this solution is clear. This is implicitly equivalent to operating the Occoquan as though the drought of record will repeat starting tomorrow, irrespective of current hydrologic conditions.

Figure 1 shows a graphical representation of current seasonally varying reliable refill recommendations based on reservoir storage levels.

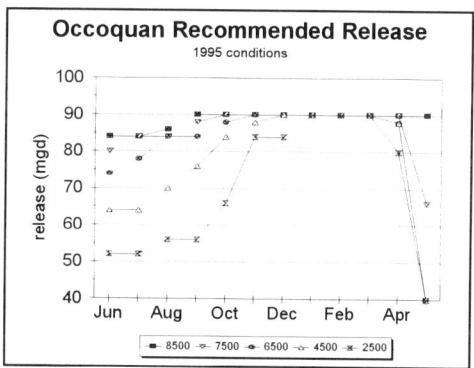

**Figure 1** Current maximum recommended withdrawal rates for reliable refill

Managing hydrologic risks in real-time

In real-time operation there is a clear and critical need to continuously hedge releases against the possibility of drought conditions developing. Drought conditions develop slowly and reveal themselves gradually through the cumulative development of precipitation and soil moisture deficits. For the majority of non-drought conditions, the extremely conservative operating recommendations based on drought of record hydrology can be relaxed without compromising water supply reliability. Hydrologic forecast tools that allow us to quantify the current drought risk can be effectively utilized in dynamic real-time operating rules to maintain reliable assurance against drought conditions.

In addition to lacking current hydrologic information, the current operating recommendations are based on the assumption that the release implemented today *must* be maintained until June 1. For this reason the current release must hedge against the worst conditions that could be envisioned for the remainder of the operating horizon. This open loop, or "once-and-for-all" control philosophy similarly imposes an extremely conservative limit on the operating recommendation; the decision must be appropriate for both current conditions, as well as all possible conditions that could occur between today and June 1. In practice the operating recommendation is revised over the operating horizon, based on the feedback provided by observations of future reservoir storage. The resulting series of operating recommendations nevertheless consists of sequential open-loop solutions to problem P1.

As an alternative, real-time operation can be enhanced through the use of anticipative, or closed-loop operating decisions. A closed loop operating rule can recommend relatively high releases when conditions do not indicate significant drought risk. These higher rates can be supported because the closed-loop operating rules explicitly incorporate future conservative releases, contingent on the development of drought conditions. The difference between current operations and a general closed loop operating rule can be seen by contrasting problem P1 to the recursive problem:

**P2**

$$Max \sum_{t=t_0}^{\tau} R_t = f_t(S_t)$$

subject to:

$$f_t(S_t) = [B_t(R_t) + f_{t-1}(S_{t-1})]$$

where $B_t(R_t)$ is the immediate benefit of releasing $R_t$ in the current time period, and $f_{t+1}(S_{t+1})$ is the future benefit from the sequence of optimal releases that will be made from time $t+1$ through the end of the operating horizon, on June 1.

The incorporation of real-time forecast skill with the anticipative, closed loop solution of problem P2 provides the opportunity to operate the Occoquan reservoir at much higher production rates during non-drought conditions without compromising system reliability. Figure 2 shows the resulting operating curves.

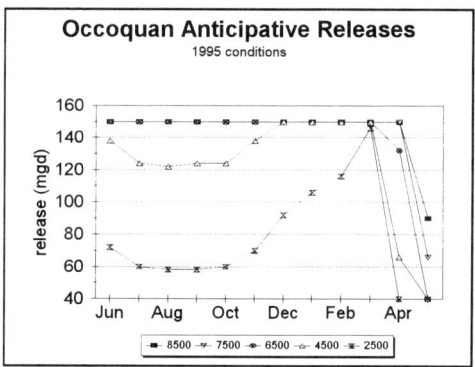

**Figure 2** Anticipative operating rules: 1995 design conditions

Risk-based Operation

The operational consequence of forecast-based anticipative operation can be quantified through operational simulation of the operating rules shown in Figures 1 and 2. The operational differences can be summarized by the cumulative density function of daily Occoquan treatment rates. These cumulative distributions show the load factor or operating capacity that can be supported by the Occoquan under alternate operating rules. The comparative CDFs are shown in Figure 3. The impact of using an anticipative operating rule is dramatic. While current operating rules are able to sustain an operating rate of 80 mgd with a load factor of 80%, the anticipative operating rules support an 85 mgd rate, with 140 mgd achievable with a load factor of roughly 75%.

This analysis was part of a comprehensive reevaluation of treatment capacity at Occoquan reservoir. With appropriate risk-based tools the Occoquan is clearly able to support sustained withdrawals exceeding 130 mgd - in stark contrast to the project firm yield of only 64 mgd. Risk-based operation supports reliable utilization of this valuable resource at substantially higher load factors.

It is important to recognize that the high load factor at which a 150 mgd plant could be operated results from utilizing this resource under non-drought conditions. Under drought conditions both current and anticipative operating rules severely curtail the use of the Occoquan, resulting in heavy reliance on the Potomac for primary supply. For this reason increased treatment capacity on the Occoquan is not a direct substitute for capacity expansion on the Potomac.

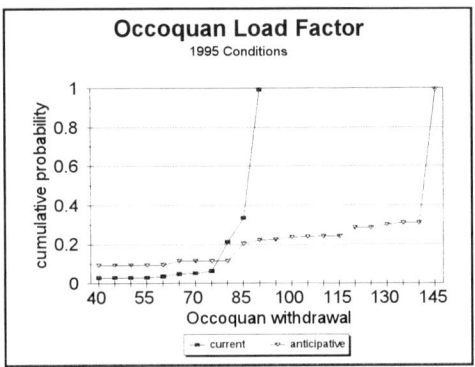

**Figure 3** Release CDF: 1995 design conditions

## Conclusion

Current drought operation of Occoquan Reservoir imposes significant limits on the rate and frequency with which water can be withdrawn. These limitations result from conservative operating practices that are intended to assure reliable refill, consistent with regional commitments for cooperative drought operations. Operating rules that utilize forecast skill in an anticipative manner can support substantially higher production rates from the Occoquan without compromising drought reliability goals.

Detailed period-of-record operating simulations demonstrate that risk-based operating rules will allow the full potential of the resource to be exploited by dynamically managing hydrologic risk in real-time. The ability to more fully utilize the Occoquan enhances both operational flexibility under normal operating conditions as well as system redundancy in the event of a supply disruption on the Potomac River. If the Potomac became unavailable, reservoir storage on the Occoquan represents a vital raw water supply for emergency operations. The value of this redundant supply is limited by the ability to treat and distribute this water during a Potomac outage. For this reason redundancy planning on the Occoquan must consider both raw water availability, as well as treatment and distribution system capacity. The results presented here demonstrate that significantly increased treatment rates can be supported from Occoquan Reservoir without compromising water supply reliability. The expanded treatment capacity that can be reliably supported provides a valuable option for system redundancy planning. The efficient utilization of real-time forecast information supports the non-structural management of hydrologic risk, effectively creating "new" storage in existing facilities.

# Multimedia Network System for Landfill Siting

Jehng-Jung Kao[1], Associate Member

*Abstract*

This paper describes a series of studies for developing a computerized multimedia network system for landfill siting. Siting analysis for a municipal solid waste landfill is a complex process. Various siting factors and criteria must be carefully analyzed. The landfill capacity crisis is a critical environmental issue to be resolved in many local communities in Taiwan, R.O.C. Appropriate inexpensive computerized tools capable of facilitating the siting analyses are therefore explored in this work. Several sub-systems including a computerized guideline, a geographical information system, a prototypical expert system, spatial optimization and multifactor models and a multimedia network interface had been developed. The entire structure of the system is described in this article, with discussions for the design and development of all sub-systems. The system is made available for 24-hour access from anywhere on the Internet. This networking capability allows a user without tools to utilize the system and to avoid the cost of system installation and training appropriate personnel to manage the system.

*Introduction*

Landfill siting is a relatively difficult, complex, tedious and protracted process (Allanach 1992). Various siting data, information, issues, rules, criteria, constraints, and factors must be carefully processed and analyzed. An initially selected candidate site may be later abandoned because opposition arises on the basis of previously neglected but important factors; costs are subsequently increased and the final decision of a landfill site postponed. Moreover, such a

---

[1] Professor, Environmental Systems Analysis Laboratory, Institute of Environmental Engineering, National Chiao Tung University, 75 Po-Ai Street, Hsinchu, Taiwan 30090, R.O.C. E-mail: jjkao@green.ev.nctu.edu.tw.

process may have to be repeated several times as new factors are introduced or as siting constraints alter.

For local governments in Taiwan, inappropriate sites may be chosen because critical factors are not fully analyzed owing to limited budget and manpower, coupled with tedious procedures required to process siting data and information, for enforcement verification of related constraints. On the other hand, A local landfill siting decision frequently becomes a controversial issue receiving national attention because of strong local opposition. Such a disputed situation typically derives from either, as aforementioned, an inappropriate or incomplete siting analysis or the public's misunderstanding of the siting procedure.

The central environmental protection agency in Taiwan is therefore looking for an inexpensive computerized tool capable of facilitating the siting procedure, opening easily related information to the general public, conveniently instructing local agencies to implement the siting analysis, and managing and monitoring the progress of local siting tasks. A series of studies (Kao et al. 1996a; Kao et al. 1996b; Kao 1996; Kao et al. 1997; Chen and Kao 1997) were therefore initiated to develop a multimedia network system for landfill siting analyses.

*System Structure*

The final expected system will include a rule database, a spatial database, an expert system, spatial optimization and multifactor models, an integrated decision support system, and a multimedia network interface. Currently, a prototypical system with a rule database, an expert system, a geographical information for storing spatial data, a spatial optimization model, a multifactor model, and a friendly World Wide Web (WWW) network interface had been developed, although these sub-systems are currently under modification by several up-to-date WWW-based tools. In the following sections, each sub-system is briefly described with discussion for its current status and future development goals.

*Siting Rule Database*

The first product of this research was a knowledge database (Kao et al. 1996a). The knowledge database, extracted from various local and foreign regulations, case studies, reports, and other literature, provides siting guidance and related information. All collected rules are divided into four major types: essential, secondary, recommended, and particular rules. Each type of rules is further classified into three groups: environmental, sociocultural, and engineering and economic groups. The previous version of the system with a demonstrated English version can be accessed via Internet from the home page http://ev004.ev.nctu.edu.tw/ENGLISH.

## Spatial Database

A spatial database was built by GRASS (1993), a raster-based geographical information system (GIS). GIS is an information management system capable of providing spatial analysis tools for sorting, retrieving, and manipulating georeferenced computerized maps. Various map layers were collected, e.g., ground-water level, road network, land slope, land cost, and others. Again, a demonstrated system is accessible via Internet from the home page provided above.

## Expert GIS System

The rule database, other than serving as a siting guidance system, was also used to set up an expert system. CLIPS, a forward-chaining expert shell (Giarratano and Riley 1991), was applied to build the expert system. The expert system was further integrated with the spatial database to form an expert-GIS system (Kao et at. 1996b). Computerized expertise can be provided by the expert system based on user-provided facts; spatial information can be effectively processed by the GIS. Clearly defined rules triggered by the expert system are expressed by a formatted text file that is processed by developed interface programs to produce GRASS commands for implementing desired map analyses. Qualitative rules are, however, not implemented with the GIS; a fuzzy expert system is currently being developed for processing such rules with the GIS.

## Spatial Optimization Model

Although a GIS is useful in siting applications, the algorithm for obtaining the optimal site, with simultaneous consideration of site compactness and other factors, is generally unavailable. Compactness represents the nature of the site and the extent to which it can be regarded as integrated tightly. The lower the level of compactness, the less likely the solution is to satisfy siting requirement, subsequently making general land planning difficult. A raster-based compactness optimization model was therefore developed (Kao and Lin 1997) a C program (Kao 1996) was written to efficiently solve the model.

## Multifactor Model

In addition to the site compactness, numerous factors must be evaluated when siting a landfill. An appropriate landfill should have minimum impact on environment, society, and economy, comply with regulations, and be generally acceptable to the public (Zyma 1990). A multifactor model was therefore developed (Kao and Lin 1997). A multifactor model with different scores assigned to varied levels of a factor and a weight is assigned to each factor to express its relative importance when compared to other considered factors. A case study for applying the model had been presented by Kao and Lin (1997).

*Decision Support System*

A landfill sting decision making process is generally complex. Before a compromise solution is accepted, the siting procedure is probably iterated several times. A comprehensive system is therefore desired for convenience and effectiveness in implementing related siting tasks under a fully integrated working environment. Furthermore, some important siting issues may not be easily numerically specified and therefore unsolvable by a mathematical approach. In addition to extending the system to be able to answer "What-if" questions with a friendly network interface described in the next section, several other sub-systems are planed to include in near future:

(1) A fuzzy DRASTIC system (Chen and Kao, 1997): DRASTIC (Aller et al. 1985) is an index system originally developed for evaluating general groundwater pollution potential and weights assigned for each factor may not be appropriate for assessing impact from a landfill. The single weight assigned to each hydrogeologic parameter is not appropriate because of the uncertainties of the parameter itself and its effects on pollution potential. A Fuzzy set approach was therefore applied to develop a Fuzzy DRASTIC.

(2) A fuzzy GIS system (Lin et al. 1997): the siting GIS classifies a siting area into unacceptable and candidate regions based on some siting criteria or factors. However, such a binary division is not appropriate to evaluate the suitability of possible candidate sites. Fuzzy algorithms are therefore explored to deal with this issue. Several fuzzy membership functions were developed and integrated with the GIS and the spatial model.

(3) A questionnaire analysis system: a landfill siting decision is a difficult environmental decision to make because some unquantifiable issues may play important roles and frequently a questionnaire survey is implemented to obtain the opinion and evaluation from varied sectors.

(4) A landfill evaluation model: when a siting process reach the very last stage, few possible sites may be left on the candidate list, if no obvious best site exists. In this stage, a detail comparison among them is necessary. A simulation model is under development based on several existing models.

*Multimedia Network Interface*

The user interface, as the homepage shown in Fig. 1, of the system is mainly developed with a WWW server that can transfer textural, audio, graphical, and video data. With the friendly interface, a general user does not require previous knowledge of developed complex tools for using the system. Although the system is mainly constructed on a UNIX platform, a personal computer with a WWW browser and connected to the Internet can access the system. This advantage minimizes the distribution cost and maximizes the accessibility of the system. The proposed system, if desired, allows the public to access the siting system entirely or partially.

*Conclusion*

Landfill siting is a complex procedure that is generally multidisciplinary and requires extensive effort to assess numerous factors. Implementing such a complicated procedure in a conventional information processing approach would be expensive and tedious. A low-cost computerized tool is therefore to be developed by integrating a rule database, a spatial database, an expert GIS system, spatial optimization and multifactor models, and other decision support tools. Although the entire system is currently still under development, several prototypical sub-systems had been developed and demonstrated for their effectiveness to facilitate landfill siting analyses.

*Acknowledgement*

The author would like to thank Mr. H.-Y. Lin, Mr. W.-E. Chen, and Mr. S.-J. Guo for their assistance in developing several previous prototypical sub-systems. Financial support provided by National Science Council, Taiwan, R.O.C., for this work under Grant NO. NSC 87-2211-E-009-007 is also greatly appreciated.

*References*

Aller, L., Bennett, T., Lehr, J. H., and Petty, R. J. (1985). *DRASTIC: a standardized system for evaluating ground water pollution potential using hydrogeologic settings.* U.S. Environmental Protection Agency. EPA-600/2-85/018.
Allanach, W. C. (1992). "Regional landfill planning and siting." *Public Works*, December, 48-50.
Kao, J.-J., Li, K.-C., and Guo, S.-J. (1996a). "Computer assisted System for landfill siting." *J. of the Chinese Institute of Environmental Engineering*, 6(2), 117-130. (in Chinese).
Kao, J.-J., Chen, W.-Y., Lin, H.-Y., and Guo, S.-J. (1996b). "Multifactor spatial analysis for landfill Siting." *J. of Envir. Engrg.*, ASCE, 122(10), 902-908.
Kao, J.-J. (1996). "A raster-based C program for siting a landfill with optimal compactness." *Computers & Geosciences*, 22(8), 837-847.
Kao, J.-J., Lin, H.-Y., and Chen, W.-Y. (1997). "Network geographic

information system for landfill siting." *Waste Management & Research*, 15, 239-253.

Chen, W.-Y. and Kao, J.-J. (1997). "Fuzzy DRASTIC for landfill siting." *Proceeding of the 13th International Conference on Solid Waste Technology and Management*, Philadelphia, PA, U.S.A.

Lin, H.-Y., Kao, J.-J., Lee, K.-J., and Huang, H.-Y. (1997). "Fuzzy GIS assisted landfill siting analysis." *Proceeding of the 11th International Conference on Solid Waste Technology and Management*, Philadelphia, PA, U.S.A.

*Grass 4.1 user's reference manual.* (1993). U.S. Army Constr. Engrg. Res. Lab. (USACERL), Champaign, Ill.

Giarratano, C. J., and Riley, G. (1991). *CLIPS user's guide.* Nat. Aeronautics and Space Admin. Lyndon B. Johnson Space Ctr. Information Sys. Directorate Software Technol. Branch.

Zyma, R. (1990). "Siting considerations for resource recovery facilities." *Public Works*, 121(Sept.), 84-86.

Figure 1. Homepage of the System.

Protecting Drinking Water Supply Sources:
San Francisco's Water Quality Vulnerability Zones

Karen E. Johnson [1]
and Edward H. Stewart [2]

Abstract

Water utilities are facing increasing challenges in meeting drinking water treatment regulations because of the discovery of new pathogens in raw waters and the difficulties in balancing disinfectant usage. A greater emphasis is therefore placed on source protection of drinking water supplies. The U.S. EPA's Source Water Assessment Program (Safe Drinking Water Act Amendments of 1996) reinforces the importance of source protection. A unique methodology was developed for the City and County of San Francisco Public Utilities Commission to aid in the delineation of its local watersheds and spatially define water quality vulnerability zones within the watersheds.

Introduction

Water utilities providing drinking water to their customers need management tools to enable them to strengthen their first barrier to water quality degradation -- source protection of water supplies. From a water quality perspective, algal blooms and taste and odor concerns have driven the use of in-reservoir treatment controls, but water purveyors are now looking at ways of controlling various parameters of concern within the watersheds before they reach the reservoirs or lakes.

A unique methodology was developed for the City and County of San Francisco to aid in the delineation of its 63,000-acre local watersheds and spatially define water quality vulnerability zones within the watersheds. This analysis provided a clear link between the natural characteristics of lands and their potential impacts to water quality from a source and transport perspective. This project was initiated as a source water protection tool for use in the overall watershed management planning process undertaken in response to growing public pressure to increase access in the watersheds. Areas with a higher vulnerability to water quality impacts which required different watershed management policies, restrictions, and maintenance procedures were delineated. These vulnerability zones are also continually being used in maintaining and protecting water quality through the review of development proposals.

---

[1] Principal Planner, Montgomery Watson Americas, Inc. 1340 Treat Blvd. #300, Walnut Creek, CA 94596

[2] Watershed Manager, currently with Contra Costa Water District, P.O.Box H2O, Concord, CA 94524; previously with San Francisco Public Utilities Commission

The vulnerability zone development methodology can be easily tailored by utilities to meet the goals of EPA's source water assessment and protection programs being developed in conformance with the Safe Drinking Water Act (SDWA) Amendments of 1996. The primary components of the source water assessment include the delineation of watersheds and recharge areas, an inventory of contaminant sources, and a determination of the susceptibility to contamination by sources inventoried within the source water protection areas. San Francisco's Water Quality Vulnerability Zone methodology is based on spatial attributes managed in a geographic information system (GIS), which can include the specific inventories of individual watershed and recharge areas, and is a precursor to the development of drinking water source protection programs.

Methodology

The methodology developed to produce the vulnerability zones generally consists of the following key steps:
- Determine relationships between the individual watershed characteristics of importance and each water quality parameter group of concern;
- Develop a database of these relationships;
- Composite the database relationship attributes into vulnerability zones for each water quality parameter group; and
- Combine the individual water quality parameter group vulnerability zones into the composite water quality vulnerability zones.

Physical characteristics (soils, slope, vegetation, proximity to waterbodies, and wildlife concentrations) of the five local watersheds were analyzed to determine which attributes 1) were of concern from a contaminant source perspective, or 2) exacerbated the transport of contaminants. Five groups of water quality parameters represent the prioritized drinking water parameters of concern: particulates (e.g., turbidity); disinfection by-product precursors (e.g., organic carbon); microorganisms (e.g., *Cryptosporidium*); nutrients (e.g., nitrogen); and synthetic organic compounds and pesticides (e.g., chemicals).

Watershed physical characteristics of source or transport vulnerability specifically for the water quality parameter groups were determined and relationships developed. For example, soils were evaluated according to their density, organic carbon content, moisture content, and adsorption capacity for the parameter groups listed above, respectively. Slope ranges were evaluated for their aqueous or particle transport mechanisms. Vegetation communities were evaluated for their ability to either provide a protective layer between rainfall and soil and to stabilize the soils with the presence of leaf debris and root systems, or provide a protective layer from the sun, thereby providing for higher moisture conditions which allow some microorganisms to survive longer.

The proximity to waterbody relationship to the water quality parameter groups was more challenging to develop. Because the movement of particulates, for example, is a function of overland flow characteristics, rainfall attributes of intensity and duration of the rainfall event are important. A range, starting with 300 feet, was ascertained as the distance required to adequately impede the movement of particulates into the watershed waterbodies. This baseline boundary was then adjusted to reflect watersheds with higher rainfall intensity and duration. The wildlife concentration relationship to water quality parameter groups was found to coincide with the proximity to waterbody characteristics because mammals in the watershed must travel to and from the reservoirs and streams for water. Impacts associated with wildlife such as soil disturbance and compaction and deposition of feces will be concentrated near the waterbodies. In addition to the above discussion of surface water, for the proximity to water relationship to synthetic organic compounds (SOCs), the assessment included the recharge area of a groundwater basin. The ease of transport of SOCs through alluvial materials into the upper aquifer influenced this decision.

The San Francisco watershed management plan GIS database was then utilized to apply these relationships to develop baseline vulnerability data for each physical characteristic which reflects each water quality parameter group. The data were mapped to reflect high, medium, and low vulnerabilities within each watershed which could adversely affect the water quality of the corresponding receiving waterbody. For example, forested vegetation communities were mapped to reflect a low level of vulnerability of particulate transport due to their ability to intercept the kinetic energy between rainfall and soil; areas lacking vegetation were considered highly vulnerable to particulate transport.

A compositing approach was defined to combine the five relationship maps for each parameter group into one vulnerability zone map for each parameter group. Using the example started above, the highly vulnerable vegetation areas, in combination with a steep slope range creates a high vulnerability zone for the transport of particulates. If the highly vulnerable vegetation areas were combined with soil types reflecting a low dry density, this creates a high vulnerability zone for a source of particulates of concern to drinking water treatment.

This compositing approach is summarized as follows. The high vulnerability areas defined for the "proximity to water" layer take precedence and therefore are defined as high vulnerability zones on all water quality parameter maps. For other areas to be defined as high, slope must be high and either soils or vegetation must be high. For areas to be defined as a "low", two of either slope, soils, or vegetation must be low. All other areas not already defined based on the three points above are defined as medium.

Once verified using the universal soil loss equation, the high, medium, and low vulnerability zones were defined and mapped for each water quality parameter group. These parameter group zones were then composited further into one overall water quality vulnerability zone map. Where differences between the zones occurred, the more restrictive zone rating was used.

Conclusions

The water quality vulnerability zones were used as a management tool in San Francisco's comprehensive watershed management planning process. The vulnerability zones aided in the development of overall management policies and strategies to maintain or enhance the reservoirs' high water quality. For example, if it is determined that an area in the watershed is a high vulnerability zone for an increased particulate contribution, this targets that area for "no or minimal" soil disturbance activities and would result in a requirement for erosion control practices. The vulnerability zone maps and associated database are currently used along with the rest of the GIS tools for overseeing the watershed maintenance program, evaluating potential recreational activities, and developing corresponding mitigation measures in the watersheds.

In one recent case, an alignment for a public access trail proposed by a local group was reviewed by San Francisco and consultant staff using the vulnerability zone maps to identify the trail's location in relation to the high vulnerability zones of water quality parameters of concern to trail usage. Recommendations to minimize risks associated with trail water quality impacts were easily tailored to reflect the level of corrective actions needed based on the vulnerability zones of the lands proximate to the trail.

The development of vulnerability zones highlights the importance of having available detailed natural resource data and quantitative data regarding physical characteristics to assess watershed conditions. The vulnerability zones can be used in establishing water quality monitoring programs to both monitor land uses and activities in these vulnerability zones, as well as to fill any quantitative data gaps.

The steps taken to develop the vulnerability zones can be adapted to other watersheds and to groundwater recharge areas for support in developing and implementing source protection strategies, or for other individual agency needs. Vulnerability zone GIS maps are also a useful public education tool used in explaining the susceptibility of lands to increased contamination risk, in devising volunteer activities, and in working with local land use planning authorities in delineating protection zones needing stricter development standards.

Another use of the vulnerability zones relates to the new SDWA source water assessment programs. These assessments require delineation of the area around a drinking water source through which contaminants move and reach the source. It also requires an identification of contaminants and their sources within the delineated area in order to assess the susceptibility of the water source to these contaminants. With an inventory of human land uses and activities included in the database, the water quality vulnerability zone approach mirrors the surface water requirements. Additional spatial and temporal data are required for the groundwater assessment, particularly for vertical transport in the recharge areas, and subsurface horizontal transport within the aquifer. Water quality vulnerability zones can serve as a basis for a water provider to effectively establish watershed source controls as the first barrier in the protection of its water supplies, and as an opportunity to improve its drinking water quality.

Acknowledgments

The authors would like to thank participants in the development of this project including the San Francisco Public Utilities Commission staff, C. R. James, Systech, EDAW, and ESA.

Approaches and Levels of Service Analysis
for an Area Subject to Interbasin Flow Transfers

L. Moris Cabezas, Ph.D., P.E.[1]

Abstract

Parsons Engineering Science developed the Basin Master Plan for South Creek in Sarasota County, Florida. The study addressed watershed management issues in the basin. Currently, minor flooding concerns exist within the study area. However, many areas in adjacent drainage basins have been largely developed and, during large rainfall events, interbasin flow transfers from South Creek contribute to flooding and violations of levels of service in those adjacent basins. This condition presented special watershed management challenges from the technical standpoint, as well as from the assessment of levels of service for flood protection, water quality control, and ecological protection.

General

In an effort to address area watershed management issues, Sarasota County is implementing its Basin Master Planning Program. The Basin Master Plan for South Creek is an element of that Program. The Plan will be used to identify the stormwater system improvements needed to provide acceptable levels of service (LOS) conditions for flood control, water quality and environmental protection in the basin. Furthermore, an objective of this project is to determine the impact of the transfer of flood flows from South Creek to adjacent basins. The local community's perception is that flows from South Creek contribute to create flooding conditions in those adjacent basins, particularly in the North Creek basin which is largely developed. Numerous cases of house and road flooding have been reported in the North Creek basin in the past.

The South Creek basin is located in the central portion of Sarasota County. It encompasses approximately 20 square miles and discharges into Dryman Bay, which is part of the Sarasota Bay estuarine system. South Creek is categorized as Class III surface waters, that is "waters used for recreation,

---

[1]Manager, Water Resources, Parsons Engineering Science, Inc., 2901 West Busch Boulevard, Suite 905, Tampa, FL 33618

propagation, and maintenance of a healthy, well-balanced population of fish and wildlife". Typical of southern Florida conditions, the South Creek drainage basin is characterized by a flat topography, high water table, and many natural depressions and wetlands. Historically, the basin consisted of an aggregation of wetland systems overflowing through natural sloughs. However, due to agricultural activities, most of the wetlands have been connected by man-made ditches. These actions have drained numerous wetland systems and have resulted in runoff being directly conveyed to the main creek channel, which has also been extensively ditched.

Characteristics of the Drainage System

Most of the basin is currently underdeveloped. However, the area is experiencing rapid development. The existing land use map indicates that about 76 percent of the area is categorized as open land and about 27 percent of the open land area is categorized as wetlands. Developments consists mainly of low and medium density residential with pockets of commercial land uses. For this study, the study area was subdivided in 203 subbasins, that varied in size from less than 2 acres to 436 acres. The median size is 46 acres.

South Creek has a main branch and five secondary branches. A total of nine potential areas of interbasin flow transfers were identified. Data on drainage, storage, and control structures were available from existing sources and analysis of aerial photographs. In addition, a significant amount of field survey work was conducted as part of this study to supplement that information.

Development of the Hydrologic/Hydraulic Model

**Model Development.** The Advanced Interconnected Pond Model (AdICPR) was used for this application. All hydrologic analyses were conducted using the Soil Conservation Service (SCS) Curve Number (CN) method. The SCS synthetic unit hydrograph with a shape factor of 100 was used in this application to convert precipitation excess into a runoff hydrograph.

The development of the hydraulic routing model included determining the geometric characteristics of the conveyance system, including junctions and conduits. An extensive amount of effort and detail was devoted to quantification and definition of storage facilities within the basin such as wetlands, lakes, borrow pits, detention ponds, and floodplains. The interbasin transfer connections were defined based on available data. Rating curves were input to the model for the Catfish Creek and North Creek interconnecting points using information from modeling studies in these basins to define stage-discharge relationships. The interconnects with Cow Pen Slough and Fox Creek were defined as broad-crested overflow weirs to represent sheet flow across drainage boundaries.

**Model Calibration Process.** Model calibration and verification was conducted using measured hydrologic data provided by the United States Geological Survey (USGS). The South Creek gauging station consisted of a continuous water level gauge and a rainfall gauge. Currently, the station has been removed and is no longer in operation. An important added parameter of the calibration process in this case was the interbasin flow transfer characteristics. The loss of runoff volume due to those transfers was initially assessed by back-calculating and comparing the average curve number resulting from a number of recorded hydrographs. As shown below, the August 1992 and April 1993 storms were relatively small and it is likely that no flow transfers occurred during those events. The difference in the calculated average CN between the two storms is due to antecedent moisture conditions and natural variations in soil infiltration characteristics. However, the CNs are within values reported in the literature for a mostly undeveloped basin such as South Creek.

| Storm Event | Total Precipitation (inches) | Recorded Streamflow Volume (inch) | Calculated Average CN |
|---|---|---|---|
| June '92 | 17.82 | 4.3 | 28 |
| March '93 | 1.83 | .39 | 78 |
| April '93 | 3.75 | .69 | 72 |

On the other hand, the low back-calculated CN value for the extremely large June 1992 storm indicates runoff volume transfers into other basins, as all interbasin interconnects are upstream from the USGS gage. In fact, the runoff versus rainfall ratio for that storm is 31 percent. The June 1992 storm resulted in one of the most extreme flooding events on record in Sarasota County and the occurrence of flow transfers from South Creek was widely documented by local residents.

As opposed to regular calibration approaches, calibration of computer models for basins that include interbasin transfer conditions is a two-step process. The purpose of the first step is to identify the value of the calibration parameters associated with the individual subbasins without the interfering impacts of the interbasin flow transfers. Therefore, it is necessary to select an event for which no transfers have occurred. Subsequently, the second calibration step is used to assess the magnitude of flow transfers. Calibration in this case must be conducted using the data for a large storm event that resulted in basin overflows.

For this particular project, the first calibration step was conducted using the data available for the April 1993 storm. The second calibration step was developed using the data for the June 1992 storm. Finally, the model was calibrated using the data for the March 1993 storm. Figures 1 through 3 show the measured and computed stage versus time relationships for the calibration and verification events. Those results were also verified using surveyed and reported high water elevation data. The calibration and verification hydrographs also show adequate agreement between the measured and simulated values.

Typical all Sarasota County basins, the measured hydrographs show an initial period of small flow within its rising limb, which is a consequence of the high retardance effects occurring in the basin. Subsequently, both flow and stage at the creek are characterized by a rapid and substantial increase until the peak is reached, followed by a long recession period. During the June 1992 storm, the volume under the hydrograph's rising limb amounted to about 19 percent of the total storm runoff.

## Standard Storms and Levels of Service Analysis

Once the hydrologic/hydraulic model was calibrated, existing conditions in the basin were analyzed on a preliminary basis using the SWFWMD 24-hour design storms with return periods of 2, 5, 10, 25, and 100 years. Simulations were conducted assuming average antecedent rainfall conditions. This type of analysis is necessary to analyze compliance with the County's levels of service for flood control. The total rainfall values per return period were as specified by SWFWMD. The accumulated rainfall volume for the 100-year design storm showed lower water elevations than those observed during both the June 1992 and July 1995 storm events. This is probably due to the differences in accumulated rainfall between the theoretical and actual events.

**Levels of Service for Flood Control.** LOS objectives for flood protection throughout Sarasota County have been defined in the County's Land Development Regulations. The concept, based on expected flood damages, is well established and defines flooding conditions associated with several design storms for structures and road access. For example, the LOS establish that structures and evacuation routes should not be flooded during the 100-year, 24-hour storm and neighborhood streets should not experience over one foot of flood depth during that same storm.

The analysis consisted of listing the modeled junctions along the drainage system and the corresponding model predictions of flood elevation. In addition, the list included descriptions of the road crossing structures along the drainage system and the corresponding low road elevation. The difference between peak flood elevation and low road elevation indicates the degree of the flooding condition at any particular location. This process allows determination of whether or not the LOS are being met. Results showed that LOS for existing and future condition are being met, except for one location on the northern part of the basin.

As indicated previously, a major concern in this study was to assess the impacts of interbasin flow transfers from South Creek. For this analysis, results of the model simulations for the various design storms were provided for input to the neighboring basin's hydraulic models. Results indicated that, because of timing differences, the South Creek overflows are the main cause of flooding in North Creek. Figure 4 shows the stage versus time relationship for the June 1992 storm at one of the North Creek stations. The graphs show that the flow and stage peaks at that location are reached long before South Creek flows have reached their peak. In fact, the North Creek peaks are reached when water elevations in South Creek are well below the channel's overtop elevation at the basin divide. The net effect of the interbasin flow is not to increase the peaks but to sustain them for a relatively long time. This illustrates the importance of carefully analyzing hydraulic timing when identifying stormwater system improvements.

**Levels of Service for Water Quality.** LOS for water quality impacts due to stormwater discharges have not been as well defined as those for flood control, but significant progress has been made over the last years toward the adoption of a generalized approach. These LOS are currently being established at all levels of government and are taking the form of discharge limitations of flow and/or pollutant concentration, pollutant load reduction goals, or even extent of treatment coverage. In addition, because water quality considerations are relevant primarily during periods of low streamflows, interbasin transfers are not an issue of importance.

For this study, a methodology to determine water quality LOS deficiencies and objectives was developed based on both a water quality index (WQI) and a watershed assessment procedure. The Florida Department of Environmental Protection (FDEP) has establish a procedure to calculate a stream's WQI based on concentrations of six chemical indicators. The cutoff values for the WQI are as follows: 0 to less than 45 represents good quality, 45 to less than 60 represents fair quality, and 60 to 99 represents poor quality. It was agreed that the water quality goal for South Creek should be to maintain the WQI at or below 60. Results showed that the Creek is meeting water quality LOS in the downstream portions, but improvements are needed in the upper portions of the watershed.

In terms of the watershed assessment procedure, FDEP has established criteria that measures watershed condition based on both land use composition and extent of BMP coverage within a watershed. The condition of a watershed is represented by an index value between 0 and 5. A value of 5 is assigned to conservation and natural wetland areas whereas zero applies to developed areas with no runoff treatment capabilities. The index is computed as the weighted average of all areas in the watershed. For this study, it was established that an acceptable watershed index should be of at least 3. A calculated value of 2.69 indicated the need to improve runoff quality control.

**Levels of Service for Ecological Protection.** The approach applied in this study incorporated the concept of BIOLOGICAL LEVELS OF SERVICE (BLOS). The BLOS are defined for each habitat area and reflect the overall habitat value of an ecological community. Criteria for the BLOS included community type; plant diversity, density, and cover; canopy coverage; level of impact to native vegetation; level of disturbance, and other factors. In addition, The BLOS required that target conditions be established for each individual community to identify such features as habitat improvement, invasive species removal, buffer zone establishment, and other considerations.

458 WATER RESOURCES AND THE URBAN ENVIRONMENT

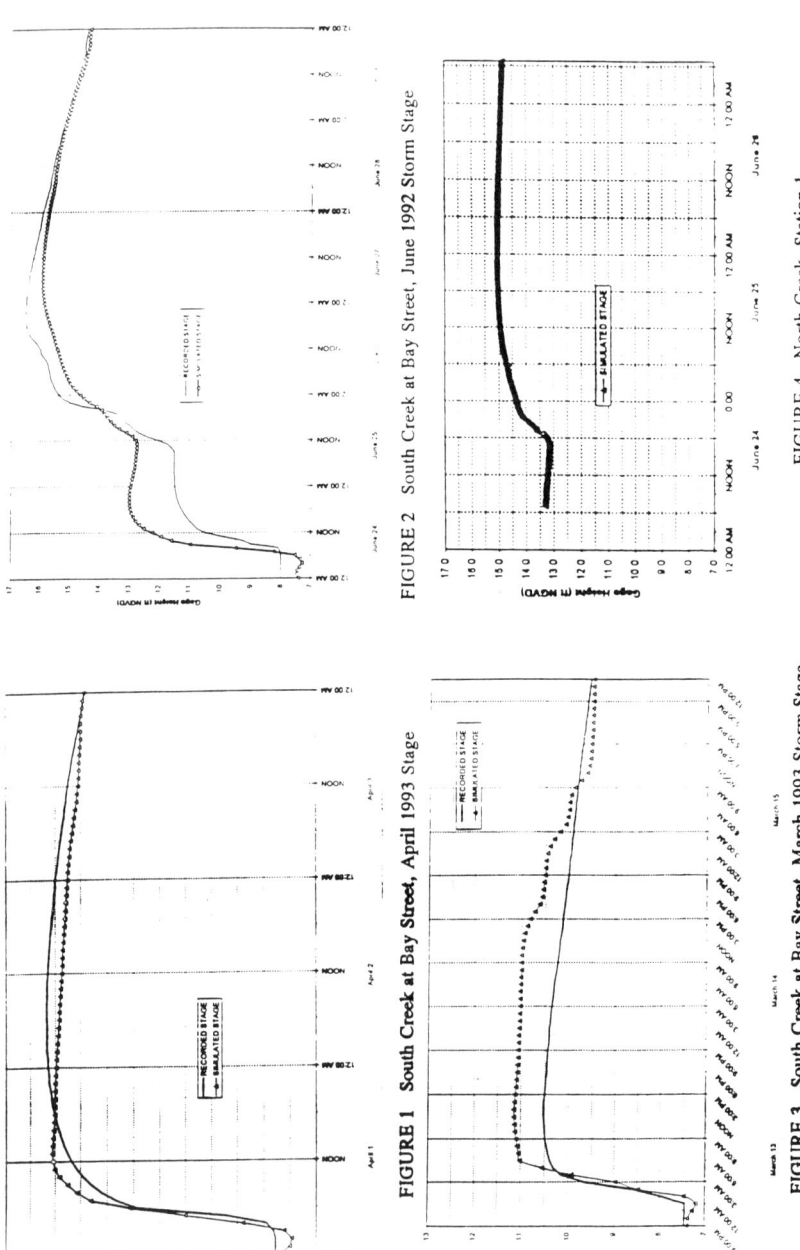

FIGURE 1  South Creek at Bay Street, April 1993 Stage

FIGURE 2  South Creek at Bay Street, June 1992 Storm Stage

FIGURE 3  South Creek at Bay Street, March 1993 Storm Stage

FIGURE 4  North Creek, Station 1

Kazakstan Irrigation and Environmental Regulations

D. M. Manbeck, M. R. Headrick and P. L. Ames[1]

Abstract

The Republic of Kazakstan is a steppe nation with about 2.5 million hectares of irrigated agriculture. Annual precipitation in the agricultural areas varies from less than 150 mm to nearly 450 mm per year. Snowmelt is the source of most surface water. Due to the highly irregular flow of most rivers, their development for economic uses is only possible through streamflow regulation by storage reservoirs. Active interest in environmental issues has developed in recent years. The Government of Kazakstan has promulgated laws, regulations and decrees which, if strictly enforced, will seriously affect irrigated agriculture. Governmental and economic changes following national independence have led to the establishment of new institutions to control the distribution of water and the regulation of its quality. With limited surface water available, conflicts are developing between agricultural, industrial, and urban users. The commonly perceived problems of chemical, sediment and salinity runoff from irrigated lands into natural lakes and rivers are prevalent. Improved water management by all users, realistic compromises between environmental and agricultural interests, and economic reuse or disposal of irrigation return flow are needed.

---

[1]Senior Agricultural Engineer, Senior Environmental Scientist and Principal Environmental Scientist, respectively. Harza Engineering Company, 233 South Wacker Drive, Chicago, IL 60606-6392

## Background

The Republic of Kazakstan is a new country established after the demise of the Union of Soviet Socialist Republics (USSR). It is located near the center of the Eurasian continent. It is bordered by three other newly independent countries in the south; China in the east; the Caspian Sea in the west; and Russia in the north. Its size is approximately equivalent to one-third of the 48 conterminous U. S. states (Figure 1). The population of Kazakstan is 17 million persons. An average density of 6.4 persons per square kilometer makes it one of the most sparsely populated countries.

Figure 1. Relative size and lattitude of the conterminous United States and Kazakstan.

Kazakstan is primarily a steppe country with large areas of short-grass plains and hills. It has a continental climate with cold, dry winters and hot, dry summers. Annual precipitation in the agricultural lands varies from less than 150 millimeters in the southwest to 450 millimeters in the foothills of the southern mountains.

There are eight major river basins but water resources are limited. Annually recurring surface water resources total about 100 cubic kilometers ($km^3$) with only 56 $km^3$ originating within the boundaries of Kazakstan. Eighty percent of the flow in Kazakstan's rivers forms in mountain regions from melt of over 2700 glaciers. Total glacier area in Kazakstan is about 2000 km$^2$. After accounting for water committed for release to Russia, domestic and ecological water supply requirements, reservoir dead storage, and conveyance losses, it is estimated that about 43 km$^3$ is available for irrigation development. Potential annual yield of usable ground water is estimated at 15 km$^3$.

Due to the highly irregular flow of most rivers, their development for economic uses is only possible through streamflow regulation by storage reservoirs which have been constructed on many rivers, especially the larger ones. At present Kazakstan has 204 reservoirs with a total capacity of 95.5 km$^3$ and a total area in excess of 10,000 km$^2$. One reservoir, Buhtarma, in eastern Kazakstan, accounts for more than half of the total volume and surface area. These reservoirs serve many purposes, e.g. hydroelectric energy, industrial and municipal water supply, and irrigation. The salinity of reservoirs varies from 0.18 deciSeimens per meter (dS/m) in East Kazakstan to about 3.1 to 4 dS/m in the central part of the country with most reservoir salinities about 1.5 dS/m.

## Environmental Issues and Regulations

Many of the major environmental issues facing the agricultural sector are concerned with the use or misuse of water in a country where precipitation is low and the demand for water is large and diverse. Most of Kazakstan's major rivers suffer from severe pollution from industrial, domestic, and agricultural wastes, due partly to excessive demand for the limited water resource. Water of various salinities is used for irrigation. About 25 percent of the total irrigated area is provided with water with a salinity of 1.5 to 3 dS/m but some irrigation water is more than 3 dS/m. This creates a critical situation, especially where drainage is poor or absent and irrigation applications are high, causing salinization and requiring additional water applications in order to leach out salts and attain acceptable soil salinities.

Some USSR environmental regulations and concepts have been retained and the Government of Kazakstan has issued a number of laws, regulations, and decrees, many of which, if strictly enforced, will seriously affect irrigated agriculture as historically practiced.

During recent decades, water levels in natural receiving water bodies, such as the Aral Sea and Lake Balkhash, have lowered excessively and the quality of the water has deteriorated. A commonly stated reason is the large amounts of water diverted upstream from the supplying rivers for inefficient irrigated agriculture. Not only has less water reached the receiving sinks but there has been less good quality mountain runoff water in the rivers to dilute the pollutants entering the downstream river sections, thus reducing the water quality in the sinks.

It is generally believed that the water resources of the Syr Darya basin, leading to the Aral Sea, are almost exhausted. However, discharge from the Chardara Reservoir on the Uzbekistan border is limited by channel encroachments on downstream settlements. Moreover, four barrages on the river in Kazakstan do not have the capacity to pass high flows. Thus, when water exceeding the limited capacity is available from upstream, it is diverted to evaporate in a blind basin in the desert in Uzbekistan. Water has not reached the mouth of the Syr Darya River (Aral Sea) since 1964, which has resulted in serious changes for delta wildlife.

Secondarily treated municipal and industrial waste water from the capital city of Almaty is conveyed in a natural waterway to a storage reservoir. It is then distributed to sprinkler systems to irrigate forage crops. Other municipal/industrial waste water in Kazakstan is often diverted into natural depressions to evaporate in order to avoid polluting natural waters. National regulations state that waste water is to be used to irrigate only forage crops. Also, international lending agencies have regulations and policies which require their most stringent Environmental Impact Assessments when waste water is proposed for irrigation, especially for edible crops. However, with well managed chemical treatment to kill pathogens, sewage effluent could even be applied to irrigate edible crops.

Kazakstan strives to avoid water pollution by regulating discharges to natural waters. Drainage from irrigated lands, waste water, and even diverted, but unused, reservoir water is prohibited by law to enter natural waters. Such waters are often diverted into natural depressions to evaporate. Obviously, this practice cannot long continue in areas that are short of water.

Runoff waters must be applied for their best beneficial use.

Environmental regulations that are likely to be enforced significantly affect irrigation planning and practices in Kazakstan. Management of water quality is the major factor. Prohibiting irrigation runoff and drainage waters from entering natural waters necessitates providing evaporation sites to dispose of the prohibited waters, which reduces the available water resources downstream. Agricultural production may decrease over time as a result of reuse of drainage water and increase of soil salinity to intolerable levels. Compliance with the prohibition may also result not irrigating at all to prevent runoff from entering natural waters.

There is a major problem of disposal of effluent from agricultural drains. Several of the major drainage systems have outlets into closed sinks whose evaporation capacity is exceeded by the drainage inflows. This causes backing up in the collector system and flooding, resulting in loss of irrigated lands. When the discharge restriction regulation is observed, irrigation planning must include provision for balancing the expected runoff with evaporation capacity.

Modification of this prohibition policy is necessary to obtain an acceptable balance of agricultural and environmental concerns, so that Kazakstan can attain sustainable agriculture production to provide for its people.

Past mismanagement of irrigation applications, the lack of effective drainage and attempts to store saline drainage flows have resulted in the creation of numerous temporary wetlands. These wetlands will either be lost as the land is redeveloped for intensive agriculture or will expand until a balance between water input and wetland evapotranspiration/evaporation is established.

The USSR banned most ecologically damaging agricultural chemicals during the 1970s. Despite this regulation, state and collective farms, under pressure to meet production quotas and with economic restrictions, continued to use stockpiles acquired in earlier years.

The selection and application of pesticides is regulated by the Interagency Commission on Sampling, which determines which chemicals may be applied to each crop for each pest, establishes application rates and methods, and sets residue tolerances for foodstuffs. Since independence, chemical use has declined due to shortages of funds, problems of supply, and lack of operable equipment. Chemicals that are used are applied at rates and frequencies well below effective levels. Frequently, the chemical applied, the method of application, or the timing is inappropriate but the chemical is used because it is available while the needed one is not. However, irrigation runoff waters with chemical products are sometimes less mineralized than the water that is in the lower reaches of many of the river systems. Discharge of runoff flows into rivers could be used as part of the strategy to control the salinization of the Aral Sea and Lake Balkhash.

The environmental assessment (EA) process is in its infancy and will take some years to develop properly. At present, each development project is required to be assessed for environmental impacts and a report included with the feasibility report when this is submitted for approval. The Ministry of Ecology and Bioresources (MEBR) reviews the EA and its comments are forwarded to those considering the project for approval. The EA is to contain a mitigation plan and a monitoring plan, which are supposed to be overseen by the regional office. Money for extensive monitoring of projects is lacking at the regional offices, making this function largely nonexistent. Regional office staff members are largely untrained in environmental assessment techniques and lack the international contacts to acquire expertise Nationally, the situation is improving due to a university level training program funded by the World Bank. The general level of expertise is substantially better than it was five years ago.

MEBR disapproval of a project can, in theory, result in cancellation but this rarely happens. There is no higher authority to which MEBR can turn for support, nor is there a precedent of court action, so the usual practice is to negotiate project modifications with the development agency. In some cases, growing public interest in the environment has resulted in opposition to a project, aiding the Ministry in imposing project mitigation.

OPERATION AND PERFORMANCE OF RESERVOIR RELEASE
IMPROVEMENTS AT 16 TVA DAMS

John M. Higgins,[1] M. ASCE, and W. Gary Brock[2]

Abstract

In 1996, the Tennessee Valley Authority (TVA) completed a five-year, $50 million program to improve low dissolved oxygen (DO) concentrations in releases from 16 dams in the Tennessee Valley. Prior to the program, over 500 km of tailwater were impacted, affecting downstream water quality, aquatic habitat, recreation, and waste assimilation. To improve the releases, TVA worked with resource agencies and private organizations to define minimum dissolved oxygen targets. Facilities were installed and operating procedures developed to meet the target conditions.

Introduction

In 1987, TVA authorized a comprehensive review of reservoir operating priorities that had been followed since 1933. The purpose was to ensure optimum operation of the reservoir system, recognizing that needs, demands, and values change over time. Navigation, flood control, and power production are primary operating objectives, but water quality, recreation, and economic development also provide substantial benefits. The review was in response to a growing number of requests from people in the Tennessee

---

[1] Water Quality Engineer, Environmental Compliance, Tennessee Valley Authority, Chattanooga, Tennessee.
[2] Program Manager, Water and Hydro Projects, Tennessee Valley Authority, Knoxville, Tennessee.

Valley for major adjustments in TVA's reservoir management policy.

Results of the review are presented in the Final Environmental Impact Statement, issued in December 1990. Dissolved oxygen improvements were recommended for 16 of the 40 major TVA dams. The 16 dams included in the program are shown in Table 1. The targets were generally set at 6.0 mg/L for coldwater fisheries and 4.0 mg/L for warmwater fisheries. In early 1991, the TVA Board of Directors accepted the report recommendations and initiated the Reservoir Releases Improvements Program.

**TABLE 1. Characteristics of Reservoir Release Dissolved Oxygen Concentrations[a]**

| Dam (1) | Dissolved Oxygen Concentrations, in milligrams per liter | | Impacted Stream Length in kilometers (4) | Mean Number of Days Concentration Below Target | |
|---|---|---|---|---|---|
| | Target (2) | Mean Minimum (3) | | 1960-1996[a] (5) | 1997 (6) |
| Apalachia | 6 | 5.0 | 3 | 64 | 0 |
| Blue Ridge | 6 | 3.4 | 24 | 83 | 24 |
| Boone | 4 | 3.9 | 16 | 46 | 0 |
| Chatuge | 4 | 1.3 | 11 | 91 | 0 |
| Cherokee | 4 | 0.2 | 80 | 122 | 3 |
| Douglas | 4 | 0.9 | 129 | 113 | 16 |
| Fontana | 6 | 2.7 | 8 | 54 | 0 |
| Ft. Loudoun | 4 | 3.7 | 68 | 17 | 0 |
| Ft. Patrick | 4 | 3.8 | 8 | 59 | 0 |
| Hiwassee | 6 | 4.1 | 5 | 82 | 83 |
| Norris | 6 | 1.0 | 21 | 120 | 72 |
| Nottely | 4 | 1.0 | 5 | 81 | 1 |
| S. Holston | 6 | 0.8 | 10 | 122 | 0 |
| Tims Ford | 6 | 0.4 | 64 | 199 | 47 |
| Watauga | 6 | 3.8 | 3 | 66 | 21 |
| Watts Bar | 4 | 4.5 | 48 | 27 | 0 |
| Total | | | 503 | 1346 | 267 |

[a]Based on weekly dissolved oxygen data from 1960-1996. Data affected by aeration are omitted.

Improvement Measures

Five basic techniques are used to increase dissolved oxygen concentrations: 1) turbine venting, 2) turbine air injection, 3) surface water pumps, 4) oxygen injection, and 5) aerating weir. These techniques offer a range of aeration alternatives with varying costs and operational requirements. Each project was evaluated separately to determine the most appropriate approach. The primary factors included aeration capacity, physical characteristics, and cost.

Turbine venting takes advantage of subatmospheric pressure that exists at the vacuum breaker outlet of some turbines (Carter, 1995). This approach is used for 22 turbine units at nine TVA dams (Apalachia, Boone, Cherokee, Douglas, Fontana, Hiwassee, Norris, South Holston, and Watauga). Dissolved oxygen uptake ranges from 0.7 mg/L at Boone Dam to over 3 mg/L at Norris Dam. The loss in power efficiency is less than 1.3 percent at all projects. Turbine venting is often the least costly aeration alternative.

Turbine air injection (i.e., blowers and compressors) is used at Nottely and Tims Ford Dams where draft tube pressures are not low enough to aspirate sufficient air. The air injection systems increase DO approximately 5 mg/L at Nottely and 2 mg/L at Tims Ford.

Nine surface water pumps are used at Cherokee and at Douglas Dams when the DO falls below the turbine venting capacity (Mobley, et al., 1995). The pumps are designed to move 15 $m^3$/s of water and make up about 30 percent of the total discharge. The pumps increase DO from 1.5 to 2.0 mg/L. Costs are typically more than turbine venting but less than oxygen injection.

Oxygen injection is used at seven dams where other, less costly, alternatives cannot meet the DO target (Blue Ridge, Cherokee, Douglas, Fort Loudoun, Hiwassee, Tims Ford, and Watts Bar). The basic system consists of an oxygen supply facility (using commercial liquid oxygen at most projects) and a porous hose diffuser system for transferring the oxygen to the reservoir water just before it enters the turbine intake (Mobley, et al., 1996). The DO increase ranges from just over 1 mg/L at Fort Loudoun and Watts Bar to over 3 mg/L at Blue Ridge. Capital costs typically exceed $1 million.

An aerating weir is the only facility used at Chatuge Dam and the primary facility at South Holston Dam. The Chatuge infuser weir increases DO by 2 to 4 mg/L. The South Holston labyrinth weir increases DO by 2 to 3 mg/L (Hauser and Morris, 1995). These passive aeration facilities function automatically with few operational requirements.

Operations

The basic requirements include: 1) equipment operation and maintenance, 2) monitoring of reservoir releases and equipment performance, and 3) operational decisions and coordination. Existing organizations and personnel perform these functions with additional training, communication, and operating procedures. Operation and maintenance manuals are provided for each plant. Aeration orders are issued during the season for the operation of each system (i.e., initiating, terminating, or modifying aeration). Technical support is provided by the design engineers for nonroutine operation and maintenance problems.

Three types of performance and operational monitoring are conducted. The first involves visual observations by on-site plant personnel. The second is remote observation, recording, and electronic transfer of equipment performance parameters. The third involves monitoring of downstream river conditions to determine aeration results and to modify operations.

Minor adjustments, based on real time observations, are the responsibility of hydro plant personnel. Decisions regarding deployment or major change in aeration is based on a broader view of river conditions and system operations. These actions are centralized among a water quality specialist, an aeration specialist, and a reservoir operations specialist. The specialists have access to the expertise and information needed for three basic operational activities: 1) aeration orders to start, stop, or modify aeration; 2) repair or replacement of malfunctioning equipment; and 3) material procurement, primarily liquid oxygen.

Performance

Dissolved oxygen performance is measured by comparing actual river concentrations with the target concentration. Figure 1 shows the 1997 (with aeration) and the historical (without aeration) DO concentration of releases from Cherokee Reservoir. Historically, the

DO level fell below the target between mid-May and mid-June. Turbine venting was initiated in May, the surface water pumps started in July, and oxygen injection initiated in August. With few exceptions, the aeration systems maintained DO above the target until lake turnover (thermal mixing) restored natural aeration in October.

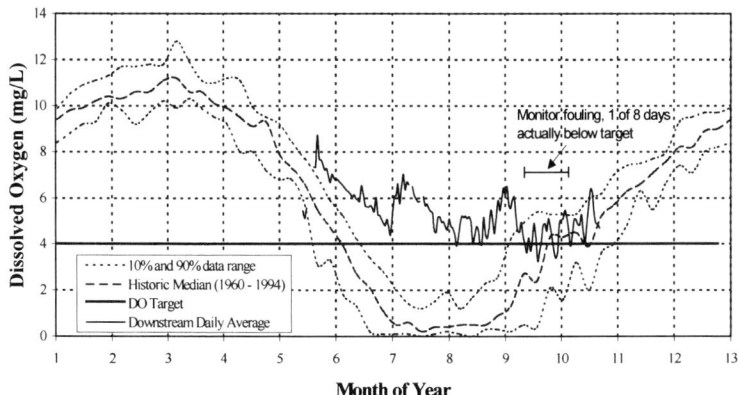

**Figure 1. Dissolved Oxygen Concentration in Releases from Cherokee Dam During 1997**

In 1997, the DO of reservoir releases from Cherokee Dam was below the target for three days (based on the daily average of 15-minute observations and excluding low readings due to probe fouling). Table 1 compares 1997 results for the 16 aeration projects with the historical record. The total number of days for all projects was reduced from an historical average of 1346 days per year to less than 267 days in 1997. Approximately 70 percent of the remaining days below the target are from Hiwassee, Norris, and Tims Ford where additional work is needed to improve aeration capacity.

Fish and macroinvertebrate communities have generally responded positively to the improved DO conditions. Data from 1971 through 1995 indicate improvements in both biological communities and fishing activity (Scott, et al., 1996).

Conclusions

1. Historic reservoir operating policies can be reviewed and updated with proper public participation and where consistent with statutory requirements.

2. Sufficient technology and operating experience is available for economically improving the dissolved oxygen concentration of reservoir releases.

3. Operation and maintenance requirements of aeration facilities are significant, requiring integration of diverse expertise, personnel, and responsibilities.

Appendix 1. References

1. Carter, J., 1995. "Recent Experience with Turbine Venting at TVA," Waterpower '95, ASCE.

2. Hauser, G. E., and Morris, D. I., 1995. "High Performance Aerating Weirs for Dissolved Oxygen Improvement," Proceedings of Waterpower '95, San Francisco, CA.

3. Mobley, M. H., and Brock W. G., 1996. "Aeration of Reservoirs and Releases Using TVA Porous Hose Line Diffusers," ASCE North American Congress on Water and the Environment, Anaheim, CA.

4. Mobley, M., Tyson, W., Webb, J., and Brock, G., 1995. "Surface Water Pumps to Improve Dissolved Oxygen Content of Hydropower Releases," Proceedings of the International Conference on Hydropower, Vol. 1.

5. Scott, E. M., Gardner, K. D., Baxter, D. S., and Yeager B. L., 1996. "Biological and Water Quality Responses in Tributary Tailwaters to Dissolved Oxygen and Minimum Flow Improvements," Tennessee Valley Authority.

# DECISION SUPPORT SYSTEM FOR RIVER BASIN WATER MANAGEMENT

Ick Hwan Ko[1], Darrell G. Fontane[2], M ASCE, and John W. Labadie[2], M ASCE

## ABSTRACT

One of the major concerns for environmentally sensitive water management is how to plan and operate large-scale upstream reservoir systems within a river basin to mitigate downstream water quality problems, while satisfying other system objectives such as water supply and hydropower generation. A prototype decision support system (DSS) is presented to assist in operational planning for the Han river basin in Korea, considering both water quantity and water quality objectives. Three models were developed for the DSS. The models are: a multi-objective, multi-reservoir, dynamic programming model; a river water quality simulation model and; a linear programming, optimal waste load allocation model. The DSS also contains geographic information system software for display of spatial data and a spreadsheet software based user interface, database, and graphical display system.

## INTRODUCTION

Planning and operating river basin water resources in a sustainable way requires meeting water quantity and quality objectives over time and space. This is one of the most difficult problems in water management. Since river management involves the coordination of the uses of water within a basin, achieving geographic and ecological integration is a key issue. However, due to the differing objectives for the operation of river basin water resources systems, difficult problems may arise for water resources managers since multi-objectives may be in direct conflict or competition with each other. Furthermore, more complicated relationships among multiple water users and the water system,

---

1 Principal Researcher at Water Resources Research Institute, Korea Water Resources Corporation, 462-1 Jeonmin-Dong, Taejon, 305-390, South Korea.
2 Professors, Department of Civil Engineering, Colorado State University, Ft. Collins, Colorado 80523.

decision making considering large numbers of planning and management alternatives, and the far ranging effects of management policies are dramatically increasing difficulties in planning and managing river basin water resources. Innovative and holistic approaches are urgently required. A decision support system (DSS) is a computer-based, advisory system for management that employs databases, models, and communication/dialogue systems to provide decision makers with management information (Grigg, 1996). In this study, a DSS, integrating water quantity and quality analysis in the context of river basin water resources planning and management, is presented for the Han River basin in Korea.

## HAN RIVER BASIN

The Han river basin, located in the middle part of the Korean peninsula, has a drainage area of 26,200 squared kilometers and discharges about 18,000 million cubic meters (MCM) of flow annually. The river consists of two main branches, the North Han and the South Han that confluence about 35 km upstream of the capital city of Seoul. From July to September the weather is influenced by the monsoon characteristics of the South Pacific Ocean. More than two-thirds of the annual precipitation, ranging from 500 to 1,500 millimeters, generally occurs during these three months. There are nine reservoirs with hydropower plants in the basin. Two multipurpose storage reservoirs, Soyang and Chungju-1, and one flow-through reservoir, Chungju-2, are managed by a government-run-agency, the Korea Water Resources Corporation (KOWACO). The other storage reservoir, Hwachon, and five other smaller, flow-through reservoirs are managed, for the single purpose of hydro-energy generation, by a different government-run-agency, the Korea Electric Power Corporation (KEPCO). The need for integrated and sustainable water management in the basin is becoming more crucial since water supply demand is rapidly increasing and water quality is deteriorating. Due to the low potential storage and the limitation of feasible sites for ground water development, surface storage reservoirs play an important role in the region as a main source of the basin water supply. Especially, the operation of the three headwater storage projects, Hwachon, Soyang, and Chungju-1, is fundamental to meet the municipal and industrial water supply of the Seoul metropolitan area. Releases from these three reservoirs also have a direct impact on water quality throughout the basin.

## HANDSS DEVELOPMENT

A decision support system was designed to organize and automate the process of alternative evaluation and selection into a flexible, user-friendly computer environment. The design of the DSS is intended to be general and applicable to various river basins. However, in developing the general DSS, the Han river basin was used as a guide to design the components. The developed system, HANDSS, was focused on answering two fundamental questions which should be raised for integrated basin water quantity and quality control problems: (1) How much water should be released from each upstream reservoir, at each time period, to meet the downstream water quality objectives, while simultaneously maximizing various water quantity objectives for the best utilization of the limited basin water resources?; and (2) To what degree should the wastewater be treated, for the areas between the headwater storage reservoirs and the downstream water quality control point, based upon optimal waste load allocation strategies determined in conjunction with optimal reservoir system operation? The developed prototype system integrates water quality simulation and water quantity optimization models, a geographical information system (GIS) module, a database, and a graphical user interface.

Data Subsystem: The general data subsystem consists of various "Notebook Tables", or pages that contain data for a specific purpose, developed using the Windows Version of the Quattro Pro spreadsheet (Novell, 1995). Since the purpose of this effort was to assess operational planning strategies, a monthly time step for the models was used. Therefore, the data requirements are small and can easily handled by spreadsheet software. Specific "Notebook Data Tables" included in the DSS are: (a) Historic monthly inflow records; (b) Stochastically generated monthly inflow sequences; (c) Input control data for the reservoir system optimization model; (d) Physical data for reservoirs and hydro-power plants (static data); (e) Hydrologic and operational data for reservoirs/hydro-power plants data (dynamic data); (f) Results from the reservoir system optimization model; (g) Monthly headwater flows and other input data for the monthly water quality simulation model; (h) Results from the monthly water quality simulation model; (i) Project and DB files for the GIS module; (j) Monthly input data and results from the waste load allocation module. Since the DSS was designed to allow flexibility to change components of the software, a modular approach based upon individual files was created.

Model Subsystem: Three main models have been developed to answer the two fundamental questions posed for basin water management. The first model, H3DP, is a multi-dimensional dynamic programming optimization model for reservoir system analysis considering both downstream water quantity and quality objectives. The developed model is based on the incremental dynamic programming (IDP) algorithm (Larson, 1968) imbedded within generalized dynamic programming software, CSUDP (Labadie, 1990). The model is used to optimize the monthly operation of the Hwachon and Soyang reservoirs on the North Han river and Chungju-1 reservoir on the South Han river. A flow-through reservoir, Paldang, is located at the confluence of the North and the South Han rivers. Because this reservoir is the principal water depletion site for the Seoul metropolitan area, it is considered as both a primary water quantity (basin firm water supply) and water quality control point. The main objectives considered in the H3DP optimization model include hydro-energy generation, water supply, and water quality (BOD) control. To simulate the water quality in the river system, the QUAL2E model (Brown and Barnwell, 1987) developed by the United States Environmental Protection Agency (USEPA) was used as the basis for the second main model. Reservoir water quality was not explicitly modeling in this study, although it is anticipated as a future system enhancement. Historical data was used to estimate the likely quality of the reservoir release and this was included as an input to the QUAL2E model. Monthly flow data and water quality data from 1993 were used for the QUAL2E model calibration. To effectively link the results of the water quality and water quantity models, it is necessary to include constraints on water quality within the H3DP optimization model. Since the models are separate, this must be done in an iterative manner. For example, the water quality model might be run with flows generated from the water quantity model. If the predicted water quality does not satisfy desired limits, then the input for water quantity model must be adjusted to produce a more desirable distribution of flows. This process can sometimes be very time consuming. In an attempt to minimize the number of iterations and time required for this model adjustment process, least squares regression analysis was used to develop empirical water quality equations for each month for the basin. These equations related headwater releases with predicted DO and BOD at downstream control points. The data for the regression analysis were developed by running the QUAL2E model using 150 to 350 sets of potential reservoir releases for each month. The empirical equations fit the data extremely well. The developed empirical equations for BOD, which evolved as the primary variable of interest in this study, were then directly incorporated into the H3DP model to evaluate the water quality constraints. The third main model was the optimal waste load allocation (WLA) model. This model was used to determine the optimal amount of waste treatment to be provided at various locations in the basin. This model was developed for the Han river basin based upon a linear programming formulation proposed by Deninger (1965). The linear programming WLA model was solved using the optimization module within the Quattro Pro spreadsheet software. Details of the WLA model are provided by Ko (1997).

GUI/GDS Subsystem: The graphical user interface (GUI) for the system consists of a menu driven control program and a graphical display system (GDS) developed using the Quattro Pro spreadsheet software. The macro-based control program organizes the process of querying the data tables for the required model inputs, running the models, importing the model output into "Notebook Data Tables", and finally calling either the GDS or the GIS module. The program also provides the interactive feedback process for the DSS. Flexible input data selection and editing options were prepared for each module using dialog boxes. For example, within this GUI environment, a user can run the H3DP optimization model using either historic and stochastic inflow series, specify the number of peak generation hours for each power plant, set minimum monthly downstream water supply requirements at Paldang reservoir, and change the weighting factors for the various purposes contained in the objective function. The user can also select required water quality standards for the basin.

GIS Module: One of the most visible and impressive aspects of GIS technology is in its output and display capability which provides the user with a visual image of the area. GIS modules for the HANDSS were developed to display the output using ARC-VIEW software (ESRI, 1994). One module displays spatial and temporal water quality related information in the basin. Another module integrates GIS with the Q2LF basin water quality simulation module, to display the predicted water quality along the Han river reaches.

## CONCLUSIONS

With available computing technology, it is possible to effectively combine water quantity and water quality models within a DSS framework. The prototype HANDSS has potential for use by basin water and environmental managers as a common tool to assist joint and coordinated decision making for evaluating Han river basin operational strategies. The key advantage of utilizing the developed HANDSS is that the user can analyze the combination of streamflow regulation, through optimal reservoir system operation, and cost-effective regional waste treatment, in order to meet specified basin water objectives. Generating viable options leads to a foundation for maximizing the productivity and reliability of complex water resources systems while minimizing adverse environmental impacts, and minimizing conflicts between multiple water uses and users.

## REFERENCES
1. Brown, L.C., and Barnwell, T.O. (1987). "The Enhanced Stream Water Quality Models QUAL2E and QUAL2E-UNCAS: Documentation and User Manual," Report # EPA/600/3-87/007, Athens, GA.
2. Deininger, R.A. (1965). "Water Quality Management - The Planning of Economically Optimal Pollution Control Systems", Proceedings of the 1st Annual Meeting, AWRA, University of Chicago, pp. 254-282.
3. ESRI (1994). **ArcView** Version 2.0.
4. Grigg, N.S. (1996). **Water Management - Principal and Cases**, McGraw Hill, New York, NY.
5. Ko, I.H. (1997). " Integrated River Basin Operational Planning Considering Water Quantity and Water Quality", Ph.D. Dissertation, Department of Civil Engineering, Colorado State University, Fort Collins, CO.
6. Labadie, J.W. (1990). "Dynamic Programming with the Microcomputer," **Encyclopedia of Microcomputers**, A. Kent and J. Williams, eds., Marcel Dekker, Inc., New York, NY.
7. Larson, R.E. (1968). **State Incremental Dynamic Programming**, American Elsevier Publishing Co., Inc., New York, NY.
8. Novell (1995). Perfect Office for Windows, Quattro Pro Spreadsheet and Graphics, User's Guide, Version 6.1.

Flood Control Effects of Hwachon Dam
in Connection with Peace Dam

Myung Pil Shim[1], Oh Ig Kwon[2], Kyung Tak Kim[3]

Abstract

The Peace Dam was built primarily to protect flooding from Kumgangsan Dam currently under construction in North Korea and has direct effects on hydrological characteristics of the Hwacheon Dam(H-dam) located downstream in South Korea. This paper is a review of the role of the Peace Dam and its use in connection with H-dam for the purpose of both flood control and water conservation. This study suggests the most efficient policies for using the unused capacity of the Peace Dam. Also, a possible scenario with severe flood inflows from North Korea is considered.

Introduction

The Peace Dam(P-dam) in Korea might be a unique type of dam because its purpose is to protect flooding from the Kumgangsan Dam(K-dam, called Imnam Dam in North Korea). The K-dam, identified as 121.5 m high and 2.6 billion $m^3$ of storage capacity, is under construction in North Korea, across the Demilitarized Zone(DMZ). In 1988 the P-dam, a rockfilled dam of 420 m in length and 80 m in height, was built primarily as a buffer to cope with failure in reservoir operation or destruction of K-dam along the Han River 11 km downstream to the south of the DMZ. It has a storage capacity of 590 million $m^3$ with four drainage tunnels of 10 meters in diameter on the bottom side of the reservoir instead of usual spillway over the dam to mitigate the flood damage(Shim et al., 1995).

But now, it is recommended to use the dam in connection with

---

[1]Professor, [3]Graduate Student, Dept. of Civil Eng., Inha University, Inchon 402-751, Seoul, Korea
[2]Senior Researcher, Korea Institute of Construction and Technology, GoYang 411-410, Korea

H-dam, located 27 km downstream the P-dam, for the purposes of both flood control and water conservation. During flood period from June 21 until September 20 in Korea, normal high water level(NHWL) of reservoir is required to come down to temporary high water level(THWL) or restricted water level(RWL) to ensure more space of inflows expected in the reservoir. It might be an opportunity to raise THWL of H-dam a little higher than current level when the capacity of P-dam can be practically used for flood control until completion of K-dam.

This study presents and compares the simulation results between single operation of H-dam only and combined operation in connection with P-dam through separation of the basin situations in the past, present and future. Also, a possible scenario with severe flood inflows from North Korea is considered. This paper analyzes the reallocation of THWL of H-dam and suggests the most efficient policies for using the unused capacity of P-dam.

Definition of subbasin area and its flood inflows

The inflow of flood to H-dam will be a sum of inflows from H-dam basin and emergency tunnels of P-dam and controlled inflows from K-dam after K-dam is built completely. So, the reservoir water level of H-dam depends on the various situations of the total inflows release scheme, capacity of flood control and so on. Fig.1(a) illustrates the location of each dam site and its associated area of subbasin in the past, present and future. Fig.1(b) shows the whole basin before the construction of P-dam in 1988, and its inflow hydrograph can be expressed as one hydrograph. Fig.1(c) illustrates the present situation between after construction of P-dam and before completion of K-dam. For this case, two inflows hydrograph must be used respectively and flooding into P-dam reservoir will raise the water level of P-dam and will be released through the four drainage tunnels on the bottom side of P-dam. So, P-dam has a role as storage reservoir due to the effect of delay through the tunnels although it has not gates. Fig.1(d) shows that the area will be composed of three divided subbasin in future situation.

Formulation of simulation models

Many models have been developed in the past to derive the best policy for reservoir operations during rainy season. Application of the systems approach to reservoir management and operation can be classified into four categories: simulation, optimization, multi-objective analysis and combinations of the previous techniques(HEC-5, 1985; Feldman, 1992).

Reservoir water level of H-dam depends on the various situations of the total inflows and release schemes and so on. This study uses release schemes of H-dam with three cases of constant rate, constant amount of inflow discharge and both respectively. Fig.2 shows the diagram for the whole system of inflows and releases of each dam. The first character and subscript of each variable means the initial of each dam. Small letters as *qin* and *qouts* mean inflow and releases.

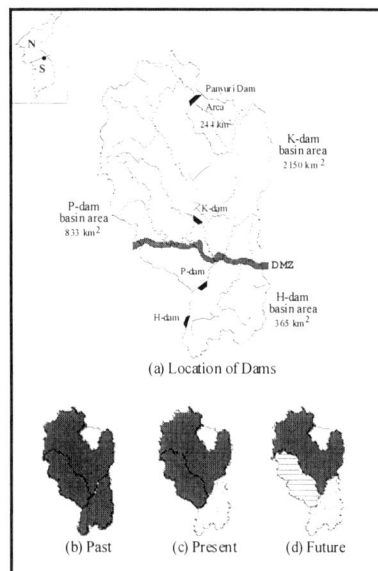

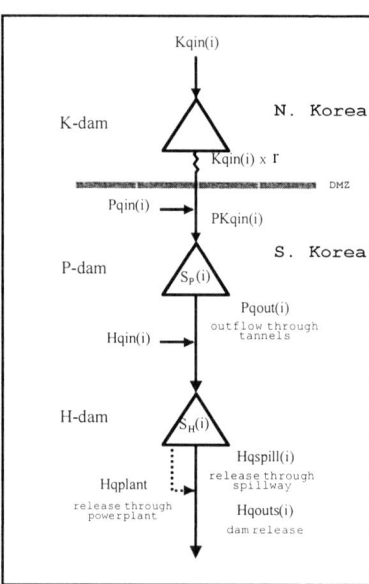

Fig.1 Location of dam site and its classification of subbasin

Fig.2 Diagram for the whole system.

## Results of simulation and analysis

(1) Single operation of H-dam before construction of P-dam

Simulation is carried out based on historical flood of 50,100 and 200 years respectively. Table 1 shows the results of simulation based 50 years flood with RWL of El. 175 m. Release rate is the ratio of maximum release to peak inflows and controlled rate is equal to one minus the release rate which means the ratio of storage in the reservoir to the peak inflow. Constant rate-amount type means to release constant rate of inflows up to peak time and constant amount there after. For the case of 100 and 200 years flood, the highest water level during operation goes over the FWL of H-dam with any type of release scheme because H-dam was designed basically for 50 years flood.

(2) Combined operation after construction of P-dam

Table 2 shows the results of simulation after construction of P-dam with release scheme of constant rate based on historical floods. Peak inflow of H-dam is much less than the one obtained with single operation of H-dam due to the effect of storage of P-dam.

Table.1 Results of single operation of H-dam based 50 years flood before construction of P-dam

| release type | peak inflow | max. release | total volume | | | release rate | control rate | Hwanchon Dam | |
|---|---|---|---|---|---|---|---|---|---|
| | | | inflow | release | storage | | | highest level | time |
| | m³/sec | | 10⁶ m³ | | | % | | EL. m | hours |
| constant rate | 11,500 | 8,620 | 1,046.81 | 768.57 | 278.24 | 75 | 25 | 180.00 | 119 |
| constant amount | 11,500 | 7,510 | 1,046.81 | 768.96 | 277.85 | 65 | 35 | 183.00 | 23 |

rem. Shadowed area means the overflow of flood water level of H-dam.

Table.2 Results of single operation of H-dam with release type of constant rate after construction of P-dam.

| return period (years) | peak inflow | max. release | total volume | | | | release rate | control rate | highest water level of P-dam |
|---|---|---|---|---|---|---|---|---|---|
| | | | inflow | release | storage | | | | |
| | | | | | H-dam | P-dam | | | |
| | m³/sec | | 10⁶ m³ | | | | % | | EL. m |
| 50 | 7,327 | 4,922 | 1,059.79 | 782.48 | 231.54 | 45.78 | 67 | 33 | 205.93 |
| 100 | 8,106 | 5,414 | 1,175.41 | 902.28 | 227.95 | 45.18 | 67 | 33 | 209.19 |
| 200 | 8,926 | 5,916 | 1,303.49 | 1,028.66 | 229.16 | 45.66 | 66 | 34 | 212.44 |

Table.3 Comparisons of results before and after construction of P-dam.(50 years flood with RWL of H-dam, El. 175 m)

| release type | peak inflows(m³/sec) | | | max. releases (m³/sec) | | | highest water level of P-dam (EL. m) |
|---|---|---|---|---|---|---|---|
| | before | after | reduction (%) | before | after | reduction (%) | |
| constant rate | 11,500 | 7,327 | 36.3 | 8,620 | 4,922 | 42.9 | 205.93 |
| constant amount | 11,500 | 7,326 | 36.3 | 7,510 | 4,000 | 46.7 | 206.05 |

Table 3 compares the simulation results of H-dam before and after construction of P-dam with each type of release scheme to show the effect of reduction for peak inflows and maximum releases. Table 3 illustrates that 36.3% of peak inflow was reduced and 42.9% and 46.7% of maximum release were reduced respectively after construction of P-dam. This means storage capacity of P-dam was used to mitigate the flood. Fig.3 shows the variation of inflows and releases to know the effect of P-dam. It can be concluded that the P-dam has a role to reduce peak inflows and releases of H-dam.

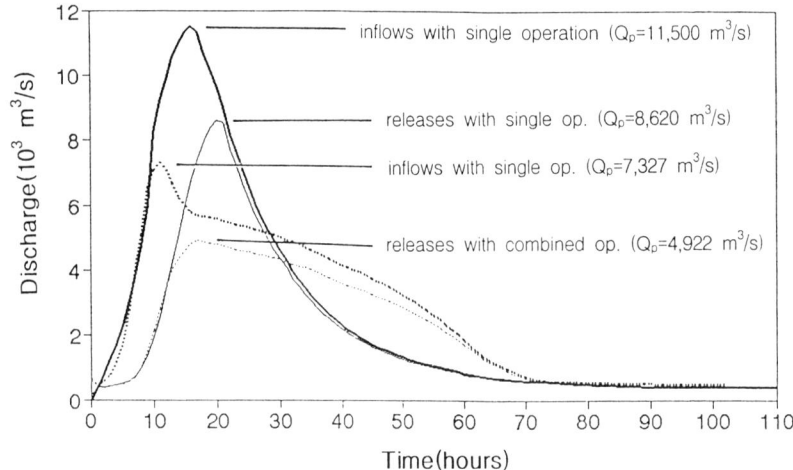

Fig.3 Variation of inflows and releases

(3) Simulated operation after completion of K-dam

After completion of K-dam, releases from the dam will depend on operation rules of K-dam in North Korea. Unfortunately, it is hard to anticipate the amount, time and duration of releases from the K-dam. Therefore, it should be assumed that the release is occurred during either around the peak time of discharge or at the time of the highest water level of H-dam to consider severe flood inflows. Table 4 compares the results of single operation with ones obtained with/without inflows from the K-dam while maintaining current RWL of H-dam as El. 175 m. If floods are intercepted completely by K-dam, total volume of inflow are reduced to 41% than before. The release rates of H-dam are 47% and 67% respectively depending on the situation of inflow from K-dam.

(4) Reallocation of Restricted Water Level of Hwachon Dam

The study was carried out for the simulation with different release type according to different RWL of H-dam. This study suggests the efficient policies for using the unused capacity of

P-dam. The benefits of the approach are illustrated through an application of floods with exceedable probability according to different THWL of H-dam. The study suggests to raise THWL of H-dam than the current level. If RWL of H-dam can be raised from El. 175 m to 180 m during flood period, the capacity of 167 million m$^3$ can be utilized for water conservation.

Table.4 Results of simulation with and without inflows from K-dam.

| 50-years flood with RWL of El. 175 m | single operation (before construction of K-dam) | combined operation (inflow of K-dam) | |
|---|---|---|---|
| | | without | with |
| total volume of inflows (10$^6$ m$^3$) | 1,046.81 | 424.52 | 1,014.02 |
| peak inflow (m$^3$/sec) | 11,500 | 6,597 | 7,327 |
| max. release(m$^3$/sec) | 8,620 | 3,092 | 4,922 |
| release rate (%) | 75 | 47 | 67 |
| control rate (%) | 25 | 53 | 33 |

Conclusions

From the study, it is found that Peace Dam stores incoming flood water and improves the flood control capacity of Hwachon Dam although it has no gate control. This study suggests the efficient policies for using the unused capacity of P-dam. The study suggests raising THWL of H-dam than the current level while the K-dam is under construction. If RWL of H-dam can be raised from El. 175 m to 180 m, the capacity of 167 million m$^3$ can be utilized for water conservation. After completion of K-dam, inflows with all possible scenario must be considered including failure in reservoir operation or destruction of the dam itself.

References

Feldman, A. D.(1992). "Systems analysis applications at the Hydrologic Engineering Center." J. Water Resour. Plng. Mgmt.,ASCE,118(3), pp.249-261.
HEC-5(1985). "HEC-5 simulation of flood and conservation systems, users manual." U.S. Army Corps of Engineers, Hydrologic Engineering Center.
Shim et al.(1995), "A study on combined operation of Hwachon Dam in Connection with Peace Dam and its management", final report prepared to Korea Water Resources Corporation, Korea.

# MAINTENANCE OF STORMWATER RETROFIT PROJECTS

*Gordon England, P.E.- A.S.C.E. Member*[1]

## ABSTRACT

Over the last few years stormwater retrofitting has become widely accepted terminology and a specialty in engineering design. While the rush to address retrofitting needs has led to the development of new BMP's. After the engineers, contractors and permitting agencies have left, it is usually the responsibility of a municipality's Public Works or Stormwater Utility to properly maintain the BMP's to ensure that stormwater treatment will not decline with the age of the project. While intentions are often noble, the long term costs for maintenance are often underestimated and/or misunderstood.

## INTRODUCTION

Brevard County's Stormwater Utility has been constructing stormwater retrofit projects since 1991. While not as glamorous as construction, long term maintenance is just as, if not more, important than the initial design in order to provide for continued pollutant removal from stormwater runoff. This paper will explore the types and costs of maintenance for the following retrofit technologies:
1. Wet Ponds
2. Dry Ponds
3. Exfiltration/Infiltration Trenches
4. Porous Pavement
5. Baffle Boxes
6. Inlet Weirs
7. Grate Inlet Baskets
8. Curb Inlet Baskets
9. Sediment Sumps

---

[1] Brevard County Surface Water Improvement
2725 Judge Fran Jamieson Way, Suite 203A
Viera, FL  32940

## DISCUSSION

### Wet Ponds

Wet Ponds are utilized for water quality as well as flooding benefits. Side slopes should be designed no steeper than 4:1 if driving mowers are to be used. A 20 foot maintenance berm around the perimeter will allow for mowing and truck or grader access. Use drought resistant grass such as Bahia to avoid irrigation demands. Mowing intervals should be more frequent in the rainy season than the dry season. Due to citizen expectations, ponds in residential areas require more frequent mowing than non-residential areas; typically monthly in the wet season and bimonthly in the dry season. Utilizing a private lawn service it costs $100.00 to $500.00 per mowing per pond, depending on the size.

Skimmers work well to keep floating debris from obstructing overflow weirs and orifices. We have also found that fiberglass skimmers are less likely to be stolen than metal skimmers. Cattails and vegetation should be kept clear of skimmers and pipes. While perimeter cattails do not affect the functions of the pond, ponds in residential areas are under higher scrutiny and should be cleaned of cattails (by non-hazardous spraying) yearly.

While not aesthetically pleasing, aquatic vegetation such as hydrilla does not affect the functioning of the pond unless they block the control structure. As a matter of fact, they increase pollutant removal efficiency through nutrient uptake. Two ways to control aquatic vegetation are through chemical spraying and introducing plant-eating fish such as carp. Where wetland and littoral vegetation is used in a pond, make sure the maintenance personnel know not to spray or mow these plants.

### Dry Ponds

Dry ponds are similar to wet ponds with respect to mowing and skimmers. Cattail growth is more of an issue with dry ponds because they will rapidly reduce infiltration rates through root mass growth. Inspections should be made 3-4 times a year and cattails removed when present. Where ground water levels are high or percolation slow, these ponds will have functionality loss through decreased retention volumes and will have soggy bottoms restricting mowing. This situation can be alleviated by:
a) Converting the dry pond to a wet pond via excavation.
b) Excavating the top layer of soil and replacing it with sand.
c) Filling the pond with well-drained soil to an elevation at least 1 foot above the seasonal high ground water level.
d) Grass should be seeded not sodded since the root mass on sod and organic dirt on the back of the sod will reduce percolation considerably.

Dry ponds do not have the sediment storage depth of wet ponds and will require excavation at much more frequent intervals; 5-10 years, depending on depth.

Dry retention swales will rapidly fill with sediment and grass and often require excavation even more frequently. Where earthen ditch blocks are used in swales they are prone to erosion and should be hardened and tied into sidebanks.

Exfiltration/Infiltration Trenches

Exfiltration/Infiltration Trenches are high maintenance items. The sand or gravel beds rapidly collect sediment and debris, which is their function, and lose percolation ability. With exfiltration trenches, 2 foot-3 foot sumps should be placed at all inlets and skimmers placed over the pipe entrances to minimize dirt and trash entering the system. These sumps should be cleaned monthly to minimize resuspension.

Vegetation over the top of infiltration trenches will rapidly clog the trenches. Any vegetation should be removed at least twice a year. Vegetation growth can be reduced by using gravel in the top 6" of the trench instead of sand

Either type of system will require flushing of the pipes when infiltration capacity is observed, to be reduced. Infiltration systems have short lives, sometimes 2-3 years. Exfiltration systems will require complete replacement periodically to stay functional. They should not be placed under pavement!

Porous Pavement

There are several types of porous pavement. Porous concrete and asphalt can easily become clogged with dirt and oil. They should be cleaned yearly with a pavement vacuum truck (not a sweeper). In parking lots and areas with high oil loadings from vehicles, the pavement may need periodic pressure washing to keep the pores from clogging with oil.

Paver blocks are prone to losing their sand fill between blocks, causing them to settle. The sand may need periodic replacement. Where grass is grows between the blocks it may be necessary to periodically mow, if traffic is not frequent enough to keep the grass down.

Baffle Boxes

Baffle Boxes are large sediment traps. The sediment accumulation rates vary from site to site depending on many factors such as drainage basin size, land use, soil type, ground slope, mowing frequency, etc. These boxes may accumulate from 227 kg (500 lbs) to 22,680 kg (50,000 lbs) per month. These devices should be cleaned monthly during the wet season and every 2-3 months during the dry season. Research has shown that grass will leach much of their nutrients out after 20-30 days in stormwater, which may then be flushed out with the next rainfall or base flows. Baffle boxes require vacuum trucks to clean the compartments. A manhole should be centered over each compartment to facilitate cleaning. This necessitates the manholes be within 15 feet of the paved road to allow the heavy vacuum trucks to reach them. The cost to vacuum a baffle box may vary from $2,500 - $5,000 per year if a private service is utilized, depending on the box size and loading rate.

Another factor influencing cleanout cost is the presence of a base flow in the pipes. The box cannot be vacuumed dry if there is a constant flow of incoming water. Therefore inflatable plugs or sandbags are placed in the incoming pipe (difficult) or the next upstream manhole (easier). Also, if the box is below the outfall level then there will be backflows and another plug is needed downstream. Inflatable plugs will cost about $1,500 up to 24" pipe and $4,000 for a plug to fit pipes 24" to 42" in diameter. These plugs are rubber and have a useful life of perhaps 2 years. A source of compressed air, often available on a vacuum truck, is needed to inflate them.

We initially used the vacuum trucks from the county's Sewer Department. This led to erratic cleanout schedules since cleaning broken lift stations and sewer lines were always a higher priory than storm drains. Next we tried a private vacuum truck service that charged $800 per day. There was an initial learning curve to train them to use the plugs we supplied. This company performed very well but after the first year fell prey to the County's low bid process.

The second private contractor initially performed the work of cleaning baffle boxes and inlet devices in an acceptable manner. In 1995 to 1996 we undertook an aggressive program to install over 120 baffle boxes and inlet devices. This magnitude of work soon overwhelmed the lone vacuum truck our contractor had allocated for this work. They fell behind schedule and we were unable to verify the quantity of work being performed. After finding numerous discrepancies between our schedules and the contractors invoices it was decided not to renew the contract at year's end.

Due to the large number of devices to be cleaned, the County decided to purchase a vacuum truck and allocate personnel to man the truck. Since there was not enough demand for full time cleaning of stormwater retrofit projects, the cost of these operations was split between 2 County departments. The stormwater Utility program paid for the vacuum truck ($160,000) and Public Works personnel from Road and Bridge Department man the truck and pay for maintenance. When the truck is not being used for stormwater projects it is available for pipe cleaning duties with the Road and Bridge Department. What has resulted is that Road and Bridge Department uses the truck the majority of the dry season when there are few cleanouts scheduled and during the wet season the truck is mostly used for our Stormwater Utility project cleaning. This arrangement works pretty well but there are occasional conflicts with the priority of schedules between the two organizations.

Disposal of the sediments removed from the baffle boxes has not been a problem. Clean spoil is stockpiled at a Public Works yard, dried, and used on road projects.

Unsuitable material is taken to the landfill. The vacuum truck operator measures the sediment depth in each chamber with a probe. We later convert this to volumes and weights. This documentation has been challenging to the Public Works personnel who are more comfortable operating a shovel than a pencil.

Inlet Weirs

Inlet Weirs are weirs with trash baskets placed in existing inlets and manholes to trap sediment and trash. The volumes trapped are relatively small and cleanout intervals are more critical to efficiency levels than baffle boxes. These inlet weirs also require a vacuum truck to clean dirt trapped in the inlets and upstream pipes.

Some inlets fill with trash after any significant rainfall event while others fill up more slowly, again depending on many factors. Our approach has been to clean these devices once a month during the wet season and every 2-3 months (or after a large rainfall) during dry season. The depth of sediment is estimated and the weight of debris in the basket is weighed using a hand scale.

Grate Inlet Basket

Grate Inlet Baskets are boxes put in grated inlets to trap sediment and trash. Due to their depth and weight of accumulated sediments vacuum trucks are also required to clean these devices. Again the trapped volume is small and the clean out schedule is once a month during wet season and every 2-3 months in the dry season. The depth of debris in the basket is measured and converted to volumes and weights.

Curb Inlet Basket

Curb Inlet Baskets are placed in curb inlets and have a removable basket to trap trash (no sediment). These baskets trap a significant amount of grass, leaves, papers, bottles, cans and other debris not normally quantified when evaluating pollutant removal efficiencies. Ideally these devices should be cleaned after every rain to maximize their efficiency but economics restricts the cleaning to once a month during wet season and every 2-3 months during the dry season. For these devices a vacuum truck is not required and a pickup truck, shovel, barrel and 1 employee is used for cleaning these devices. This is much more economical than a vacuum truck. An average of 10-15 inlets per day can be cleaned this way.

Sediment Trap

Sediment traps are typically sumps in inlets or manholes to trap sediment. Sometimes these are grated inlets constructed with no outfall pipes. When installed along dirt roads, they may fill after every rain. Other installations are sumps in inlets upstream of culverts or exfiltration pipes. The maintenance schedule is again entirely location dependent.

## RESULTS

Brevard County has installed a large number of small inlet devices and baffle boxes due to the large number of small outfalls to the Indian River. Table 1 shows the types and numbers of devices currently installed and a summary of average debris removal. Baffle boxes, inlet weirs and sediment traps principally trap dirt while the other devices trap trash, grass and leaves. The different devices target different pollutants.

| Type | Number | Average Weight Cleaned | Average Cost/Cleaning |
| --- | --- | --- | --- |
| Baffle Box | 24 | 1925 kg | $450 |
| Inlet Weir | 40 | 2.76 kg | $3.50 |
| Grate Inlet Basket | 30 | 16.3 kg | $45 |
| Curb Inlet Basket | 68 | 4.6 kg | $3.50 |
| Sediment Traps | 13 | 101.6 kg | $6 |

Table 1

It rapidly became apparent that scheduling and tracking maintenance of these devices would be a demanding job. We have one employee who devotes up to 60% of his time to this effort.

A database has been set up in Microsoft Access for tracking these cleanouts. It is important to keep the paperwork simple for the cleanout personnel or accurate records will not be kept. The database will provide a form listing projects to be cleaned each month on which the operator fills in the pertinent information such as date and amount removed and returns the forms to us. This avoids the confusion of writing down wrong or adjacent address and misnaming the type of device. This information is then entered into the database and reports are generated. The data entry must be kept simple, but organized, at this point also. Every time a contractor, cleaning personnel, scheduler, or data entry person changes, there is a loss of consistency in the records and cleaning. Complete, routine, defined procedures and database management will help minimize these problems.

## CONCLUSION

In order for stormwater retrofit projects to successfully remove pollutants on a long term basis, the type of maintenance needs should be considered in their design and construction. If adequate personnel, equipment, and funding are not allocated, the projects will be useless. Retrofit projects need maintenance just as roads and wastewater plants need maintenance.

# A COMPARISON OF BMP REQUIREMENTS - A CASE STUDY
Robert A. Sherman[1], David D. Dee, Jr.[2], PE, M. ASCE

## Introduction

When dealing with environmental issues such as maintaining the quality of the nation's waters, sometimes it is useful to step back and examine the big picture. This paper considers the variations in stormwater management approaches and regulations along with Best Management Practices (BMP) for several states in the Chesapeake Bay watershed.

The Chesapeake Bay is the United States' largest estuary. The Bay's watershed covers approximately 64,000 square miles of land encompassing portions of Maryland, Virginia, West Virginia, Pennsylvania, the District of Columbia, and New York. Approximately 15 million people live within the watershed and their activities and surrounding land use are reflected in the bay's water quality. The three main tributaries that contribute the majority of the bay's water are the Susquehanna (50 percent), the Potomac (19 percent), and the James (10 percent).

In 1987, the Chesapeake Bay Agreement was signed in response to an increased awareness of the Bay's decline in water quality. The signers of the agreement were Virginia, Maryland, Pennsylvania, the District of Columbia, the Environmental Protection Agency, and the Chesapeake Bay Commission. The purpose of the agreement was to promote cooperation between state and local governments to reduce point and nonpoint source pollution. The Agreement recognizes that land use decision making is the responsibility of local and state governments. Under this approach treatment for water quality is then also defined by political boundaries. What follows is a comparison of the stormwater management regulations and guidelines of the signers of the Chesapeake Bay Agreement Act, showing the similarities and dissimilarities of each of their programs.

---

[1] Water Resources Engineer, Parsons Brinckerhoff Quade & Douglas, Inc., 6161 Kempville Circle, Suite 110, Norfolk, Virginia 23502.
[2] Professional Associate, Senior Water Resources Engineer, Parsons Brinckerhoff Quade & Douglas, Inc., 301 N. Charles Street, Suite 200, Baltimore, Maryland 21201.

## The State of Maryland

As a result of the Chesapeake Bay Agreement, the Maryland General Assembly passed the Chesapeake Bay Critical Area Protection Act in 1984. With about 5 million people living in Maryland alone, the Act was passed in an effort to reduce the destruction of natural habitat and the deterioration of water quality associated with this many people. The Act designates all lands within 1000 feet of the edge of tidal waters as "Critical Areas" and requires each of the 60 local jurisdictions surrounding the Bay to establish a Critical Area Program. One problem is that many pre-existing uses were grandfathered in, and thus, allowed to continue despite being inconsistent with the law.

### Stormwater Management Regulations

Maryland's Regulations specify that stormwater management measures are necessary for all land development projects involving over 5000 square feet of disturbed land area. The stormwater quantity requirement for most of the state is to require attenuation of the 2- and 10-year post-development peak discharges for a project area such that they are equal to or less than the pre-development peak discharges. This is accomplished through stormwater measures that control the volume, timing, and rate of runoff. For those areas of the state that are in flood hazard zones, attenuation of the 100-year event is also required. There are also exceptions and additions to the regulations depending on the location of the project. Calculation of the runoff design events is based on accepted Soil Conservation Service (SCS) methodologies such as TR-55 and TR-20 using drainage area, runoff coefficients, time of concentration and rainfall intensity. The stormwater quality requirement is basically a rule of thumb that states that the first half-inch of runoff otherwise known as the "first flush" must be treated.

### BMP Practices

Maryland has a preferred order of BMP practices to use in complying with the stormwater management quantity and quality control requirements. The first preferred BMP practice is infiltration of stormwater runoff. If this is not possible because of site or soil limitations, the next preferred practice is flow attenuation using open swales. The third preferred practice is a stormwater retention "wet pond" facility, and the final preferred practice is a stormwater detention "dry pond" facility.

## The Commonwealth of Virginia

As a result of the Chesapeake Bay Agreement, the Virginia General Assembly enacted the Chesapeake Bay Preservation Act in 1988. The Bay Act established a cooperative program between state and local governments aimed specifically at reducing nonpoint source pollution. The program attempts to improve water quality in the Bay by requiring wise resource management practices for environmentally sensitive lands being utilized or developed; the idea being that land can be used and developed in a manner that will minimize impact on water quality. These areas are designated Chesapeake Bay Preservation Areas and include a 100-foot wide buffer around wetlands, tidal shores, tributary streams, and floodplains and areas with highly erodible soils.

The Preservation Act is an additional requirement to those specified in the Virginia Stormwater Management Regulations and the Erosion and Sediment Control Regulations. It sets limits on non-point source pollution for new and redevelopment by considering the extent of the development and then arbitrarily monitoring pre- and post-development phosphorus loadings.

Stormwater Management Regulations

Virginia's regulations specify that stormwater management measures are necessary for all land disturbing projects greater than one acre that significantly alter the hydrologic characteristics of the existing topography. The requirement for stormwater management quantity control is the attenuation of the 2- and 10-year post-development peak discharges to levels less than or equal to the pre-development discharges. Land disturbing projects located in a flood hazard zone also have to attenuate the 100-year design event. Otherwise, this event must safely be conveyed through the facility. Calculation of the runoff design events is based on accepted Soil Conservation Service (SCS) methodologies such as TR-55 and TR-20 using drainage area, runoff coefficients, time of concentration and rainfall intensity. However, storms of like frequency and longer duration than the watershed time of concentration must be analyzed, because the inflow hydrograph volume is critical to detention basin design. The critical storm duration is based on a rainfall intensity-duration relationship. If the critical storm duration is greater than the time of concentration after development, than the basin volume required is the maximum volume based on the critical storm event. The stormwater quality requirement is a rule of thumb that states that the first half-inch of runoff, otherwise known as the "first flush", must be treated. This first half-inch of runoff is known as the water quality volume. For detention basins, the quality requirement is to treat the water quality volume. For retention basins, the water quality requirement is to provide storage for three times the water quality volume.

BMP Practices

Virginia has no particular order of preferences for BMP practices for addressing stormwater management quantity and quality regulations; however, different localities may have a preferred order. These practices consist of retention/detention facilities, infiltration trenches and filter strips, alone or in any combination.

## The Commonwealth of Pennsylvania

The Pennsylvania legislature passed the Storm Water Management Act in 1978. The Act establishes a program of comprehensive watershed stormwater management with the authority at the local level for implementation and enforcement of stormwater ordinances. Pennsylvania differs from Maryland and Virginia in the sense that state statutes place the authority in the hands of Pennsylvania cities, boroughs, townships and counties to prepare comprehensive plans for community development, zoning, and land development ordinances which may include provisions for drainage and stormwater management. However, there is no requirement that requires that the local governments must adopt these plans or that they must consider the effects of runoff beyond their boundaries.

Pennsylvania does have Stormwater Management Guidelines and Model Ordinances, which give recommendations for specific watersheds based on specific studies where current conditions need to be improved. In addition, the Department of Environmental Protection will assist counties in developing stormwater management plans for designated watersheds by providing grant money.

Pennsylvania does have programs that support the State's commitment to the Bay restoration effort such as the Nutrient Management Act which deals with runoff from the agricultural community. Also, the Pollution Prevention and Compliance Assistance Program assists companies, communities and individuals with environmental compliance.

## The District of Columbia

The responsibility for ensuring compliance with stormwater management regulations and erosion and sediment control regulations falls within the District of Columbia Department of Consumer and Regulatory Affairs

(DCRA) Stormwater Management and Erosion and Sediment Control Divisions.

Stormwater Management Regulations

DCRA's Regulations specify that stormwater management measures are necessary for all land disturbing projects greater than one acre. The requirement for stormwater management quantity control is the attenuation of the 2- and 15-year post-development peak discharges to levels less than or equal to the pre-development discharges. Land disturbing projects located in a flood hazard zone also have to attenuate the 100-year design event. Otherwise, this event must safely be conveyed through the facility. Calculation of the runoff design events is based on accepted Soil Conservation Service (SCS) methodologies such as TR-55 and TR-20 using drainage area, runoff coefficients, time of concentration and rainfall intensity. The stormwater quality requirement is basically a rule of thumb that states that the first half-inch of runoff otherwise known as the "first flush" must be treated. There are also exceptions and waivers to the requirements.

BMP Practices

DCRA's order of preference for complying with stormwater management quantity and quality control requirements is to use detention or retention ponds wherever possible. Because of its largely urban nature, many development sites within the district have severe space or land limitations for providing a pond design. For these situations, along with projects where a waiver of quantity requirements is granted, DCRA has been requiring the installation of sand filters as the preferred BMP practice. Because the effectiveness of sand filters depends on accessibility and maintenance issues, DCRA has implemented a monitoring and sampling program for all sand filters in the District.

**Conclusion**

Ideally, the National Pollutant Discharge Elimination System (NPDES) permit program will expand to include all point and non-point discharges, including agricultural uses. Perhaps an umbrella organization could be formed with the authority to cross political boundaries that would be able to analyze the various BMP approaches used by the various states in the Chesapeake Bay watershed. Then, by taking the best approaches to curbing urban and agricultural runoff pollution, either a regional design manual could be established, or existing design guidelines that are working for one state could be adopted to apply to the entire region. In either case, additional commitment, funding, and cooperation are needed to restore the Bay's health.

## References

Division of Soil and Water Conservation, *Virginia Erosion and Sediment Control Handbook, Third Edition*, Virginia Department of Conservation and Recreation, Richmond, Virginia 1992

Karikari, T.J., H.V. Troung, and M.K. Mitchell, *Stormwater Management Guidebook*, District of Columbia Department of Consumer and Regulatory Affairs Environmental Regulation Administration Soil Resources Management Division Stormwater Management Branch, Washington, D.C., February 1988.

Location and Design Division, Hydraulics Section, *Drainage Manual*, Virginia Department of Highways and Transportation, Richmond, Virginia 1980.

Maryland Department of Transportation, State Highway Administration, Highway Drainage Manual, December 1981.

Maryland Water Resources Administration, Soil Conservation Service, State Soil Conservation Committee, 1983 Maryland Standards and Specifications for Soil Erosion and Sediment Control, 1983.

"*Renewing A Pledge To Restore The Chesapeake Bay*", Remarks Of Pennsylvania Governor Tom Ridge, Chesapeake Bay Executive Council Meeting, Reston, Virginia November 1995.

Shueler, T.R., *A Current Assessment of Urban Best Management Practices*, Metropolitan Council of Governments, Washington, D.C., March 1992.

Shueler, T.R., *Controlling Urban Runoff: A Practical Manual for Planning and Designing Urban BMP's*, Metropolitan Council of Governments, Washington, D.C., 1987.

West Virginia Department of Highways, *Drainage Manual*, Roadway Design Division, 1984.

West Virginia Department of Highways, *Erosion and Sediment Control*, April, 1972.

Yu, Shaw L. & Robert J. Kaighn, Jr., *VDOT Manual of Practice for Planning Stormwater Management*, Virginia Transportation Research Council, Charlottesville, Virginia, 1992.

## BMP CONSIDERATIONS FOR A ROADWAY PROJECT LOCATED WITHIN A WATER QUALITY RESERVOIR WATERSHED

David D. Dee, Jr.[1], PE, M. ASCE, Glenn Bottomley[2], PE, M. ASCE, John Olenik[3]

### Introduction

The selected alignment for the Virginia Department of Transportation's (VDOT) Route 29 Bypass Project is located in close proximity to the Rivanna River Reservoir. Approximately four miles of the proposed four-lane divided limited access highway lies within the reservoir's watershed. The reservoir provides the main source of drinking water for the city of Charlottesville, Virginia.

The proposed highway provides a bypass to the existing Route 29 Corridor as well as a direct access to the University of Virginia's North Grounds. Access to the proposed roadway from the north is via an interchange which ties to existing Route 29 and includes a proposed bridge crossing over the South Fork Rivanna River downstream from the reservoir. Access on the south end is from the existing Route 250 bypass and includes a southern interchange with a direct connection to the North Grounds by way of a connector roadway.

Public involvement and participation has played a major role in the planning and design process of the VDOT roadway. Because of the proposed roadway's proximity to the reservoir, the designers recognized early on the importance of protecting the reservoir watershed from any potential impacts from the highway runoff. During numerous and well attended public meetings, the citizens expressed concern about drinking water contamination and increased sedimentation of the reservoir which they believed could occur due to runoff from the proposed roadway. A significant component of the public information process was to educate

---

[1] Professional Associate, Senior Water Resources Engineer, Parsons Brinckerhoff Quade & Douglas, Inc., 301 N. Charles Street, Suite 200, Baltimore, Maryland 21201
[2] Professional Associate, Senior Water Resources Engineer, Parsons Brinckerhoff Quade & Douglas, Inc., 6161 Kempsville Circle, Suite 110, Norfolk, Virginia 23502
[3] Hydraulics Engineer, Virginia Department of Transportation, 1401 East Broad Street, Richmond, Virginia 23219

citizens about established VDOT criteria and requirements, already utilized statewide, to protect the water quality of the aquatic environment and to ensure minimal impact from land disturbing activities.

Additional measures and practices are being implemented into the roadway and stormwater management design by VDOT to help protect the water quality of the reservoir before, during and after construction of the roadway project. VDOT's Hydraulic Section is the reviewing agency responsible for acceptance and approval of the stormwater management and best management practices (BMP) design. Many of the studies and design measures being implemented are additional measures beyond those normally required by VDOT's Stormwater Management and Drainage Design Guidelines.

### Stormwater Management Strategy

Virginia's Stormwater Management Regulations contain requirements for both stormwater quantity and quality control. The stormwater quantity requirements are to reduce the risk of flooding by ensuring the peak flows from the developed area do not exceed the predevelopment flows resulting from a two- and a ten-year storm. The stormwater quality requirements are to detain the first half inch of runoff for an extended period of time, thus allowing pollutants to settle out.

The use of retention "wet" ponds within the reservoir's watershed was selected over the other forms of BMPs, such as detention "dry" ponds, and infiltration facilities, because of the higher pollutant removal efficiencies associated with wet ponds. The water quality volume requirement for a wet pond is three times the volume for the first 0.5 inches of runoff from the developed area. The additional volume allows sufficient time for the pollutants to settle out. The incoming runoff to the pond displaces the old water already in the pond.

The design of the wet ponds incorporates the following features in order to improve pollutant-removal efficiencies and increase safety. These include:

- using a 3:1 length to width ratio for shape,
- making the shape at the outlet wider than at the inlet,
- using 3:1 side slopes for easy maintenance access,
- providing a shallow safety ledge around the perimeter of the wet pond area,
- maintaining a three foot maximum wet pond depth for safety reasons,
- providing fencing around the perimeter,
- providing a dry sump pretreatment area for hazardous spill storage,

- providing a shallow sediment forebay at the entrance to the pond with rock riprap protection, and
- using the pond as a temporary sediment control basin during construction.

The ponds are designed with riser or weir outlet structures that provide for the water quality volume as well as manage the two- and ten-year events and safely convey the 100-year event. However, because of the steep terrain, the presence of rock, and the limited amount of available space, several of the wet ponds are being created in existing swales located on the upstream and downstream sides of the roadway embankment. At these locations culverts are normally used to convey drainage from the upstream drainage area. These pond locations are only proposed where the upstream drainage areas are relatively small.

At locations on the downstream side of the roadway, a wet pond is created by placing a large embankment across the swale, excavating the sides out, and providing a riser outlet structure and emergency spillway. At locations on the upstream side of the roadway, the wet ponds are created by constructing a weirwall in front of the culvert headwall and attaching it to the culvert's wingwalls. The required volume can then be created upstream in the existing swale through excavation. The weirwall is sized to provide the required water quality volume as well as manage the two- and the ten-year events, while having with very little effect on the conveyance design of the culvert for higher flows. Figure 1 provides an example of a weirwall outlet structure for a wet pond:

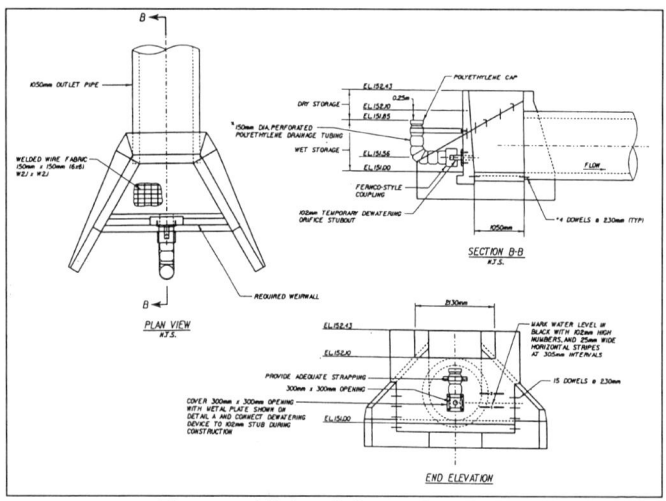

Figure 1 -Weirwall Outlet Structure

## Drainage Design

Because the wet ponds can only provide pollutant removal for runoff conveyed from the roadway, it became necessary to ensure the capture of 100 percent of the runoff from the roadway. VDOT's typical roadway section is usually an open section that allows for runoff to flow over the shoulder to be conveyed in ditches. However, for this project, VDOT is mandating the use of concrete curb along the entire length of roadway for fill sections within the reservoir's watershed. Drainage from the roadway is then collected through a series of curb, median, and ditch inlets and conveyed to the stormwater management facilities through concrete pipe systems.

There are also several locations within the watershed where, through the strategic placement of inlets and drainage systems, runoff from existing developed areas is collected and taken to the proposed ponds for treatment. These areas, which drain directly into the reservoir, are untreated in the existing condition and include several commercial businesses and existing roadways. The priority for collecting impervious areas for quality treatment in order of preference is:

- impervious area proposed by the project,
- existing roadway and parking facilities, and
- existing residential impervious areas.

## Hazardous Spill and Contamination Concerns

Another public concern over the roadway alignment's proximity to the reservoir is the potential for contamination of the reservoir in the event of a hazardous material spill on the roadway. There are several measures being applied to minimize the chance for this occurrence. The use of concrete curb along the length of the roadway with the reservoir watershed will help to convey any spills into the drainage system. Also, a dry sump area is being created at the outfall of each drainage system where runoff is conveyed to a wet pond. The sump area is sized for the volume of a tanker truck, approximately 1,100 cubic feet. Runoff from the roadway must first fill the dry sump area before it is conveyed into the shallow sediment forebay of the pond. This pretreatment area provides additional valuable time while the cleanup crews respond, capture, contain, and cleanup any spill before it enters the wet pond.

In addition, rock check dams are being used in all the fill ditches of the proposed roadway within the reservoir's watershed. The purpose of the rock check dams is to increase the travel time for runoff to reach the reservoir, to improve the pollutant removal capability of the ditch, and to provide additional time for cleanup crews to capture the spill if it should enter into a ditch.

The rock check dams, along with the dry sump areas, provide additional time only if the spill occurs during a dry period. The possibility that a spill might occur during a rain event mandates the creation of an emergency hazardous plan and response team. Part of the response team's initiative may be to carry an inflatable balloon device in order to plug the wet pond's outfall pipe should the pond become contaminated from the spill. Figure 2 illustrates the wet pond design features:

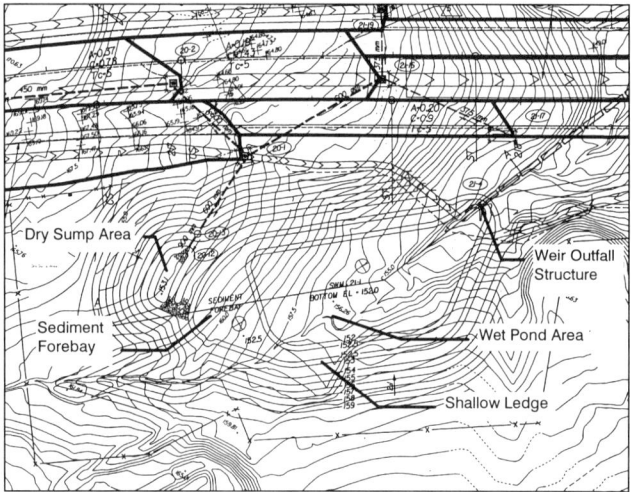

Figure 2 - Wet pond design features

**Additional Measures**

As part of its on-going efforts to monitor the effectiveness of BMPs, VDOT has hired an additional consultant to evaluate the design of the wet ponds. Their efforts will be to try to incorporate additional measures and practices into the wet pond designs in an effort to improve the pollutant removal efficiencies. This will be accomplished first through the use of a monitoring program of several outfall locations from the project area for pollutant runoff before, during, and after construction. This will help in determining the percentage of pollutant runoff attributable to the proposed roadway. Then, additional measures to be evaluated may include the use of perimeter vegetation in the wet ponds to increase biological uptake, or the use of oil/grit separator chambers in the roadway drainage system. The chambers could act to not only remove oil and gas pollution from the roadway runoff but also to aid in capturing hazardous spill material.

Another design measure involves the use of turbidity curtains during construction at three locations along the reservoir. The turbidity curtains are located where existing drainage swales convey runoff that discharges from the stormwater management facilities and eventually reaches the reservoir. This would be one last effort to capture sediment that escapes from the project before it enters the reservoir. VDOT is purchasing permanent drainage easements along these existing swales and proposing the construction of rock check dams in the swales. The easements will allow VDOT to access the swale before, during and after construction should the need arise.

These additional measures along with the proposed BMP designs for the Route 29 Bypass project address the public's concerns as well as help protect and maintain the water quality of the Rivanna River Reservoir. VDOT's commitment to the project is demonstrated through:

- the use of wet ponds over dry ponds for the moderate to high pollutant removal efficiencies,
- the use of a monitoring program of runoff from the roadway project before, during, and after construction,
- the utilization of significant additional concrete curb and storm drain collection systems,
- the implementation of additional sediment control measures during construction such as turbidity curtains and rock check dams,
- the implementation of hazardous spill containment measures,
- the implementation of a regular inspection and maintenance program during and after construction. and,
- the implementation of a proactive public information/participation program.

## References

Division of Soil and Water Conservation, *Virginia Erosion and Sediment Control Handbook*, *Third Edition*, Virginia Department of Conservation and Recreation, Richmond, Virginia 1992

Location and Design Division, Hydraulics Section, *Drainage Manual*, Virginia Department of Highways and Transportation, Richmond, Virginia 1980.

Shueler, T.R., *Controlling Urban Runoff: A Practical Manual for Planning and Designing Urban BMP's*, Metropolitan Council of Governments, Washington, D.C. 1987.

Yu, Shaw L. & Robert J. Kaighn, Jr., *VDOT Manual of Practice for Planning Stormwater Management*, Virginia Transportation Research Council, Charlottesville, Virginia 1992

# WALKING A TIGHT ROPE BETWEEN DEVELOPERS AND BANANAS[1]

GREG BOEHM, P.E., ASCE MEMBER[2]

## ABSTRACT

The Lake County Stormwater Management Commission (SMC) regulatory program regulates development on a watershed basis with respect to preserving natural areas and natural functions of watersheds, reducing potential flood damages, and attention to water quality. Floods of 1986 and 1987 motivated the State of Illinois "Part 708 Floodway Regulations" which were at the time one of the most restrictive sets of regulations over water resources of any jurisdiction in the country and included a short list of appropriate (or permissible) uses within the Regulatory Floodway. The Lake County Watershed Development Ordinance (WDO) was adopted October 18, 1992 by Lake County and was designed to incorporate comprehensive watershed management of a rapidly developing county. The Lake County Comprehensive Stormwater Management Plan provides for consistent countywide regulation on a watershed basis by empowering municipal engineering offices with appropriate local authority and retaining more technical and general authority for the SMC staff. The municipal offices coordinate with the county staff to regulate development according to the WDO.

## INTRODUCTION

Lake County, Illinois at the most northeast corner of Illinois was one of the six fastest developing counties in the country in 1990. Pressure from property owners and investors wishing to capitalize on the development clashes directly with the desire of property owners who want no further development. Nearly one-fifth of Lake County consists of streams, ponds, lakes, wetlands, or floodplains. Much of the development is dense and the topography is relatively flat. Historically as development has increased the imperviousness and resulting runoff volumes during

---

[1]  Build Anything Not Anywhere Near Anyone.
[2]  Permit Engineer, Lake County Stormwater Management Commission, 333 Peterson Road, Libertyville, Illinois 60048

rainfall events, it has also constricted floodplains and increased flood profiles. Wetlands have been filled or drained for agricultural purposes. Water quality and habitat in some lakes and streams has been reduced drastically in the course of the past 75 years. Many lakes, streams, wetlands, and floodplains continue to serve plant and animal life in much the same way as they have for ages. If development can minimize its impacts to mimic the drainage and hydrology of the existing or natural watersheds, existing natural resources can sustain themselves and the current risk of flood damage will not be increased. Property owners, developers, and future investors expect existing natural resources and flood damage risks to be maintained. Development standards should be based on maintaining natural resources and existing flood damage risks.

## EVALUATING THE EFFECTIVENESS OF THE REGULATORY PROGRAM

The earliest development within a watershed will tend to be on land that costs the least to develop and maintain and which is accessible to water. Later development within a watershed will tend to involve the less desirable parcels of land for development: land that will cost more to develop and maintain due to factors such as flood frequency, soil saturation, and stability. In order to develop land remaining in an urbanizing environment, it is often necessary to effect a greater change on pre-construction drainage conditions than had been necessary to construct older developments. In an area that had been concentrated as wetlands, development may require significant cost in draining, filling, and compacting of earth and building bridges and drainage facilities. How best to require developers to spend this additional cost is the purpose of stormwater and watershed development regulation.

In a sense regulatory control of development is a simple matter of trying to define the most significant existing conditions and maintain them. The most significant problems and the most significant benefits of the current condition will ultimately be affected by development unless it is regulated. Watershed development regulation will be most greatly affect by those developments that desire the greatest impact in terms of drainage and density. A regulatory program may be evaluated on the impact it has on development compared to the impact it has on the watershed. Major issues include:
- minimizing the increase of damages due to flooding and drainage problems created by development,
- preserving the hydrology of the natural environment,
- mitigating water quality impacts from detrimental long-term effects of developments,
- minimizing sedimentation and erosion during development activities,

In order to measure the utility of regulation, regulatory ideals should be considered in regard to immediacy, certainty, duration, magnitude of cost or effort, magnitude of impact, scope of impact, and number of persons affected.

### IMMEDIACY

Immediacy can be evaluated in terms of the impact to the environment or by the adjoining property owners' frequency, depth, or duration of flooding. Neither of these two effects is the most critical issue to a regulatory process. The most critical issue to a regulatory process is the immediacy of the regulation. Delay of development due to excessively lengthy regulatory review processes results in inconsistency and excessive development cost. No amount of effort to reach regulatory ideals and consistency is of any consequence if the regulatory agency is not timely and complete in its response to permit applications. A regulatory program must strive regulate its ideals while streamlining its process and minimizing the time required to obtain a permit. As much as possible simpler requirements must be made clear immediately and the cost of lengthy analyses, reviews, and studies should be weighed against the value of the results.

### CERTAINTY

The more significant the change from the pre-construction conditions to the with-project conditions, the more certain the impacts will be. A dam at the upstream limit of a development will certainly cause increased inundation to the adjacent upstream property. Increased velocity or dramatically reduced sediment load at the downstream limit of development will certainly increase erosion downstream of the development. Significant effects to the watershed such as increases in peak discharge, reduction of storage, reduced conveyance capacity, reduced normal water flow, eliminating floodplain and wetlands areas should be evaluated from a broad perspective. A significant impact such as the draining and filling of a large wetlands and overbank floodplain area may affect normal and extreme drainage conditions and natural habitat well beyond the upstream and downstream limits of a development.

### DURATION

The notion of self sustaining development is one that will not only enable continued similar development within a watershed but will enable surrounding natural resources to endure. Regulatory standards should minimize environmental impacts to groundwater, surface water quality, and normal water conditions between storm events as well as severe flooding conditions. Where developmental impacts are most certain and enduring to valuable natural resources, the resources should be permanently preserved from development.

## MAGNITUDE OF COST OR EFFORT

The cost of sedimentation and erosion control to a development during construction can provide reasonably clear runoff discharge all runoff for 90% or 95% of all storm events. The cost of providing sedimentation and erosion control to provide reasonably clear discharge for a worst case 100-year frequency flood event may require settling basins to provide two or three times the detention volume. While the 100-year event has been settled upon nationally as a standard for development and flood insurance requirements of completed projects, it may be more reasonable to address only the much more frequent events for sedimentation and erosion control. It is the more frequent events, the ones that will occur 90% to 95% of the time, that will carry most of the sediment from development sites to the floodplains and streams. To protect silted runoff from more frequent storm events is more of an issue of thorough construction planning than construction cost. Construction activities should therefore be closely monitored with regular inspections.

## MAGNITUDE OF IMPACT

It is often clear during the regulatory process what the project qualities are that are motivated entirely by regulations. For most developments, detention basins, wetlands buffers, water quality treatment, and infiltration are included only to facilitate permitting per regulatory requirements and would not otherwise be included in the development plans. It is the regulations motivating such a great magnitude of impact to development that most need to be justified in terms of the greatest good. By their nature these also are the characteristics of development that should be most closely monitored over time and retrofitted as necessary to ensure that their desired benefit is realized. Detention ponds should hold water during storm events and buffers and swales should be suitably maintained.

## SCOPE OF IMPACT

Whether a regulatory requirement should be required of all developments or whether there should be a threshold below which a development is not subject to specific regulations is a matter that becomes an issue of regulatory work load as well as an unreasonable burden to the smallest developments. Once target natural resources have been protected with specified regulatory ideals, thresholds of applicability should be applied to development outside of the natural resource areas. Complete maintenance of base flood conveyance, wetlands, wetlands buffers, and water quality may be required for all developments within a watershed. Detention during 100-year frequency storm events may be a development standard that should apply only to developments larger than a specified threshold.

## NUMBER OF PERSONS AFFECTED

Quantifiable impacts from development should be restricted to the owner of the subject property and any parties that have provided explicit and permanent authority to allow a specified impact. The regulatory program should first enable existing development to continue to enjoy the same potential flood damages, drainage conditions, and surrounding natural resources. Enabling development according to regulatory standards should be secondary to maintaining existing flood damage risk and maintaining critical natural resources.

## SUMMARY

From a regulatory perspective, sustainable development requires close monitoring of the cost regulation has on development to the impact the development has on the watershed. The ideals of the regulatory program should be investigated and a comparison evaluated of the costs and benefits in terms of the greatest good.

## REFERENCES

Lake County Department of Planning, Zoning and Environmental Quality, Lake County Comprehensive Stormwater Management Plan (1990).

Lake County Stormwater Management Commission, Lake County Watershed Development Ordinance (1994).

Flood Management that Revitalizes a Community

Kevin Shafer, Associate Member[1]
Bob Wolf,[2]

Abstract
The City of Lincoln, Nebraska, has joined forces with the University of Nebraska-Lincoln (UNL) and the Lower Platte South Natural Resources District to sponsor the Antelope Valley Major Investment Study (AVMIS). The purpose of the study is to develop a Draft Single Package that addresses the community, transportation, and storm water problems in the central business district and UNL. These three components are referred to as the "three legged stool" that supports the project. The storm water component of the AVMIS is centered on the Antelope Creek watershed, which flows through the central business district and the university. The goal of the storm water task is to develop a 100 year flood management program that works in conjunction with the other two legs of the stool.

The AVMIS study team analyzed detention storage and channelization in this urban setting to achieve 100 year flood management. Using the UNET and HEC-2 computer programs, a series of alternatives were analyzed. Each was developed not only for 100 year flood management, but also with the hope of converting an underground concrete-lined, forgotten creek into a focal point for the surrounding community.

Introduction
The City of Lincoln, Nebraska, joined forces with the University of Nebraska-Lincoln (UNL) and the Lower Platte South Natural Resources District to consider transportation, storm water and community revitalization issues in and near the Antelope Creek valley. The valley cuts through the city's core from southeast to northwest, flowing into Salt Creek just north of Downtown

---

[1] Area Manager, Parsons Brinckerhoff Quade & Douglas, Inc., 135 W. Wells, The Germania Building, Suite 222, Milwaukee, Wisconsin 53203
[2] Managing Principal, Olsson Environmental Sciences, 1111 Lincoln Mall, P.O. Box 81307, Lincoln, Nebraska 68501

Lincoln and the UNL's City Campus. By virtue of location, jurisdiction and interest, all three partners have a stake in protecting and enhancing the area. Because problems and solutions regarding transportation, storm water and community revitalization are profoundly interrelated, the partners recognized the wisdom of considering the topics in a unified, comprehensive and systematic manner.

Basin Description
Antelope Creek in Lincoln, Nebraska has a 33.15 square kilometer drainage area of which runoff from the upper 13.99 square kilometers is controlled by Holmes Dam. The lower 19.17 square kilometers is totally developed and drains through an open channel to just south of "N" Street. From this point north to Vine Street, the creek is enclosed in 1,250 meter of concrete conduit. This 3 meter by 6 meter conduit was built around 1915 to convey the flow from Antelope Creek from south of "N" Street to north of Vine Street. The creek channel was obliterated and overtime, the general public has forgotten that a creek channel historically existed in that area. The creek valley was fully developed with businesses and residences, creating a substantial flood hazard that was largely unrecognized.

In 1994, the conduit was lined with twin 2.44 meter by 2.74 meter conduits in order to restore structural integrity and avoid collapse. The restoration substantially extends the service life of the conduit, but reduces its hydraulic capacity to less than the fully urbanized 5 year event which is substantially less than the upstream and downstream channels. Consequently, excess flood flows bypass the conduit and flow overland through downtown Lincoln and portions of the UNL.

In heavy rainfalls, the water collected in the drainage basin fills the conduit quickly and excess water would spread over residential, commercial, industrial and UNL land. The overflow area during the 100 year storm would be approximately 3 to 5 blocks wide and up to 1.68 meters deep. Following such a storm, approximately $20 million in damage may occur, most not covered by insurance. The potential for tragic loss of life is also present.

Typically, storm water management designs provide for handling significantly more water efficiently than a five-year storm. The AVMIS examined options for handling water resulting from the 100 year storm event.

Flood Characteristics
Most precipitation events in Lincoln are high intensity and short duration thunderstorms. Flooding from Antelope Creek generally results from accelerated runoff due to urban development and the choking

characteristics of the drainage system, particularly the "N" Street to Vine Street conduit. Prior to construction of Holmes Lake Dam in 1962, floods are known to have occurred in 1908, 1910, 1940, 1950, 1951, 1952, 1957, and 1958. In the year following the completion of Holmes Lake dam, there has been no widespread flooding, although storms in 1963, 1966, 1967, 1973, 1989, and 1993 produced substantial runoff volume and peak discharges. At least three of those storms produced runoff that exceeded the rehabilitated conduit's capacity.

Flood Hazard Limitations on Land Use
The existing floodplain severely limits potential high value land use in the lower reaches of Antelope Valley, especially where no designated floodway exists along the conduit enclosed portion of the creek. Where the floodway has been designated by the Federal Emergency Management Agency (FEMA), new construction in the flood fringe area outside the floodway can receive a floodplain permit if the first floor is elevated to 300 millimeter (one foot) above the base 100 year flood level. Within the floodway, construction can only be permitted if a hydraulic analysis proves the construction will not cause any rise in the regulatory hydraulic profile of the floodway.

Where there is no designated FEMA floodway, between "N" and Vine Streets, construction anywhere within the entire 3 to 5 block wide floodplain can only be permitted if a hydraulic analysis proves that the cumulative effect of proposed construction and all previously permitted floodplain construction in the area would not cause more than a 300 millimeter (1.00 foot) rise in the FEMA base flood profile. The uncertainties and costs of dealing with the flood hazard to the area currently limit and deter reinvestment by potential and existing property owners. Long term negative impacts of the flood hazard to area neighborhoods, the central part of the City, and the UNL are forecasted to be substantial.

Development of Storm Water Elements
The characteristics of the watershed (i.e. urbanized, limited open space, etc.) led the study team to the conclusion that two means of storm water management were most appropriate for this project; detention storage and/or conveyance channels.

Detention storage facilities were analyzed to be placed along side Antelope Creek. Inflows to the detention facilities would be controlled by a side-flow weir that would be placed between the creek and the storage facility. Additionally, a low flow, flap gated outlet structure was analyzed to provide low flow drainage from the facility back into the creek. The detention operation would be such that as flood event water surface elevations rise in the creek they would begin to overtop the side-weir levee and spill into the detention area. This would allow for the more frequent flows to continue downstream and only divert the larger flood flows into the detention areas. Once the facilities were full, the water surface elevations on both sides of the weir would equalize. Then as the water surface elevation on the creek side of the weir declines, the detention facility would begin to drain back into the creek.

As alternatives, conveyance channels were analyzed between "N" Street and Vine Street to contain the conduit's 100 year bypass flows within the channel limits. A grass-lined, trapezoidal shaped channel was modeled. Enhancements to the channel for community and recreational considerations were also modeled.

Storm Water Analysis
The analysis of each storm water solution was based on the Corps of Engineers HEC-1, HEC-2 and UNET computer programs. The Omaha District of the Corps of Engineers had developed the Antelope Creek computer models and the study team modified these models to reflect the proposed alternatives.

The HEC-1 computer model was developed to generate the flood hydrographs for the various subbasins within the watershed. These hydrographs were stored in a Data Storage System (DSS) database for further use by the UNET computer program.

The UNET model combined the HEC-1 flood hydrographs into a riverine model that simulates the creek two-dimensionally. This program develops both peak discharges as well as water surface elevations along the creek for the design 100 year storm event. This UNET model was used to size the detention facilities because the two-dimensional flow could be modeled more accurately.

The HEC-2 computer model was used to calibrate the UNET model and to develop water surface elevations for the conveyance channel options. Peak discharges generated by the UNET program were used as input to HEC-2 during the alternatives analysis.

Storm Water Elements – Detention versus Conveyance
While successful elsewhere, a number of obstacles make implementing detention areas in the Antelope Creek basin very difficult. What seemed on the surface an attractive storm water solution quickly became burdened with perplexing questions about costs, responsibilities and effectiveness.

The most effective location for a detention facility is near Lincoln High School, just upstream of the conduit's inlet. Even with 387 ac-ft of detention storage in place at this location, additional upstream detention facilities would also be needed to eliminate the bypass flows around the conduit. Because the significant redevelopment of the site and existing use mitigation would be required at the multiple locations, detention was dropped from consideration as a stand-alone solution to flooding.

The conveyance option looked at providing a bypass open channel for the flood flows that exceeded the capacity of the conduit. This solution offered superior flood management reliability, i.e. the 100 year bypass flows could be limited to within the channel, and since it would share the transportation corridor, it proved to be more cost effective. Both the expensive concrete-channel conveyance and the less costly wide, grassy option provided effective 100 year flood management.

Providing a combination of detention ponds and conveyance channels were also considered. When examined as part of a joint system, however, the costs were seen as prohibitive, requiring the more expensive parts of both conveyance and detention. In addition, the conveyance channel option offered some potential aesthetic enhancements, adding support for the selection of grassy, open-channel conveyance as the storm water solution. Another benefit is that the conveyance channel option provides the ability to provide a continuous bicycle path from Holmes Lake to Salt Creek. Rather than loosely linking several detention sites, conveyance presents the opportunity to provide the community with a lengthy off-street trail.

While both sets of solutions were considered in the alternative analysis, conveyance channels were favored because they could occupy the same limited right of way as the transportation solution and because there was limited open space upstream of the conduit that was available for the large detention area that was required for the 100 year storm events.

<u>Draft Single Package</u>
The storm water management component of the Draft Single Package is focused on providing an Antelope Creek conveyance system which has adequate channel and bridge capacity to shrink the 100 year floodplain to within the limits of a planned conveyance corridor.

The Draft Single Package is comprised of three storm water elements that improve the conveyance capacity along the creek:

1. Remove 38$^{th}$ Street bridge and replace South Street bridge with a bridge with a larger hydraulic capacity - removal and replacement at these two

locations will remove the bridge constrictions allowing the flood flows to continue downstream without overflowing the creek corridor at these locations.

2. Provide an open creek bypass channel to supplement the Antelope Creek Conduit - re-establish the Antelope Creek channel through this area, providing adequate conveyance to limit the 100 year floodplain to the width of the channel, approximately one-half of a block. A flat sloped, vegetated, landscaped channel and bridge system is envisioned to provide a park-like setting. The resulting multi-purpose corridor would include a small, aesthetically designed stream at the bottom of the flood channel and a trail to facilitate public use and access when the flood carrying capacity of the channel is not in use.

3. Reconstruct the Antelope Creek channel downstream of the conduit - a landscaped flood channel which includes a trail and architecturally designed open stream will be constructed along side of the proposed roadway. This minimizes the required right of way and also provides a visual enhancement/buffer for the roadway and adjacent properties.

Conclusion

The focus of this study was to develop a 100 year flood management package that supports and enhances the other two components of the study, transportation improvements and community revitalization. The results of the alternatives analysis phase of the study showed that detention storage facilities were not feasible due to financial and land acquisition issues.

Conveyance channels were determined to be the best solution to the problem due to their:
- ability to provide adequate 100 year flood management at a feasible cost,
- capability to share the right of way with the transportation component,
- ability to extend the extensive trail network in Lincoln to the downtown area,
- ability to provide a sense of "place" for the community

The "greening" of Antelope Creek will provide a focal point that will allow this community to rebuild.

# BAFFLE BOXES AND INLET DEVICES FOR STORMWATER BMP's

*Gordon England, P.E.- A.S.C.E. Member[1]*

## ABSTRACT

With the advent of NPDES Stormwater Permits and increased environmental awareness, many municipalities are confronted with the daunting task of retrofitting existing developed areas, which provide little or no water quality treatment for stormwater runoff. While retention ponds are the traditional method used for treating stormwater, they are often not feasible for retrofit projects due to available land constraints. This study will present several new types of treatment methods utilizing existing inlets and manholes.

## BACKGROUND

While there are numerous pollutants which may be in stormwater runoff, the most commonly targeted are: suspended solids, floating trash, hydrocarbons, nutrients (such as nitrates and phosphates), and heavy metals. The suspended solids and floating trash are the most visible and easiest components to treat. Suspended solids removal will also address heavy metals and phosphates, to a limited degree, since they will bind to particulates. Although floating trash is invisible to most pollutant testing techniques, a significant amount of grass clippings, leaves, and yard debris often can contribute to nutrient loadings of receiving water bodies.

To date, the most effective methods for pollutant removal are retention/detention ponds or alum treatment. These methods usually require significant land area or high maintenance costs, which are not often economically feasible in developed areas. Brevard County's Stormwater Utility used two EPA grants to install and test three (3) treatment devices which may be constructed in existing manholes, inlets, or pipes for sediment and trash collection.

---

[1] Engineer II, Brevard County, Surface Water Improvement
2725 Judge Fran Jamieson Way, Suite 203A
Viera, FL  32940

## METHODS

### Baffle Boxes

Baffle Boxes have been pioneered by Brevard County to provide an end of pipe treatment method. These are concrete or fiberglass sediment boxes constructed in line, or at the end of, existing storm drain pipes. They are typically 10-15 feet long, 2 feet wider than the pipe, and 6-8 feet high. The box is divided into 2 or 3 chambers by weirs set at the same level as the pipe invert in order to minimize hydraulic losses. There are trash screens or skimmers to trap floating trash and yard debris. Manholes are set over each chamber to allow access for cleaning with vacuum trucks. Baffle Boxes are principally designed for sediment removal. The trash screens will trap floatables but will swivel up in high flows losing the trash. Clean out records are only kept on sediment removal.

Scale model testing by Dr. Ashok Pandit at Florida Institute of Technology has shown that baffle boxes have a removal efficiency of 90% for "baseball field" sand or sandy clay at entrance velocities up to 6 feet/sec. For fly ash, smaller than most clays and silts, the removal efficiency is 28% at velocities up to 6 feet/sec.

While it is intuitively obvious that larger boxes will be more efficient than smaller boxes due to slower velocities, construction and economic constraints limit their size. A weir height of 3 feet is generally used. Experiments showed that the deeper the chamber the more efficient the box is and the time between cleanouts increases. Vacuum trucks are used for cleaning the chambers which necessitates the box be within 15 feet of a paved or stabilized road. Baffle boxes for pipes up to 48" in diameter can usually be precast making for fast installation. Boxes for pipes up to 60" in diameter have been used, but they must be poured in place and are more expensive and time consuming. The installation cost for most baffle boxes is $20,000-$30,000, depending on the utilities to be relocated.

### Grate Inlet Basket

The grate inlet basket is a fiberglass box resembling a trash can, which is placed inside a grate inlet. The unit has several filter holes at different elevations in the box, which are covered with stainless steel filter screen. The filter screens allow low to medium flows to pass through while trapping sediment and floating debris. An insert in the top of the box contains an absorbent pad, which provides for removal of oil, grease and other hydrocarbons from the stormwater as they pass through the box. The insert also serves as a skimmer to prevent floatables such as leaves and trash from leaving the box through the overflow holes located around the top of the container. These holes are designed to act as orifices with the same hydraulic capacity (maximum flow rate) as the grated inlet.

The grate inlet baskets are custom designed for each inlet. Installation is easy; simply remove the grate, drop in the basket, and replace the grate. A lip around the top of the basket fits under the grate, securing it in place. To prevent hydraulic losses the depth of the basket should be kept above the pipes leaving the inlet.

## Curb Inlet Baskets

The third configuration of pollution control device utilized is the curb inlet basket, which is a fiberglass form that fits inside curb inlets. It attaches to and extends the throat of the inlet while dropping 5 inches vertically. At the end of the fiberglass throat is a 4-inch-high weir with a notch that funnels incoming water into a basket hanging on the backside of the weir. The weir is configured so the elevation of the top of the weir is below the bottom of the gutter inlet elevation to prevent any reduction in inlet flow rate capacity. The trash basket catches floating trash, grass clippings, and yard debris. It is located under the manhole cover where it can be easily removed and emptied without human entry into confined spaces. The unit has unique design features that allow it to expand to fit various curb widths. The back side of the weir is supported by fiberglass legs designed so as not to impede the flow from any other pipes that may be connected to the inlet box. Installation consists of easily attaching the unit to the walls and floor of the inlet using concrete anchors. The inlet devices are made of fiberglass or corrosion-resistant materials to extend their service life. This also allows them to be reused at other locations, should they no longer be required where initially installed.

A total of 117 inlet devices were installed along Merritt Island and the barrier island in Brevard County as part of the EPA grant. The distribution was 39 grate inlet baskets and 68 curb inlet baskets. Suntree Isles of Cape Canaveral manufactures these devices.

## RESULTS

### Baffle Boxes

Cleanout frequency is dependent upon rainfall frequency, land use, and drainage basin size. The typical cleanout schedule was once a month in the rainy season and every other month in the dry season. There was an average of 6 cleanouts per baffle box per year. If boxes with no base flows which are left for longer periods, they tend to become septic and odorous. Maintenance records for 24 baffle boxes show that 202 cubic meters of sediment were removed from these boxes over the last 3 years. Sampling has indicated that the average wet density of sediment removed is 88 lb. per cu. ft. The sediment removal rate ranged between 5.9 kg/cleanout and 17,796 kg/cleanout with an average of 1925 kg/cleanout. The highest yearly removal record from a box was 100,633 kg; which had a 43" x 68" pipe and drainage basin of 63 acres. Upstream of this box was a highly eroded open channel.

The lowest yearly removal record was 18 kg from a baffle box with a residential drainage basin of .1 acres. A vacuum truck will typically clean 2 baffle boxes per day which runs about $450 per box. This averages about $0.24 per kg of sediment removed.

### Grate Inlet Devices

Observations during several storms did not show any loss of hydraulic capacity of the inlet due to the presence of the grate inlet baskets. They proved effective in collecting significant quantities of leaves, cigarettes, and floating trash. The grates prevent large items from entering the baskets.

The hydrocarbon absorbing booms turned gray and black as they absorbed oils. However a method to measure their efficiency could not be determined. The replacement frequency depends upon the hydrocarbon-loading rate. The booms were changed at 3-4 month intervals. If left in longer, the booms turned brown from organic growth and probably become ineffective. Inlets in grassed swales had virtually no oil loadings due to filtering by grass. Inlets along pavement had minor oil loadings and parking lots had the highest loadings. It was concluded that oil booms were only cost effective in parking lot and commercial areas with high hydrocarbon loadings. In high loading areas considerable oil and grease also collected on the walls and filters in the box.

### Curb Inlet Devices

Curb inlet baskets were successfully used in curb inlets of various sizes and shapes. Adapters are available for round inlets and multiple devices can be placed side by side to accommodate wide curb openings. The trash basket hanging on the backside of the weir is effective in trapping grass clippings, leaves, paper, bottles, and other trash. The constituent loading was approximately 50% grass and leaves by weight. There was frequently observed to be several inches of sediment in the bottom of these inlets even though not designed to do so. This dirt was cleaned out but not weighed. The grass and leaves dry out and through oxidation release nutrients to the air. Nutrients not oxidized will remain in the plant material until they are cleaned from the inlet.

### SUMMARY

The cleanout frequency depends upon rainfall frequency and loading rates. Streets with trees obviously had higher leaf loadings. Residents often utilized the inlet as a convenient place to pile yard waste for pick-up and disposal and these inlets consistently showed higher loadings. This points out the importance of public education as a component of pollution reduction. Baffle Boxes are very effective at removing sediments and some floating debris. They have limited effectiveness at nutrient removal. The grate inlet basket is most effective at trapping sediment, greases, and oils. It traps finer sediments than those collected by the curb inlet basket. The curb inlet basket traps primarily grass, leaves, floating trash, and small amounts of sand. Ideally these devices would be cleaned after every rain but this is not feasible due to manpower and cost constraints. It is recommended they be cleaned once a month in the wet season and every 2-3 months in the dry season.

Records show that grate inlet baskets average 16.3 kg per clean out and the curb inlet baskets 4.6 kg per clean out. Clean out costs for grate inlet baskets have averaged $45 per cleanout using a private service with a vacuum truck at $900 per day. Curb inlet baskets are cleaned with County employees. One man in a pickup truck with a barrel can clean 20 per day. This averages $3.50 per cleanout. See Table 1

## CONCLUSION

There are no universal fixes for stormwater pollution control. Each outfall and drainage basin must be analyzed to determine types of pollutant loadings, size of drainage basin, type of conveyance system, and pollutants targeted for removal. Then a BMP or series of BMP's should be selected. See Table 2.

Baffle boxes are effective BMP's for sediment removal in small to medium size drainage basins. They are installed inline with existing pipes requiring minimal easements and utility relocations.

For small flows and drainage basins, grate inlet baskets, and curb inlet baskets are affordable alternatives for providing stormwater treatment. Installation into existing inlets and manholes avoids disruptive and expensive conventional construction. Inlet devices trap small amounts of sediment and larger volumes of yard debris and trash. Research is being continued to quantify nutrient loading rates from grass clippings.

The tradeoff for these low cost treatment methods is the perpetual maintenance expense. It is important to note that if the devices are not going to be frequently maintained they will not be effective. A dedicated source of manpower and equipment is needed to remove the pollutants from these BMP's.

| Type of BMP | Number Installed | Average Weight Cleaned | Average Cost per Cleaning | Average Cost/lb Sediment Removal | Average Cost/lb Phosphorus Removal | Average Cost/lb Nitrate Removal |
|---|---|---|---|---|---|---|
| BB | 31 | 5054 | $267 | $0.11 | | |
| CIB | 50 | 8.9 | $7 | | $8.54 | $3.33 |
| GIB | 39 | 21 | $40 | $1.25 | | |

Table 1

## BMP SELECTION

|  |  |  | Pollutant Type | |
| --- | --- | --- | --- | --- |
| Basin Size | Sediment | Trash | Nutrients/Grass and Leaves | Nutrients from Other Sources |
| Small | GIB, BB | GIB, CIB | GIB, CIB | other |
| Medium | GIB, BB | GIB, CIB, BB | GIB, CIB | other |
| Large | other | other | GIB, CIB | other |

BB – Baffle Box   GIB - Grate Inlet Basket   CIB – Curb Inlet Basket

### Table 2

**APPENDIX** Dr. Ashok Pandit 1996

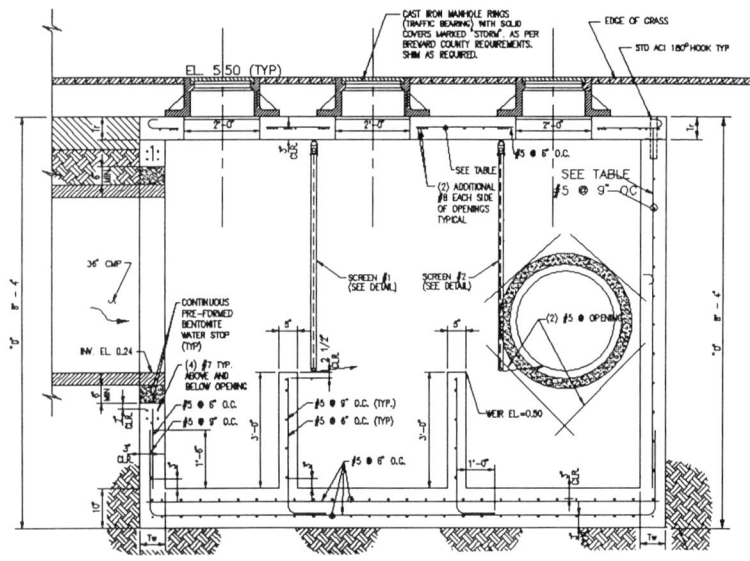

**BAFFLE BOX**

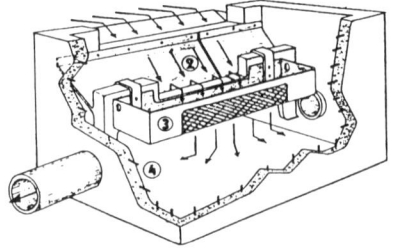

**CURB INLET BASKET**

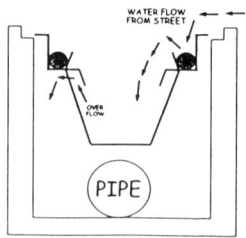

**GRATE INLET BASKET**

Implementing Municipal Stormwater Management Program:
An Overview of Planning and Administration

Shih-Long Liao[1], Richard Field[2], Daniel Sullivan[3], and Chi-Yuan Fan[4]

Abstract

According to Clean Water Act's mandate, municipalities regulated under the National Pollutant Discharge Elimination System (NPDES) program must, at a minimum, achieve technology-based requirements as a first step toward achieving loading reductions consistent with applicable water quality standards. In order to help municipalities implement their stormwater management programs (SWMPs) in a practical and cost-effective manner, the U.S. Environmental Protection Agency (EPA) has been developing a number of guidance manuals for assisting municipalities to implement their SWMPs. This paper summarizes the contents of the first volume of a two-volume manual currently being developed by EPA's Office of Wastewater Management and Office of Research and Development to: (1) provide a basic seven-step process for effective decision-making and long-term planning; (2) suggest practical guidance for municipalities to best implement their SWMPs; and (3) provide a watershed protection approach (an integrated, holistic strategy) for more effectively restoring and protecting aquatic ecosystems and protecting human health.

Background

Studies, e.g., the Nationwide Urban Runoff Program (NURP) study (Athayde et al. 1983), have shown that stormwater (SW) from urban and industrial areas is commonly contaminated by heavy metals, synthetic organics, pesticides, fuels, waste oils, and pathogens. Congress, recognizing the importance of controlling these discharges, passed amendments to the Clean Water Act (CWA) in 1987 requiring that the U.S. Environmental Protection Agency (EPA) issue regulations addressing SW discharges under the National Pollutant Discharge Elimination System (NPDES) program. These

---

[1]ORISE Research Engineer, [2]Leader, [3]Branch Chief, and [4]Environmental Engineer, Wet Weather Flow Research Program, Urban Watershed Management Branch, Water Supply and Water Resources Division, National Risk Management Research Laboratory, U.S. Environmental Protection Agency, Edison, NJ 08837

programs are required to address a number of SW control measures, including methods to detect and remove illicit discharges entering municipal storm sewer systems, as well as appropriate best management practices (BMPs) to address discharges from industrial, commercial, and development activities.

To help municipalities implement their SW management programs (SWMPs) in a practical and cost-effective manner, EPA is developing the "Guidance Manual For Implementing Municipal Stormwater Management Program" (EPA 1997). This is a two-volume manual: Volume I addresses policy and institutional issues and Volume II will present technical information on control/treatment, design, and operation for SW management. Volume I of the manual is being published and Volume II is expected to be completed next year. This paper presents an overview of Volume I.

Chapter 1. Introduction

This chapter outlines the SWMP planning process (a basic seven-step planning process) and examines the relationship between the NPDES program and other urban SWMPs. This manual differs from most of the other publications because rather than focusing on completing municipal permit application requirements, it provides guidance on how to develop and implement a long-term, cost-effective SW management program. Specifically, this document will help municipalities to set priorities for successful program implementation.

*Developing a Water Management Program: the Planning Process* This section delineates a basic seven-step planning process that will help municipalities design cost-effective and sensible SWMPs. For municipalities that have already completed Parts 1 and 2 of the NPDES municipal permit application, this planning process may suggest ways to improve or enhance the proposed SWMP. A flow chart (Figure 1) has been developed to give municipalities a sense of how each step in the planning process logically leads to the next and ultimately of how the process feeds back into itself, thereby forming a cycle.

*Related Regulations/Statutes and Programs That Address Municipal SW Runoff* While this manual focuses on providing guidance for NPDES SW program implementation, municipalities should carefully consider other related watershed protection programs. The related Federal statutes, regulations, and programs that address municipal SW runoff, pollution prevention and control include: (1) The National Combined Sewer Overflow Control Policy (59 FR 18688); (2) Nonpoint Source Program (under CWA § 319); (3) Coastal Zone Nonpoint Source Pollution Control (under the Costal Zone Act Preauthorization Amendment of 1990, § 6217); (4) Safe Drinking Water Act; (5) Clean Lakes Program (under CWA § 314); (6) Regulations/Wetlands Program (under CWA § 404); (7) National Estuary Program; (8) Federal Emergency Management Agency Regulations; and (9) Pollution Prevention Act of 1990.

# WATER RESOURCES AND THE URBAN ENVIRONMENT

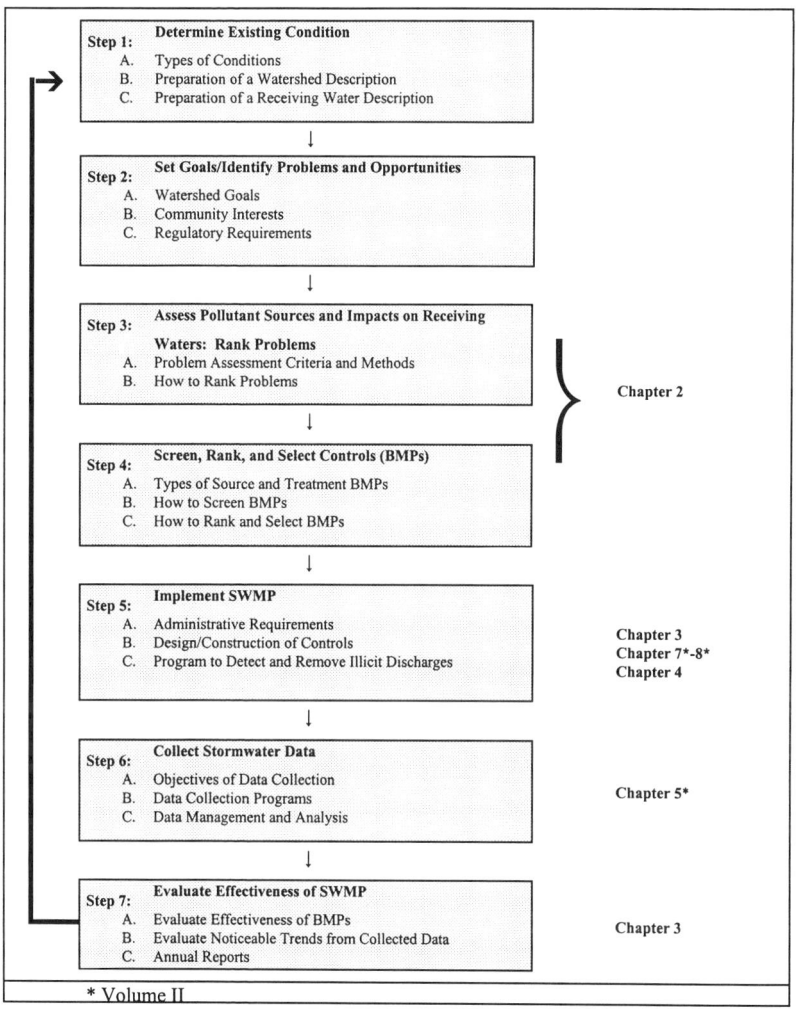

FIGURE 1. THE SEVEN-STEP STORMWATER MANAGEMENT PROGRAM PLANNING PROCESS

Chapter 2. Assessing SW Runoff Problems and Developing Solutions: How to Set Priority

This chapter consists of three primary sections. The first addresses methods for assessing problems and ranking them in order of importance. The second offers methods for evaluating and selecting controls to solve these problems. The third section includes case studies of municipalities assessing SW runoff problems and evaluating/selecting appropriate BMPs.

*Assessing Pollutant Sources and Impacts on Receiving Waters: Ranking Problems* To determine the need for, and appropriate level of, pollution prevention and control measures under their SWMPs, municipalities need to assess and rank existing watershed conditions. While many different types of problem assessments may be conducted as part of the SWMP, this chapter focuses on the following four major types: (1) resource assessments; (2) pollutant source assessments; (3) institutional assessments; and (4) goals and objectives assessments. A numerical ranking system for prioritizing pollution sources is included in this chapter. The hypothetical application of this weighted ranking methodology uses criteria such as water body, type of use, status of use, level of use, pollutant loads, and implementability of controls. Finally, methods for ranking problems using both quantitative and qualitative measures are explained. Once SW runoff problems have been fully assessed and ordered, municipalities will begin to screen and select BMPs.

*BMP Selection Process* Selecting BMPs for preventing and controlling SW runoff pollution is a two-step process. First, a comprehensive list of BMPs should be compiled and screened to eliminate those that are inappropriate for the program. The appropriate BMPs are then assessed to select those that will ultimately be implemented in the SWMP. The construction of facilities to collect and treat urban SW runoff may be prohibitively expensive. Therefore, the emphasis of SW pollution control should be on developing a cost-effective approach that includes nonstructural controls and low-cost structural controls.

The goal of the *BMP screening* process is to reduce the list of BMPs to a more manageable number to be considered for implementation. For the purposes of screening, BMPs are divided into two general categories: structural and nonstructural. Structural BMPs, e.g., detention ponds and infiltration practices, are designed to address specific pollutants from known sources. In contrast, nonstructural BMPs, which include regulatory practices (e.g., those that limit impervious areas or protect natural resources) and source controls (such as street sweeping or solid waste management) are typically implemented throughout an entire community, watershed, or special area to be protected. Municipal SWMPs will, in most cases, rely on a combination of both structural and nonstructural practices. Methods for screening these two types of BMPs are outlined in Chapter 2.

The process outlined acts as a guide for decision making and cannot account for all of the circumstances that might be encountered. Professional judgment and care is needed at each step along the way. Once these choices have been made and BMPs have been selected, the SWMP is ready to be implemented.

### Chapters 3. Guidance on Completing Administrative Requirements

This chapter discusses the procedures on implementing specific administrative requirements, which include public participation and public information programs, fiscal analysis, and annual reports.

*Public Information and Public Participation Programs* are essential to the implementation of an effective municipal SWMP. Public participation and public information programs solicit public support by informing individuals of the importance of good SW management and its effect on water quality. The key points to consider in developing this component of the program include creating appropriate goals and objectives, targeting the proper audience, explaining and selling the program to the audience, and having the necessary equipment and staff for proper program implementation. Case studies of selected municipalities are provided in this section.

The availability of *fiscal resources* is another essential component of municipal SWMPs. Several funding options are available to municipalities: local funding mechanisms, matching fund programs, and grant programs. By conducting a thorough fiscal analysis program, a municipality examines all of the available sources of funding and selects the funding option(s) according to its specific needs. Selected Federal grant programs that can assist in the financing of SW management needs are also included.

An *annual report* assesses the effectiveness of the management program and allows the municipality to revise the program based on the results of the assessment. An excerpt from an annual report on the Santa Clara County program and pamphlets and booklets of public education materials that attract attention, are easy to read, and provide steps that the public can take to help improve water quality are provided at the end of this chapter.

### Chapter 4. Procedures for Implementing a Program to Identify and Remove Illicit Discharges from Storm Sewer Systems

This chapter describes the procedures for identifying illicit discharges and implementing illicit discharge control programs. Specifically, it describes the components of an effective illicit discharge detection program, EPA's method for identifying illicit discharges, and examples of illicit discharge programs that have been or will be implemented in different municipalities.

*Required Components of an Illicit Discharge Detection and Removal Program* The regulations (40 CFR 122.27) require that the SWMPs include "a description of a program to detect and remove . . . illicit discharges into the storm sewer." The components of an effective program include a mechanism for prohibiting illicit discharges, field screening, investigation of potential illicit discharges, spill response and prevention procedures, public awareness and reporting program, used oil/toxic materials management and disposal procedures, and methods to control infiltration from sanitary sewers to storm sewers. Within these components, the use of GIS for mapping illicit discharge connections and for maintaining a database of information on illicit discharges throughout the municipality is essential.

*EPA's Suggested Method for Detecting Illicit Connections* EPA's suggested method for detecting illicit discharge connections, developed by the Office of Research and Development, is described in "Investigation of Inappropriate Pollutant Entries into Storm Drainage Systems: A User's Guide" (Pitt et al. 1993). This method focuses on data collection and quantitative analysis to implement a proper illicit discharge identification program. The procedure of this method can be briefly described in the following investigation steps: (1) drainage area mapping; (2) identification using tracers; (3) field survey and data collection; (4) analyses of data collected; (5) categorization of outfalls; (6) investigation and remediation; and (7) pollution prevention program. This method relies on the quantitative analysis of dry weather flows to identify the pollutants within illicit discharges. This information is then used to locate the potential source(s) of the discharges.

Several case studies on the various ways illicit discharge programs can be developed and implemented are provided at the end of the chapter. These municipalities have incorporated the components of an effective program in ways that are most effective to their specific needs.

References

Athayde, D.N., Shelley, P. E., Driscoll, E.D., Gaboury, D., and Boyd, G. (1983). "Results of the Nationwide Urban Runoff Program, Volume 1, Final Report", *(NTIS PB 84-185 552)*, U.S. Environmental Protection Agency, Washington, DC.

Pitt, R., Field, R., Lalor, M., Adrian, D., and Barbe', D. (1993). "Investigation of Inappropriate Pollutant Entries into Stormwater Drainage Systems: A User's Guide", EPA 600/R-92/238 *(NTIS PB 93-131 472)*, U.S. Environmental Protection Agency, Washington, DC.

U.S. Environmental Protection Agency (1997). "Guidance Manual for Implementing Municipal Stormwater Management Programs: Volume I - Planning and Administration.", Office of Wastewater Management and Office of Research and Development, Washington, D.C. (Final Draft)

Constructed Wetlands for Stormwater Management: Applications, Design, and Evaluation

Shih-Long Liao[1], Richard Field[2], and Shaw L. Yu[3]

Abstract

An analysis model, VirginiA Stormwater WETland Simulation Model (VASWETS) based on a double-layer (water column and substratum) box approach, was successfully developed to model the fate of pollutant transport in bucket wetlands. Results obtained indicate that the model has the ability to predict and optimize performance and to compare design criteria of stormwater (SW)-wetland systems. Design graphs for system-performance prediction and a comparative pollutant removal analysis for different detention times and pond/wetland (P/W) volume ratios for SW-P/W-management applications, both developed from the model calculations, are also included. A newly funded EPA bench-scale (continuous) and field-scale wetland/disinfection channel study will further examine and verify the model's algorithm for nutrient and pollutant dynamics at a variety of detention times and nutrient and metal loadings for three plant species.

Introduction

Fairly extensive literature is available on the performance of wastewater-wetland systems; however, information on the use of natural or constructed wetlands for controlling stormwater (SW) pollution is scarce as it is a new concept. Thus, there is no general agreement on design criteria. To address this scarcity, a steady-state model, VirginiA Stormwater WETland Simulation Model (VASWETS) was developed to compare design alternatives, predict, and optimize the performance of SW-wetland systems. VASWETS was verified with experimental data.

Experimental Design and Results

Four experimental bucket, batch-type, wetland systems were installed in the University of Virginia's Environmental Engineering Laboratory. Each of these four systems was comprised of a 15-liter plastic bucket filled with 12 kg of washed gravel

[1]ORISE Research Engineer and [2]Leader, Wet Weather Flow Program, Urban Watershed Management Branch, Water Supply and Water Resource Division, National Risk Management Research Laboratory, U.S. Environmental Protection Agency, Edison, NJ 08837 and [3]Professor, Department of Civil Engineering, Charlottesville, VA 22903

(3—7 mm dia.) substratum. Cattail (*Typha latifolia*), reed (*Phragmites sp.*), and bulrush (*Scirpus*) were planted directly into the substratum with one bucket not planted to act as the control. Water quality parameters including suspended solids (SS), chemical oxygen demand (COD), total phosphorus (TP), orthophosphate (OP), and Zinc (Zn) in the water column and the substratum were monitored at 1, 5, 7, 14, and 21 days (d) detention. Two SW-runoff (SWR) tests and four synthesized-runoffs (SR) tests, spiked with stock Zn and TP solutions with initial concentrations of 1000 mg/L as Zn and 100 mg/L as $PO_4^{-3}$, respectively, were conducted. The SWR was considered to be low in COD, TP, OP, and Zn with average concentrations of 37, 3.6, 2.8, and 1.8 mg/L, respectively and the SR was considered to be high in COD, TP, OP, and Zn with average concentrations of 96, 16, 13, and 4.1 mg/L, respectively.

The results suggested that of the three plant species, bulrush is the most effective for TP (average 62 %) and OP (average 68 %) removal (based on the detention times of 1—21 d). Cattail and reed were very effective for Zn (average 78 %) and COD (average 42 %) removal, respectively. The removal-efficiency differentials between the vegetated buckets and control bucket is highest for TP (18 %) and is lowest for COD (5 %) (Yu and Liao 1996). This study demonstrated that bucket wetlands have the ability to remove nutrients. In a subsequent field study, close agreement between the laboratory experimental and field study was found (Yu et al. 1996).

Modeling Approach

Using data collected (pollutant concentration versus detention time) from this study, VASWETS, based on a double-layer (water column and substratum) box approach, was successfully developed to model pollutant fate and transport in bucket wetlands. In this model, five transport mechanisms, i.e., diffusion, settling/sedimentation, sorption/filtration (both to plant species and substratum), and plant species uptake are used to examine pollutant concentration with respect to time. The algorithm for the experimental-wetland system is described in the following equations:

*Pollutant in the water column*

$$V_{(1)}\frac{dC_1}{dt} = - [v_{s(1)}Afp_1 + (K_2 + K_3)V_{(1)}fd_1]C_1 + K_1A[\frac{fd_2C_2}{\phi} - fd_1C_1] \quad [1]$$

*Pollutant in the substratum*

$$V_{(2)}\frac{dC_2}{dt} = - [v_{s(2)}Afp_2 + (K_2 + K_4)V_{(2)}fd_2]C_2 + v_{s(1)}Afp_1C_1 - K_1A[\frac{fd_2C_2}{\phi} - fd_1C_1] \quad [2]$$

where:

$A$=interfacial area of the segments (m$^2$); $C_1$=concentration of pollutant in water column (mg/L); $C_2$=concentration of pollutant in substratum (mg/L); $E$=dispersion coefficient (m$^2$/d); $K_1$=diffusion coefficient between water column and the substratum (m$^2$/d) (usually ranges from 2×10$^{-3}$ —2×10$^{-5}$ m$^2$/d); $K_2$=sorption/filtration coefficient (to substratum) (d$^{-1}$); $K_3$= sorption/filtration coefficient (to plant species) (d$^{-1}$); $K_4$=plant species uptake coefficient (d$^{-1}$); $K_p$= partition coefficient (L/mg); $V_{(1)}$=volume of water column in the segment (m$^3$); $V_{(2)}$=volume of substratum (m$^3$); $v_{s(1)}$=settling velocity (m/d) in water column (usually ranges from 0.1—0.5 m/d); $v_{s(2)}$=settling velocity (m/d) in substratum (usually range from 0.1—2.0 m/d); $fd$=fraction of dissolved pollutant=$(1 + K_p \times (SS))^{-1}$; $fp$=fraction of particulate pollutant=$K_p \times (SS) \times (1 + K_p \times (SS))^{-1}$; $SS$=concentration of suspended solids (mg/L); $fp_1$=fraction of particulate pollutant in water column; $fp_2$=fraction of particulate pollutant in substratum; $fd_1$=fraction of dissolved pollutant in water column; $fd_2$=fraction of dissolved pollutant in substratum; $t$=time (d); and $\phi$= porosity of substratum (unitless).

## Modeling Results

Close agreement between the model calculations and the laboratory data was obtained (Liao and Yu 1997, Liao 1996, and Liao and Yu 1996). With the same input parameters, a comparison of simulated results made between EPA's Water Quality Simulation Program WASP (DiToro and Chick 1983) and VASWETS shows similar predictions for the fate of TP, COD, and Zn in both water column and substratum (Liao and Yu 1998 and Liao et al. 1998). Sensitivity analyses for the model further substantiate the model calculations (Liao 1996). In order to apply the model to a SW-wetland design, the parameter's trends were analyzed with respect to the initial pollutant concentrations of all runoff. A trend analysis example of K rate constants for TP, OP, and Zn was presented elsewhere (Liao and Yu 1998).

## Model Application

Design graphs (Figure 1) based on system-performance predictions was developed using VASWETS assuming the proposed pond/wetland (P/W) system is part of a 30-hectare (≈ 75-acre) drainage basin. The basin contains a large open-air shopping mall and a parking lot. It was further assumed that the runoff into the P/W was generated by a 13 mm (≈ 0.5 in.) one-hour storm having six antecedent dry days along with a runoff coefficient of 4.6 and pollutant washoff coefficients for TP, OP, and Zn of 5, 5, and 5, respectively. The antecedent-dry day pollutant-buildup parameters for TP, OP, and Zn are 0.02, 0.015, and 0.02 lb/acre/d, respectively. From simulation, the total pollutant loads are 2.18, 1.16, and 2.18 kg for TP, OP, and Zn, respectively. Also the ratio of cattail/bulrush/reed was assumed to be 1:1:1 and the pollutant removal mechanisms in the control bucket were sedimentation and sorption/filtration, analogous to a detention pond having a gravel substratum (with free-water and subsurface flow) without vegetation. The design graphs (Figure 1) containing removal efficiencies for different P/W volume ratios at various detention times were developed from VASWATS's calculations. Based on

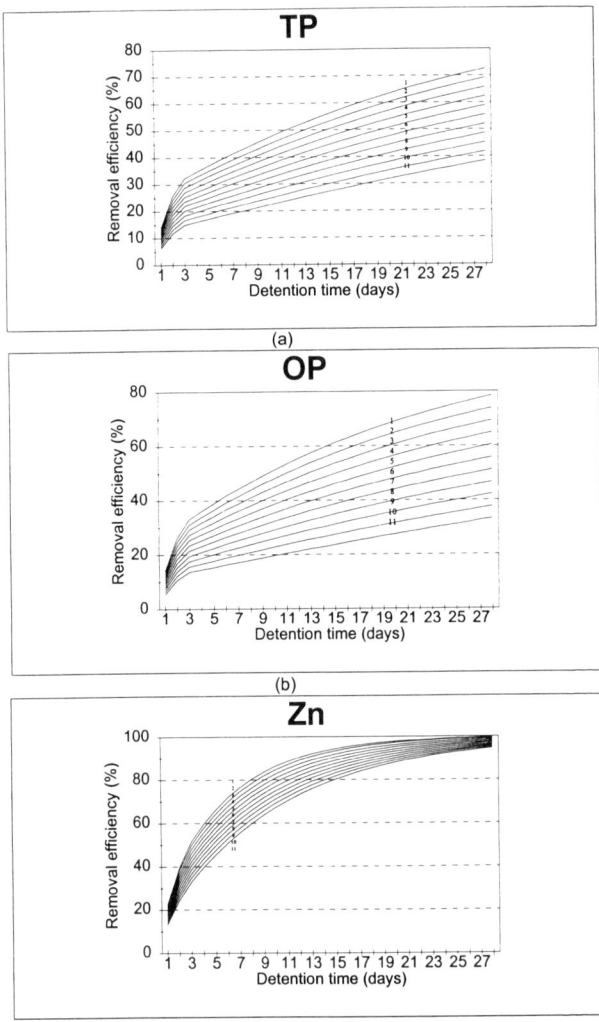

Figure 1. Design Graphs: Removal Efficiency Versus Time with Respect to Pond /Wetland (P/W) Volume Ratio at Bulrush:Cattail:Reed Ratio =1:1:1 for (a) TP, (b) OP, and (c) Zn
1: All Wetland; 2: P/W=1:9; 3: P/W=2:8; 4: P/W=3:7; 5: P/W=4:6; 6: P/W=5:5; 7: P/W=6:4; 8: P/W=7:3; 9: P/W=8:2; 10: P/W=9:1; 11: All Pond

these graphs, the engineer can select the proper P/W volume ratio and detention time to achieve a desired removal for each pollutant (TP, OP, and Zn). For example, if a detention time and a P/W volume ratio of 10 d and 3/7, respectively are selected, then 41, 43, and 82 % removal for TP, OP, and Zn, respectively will be achieved. The results suggest that the ideal P/W ratios range from 3/7—7/3. For P/W ratios between 3/7 and 7/3, the ideal detention times for a 40 % removal of TP, OP, Zn are 11—14, 10—12, and > 4 d, respectively.

As previously indicated, there is no general agreement on design criteria because the use of natural or constructed wetlands for controlling SW pollution is a new concept. However, several existing design guidelines suggest detention times ranging from 4—14 d. Using the example described above, a comparative-pollutant-removal analysis for different detention times (4—14 d) and different P/W volume ratios derived from Figure 1, is shown in Table 1. The pollutant removal increases as the P/W volume ratio decreases. The results also show that the pollutant removal increases as the detention time increases. This is reasonable since the detention time is proportional to the treatment volume. A wetland requires a shorter detention time than a pond without vegetation for an equivalent pollutant removal; accordingly, a smaller treatment volume is required for the wetland.

Table 1: Comparative Pollutant Removal for Different Detention Times*

| Detention time (d) | 4 | 6 | 8 | 10 | 12 | 14 |
|---|---|---|---|---|---|---|
| Pollutant removal (%) | | | | | | |
| Pollutant Parameter | Pond/wetland (P/W) volume ratio (7/3) | | | | | |
| TP | 24 | 25 | 27 | 30 | 32 | 35 |
| OP | 21 | 25 | 27 | 30 | 32 | 35 |
| Zn | 50 | 53 | 60 | 66 | 75 | 82 |
| All pond (without vegetation) | | | | | | |
| TP | 15 | 16 | 19 | 21 | 25 | 25 |
| OP | 15 | 16 | 17 | 18 | 20 | 22 |
| Zn | 40 | 51 | 58 | 64 | 72 | 75 |
| All wetland (cattail/bulrush/reed of 1:1:1) | | | | | | |
| TP | 35 | 39 | 43 | 46 | 51 | 54 |
| OP | 36 | 41 | 45 | 48 | 55 | 58 |
| Zn | 60 | 75 | 81 | 85 | 90 | 95 |

* Liao 1996

## Model Improvement

VASWETS has been successfully improved to accommodate flood routing and both surface-, and subsurface- flow algorithms for SW-wetland simulation. A newly funded EPA bench-scale (continuous) and field-scale wetland/disinfection channel study will further examine and verify the these algorithms for nutrient and pollutant dynamics at a variety of detention times and nutrient and metal loadings for the aforementioned three plant species.

## Conclusions and Recommendations

Results obtained indicate that VASWETS has the ability to select SW-P/W design criteria (P/W volume ratio and detention time) based on comparative pollutant removal analyses of SW-P/W systems.

## References

DiToro and Chick 1983. "Water Quality Simulation Program (WASP) and Model Verification Program (MVP) Documentation", *EPA 600/3-81-044 (NTIS 83-225573)*, US Environmental Protection Agency, 135 pages.

Liao, S.L.,1996. "An Experimental Evaluation of Using Constructed Wetlands for Controlling Stormwater Pollution", Ph.D. dissertation, Department of Civil Engineering, University of Virginia, Charlottesville, VA, 273 pages.

Liao, S.L. and Yu, S.L., and Long, S.J., 1996. "An Experimental Study on the Use of Constructed Wetlands for Controlling Stormwater Pollution", 1996 ASCE North American Water and Environment Congress, Anaheim, CA.

Liao, S.L. and Yu, S.L., 1997 "The Use of Constructed Wetlands for Controlling Stormwater Pollution", *Journal of Chinese Institute of Environmental Engineering Vol. 7, No 4, pp 20-26.*

Liao, S.L. and Yu, S.L., 1998 "The Use of Constructed Wetlands for Stormwater Management", Submitted to the ASCE Journal of Water Resources Planning and Management.

Liao, S.L., Field R., Fan C.Y, and Yu, S.L. 1998 "The Use of Three Wetland Plant Species for Stormwater Management" -IAWQ Conference (accepted)

Yu, S.L., Liao, S.L., Fitch, M., and Long, S.J., 1996. "Field Testing of Stormwater Wetland and GIS application", National Conference on Environmental Engineering, WEF, Dallas, TX.

Yu, S.L. and Liao, S.L., 1996. "The Control of pollution in Highway Runoff through Biofiltration, Volume III: Laboratory Test of Roadside Vegetation", Virginia Department of Transportation, Charlottesville, VA, 66 pages, *FHWA/VTRC 95-R30*.

Optimization of Storage and Treatment Systems for CSO Pollution Control

Richard Field[1], Member, ASCE, and Thomas P. O'Connor[2]

Abstract

Combined-sewer overflow (CSO) must be controlled by a storage-treatment system because storm flow in the combined sewerage system is intermittent and highly variable in both pollutant concentration and flowrate. A treatment facility operating without the benefit of upstream storage would need to be very large and costly in order to handle the relatively high volume and flowrate of a CSO. Similarly, if storage is used without treatment, the storage volume required would be very large and also expensive. This paper describes a strategy to optimize the CSO-control system. This optimized system maximizes the use of the existing system before new construction and sizes the storage volume in concert with the wastewater treatment plant (WWTP) capacity to obtain the lowest-cost storage-treatment system.

Introduction

One commonly used approach to combined-sewer overflow (CSO)-pollution abatement is to rely on arbitrarily sized storage tanks to capture CSO for pump-back or bleed-back (gravity-flow) to the existing wastewater treatment plant WWTP) for treatment based on a design-storm. However, this is not the most economical approach because of the high costs involved in the construction of storage and/or expanded WWTP capacity. Therefore, a strategy should be adopted to create an optimized CSO-control system which maximizes the use of the existing system before

---

[1]Leader, Wet-Weather Flow Research Program and [2]Environmental Engineer, U.S. EPA, National Risk Management Research Laboratory, Water Supply and Water Resources Division, Urban Watershed Management Branch

constructing new facilities and sizes the storage volume in concert with the treatment rate to obtain the lowest-cost storage-treatment system. An integraged optimization strategy will minimize the construction of new storage and treatment facilities by: (1) making operational and low-cost, inline improvements in the existing combined-sewerage (collection-treatment) system to maximize storage and treatment before new storage or treatment unit construction, (2) accounting for treatment by sedimentation when storage tanks overflow, (3) selecting design capacity of the CSO-control system to reflect the most economical approach by choosing the point of diminishing returns on the curve of pollution control vs. cost, and (4) sizing of the storage-treatment system according to the most economical break-even point between the amount of storage vs. the treatment capacity (Field and O'Connor, 1997).

Optimization

The CSO-pollution-abatement system must integrate storage and treatment because storm flow is intermittent and highly variable in pollutant concentration and flowrate.

Optimization should begin with operational changes that increase the use of the existing system. Long-term urban-hydrological and CSO-quality-monitoring data should then be used to model the existing system. This analysis should include treatment by sedimentation in the existing storage facilities. Analyses should then be made of comparatively lower-cost modifications to the collection system followed by higher cost modifications to the WWTP, i.e., retrofitting in the existing primary-settling tanks (PSTs) with lamellae settling plates/tubes, chemical coagulation, microsand, or combinations thereof; and dissolved air flotation (DAF). The more expensive options of constructing additional storage facilities, parallel interceptors or facilities at the WWTP, and satellite-treatment plants at the outfalls should be considered last.

*Economic Analysis.* A complete and thorough storage-treatment-system analysis for the entire drainage area should be conducted using break-even economics for cost optimization of the overall pollution control requirements of the system. Storage-treatment strategies should identify the point of diminishing returns based on long-term rainfall-CSO-quantity and quality analyses, identified by the increasing cost without increased benefits to treatment. Figure 1 contains site-specific data for

Atlanta, Georgia, and shows that the rate of increase in the overall percent precipitation control drops off dramatically after the 90% control point. This point should be selected for design of the system unless regulations or receiving-water impacts require a higher capacity system. The exact shape of this curve will vary from site to site.

Another aspect of system optimization is illustrated by the simple hypothetical example in Figure 2. It assumes a design for a 3-h CSO event as characterized by a triangular hydrograph with a maximum flowrate of 25 MGD at end of the first hour. The construction cost of a *separate* approach for a storage tank of 1.56 Mgal or a treatment facility able to handle this flow would be $1,170,000 (@ $0.75/gal) or $1,250,000 (@ $50,000/MGD), respectively. However, an *optimized* approach for storage-treatment system design for this event, where the storage facility is designed for 0.45 Mgal @ $340,000 and the treatment facility is designed for 11.6 MGD @ $580,000 could handle this CSO event at a total cost of $920,000 -- a savings of $250,000 compared to the *separate* storage and treatment approach.

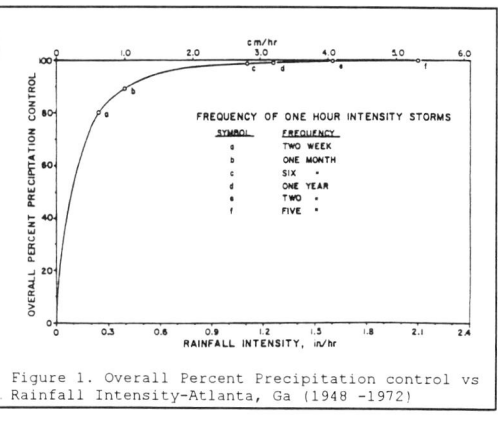

Figure 1. Overall Percent Precipitation control vs Rainfall Intensity-Atlanta, Ga (1948 -1972)

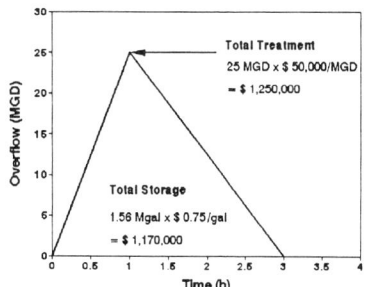 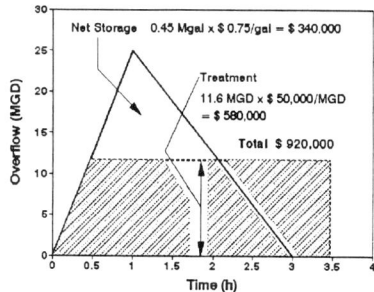

Figure 2. Storage Treatment Example: *Separate* and *Optimized* Approach

*Operational Changes.* Advantage should be taken of the storage available in the existing collection system. Storage capacity exists in most combined sewers because they were designed for the very infrequent storms, e.g., 1 in 5 and 1 in 10 year storms for flood control.

At capacity, overflowing storage tanks behave like, PSTs as previously mentioned. This dual-function of sedimentation treatment obtained during overflows should be taken into account before any further control designs. Removal by settling can be estimated by continuous long-term modeling of storm-flow influent to the storage tank using the settling-velocity distribution of the influent particles and the overflow rate of the tank.

In optimizing the WWTP, the first and cheapest of all optimization alternatives is to direct more flow to the WWTP and through the PSTs. This entails maximizing the treatment at the WWTP during storm-flow conditions by effectively treating as much flow as possible during the storm-flow occurrence.

This practice is well worth the effort. It reduces the release of raw-sanitary sewage and polluted stormwater at upstream outfalls by redirecting the CSO to the WWTP where at least some measure of primary treatment can be accomplished. In some systems for certain portions of the storm-flow period, the particle-settling velocities are relatively high in comparison to the dry-weather flow (DWF) particle-settling velocities due to the resuspension of antecedent DWF deposits in the sewer system. Therefore, during storm-flow periods, higher flowrates can be directed through the PSTs.

However decreases in settling-tank efficiency as a function of the higher influent-hydraulic loading should be taken into account. As the PST-hydraulic loading increases, its efficiency will decrease until a point is reached whereby diverting more flow to the WWTP will have a negligible effect on the increase of CSO-pollution removal.

*Lower-Cost Modifications.* Treatment maximization at the WWTP can also be enhanced by: 1) using advanced signals which indicate relatively high storm flows from monitoring systems (flowmeters, raingauges and/or radar) upstream of the WWTP; 2) maximizing in-sewer storage by simple CSO flow regulator adjustments; and 3) installation of dynamic regulators that respond to remote sensing devices to increase in-sewer storage.

In conjunction with directing more flow to the WWTP, an evaluation should be made to increase the interceptor carrying capacity from storage to the WWTP. This evaluation should include the use of: 1) increased pumping capacity for surcharged interceptors, which is the easiest and cheapest, as it is an operational change not requiring additional capital cost; 2) polymer injection, which has been effectively demonstrated to reduce sewer roughness and increase carrying capacity in both open-channel and surcharged flows; and 3) parallel interceptors.

*Higher-Cost Modifications.* Storage tanks are a basic component of the CSO- control system that require relatively little operation and maintenance (O&M). Storage-tank design should incorporate the hydrology of the entire drainage system and the withdrawal rate from the WWTP and/or satellite treatment facilities. The design should be based on the historic rainfall records using a continuous-modeling approach to simulate long-term effects and dual function of sedimentation treatment (as previously discussed). A reliable answer to the required storage volume can only be obtained through the use of continuous modeling, as it does not assume that the storage tank is empty for every storm event and incorporates the dependence on the storm interval. This is sometimes referred to as a "mass-flow" analysis.

The performance of PSTs can be enhanced by *retrofitting* with appurtenances that improve settling or separation of suspended solids so that higher overflow rates and shorter detention times can be used. Three proven methods are: (1) installing plate settlers, (2) converting to higher rate DAF, and (3) chemically-enhanced-primary treatment. Newer settling technologies rely on enhancement by coagulation/recycling systems. The microcarrier systems, e.g., Micosep® and Actiflo® use chemicals and recycled microsand, while DensaDeg® uses recycled sludge. Each system adds chemical to the recycled material to create heavier/larger floc which does not require as long a detention time.

*Parallel* facilities at the WWTP might consist of high-rate: (1) fine-mesh screens; (2) dual-media deep-bed filtration, or relatively slow-rate conventional PSTs. High-rate physical/chemical treatment processes have shown their importance in the treatment of CSOs. These physical processes, with or without chemical enhancement, can be designed for automated operation, are adaptable to intermittent and variable quantity and quality characteristics, and are resistant to shock loadings.

## Conclusions

CSO-pollution abatement must be dealt with in terms of a storage-treatment system because storm flows are intermittent and highly variable. Construction of additional treatment facilities or storage tanks to control CSOs are expensive and therefore the optimization strategy must: 1) maximize the available capacity of the existing sewerage system, 2) include the benefit of treatment by settling in overflowing storage tanks, 3) size the storage-treatment system based on continuous long-term historical, urban hydrology, and knee-of-the-curve for economical analysis (Figure 1), 4) size the storage-treatment combination using a break-even point or low-cost economical analysis; and 5) use a mass-flow analysis in the selection of control capacity, which does not assume storage tanks will be empty prior to the next storm-flow occurrence.

The existing WWTP will have treatment limitations due to the interceptor capacity, expansion costs, and space availability. Therefore, besides the enhancement of capacity for the interceptor and facilities at the WWTP, the storage-treatment-system optimization alternatives should include high-rate treatment with storage or satellite facilities at upstream CSO points.

## Appendix. Reference

Field, R. and O'Connor, T.P. (1997). "Optimization of CSO Storage and Treatment Systems." *J. of Environ. Engineering,* ASCE, 123 (3), 269-274.

# Urban Wet-Weather Flows Toxic Pollutants: Characterization and Enhanced Sedimentation Treatment

Chi-Yuan Fan[1], Richard Field[2],
Thomas P. O'Connor[1], and Mary K. Stinson[3]

## Abstract

This paper presents a brief overview of sources of toxic pollutants in the urban watershed and of a treatment approach, specifically the high-rate, microcarrier-weighted flocculation and sedimentation process, for removing these toxic pollutants, as well.

Urban wet-weather flow (WWF), which includes sanitary sewer overflow (SSO), combined sewer overflow (CSO), and stormwater discharge (SWD), contains significant quantities of toxic substances. WWF related toxic pollutants are major contributors to the degradation of receiving waters. Urban storm runoff contains a greater variety of toxic pollutants than sanitary wastewater. Field studies have identified such priority pollutants as benzene and polycyclic aromatic hydrocarbons (PAH), pesticides, and toxic heavy metals.

In urban WWF, toxic pollutants are primarily associated with the small-size particles, ranging from <50 $\mu$m to <10 $\mu$m, that are difficult to remove by conventional WWF sedimentation technologies. Advanced sedimentation systems show promise of reasonable effectiveness for removing these small-size particles containing these toxic pollutants. Importantly, these advanced sedimentation processes operate effectively at high-rates which is a needed attribute for the economical treatment of storm-generated high flowrates.

---

[1]Environmental Engineer, [2]Wet-Weather Flow Research Program Leader, and [3]Control Technologies Team Leader, U.S. Environmental Protection Agency, National Risk Management Research Laboratory, Water Supply and Water Resources Division, Urban Watershed Management Branch 2890 Woodbridge Avenue, Edison, NJ 08837.

The new sedimentation technologies use either the addition of an inert microcarrier with the addition of chemicals or the recirculation of chemically conditioned sludge. The Micosep® (International Water Solutions Corporation) and the Actiflo® (Omnium de Traitement et de Valorisation [OTV]) use recycled microcarriers (e.g., microsand), while DensaDeg®(Infilco Degremont, Inc.) uses sludge.

Introduction

Pollutants carried off urban catchments by drainage systems during wet weather emanate from many sources, e.g., commercial, industrial, and residential parking areas; roadways; automobile-service stations; sewer infiltration from leaking underground storage tanks; accidents and spills; park and residential lawns; construction sites; and active and inactive industrial sites. Past studies indicate that SSO, CSO, and urban SWD contain significant quantities of toxic substances; a number of the hazardous-waste priority pollutants have been identified. Without consideration of urban and industrial stormwater-runoff toxic-substance control, the various hazardous-substances-cleanup programs will not be effective in controlling total areawide emissions of these substances.

Toxic-organic chemicals (e.g., benzene, PAH, polychlorinated biphenyls [PCB], etc.) and heavy metals (arsenic [As], cadmium [Cd], chromium [Cr], copper [Cu], lead [Pb], mercury [Hg], and zinc [Zn]) in storm-induced discharges contribute to receiving-water degradation (Pitt, et al., 1995). The US Geological Survey reported that urban storm runoff collected from residential, commercial, and industrial areas around Phoenix, AZ was found to be toxic to fathead minnows and the water flea (Lopes and Fossum, 1995). The concentration of total petroleum hydrocarbons (TPH) and heavy metals in water and stream-bed sediments increased downstream of roadway runoff. Boudries, et al. (1996) and Estèbe, et al. (1996) reported that heavy metals and the aliphatic and aromatic hydrocarbons bound to particles in the River Seine sediments near Paris are due to urban WWF discharges. Such toxic substances in sediments create a long-term impact on ecological systems.

Origins of WWF Toxic Pollutants

Sewer sediments deposited during dry weather flow conditions and prior storms contribute significant toxic pollutants into receiving waters due to resuspension by high storm-caused flowrates, are one of the main sources of pollution emanating from CSO and SWD (Laplace, et al., 1990; McGregor, et al.,1993, Verbancu, et al.,1994, Butler, 1996). Pollutant origins in urban SWD are summarized below:

*Atmospheric Deposition.* Large amounts of toxic pollutants were associated with atmospheric deposition (Cotham and Bidleman, 1995). This has been further demonstrated in EPA's Great Lakes and Coastal Waters Program, which reported that about 90% of toxic-pollutant (e.g., As, Cr, Cu, Pb, Hg, PCB, PAH) loadings to the Great Lakes result from airborne deposition.

*Illicit Connections.* Common nonstormwater entries include the illicit connection of sanitary and industrial wastewater discharges from residential, commercial, and industrial buildings to separate storm-drainage systems that contribute large amounts of toxic and conventional pollutants. Recently, Federal and State regulatory agencies have paid much attention to the investigation of illicit connections. Baltimore, as part of its National Pollutants Elimination Systems (NPDES) permit, examined the storm-drainage system for sources of illicit entries and found approximately one third of the 344 small subcatchments are potential sources of toxic contaminants (phenols, ammonia, Cu, Pb, Zn, and TPH) from illicit entries.

*Paved Surfaces and Building Rooftops.* Major sources of toxic pollutants are wet- and dry-weather runoff from urban streets, highways, building rooftops, and parking areas. Furthermore, Table 1 (Ellis and Revitt, 1982; Vignoles and Herremans, 1995) demonstrates that the majority of heavy metals are associated with particles <10 $\mu$m that are well into the colloidal range.

Table 1. Metal distribution versus particle size

| SS Size ($\mu$m) | Cd | Co | Cr | Cu | Mn | Ni | Pb | Zn |
|---|---|---|---|---|---|---|---|---|
| >100 | 18 | 9 | 5 | 7 | 8 | 8 | 4 | 5 |
| 10 - 100 | 36 | 31 | 24 | 30 | 21 | 29 | 23 | 35 |
| <10 | 46 | 60 | 71 | 63 | 71 | 63 | 73 | 60 |

Control and Treatment

Most of the solids finer than 50 $\mu$m are the principal vector of pollution in urban WWF (Chebbo et al.,1990). Thus, methods for controlling toxicants in urban WWF must be capable of reducing these small particles. Nevertheless, very few of the usually employed WWF treatment processes are able to effectively remove colloidal solids.

The most popular method for CSO pollution control is storage (tanks) or retention (ponds). This enables overflow and associated pollutant capture for pump- or bleed-back to the regional wastewater treatment plant (WWTP) after each storm-flow event. The effectiveness of pollutant removal is based on the volume of flow captured by the storage facility and influent suspended solids (SS) settling characteristics, when the CSO volume exceeds holding capacity. Greb and Bannerman (1997) reported that SWD retention ponds are capable of removing 87% of the SS load in residential runoff. The pond's effluent SS distribution ranged from 4 $\mu m$ — 62 $\mu m$. However, in order to control toxic pollutants, effluent solids' sizes should be <2 $\mu m$.

One approach is to apply polymers to promote toxic pollutant sorption during floc formation prior to sedimentation. This physical-chemical process achieves better water quality than conventional sedimentation because it removes the smaller particles and their sorbed toxic pollutants. It can be incorporated into the existing WWTP by retrofitting the primary sedimentation tanks or as parallel processes for storm flow or set up as part of remote satellite plants in conjunction with storage at outfall sites. A high-rate sedimentation process, the Lamella angled plates settler, increases the effective settling area of the tank. The best sedimentation efficiency occurred at plate angles of $50°$ with a range of hydraulic loadings between 0.5 m/h and 3 m/h (Demir, 1995). However, the effluent particle-size distributions have not been reported.

Three new settling processes use enhanced systems to recycle either an added microcarrier (e.g., microsand with a diameter range of 50 $\mu m$ — 200 $\mu m$) or chemically-conditioned sludge. The Micosep® (International Water Solutions Corporation) and the Actiflo® (Omnium de Traitement et de Valorisation [OTV]) use microsand, while DensaDeg® (Infilco Degremont, Inc.) uses chemically conditioned return sludge. Actiflo® and DensaDeg® employ Lamella settlers whereas Micosep® does not. Each system adds polymers and metallic salts to the recycled stream to create heavier/larger floc that does not require as long a detention time as the conventional sedimentation. It has been claimed that: (1) Actiflo® removes 80% SS with effluents of 30 mg/l and operates at a 15 min detention time, a range of 10% — 100% the design flow, and a peak hydraulic loading of 130 m/h (compared to 10 m/h for conventional sedimentation); (2) Micosep® removes 90% SS, has a rapid startup (for offline installations), and operates at a 70% reduction in tank surface area (compared to conventional sedimentation tanks); and (3) DensaDeg® operates at hydraulic loadings of 17 m/h — 22 m/h and a 50% reduction in tank surface area (compared to conventional solids-contact clarifiers while also incorporating gravity-sludge thickening).

Conclusions

WWF pollution abatement must consider its highly variable nature in both flowrate and pollutant characteristics. A significant proportion of toxic pollutants in urban WWF is associated with particle sizes <10 $\mu$m. Therefore, an optimized basinwide storage-treatment system must be considered including: (1) maximization of available capacity of the existing WWTP sedimentation tanks by retrofitting or installing parallel units with high-rate (sedimentation) processes; (2) incorporation of high-rate sedimentation at the WWF outfall location either within the storage tank or in series or parallel to it; and (3) new flexible design approaches for retrofitting and/or improving operations of conventional WWTP and enhancing the control of toxic substances.

References

Boudries, H., Broguet, C., Mouchel, J-M., and Thévenot, D.R. (1996). "Urban Runoff Impact on Composition and Concentration of Hydrocarbons in River Seine Suspended Solids." *Proc. 7thInt. Conf. on Urban Storm Drainage*. Hannover, Germany, 569-574.

Butler,D., May,R.W.P., and Ackers,J.C. (1996). Sediment Transport in Sewers Part 1: Background. *Proc. Inst. Civ. Eng. Water, Maritime and Energy.* 118(2)103-112.

Chebbo, G., P. Musquere, V. Milisie and A. Bahoc (1990). Characterization of Solids Transferred into Sewer Trunks During Wet Weather. *Wat. Sci. Tech.*, 22(10/11) 231-240.

Cotham, W. and T. Bidleman (1995). Polycyclic Aromatic Hydrocarbons and Polychlorinated Biphenyls in Air at an Urban and a Rural Site near Lake Michigan, *Environ. Sci. & Technol.* **29**: 2782-2789.

Demir, A. (1995). Determination of settling Efficiency and Optimum Plate Angle for Plated Settling Tank. *Wat. Res.* 29(2)611-616.

Ellis, J.B. and Revitt, D.M. (1982). Incidence of Heavy Metals in Street Surface Sediments: Solubility and Grain Size Studies. *Water, Air, and Soil Pollution*, 17(1)87-95.

Estèbe, A., Belhomme, G., Lecomte, S., Videau, V., Mouchel, J-M., and Thévenot, D.R. (1996). Urban Runoff Impacts on Particulate Metal Concentrations in River Seine: Suspended Solid and Sediment Transport. *Proc. 7th Int. Conf. on Urban Storm Drainage*, Hannover, Germany, 575-580.

Greb, S.R. and Bannerman, R.T., (1997). Influence of Particle Size on Wet Pond Effectiveness. *Water Environ. Research*, 69(6)1134-1138.

Laplace,D.,Sanchez,Y., Dartus,D., and Bachoc,A., (1990). Sediment Movement Into the Combined Trunk Sewer No. 13 in Marseille. *Wat. Sci. Tech.* 22(10/11)259-267.

Lopes, T. J. and Fossum, K. D. (1995). Selected Chemical Characteristics and Acute Toxicity of Urban Stormwater, Streamflow, and Bed Material, Maricopa County, Arizona. *Water-Resources Investigations Report 95-4074*, U.S. Geological Survey, Denver, Colorado.

McGregor, I., Ashley, R.M., and Oduyemi, K.O.K., (1993). Pollutant Release from Sediments in Sewer Systems and Their Potential for Release into Receiving Waters. *Wat. Sci. Tech.* 28(8/9)161-168.

Pitt, R., Field,R. Lalor,M., and Brown,M. (1995). Urban Stormwater Toxic Pollutant Assessment, Sources, and Treatability. *Water Environ. Research*. 67(3) 260-275.

Verbanck, M.A., Ashley,R.M., and Bachoc,A. (1994). International Workshop on Origin, Occurrence and Behaviour of Sediments in Sewer Systems: Summary of Conclusions. *Water Environ. Research*. 28(1)187-194.

Vignoles, M. And Herremans, L. (1995). Metal Pollution of Sediments Contained in Runoff Water in the Toulouse City. *NOVATECH 95, Second International Conference on Innovative Technologies in Urban Storm Drainage*, May 30-June 1, 1995. Lyon, France, Organized by Eurydice 92 and GRAIE. 611-618.

**Key Words:**

Stormwater runoff, water quality, water pollution control, storage tanks, combined sewers, combined sewer overflows, separated sewer overflow, urban hydrology, drainage systems.

Planned EPA Research in Urban Watershed Modeling

Michael Borst[1]

Abstract

Many mathematical watershed planning and management models exist. Various organizations, public and private, are using and developing these models. This paper outlines the EPA research direction for modeling efforts supporting the urban watershed over the next few years. EPA's modeling research plan relies heavily on past efforts, and following Agency policy, recognizes the ultimate user is the local community. The peer-reviewed EPA Wet Weather Flow Research Plan[1] broadly outlines the plan and it was the topic of a targeted three-day discussion in Edison, NJ.

Background

As an Agency, EPA needs to identify a small collection of models that encompass most watershed-scale water quality and ecosystem needs. EPA cannot reasonably support many single-application models. Alternately, EPA should support a single package, assembling proven, well-recognized models with transparent translation software assuring the component models share databases and exchange results. This approach concentrates efforts and simplifies the needed support. The urban watershed modeling effort fully supports the approach selected for the Better Assessment Science Integrating Point and Nonpoint Sources (BASINS) system and plans to build on this approach of coupling existing models with common databases.

The Urban Watershed Management Branch (UWMB) is undertaking to develop standard operating procedures (SOPs)

---

[1] Chemical Engineer, US EPA, Urban Watershed Management Branch, Water Supply and Water Resources Division, National Risk Management Research Laboratory, Edison, NJ 08837-3679

to help assure the user community will efficiently collect the correct information using EPA-defined and EPA-accepted quality assurance and quality control procedures in a uniform data format. Building from the SOPs and past efforts, this research can generate the required minima to assure integrated, technically sound, and reliable models. UWMB is also developing an interface between the Stormwater Management Model (SWMM) and a BASINS-compatible geographic information system (GIS).

EPA has long supported watershed and other computer models. Separate EPA airshed and groundwater models support well-known watershed application tools. This history yielded many well-documented models that provide the engineering and scientific community with reliable computational capabilities and have advanced the national ability to protect water quality.

As EPA continues shifting emphasis and strategy to stress community-based environmental protection (CBEP) and a holistic watershed protection process, models must play an increasingly important role in watershed planning. Models are the only practical tools providing the time compression required to evaluate alternatives when the results of a given action or approach are likely to be visible years, and commonly decades, in the future. Models are the only realistic tools that can give users the ability to evaluate progress by comparing monitoring against forecasted milestones and integrate collected data fully into the management plans. Models are the only tools that allow decision makers to evaluate forecasted future conditions under alternative actions. Without the ability to forecast future conditions, the regulatory bodies with the delegated authority and responsibility lack a systematic approach for maintaining or improving the watershed. For models to be fully effective, they must integrate smoothly, share common data sets, fit the expectations of today's computer user, and present results in a format rapidly understood by the lay community.

The reclamations accomplished by community groups are noteworthy. Twenty-five years of Clean Water Act (CWA) ventures have handled the affordable, obvious, and simple requirements leaving subtler, more costly, and more complicated decisions. Well-defined, readily-implemented, linked models that can be used by CBEP decision makers will enable faster, less costly restoration. Fortunately, the foundation laid by the Agency's long history of model support, the lessons learned in large, well-documented studies, and the advances in computer hardware position us to advance the state of the practice. This foundation coincides other events that will accelerate the modeling process such as the maturing of several monitoring programs, new software products, and the transition of significant computational capability from the reserved domain of scientists to the average desktop. The combined effects place EPA in an excellent position to revisit watershed modeling needs.

Agency resources, measured in both available staff time and funding, have decreased over past years. The Congressional mandate to "protect human health and the environment," remains active and the Administrations' interpretation of the water mission continues to broaden from the "fishable and swimmable" vision of the CWA to broad ecologic protection[2].

As the broadening interpretation of the Agency mission collides with the reduced resource allocation, the need for a cohesive Agency modeling strategy becomes paramount. The Agency Task Force on Environmental Regulatory Modeling[3] (AFTERM) is addressing this issue. EPA is forming a separate modeling Peer Review and Quality Assurance structure based on ATFERM recommendations[4]. EPA drafted guidance on Monte Carlo applications[5], and extending the probabilistic analysis to other arenas appears likely.

The 1995 reorganization in EPA's Office of Research and Development created a structure paralleling the risk paradigm and assigned the responsibility of risk management research to the National Risk Management Research Laboratory (NRMRL). Within NRMRL, urban watershed research centers in UWMB. With the creation of UWMB the research budget measured in both extramural resources and staff assigned to the urban watershed increased significantly.

Near-term Research

With the shift of watershed management to local decision makers comes a need for EPA to provide simple multiple-use models that users can apply nearly universally. EPA must generate the object-oriented packages that make each application a special circumstance of a general condition. This object-oriented approach de-emphasizes situational uniqueness and emphasizes cross-situational similarities.

An appropriate watershed model under this object-oriented approach must allow an arbitrary mixture of water elements and land uses. EPA's Office of Science and Technology developed BASINS largely from existing models[6], including Hydrologic Simulation Program-FORTRAN (HSPF), QUAL2E, and TOXIROUTE. This backbone is a strong starting point. The BASINS approach, bundling several existing models with a common set of software-provided databases, a GIS overlay, and data translation tools on a single CD ROM, is practical, well crafted and well considered. The GIS overlay provides not only for common data needs, but, at least as important, provides an information transfer mechanism to nontechnical users.

UWMB believes the next obvious step for this effort is incorporating SWMM into the BASINS system. The long history of SWMM use, the strong private sector following, SWMM's selection as a teaching tool in many universities, and the proven, flexible capabilities of this tool clearly supports continued EPA backing. Although essentially unsupported by EPA for many years, the SWMM users group has sustained the model through several updates.

With the addition of SWMM, the urban environment's important contribution to watershed health can be added to the BASINS model collection. Urban decision makers require a proven model to evaluate the effects of the urban environment on the watershed while continuing to assure the needed urban infrastructure. This addition is not simply adding a Graphical User Interface (GUI) to SWMM. This addition is the next step to including urban contribution as part of the watershed considerations and not as a separate entity, independent of, and isolated from the remaining watershed.

As part of linking SWMM to BASINS, we similarly need ancillary programing support to provide context-sensitive help and rudimentary error checking. This coding will, for example, advise users when entered data is outside a reasonable or possible range of values. An analogous check will compare computed values against audit values to identify questionable results. These tools cannot replace a knowledgeable user. Expertise will remain necessary for a full interpretation of the results.

In defining the UWMB contribution, we face an unclear vision of what to include in the redefined "research" arena under the revised ORD strategy[7]. The lines among modeling tasks, such as engine development, maintenance, and interfaces shift. Are code revisions and maintenance research? Is user technical support research? While these are clearly necessary parts of a cohesive modeling strategy, neither model maintenance nor user support is part of the current research vision.

Modeling is an important part of the watershed research needs outlined in UWMB's Wet Weather Flow Research Plan. NRMRL similarly emphasized these needs in its ecosystem restoration research plan[8]. These tools are critical to reaching the Agency's announced approach to environmental protection. Modeling is not, however, the full research strategy and comprises only part of a larger mission. Alone, models do not suffice as a decision support system and the urban watershed modeling effort must interlace with other watershed research efforts.

Surface source water protection follows the CBEP paradigm encouraging local decision makers to assume greater responsibility for drinking water quality. The Safe Drinking Water Act (SDWA) requires the states to submit plans delineating the source water watershed, identifying and managing the potential contamination sources. The analogy to watershed modeling is clear. When part of the surface water is a drinking water supply, the watershed model is partially supporting the SDWA needs although the regulations do not explicitly require modeling to comply with the provisions of the SDWA.

Appendix. References

1. USEPA, "Risk Management Research Plans for Wet Weather Flows," November 1966, EPA 600/R-96/140.

2. USEPA, Office of the Administrator, "The New Generation of Environmental Protection," July 1994, EPA 200-B-94-002.

3. USEPA, Office of Solid Waste and Emergency Response, "Report of the Agency Task Force on Environmental Regulatory Modeling Guidance, Support Needs, Draft Criteria and Charter," March 1994, EPA 500-R-94-001.

4. USEPA, Office of the Administrator, "Guidance for Conducting External Peer Review of Environmental Models," July 1994, EPA 100-B-94-001

5. USEPA, Office of Research and Development, "Guiding Principles of Monte Carlo Analysis," March 1997, EPA 630/R-97-001.

6. USEPA, Office of Science and Technology, "Better Assessment Science Integrating Point and Nonpoint Sources," BASINS Version 1.0, May 1996, EPA 823-R-96-001.

7. USEPA, Office of Research and Development, "Strategic Plan for the Office of Research and Development," May 1996, EPA 600/R-96/059

8. USEPA, National Risk Management Research Laboratory, "DRAFT Risk Management Research Plan for Ecosystem Restoration in Watersheds," September 1996, (In peer review).

# PERFORMANCE MEASURES FOR CSO CONTROL

Michael P. Sullivan[1], Karen T. Gontasz[2] and Mark P. Hoeke[3]

**Background and Objective**

Hundreds of communities across the nation are making substantial investments in staffing and infrastructure in order to meet the expectations of the National Combined Sewer Overflow (CSO) Control Policy of 1994 (EPA, 1994). These investments are directed toward minimum control technologies focused on improved operation and maintenance of collection systems and, where necessary, the design and construction of new CSO control facilities. The need to measure and track the results of this investment with regard to environmental goals and benefits is widely recognized. Sewerage agencies must be accountable to rate payers, and must be able to demonstrate results in terms of the environmental improvement associated with CSO control.

In a cooperative project with the U.S. Environmental Protection Agency, AMSA (the Association of Metropolitan Sewerage Agencies) implemented a study to develop a series of performance measures for use by utilities, sanitation districts and local government agencies to track improvements and benefits associated with CSO control. The study involved "stakeholders" from all of the parties interested in CSO control, and focused on the needs of both small and large CSO communities. The principal objectives of the study were to explore a broad range of performance measures, and to identify and recommend the most important and practical measures for local use. The results of this study included a set of twenty-four performance measures for CSO control, and recommendations on their application and use.

---

[1] Regional Manager, Limno-Tech, Inc., 1150 17$^{th}$ St., NW, Suite 705, Washington, DC 20036
[2] Design Engineer, Johnson, Mirmiran, and Thompson, 72 Loveton Circle, Sparks, MD 21152 (formerly Limno-Tech, Inc.)
[3] Manager, Regulatory Affairs and Technical Services, AMSA, 1000 Conn. Ave., NW, Suite 410, Washington, DC 20036

## Methodology

The study built upon the cooperative working relationships among regulatory authorities, the regulated community, and citizen and environmental interests established during formulation of the National CSO Control Policy. It was designed to maximize the number of participants involved and the amount of input received from parties having a substantial interest in CSO control. As shown in Figure 1, the study process consisted of several major elements. AMSA and EPA provided the funding and administrative support. A Work Group comprised of representatives from CSO communities, environmental organizations, and state and federal agencies was formed to guide the study. The Work Group members had substantial experience and expertise with CSOs, monitoring, and urban wet weather conditions, and their extensive discussions over the course of one year were central to the development of appropriate performance measures. In addition, a determined effort was undertaken to get as many other interested parties involved in the process as possible.

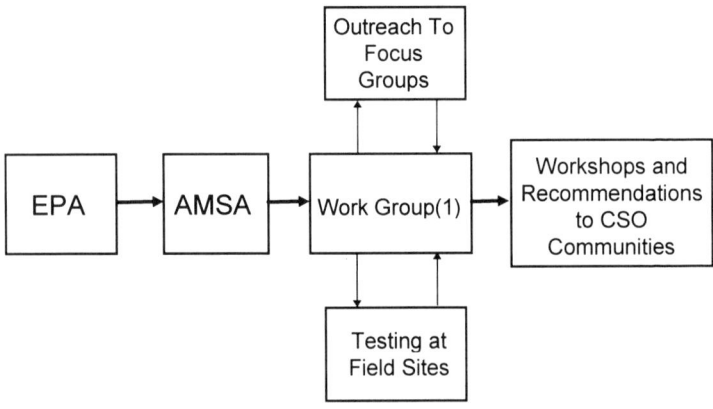

**Figure 1. Study Process**

(1)Columbus Water Works, GA; EPA OW, DC; Narragansett Bay Comm., RI; Save the Bay, RI; Detroit Water and Sewer Dept., MI; EPA Region II, NY; Center for Marine Conservation, DC; NLC, DC; City of S. Portland, ME; American Rivers, DC; City of Philadelphia Water Dept., PA; San Francisco Dept. of Public Works, CA; WV DEP; EPA Region I, MA; IL EPA; PA DER.

The Work Group initiated the following activities to garner input:

- Focus group meetings targeted at CSO communities, environmental organizations, and state and federal regulatory and resource agencies were used to obtain ideas and to hear concerns about CSO control and performance measures. Focus groups reinforced the need for performance measures, cautioned against separating out CSO impacts from other pollutant sources, ranked performance measures according to their importance, and identified key performance measures.
- Field interviews with CSO communities and regulatory agencies were utilized to determine the applicability and usefulness of a preliminary group of performance

measures developed by the Work Group. This input allowed the Work Group to eliminate several performance measures as "not applicable" or "not useful", and to separate those with broad applicability from those with limited, site specific applicability.

Following receipt and assimilation of input from the focus groups and field interviews, the Work Group developed and applied criteria for the final selection of performance measures. This process was integrated with extensive discussions on related subjects that would influence enthusiasm for the performance measures and their use. These subjects included the attainment of designated use in receiving waters; the importance of linking performance measures to locally defined objectives such as increased waterfront access or other planning initiatives; and the value of incorporating CSO control into a broader watershed approach for the restoration of impaired urban waterways. The Work Group also recognized that cost is an important factor in the consideration of performance measures, and that the cost for data collection and management should be commensurate with the usefulness of the data. In addition, the Work Group recommended that the cost of obtaining data for performance measures should not represent an extra burden on CSO communities beyond what was intended in the CSO Policy.

The Work Group produced a study report for AMSA entitled *Performance Measures for the National CSO Control Program* in January, 1996 (AMSA, 1996). Subsequently, a series of workshops were held during the fall of 1996 to promote and encourage the use of performance measures. The workshops included presentations on performance measures and their use, panel discussions on the relationship between performance measures and NPDES permits, and round-table discussions on implementation issues. Case studies on the use of performance measures from laarge and small CSO communities were also prepared and discussed. The workshops were attended by nearly 300 individuals representing sewerage agencies, city and county government, state regulatory agencies, consultants and environmental interests.

**Findings and Recommendations**
The recommended performance measures are presented in a menu-like format in Table 1. They are arranged within four categories: Administrative Measures; End-of-Pipe Measures; Receiving Water Measures; and a category encompassing Ecological, Human Health and Resource Use Measures. All twenty-four performance measures were considered to be *Appropriate for General Use* by CSO communities. Four of the performance measures were also recommended as being *Appropriate for Tracking on a National Basis*, with the expectation that regulatory authorities might use these measures to assess the programmatic effectiveness of CSO control at the state or national level.

| ADMINISTRATIVE | END-OF-PIPE | |
|---|---|---|
| ♦ Documented Implementation of NMCs (1)<br>♦ Status of LTCP(1)<br>♦ Waste Reduction | Flow Measurement<br>♦ Wet Weather Flow Budget<br>♦ CSO Frequency(1)<br>♦ CSO Frequency in Sensitive Areas(1)<br>♦ CSO Volume<br>♦ CSO Volume in Sensitive Areas<br>♦ Dry Weather Overflows | Pollutant Load Reduction<br>♦ BOD Load<br>♦ TSS Load<br>♦ Nutrient Load<br>♦ Floatables |
| RECEIVING WATER | ECOLOGICAL/HUMAN HEALTH/RESOURCE USE | |
| ♦ Dissolved Oxygen Trend<br>♦ Fecal Coliform Trend<br>♦ Floatables Trend<br>♦ Sediment Oxygen Demand Trend<br>♦ Trend of Metals in Bottom Sediments | ♦ Shellfish Bed Closures<br>♦ Benthic Organism Index<br>♦ Biological Diversity Index<br>♦ Recreational Activities<br>♦ Beach Closures<br>♦ Commercial Activities | |

(1) Appropriate for National Tracking

**Table 1. Summary of Recommended Performance Measures**

Substantial effort throughout the study was directed toward identifying performance measures that would quantify benefits in terms of the resources that CSO control is trying to protect. The identified receiving water, ecological, human health and resource use measures represent this effort. However, it was found that the performance measures in these categories (e.g., dissolved oxygen trend, beach closures, etc.) track conditions that are typically influenced by many pollutant sources in addition to CSOs. In fact, quantified improvements in most of the measures in these categories would only be expected if all of the contributing sources of pollutants and environmental degradation were taken into account. In contrast, the performance measures in the administrative and end-of-pipe categories (e.g., documented compliance with the nine minimum controls, CSO frequency, etc.) do not directly quantify environmental results. However, these performance measures are directly related to CSOs and CSO control efforts, and are quite valuable in their own right.

Several examples that illustrate how CSO communities can use performance measures to track improvements due to CSO control are presented in Figures 2, 3, and 4.

Other major findings and recommendations are as follows:
- The operation of combined sewer systems and CSO impacts are very site specific.
- Urban receiving waters are very complex. A watershed approach that includes CSO control in conjunction with the control of other pollution sources is recommended as the most appropriate approach to address urban wet weather water quality problems.

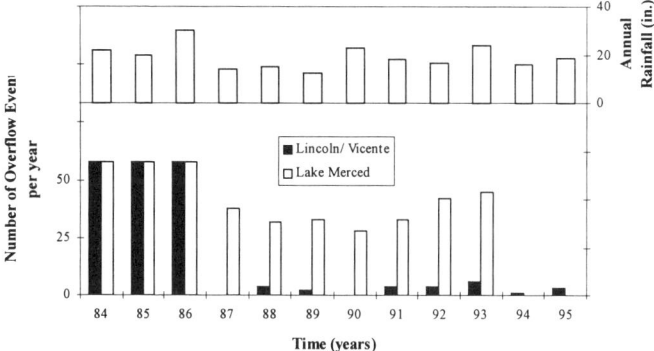

Source: San Francisco Department of Public Works, San Francisco, CA.

**Figure 2. CSO Frequency at Lincoln/Vicente and Lake Merced Overflow Structures in San Francisco From 1984-1995**

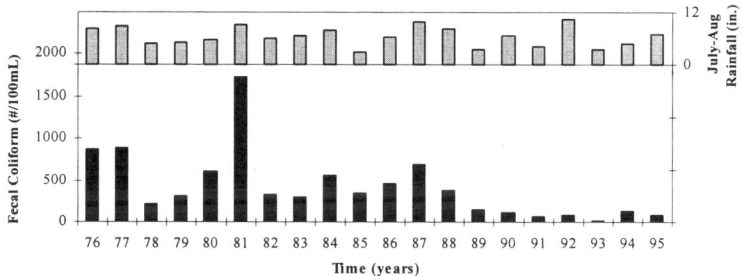

Source: Monroe County Health Department, Environmental Health Laboratory, NY.

**Figure 3. Summer Fecal Coliform Trend in the Lower Genesee River, New York: 1976-1995 (Geometric Mean)**

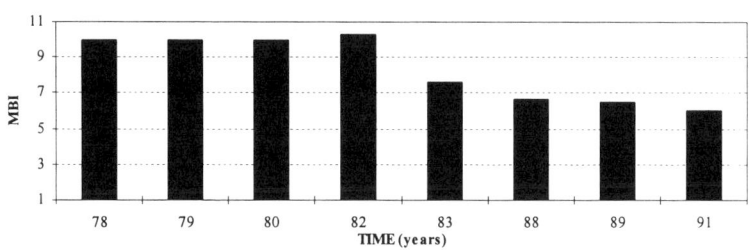

Source: Illinois EPA and the Metropolitan Water Reclamation District of Greater Chicago, Chicago, IL.

**Figure 4. Summary of Macroinvertebrate Data in the Chicago Sanitary and Ship Canal at Lockport: 1978-1991**

- In selecting appropriate performance measures, CSO communities should consider locally defined objectives including potential recreational or commercial development benefits that go beyond regulatory requirements.
- Most of the participants at study-sponsored focus groups, field interviews and workshops were supportive of the use of performance measures and recognized the need to track results and benefits. Many CSO communities are already using several of the performance measures or collecting data that would support their use.
- State and federal permitting authorities generally expect several of the administrative and end-of-pipe performance measures to be required either in permits or within other more flexible agreements. In particular, documented implementation of the nine minimum controls, the status of Long Term Control Plans, and CSO frequency and volume have broad appeal with regulatory authorities.
- CSO communities generally recognize the importance of using performance measures on a voluntary or cooperative basis, but have serious reservations about the incorporation of performance measures in permits or other legally binding instruments.
- Negotiation between CSO communities and permit writers on monitoring requirements and performance measures have been successful.
- Costs associated with data collection for performance measures should be commensurate with the usefulness of the data. While the cost of CSO controls is not a performance measure per se, cost should be tracked to put the financial investment in CSO control into context.

## Conclusions

The study has been significant because the process employed and the report and workshops were designed to reach out to all of the nearly one thousand CSO communities across the nation. This outreach identified the use and benefits of performance measures, and recommended a process for selecting among twenty-four practical and useful performance measures.

A major benefit of the study was increased awareness of the need to be accountable and to communicate the benefits of CSO control to the public and rate payers through the use of performance measures. Other benefits included use of a consistent set of performance measures on a national basis; development of a better understanding between CSO control and water quality, ecological and resource use benefits; and the establishment of a dialog between regulatory entities and permitting authorities.

## References

AMSA, 1996. "Performance Measures for the National CSO Control Program," Washington, DC.

EPA, 1994. "Combined Sewer Overflow (CSO) Control Policy," EPA 830-B-94-001, Washington, DC.

# STRETCHING THE LIFE OF AN OLD SEWER SYSTEM

David J. Anderson[1], P.E.

James P. Pistilli[2], P.E.

## Abstract

Designed in the late 1960's and built in the early 1970's, the Clinton Oakland Sewage Disposal District reached the end of its original demographic planning period in 1995. To ensure adequate service to this District of twelve rapidly growing communities, a new planning effort examined the system needs to the year 2020. This work involved demographic planning, flow metering, and computer modeling to identify the system needs. The most important results showed that with good maintenance practices the existing system could serve the needs of every community. Increased cooperation between member communities utilizing new knowledge of the system behavior gained in this study would also be required.

## Introduction

Twelve communities in the northeastern part of Oakland County, Michigan discharge sewage to the Clinton-Oakland Interceptor Sewer. Built in the early 1970s, the 90 km (57 mi.) interceptor serves over 160,000 people in a 650 km$^2$ (250 mi$^2$) sewage disposal district. It reached the end of its 25-year "design life" in 1997. Monitoring, modeling, and population forecasting, coupled with an interactive computerized <u>M</u>anagement <u>A</u>nd <u>R</u>eporting <u>S</u>ystem **(MARS)**, will stretch the usefulness of this pipe network into the next century. These engineering tools

---

[1] Principal Engineer, Montgomery Watson, 505 Highway 169, Suite 555, Minneapolis, MN 55441.
[2] Assistant Chief Engineer, Oakland County Drain Commissioner, One Public Works Drive, Waterford, MI 48328.

have been complemented by a cooperative approach to infrastructure use which results in a fiscally and environmentally responsible plan.

## Monitoring - Continuous

A network of 28 flumes and 15 rain gauges provides the data needed to understand sewer system performance. Most of these flumes were constructed shortly after the interceptor was built. The location of the flumes isolates flow from each member community.

Depth of flow in the flumes was originally measured by bubblers and recorded on circular charts. In the early 1990's all flumes were updated to measure depth with ultrasonic level sensors. A datalogger records the 4 to 20 milliamp level proportional signal every five minutes. Twice a week, dataloggers transmit stored information to office computers. This data is reviewed and converted from level to flow using appropriate flume equations.

Historically, rainfall was measured using weighing bucket gauges with data recorded on strip charts. This network of gauges provided good long-term averages for rainfall. However, its density was only about one gauge for every 260 km$^2$ (100 mi$^2$). To improve the spatial resolution and accessibility of the data, a new network of 15 tipping bucket rain gauges was established. This network has a density of approximately one gauge for every 50 km$^2$ (20 mi$^2$). These gauges were connected to the same type of dataloggers as the flow meters. With the rainfall now in digital form and recorded at the same time interval as flow, the rainfall from the nearest gauge is easily plotted with each meter's flow.

## Monitoring - Short Term

As part of a comprehensive sewer system evaluation, detailed infiltration and inflow (I/I) studies were undertaken. This effort utilized an additional 35 temporary flow meters (area/velocity type). These meters were placed on major connections to the interceptor and along the interceptor in an effort to identify and isolate extraneous flow. From this work, subbasins with high inflow and/or infiltration were identified for further analysis by the local municipality.

## Example of Monitoring Data Use - Dry Weather Flow

Computer routines were developed to calculate weekly dry weather flow patterns without the need for manual intervention to mark wet periods. Long-term (>2 years) data recorded digitally on a constant interval makes this possible. Each year of data represents 52 data points for every five minute recording interval over an entire week. By averaging the values for each five minute interval, the weekly flow pattern was determined. After excluding data outside two standard deviations from the initially calculated mean, data was reaveraged. This refinement removed most wet weather data, and the vast majority of bad metering data (See Figure 1).

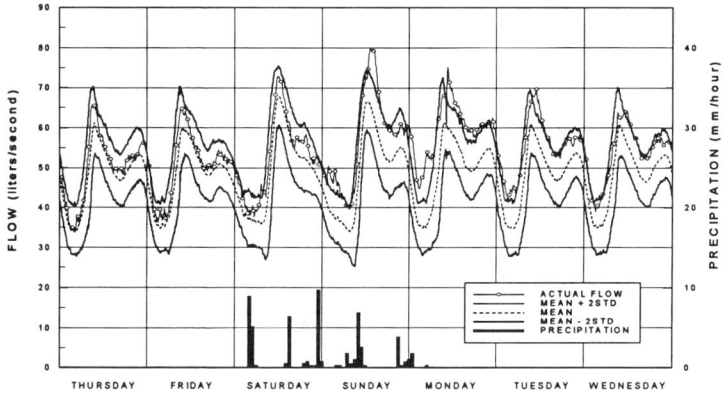

**Figure 1  Actual Flow and Historic Bounds**

These curves were compared to manually calculated average dry weather flow (DWF) curves produced by local engineering firms as part of the original I/I study. In every case, these automated DWF curves matched those produced from the rigorous extraction of dry weather flow periods (See Figure 2). This automated technique allows the annual recalculation of DWF curves quickly and inexpensively.

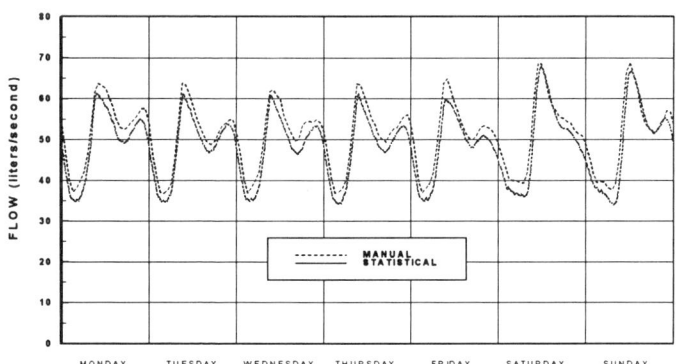

**Figure 2  Manual and Statistical Calculation of Dry Weather Flow**

Tracking changes in the annual DWF curves will demonstrate the effect of development on the tributary basin as represented by increases in base flow. Comparison of wet flows due to rainfall versus dry flows establishes wet weather inflow. Seasonal groundwater infiltration is clearly demonstrated by comparison of dry weather flow in the spring with long-term average weekly flow curves (See Figure 3).

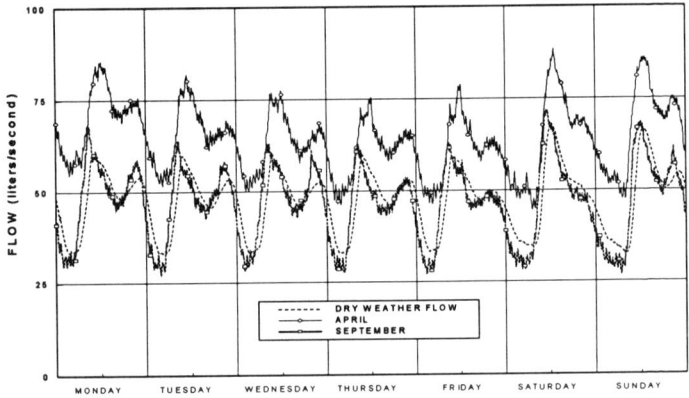

Figure 3 Seasonal Flow Comparison to Dry Weather Flow

## Modeling and Population Forecasting

Using the EPA Storm Water Management Model (SWMM) to examine system hydraulics, hydrographs (wet or dry weather) were routed from upstream meters to downstream (Pistilli, Anderson, Nelson, 1993). The routed flow is then subtracted from the downstream meter flows to produce the incremental flow the local community contributes to the interceptor. This produced a continuous graph of local community flows to the interceptor. For the first time the county was able to accurately establish community flow rates. Member community flows are now reasonably monitored for compliance with contractual peak flow rates.

A demographic planning model estimated and distributed future population over a projected service area. Future community flows were estimated based on these projections. These tools demonstrated the interceptor sewer system can easily serve the dry weather flow needs of all member communities through buildout.

This effort also showed that collection system configuration, typical storm tracks, and peak flow attenuation allow the system to safely handle wet weather events. With this system understanding, local communities realized improved maintenance, cooperation and sharing of existing capacity would eliminate the need to construct addition pipe. The communities agreed to "rent" wet weather capacity from each other on an "as-utilized" basis when their flows exceed their peak contractual flow. This arrangement compensates communities who control their flows to contract limits and provides an incentive to those who exceed their limits to reduce extraneous flows.

## Management And Reporting System (MARS)

Stretching the useful life of the interceptor requires constant vigilance. That process depends on converting data into information. Using an interactive, graphic environment, a map based routine allows staff to select flow meters, rain gauges or communities and pick various graphics or reports describing performance. Data "repair," eliminating the influence of lift station operation on community flows and accounting for I/I to the interceptor proved challenging.

MARS also produces hard copy reports which are provided to the 12 communities. These are produced quarterly and provide back-up information for any peak flow changes/credits. Community flow performance is summarized in tables and graphs, similar to Figure 4. In this way, the communities are regularly appraised of their interceptor sewer usage. They can also request detailed plots for any meter, etc., from the county. Over time, it is hoped that this information can be made available on a Web page.

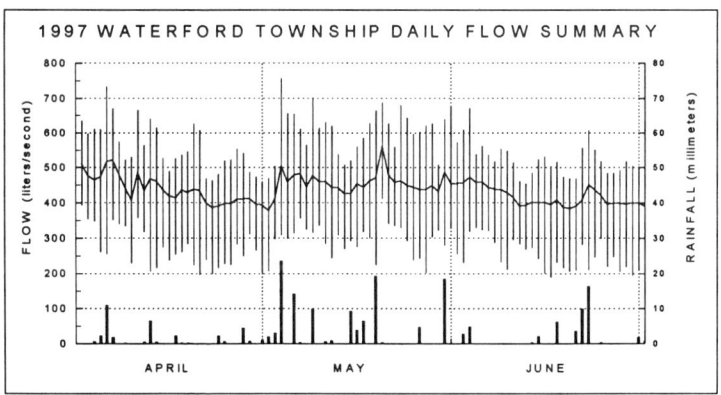

Figure 4 MARS Quarterly Flow Summary

Flow routing in the MARS routine uses the Progressive Average Log (PAL) method (Dooge, Harley 1967). SWMM and recorded data were used to select the two variables used by PAL (the number of points to average, which accounts for attenuation and the number of time steps to lag the flow, which accounts for travel time (Harris, 1968)). Except in unusually rapid flow change situations, PAL works quite well. In fact, when there are major pump station impacts, the method can be tuned to closely match SWMM results (See Figure 5). Using PAL simplifies the MARS process, which involves some 1.25 million data points per quarter.

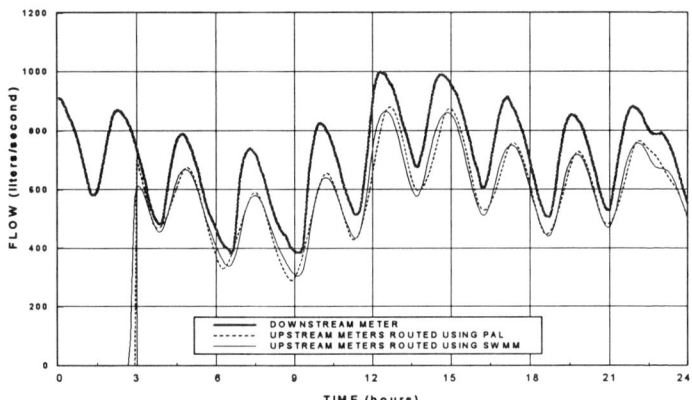

Figure 5 Comparison of SWMM and PAL Flow Routing

## Cooperative Approach

A task force of county, community, and engineering members continues to guide this effort. Better management and a sharing of interceptor capacity eliminated the need for additional pipe construction, with an estimated cost of $5 to $20 million. Communities may temporarily rent peak flow capacity above their purchased capacity from other members. MARS and continuous monitoring provide the mechanism for implementing this novel infrastructure management system.

## Appendix I. Reference

Dooge, J. C. and Harley, B. M. (1967) "Linear Routing in Uniform Open Channels", International Hydrology Symposium Proceedings. Sept. 6-8, 1967, Fort Collins, Colorado, pp. 8.1 - 8.7.

Harris, G. S. (1968) "Development of a Computer Program to Route Runoff in the Minneapolis-St. Paul Interceptor Sewers." Memorandum No. 121, University of Minnesota., St. Anthony Falls Hydraulics Laboratory.

Pistilli, J. P., Anderson, D. J., and Nelson, W. G. (1993). "Modeling and Metering Flow." Water Environment & Technology, Vol. 5, No. 3, pp. 34 - 36.

# THE MANY FACES OF COMBINED SEWER SYSTEM PERFORMANCE
Kenneth A. Pew[1]

## Abstract

The Combined Sewer Overflow (CSO) Control Policy issued by the United States Environmental Protection Agency in April 1994 "establishes a consistent national approach for controlling discharges from CSOs to the Nation's waters through the National Pollutant Discharge Elimination System (NPDES) permit program." Measurement of combined sewer system performance is key to complying with this policy. This paper addresses the various techniques utilized by the Northeast Ohio Regional Sewer District to measure combined sewer system performance.

## Background

The Northeast Ohio Regional Sewer District has a 764 square kilometer (295 square-mile) service area, which includes 194 square kilometers (75 square miles) served by combined sewers. The service area encompasses all or part of 53 communities, each responsible for their own local sewer system. The District serves over one million people and treats over 330 million gallons of wastewater per day on average.

The District owns and operates three wastewater treatment plants, six pump stations, 343 kilometers (213 miles) of interceptors, 462 static combined sewer regulators, and 29 automated combined sewer regulators. The 491 combined sewer regulators discharge to the environment at 126 locations permitted by the National Pollutant Discharge Elimination System (NPDES). Discharges at these 126 locations impact many receiving waters:

- Lake Erie
- Four streams tributary to Lake Erie
- Cuyahoga River
- Four streams tributary to the Cuyahoga River
- the Ohio Canal

[1]Deputy Executive Director, Northeast Ohio Regional Sewer District, 3826 Euclid Avenue, Cleveland, OH 44115-2504

The District utilizes several programs to measure combined sewer system performance:

- Visual inspection, cleaning, and maintenance
- Permanent real-time monitoring and control
- Temporary monitoring
- Water quality monitoring
- Long-Term Control Plan development

**Visual Inspection, Cleaning, and Maintenance**

Each of the 462 static combined sewer regulators, along with 29 sanitary sewer overflow locations, are visually inspected every other week. The 143 regulators that are more prone to block and cause a dry weather discharge are inspected every week, and sometimes twice a week. During 1996, 100 percent of the dry weather discharges caused by the regulator itself were eliminated within 24 hours of discovery, with 87 percent eliminated within two hours. Inspection time is currently being decreased through the use of portable bar code readers.

**Permanent Real-Time Monitoring and Control**

A Central Sewer Control Facility monitors 25 rain gauges and 41 sewer level and/or flow gauges. It also monitors and controls 29 automated combined sewer regulators and provides real-time rainfall and sewer flow data to two of the three wastewater treatment plants. Rainfall data dates back to the early 1970s, with weighing rain gauges being used until they were replaced with tipping bucket gauges in 1994. Various types of sewer level and flow gauges are used. Data is transmitted via leased telephone lines and radio telemetry. Control devices include slide gates, sluice gates, knife gates, bascule gates, and inflatable dams. Most control devices date back to the early 1980s.

**Temporary Monitoring**

The NPDES Permit in effect from early 1988 until early 1997 required rotating monitoring of five discharge locations at a time. The required daily reporting parameters were occurrences, duration, average flow rate, and biochemical oxygen demand and suspended solids during the first 30 minutes of discharge. With moving monitoring equipment about every two months, each of the 126 discharge locations have been monitored twice during this time period.

**Water Quality Monitoring**

Water quality monitoring has substantially increased since it began in 1987. It currently encompasses all Greater Cleveland Area streams, two beaches, and 15 sites in the near shore area of Lake Erie. This program includes:

- Visual surveys of streams and follow-up investigations for discovered environmental disruptions
- In-field measurements of temperature, dissolved oxygen, pH, turbidity, and flow rates
- Collection of samples for chemical and bacteriological analyses
- Qualitative, quantitative, and semi-quantitative analyses of benthic macro invertebrates and the corresponding use of the Ohio Environmental Protection Agency's (EPA) Invertebrate Community Index (ICI) and the Hilsenhoff Biotic Index (HBI)
- Quantitative analyses of fish using long-line and boat electroshocking techniques and the corresponding use of Ohio EPA's modified Index of Well-Being (MIwb) and Index of Biotic Integrity (IBI)
- The evaluation of aquatic habitat using Ohio EPA's Qualitative Habitat Evaluation Index (QHEI)

**Long-Term Control Plan Development**

The District must complete five Combined Sewer System Long-Term Control Plans by early 2002. The first (Mill Creek Watershed) is essentially complete. The second (Westerly Tributary Area) began late last year. The third, fourth, and fifth will begin in 1997, 1998, and 1999, respectively. Each plan will include extensive combined sewer system monitoring and modeling and stream monitoring and modeling.

**Scope**

More details, experiences, and results of these programs will be presented. Applicability to the report entitled *Performance Measures for the National CSO Control Program* published by the Association of Metropolitan Sewerage Agencies in January 1996 will also be discussed.

Sheetflow Water Quality Monitoring Device

Stuart M. Stein, Member[1], Frank R. Graziano[2],
G. Kenneth Young, Member[3], Pat Cazenas[4]

Abstract

This Small Business Innovative Research (SBIR) project funded by the Federal Highway Administration (FHWA) involves the development of a sheetflow sampler to collect highway runoff. The need for a simple and inexpensive sampling device to collect sheetflow from highway sections is evident with stormwater runoff monitoring requirements often stipulated in permits required to construct, maintain, and operate highway facilities. The sampler will be an expendable, simple, and cost-effective device which will operate in the highway roadside environment. It is being designed to collect a predetermined depth of runoff (most commonly the first flush), provide storage, and to be easily removed and replaced by maintenance workers.

Introduction

Stormwater monitoring is currently underfunded. Such monitoring is necessary to protect our natural resources and surface waters. Monitoring of non-point source (NPS) stormwater runoff to support US EPA regulatory requirements

---

[1] Executive Vice President, GKY and Associates, Inc., 5411-E Backlick Road, Springfield, VA, 22151, (703) 642-5080 (ph), (703) 642-5367 (fax), sstein@gky.com.

[2] Senior Engineer, GKY and Associates, Inc., 5411-E Backlick Road, Springfield, VA, 22151, (703) 642-5080 (ph), (703) 642-5367 (fax), fgraziano@gky.com.

[3] President, GKY and Associates, Inc., 5411-E Backlick Road, Springfield, VA, 22151, (703) 642-5080 (ph), (703) 642-5367 (fax), kyoung@gky.com.

[4] Federal Highway Administration, HEP-40, 400 7$^{th}$ Street, SW, Washington, DC, 20590, (202) 366-4085 (ph).

and to determine pollutant export rates by land use and by intensity of use, or other factors, is underfunded and "end-of-pipe" equipment is expensive. The sheetflow sampler device achieves more monitoring at less cost–the project's driving force.

The sheetflow runoff from impervious surfaces contains nearly all the NPS pollution associated with highway and urban non-point sources. This pollution is traditionally measured at "end-of-pipe" sampling sites where the sheetflows have been collected and concentrated and the measurement of pollution is confounded by the introduction of non-impervious surface runoff, off-site runoff, and by effects introduced in the collection pathways: settling, infiltration, and interactions with vegetation. With these shortcomings, the present systems are also saddled with high costs ($5,000 to $10,000) to provide storage space, servo-mechanisms, and electro-mechanical systems that only sample portions of the flow.

The sheetflow sampler, whose feasibility was demonstrated in the successful completion of Phase I (prototype development and testing), has none of the shortcomings of present measurement systems and has the virtue of simplicity:

1. It directly measures the source of the pollution with no confounding factors.
2. It collects the sample for chemical testing and enables direct estimates of the associated rainfall runoff.
3. It does not sample the flow, rather it collects a representative aliquot of all the flow.
4. It is passive with no moving parts, nor need for energy.
5. Its projected cost is much less than present samplers.
6. It enables a set-up that is nonobtrusive and will not suggest a target for vandalism and which is safe in conjunction with vehicles.

Basic Design Concept

Figure 1 shows the basic design of the GKY Sheetflow Water Quality Monitoring Device. The device consists of a grate, a sample receptacle, O-rings, an insert, and a protective plate. The figure 1 exploded view further shows how the main components combine to form the entire sheetflow water quality monitoring device. The top of the grate section contains one or more sample-port(s) that are rectangular in nature. The sample-port(s) have a rounded leading edge and square side and back edges. The grate section also has an internal baffle designed to trap floating pollutants in the event the device is filled by a rain event. Figure 2 is a cross-section through the grate showing the details of a typical sample-port and the internal baffle. Figure 3 shows a typical application, with the sampler located in the shoulder. The current thought is that the grate section would be one piece and made of a chemically inert plastic, such as glass-filled polycarbonate, using plastic injection mold techniques. This would make the grate section economical.

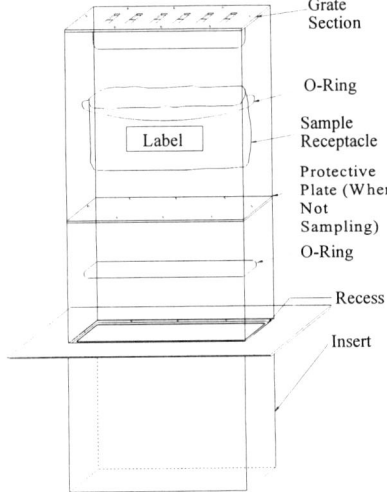

Figure 1. Schematic of the GKY Sheetflow Water Quality Monitoring Device.

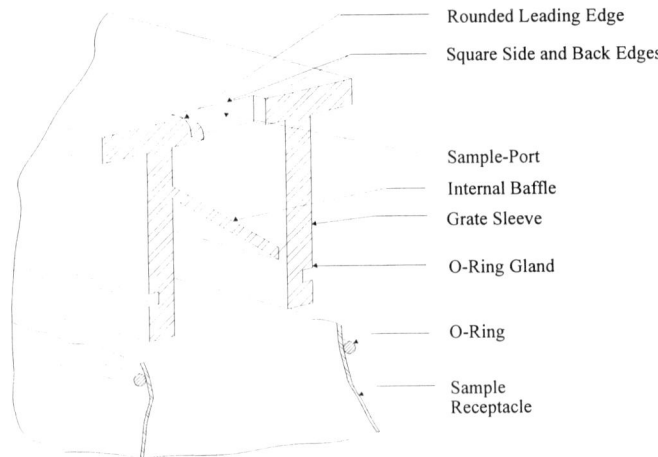

Figure 2. Cross-section view through the grate section.

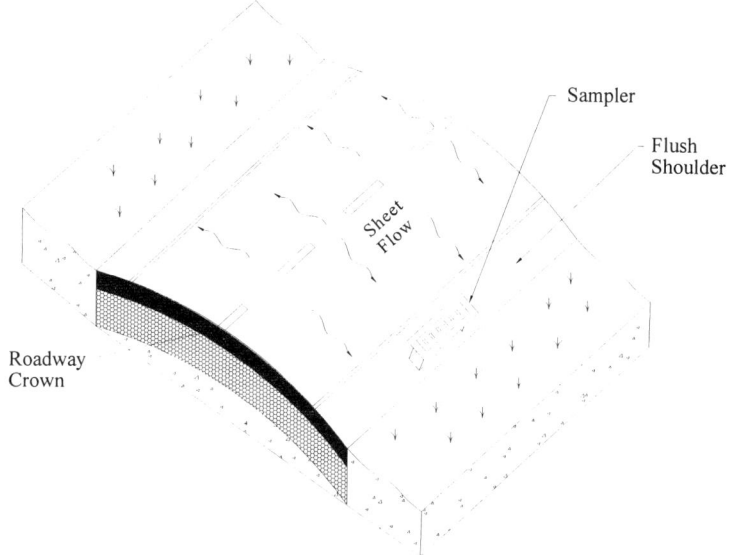

Figure 3. Typical application of the GKY Sheetflow Water Quality Monitoring Device

Theory of Operation

The GKY Sheetflow Water Quality Monitoring Device can be easily configured to capture various runoff volumes over various pavement widths. The runoff volume is defined as the depth of runoff over a defined surface area. The first 12.5 mm of runoff is often considered to be the "water quality volume" as it represents the "first flush" of a rain event.

The theory of how the sample-port(s) and sample receptacle combine to capture a desired runoff volume is very simple. The number of the sample-port(s) can be configured so that the water quality monitoring device is filled by the desired runoff volume. When full, any runoff in excess of the desired runoff volume will overtop the device and will not be captured. The internal baffle traps floating pollutants within the sampler body, preventing them from spilling out.

In order for the above theory to function properly, the sample-port(s) must capture a predictable percentage of the sheetflow runoff approaching the sample-port leading edge regardless of runoff depth. This leads to the importance of the design of the sample-port leading edge as well as the sample-port side and back edges. Assuming a predictable capture efficiency for the sample-port(s), the equation to determine the required volume of the sample receptacle:

$$V_{Sampler} = 0.001 \times D_{Runoff} \times PL_{Flow} \times W_{Ports} \times N_{Ports} \times Eff_{Ports} \qquad (1)$$

Where :

    $V_{Sampler}$ = Required volume of sample receptacle, L.
    $D_{Runoff}$ = Desired runoff capture depth, mm (i.e., first 13 mm).
    $PL_{Flow}$ = Pavement flow length, m.
    $W_{Ports}$ = Width of sample-port, mm.
    $N_{Ports}$ = Number of sample-ports.
    $Eff_{Ports}$ = Capture efficiency of sample-port (fraction of frontal flow that descends into the sample receptacle, percent.
    0.001 = Conversion factor.

Laboratory Test Results

Initial testing was performed using an acrylic model. Acrylic was used because it is was convenient to fabricate. The model measured 51-mm wide by 457-mm long by 305-mm deep for a total volume of 7.11 L. The sample-port configuration was 6.4-mm wide by 25-mm long.

A series of tests were conducted on different sample-port leading edge configurations, including a square edge, a 1.5-mm bevel, a 3-mm bevel, and finally on a rounded edge configuration. Figure 4 is a plot of capture efficiency versus flow depth for each of these edge configurations. The data show that if there is a sharp edge, even when beveled, the lower flow depths tested have a much lower capture efficiency. Water was inhibited from breaching the sharp edge. The rounded edge, however, accepted water much more readily and had a narrow range of capture efficiency over the flow depths tested. A capture efficiency greater than 100 percent was achieved because water entered the sample-ports from around the front corner.

Another factor tested was the slope of the ramp. Using the rounded-edge configuration, tests were conducted with ramp slopes of one and two percent. The results indicate that slope is not a significant factor affecting capture efficiency, within the range of typical highway cross slopes.

The GKY Sheetflow Water Quality Monitoring Device is intended to be installed with the sample-port(s) oriented parallel to the direction of the sheetflow. Realizing that this may not always occur, a test was performed with the sample-ports skewed 10 degrees from parallel to see what effect this may have on capture efficiency. Tests show that a skew of 10 degrees had no affect on capture efficiency. Visual observation during the test revealed that water would not breech the square side edges, even at the higher flow depths. By placing small particles in the flow, it was evident that flow would orient parallel to the sample-ports while between them. The square side and back edges are therefore an important design consideration.

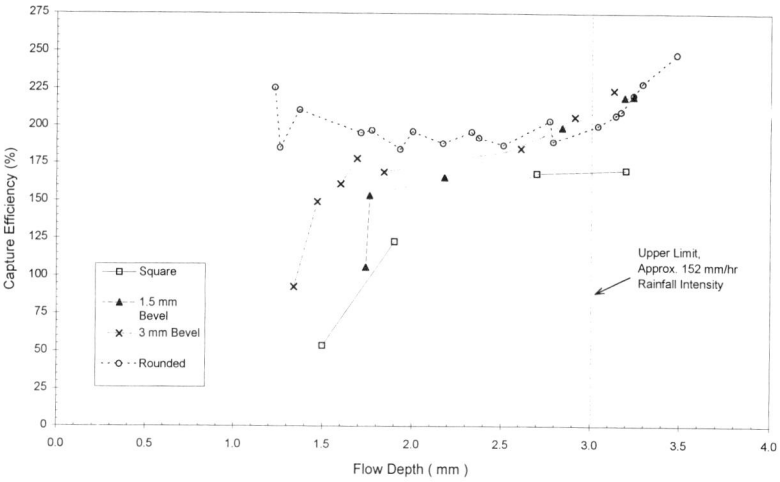

Figure 4. Effect of leading-edge configuration, 6.4 mm sample-ports, acrylic model.

Conclusions

Laboratory results indicate that this simple and inexpensive monitoring device can effectively sample pavement sheetflow runoff. Very limited field testing supports this conclusion and convinced FHWA to fund Phase II of this SBIR project. Phase II will include further testing and specifications for the final design. Testing will involve various state highway departments to thoroughly evaluate design, installation, and sampling protocol robustness. After testing is complete and the design is finalized, manufacturing of plastic injection molded copies will be initiated, as will development of channels of distribution and development of operators' field manuals for installation and sample handling.

# Effectiveness of a Combined Urban Wetland Detention Basin - First Results

Robert G. Traver, M. ASCE[1]

Abstract
　The author presents initial results from an ongoing research project into the effectiveness of two wetland Best Management Practices. Hydrographs and Pollutographs from one of the 61 collected storms are described in detail. The project scope was previously described in "A Study into Effectiveness of Urban Best Management Practices" (Traver 96) presented at the 1996 ASCE North American Water and Environment Congress.

Objective
　The research objective of the project is to quantify the hydrological and biogeochemical effects of a combined wetland and detention pond "Best Management Practice (BMP)" on urban stormwater quality and quantity.

Introduction
　Stormwater, with its rapidly changing and widely fluctuating flows, has the potential to deliver large pollutant loads to receiving waters during short time intervals (Schueler 92). Current practice is to use wetland detention ponds and other BMP's to reduce nonpoint source pollutants. This study was initiated to closely scrutinize multiple storm events over a 21 month period in order to determine the pollutant removal effectiveness of a wetland detention pond and small wetland stream system. What follows is a follow up to a previous conference proceedings (Traver 96) and is based upon the first year end report to the Nutrient Subcommittee of the Chesapeake Bay Program (Traver Et Al. 97).

Project Description
　The study is based at a newly constructed stormwater detention basin and adjacent riparian wetland literally "across the street" from Villanova University in southeastern Pennsylvania. Constructed under the direction of PennDOT in 1995, the basin quickly took on the form of a stormwater wetlands BMP due to its hydrology and the migration of nearby wetland plants. The wetland stream was existing, though a small mitigation site was included as part of the basin construction (Fig. 1). The BMP and wetland stream each

---

[1] Associate Prof., Department of Civil and Env. Engineering, Villanova University, Villanova PA 19085

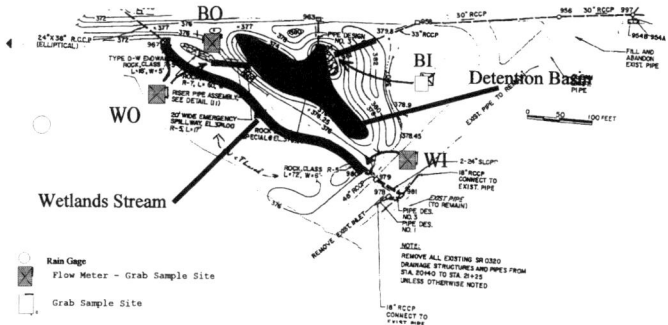

**Figure 1** Culvert and Instrumentation Plan - Source: PaDOT Construction Drawings - State Route 30 Section 630, Delaware County. Sheet 23/28.

receives stormwater runoff from a combined watershed area of 33.7 ha (.13 mi$^2$). The two contributing drainage areas are highly developed, being over 35% impervious. All flow is piped in and out of the two BMP's due to the surrounding road system.

Instrumentation was installed upon completion of the basin. Instrumentation included a rain gage, three flowmeters, two water level recorders, and four automated samplers. The positioning of the sampling equipment is shown in Figure 1. A routine schedule of weekly and storm-event water sampling was implemented with analysis for pH, specific conductance, suspended solids, dissolved and particulate forms of N, P, Cd, Cu, Cr, Ni, Pb, and Zn.

In order to characterize the site, the equipment was grouped into four flow and quality sampling stations. These were identified as Wetlands In (WI), Wetlands Out (WO), Basin In (BI), and Basin Out (BO). This nomenclature will be used throughout the paper. Note that wetland out includes the basin outflow, and that the only location without a flowmeter is basin in. Basin in flow data is developed using the Storage Indication method of hydraulic routing using the basin out flow and surveyed site data. Water quality samples were collected from the locations where Brown's Run enters (WI) and exits the wetland (WO), the point of inflow into the detention basin (BI), and from the exit point of the stormwater basin (BO).

Project Data
Over the course of the study, 21 months of weekly data and 61 storm events were collected. Due to high rainfall and shear quantity of data collected over the study period, chemical analysis is still underway. Currently efforts are underway to complete the quality analyses, and to create pollutographs for each storm event and pollutant.

Pollutograph Formulation
The water quality and quantity data are in the process of being combined to form pollutographs representing both the pollutant rate

and mass quantity entering and exiting the site for each of the 61 storms. To incorporate the disparate time increments, a Lagrange scheme was used to interpolate the quality data. The basin outflow quality and quantity parameters were subtracted from the wetland outflow to separate the mechanisms of the two sites.

Event 5

The first significant rainfall event recorded occurred on the 25th of September 1995. A drought had occurred over the previous summer, with only four small rainfall events occurring after the project start in August. While the banks of the detention pond were still mostly bare, wetland vegetation had migrated in force, covering at least half of the pond surface. A light rain had occurred six hours prior to the start of Event 5. The event itself lasted ap proximately 6 hours totaling 2.5 cm (1").

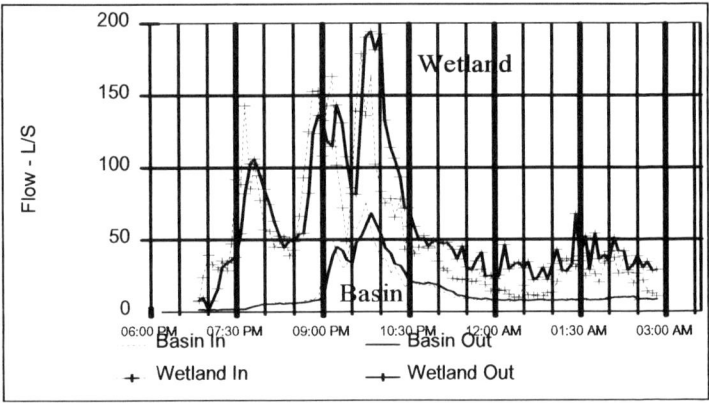

**Figure 2** - Flow Hydrographs

Flow Results

As shown in Figure 2, the flowmeters were able to record the effects of this modest storm within both the wetland stream and basin. When the rainfall is added to Figure 2 (not shown) the quick response of the watershed can be seen in the wetland stream system. As expected, the stream dampens some of the initial flow peaks but the flow eventually was increased in the latter parts of the storm. The storm was too small to inundate the flood plain. When comparing the wetland stream flows to the basin, it appears that the beginning part of the rainfall is missing. This was caused by two underground storage facilities within the contributing basin watershed. The detention basin does decrease the outflow peak of the system but not as much as expected. Part of the poor performance may be that the perforated circular riser designed to control this type of storm was clogged early in the life of the structure. As a result, flow was controlled by the weir created by the top of the riser pipe, not the riser perforations.

Quality Data - TSS and Conductance

Figures 3 and 4 display the total Suspended Solids mass pollutograph for the basin and wetlands site. Both the basin and wetland appear to reduce the suspended sediment load. The wetland

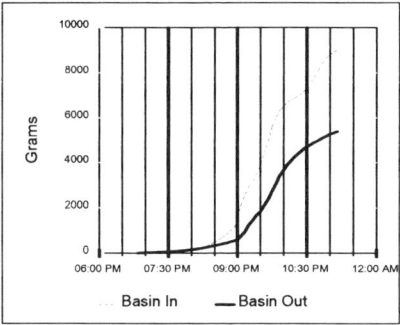

**Figure 3** TSS - Basin

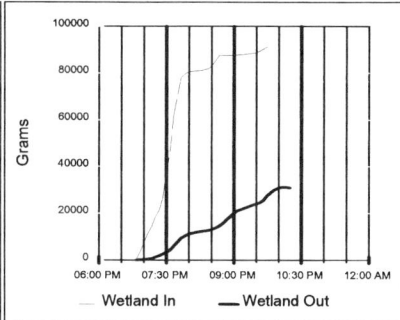

**Figure 4** TSS - Wetland

load is about 10 times higher than the basin, and appears to be more efficient. Note that the suspended sediment load arrives at the basin much later than the wetland, again illustrating the effect of the underground detention structure. Figure 5 displays the flow averaged conductivity values for both the basin and the wetland stream. This can be used to estimate the concentration of soluble solids. Note that the conductivity of the basin outflow starts off high and drops over the course of the event.

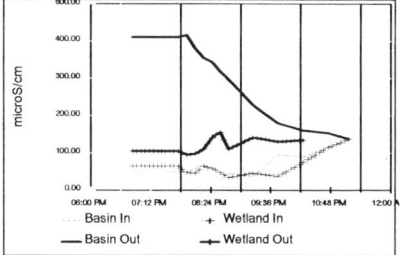

**Figure 5** - Conductance

Quality Data - Metals
Figures 6 and 7 display the cadmium mass curves for soluble and particulate forms of this metal. In the soluble form, both the wetland and basin remove a portion of cadmium. While the basin also removes total cadmium, this is not the case for the wetlands. In this storm the wetlands actually was a source of cadmium over the course of the storm. The increase therefore is in the particulate form. All metals did not act identical in all storms. For example the amount of soluble copper was higher leaving both the basin and wetlands for event 5, while the particulate form of both was reduced.

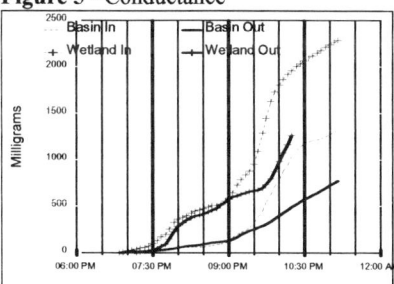

**Figure 6** Cadmium - Soluble

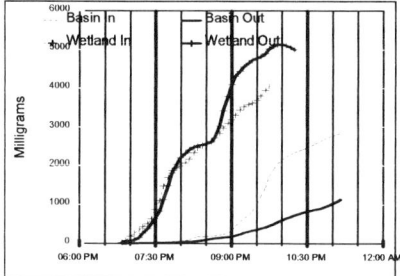

**Figure 7** Cadmium - Particulate

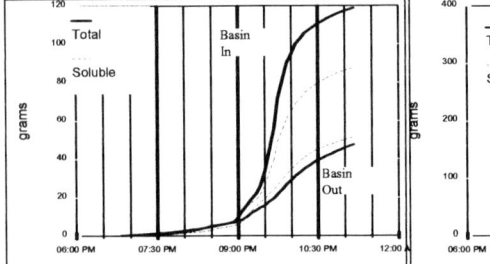

**Figure 8** - Basin Phosphorus

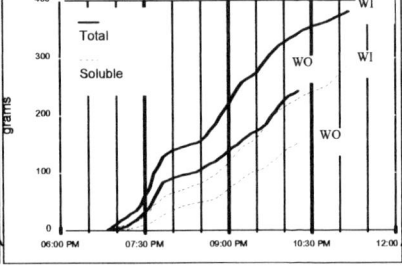

**Figure 9** - Wetland Phosphorus

### Quality Data - Nutrients

Figures 8 and 9 display the mass curves for total and soluble phosphorus. Both forms are reduced by the two BMP's, and it appears that all particulate phosphorus was removed by the basin. Some error is shown by the higher soluble versus total levels of phosphorus leaving the basin. Figure 10 displays total organic phosphorus. Note that this pollutant is removed by both systems.

Total and Soluble Nitrogen Pollutographs are displayed in Figures 11 and 12. Again the basin reduces the outflow of both forms of nitrogen. The wetland stream produces different results. Though total nitrogen is reduced, it appears to have no effect on the soluble form. Note that it appears that the wetland outlet nitrogen is totally soluble. Total organic nitrogen has slightly different effects then that of organic phosphorus. While space precludes the inclusion of this graph, it shows that while the basin again removes some organic nitrogen, the wetland does not have this effect and in fact raises the level slightly.

### Quality Data Results

What has been presented is one of 61 events. When observing some of the other early storm events it is apparent that these results are not all the same for each storm. Some storm results were diametrically opposite to those presented here. Storm duration, and intensity as

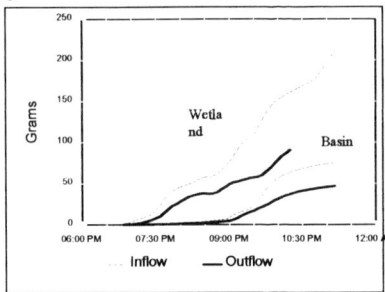

**Figure 10** Total Organic Phosphorus

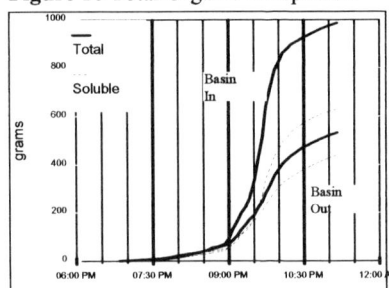

**Figure 11** - Basin Nitrogen

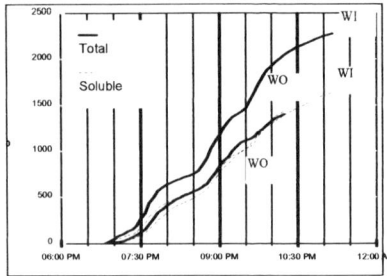

**Figure 12** - Wetland Nitrogen

well as seasonal factors will be studied as work on each of the 61 events is concluded.

Summary
The goal of this research project is to quantify the hydrological and biogeochemical effects of a combined wetland and detention pond system on stormwater runoff quality and quantity. The emphasis of the project to date has been collecting data to quantify changes in flow and in the concentrations of suspended solids, N, P, and selected metals over different temporal scales (seasonal, weekly, and in association with rain storm events). Proximity of the field site to the physical facilities of Villanova University and the use of graduate students conducting thesis research has allowed an intensity of sampling that typically is precluded because of logistical and financial constraints.

While the results of this study are an evaluation of only one wetland basin and stream, they will be more generally relevant to the implementation of extended detention ponds and wetland systems as BMPs for reducing and controlling nonpoint source pollution. In addition to the scientific and technological benefits, a person-power benefit of this project is the training of graduate students in field and laboratory approaches to environmental assessment.

We expect to continue to study sources and solutions to urban nonpoint source pollution. Villanova's location in a predominantly urban - suburban area provides ample field research opportunities. Three unique BMP sites have been identified for future study. Additional topics of interest include contaminant fate, harvesting, and pollutant buildup. We expect that the return on this initial investment will be extraordinary.

Project Sponsors and Acknowledgements
The primary project sponsors are the Nutrient Subcommittee of the Chesapeake Bay Program, Bureau of Environmental Quality - Pennsylvania Department of Transportation, and EPA Region III.
The author wishes to acknowledge his co-principal investigators, R. A. Chadderton, Professor of Civil Engineering, and R. K. Wieder Professor of Biology. As with any team project it is impossible to single out individual credit.

Bibliography
Schueler, T.R., P.A. Kumble, and M.A. Heraty. 1992. A Current Assessment of Urban Best Management Practices - Techniques for Reducing Non-point Source Pollution in the Coastal Zone. Information Center, Metropolitan Washington Council of Governments, Washington, D.C. 127 pp.

Traver, R.G. 1996 "A Study into Effectiveness of Urban Best Management Practices," 1996 ASCE North American Water and Environment Congress. ASCE

Traver, R.G., Chadderton, R.A., and Wieder R.K., 1997 - Year End Report to the Nutrient Subcommittee of the Chesapeake Bay Program.

# Economic Analysis for Stormwater Quality Management

Orit Wilchfort-Kalman[1], Jay R. Lund, member ASCE[2],
Dan Lew[3], and Douglas M. Larson[4]

## Abstract

In evaluating stormwater quality management options and alternatives, their feasibility, removal effectiveness, and economic efficiency should be considered. Commonly, stormwater quality management decisions are based on expert advice and minimizing total expenditures. Because funds are limited and the magnitude of stormwater quality management problems is often great, it is important to use funds in an efficient manner to maximize environmental benefits and to justify environmental expenditures. Benefit cost analysis (BCA) can help in stormwater quality management decisions. Ballona Creek, a case study in urban Los Angeles, provides some lessons to the applicability of BCA for stormwater quality management.

## Introduction

Many receiving waters are compromised by human activities, particularly due to nonpoint source and stormwater runoff pollution. Jeopardizing water quality reduces recreational opportunities, degrades scarce habitats, and limits water supply availability. Under the Clean Water Act, states have the primary authority for establishing designated uses for waterbodies and for developing numerical and narrative water quality criteria for nutrients, minerals, physical characteristics, and biological indicators to protect those uses. Federal, state, and local agencies are struggling to satisfy these stringent regulatory requirements, as required by the Clean Water Act, to improve degraded receiving waters.

---

[1] Ph.D. Student, Dept. of Civil and Envir. Eng., University of California, Davis, CA 95616-5294.
[2] Professor, Dept. of Civil and Envir. Eng., University of California, Davis, CA 95616-5294.
[3] Ph.D. Student, Dept. of Agriculture and Resource Econ., University of California, Davis, CA 95616-5294.
[4] Professor, Dept. of Agriculture and Resource Econ., University of California, Davis, CA 95616-5294.

Scientific and financial resources for protecting and enhancing receiving waters are limited. Agencies and municipalities generally develop stormwater management programs based on expert advice, familiarity with management practices, availability of management practices, and by selecting the lowest cost stormwater quality management options available. The agencies setting the standards and the agencies implementing stormwater pollution management are confined by the many uncertainties associated with the characteristics of stormwater pollution and its impact on receiving waters. Unlike point source pollution that discharges into receiving waters, stormwater runoff is highly variable both in quantity and quality and is therefore difficult to address with traditional treatment methods.

The effectiveness of stormwater management is highly uncertain and greatly dependent on the environment in which it is placed. Because funds are limited, most stormwater management practices are not revisited and monitored for effectiveness by either the agencies that install them or regulatory agencies. The effectiveness of stormwater management in removing pertinent pollutants and improving receiving waters is therefore not well understood or documented. Due to these limitations, funds spent on stormwater management do not always lead to the desired or perceivable improvements in the receiving water quality.

Since the protection and enhancement of receiving waters affected by stormwater runoff is important to the preservation of the environment, the regional economy, and the numerous uses of receiving waters, stormwater quality management evaluation should be based on the benefits realized from expenditures. Benefit cost analysis is proposed for helping to evaluate and develop stormwater quality management alternatives and plans.

Environmental applications of benefit-cost analysis

Benefit Cost Analysis (BCA) compares the likely costs and benefits associated with a project or policy to assess the relative desirability of implementing alternatives, plans, or designs. BCA has been widely used to evaluate the economic soundness of projects and policies, but remains controversial in particular because of the benefits estimation. In most cases relating to natural resources, it has been easier to estimate costs (such as wastewater treatment) than benefits associated with implementing a policy (i.e. the value of treating effluent discharging into the receiving waters). Economic analyses of public policies and projects have evolved from limited consideration of salient costs and benefits to more comprehensive analyses that incorporate less tangible costs and benefits associated with the environment, public health, and safety (Arrow, et al., 1996; Howe, 1971; Freeman, 1993; Lew et al., 1997). Even though there are recognizable difficulties in quantifying benefits and costs related to natural resource management projects, numerous examples are available to demonstrate the usefulness of such analyses in policy decision making.

Loomis (1987) used an economic approach to assess the balance between water rights and public trust resources in Mono Lake, California. The City of Los Angeles' diversion of 100 thousand acre feet/year of water from rivers tributary to Mono Lake adversely impacted Mono Lake's ecology. Benefits and costs were estimated for three different lake levels. Lake levels were used to represent recreational opportunity (ability to access the lake) and the ecological health of the lake. A Contingent Valuation Method (CVM) was used to assess the benefits of each lake level. Costs included the replacement of water and power generated by water diversions and benefits were based on California residents' willingness to pay (WTP) for an incremental gain in water level of the lake. In this analysis, costs were estimated at $26.2 million annually and benefits at $1.5 billion annually. The results suggested that lake levels should be restored. This conclusion was found to be insensitive to major assumptions made in estimating benefit values.

More recently, an economic analysis was prepared by EPA to evaluate the proposed California Toxics Rule (CTR), which establishes ambient water quality criteria. The economic analysis estimated the costs accrued by point source dischargers that may be subject to water quality based effluent limits calculated using the CTR criteria, and the benefits attributable to those point sources. The benefits included direct and indirect uses and passive use of the impacted waters. Contingent valuation and travel cost methods were used to estimate the value of human health, recreation, and passive use. Annual costs were estimated between $14.9 million and $86.6 million and annual benefits were estimated between $1.5 million and $51.7 million. Comparison results showed that benefits and costs were similar in magnitude. Based on the results and given that costs were felt to be generally over stated and benefits not inclusive, the CTR was found to be economically acceptable (USEPA, 1997).

Benefit cost analysis for stormwater management at Ballona Creek

Benefit costs analysis was used to evaluate stormwater quality management at Ballona Creek (Wilchfort et al., 1997). Ballona Creek, originally a natural creek draining to the Santa Monica Bay in Los Angeles, California, is now a concrete lined channel and an extension of an extensive underground system of storm drains (US Army Corps of Engineers, 1995). The Ballona Creek watershed is approximately 130 square miles of which 80 percent is urbanized area with the partially developed foothills and mountains making up the remaining 20 percent. Pollutants from industrial and municipal effluents as well as urban and stormwater runoff have degraded Ballona Creek water quality. Pollutants such as debris, dissolved metals, and fecal coliform impair designated the beneficial uses of Ballona Creek, Santa Monica Bay, and the marina adjacent to the creek.

Stormwater management practices examined included three levels of stormwater treatment: detention and screening, filtration and disinfection, and advanced treatment by reverse osmosis. These treatment levels were selected because of their range in costs and reported pollutant removal efficiencies. Detention and screening provides aesthetic improvement and partial reduction in mass loading of some pollutants. Filtration and disinfection provides improvements in stormwater quality to meet recreation use standards. Reverse osmosis treatment improves stormwater quality to meet all receiving waters objectives as identified in the Regional Water Quality Control Board (RWQCB) basin plan.

Costs of stormwater management practices are relatively straightforward and include capital expenditure as well as annual operation and maintenance costs. Benefits identified with environmental improvements attained through these stormwater management practices are more difficult to quantify. Environmental improvements do not have a market and are considered a public good. Additionally, it is generally accepted that accounting for use values does not always sufficiently account for the value of environmental improvements, since nonuse values are often important. Stormwater management economic benefits can be inferred from the incremental improvements in receiving water quality that support or improve designated and perceived beneficial uses. In addition to the challenges of estimating the value of beneficial uses, the effectiveness of management practices in removing pollutants and improving water quality are not completely understood and are dependent on the environment in which they are placed.

Benefits estimated for the three stormwater management practices represent the economic value associated with incremental beneficial changes in uses of Ballona Creek. The incremental benefits were developed from the estimated changes in present water quality conditions due to future water quality improvements and the unimpaired value of the beneficial uses. The unimpaired values of beneficial uses were estimated from field research and an extensive economics literature review (Lew, et al., 1997). Costs for the three stormwater management practices were obtained from an engineering firm. Benefits and costs were compared for two scenarios. One scenario assumed that only one polluter treats its runoff while the second scenario assumes basinwide treatment of all stormwater runoff discharging to Ballona Creek (Wilchfort et al, 1997).

The results of the benefit cost analysis (presented in Table 1) indicated that the costs of implementing stormwater management practices were at least 110 times higher than the benefits. Ballona Creek is situated in a highly urbanized area and its water quality is affected not only by stormwater runoff but also by factors such as stormdrain system design, neighborhood land uses, sewage spills, and the behavior of people upstream (debris, oil changes and leaks). Because of its urban nature, beneficial use values tend to be low and water quality response to stormwater treatment is very limited. The combination of limited beneficial uses and limited response to stormwater management lead to the huge difference in the magnitude of costs and benefits and the conclusion that stormwater quality management is not economically desirable. BCA results also revealed that basinwide treatment of runoff is much more cost effective than a single polluter treatment, since benefits were much higher for basinwide treatment than single polluter treatment. Nevertheless, for the case of Ballona Creek, basinwide treatment still was not shown to be economically efficient.

Table 1: Comparison of Annualized Benefits and Costs ($1,000/year)

| Scenario | Existing condition | | Detention and screening | | filtration and disinfection | | reverse osmosis | |
|---|---|---|---|---|---|---|---|---|
| | Costs | Benefits | Costs | Benefits | Costs | Benefits | Costs | Benefits |
| One Polluter | 0 | 0 | 15,062 | 20 | 17,116 | 36 | 24,873 | 38 |
| Basinwide Treatment | 0 | 0 | 168,166 | 1,492 | 262,170 | 1,613 | 418,672 | 1,644 |

Numerous assumptions and simplifications were made in estimating the impact of stormwater management practices on water quality, costs, and benefits for the BCA. These assumptions were revisited with a sensitivity analysis to evaluate the robustness of the analysis results. In the case of Ballona Creek, the estimated benefits and costs were so wide apart that even gross underestimation of benefits or overestimating of costs would not have changed the results and the conclusion that stormwater treatment at Ballona Creek is economically undesirable.

Lessons from the benefit cost analysis

*1. Evaluation of stormwater quality management with BCA:* Despite the limitations of BCA and estimating costs and benefits of stormwater management, the Ballona Creek case study demonstrates the usefulness of benefit-cost analysis for environmental problem solving. BCA was able to demonstrate that treatment methods were far from economically justified, and would have imposed a tremendous cost without major environmental benefits. However, BCA should not be the sole approach used for environmental planning and policy analysis. BCA can be helpful in developing and evaluating stormwater management decisions, provide a basis for deliberations, and help planning become more focused and productive (Arrow, et al., 1996).

*2. Identifying limits of stormwater quality management:* In cases where receiving waters are highly polluted and beneficial uses are either eliminated or greatly jeopardized, it may not be economically prudent to invest limited resources in stormwater quality management. Investing resources will not improve or revitalize destroyed environments and limited funds may be better spent in other areas where beneficial uses can be improved (Mar, 1981).

In extreme cases such as Ballona Creek, where the costs are orders of magnitude higher than the benefits estimated, BCA can be used as a preliminary economic analysis to reveal economic inefficiency quickly. Identification of uneconomical alternatives can lead to focusing on more promising planning and policy alternatives and the dedication of scarce economic resources to more environmentally productive investments (Arrow *et al.*, 1996; Howe, 1971).

*3. Research needs for economic stormwater quality management:* In evaluating stormwater management options for Ballona Creek, many assumptions and simplifications were applied. The relationship between beneficial uses and water quality was assumed linear, the removal efficiencies of the management practices was based on limited knowledge, and the unimpaired values of beneficial uses were obtained from the literature and were not location specific. These uncertainties and assumptions were not found to be critical in the Ballona Creek analysis because of the clear difference in costs and benefits magnitudes (Wilchfort, et al., 1997). However, in studying other waters, these assumptions may become critical to the results of the economic evaluation. To improve the process of stormwater management evaluation and planning, additional research is needed in identifying the controlling physical processes of pollution, beneficial uses, and stormwater quality management, their variables, and uncertainties. It is particularly important to revisit constructed and implemented stormwater management practices and evaluate their removal efficiency and engineering limitations. BCA also would benefit from improving the understanding of the relationship between pollutants and beneficial uses at the receiving waters and the effects of water quality and perception on nonuse values.

References

Arrow, K.J., M.L. Cropper, G.C. Eads, R.W. Hahn, and others (1996), "Is there a role for benefit-cost analysis in environmental, health and safety regulation?", *Science* 272(5259): 221.

Freeman III, M.A. (1993), The Measurement of Environmental and Resource Values. Washington D.C.: Resources for the Future.

Howe, C.W. (1971), Benefit-Cost Analysis for Water System Planning, American Geophysical Union, Washington, DC.

Lew, D, O. Wilchfort, J.R. Lund, and D.M. Larson (1997), Approaches to Economic Valuation of Changes in Receiving Water Quality: A Critical Review. CEWRE Report 97-1, Department of Civil & Env. Engr., UC Davis, May.

Loomis, J.B. (1987), "Balancing Public Trust Resources of Mono Lake and Los Angeles' Water Right: An Economic Approach" *Water Resources Research*, 23(8): 1449-1459.

Mar, Brian M. (1981), "Dead is Dead - An Alternative Strategy for Urban Water Management" *Urban Ecology*, 5(1980/1):103-112.

US Army Corps of Engineers, Los Angeles District (1995), Marina del Rey and Ballona Creek, California Final Reconnaissance Report.

US Environment Protection Agency (1997), Economic Analysis of the Proposed California Water Quality Toxics Rule, EPA Contract no. 68-C4-0046.

Wilchfort, O, J.R. Lund, D. Lew, and D.M. Larson (1997), An Economic Valuation of Stormwater Quality Improvement for Ballona Creek, California. CEWRE Report 97-2, Department of Civil & Env. Engr., UC Davis, May.

Alternative Methods in Stormwater Management

Isabel C. Escobar, M.S., E.I.[1], Andrew A. Randall, Ph.D., P.E.[2], Frank E. Marshall, III, M.S., P.E.[3]

Abstract

In the Mosquito Lagoon located in New Smyrna Beach, Florida, stormwater runoff was discharged without any treatment, and this runoff, containing suspended material, has caused a 4-5 ft accumulation of muck. Coagulation was proposed as a treatment method for the runoff in a partially hydraulically isolated basin from the lagoon. In addition, recycle of saline water from the lagoon upstream of the storm sewer outfall to act as a settling aid in the basin was proposed. Laboratory experiments were conducted to determine the effectiveness of the use of two known coagulants, alum ($Al_2(SO_4)_3 \cdot 18H_2O$) and ferric chloride ($FeCl_3 \cdot 6H_2O$). The removal efficiencies obtained by the coagulants were compared with the effect of salinity on suspended solids (TSS) removal. The results of the study were that alum was the best coagulant for the stormwater runoff composition while salinity could not be shown to be an effective settling rate enhancer.

Introduction

In Florida as well as the US, the primary cause of water pollution on a mass and volume basis is typically from non-point sources (Randall et al., 1983). This phenomenon is associated with changes in the watershed both from urban development as well as agricultural development. The bulk of pollutant and

---

[1] Research Assistant, University of Central Florida, P.O.Box 162450, Orlando, FL 32816-2450
[2] Assistant Professor, University of Central Florida, P.O.Box 162450, Orlando, FL 32816-2450
[3] President, Cetacean Logic Foundation, Inc., 340 North Causeway, New Smyrna Beach, FL 32169

nutrient mass from stormwater pollution may be found mostly in the first flush since, during dry periods, suspended solids accumulate on impervious surfaces and land. Typically, suspended solids are removed from stormwater runoff in detention basins, which provide sufficient detention times to allow sedimentation. However, along the boundaries of the Mosquito Lagoon, there are a number of outfalls in urban areas, where there is no possibility of constructing detention basins with hydraulic retention times (HRTs) adequate to remove most suspended solids without some settling aid or coagulant being added. In using coagulation as a settling aid, the coagulant may be added upstream in the pipes to provide rapid mixing, and when the runoff reaches a detention pond, flocculation and settling are accomplished more efficiently.

Because there has been little research into the effect of salinity on coagulation processes, one of the objectives of this research was to investigate the use of saline water as a settling aid. Therefore, this study investigated the feasibility of increasing the salinity of the stormwater runoff to enhance the settling characteristics of suspended particles, and to compare this removal to that obtained by using alum or ferric chloride. The concept of using salinity to intensify particle removal corresponds to double layer compression by the addition of indifferent electrolytes. Seawater was considered as a settling aid for stormwater runoff since it can be relatively inexpensive in coastal areas.

Literature Review

Surface waters generally contain a wide variety of colloidal impurities that cause the water to appear turbid or impart color. Benefield *et al.* (1982) explains that turbidity is usually caused by colloidal clay particles produced by soil erosion while color results from colloidal forms of iron and manganese, or, more commonly, from organic compounds contributed by decaying vegetation. Common unit operations to reduce turbidity and color are filtration, flotation, and sedimentation, which have their effectiveness increased as the size of the colloids increases. However, if the particles lack the size and density to be removed, the effective size of the colloids can be increased by causing them to aggregate. This aggregation process is known as coagulation and flocculation. According to Gregory (1989), particles, which are to be flocculated, may initially be very small and fall in the colloidal size range, which is conventionally assumed to cover sizes between $10^{-9}$ m and $10^{-6}$ m. Moreover, colloidal particles are said to be stable if they are resistant to aggregation. Coagulation and flocculation only occur if particles (1) collide with each other, and (2) can adhere when brought together.

In order to induce particles to aggregate through coagulation and flocculation, Benefield *et al.* (1982) writes that two distinctive steps must occur: (1) the repulsion forces must be reduced, or the particle destabilized, and (2) particle transport must be achieved to provide contacts between the destabilized particles. At this point, it is important to define coagulation as a typically rapidly mixed process in which the energy barriers are eliminated, and flocculation as a slowly mixed procedure during which particles collide and aggregate. Particle removal can be achieved by four mechanisms: (1) double layer compression, and (2) adsorption and charge neutralization, during coagulation; (3) adsorption and interparticle bridging, during flocculation; and (4) enmeshment or entrapment into a precipitate, or sweep floc, during precipitation or gravity settling.

Concerning the use of salinity as a settling aid, the controlling coagulation process is double layer compression. Double layer compression involves the addition of large quantities of indifferent electrolytes that retain their identity and do not absorb to the colloid, such as sodium ions. When an indifferent electrolyte is added to a colloidal dispersion, the added electrolyte increases the charge density in the diffuse layer and results in less volume of the diffuse layer being required to neutralize the surface charge of the colloid, and, thus, the diffuse layer is compressed (Benefield *et al.*, 1982). The effect of this compression is to change the distribution of double layer repulsion forces, which allow the Van der Waals' attractive forces to be more dominant and enhance aggregation. Seawater contains indifferent electrolytes, such as sodium and chlorine, that can behave as settling aids to remove suspended solids from water by charge neutralization.

Ayoub *et al.* (1992) performed batch studies with the aid of 2-liter jar testers for a synthetic oily wastewater with a solubility in water > 10%, specific gravity of 1.006, and pH of 8.7. The solids' concentration in the tests before seawater addition was 1000 mg/L. The experiment performed by Ayoub *et al.* (1992) indicated that a total suspended solids' removal of 70% or more could be achieved by an addition of 10% seawater by weight. Seawater addition to induce settling of colloids is an inexpensive treatment compared to the addition of coagulants, and the increased salinity of the final effluent does not pose a problem for ocean or estuarine disposal. However, since the suspended solids were observed to be oily, they were probably more hydrophobic, so the process of seawater addition for removal may have been more efficient than it would be for suspended solids without this characteristic.

Methods and Materials

A synthetic stormwater was used in the experiments due to the difficulty of transporting the actual stormwater runoff from New Smyrna Beach to Orlando. The source of suspended solids for the experiments were street sweepings from the city of New Smyrna, which were oven dried at 103 - 108 °C over night, and sieved to an average size less than 250 µm. Characteristics of the synthetic stormwater were based off of sample data from a storm outfall in New Smyrna Beach and historical stormwater data for central Florida (Wanielista et al., 1993). In order to model settling of suspended particles due to the addition of a settling aid, a pragmatic setup was a tall upflow column with the stormwater and settling aid mixing in pipes immediately before entering the column. The settling was analogous to the vertical settling of particles in detention and retention basins. The column used was designed to closely approach the characteristics of the outfall basin in New Smyrna Beach, FL. The average overflow rate measured for the basin in New Smyrna Beach was determined to be $V_0 = 27.432 \; \frac{cm}{min}$ for a 10-year storm event with a hydraulic retention time (HRT) of 10 minutes. The available column used was 188.10 cm in height, which gave a very conservative estimate of the available height in the basin, and 9.525 cm diameter. However, the sampling port was built on the column at a height equal to 177.80 cm from the bottom to allow the column to operate as a continuous upflow system, so 177.80 cm was considered to be the effective height of the column.

Two columns were run in parallel at all times, one as a control and one with a constant settling aid addition, to allow quantitative comparisons while the parameter manipulated was the length of the hydraulic retention time. The column studies ran a minimum of four HRTs during each experiment, so that the system was allowed to reach steady state. The stormwater runoff was input from the bottom of the column while the effluent was taken from the top of the column. The flow rate for the coagulants or the settling aid was also calibrated daily with tap water, and it was kept below 10% of the synthetic stormwater flow rate.

A synthetic seawater was employed in the experiments using White Sea salt coarsed crystals/ionized, made by La Baleine des Salins du Midi. The concentrations of alum, ferric chloride and synthetic seawater to be added to the synthetic stormwater runoff during the column studies were obtained from jar tests following procedures outlined by Hudson (1981). Six acrylic jars with square bottoms with 11.5 cm on each side and 21 cm high were filled with 1 L of synthetic stormwater runoff, and a different concentration of alum, ferric chloride or seawater was added to each. Dye tracer tests using methylene blue were performed to determine that there was no short-circuiting in the column according to procedures described by Schulz (1994) and Dietz (1982).

Results and Discussion

Batch experiments indicated the optimum concentrations for each coagulant/settling aid to be used. These values were 10 mg/L alum, 10 mg/L ferric chloride, and 10% seawater by weight (3500 mg/L). Figure 1 shows the comparison between alum addition, ferric chloride addition and salt-water addition. TSS removal was the target parameter because its removal was the main objective of the research. The experiments shown in Figure 2 were performed with hydraulic residence times ranging from 8 minutes to 20 minutes. The actual influent TSS ranged from 150 to 190 mg/L when 200 mg/L TSS was added to the influent basin since even with a rotor to provide complete mixing some variability was unavoidable due to the inhomogeneous nature of the solids added and the large volume of the stormwater container (100 L of stormwater).

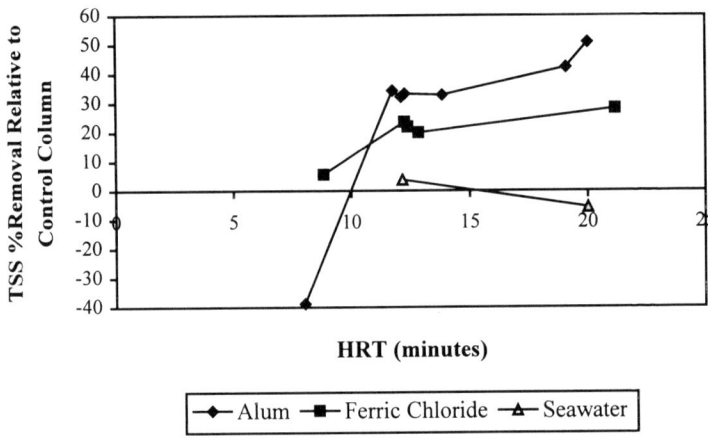

Figure 1. Comparison of Settling Aids Efficiencies.

From Figure 1, alum was the most effective coagulant for improving TSS removals in the range of HRTs analyzed. At its optimum HRT, 20 minutes, an alum concentration of 10 mg/L removed 51% of the suspended solids in the experimental column in comparison to the control column. Ferric chloride trailed alum, removing 28% of the suspended particles in the stormwater runoff after 21 minutes of residence time. Conversely, the removal of suspended solids obtained by the addition of salinity was insignificant and ranged between +/- 5%.

TSS removal was a function of HRT for the coagulants, alum and ferric chloride, but not for salinity. This was probably the result of the fact that double layer compression was a very rapid phenomenon and additional time did not change the level of compression achieved. Charge neutralization, flocculation, and precipitation would all have much slower kinetic rates, and, thus, they were probably helped by the additional time at higher HRTs. For alum, TSS removal efficiencies increased from -38.8%, at a HRT of 8 minutes, to 50.8%, at a HRT of 20 minutes. Similarly, TSS removal due to ferric chloride addition increased from 5.6% when HRT was 9 minutes, to 28.2% when HRT was 21 minutes.

Poor correlation between turbidity and TSS removals was observed throughout the study for both alum and ferric chloride. This may have been due to small light scattering particles remaining in suspension when coagulants were used, and also to the development of color with ferric chloride addition. Seawater provided significant reduction of turbidity (about 15% at HRTs of 8 and 20 minutes) which may indicate that some aggregation of colloids occurred. However, this aggregation was insufficient to result in a significant increase in settling velocities and, thus, increase TSS removals.

From the data obtained, it was determined that salt water addition did not significantly improve settling of suspended solids representative of those likely to be found in the stormwater runoff from the city of New Smyrna Beach. It may be that salt-water recirculation might work better if runoff characteristics were different as with a runoff with oily sediments, similar to the TSS studied by Ayoub (1992). This study strongly implied that the observed phenomenon of particle settling when the runoff came in contact with brackish water at New Smyrna was largely due to the sudden decrease in flow velocity rather than the increase in salinity as the runoff entered the lagoon from the outfall.

References:

Ayoub, G. M.; Lee, S. I.; Mazidji, C. N.; Seo, I. S.; Cho, H. M.; and Koopman, B. "Seawater Flocculation of Emulsified Oil and Alkaline Wastewaters." Water Resources 26 (6) 1992: 817-823.

Benefield, L. D.; Judkins, J. F.; Weand, B. L. Process Chemistry for Water and Wastewater Treatment. New Jersey: Prentice-Hall, Inc., 1982.

Dietz, J. D. "Clarification Mechanisms for a Flocculent Slurry." Ph.D. Thesis, Clemson University, Clemson, SC, 1982.

Hudson, H. E. Water Clarification Processes Practical Design and Evaluation. New York: Litton Educational Publishing, Inc., 1981.

Randall, C. W.; Whipple, W.; Grigg, N. S.; Shubinski, R. P.; and Grizzard, T. J. Stormwater Management in Urbanizing Areas. New Jersey: Prentice-Hall, Inc., 1983.

Schulz, C. E.; Singer, P. C.; Gandley, R.; and Nix, J. E. "Evaluating Buoyant Coarse Media Flocculation." Journal AWWA 86 Aug. 1994: 51-63.

Wanielista, M. P.; and Yousef, Y. A. Stormwater Management. New York: John Wiley & Sons, Inc., 1993.

# Improved Estimation of Urban Non-Point Pollution
Stuart S. Schwartz, Ph.D.[1]

## ABSTRACT

Planning level techniques are typically used to estimate mean annual contaminant loads when monitoring data are inadequate or unavailable. The computational ease and modest data required for most planning level load estimation techniques has encouraged their regular use in planning, watershed assessment, and resource allocation decisions. While planning level load estimates are readily calculated, the quality of management decisions based on these estimates will be critically linked to their accuracy and variability.

Common planning-level techniques estimate pollutant loads as the product of an annual or seasonal runoff volume and a characteristic concentration. The common use of the mean, median or geometric mean of event concentrations results, in general, in biased estimates of the mean annual load. This motivates the need to consider both the relative variance as well as the correlation of discharge and characteristic concentration in planning level load estimates. While primarily focussed on nonpoint pollution, these results are equally applicable to planning level estimates of mean annual contaminant loads in rivers and point source discharges.

These findings have substantive implications for both the design of monitoring programs as well as planning, policy making, and watershed management. A simple example illustrates the importance of quantifying and incorporating both the accuracy and the inherent variability in planning level load estimates, in planning and management decisions. The results presented here support risk-based planning and watershed management by extending the accuracy and quantifying the variability in planning level estimates of mean annual contaminant loads.

---

[1]

Associate Director for Water Resources - Interstate Commission on the Potomac River Basin. **Current Address**: Hydrologic Research Center; 12780 High Bluff Drive, Suite 250; San Diego CA 92130-2069 (V) 619.794.2726 (FAX) 619.792.2519

## Introduction

The estimation and control of nonpoint contaminant loads has come to be recognized as a crucial part of the national strategy to attain water quality goals. Consistent with this need, Schwartz and Naiman (1996) examined the estimation of mean annual contaminant loads, emphasizing sources of bias and the quantitative characterization of variability. Their analysis focused on planning level techniques based on the product of annual runoff and a characteristic contaminant concentration. In this context, "planning level" refers to computationally simple techniques that are primarily intended to provide estimates of mean annual contaminant loads for screening and assessment purposes, by rationally extrapolating nonpoint loading characteristics from monitored locations to ungauged catchments.

The computational ease and modest data required for most planning level load estimation techniques allows their application over spatial and temporal scales that can dramatically exceed the domain over which these methods were developed or are reasonably supported. Common among planning level techniques is their development and intended application on small drainage areas, where the fate and transport of contaminants can be ignored. When extrapolated to watershed scale estimation, particularly for hydrocarbons, metals, and other sediment associated contaminants, the inherent limitations in planning level load estimation techniques restrict their applicability. The ease with which simple load estimates can be calculated belies the uncertainty associated with these estimates.

Many planning level techniques have been developed, ranging from the use of average loading factors and regression relationships, to techniques extrapolating process-based event statistics to annual time scales. Quantifying both bias and uncertainty in planning-level load estimates extends their utility in risk-based planning, supporting management and resource allocation decisions by allowing explicit consideration of uncertainty and expected cost-effectiveness.

### Planing Level Estimation: An example

To illustrate the importance in quantifying bias and uncertainty in planning level estimates, the moment estimators in Schwartz and Naiman (1996) are used to analyze a hypothetical planning problem. Consider a budget-limited planner evaluating watershed-scale copper control options for both point and nonpoint sources. In this example, we assume routine monitoring data for copper has not been collected, so the simple flow-concentration estimator is used to develop preliminary watershed estimates for the mean annual copper load from both point and nonpoint sources. For illustrative purposes the characteristic point source copper concentration is taken to be identical to the nonpoint concentration, $\mu_C$, and the annual average point source discharge, normalized for area served, is also assumed to be identical to the annual nonpoint runoff depth $\mu_Q$. The point source discharges are assumed to be inherently less variable than nonpoint discharges. Both point source and nonpoint source loads are expressed as areal loading rates, in order to support the planning level comparison and assessment of their relative magnitudes.

This example is constructed so that the naive estimator, taken as the product of annual runoff and characteristic concentration, yields equal load estimates of the point and nonpoint source loads. These estimates will only be unbiased if flow and concentration are uncorrelated. Lacking correlation data from either source, the planner might consider the uncertainty in these estimates, by examining the sensitivity of the mean and standard deviation to the correlation coefficient between annual runoff and annual flow-weighted EMC. The expected load is plotted with error bounds $\pm 1$ standard deviation for a range of correlations between concentration and discharge, $\rho_{CQ}$. Although the expected load is sensitive to the value of $\rho_{CQ}$, the dramatic difference between the two hypothetical load estimates is clearly in the variability or confidence in the estimated load, reflecting the greater *inherent variability* in nonpoint source runoff and concentration assumed in the values of the coefficient of variation for concentration and runoff, $v_C$ and $v_Q$. In addition to relative magnitude, the relative confidence in the load estimate should significantly influence planning-level efforts to develop a watershed control strategy.

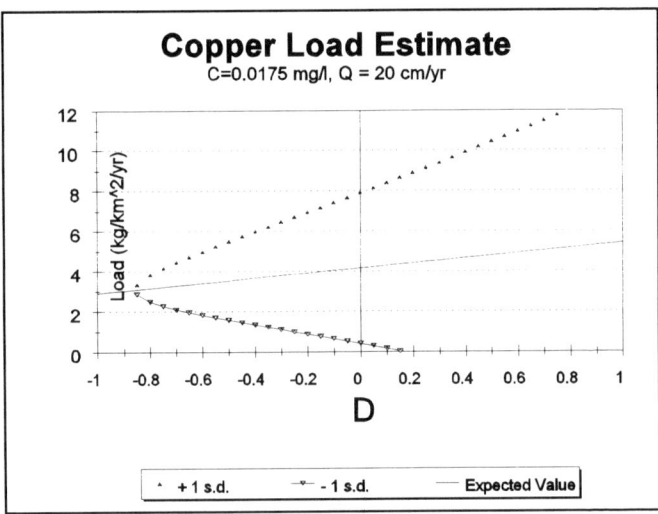

**Figure 1** Copper Load Error Bars $v_C = .75$, $v_Q = .4$

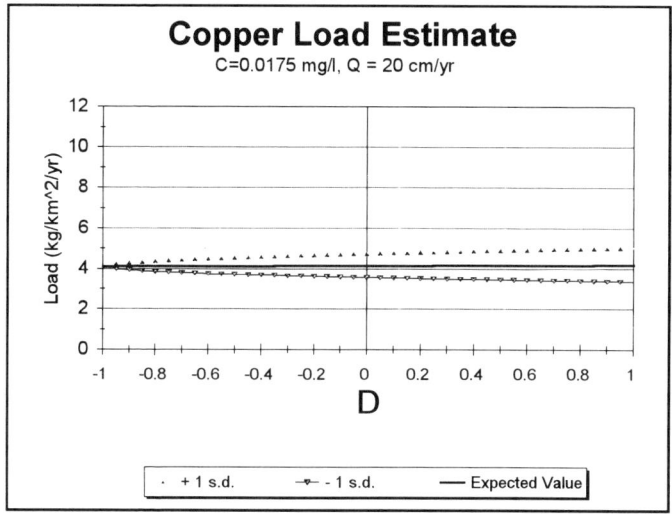

**Figure 2** Point Source Copper Load Error Bars $v_C = 0.1$ $v_Q = 0.1$

In developing a watershed plan to control copper loading, the planner's optimal strategy would allocate limited resources between point source and nonpoint source control options in order to equate marginal removal costs over all alternatives. If expected marginal reductions in copper loading from nonpoint runoff controls (such as retrofitting stormwater management ponds or street sweeping) are less expensive than point source process controls, watershed strategies based only on the expected value of the mean annual load will result in planning level strategies favoring low cost nonpoint source controls. When the variability of the load is considered, the likelihood that nonpoint controls could be expected to reliably reduce the mean annual load becomes more problematic.

The example is intended to illustrate the need to balance the high degree of confidence in the point source load against the lower (presumed) unit cost for non-point source control of a much more variable load. Higher unit cost point source controls may be more cost-effective when the greater likelihood that the expenditure will result in reduced loading is considered. A more robust optimal watershed policy with uncertainty would therefore allocate limited resources in order to maximize the *probability* of reducing the mean annual load. Quantifying the variability or distribution of planning level load estimates significantly expands their utility in supporting this risk-based decision.

## Conclusions

Planning-level procedures for contaminant load estimation offer convenient and useful tools for screening-level evaluations of the relative magnitude of contaminant loads. While many planning level techniques are readily available and widely used, the uncertainty in these estimates has heretofore been poorly studied. The computational ease with which these procedures can be applied creates the potential for misapplication across inconsistent spatial and temporal scales, and may lead to dramatically inaccurate estimates. While planning-level techniques are useful in reconnaissance level studies, their use in developing watershed strategies and targeting resource allocation decisions requires a quantitative characterization of both the expected load, as well as the variability and uncertainty in that load estimate.

Planning-level load estimates derived as the product of a runoff volume and characteristic concentration have been analyzed in order to identify and quantify sources of bias and variability in mean annual load estimates. The use of annual flow-weighted event mean concentration as the characteristic concentration for planning-level load estimation provides an unbiased estimate of the mean annual load. The common practice of using either the arithmetic mean, median, or geometric mean of EMCs to characterize mean annual load results, in general, in biased load estimates. This finding is significant and implies that monitoring programs designed to support the estimation of mean annual loads should sample both flow and concentration, in order to support the estimation of flow weighted EMC, as well as the correlation between flow and concentration.

For planning and policy purposes, the allocation of limited mitigation resources will be most cost-effectively targeted by accounting for estimates of both expected load, as well as the variability in the load estimate. Quantitative estimates support risk-based allocation of pollution abatement resources, by extending the accuracy and quantifying the variability in planning level estimates of mean annual contaminant loads.

## References

Schwartz, S. S. and D.Q. Naiman, 1996; Planing Level Estimates of Annual Contaminant Loads: Bias and Confidence. Submitted to Water Resources Research; under revision.

# A Postaudit of Optimal Conjunctive Use Policies
Tracy Nishikawa, A. M. ASCE and Peter Martin[1]

## Abstract

A simulation-optimization model was developed for the optimal management of the city of Santa Barbara's water resources during a drought; however, this model addressed only groundwater flow and not the advective-dispersive, density-dependent transport of seawater. Zero-m freshwater head constraints at the coastal boundary were used as surrogates for the control of seawater intrusion. In this study, the strategies derived from the simulation-optimization model using two surface water supply scenarios are evaluated using a two-dimensional, density-dependent groundwater flow and transport model. Comparisons of simulated chloride mass fractions are made between maintaining the actual pumping policies of the 1987-91 drought and implementing the optimal pumping strategies for each scenario. The results indicate that using 0-m freshwater head constraints allowed no more seawater intrusion than under actual 1987-91 drought conditions and that the simulation-optimization model yields least-cost strategies that deliver more water than under actual drought conditions while controlling seawater intrusion.

## Introduction

During non-drought years, the city of Santa Barbara (the City) obtains adequate water supply from local surface-water reservoirs and ground water--most of the water coming from the reservoirs. During times of drought, however, these local surface-water supplies are insufficient to meet demand. In addition, the City is aware that the overpumping of ground water may induce seawater intrusion and other water-quality problems as well as excessive drawdown in sensitive areas. In the Santa Barbara area, the primary management issues during a drought are minimizing the cost of water supply while satisfying water demand and controlling seawater intrusion.

The U.S. Geological Survey (USGS), in cooperation with the City, developed a simulation-optimization model as an aid to conjunctively manage the City's water resources, assuming the City's 5-year design drought (Nishikawa, in press). This model did not explicitly address the density differences between seawater and

---

[1] both authors: Hydrologist, U.S. Geological Survey, 5735 Kearny Villa Rd., Suite O, San Diego, CA 92123

freshwater, but used 0-m fresh-water head constraints at the coastal boundary as surrogates for the control of seawater intrusion.

In this study, selected optimal pumping strategies from Nishikawa (in press) are used as input for a density-dependent solute-transport model to determine the effect on seawater intrusion by using the 0-m fresh-water head constraints over the model's 5-year management period. In addition, the water-delivery strategies are compared with the actual strategy implemented by the City during the 1987-91 drought.

## Description and Available Water Resources of the Study Area

The city of Santa Barbara obtains its ground-water supplies from the Santa Barbara and Foothill ground-water basins, which are on the south coast of Santa Barbara County about 193 km northwest of Los Angeles (fig. 1). Hydrologically, the Santa Barbara ground-water basin is divided into two subbasins--Storage Unit I and Storage Unit III--by the Mesa Fault and is separated from the Foothill ground-water basin by the Mission Ridge Fault (Freckleton, 1989). The combined area of the two ground-water basins is about 36 km$^2$.

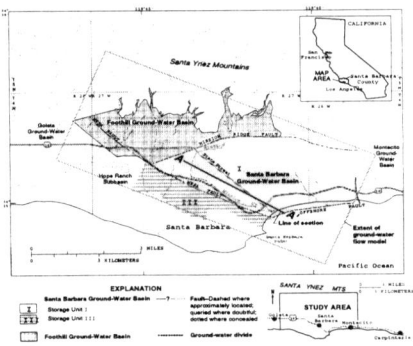

**Figure 1**: Location map showing ground-water flow model domain and solute-transport model trace.

On the basis of data from geophysical and geologic logs of selected wells, Martin (1984) subdivided the unconsolidated deposits underlying the Santa Barbara ground-water basin into five main zones: (1) the shallow zone, (2) the upper producing zone, (3) the middle zone, (4) the lower producing zone, and (5) the deep zone (fig. 2). The upper and lower producing zones are the two main water-bearing zones tapped by wells in the Santa Barbara area.

Groundwater is pumped from 13 wells most of which are screened within the upper and/or lower producing zones. Most of the largest producing wells are within 2 km of the coast. Surface water sources are the Cachuma and Gibraltar Reservoirs, infiltration into Mission Tunnel (conveys water from Gibraltar Reservoir to the City), desalinated water, and imported water from the California State Water Project (SWP).

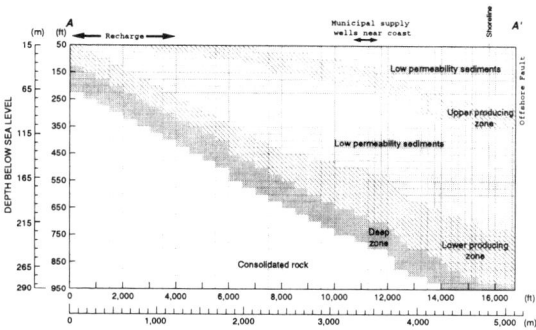

**Figure 2:** Solute-transport model grid showing aquifer layers.

## Models

Nishikawa (in press) applied a simulation-optimization approach as an aid to manage conjunctively the City's water resources while controlling seawater intrusion. His simulation model only addressed groundwater flow. In this study, the strategies derived from the simulation-optimization model using two surface-water supply scenarios are evaluated using a solute-transport model.

### Simulation-Optimization Model

The ground-water system was simulated as two horizontal layers (the upper and lower producing zones) separated by a confining bed (the middle zone) throughout most of the modeled area. The model grid consists of blocks that are about 300 m on a side. Additional specifics regarding the ground-water flow model of the Santa Barbara and Foothill basins are presented by Martin and Berenbrock (1986), Freckleton (1989), and Freckleton et al. (in press).

The optimization model was formulated as a linear programming problem with the objective of minimizing the cost of water supply subject to: (1) water-supply capacity constraints; (2) maximum and minimum heads at the coastal boundary to control seawater intrusion; (3) constraints maintaining pumping distributions between upper and lower producing zones; and (4) satisfying water-supply demands specified by the City. The decision variables are the monthly amounts of water (in $m^3/s$) supplied from surface water and groundwater from 13 production wells. The state variables are hydraulic heads. Cost coefficients were about $50,000/$10^6$ $m^3$ ($10^6$ $m^3$ = MCM) for reservoir water, $370,000/MCM for SWP water, $890,000/MCM for desalinated water, and $100,000 /MCM for ground water.

Basic assumptions for the simulation-optimization model include:
- The surface-water supply conditions for the drought of 1947-51, the City's design drought, were used for all simulations.
- The minimum freshwater head constraints at the seawater boundary were set to 0 m (sea level) in both layers (allowing a landward hydraulic gradient). Specifying a

0 gradient was infeasible. The maximum heads were the land surface elevations.
- Annual water demand for the simulation-optimization model are shown in fig. 3. Note the 7% decrease in demand in year 5 reflecting water conservation efforts.

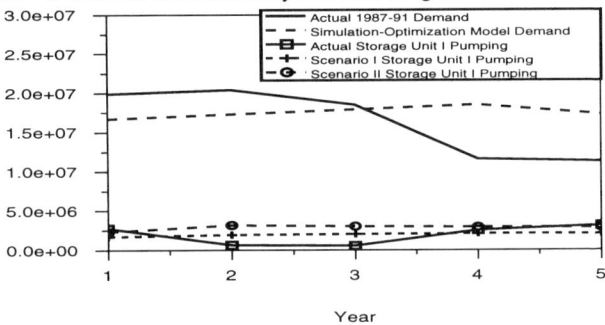

**Figure 3**: Annually varying water demand and total Storage Unit I pumpage.

## Solute-Transport Model

For comparative purposes only, a preliminary solute-transport model was developed for this study. The numerical solute-transport model used in this study was SUTRA, a two-dimensional, density-dependent, finite-element model (Voss, 1984). The trace for the vertical section simulated is shown in fig. 1 and the conceptual model is shown in fig. 2. The model domain extends from the offshore fault inland to the aquifer recharge area of Storage Unit I. The model grid has a total length of about 5,075 m, a height of about 275 m, variable horizontal grid spacing ranging from about 40 m at the coast to 150 m at the recharge area, and uniform vertical spacing of about 8 m. The model grid has a uniform thickness of about 1800 m. The dependent variables are pressure and chloride (Cl) mass fractions (100% Cl = 19,000 mg/L Cl).

Boundary conditions for the model include a hydrostatic pressure boundary at the offshore fault, drains to simulate leakage from the top layer of nodes near the coast, constant-flux boundary in the recharge area, and an impermeable basal boundary at an altitude of about 290 m below sea level (fig. 2). The drain boundaries are similar in concept to the hydrostatic pressure boundary, except they only allow water to leave the aquifer. Hydraulic properties used in the solute-transport model approximate the values used in the calibrated flow model.

Steady-state conditions were simulated by setting the initial pressures and Cl mass fractions equal to zero and running the model until the seawater front stabilized. The model was then used to simulate the transport of Cl in response to measured monthly pumpage during 1947-86. The transport parameters were based on trial and error parameter estimation using observed head and Cl data from 1976-92 at USGS monitoring well clusters.

The initial conditions for this study are the simulated 1986 pressures and Cl mass fractions. The optimal Storage Unit I pumping values were then used as input to the solute-transport model. All other parameters were held constant.

## Results

Two operating scenarios for the release schedule of water from Gibraltar and Cachuma Reservoirs were tested using the simulation-optimization model: (1) uniform monthly releases from both reservoirs and (2) current, variable monthly release percentages from Gibraltar Reservoir and variable monthly releases from Cachuma Reservoir. The use of variable releases from Cachuma Reservoir allows the monthly water deliveries to vary within a year such that the total water delivered within that year is less than or equal to the maximum annual volume of water available to the City [see Nishikawa (in press) for additional details]. The resulting optimal pumping schedules are used as input for the solute-transport model and the simulated Cl mass-fraction distributions are compared to determine the effect on seawater intrusion by using the equivalent fresh-water head constraints.

As stated above, the water supply conditions for the City's design drought of 1947-51 were used in both scenarios. The City also experienced drought conditions in 1987-91 during which water demand differed somewhat from that assumed in the simulation-optimization model (Steven Mack, city of Santa Barbara Public Works Dept., written commun., 1997) (fig. 3). The actual pumping from Storage Unit I for this drought was used as input for the solute-transport model (fig. 3). The simulated Cl mass fractions were compared with the Cl mass fractions under optimal pumping conditions. In addition, the actual and optimal water delivery strategies were compared.

The optimal total ground-water deliveries for the 1947-51 design drought from Storage Unit I for Scenarios I and II are shown in fig. 3. The simulated 10% Cl contours for the actual pumping and 2 simulation-optimization scenarios at the end of the 5-year management period are shown in fig. 4. The cost of delivering water over the 5-year design period for Scenarios I and II are $5.56 and $4.53 million, respectively illustrating the value of variable reservoir releases.

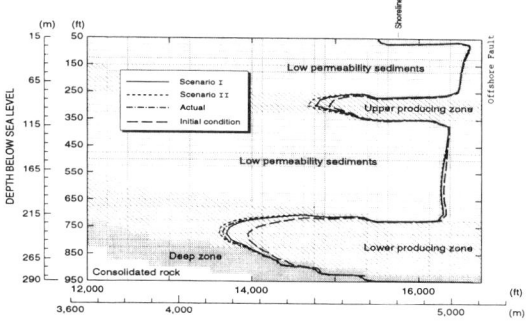

Figure 4: Simulated 10% Cl contours.

There is little difference between the location of the 10% Cl contour for the actual pumping and the 2 simulation-optimization scenarios (fig. 4). In addition, all the 10% contours have moved less than 250 m inland from the initial conditions. The total actual 1987-91 pumping from Storage Unit I was approximately equal to the total optimal pumping from Storage Unit I in Scenario I (9.60 MCM and 9.85 MCM, respectively). However, the total optimal pumpage from Storage Unit I in Scenario II was about 50% greater than under the 1987-91 drought (14.3 MCM); explaining the slightly greater amount of seawater intrusion in Scenario II (fig. 4). Note that seawater intrusion is a slow process and that the impact of all pumping strategies may induce greater seawater intrusion over a longer management period.

The total volume of water delivered in the 2 scenarios was 7% greater than the actual volume of water delivered during the 1987-91 drought (86.1 MCM vs. 80.4 MCM). However, the amount of seawater intrusion under actual and optimal conditions were approximately equal.

## Conclusions

The results from the 2 optimization scenarios indicate that the surrogate 0-m freshwater head constraints allowed no increase in seawater intrusion in comparison with results from the actual drought conditions. Specifically, Scenario II delivered about 50% more water from Storage Unit I with no increase in seawater intrusion. In addition, the simulation-optimization model identified least-cost strategies that delivered 7% more water for the design drought of 1947-51 than under actual 1987-91 drought conditions with no increase in seawater intrusion.

## References

Freckleton, J.R. (1989). "Geohydrology of the Foothill Ground-water Basin near Santa Barbara, California." *U.S.G.S. Water-Resources Investigations Report 89-4017*, Reston, Va.

Freckleton, J.R., and Martin, P., and Nishikawa, T. (in press). "Geohydrology of Storage Unit III and a combined flow model of the Santa Barbara and Foothill ground-water basins, California." *U.S.G.S. Water Resources-Investigations Report 97-4121*.

Martin, P. (1984). "Ground-water monitoring at Santa Barbara, California, California: Phase 2--Effects of pumping on water levels and on water quality in the Santa Barbara ground-water basin." *U.S.G.S. Water-Supply Paper 2197*, Reston, Va.

Martin, P. and Berenbrock, C. (1986). "Ground-water monitoring at Santa Barbara, California: Phase 3--Development of a three-dimensional digital ground-water flow model for Storage Unit I of the Santa Barbara ground-water basin." *U.S.G.S. Water Resources-Investigations Report 86-4103*, Reston, Va.

Nishikawa, T. (in press). "A simulation-optimization model for water resources management, Santa Barbara, California." *U.S.G.S. Water Resources-Investigations Report 98-4246*.

Voss, C.I. (1984). "SUTRA: A finite-element simulation model for saturated-unsaturated, fluid-density-dependent ground-water flow with energy or chemically-reactive single-species solute transport." *U.S.G.S. Water-Resources Investigations Report 84-4369*, Reston, Va.

## Acknowledgments

The authors acknowledge the city of Santa Barbara.

## Model Development for Conjunctive Use Planning in Taiwan

Shu-li Yang[1], Associate Member, ASCE, Nien-Sheng Hsu[2], Associate Member, ASCE, Shiang-Kueen Hsu[3], and William W-G. Yeh[4], Honorary Member, ASCE

### Abstract

To mitigate the groundwater overdraft from a coastal alluvial aquifer, thereby reducing land subsidence and preserving the groundwater resources, a management model that links a simulation model and an optimization model for conjunctive use of groundwater and surface water was developed. The final objective of the management model is to reduce the area in which the groundwater level is below the mean sea level, subject to the safe yield of the groundwater aquifer and the availability of surface water. The groundwater simulation model is linked with the optimization model to ascertain the optimal pumping strategy.

### Introduction

Rapid growth of population in Taiwan, along with the attendant increased urbanization and developing industry, has made strong demands on groundwater resources. The resulting groundwater overdraft has caused severe land subsidence, especially in the coastal areas. In response, the Executive Yuan of Taiwan, in 1995, developed the Land Subsidence Prevention Plan, which has as its goal the complete elimination of all land subsidence by the year 2000. To achieve this goal, the Plan sets out the extent to which the groundwater use is to be reduced by and the percentage of wells that are to be gradually shut down. Local and central government agencies, as they struggle to comply with requirements of the Plan, are seeking means to mitigate the groundwater overdraft, particularly in the fragile coastal areas.

To mitigate the overdraft in Taiwan's coastal areas, this study focuses on

---

[1] Postdoctoral Fellow, Civil & Env. Eng. Dept., UCLA, Los Angeles, CA 90095.
[2] Associate Professor, Dept. of Civil Eng., National Taiwan Univ., Taipei, Taiwan, R.O.C.
[3] General Director, Water Resources Bureau, Ministry of Economic Affairs, Taipei, Taiwan, R. O. C.
[4] Professor, Civil & Env. Eng. Dept., UCLA, Los Angeles, CA 90095.

developing a model for conjunctive use of groundwater and surface water. Taking the aquifer of the Choushui Creek Alluvial Fan as an example, application of the management model is illustrated.

## System Description

The Choushui Creek Alluvial Fan is located in the semi-tropical, central west part of Taiwan, where virtually all precipitation occurs during the period from May to September. The long-term average annual rainfall in the area is approximately 2,197 mm and annual evaporation is approximately 1,263 mm. Earlier, up to 1994, the Central Geological Survey (CGS), Ministry of Economic Affairs, R.O.C., had established 53 well logs on the Fan. These well logs have provided the data for analyzing the aquifer stratigraphy and for determining the system boundaries—information necessary to characterize the geologic and hydrogeologic conditions of the area. The groundwater system of the Choushui Creek Alluvial Fan is divided into seven layers, with four aquifers and three aquitards. Aquifer 2, at some points, is divided by a thin and less permeable layer that further divides aquifer 2 into aquifers 2-1 and 2-2 (Fig. 1). The system overall is heterogeneous and multi-layered. In general, hydraulic conductivity is higher in the upstream area of the Fan, where most of the interaction between the layers occurs. Silt and clay flood plain deposits are present in the downstream area. Because materials with a lower hydraulic conductivity, such as silt and clay, cumulate toward the ocean, the aquifer system is almost sealed at the west boundary; consequently, no water intrusion is present. The groundwater flow direction, in general, moves westward toward the ocean, as does the topology of the Fan.

The top aquifer of the aquifer system is an unconfined aquifer, while the other three aquifers are confined. The groundwater system is confined by no-flow boundaries. The unconfined aquifer recharged through areal irrigation and the seepage from the riverbed of the Choushui Creek.

Although the groundwater levels in the Choushui Creek Alluvial Fan have been periodically recorded since 1968, only recently have layered groundwater-level data been available. Data on groundwater being extracted from the Choushui Creek Alluvial Fan is limited and the actual amount of extraction is unknown. Because current control measures are inadequate, many illegal pumping wells are operating without permits, a response to the increasing demand for water in the face of insufficient surface water supply. Nevertheless, both local and central government agencies are attempting to control the amount of pumping and thus to prevent further land subsidence in the Fan. One such attempt is seen in the construction of a major water supply project that will divert water from upstream of the Choushui Creek, thus reducing the prevailing groundwater use.

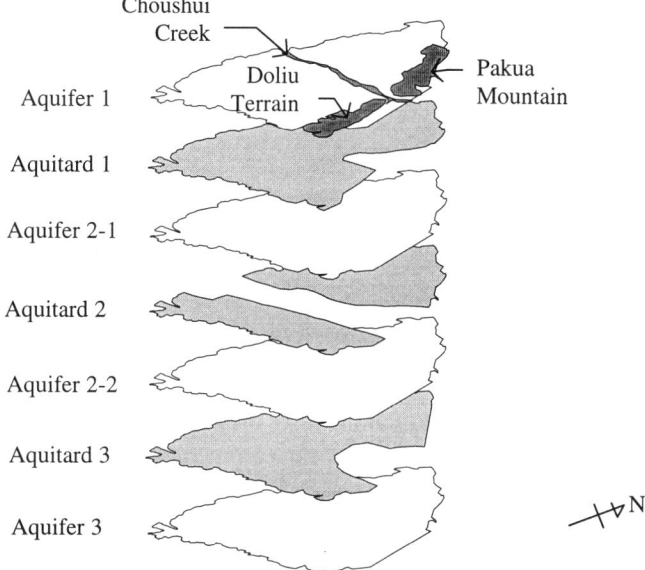

Figure 1. Layered structure of the Choushui Creek Alluvial Fan

## **Optimization Model**

The optimization model is used to find the optimal pumping rates that will allow compliance with the prescribed objectives of the Plan. The model is provided with the location of the pumping wells, data on the available groundwater and surface water, and data on the current demand. Taking as an indicator of the groundwater recovery, the extent of the area in which the groundwater level is lower than the mean sea water level is examined. If the size of the area decreases, the groundwater condition is considered improved. Therefore, the ultimate objective of the optimization model is to reduce the extent of the area in which the groundwater level is lower than the mean sea water level. The mathematical form of this objective is expressed in terms of the number of model cells with negative groundwater level. That is,

$$\min Z_1 = \sum_{i=1}^{N} \delta_i$$

$$\delta_i = \begin{cases} 1, & h_i < 0 \\ 0, & h_i \geq 0 \end{cases} \quad i = 1, 2, \ldots, N$$

$N$ = total number of model cells in the simulation model.

The second objective of the optimization model is to minimize the total cost of the water supply:

$$\min Z_2 = \sum_{i=1}^{2} \sum_{j=1}^{N_i} c_{i,j} q_{i,j}$$

where  $N_i$ = number of supply nodes for source $i$
  $c_{i,j}$ = unit cost of supply node $j$ of source $i$
  $q_{i,j}$ = supply rate of supply node $j$ of source $i$
  $i$ = source index, 1: groundwater, 2: surface water
  $j$ = supply node index

As the unit cost of groundwater use is lower than that of surface water use for the study area and since groundwater use usually increases the number of model cells with negative groundwater level, the two objectives conflict with each other. A weighting method is used to generate the non-inferior solutions, and a trade-off table is constructed.

Here, the first objective contains a binary variable, whose value is derived from the groundwater simulation results. Therefore, the optimization model has to work with a simulation model to determine the response of the aquifer to the current estimates of the pumping rates. A variety of groundwater models with similar mechanism have been developed by previous researchers. Gorelick (1983) classified such groundwater optimization models as based upon either the embedding method or the response matrix approach. Embedding methods incorporate the response equations of the aquifer system as constraints in the optimization model. Examples include those of Willis and Liu (1984) and Willis and Finney (1985). This embedding method is computationally expensive because it embeds the entire system of finite difference equations in the constraint set. On the other hand, the response matrix approach yields incomplete information regarding system functioning, though it is computationally much less demanding. However, iteration is generally required for convergence. In the methodology developed in the present study, the optimization procedure is linked directly to the flow simulation. The optimal search proceeds in a manner similar to that of the response matrix approach. In every iteration, the responses of the objective function values to the current estimates of the decision variables are evaluated, and the information is used to guide the direction of the search. Since the decision variable pertaining to objective function $Z_1$ is nonlinearly related to the decision variables, the optimization problem is a nonlinearly constrained problem with linear objective function. Various mathematical techniques have been developed to solve nonlinear optimization problems (Gill et al., 1981; Willis and Yeh, 1987). MINOS (Murtagh and Saunders, 1987) is used to solve the optimization portion of the management model.

## Simulation Model

A groundwater simulation model has been developed by Yang and Yeh (1997) for the aquifer system of the Choushui Creek Alluvial Fan. The groundwater system of the Choushui Creek Alluvial Fan is divided into 4 horizontal aquifers and 3 discontinuous aquitards, and the flow is simulated by MODFLOW developed by McDonald and Harbaugh (1988).

The basic assumptions for the groundwater system are as follows: 1) The system has a layered structure, and the vertical velocity component within each layer is small. 2) All aquifers are confined except for the top layer, which is unconfined. 3) The aquifers are heterogeneous but assumed to be isotropic. 4) The groundwater system has no-flow boundaries.

Pumping contributes to the majority of the outflow from the system. Because of insufficient data, the simulation model is calibrated against pumping and recharge rates, as well as against the aquifer parameters. Prior information from borehole data is incorporated in the parameter estimations. The error criterion used for the model calibration is the sum of the squared errors between observed and simulated head values.

The 3-dimensional movement of groundwater with a constant density through porous media in a saturated zone is described by the following partial differential equation:

$$\frac{\partial}{\partial x}\left(K_{xx}\frac{\partial h}{\partial x}\right) + \frac{\partial}{\partial y}\left(K_{yy}\frac{\partial h}{\partial y}\right) + \frac{\partial}{\partial z}\left(K_{zz}\frac{\partial h}{\partial z}\right) - W = S_s\frac{\partial h}{\partial t}$$

where $K_{xx}$, $K_{yy}$, and $K_{zz}$ are hydraulic conductivity values along the $x$, $y$, and $z$ coordinate axes, which are assumed to be parallel to the major axes of hydraulic conductivity [L/T]; h is the potentiometric head [L]; $W$ is a volumetric flux per unit volume, representing the source and/or sink of water [$T^{-1}$]; $S_s$ is the specific storage of the porous media [$L^{-1}$]; and $t$ is time [T]. The flow domain is discretized in both horizontal and vertical directions. A nodal spacing of 2 km is deemed sufficient to account for the spatial variation of the various model parameters, resulting in 8,000 model cells in total.

To estimate aquifer parameters, this study adopts the zonation method, in which the flow region is divided into a number of sub-regions or zones and a constant parameter value characterizes each zone. The calibrated storage coefficients for the confined aquifers range from 0.0001 to 0.005, and transmissivity values range from $10^{-6}$ to $10^{-2}$ m$^2$/s. The total pumpage after calibration, which is approximately $10.42 \times 10^8$ m$^3$/year, exceeds the annual recharge by $1.45 \times 10^8$ m$^3$/year. It is assumed that the safe yield of the aquifer equals the annual natural recharge and will be used as the upper bound of groundwater use in the constraint set of the optimization model.

## Summary

With the goal of mitigating the groundwater overdraft, thereby preserving groundwater resources and reducing land subsidence in the Choushui Creek Alluvial Fan in Taiwan, a management model for conjunctive use of groundwater and surface water was developed for a coastal alluvial groundwater basin with extensive groundwater overdraft. The objective of the model is to minimize the area where the groundwater level is lower than the mean sea level and to minimize the water supply cost, while meeting water demand, subject to the availability of the surface water and the groundwater.

The groundwater simulation model is linked with the optimization model, which is solved using MINOS. The groundwater system of the study area is heterogeneous and multi-layered. The 3-D groundwater flow equation is solved by MODFLOW. The model calibration results show that the annual groundwater use exceeds the recharge by $1.45 \times 10^8$ m$^3$/year. Through the conjunctive use of groundwater and surface water, the overdraft problem can be mitigated to achieve the goals set out in the Land Subsidence Prevention Plan in Taiwan.

## References

1. Gill, P. E., Murray, W., and Wright, M. H. (1981). *Practical Optimization*, Academic Press, London, 401 pp.

2. Gorelick, S. M. (1983). "A review of distributed parameter groundwater management modeling methods," *Water Resour. Res.*, 19(2), 305-319.

3. McDonald, M. G., Harbaugh, A. W., (1988). *A Modular Three-Dimensional Finite Difference Ground-Water Flow Model.* USGS Open-File Report 83-875

4. Murtagh, B. A., and Saunders, M. A. (1987). *MINOS 5.1 User's Guide, Technical Report 83-20R,* Systems Optimization Laboratory, Department of Operations Research, Stanford University, Stanford, California, 118 pp.

5. Willis, R., and Liu, P. (1984). "Optimization model for ground-water planning," *J. Water Resour. Plng. and Mgmt.*, ASCE, 110(3), 333-347.

6. Willis, R., and Finney, B. A. (1985). "Optimal control of nonlinear groundwater hydraulics: Theoretical development and numerical experiments," *Water Resour. Res.*, 21(10), 1476-1482.

7. Willis, R., and Yeh, W. W-G. (1987). *Groundwater Systems Planning and Management,* Prentice-Hall, Englewood Cliffs, New Jersey, 416 pp.

8. Yang, S., and Yeh, W. W-G., *Formulation and Evaluation of Alternatives for Mitigating Groundwater Overdraft in Taiwan's Coastal Area (I)*, Water Resources Bureau, Taiwan, R.O.C. May, 1997.

# APPLICATION OF GROUNDWATER FLOW AND SOLUTE TRANSPORT MODELS FOR GROUNDWATER REMEDIAL DESIGN

Zafar Adeel[1], Charles Faust[1], Barry Lester[1], Todd Hagemeyer[2], and Ron Lantzy[3]

## ABSTRACT

Conventional approaches to remediation typically comprise source control and pump-and-treat, and are designed for aquifer restoration and plume containment. In this context, mathematical models are useful tools for setting goals and for optimizing design/performance of conventional groundwater remedies. Based on available data, three-dimensional groundwater flow and solute transport models can account for complex hydrogeologic conditions. The reliability of predictions from these models is dependent on level of hydrogeologic detail incorporated into the conceptual model, confidence in available site-specific data, and availability of system performance data. During implementation of remedial action, sensitivity analyses or formal models (e.g., linear optimization) can be used for system optimization. Case studies for two sites with fairly complex conditions are presented to illustrate these modeling capabilities and their use in goal setting, remedial design, and post-design optimization.

## INTRODUCTION

Remediation of groundwater contamination is often a complex problem which requires in-depth understanding of the hydrogeologic and subsurface conditions and the nature of contamination (Bouwer et al., 1988). This is true for conventional remediation technologies, such as pump-and-treat recovery. In the regulatory context, mathematical models are useful tools for setting goals (remedial action objectives). Additionally, modeling can play a key role in risk assessment by identifying contaminant migration pathways and potential receptors.

Comparison of various remediation strategies and design of remedial action often requires prediction of system performance. Greater confidence in flow and transport models has led to their increased use as such predictive tools. Based on remedial objectives and site conditions, model predictions can be used to select either plume containment or aquifer restoration, or both. Further, these models have also been used

---
[1] HSI GeoTrans, Inc., 46050 Manekin Plaza, Suite 100, Sterling, VA 20166.
[2] HSI GeoTrans, Inc., 1080 Holcomb Bridge Road, Bldg. 200, Roswell, GA 30076
[3] Rohm and Haas Company, Inc., Bristol, PA 19007

for optimization of remedial design to effectively meet the remediation goals. The objective of this paper is to demonstrate the applicability of three-dimensional models to the remedial design process and to show their effectiveness in optimization.

### REMEDIAL DESIGN USING MODELS

Once the remedial action objectives have been identified and accepted by the stake holders, numerical modeling can be used for system optimization for both the remedial design and the post-design analysis. In general, these steps may be followed in application of flow and transport models to remedial design:

- Model development and calibration against transient and steady-state data
- Risk assessment; identifying exposure pathways and potential receptors
- Goal-setting; identifying source areas and containment/restoration objectives
- Remedial design; using model predictions to define key design elements such as:
  - depth and location of recovery wells,
  - practicable recovery rates,
  - concentrations in pumped water that must be treated, recharged, or discharged,
  - monitoring necessary for performance evaluation, and
  - long-term contaminant mass recovery.
- Post-design optimization and cost minimization, based on the preliminary system performance and new hydrogeologic information.

### RELIABILITY OF MODELS FOR COMPLEX HYDROGEOLOGIC SCENARIOS

The level of confidence in model predictions is generally correlated to the degree of knowledge about site hydrology and hydrogeology. The quality of site-specific data is also a concern; good-quality data can minimize the uncertainty associated with model predictions. In general, the longer the time-span of the available data the greater the confidence in modeling results. Typical sources of uncertainty are:

- Lack of site-specific flow and transport parameters
- Too few data points associated with the key hydrogeologic parameters
- Improper conceptualization of boundary conditions and sources/sinks
- Coarse delineation with insufficient detail of geologic formations
- Inability to identify or include local subsurface features like fractures and faults
- Improper description of heterogeneity and anisotropy in subsurface properties
- Insufficient historical information regarding seasonal aquifer behavior
- Lack of appropriate calibration data (steady-state and transient) for the model

Two success stories are presented here in which good quality, site-specific data were used to develop reliable three-dimensional models for sites with fairly complex hydrogeology. In the first case, groundwater flow and solute transport modeling were performed for an arsenic-contaminated site to obtain the optimum well locations and pumping rates. In the second case, three strategies for plume containment at a site were rigorously evaluated using a calibrated flow model and a linear optimization package.

## CASE STUDY 1. Arsenic Removal at a Superfund Site in Pennsylvania.

OVERVIEW

The Remedial Action (RA) program at the Whitmoyer Laboratories Superfund site (see Figure 1) includes a groundwater extraction and treatment system for arsenic (As) removal. The proposed design is based on aggressive pumping for mass reduction and containment of the central plume (As > 1,000 ppb) and relies on natural attenuation for the exterior plume (As > 50 ppb). The effectiveness of the extraction system can be evaluated in terms of arsenic mass removal and decrease in arsenic levels to the remedial goal of 50 ppb within the central plume. A numerical, three- dimensional groundwater flow model was developed as part of the Remedial Design (GeoTrans, 1996a); the model was critically evaluated and approved by the U.S. EPA. Arsenic transport modeling was used to quantitatively evaluate the RA extraction system.

HYDROGEOLOGIC FRAMEWORK

The site is located in the Lebanon Valley in southeastern Pennsylvania, which is underlain by Martinsburg shale in the north and alternating carbonate beds of the Cambrian and Ordovician Ages (Meisler, 1963). The site is underlain mostly by the Ontelaunee Formation. Structurally, the Lebanon Valley comprises a complex system of folds and faults, which along with secondary openings in rock formations govern groundwater flow. The transmissivity of the formations is highly variant, reflecting the heterogeneities within the lithologic units and the variability from one unit to another.

GROUNDWATER FLOW MODEL

The subsurface formations were represented in the model as a highly complex, heterogenous, unconfined aquifer. A porous-medium continuum model was considered appropriate for the fractured rock because: (a) the secondary openings are much smaller than the model scale and are sufficiently well-connected to transmit water in a manner analogous to a porous medium, and (b) delineation of all the subsurface fractures was neither feasible nor necessary, given the reasonable performance of the porous-medium analogous model.

The hydrogeologic complexity necessitated that each lithologic unit and subsurface feature (e.g. fractures, fault zones) be assigned a distinct zone and unique properties. The groundwater flow system was approximated using a grid with 146 columns and 129 rows. Non-uniform areal discretization allowed greater detail to be incorporated for the site area. The model was vertically discretized into 11 layers (top five layers: 25-ft thick; bottom six layers: 50-ft thick). Calibration against existing groundwater levels and on-site pump tests reduced the chance of a non-unique solution.

ARSENIC TRANSPORT MODEL DEVELOPMENT

The initial aqueous arsenic distribution for the transport model was based on the site-specific arsenic data and the known extent of the arsenic plume. The Triangulated Irregular Network (TIN) algorithm was used for this purpose. Vertically, the arsenic distribution was kept uniform within the shallow zone (layers one to five) and the mid-depth zone (layers six to ten). A numerical code, MODFLOWT (GeoTrans, 1996b), was selected for simulating arsenic transport. This model was calibrated against historical data from a pump-and-treat operation (1965-71), during which approximately 450,000 lbs arsenic was removed from groundwater.

KEY MODELING RESULTS

The calibrated transport model was used to simulate the RA extraction system. It predicted that the RA system will be successful in capturing greater than 95% of the arsenic discharge to the creek. The baseline simulation predicted recovery of about 31,000 lbs of arsenic from groundwater over a five-year period. After an initial rapid removal of arsenic in the first two years, the removal rate is predicted to decline asymptotically. Initially, the average arsenic concentration is about 16 ppm, which reduces to 4 ppm after five years.

The model also predicts that the central plume will be somewhat reduced in extent after five years, while exterior plume remains relatively unchanged. Further, the exterior plume is predicted to remain stable even after 30 years. It can, therefore, be surmised that the intrinsic attenuation processes are sufficient for containment of the exterior plume. Because the arsenic removal rate appears to be asymptotic, it is unlikely that the remedial goal for the central plume (As < 50 ppb) will be achieved within five years. Essentially consistent results were obtained in a sensitivity analysis in which the retardation factor was varied from 1 to 6.

In the post-design stage, a variable sequence of pumping to optimize extraction performance may be evaluated, in which wells are turned off as the arsenic levels approach the asymptote and re-started for pumping when arsenic rebounds.

**CASE STUDY 2. Plume containment at Paducah Gaseous Diffusion Plant.**

OVERVIEW

A groundwater plume of dissolved trichloroethylene (TCE) and technetium ($Tc^{99}$) was identified at the Paducah Gaseous Diffusion Plant (PGDP), Kentucky (Hagemeyer et al., 1993). Groundwater extraction was selected as an interim remedial measure to prevent offsite migration of the 500-ft wide plume centroid, which had impacted the Ohio River and some residential wells (Figure 2). Three different plume containment alternatives were evaluated using a groundwater management code called MODMAN (GeoTrans, 1992): (a) groundwater extraction, (b) groundwater extraction with injection, and (c) groundwater extraction in conjunction with a barrier wall.

HYDROGEOLOGIC FRAMEWORK

PGDP is located at the northern apex of the Mississippian embayment, a gentle syncline of Paleozoic bedrock infilled with unconsolidated sedimentary deposits. These deposits are differentiated into an upper silty-clay facies, called the Upper Continental Recharge System (UCRS), and a lower gravel-sand facies, called the Regional Gravel Aquifer (RGA). The RGA forms the principal water-transmitting zone at PGDP; its hydraulic conductivity (100-400 ft/day) is greater by at least two orders of magnitude than the UCRS or the unit below it (McNairy formation). Groundwater flow direction in the RGA is north to northeasterly towards the Ohio River, which is consistent with the observed plume configuration.

## Groundwater Flow Model

A three-dimensional flow model was developed using MODFLOW (McDonald and Harbaugh, 1988) and calibrated against data from more than 100 monitor wells. The model domain is shown in Figure 2. The model grid has 91 columns and 117 rows with 50-ft grid spacing in the area chosen for plume containment measures. Vertically, the model has three layers which represent the UCRS, the RGA, and the McNairy formation. Unique boundary conditions and aquifer parameters were assigned to each layer. An infiltration recharge value of 4.7 in/yr was assigned to the UCRS.

## Design Optimization

The remedial goal was to control the plume centroid at a proposed containment line (Figure 2). The alternatives listed in Table 1 were simulated using MODMAN. The barrier wall design for Alternative **c** contained a 50-ft wide "gate" at its center as part of a "funnel-and-gate" system. For all the alternatives, the maximum yield for each well was limited to 250 gpm. The imposed gradient was approximately the same magnitude as the natural gradient (0.0007) but opposite in direction.

**Table 1.** Total groundwater extraction rates (gpm) required for plume containment.

| Alternative Description | | Number of Extraction Wells | | | |
|---|---|---|---|---|---|
| | | 1 | 2 | 3 | 4 |
| a | Extraction only | infeasible | 245 | 193 | 187 |
| b | Extraction with injection[1] | 196 | 125 | 114 | 111 |
| c | Extraction with barrier wall | 45 | 44 | 43 | 43 |

[1] Assumes number of injection wells equals the number of extraction wells.

Increasing the number of wells indicates a tradeoff between installation/maintenance costs and water treatment costs, i.e., the extraction rate decreases as the number of wells increases. Supplementing extraction with injection can result in a substantial increase in efficiency over Alternative **a**. Alternative **c** is clearly the most efficient plume capture method, in which containment is achieved with just one well and a 60% flow reduction over the next best alternative (**b-4**). For Alternative **c**, efficiency is insensitive to the number of wells and the flow field is least impacted. However, after performing a rigorous cost comparison, Alternative **a-2** was selected.

## Conclusions

For both the sites, modeling helped in defining the remediation goals and selecting the optimum remedial design (based on performance <u>and</u> cost). In general, the model predictions were found to be reliable when compared against system performance. For the Whitmoyer site, modeling provided further insights into post-design optimization.

## References

Bouwer, E., Mercer, J., Kavanaugh, M., DiGiano, F. (1988) "Coping with groundwater contamination." *J. Water Poll. Cont. Fed*, 60(8), 1415-1427.

GeoTrans, (1992) *MODMAN: An Optimization Module for MODFLOW, Version 2.1.*

GeoTrans (1996a) *Remedial Design Groundwater Flow Model*. Submitted to U.S.

Environmental Protection Agency, August 29, 1996.

GeoTrans (1996b) *MODFLOWT: A Modular Three Dimensional Flow and Transport Model.*

Hagemeyer, T.R., Andersen, P.F., Greenwald, R.M., and Clausen, J.L. (1993) "Evaluation of alternative plume containment designs at the Paducah Gaseous Diffusion Plant using MODMAN, a well pumpage optimization module for MODFLOW." 1993 Ground Water Modeling Conference, Golden, Colorado.

McDonald, M.G., and Harbaugh, A.W. (1988) "A modular three-dimensional finite-difference groundwater flow model." Techniques of Water Resources Investigations 06-A1, USGS.

Meisler, H. (1963) *Hydrogeology of the carbonate rocks of the Lebanon Valley, Pennsylvania*, Pennsylvania Geological Survey, 4th Ser. Water Resources Report 18.

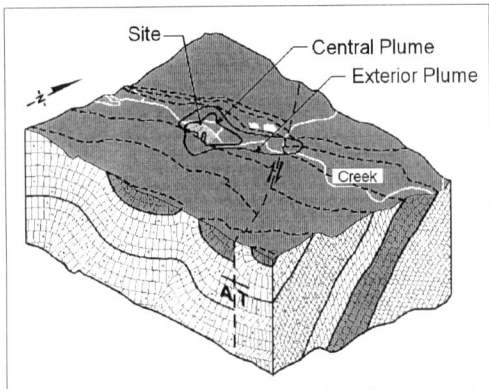

**Figure 1.** The area included in the Whitmoyer model, showing varying lithologic units (vertical exaggeration is approximately 1:3)

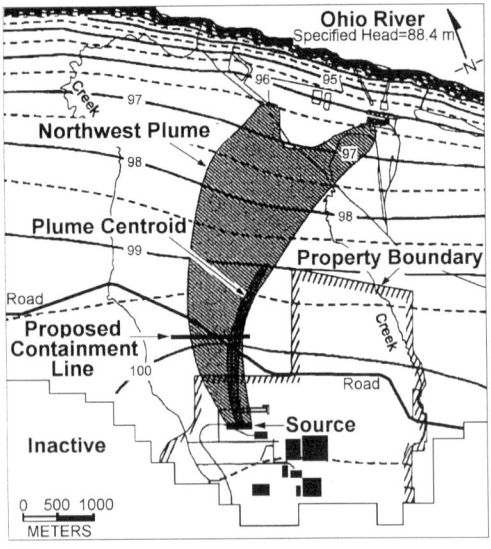

**Figure 2.** The modeled area for the PGDP site, showing the Northwest plume.

Selection of Remedial Action for a DNAPL Spill

Sehayek Lily[1], Terry Vandell[2], Brent Sleep[3], Calvin Chien[4]

Abstract

Field investigations, laboratory studies, and three models were used to evaluate remedial alternatives at a Gulf Coast site with a 1,2-Dichloroethane (EDC) dense non aqueous phase (DNAPL) spill. The spill resulted in contamination of a ditch, river sediments, and a shallow water bearing zone. Immediate remediation of shallow ditch and river sediments, followed by intensive field investigation and laboratory studies, and comprehensive evaluation of results, led to environmentally sound, cost effective remediation. Active remedial action focused on recovery of EDC DNAPL, and containment of dissolved phase EDC in the shallow water bearing zone. Results of multiphase modeling demonstrated that under the worst case scenario, vertical migration of EDC DNAPL would not extend to the deep regional aquifer. Natural attenuation through intrinsic biodegradation was demonstrated and used to evaluate the fate of dissolved phase EDC not contained under the worst case scenario. Results of field investigations 4 years after the release support modeling results, and demonstrate that the remedial actions selected for the site are effective.

Introduction

In March 1994, approximately 570 m$^3$ (150,000 gallons) of EDC DNAPL were spilled from a buried pipe line leak. This resulted in ditch and shallow soil and groundwater (< 13 meter) subsurface contamination of a Gulf Coast site.

---

[1]Assistant Professor, Penn State Great Valley, 30 East Swedesford Road, Malvern, PA 19355.
[2]Senior Environmental Consultant, Dupont, 1000 S. Pine, Ponca City, OK 74602.
[3]Associate Professor, University of Toronto, 35 St. George Street, Toronto, Ontario, Canada M551A4.
[4]Environmental Fellow, Dupont Corporate Remediation, Barley Mill Plaza, Bldg 27, Routes 141/48, Wilmington DE 19805.

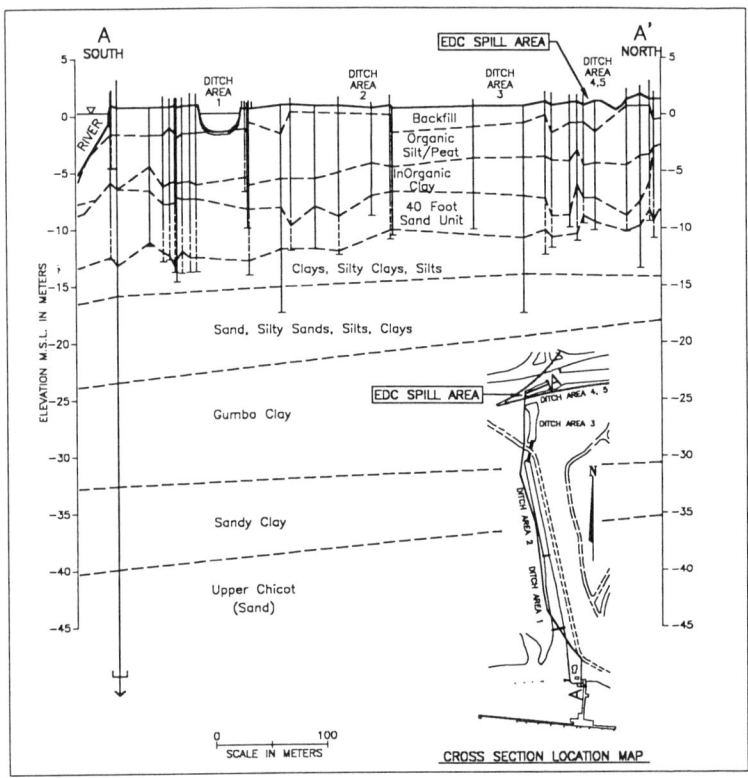

Figure 1. Geologic Cross Section.

There are three distinct water-bearing zones at the site. These zones include the 13 - meter Sand Unit (SU), the Interbedded Sand Unit (ISU), and the Upper Chicot Aquifer (UCA, a regional aquifer). The Upper Interbedded Unit (UIU) comprises the aquitard that separates the SU from ISU. The Gumbo Clay Unit (GCU) and the Sandy Clay (SC) comprise the aquitard that separates the ISU from the UCA.

Figure 1 shows the spill and ditch locations and geologic cross section in the EDC DNAPL contaminated area. Approximately 240 $m^3$ (63,000 gallons) of EDC DNAPL were recovered by excavating and vacuuming the ditch within six months of the spill. River dredging only recovered 10 $m^3$ (3000 gallons) of EDC DNAPL.

Subsequent to these interim remedial measures, the regulatory agency suggested removal of Organic Silt/Peat Unit (OSPU) and Inorganic Clay Unit (ICU) sediments located beneath the impacted ditch. EDC concentrations in impacted soil collected six months after the spill (from the OSPU and ICU) were analyzed to determine if additional excavation was necessary to minimize the potential vertical migration of EDC DNAPL. Results of these analyses indicated that although EDC concentration in the soils ranged from approximately 7,000 to 60,000 mg/kg, EDC DNAPL saturation in these units was below 0.028. Laboratory studies conducted at the University of Toronto demonstrated that residual saturation for these units ranged from 0.07 to 0.14 percent. Furthermore, pilot scale vacuum enhanced pumping for recovery of EDC was attempted in these units. Results indicated that even with the application of 0.5 meters (20 inches) of water vacuum, EDC DNAPL was immobile in these units.

Hydraulic containment in the SU is comprised of groundwater recovery from twelve SU wells with recovery rates of approximately 87 $m^3/d$ (cubic meters per day, 16 gpm (gallons per minute)). The quantity of EDC DNAPL recovery varied from about 0.8 $m^3/d$ (200 gpd (gallons per day)) during early recovery to about 0.04 $m^3/d$ (10 gpd) four years after the spill. To date approximately 150 $m^3$ (40,000 gallons) of EDC DNAPL and 150 $m^3$ (40,000 gallons) of dissolved phase EDC have been recovered in pumping from the shallow SU. Neither EDC DNAPL or dissolved phase EDC are encountered in the deeper water bearing units.

Field and Laboratory Investigation, Modeling and Results

Field investigation at the site included 29 cone penetrometer soil and groundwater sampling tests, installation of 56 monitoring wells, pump tests, high vacuum tests, slug tests, groundwater elevation and tidal - groundwater influence studies, measurement of product thickness in the SU, and groundwater and soil analyses for EDC and EDC degradation products. Results of these analyses were used to generate geologic cross sections, plot piezometric contours for each water bearing unit, plot dissolved phase EDC isoconcentration maps, determine the horizontal and vertical extent of EDC DNAPL, and calculate the hydraulic conductivity values for each formation.

Laboratory studies included: tri-axial permeability tests to determine hydraulic conductivity, batch adsorption tests with dialysis tubing (Allen-King et al, 1995) to determine adsorption coefficients, and a modified version of Salehzadeh and Demond (1994) dynamic method to generate capillary pressure curves.

Aperture and frequency characterization of micro fractures encountered in the UIU and GCU were determined based on laboratory studies (mercury injection porosimetry, x-ray tomography, and x-ray radiograph) conducted on core samples obtained from a nearby site.

Results of field investigation and laboratory studies were used to develop conceptual models for groundwater flow, transport of EDC DNAPL, and transport of dissolved phase EDC. Natural attenuation and intrinsic biodegradation were demonstrated through laboratory microcosm studies, EDC concentration reduction over time, and documentation of EDC degradation products. Numerical and analytical models were used to: confirm the conceptualization of the hydrogeologic system in three dimensions; simulate groundwater flow in the SU, ISU, and the UCA; determine sensitive aquifer parameters; determine average hydrogeologic properties for water-bearing zones and intermediate aquitards; assist in the conceptual design of the containment system for dissolved phase EDC in the SU; evaluate the potential for EDC DNAPL to reach the UCA under the worst case scenario; provide worst case scenario input for modeling dissolved-phase EDC migration in the UCA; and evaluate the worst case scenario risk posed by dissolved phase EDC at potential receptor points located in the UCA.

Groundwater flow was analyzed in three dimensions using MODFLOW. The groundwater flow model was calibrated and verified. Results of the model supported the conceptualized hydrogeologic system, the average values of aquifer parameters determined in field investigations and laboratory studies, and provided flow rate of 87 $m^3/d$ (16 gpm) for the conceptual design of the groundwater containment system in the SU. Results of the groundwater flow model were confirmed subsequent to implementation of the groundwater containment system in the SU.

Field investigations showed that EDC DNAPL in the subsurface was confined to the ditch area, and that the majority of the EDC DNAPL plume was oriented along the ditch. To minimize simulation time and memory, the conceptual model was simplified from a three-dimensional system to a two-dimensional vertical cross section system, using worst case assumptions with respect to vertical migration of EDC DNAPL. This conceptualization assumed a horizontal x-axis oriented along the ditch (section A-A' in Figure 1) and a vertical z-axis oriented along gravity. This simplification is conservative with respect to vertical migration of EDC DNAPL, since it assumes symmetry along the Y-axis (along the width of the ditch). Hence, the EDC DNAPL model assumes that there is no lateral flow of EDC DNAPL or dissolved-phase EDC across the ditch (y-axis). Vertical migration of EDC DNAPL in the aquitards which separate the ditch from the SU (see Figure 1) were not simulated since EDC DNAPL in these formation was found to be immobilized six months after the spill. EDC DNAPL was infiltrated directly into the SU using conservative assumptions with respect to quantity and duration of the spill. No EDC groundwater recovery was assumed, and the total EDC volume after six months was used, 400 $m^3$ (150,000 gallons). Other conservative assumptions in the EDC DNAPL model included orientation of formations and boundary conditions. The EDC DNAPL (Sleep and Sykes, 1993) model simulated vertical migration of EDC DNAPL using the equivalent porous media approach for the SU, ISU, SC and UCA, and the dual porosity approach for the UIU and GCU. Input data for the EDC DNAPL model included soil properties, fluid properties and reaction parameters which were based on results from field and laboratory investigations, the calibrated groundwater flow model, and literature values.

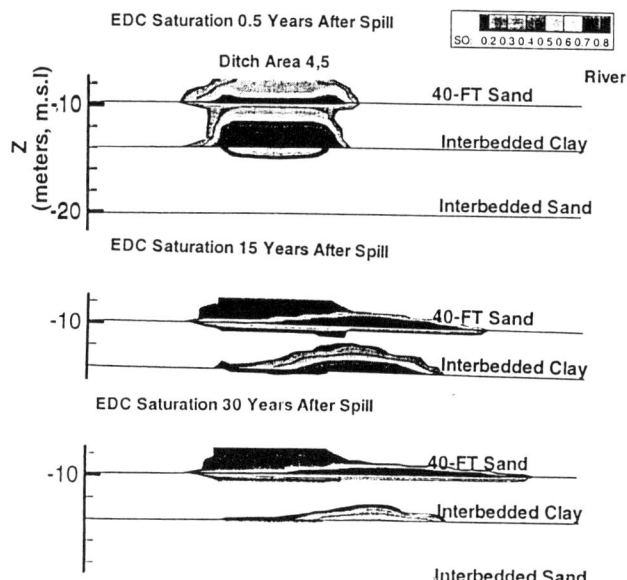

Figure 2 EDC Saturation in Matrix and Fracture 0.5, 15, and 30 Years After the Spill.

Fracture porosity in the UIU and GCU were computed using the Gerke and Van Genuchten's (1993) expression with average hydraulic conductivity (obtained from calibrated groundwater flow model), fracture hydraulic conductivity (calculated from fracture aperture and frequency), and matrix hydraulic conductivity (determined from tri-axial laboratory studies using ASTM methods). This expression yielded fracture porosity in the range of 0.0035 to 0.0088 for the UIU and GCU. A fracture porosity value of 0.0078 (calculated from fracture aperture and frequency) used in the EDC DNAPL model fall within the range calculated from Gerke and Van Genuchten's (1993) expression. Figure 2 shows the distribution of EDC DNAPL six months, five years, and thirty years after the spill. The length of the EDC DNAPL plume in the SU six months after the spill is within 10 % of the total length measured in the field at six months. Figure 2 shows that EDC DNAPL will not extend beyond the UIU. Field investigation four years after the EDC DNAPL pipeline spill confirm that EDC DNAPL is not encountered beyond the UIU. Figure 2 shows that although EDC DNAPL pool thickness exceeded 7 meters (23 feet), this thickness is insufficient to initiate EDC DNAPL entry into the ISU. This value is one order of magnitude greater than the minimum pool thickness calculated from the simple expression provided in the literature. This is attributed to dissolution of the small quantity of EDC mass that enters the ISU from the fractures in the UIU.

Worst case scenario horizontal transport of dissolved phase EDC in the UCA was modeled using the analytical solution of the two dimensional mass transport equation for a strip source in an infinite aquifer (Model 4, Cleary and Ung, 1994). Results of this modeling effort demonstrated that under the worst case scenario, no risk is posed at potential exposure points in the UCA.

Summary and Conclusion

Immediate response to a spill of approximately 150,000 gallons of EDC DNAPL from a buried pipe line, and a well planned long-term remediation approach resulted in effective recovery and containment of EDC DNAPL and dissolved phase EDC. Effectiveness of remediation was demonstrated through documented total recovery of about 540 m$^3$ (143,000 gallons) of EDC DNAPL, and groundwater monitoring results which show that both EDC DNAPL and dissolved phase EDC are confined to the shallow (< 13 meter) water bearing zone. These studies resulted in cost savings in excess of $ 20-30 million due to avoidance of unnecessary excavation of sediments with EDC DNAPL concentrations below residual saturation, and due to optimization of appropriate field and laboratory investigations and site specific remediation.

References

Allen-King, R. M., Groenevelt, H., and Mackay, D. M., Analytical method for sorption of hydrophobic organic pollutants in clay-rich materials, *Environ. Sci. Tech.*, 29, 148-153, 1995.

Cleary, R. W., and Ungs, M. J., *Princeton analytical models of flow and mass transport*, Waterloo, Ontario, Canada, Waterloo Hydrologic Software, 1994.

Gerke, H. H., and Van Genuchten M. T., A dual-porosity model for simulating the preferential movement of water and solutes in structured porous media, *Water Resour. Res.*, 22 (2), 305-315, 1993.

Salehzadeh, A., and Demond A. H, Apparatus for rapid automated measurement of unsaturated soil transport properties, *Water Resour. Res.*, 30(10), 2679-2690., 1994.

Sleep, B. E., and Sykes, J. F., Compositional simulation of groundwater contamination by organic compounds, 1: model development and verification, *Water Resour. Res.*, 29(6), pp 1697-1708, 1993.

Sleep, B. E., and Sykes, J. F., Compositional simulation of groundwater contamination by organic compounds, 2: model applications, *Water Resour. Res.*, 26(9), pp 1709-1718, 1993.

# Groundwater Modeling for Contaminated Site Closures - Successful Case Studies for Site Assessment, Remediation Design and Remediation Performance Evaluation

C.C. Chang[1]

## Abstract

Groundwater modeling is a proven technology used to support the closure of contaminated sites. Normal applications of this technology include site assessment, remediation design and remediation performance evaluation for a site closure. This document summarizes practical examples of applying groundwater modeling in conjunction with engineering calculations. These examples include a site assessment for an industrial site and remediation design for both an industrial site and a superfund site. At these project sites, only the dissolved-phase contaminants and limited amounts of residual Non-Aqueous Phase Liquids (NAPLs) were present. Based on the objectives of site closure, site characteristics and cost consideration, both analytical and numerical models were used in supporting these site closures. The objectives of site closure included 1) the removal of dissolved-phase contaminants in groundwater to required levels, and 2) at the same time, containment of the plume.

## Introduction

Groundwater is a valuable water resource in the United States and the world. To protect this valuable resource, assessment of the site conditions and corrective measures for a contaminated site and waste management facilities are required. However, most site assessments and corrective measures are costly. Therefore, the waste management and remediation strategy for a site needs to be formulated, and effective corrective measures need to be selected. The strategy formulation and corrective measures selection should significantly reduce the time and cost to manage and remediate a site.

The focus of this document is to present the practical application of groundwater modeling to formulate solutions to achieve the primary objectives of site closure. Discussions of closure strategy formulation, design parameter determination, and remedial scheme development through the use of the groundwater modeling are also presented in this document. Site conditions and available field performance data were applied to support the modeling. Based on the results of modeling and engineering calculation, the amount of groundwater to be treated and the effectiveness of contaminant removal were also determined. The results of the modeling also provided a guideline for construction, operation, maintenance, and monitoring the performance of the remedial system to complete the closure of the sites.

---
[1]Project Manager, Radian International LLC, Houston, Texas.

## Model Application for Site Assessment

**Site Conditions:** The project site is located along the south bank of a river across from an industrial facility. The site occupies approximately 200 acres and has been used for residential, cattle grazing, and farming purposes. Site stratigraphy can be divided into three transmissive zones at 40 feet, 60 feet, and 110 feet below grade. Each transmissive zone is separated by a clay stratum beneath the site.

The preliminary investigation results showed that dissolved organic constituents were present in the groundwater beneath the site in the First and Second Transmissive Zones. Based on these findings, additional site investigation and assessment were required to 1) define the lateral and vertical extent, and distribution of constituents of concern present in the subsurface media (soil and groundwater) at the site; and 2) evaluate the migration rates and directions of constituents of concern.

**Modeling and Site Assessment:** In order to increase the efficiency in conducting site investigation, a simple one-dimensional groundwater solute fate-and-transport model was used to estimate the leading edge of the plume. The results were used to determine the appropriate locations for groundwater sampling and to minimize the number of samples to be analyzed. Groundwater sampling was conducted through the installation of Cone Penetrometer Test (CPT) push-in well screen and investigation wells. This process provided significant savings on investigation cost because, traditionally, the sample locations were determined by setting a matrix with different spacing and required a large amount of samples.

The model simulated chemical migration governed by groundwater flow, mechanical dispersion, chemical retardation, and biological degradation. The governing equation of the model is as follows:

$$R_f \frac{\partial C}{\partial t} = Dx \frac{\partial^2 C}{\partial x^2} - Vx \frac{\partial C}{\partial x} - \lambda C$$

where:
- $C$ = the concentration of the contaminant in groundwater ($M/L^3$).
- $Vx$ = seepage velocity in longitudinal direction (L/T)
- $\lambda$ = biodegradation rate in groundwater (1/T)
- $t$ = time (T)  $x$ = flow distance (L)
- $Dx$ = dispersion coefficient ($L^2/T$)  $R_f$ = retardation factor (unitless)

The governing equation can be solved analytically by setting appropriate boundary and initial conditions with assumptions as follows: 1) a continuous contamination source exists; 2) no chemical reaction occurs among contaminants; and 3) transverse migration of chemicals is insignificant. The preliminary investigation results and values identified in the literature search were used for the fate-and-transport analysis to project the plume edge. Data collected by preliminary investigation were also used for model calibration. Table 1 lists all the major parameters for this analysis.

**Results:** Figure 1 shows the projected plume edge by groundwater modeling. A total of 19 CPT and 9 investigation wells were installed to define the plume boundary. Figure 2 shows the plume boundary determined by contaminants detected in groundwater samples. Based on the results of the field investigation, it was concluded that the groundwater modeling successfully projected the plume boundary at the site. Cost savings between the modeling-guided site investigation and traditional site investigation are shown in Table 2.

## Application for Corrective Measure Selection

**Site Conditions:** This 10-acre site is located near a river. The site was used for land disposal operations from the 1940's to 1960's. In 1990 a leak in an 8-inch underground fuel line was discovered approximately 130 feet inland from the river bank. The leaked oil was in fill material at the site. The leak was promptly repaired and 10 sumps were installed to help prevent migration of oil into the river. The fill material is 17 feet thick and very heterogeneous. It consists of crushed concrete, powdered lime, magnesium and organic sludges, crushed glass, black tar, carbon electrodes, shell spoils and clay soils. The base of the Fill Material Zone slopes toward the river. The groundwater flow direction within the zone is to the south, toward the river.

A Corrective Measures Study (CMS) was performed to evaluate remedial options that would protect human health and the environment from potential exposure to the constituents of concern associated with the site. This study included additional field investigative activities, a baseline Risk Assessment (RA) and a corrective measures evaluation. The RA assessed potential risks related to current site conditions and identified potential exposure pathways. The RA indicated that, with appropriate onsite restrictions, risks to human health and the environment in the site were within acceptable limits. Leak oil seepage into the river will be prevented by an impermeable barrier placed along the river bank. Leak oil would be removed using appropriately spaced recovery wells.

**Model Application:** Groundwater modeling was applied to support the configuration of corrective measures and their associated costs. An initial screening process was performed prior to applying the groundwater modeling. The initial screening process eliminates those options which obviously were not technologically feasible considering the site conditions under which they had operate, posed unnecessary health risks, or were cost prohibitive. After the initial screening process, the remaining technologies underwent a detailed evaluation based upon the following criteria: 1) technical feasibility (an in-depth evaluation), 2) effectiveness in achieving remedial objectives given the site conditions, 3) ability to achieve remedial objectives within a reasonable time-frame and 4) economic feasibility.

Corrective measures which passed the initial screening were evaluated by applying groundwater modeling. Those corrective measures included 1) hydraulic control with a free-phase recovery system and 2) subsurface physical control and free-phase recovery. A steady-state two-dimensional horizontal flow model, Flowpath 5.0 (Waterloo Hydrogeologic Software, 1994), was selected and applied for the model simulations. The governing equation for a two-dimensional, steady-state flow in heterogeneous, saturated, anisotropic, porous media, is:

$$\frac{\partial}{\partial x}\left(T_{xx}\frac{\partial h}{\partial x}\right) + \frac{\partial}{\partial y}\left(T_{yy}\frac{\partial h}{\partial y}\right) \pm Q(x,y) = 0$$

where:

$T_{xx}, T_{yy}$ = principal components of the transmissivity tensor ($L^2T^1$)
$Q(x,y)$ = volumetric fluxes of sinks (-) and sources (+) per unit surface area of aquifer ($LT^1$)
$h$ = hydraulic head (L)  $x,y$ = Cartesian coordinates

A finite difference method was employed to solve the governing equation for two-dimensional steady-state horizontal flow. A rectangular grid was superimposed over the map of the groundwater system to discretize the system into grid cells that were small compared to the spatial extent of the entire aquifer. The hydraulic conductivity of the Fill Material Zone for the modeling was 0.049 ft/day. Different scenarios were simulated to evaluate the alternatives for the corrective measures.

**Results:** Table 3 is a comparison of corrective measures based on the configurations determined by groundwater modeling. The results of modeling showed that, in order to hydraulically control and remove the hydrocarbon plume, 44 wells on 50-foot spacing or as many as 110 wells on 25-foot spacing would be required. Considering the large number of wells needed to intercept the plume, this would not be cost effective (Table 3). Therefore a hybrid system with physical control (vibrating beam slurry wall) and nine recovery wells was determined. Based on estimate cost, reliability and implementability (Table 3), this hybrid system configuration would be a cost-effective corrective measures.

**Application for Remediation Design and Performance Evaluation**

**Site Conditions:** This site was a 7.3 acre abandoned industrial waste disposal facility located east of Houston, Texas. A variety of chemicals and sludges were placed in the lagoon until 1972, when operations ceased. Groundwater contaminant plumes migrated to 600 feet south of the site in unconsolidated alluvial materials. Major groundwater contaminant plumes included dissolved benzene, vinyl chloride and naphthalene. Non-mobile DNAPL was also detected in the groundwater remediation area while implementing the groundwater remediation system. In 1983, the site was placed on the National Priority List by the U.S. Environmental Protection Agency (EPA) and became a superfund site.

Following seven years of investigation, the remedial action plan and engineering design were completed and submitted to the EPA in July 1990 and January 1991, respectively. The design included: bioremediation of sludges in the lagoon; dewatering of the lagoon and on-site treatment of waters; fixation of remaining sludges; pumpage, treatment and injection of groundwater; and groundwater remediation enhancement technologies (e.g., in-situ bioremediation). Currently, most of the active remediation activities at this site have been completed. A ten-year natural attenuation of site ground water cleanup is under the way.

**Model Application:** The focus of the application of groundwater modeling at this site was to support remediation strategy formulation and conceptual and detailed remedial system design. Remediation strategy formulation involved the design concept and consideration, the development of the remedial scheme, design parameters, estimate system cost and scenarios for upgrading the system. Modeling was also applied to evaluate the system performance and provide guidance for system upgrade.

The Modular Three-Dimensional Finite Difference Groundwater Flow (MODFLOW) model (McDonald and Harbaugh, 1988) was used to determine the configuration of remedial system. Different scenarios were applied to optimize the numbers and locations of injection and withdrawal wells and to predict withdrawal and injection rates for treatment plant design. The configuration included the following actions:

- hydraulically capture all contaminated groundwater,
- minimize the size of hydraulic dead spots,
- maximize groundwater flow velocity,
- minimize contaminant travel path lengths to points of removal,
- establish an inward hydraulic gradient across the property line, and
- minimize flexibility for future upgrade.

The Random-Walk solute transport code (Prickeft, et al., 1981) was modified to allow 1) modeling hysteresis of adsorption and desorption and 2) tracking of the distribution of the contaminants in both the aqueous and solid of a transmissive zone. After the configuration was determined, the fate-and-transport model was used to determine other design parameters listed as follows:

- optimal well/trench placements,
- the schedule and collection/reinjection rates,
- the duration of operation
- anticipated contaminant concentrations in the recovered groundwater, and
- the enhancement required.

A conceptual model which applied mass balance analysis was also developed to determine the amount of contaminant removal which can be achieved by flushing of clean water. The mass balance analyses were also used to evaluate the effectiveness and possible enhancement needed.

**Results:** The system configuration included approximately 180 injection wells and pumping wells at various rates as designed. The mass balance analyses reflected that the contaminants with high distribution coefficients (e.g., naphthalene) required a long period of time to achieve the ground water clean-up criteria at the site. The flushing test and the model simulations also suggested that an enhanced remediation technique were required to improve the efficiency of aquifer restoration at the site (Chang, et. al., 1992).

The initial phase construction for the shallow aquifer/subsoil remediation began in May-1991 and were completed in December 1991. As planned in the system design, subsequent refinements to the system (i.e., additional wells, in-situ bioremediation) were added, in order to enhance and improve aquifer remediation at the site. A total of approximately 375 recovery, injection, and monitoring wells were installed and operated for groundwater remediation at the site. The ground water in-situ bioremediation and conventional pump and treat system were operational from February 92 to December 95.

## References

Franz,T and Guiguer, N., 1994, FLOWPATH, Waterloo Hydrogeologic Software.
McDonald, M.G. and Harbaugh, A.W., 1988, "A Modular Three-Dimensional Finite Difference Groundwater Flow Model," Techniques of Water Resources Investigations of the United States Geological Survey, Book 6, Chapter A1.
Prickett, T.A., Naymik, T.G., and Lonnquist, C.G., 1981, "A Random-walk Solute Transport Model for Selected Groundwater Quility Evaluation," Illinois State Water Survey, B. 65.
Chang, C.C., Moncure, G.K., and Dorrance, D.W., 1992, "Ground Water Remedial Action Design at the French Limited Superfund Site - An Application of Ground Water Flow and Transport Models", Proceedings of The Solving Ground Water Problems With Models NWWA Conference, Dallas, Texas.

TABLE 1 ESTIMATED PARAMETERS FOR MODELING

| Corresponding CPTs ID | Indicating Chemicals | Mean Concentration (ppm) | Concentration in CPTs at Site (ppm) | Distance (feet) | *Estimated Distance to Source (feet) | *Estimated Seepage Velocity (ft/day) | Estimated Dispersivity (feet) | Estimated Retardation Factor |
|---|---|---|---|---|---|---|---|---|
| CPT-2 | EDC | 2.17 | 0.47 | 800 | 300 | 0.22 | 1.5 | 1.8 |
| CPT-3 | MC | 40.17 | 0.67 | 750 | 50 | 0.22 | 1.5 | 1.3 |
| CPT-3 | EDC | 13.23 | 0.046 | 750 | 30 | 0.2 | 1.5 | 1.8 |

EDC - 1,2 Dichloroethane
MC - Methylene Cloride
Ppm - Parts per million

* From reference wells to source

TABLE 2 COST SAVINGS FOR MODELING-GUIDED SITE INVESTIGATION

|  | Number of CPTs Required | Number of Groundwater Samples | Investigation[2] Costs (CPT) | Investigation Costs (Laboratory) | Investigation Costs (Subtotal) |
|---|---|---|---|---|---|
| Matrix Approach (100 ft spacing) | 288 | 576 | $230,400. | $201,600. | $432,000. |
| Matrix Approach (200 ft spacing) | 144 | 288 | $115,200. | $100,800. | $216,000. |
| Modeling Approach | 19 | 38 | $16,000. | $13,300. | $29,300. |
| Difference (Comparison between modeling and 200 ft spacing) | 125 | 250 | $99,200. | $87,500. | $186,700. |

1: Two samples per location (one for volatile compounds and one for semi-volatile compounds)
2: Average 60 feet deep holes (include other field costs)

Figure 1 PROJECTED PLUME BOUNDARY BY MODELING     Figure 2 DETECTED PLUME BOUNDARY

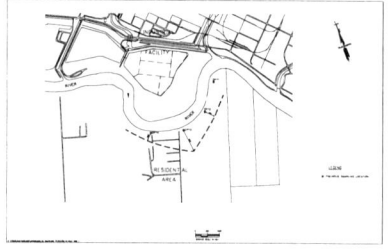

Table 3
CORRECTIVE MEASURE ALTERNATIVES COMPARISON

| Corrective Measure Alternative | Implementability | Performance | Reliability | Safety | Quantity | Installation Cost (unit price) | Pump Costs[6] (unit price) | Cumulative Cost |
|---|---|---|---|---|---|---|---|---|
| Hydraulic Control and Free-Phase Hydrocarbon Recovery | | | | | | | | |
| Vertical wells | fair[1] | poor[1] | good[1] | good | 44[4] | $1,150 | $8,625 | $570,500[5] |
| Horizontal wells | very poor[1] | poor | poor | good | 3 | $117,000 | $8,625 | $450,000[5] |
| Recovery trench | poor[1] | poor | poor | good | 1 | $215,000 | $8,625 | $305,000[5] |
| Physical Control and Free-Phase Hydrocarbon Recovery[7] | | | | | | | | |
| Sheet Pile Wall | fair | excellent | excellent | good | 17,900 SF | $19 per SF | NA | $338,300 |
| Vib. Beam Slurry wall | excellent | excellent | excellent | good | 17,900 SF | $7.35 per SF | NA | $131,500 |
| HDPE Liner[2] | poor[3] | poor | excellent | good | 17,900 SF | $2.10 per SF | NA | $30,870 |

NA Not Aapplicable
SF Square Foot
1  Very poor, poor, fair, good, excellent: describes the increasing progression of the technology's ability to meet its intended function
2  60-mil liner
3  Ability to install the liner given the soil conditions at the site
4  Average total depth of each vertical well in the vicinity of the hydrocarbon plume would be approximately 15 feet bgs; assumed LNAPL recovery only
   Vertical well spacing on 50-foot centers, vertical well diameter would be 8 inches
5  Includes estimated costs for installation of an onsite groundwater treatment system; does not include soil disposal costs
6  Includes water table depression pump and a scavenger pump
7  Same recovery system would be used with all three physical barriers, therefore recovery system costs have not been included with the costs of the different barriers
8  Number of drums of the mixture (microbes, nutrients, oxygen) used over a 6 month period, after which an evaluation must be made to determine if the
   microbes are self-sustaining

## Operational Definition of NAPL from Soil Analyses

Roger H. Page[1], Michael L. Watkins[2], Edith Sieber[3], Mohammad F.N. Mohsen[1]

### Abstract

At many industrial and disposal sites, delineating nonaqueous-phase liquid (NAPL) in the vadose and saturated zones is important in planning technically sound remedial actions. This paper describes the Threshold Ratio Summation (TRS) method for screening total concentrations in a soil sample to determine whether a mixture of NAPL was present. For each chemical, the method uses a NAPL threshold concentration, the highest total concentration that a sample could hold in the absence of NAPL. That threshold is the sorbed, dissolved, gaseous mass the sample could hold if only one contaminant were present. The TRS is a sum of the fractions of thresholds for all of the sample's analyzed constituents. Based on a Raoult's-Law-type relationship that estimates effective solubility, the TRS is less than 1 unless NAPL is present. Results are shown for two of the sources at a chemical manufacturing site.

### Introduction

Organic chemicals present as NAPL in soil are a persistent source of long-term ground water contamination. Residual NAPL in soil may be the dominant future source of dissolved contaminants and the main obstacle to a quick cleanup.

The U.S. Environmental Protection Agency (USEPA) asked a site owner to develop an "Operational Definition of NAPL" for an investigation under the Comprehensive Environmental Response, Compensation, and Liability Act (CERCLA). The operational definition was needed to enable investigators (1) to determine whether a specific soil sample likely contained NAPL, (2) to delineate NAPL-containing zones, and (3) to achieve these goals using data from substantial field work already completed. The Threshold Ratio Summation (TRS) method subsequently developed, an extension of the work of Feenstra et al. (1991), was described by Watkins et al. (1996). The TRS method identifies NAPL based on total concentrations

---
[1] ENVIRON Corporation. 214 Carnegie Center, Princeton, NJ 08540
[2] ECKENFELDER INC., 440 Franklin Turnpike, Mahwah, NJ 07430
[3] Ciba Specialty Chemicals Corporation. P.O. Box 71, Toms River, NJ 08754

of organic chemicals measured in soil samples. The TRS assumptions are consistent with the ones Mott (1995) and Mariner et al. (1997) employed to calculate NAPL saturations in soils or dissolved concentrations in NAPL-free soils. In fact, the latter authors independently proposed a method equivalent to the TRS for detecting a non-zero NAPL saturation. This paper (1) describes the method's physical basis, (2) discusses the parameters required, and (3) summarizes some trials made during the CERCLA investigation.

## Methodology

Each chemical's NAPL threshold is based on the sorption capacity of the soil particles, the dissolution capacity of the pore water, and the vapor capacity of any gas phase in the pores. The method of Feenstra et al. (1991), which is included in USEPA (1992) guidance, estimates the dissolved concentration in pore water ($C_{w,i}^{EST}$, mass of constituent i per unit volume of water) from the total concentration measured in soil ($C_{t,i}^{MEAS}$, mass of constituent i per unit mass of dry soil) by

$$C_{w,i}^{EST} = \frac{C_{t,i}^{MEAS}}{K_{T,i}} \qquad (1)$$

Here $K_{T,i}$ is a shorthand notation for $K_{d,i} + \phi_w/\rho_b + H_{c,i} \phi_a/\rho_b$, the theoretical ratio of total concentration to dissolved concentration in a NAPL-free soil sample; $K_{d,i}$ represents the soil-water partition coefficient of compound i; $\rho_b$ represents the soil's dry bulk density; $\phi_w$ represents the volume of water per unit bulk volume of soil; $\phi_a$ represents the volume of gas per unit bulk volume of soil; and $H_{c,i}$ represents the dimensionless Henry's Law constant of compound i. The ratio $\phi_w/\rho_b$ is the volume of water per unit mass of dry soil. The term $H_{c,i} \phi_a/\rho_b$ expresses the sample's capacity for mass in any vapor phase present. A simple, convenient way to check for a pure-phase NAPL is to look for $C_{t,i}^{MEAS} > K_{T,i} S_i$, where $S_i$ is the pure-phase solubility limit of constituent i. The product $K_{T,i} S_i$, which represents the NAPL threshold concentration in soil for constituent i (dry-mass basis), can be calculated for each constituent in a particular soil type.

When compound i is part of a mixed NAPL, its effective solubility limit $S_i^e$ can differ from the pure-phase solubility $S_i$. The relationship used here to calculate the effective solubility is one discussed by Shiu et al. (1988) and used by Feenstra et al. (1991), among others. It is analogous to Raoult's Law of vapor pressures.

$$\frac{S_i^e}{S_i} = X_i \qquad (2)$$

Here $X_i$ represents the mole fraction of compound i in the mixed NAPL, that is, moles of compound i per mole of NAPL. Mackay et al. (1991) describe the physical and chemical relationships and assumptions underlying that equation. It has also been evaluated using experimental results (Leinonen et al. 1973, Banerjee 1984, Feenstra 1990, Yalkowsky and Banerjee 1992, Feenstra 1997). Because $K_{T,i} S_i$ is a pure-phase NAPL threshold and $K_{T,i} S_i^e$ is an effective NAPL threshold, an effective threshold has the same ratio to a pure-phase threshold as an effective solubility has to a pure-phase solubility in Equation (2). Within the water phase, each constituent's concentration will be below its effective solubility limit given by Equation (2) unless that water phase was in equilibrium with a NAPL. If a sample of ground water has measured concentrations ($C_{w,i}^{MEAS}$) at the effective solubility limits ($S^e_i$), then Equation (2) implies that

$$\frac{C_{w,i}^{MEAS}}{S_i} = \frac{S^e_i}{S_i} = X_i \quad \text{for each constituent i, if water was in contact with NAPL} \quad (3)$$

The fraction of solubility on the left-hand side of Equation (3) offers an estimate of the mole fraction of a NAPL constituent or a lower-bound estimate when the dissolved concentrations are diluted below the effective solubility limits. The mole fractions of the various constituents of a NAPL sum to 1 by definition, so the various dissolved constituents in water in contact with that NAPL should have measured fractions of solubility that sum to 1. If only part of the water sample or none of it was in contact with NAPL, the summed fractions of solubility should be less than 1, that is,

$$\text{summed fractions of solubility} = \Sigma_i \frac{C_{w,i}^{MEAS}}{S_i} < 1 \quad \text{if NAPL is absent} \quad (4)$$

Theoretically, that sum would not rise above 1 unless the liquid tested included NAPL, a rare occurrence for ground water samples. Thus a practical screening of water samples would use a criterion lower than 1 for the summed fractions of solubility to identify samples that might indicate NAPL.

To screen total soil concentrations instead of aqueous concentrations, Equation (2) can be used to derive the following inequality:

$$\frac{C_{t,i}^{MEAS}}{K_{T,i} S_i} < \frac{K_{T,i} S^e_i}{K_{T,i} S_i} = X_i \quad \text{for each constituent i if NAPL is absent} \quad (5)$$

Estimating $X_i$ for a soil sample could be accomplished by a program such as SOILCALC (Mott 1995) or NAPLANAL (Mariner et al. 1997), but neither is necessary for simple delineation of NAPL. Summing the threshold ratios in Equation (5) for the various constituents gives the following <u>threshold ratio summation</u> for a NAPL-free soil sample.

$$\text{threshold ratio summation (TRS)} = \Sigma_i \frac{C_{t,i}^{MEAS}}{K_{T,i} S_i} < 1 \quad \text{if NAPL is absent} \quad (6)$$

Unlike a typical ground water sample, a soil sample can retain NAPL. Thus the TRS can rise well above 1 and thereby give an unequivocal indication that NAPL was present. In practice, a TRS within, say, an order of magnitude of 1 typically warrants a closer examination of the associated laboratory data to determine how to classify the sample. Considering TRS values lower than 1 helps to account for the approximate nature of laboratory analytical results and the thresholds, the possible absence of a NAPL constituent from the target list of the analysis, and the possible nonuniformity of NAPL residual in the sample.

## Calculating Pure-Phase NAPL Threshold Concentrations

The estimate of $K_{T,i}$ for the pure-phase threshold includes the capacities of up to three soil phases -- sorption capacity represented by $K_{d,i}$, dissolution capacity expressed by $\phi_w/\rho_b$, and the capacity mass in the vapor phase expressed by $H_{c,i}\ \phi_a/\rho_b$. The second and third are typically easier to estimate with satisfactory accuracy than the first; $K_{d,i}$ typically makes up the majority of $K_{T,i}$ and has the widest range of uncertainty.

The volume of water per unit mass of dry soil $\phi_w/\rho_b$ can be expressed as $(1-f_s)/(f_s\ \rho_w)$, where $f_s$ represents the fraction of solids (mass of solids per unit mass of soil, typically reported as percent solids) and $\rho_w$ represents the density of water. If $\phi_a/\rho_b$ is needed (for samples from the vadose zone with significant volatility, say, $H_c > 0.03$), it can be estimated by subtracting $\phi_w/\rho_b$ from the combined volume of liquid and gas per unit mass of dry soil, $n/[(1-n)\ \rho_s]$. Here n represents soil porosity, which may be estimated based on grading and packing, and $\rho_s$ represents the soil particle density, which is usually in the narrow range of 2.5 to 2.8 kg/L. The pore capacity of the saturated zone is roughly 0.2 to 0.3 L/kg; for the vadose zone it is typically 0.03 to 0.3 L/kg, depending on the compound's volatility.

The product $S_i\ K_{d,i}$ represents the upper limit of sorbed concentration in the case of a pure-phase NAPL. In aquifers containing a fraction of organic carbon ($f_{oc}$) greater than about 0.001 grams of carbon per gram of soil, sorption is believed to occur primarily in material containing organic carbon (Schwarzenbach and Westall 1981). Often, a constituent's $K_{d,i}$ can be estimated reasonably as the product of the constituent's organic carbon partition coefficient ($K_{oc,i}$) with the soil's fraction of organic carbon ($f_{oc}$).

If a soil sample's concentrations are expressed on a wet-mass basis, either the corresponding threshold would be reduced to $f_s\ K_{T,i}\ S_i$ or the concentrations would be scaled up by a factor of $(1/f_s)$. If a sample's concentrations are stated in terms of wet mass, $f_s$ might not have been measured; in that case, $f_s$ can be derived from the best estimate of $\phi_w/\rho_b$ using $[1 + \rho_w\ \phi_w/\rho_b]^{-1}$.

## Results

The NAPL identification method described above was applied to total soil concentrations at two former wastewater equalization basins of a chemical manufacturing facility. These sites, which have a high potential for NAPL impact, have detailed soil quality data at various depths and have no buried waste. The potential NAPL constituents were identified, and a vadose zone and a saturated zone threshold were calculated for each constituent. An initial review of analytical data identified 20 compounds commonly detected in the vicinity at elevated concentrations. A preliminary review of the tentatively identified compounds (TICs) showed that the west basin had widespread occurrences of chloromethylbenzene and its isomers, so that TIC was added to the list along with assumed chemical properties (e.g., molecular weight and solubility limit) based on the probable dominant component, 2-chlorotoluene. The $f_{oc}$ measured at the site has a geometric mean of 0.0003. For $K_{d,i}$ low, site-specific values based on column studies. At the low values of $K_{d,i}$ determined, mass in the pore space was significant and some vadose zone thresholds were less than half of their saturated zone counterparts. Arithmetic means of measured $\phi_w$ and $\rho_b$ gave $\phi_w/\rho_b$ of 0.25 and 0.062 L/kg in the vadose and saturated zones, respectively. Of the 227 samples, 85 had a TRS of 0 (no detections). The remaining TRS values included 43 below 0.1, 19 between 0.1 and 1, 25 between 1 and 10, and 55 above 10.

Figures 1 and 2 show the boring locations and distinguish the locations with at least one TRS value greater than 1 or between 0.1 and 1. The greatest potential for NAPL in the vadose zone exists in the northeast half of the west basin and the western third of the east basin. The distribution in the saturated zone is similar to that of the vadose zone but slightly larger with possible NAPL observed in the west basin's southeast corner and the east basin's northeast corner. This apparent expansion of extent of the potential residual NAPL with depth is not surprising, considering the likelihood of percolating NAPL being partially diverted laterally as it encountered the heterogeneities of the soil strata.

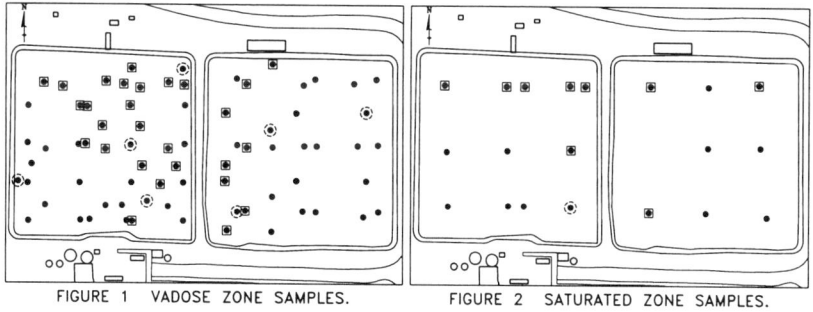

FIGURE 1   VADOSE ZONE SAMPLES.    FIGURE 2   SATURATED ZONE SAMPLES.

LEGEND:
● LOCATION OF ANALYTICAL SOIL SAMPLE(S)
□ SAMPLE LOCATION WITH A TRS VALUE GREATER THAN 1.0
◉ SAMPLE LOCATION WITH HIGHEST TRS VALUE BETWEEN 0.1 AND 1.0

Other tests (Watkins et al. 1996) found head-space measurement by HNu combined with hydrophobic dye shake testing to agree well with TRS results. Thus, samples originally analyzed for total concentrations without the benefit of field screening methods can be screened with confidence.

## Acknowledgments

The authors thank Stan Feenstra for his valuable comments. The authors also extend their appreciation to the investigators who collected the field data used in this study and to Joanna Scott whose assistance in developing and managing the information data base was so crucial to this study.

## Appendix. References

Banerjee, S. (1984). "Solubility of organic mixtures in water." *Environmental Science & Technology,* 18, 587-591.

Feenstra, S., Mackay, D.M., & Cherry, J.A. (1991). "A method for assessing the residual NAPL based on organic chemical concentrations in soil samples." *Ground Water Monitoring Review,* XI(2), 128-136.

Feenstra, S. (1990). "Evaluation of multi-component DNAPL sources by monitoring of dissolved-phase concentrations." In: *International Association of Hydrogeologists Conf. on Subsurface Contamination by Immiscible Fluids, Calgary, Alberta, April 18-20, 1990.* Weyer, K.U. ed. A.A. Balkema, Rotterdam, The Netherlands. 65-72.

Feenstra, S. (1997). *Aqueous concentration ratios to estimate mass of multi-component NAPL residual in porous media.* Doctoral thesis, University of Waterloo, Waterloo, Ontario.

Leinonen, P. J., & Mackay, D. (1973). "The multicomponent solubility of hydrocarbons in water." *Canadian J. of Chemical Engrg.,* 51, 230-233.

Mackay, D., Shiu, W.Y., Maijanen, A., & Feenstra, S. (1991). "Dissolution of non-aqueous phase liquids in groundwater." *J. of Contaminant Hydrology,* 8(1), 23-42.

Mariner, P.E., Jin, M., & Jackson, R.E. (1997). "An algorithm for the estimation of NAPL saturation and composition from typical soil chemical analyses." *Ground Water Monitoring & Remediation,* 17(2), 122-129.

Mott, H.V. (1995). "A model for determination of the phase distribution of petroleum hydrocarbons at release sites." *Ground Water Monitoring & Remediation,* 15(3), 157-167.

Schwarzenbach, R.P., & Westall, J. (1981). "Transport of nonpolar organic compounds from surface water to groundwater." *Environmental Science & Technology,* 11, 1360-1367.

Shiu, W.Y., Maijanen, A., Ng, A.L.Y., & Mackay, D. (1988). "Preparation of aqueous solutions of sparingly soluble organic substances: II. multicomponent systems - Hydrocarbon mixtures and petroleum products." *Environmental Toxicology & Chemistry,* 7, 125-137.

US Environmental Protection Agency. (1991). *Dense Nonaqueous Phase Liquids -- A Workshop Summary.* EPA/600-R-92/030. February.

US Environmental Protection Agency. (1992). *Estimating Potential Occurrence of DNAPL at Superfund Sites.* Publication 9355.4-07FS. January.

Watkins, M.L., Sieber, E., & Schulze, M.E. (1996). "Calibration of NAPL field screening methodologies for inclusion in a comprehensive NAPL investigation at a Superfund site in New Jersey." In *Proceedings of the Tenth National Outdoor Action Conf.,* NGWA.

Yalkowsky, S.H. & Banerjee, S. (1992). *Aqueous Solubility - Methods of Estimation for Organic Compounds.* Marcel Dekker, New York.

Practical Approaches for Development of Site-Wide and Local-Scale Models at Brookhaven National Laboratory: A Case Study

Douglas A. Smolensky[1], Arthur J. Zahradnik[2], Michael P. Kladias[3], Thomas W. Burke[4], William R. Dorsch[4]

Abstract

A regional groundwater flow model of the Brookhaven National Laboratory (BNL) and surrounding area provides a tool for evaluation of groundwater flow, contaminant transport, and remedial alternatives, where such processes, concerns, or actions have impacts that extend across extensive areas within the BNL site boundary or beyond the site boundary. Given, however, the distribution and scale of many of the impacted areas, the type and distribution of contaminants, the contaminant specific transport mechanisms, the hydrogeologic framework, and the human-induced impacts of water use, the regional model could not be developed to adequately address all local issues. Development of the regional (site-wide) model, therefore, proceeded with the expectation of serving as the framework and foundation for the evaluation of specific groundwater flow and contaminant migration issues at the local scale. To date, specific local-scale evaluations have been successfully completed using seven distinct sub-regional models. Three sub-regional applications are presented herein.

Introduction

Numerical groundwater modeling provides a means whereby specific hydrologic events or stresses can be simulated resulting in an approximation of the response that the natural groundwater system would have in terms of head and groundwater flow. Taken one step further, certain groundwater models may also be used to approximate the transport of contaminants within an aquifer system. If properly developed, a model will correctly represent the geometry of the groundwater system in a discrete form. A conflict often arises where the investigation of a local problem requires a detailed representation of the groundwater system near the stress (at the sub-regional scale) because accurate simulation of groundwater flow rates and directions can be obtained only by representing the entire

---
[1] Associate, ARCADIS Geraghty & Miller, 88 Duryea Road, Melville, NY 11747
[2] Project Eng/Sci, ARCADIS Geraghty & Miller, 88 Duryea Road, Melville, NY 11747
[3] Principal Hydrogeol, ARCADIS Geraghty & Miller, 1131 Benfield Blvd., Millersville, MD 21108
[4] Project Eng/Hydrogeologist, Brookhaven National Laboratory, Building 51M, Upton, NY 11973

groundwater system to its natural system boundaries (regional scale) (Buxton and Reilly 1987).

This paper describes an approach used to simulate groundwater flow and the transport of various volatile organic compounds, tritium, and strontium-90 in groundwater at Brookhaven National Laboratory (BNL). The report includes a short description of the regional flow model and three examples of sub-regional models developed for distinctly different purposes.

Background

BNL is located in Suffolk County, New York, just east of the geographic center of Long Island (Figure 1). The BNL property is a approximately 3 miles on each side, comprising an area of 5,265 acres. The principal on-site facilities occupy only approximately 20 percent of the land area of BNL and are concentrated in the center of the site; they include various research facilities, several inactive research reactors, office buildings, waste management areas, inactive landfills, etc. The remaining 80 percent of BNL is native woodlands.

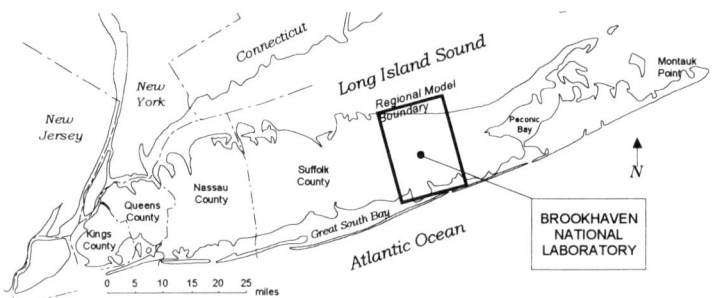

Figure 1. Regional Site Location Map and Extent of Regional Model.

BNL is directly underlain by approximately 190 feet of glacial deposits that comprise the Upper Glacial aquifer (the aquifer predominantly used for water supply at BNL). Local confining units that tend to restrict flow between the glacial aquifer and underlying aquifers may be found beneath the glacial deposits. These local units may be several feet to tens of feet in thickness. Underlying the confining units is approximately 800 feet of fine sand that comprises the Magothy aquifer. The areally extensive Raritan confining unit (200 feet thick) underlies the Magothy aquifer. Beneath the Raritan confining unit is the Lloyd aquifer (300 feet thick) which overlies bedrock (the bottom boundary of the natural hydrogeologic system).

The only source of natural recharge at BNL is recharge from precipitation (approximately 50 % of 46 in/yr.). An east-west trending groundwater divide exists just north of the BNL site boundary. Water entering the system north of the divide flows horizontally and vertically in a northerly direction and eventually discharges to the Long Island Sound (see Figure 1). Most water entering the system south of the divide flows in a southerly direction and eventually discharges to the Great South

Bay or Atlantic Ocean. Some of the water south of the divide discharges to freshwater streams and rivers whose channels intersect the water table. Natural groundwater flow velocities vary greatly across the BNL site but are generally in the range of 0.5 to 1.0 foot per day.

Regional Flow Model

The three-dimensional regional model covers an area extending from BNL to the shorelines to the north and south, and approximately 15,000 and 45,000 feet to the west and east, respectively (see Figure 1) (Geraghty & Miller, 1996). MODFLOW was the model code chosen for use as it is has the technical capabilities required for this effort and is well tested, documented, and accepted. Additionally, MODFLOW can be readily coupled with many finite-difference transport codes and has been used extensively in regional/sub-regional modeling applications. A variable-spaced grid of 189 rows by 223 columns was developed. Over the majority of the BNL site cell spacing was 100 by 100 feet. Vertically the model area was divided into 8 layers, with the most emphasis in detail placed on representation of the Upper Glacial aquifer and the local underlying confining units. A generalized stratigraphic column showing the corresponding model layering scheme for the regional model is shown on Figure 2. The resulting three-dimensional discrete representation of the

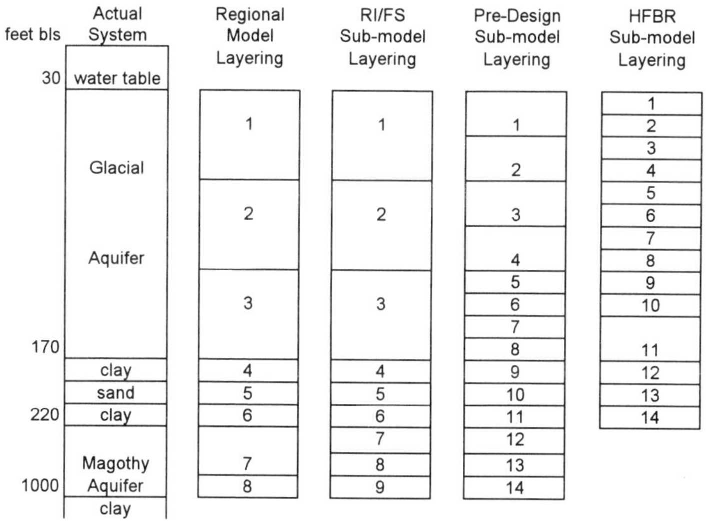

| feet bls | Actual System | Regional Model Layering | RI/FS Sub-model Layering | Pre-Design Sub-model Layering | HFBR Sub-model Layering |
|---|---|---|---|---|---|
| 30 | water table | | | | |
| | Glacial | 1 | 1 | 1 | 1 |
| | | | | | 2 |
| | | | | | 3 |
| | | | | 2 | 4 |
| | | 2 | 2 | | 5 |
| | | | | 3 | 6 |
| | Aquifer | | | | 7 |
| | | | | 4 | 8 |
| | | | | 5 | 9 |
| | | 3 | 3 | 6 | 10 |
| | | | | 7 | |
| 170 | | | | 8 | 11 |
| | clay | 4 | 4 | 9 | 12 |
| | sand | 5 | 5 | 10 | 13 |
| 220 | clay | 6 | 6 | 11 | 14 |
| | | | 7 | 12 | |
| | Magothy | 7 | 8 | 13 | |
| 1000 | Aquifer | 8 | 9 | 14 | |
| | clay | | | | |

Figure 2. Generalized Stratigraphic Column and Vertical Layering Schemes for Regional and Sub-Regional Models at Brookhaven National Laboratory.

groundwater system (337,176 nodes) provided a tool that adequately represented the complexities of the natural system, the hydraulic impacts of human use of the system, the details associated with contamination of portions of the system, and a platform from which to develop more detailed sub-regional models for specific remedial evaluation and design, or potential exposure evaluation.

RI/FS Sub-Regional Model

To meet the objectives of a fate and transport evaluation and an evaluation of remedial alternatives for a specific operable unit RI/FS, a sub-regional model was required (Geraghty & Miller, 1998). In this case, the potential transport of Volatile Organic Compounds (VOCs) downgradient of BNL or to potential receptors necessitated finer off-site grid spacing than the regional model provided. The extent of the sub-model was based on the locations of both natural and human-induced boundary conditions, the area of the existing contamination, and the expected area of potential transport over the simulation period. The latter was approximated by performing advective particle tracking (using MODPATH) and qualitatively assessing the additional effects of other transport mechanisms. Vertically, an additional model layer was added to better reflect the vertical extent of contamination in the uppermost portion of the Magothy aquifer. By dividing regional Model Layer 7 into two sub-regional layers, the vertical extent and initialized contaminant mass could be more accurately represented. The revised vertical discretization scheme is shown on Figure 2. The development of the sub-model (using MT3D for transport analysis) resulted in 273 rows, 160 columns, and 9 layers (393,120 nodes). The extent of the sub-model (Figure 3) covers approximately 18 square miles (of the approximately 175 square mile regional model).

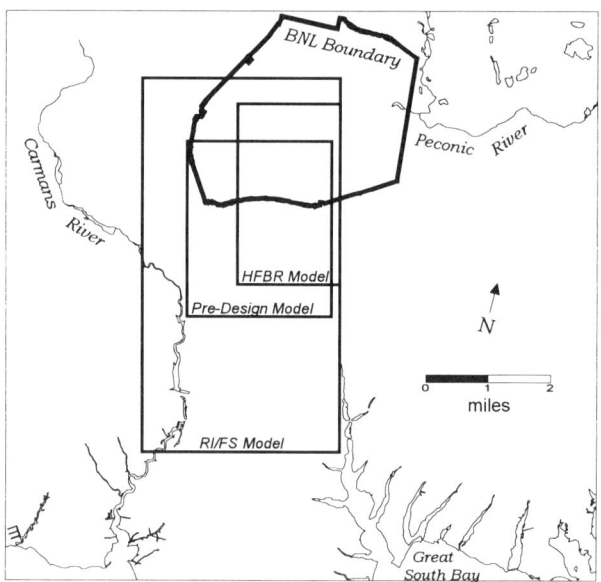

Figure 3. Extents of Sub-Regional Models at Brookhaven National Laboratory.

## Pre-Design Sub-Regional Model

The goal of this evaluation was to use existing and new modeling tools to perform advective and solute transport analyses in support of selected remedial alternatives (Geraghty & Miller, 1997). The specific objectives were to determine the optimal locations, depths, and pumping rates for the placement of a system of in-well sparging wells. The optimal in-well sparging system design must consider the degree of desired re-circulation between upper and lower screened zones, and the issues associated with subsurface discharge. Finally, the modeling tools must estimate the concentration of VOCs in the extracted water. This approximation will serve as an estimate of influent concentrations to individual in-well sparge wells and be used to estimate the operational time-frame of individual wells and the system as a whole.

Given the stated objectives, a fine-scale discrete model representation was required for a very local area. The required detail was needed not only in the horizontal direction to determine lateral hydraulic capture, but also in the vertical direction to evaluate re-circulation between screened zones and to accurately represent the distribution of VOCs (the VOC plume was characterized at 10 foot vertical intervals). Because the entire model area was within the boundaries of the RI/FS sub-regional model, the RI/FS model was the model used to develop this local-scale model from. The pre-design model was therefore developed as a sub-model of a sub-model. A variable-spaced grid was developed with horizontal grid spacing of 25 feet by 25 feet in the vicinity of the proposed remedial system. Vertically, regional Model Layers 1 and 2 (shallow and mid- glacial zones) were each divided into 2 layers, while regional Model Layer 3 (the deep glacial zone) was divided into 4 layers. The additional detail was required in the deep glacial zone because that was where the contamination was identified and where the hydraulic impacts of pumping and recharging (re-circulation) needed to be best quantified. The vertical layering scheme is shown on Figure 2. The resulting model (using MODPATH and MT3D for transport analyses) consisted of 172 rows, 167 columns, and 14 layers (402,136 nodes) over a 7 square mile area (see Figure 3).

## HFBR Sub-Regional Model

The goal of this effort was to provide a tool that could be used to evaluate historical, current, and future migration of a tritium plume emanating from a spent fuel storage pool at the on-site High Flux Beam Reactor (HFBR) (Geraghty & Miller, 1997). The model would also be used to evaluate various remedial alternatives. Although the observed plume was several thousand feet long, the level of detail achieved in the plume's vertical characterization was accomplished at the 5 and 10 foot interval levels. The expected flowpath of the core of the plume (where not significantly altered by local pumping) was clearly confirmed based on the detailed chemical data.

Given the detailed delineation of the plume and the types of remedial alternatives to be evaluated, an extremely fine-scaled sub-regional model was developed. Like the pre-design sub-regional model, the HFBR model was developed as a sub-model of a sub-model. A variable-spaced grid was developed with horizontal grid spacing of 10 feet by 10 feet throughout most of the areas occupied by and downgradient of the plume. Vertically, regional Model Layers 1 and 2 (shallow and mid-glacial zones) were each divided into 4 layers, while regional Model Layer 3 (the deep glacial zone) was divided into 3 layers. The additional detail was required throughout the glacial aquifer because of the amount of plume characterization data, the "diving" nature of the plume (a function of its advective path), and the focused remedial alternatives to be evaluated. The vertical layering scheme is shown on Figure 2. The resulting model (using MT3D for transport analysis) had 328 rows, 225 columns, and 14 layers (1,033,200 nodes) over approximately a 4 square mile area (see Figure 3).

Conclusions

The development and use of sub-regional models at Brookhaven National Laboratory has proven to be a cost-effective method of performing detailed flow, transport, and remedial alternative analyses. The effective use of the sub-models, however, is enhanced by developing the regional model with the expectation of its use as the foundation for future development of sub-regional models. At Brookhaven National Laboratory, model development and use following this philosophy has resulted in simulation tools that allow for critical evaluation of remedial alternatives for various constituents.

References

Buxton, H.T. and Reilly, T.E. 1986. "A Technique for Analysis of Ground-Water Systems at Regional and Sub-Regional Scales Applied on Long Island, New York" US Geological Survey Water-Supply Paper 2310

Geraghty & Miller, Inc. 1996. Regional Groundwater Model, Brookhaven National Laboratory, Upton, New York

Geraghty & Miller, Inc. 1997. Draft Pre-Design Report for the Operable Unit III Off-Site Removal Action, Brookhaven National Laboratory, Upton, New York

Geraghty & Miller, Inc. 1997. Draft Groundwater Flow and Tritium Transport Modeling, Brookhaven National Laboratory, Upton, New York

Geraghty & Miller, Inc. 1998. Draft Remedial Alternative Simulations in Support of OU III Feasibility Study, Brookhaven National Laboratory, Upton, New York

# Groundwater bioremediation optimization using genetic algorithms

Amy B. Chan Hilton, S.M.* and Teresa B. Culver, A.M.†

### Abstract

The cost of remediating contaminated groundwater systems is high given the transport limitations, the magnitude of the contaminated area, and the time-scale of the clean-up process. It is possible to replace trial-and-error remediation design with a mathematical optimization algorithm to determine the most cost-effective policy, resulting in potentially large cost savings. The analyst may utilize an optimization approach to explore the trade-offs between optimal remediation costs and remediation efficiency. The work utilizes a genetic algorithm to explore the sensitivity of optimal costs for in situ bioremediation given the ultimate in situ groundwater quality goals. The cost sensitivity to the remediation water quality goals will be evaluated for both static and dynamic problems.

## Introduction

Groundwater remediation systems can often cost millions of dollars at a single site (U.S. EPA 1996). Using optimization methods, rather than trial-and-error, to find groundwater remediation designs can provide significant savings. The application of optimization methods to groundwater problems has been investigated by numerous researchers. There is substantial research on optimal groundwater remediation design using linear programming (Willis 1979), nonlinear programming (McKinney and Lin 1996), and dynamic programming (Culver and Shoemaker 1992).

---

*Graduate Research Assistant, Dept. of Civil Engineering, University of Virginia, Thornton Hall, Charlottesville, VA 22903-2442. abh4b@virginia.edu
†Assistant Professor, Dept. of Civil Engineering, University of Virginia, Thornton Hall, Charlottesville, VA 22903-2442. tbc4e@virginia.edu

Genetic algorithms (GAs) are search algorithms based on concepts from natural selection. The method was developed by Holland in 1975 and inspired by the idea of "survival of the fittest." Genetic algorithms are a random search procedure which use probabilistic rather than deterministic search rules. The objective function, rather than derivative information, is used directly in the search, therefore allowing GAs to handle nonconvex, highly nonlinear, and complex problems (Goldberg 1989). The genetic algorithm procedure consists of three main operations: selection, crossover, and mutation. Previous works which applied GAs to optimal groundwater remediation design problems have found that this probabilistic search technique is able to successfully identify optimal or near-optimal solutions, with policies similar to those found by using other techniques, such as nonlinear programming (McKinney and Lin 1994) and simulated annealing (Wang and Zheng 1996). Additionally, there have been applications of GAs on determining optimal dynamic policies in which the policies change with each management period (Wang and Zheng 1997; Huang and Mayer 1995, 1997; Yoon and Shoemaker 1997).

Remedial cost sensitivity to the remediation efficiency or ultimate water quality goals has been demonstrated. For steady-state pump-and-treat remediation designs, costs were predicted to increase by more than 200% given an order of magnitude reduction in the ultimate water quality goals (Ahlfeld and Hill 1996; National Research Council 1994). This work utilizes GAs to explore the sensitivity of bioremediation designs and costs as the ultimate water quality expectations are modified. Both stead-sate design and dynamic designs, which incorporate management periods where the management approach may change over time (Culver and Shoemaker 1992), will be studied.

# Bioremediation example

A hypothetical bioremediation problem, based on one presented by Yoon (1997), was developed for a homogeneous, isotropic aquifer. This aquifer is represented by a finite element grid with 280 elements (dx=16.25 m, dy=12.5 m). The aquifer has a hydraulic conductivity of 6.1 x $10^{-6}$ m/s, longitudinal dispersivity of 10.0 m, transverse dispersivity of 2.5 m, porosity of 0.2, and saturated thickness of 30.5 m. A steady easterly flow was created by maintaining constant heads of 35 m on the west boundary and 32 m on the east boundary, with no-flow on the north and south boundaries. Also, at the east and west boundaries, the contaminant concentration was set to 0 mg/L. The bottom of the aquifer has a constant slope, with a depth of $d = 60$ m on the western boundary and $d = 63.25$ m on the eastern boundary. A contaminant plume was created with four constant point sources of toluene, with concentrations ranging from 31.7 to 100 mg/L, and simulating the contaminant movement for 3 years. The resulting plume, as well as the locations of the computational nodes and the contaminant sources, is shown in Figure 1. During the remediation stage, the constant sources were removed, and the initial biomass concentration was 0.001 mg/L. A background aqueous oxygen concentration of 3 mg/L existed at nodes where the contaminant concentration was less than 0.01 mg/L.

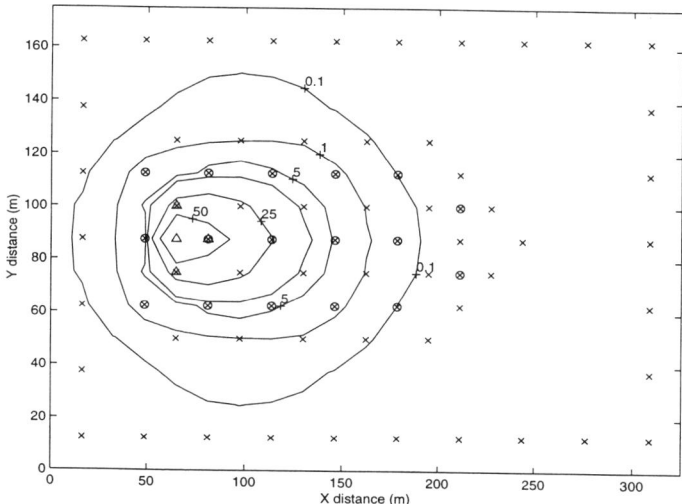

Figure 1: Initial toluene plume ($\triangle$ = point source, $\circ$ = potential well location, $\times$ = observation well location).

The objective of this pump-and-treat design problem is to minimize the total cost of remediation (capital and operating costs), given the initial plume, by varying the injection rates at 17 potential wells subject to constraints on the final contaminant concentration at 73 observation wells and the hydraulic heads in the aquifer. The locations of the potential wells and observation wells are denoted in Figure 1. At the end of the 1-year remediation period, the contaminant concentration at the observation wells must meet the clean-up standard for the contaminant. The water injected into the potential wells had an aqueous oxygen concentration of 8 mg/L

Mathematically the optimization model was described by the following:

$$\min \ Cost = \sum_{m=1}^{M} \left[ C_{treat,m} + \sum_{j=1}^{J} C_{inject,jm} \right] \quad (1)$$

subject to

$$c_{i,M} \leq c^*, \quad \forall \, i = 1, \ldots, N, \quad (2)$$
$$h_{low} \leq h_{i,m} \leq h_{up}, \quad \forall \, i = 1, \ldots, H, \quad (3)$$

where
$$C_{treat,m} = 14200\, n_{QI,m}, \quad (4)$$
$$C_{inject,jm} = 5543(d_j)^{0.299} + 5751(QI_{j,m})^{0.453} + 12227 QI_{j,m}, \quad (5)$$
where $C_{treat,m}$ and $C_{inject,m}$ are the present value of capital and operating costs of treatment and injection wells, respectively, during management period $m$; $c_{i,M}$ is contaminant concentration at the end of the final remediation period at observation well $i$ (M/L$^3$), where $N$ is the total number of observation wells and $M$ is the total number of management periods; $c^*$ is the clean-up standard for the contaminant (M/L$^3$); $h_{i,m}$ is the hydraulic head at node $i$ at the end of management period $m$, where $H$ is the total number of computational nodes (L); $h_{low}$ and $h_{up}$ are the lower and upper bounds on hydraulic head, respectively (L); $n_{QI,m}$ is the number of active injection wells during management period $m$; $d_j$ is the depth to the bottom of the aquifer at the location of injection well $j$ (L); and $QI_{j,m}$ is the injection rate at injection well $j = 1, \ldots, J$, (L$^3$/T) during period $m$. The cost functions in Equations 4 and 5 were adapted from McKinney and Lin (1996).

## Analysis

To solve the above problem, the optimization model combines the genetic algorithm (GA) approach with a groundwater simulation model. The GA model was coded using the PGAPack library, created by Levine (1996). For the steady-state case, the GA implemented had a population size of 100 strings, each with a length of 68 binary bits, replaced 80 strings each generation, implemented tournament selection and two-point crossover, and ran for 75 generations. A two-dimensional flow and transport model for in situ bioremediation, BIO2D-KE, was used to evaluate the objective function value of each string. BIO2D-KE implements the finite element method to simulate the advective and dispersive transport of a contaminant, oxygen, and biomass (Earles et al. 1996).

A base case solution for steady-state management was determined by setting $c^*$ to 1 mg/L and $M = 1$. The optimal policy found for this base case had a total cost of \$167,380, in which \$71,000 was treatment costs and \$96,380 was injection costs. This policy met all of the concentration and hydraulic head constraints and was found during generation 58 of the GA. Five potential wells were chosen to be active.

Adjustments are made to the base case bioremediation formulation to explore the design sensitivity. The water quality goal, $c^*$, itself or the total amount of violation of the water quality constraints is varied to demonstrate the design sensitivity to water quality expectations in a bioremediation system. The impact of dynamic solutions on this design sensitivity will also be presented.

# References

[1] Ahlfeld, D. P. and E. H. Hill, III. The sensitivity of remedial strategies to design criteria. Ground Water, 34(2), 341-348, 1996.

[2] Culver, T. B. and C. A. Shoemaker. Dynamic optimal control for groundwater remediation with flexible management periods. Water Resources Research, 28(3), 629-641, 1992.

[3] Earles, T. A., J. P. Gray, and T. B. Culver. BIO2D-K and BIO2D-KE User's Guide, 1996.

[4] Goldberg, D. E. Genetic Algorithms in Search, Optimization & Machine Learning. Reading, MA: Addison-Wesley Publishing Company, Inc., 1989.

[5] Huang, C. and A. S. Mayer. Development of dynamic groundwater remediation strategies for variable aquifer configurations. Integrated Water Resources Planning for the 21st Century: Proceedings of the 22nd Annual Conference, Cambridge, MA, 840-843, 1995.

[6] Huang, C. and A. S. Mayer. Pump-and-treat optimization using well locations and pumping rates as deicision variables. Water Resources Research, 33(5), 1001-1012, 1997.

[7] Levine, D. Users Guide to the PGAPack Parallel Genetic Algorithm Library. ANL-951/18, available through anonymous ftp from ftp.mcs.anl.gov/pub/pgapack/pgapack.tar.Z, 1996.

[8] McKinney, D. C. and M-D. Lin. Genetic algorithm solution of groundwater management models. Water Resources Research, 33(6), 1897-1906, 1994.

[9] McKinney, D. C. and M-D. Lin. Pump-and-treat ground-water remediation system optimization. Journal of Water Resources Planning and Management, 122(2), 128-136, 1996.

[10] National Research Council. Alternatives for Ground Water Cleanup. Washington, D.C.: National Academy Press, 1994, as cited in G. Masters, Introduction to Environmental Engineering and Science (2nd Ed.), Prentice Hall, 1998.

[11] U.S. EPA. Innovative treatment technologies: Annual status report (8th Ed.). Office of Solid Waste and Emergency Response, EPA-542-R-96-010 No. 8.

[12] Wang, M. and C. Zheng. Optimal remediation policy selection under general conditions. Groundwater, 35(5), 1997.

[13] Willis, R. L. A planning model for the management of groundwater quality. Water Resources Research, 15(6), 159-171, 1987.

[14] Yoon, J-H. Optimal Design of Groundwater in situ Bioremediation Using Evolutionaly Algorithms. Ph.D. dissertation, Cornell University, Ithaca, NY, 1997.

Use of Advanced Hydroenvironmental Modeling and Simulation in Water Resources Management

Robert F. Athow, Jr.[1], M ASCE & Jeffery P. Holland[2], AM ASCE

Abstract

The U.S. Army is the lead within the U.S. Department of Defense (DoD) modeling and simulation (M&S) tool science and technology development to support water resources development. This research and development is led for the Army by the U.S. Army Engineer Waterways Experiment Station (WES). The WES has developed a series of groundwater, surface water, and watershed modeling systems in support of water resources project design, operation, and management. These systems provide access to multiple models, analysis tools, visualization, geographic information systems, and databases from a single set of comprehensive computational environments. Applications of these systems are conducted by multiple organizations (e.g., U.S. Army Corps of Engineers field offices, U.S. Army, Navy, and Air Force installations, Department of Energy [DOE], Environmental Protection Agency [EPA], academia, etc.) and WES, particularly in support of environmental quality (EQ) resource management and water resources project design/operation. These applications involve risk and tradeoff analyses for impact assessment, compliance and regulatory issues, risk reduction, and natural/cultural resource management. All of the overviewed systems have been developed through partnerships with other DoD and Federal agencies (particularly the DOE and the EPA), and academia. The development paths for these tools over the next three years will also be presented.

[1] Hydraulic Engineer, Coastal & Hydraulics Laboratory, U.S. Army Engineer Waterways Experiment Station, 3909 Halls Ferry Rd, Vicksburg, MS 39180-6199
[2] DoD Environmental Modeling & Simulation Computational Technology Area Leader, Coastal & Hydraulics Laboratory, U.S. Army Engineer Waterways Experiment Station, 3909 Halls Ferry Rd, Vicksburg, MS 39180-6199

Introduction

The WES has developed a series of groundwater, surface water, and watershed modeling systems in support of military installation and Army civil works water resources project design, operation, and management. Applications of these systems are conducted by multiple organizations and WES in support of flood control, navigation, natural/cultural resources, and EQ management investigations. These investigations involve risk and tradeoff analyses for impact assessment, compliance and regulatory issues, risk reduction, and natural/cultural resource management. The tools overviewed below have all been developed through collaboration with other DoD and Federal agencies (particularly EPA, which often recommends and funds WES M&S tool development in the EQ area, and DOE), industry, and academia. Limited details for each of these systems are provided herein; more details are provided in Holland et al (1996).

Watershed Modeling System (WMS)

The WMS provides DoD with the ability to productively conduct lumped-parameter and multi-dimensional watershed analyses in support of groundwater cleanup (infiltration boundary conditions), runoff analyses, and resource management for DoD installations and Army civil works water resources projects. Developed through leveraging military and Army Civil Works research, the WMS has the same look and feel as the Surface Water Modeling System (SMS) and the Groundwater Modeling System (GMS). Partnering with U.S. Army Construction Engineering Research Lab and EPA has further extended the scope of the system.

Triangulated Irregular Networks (TINs) are employed within the WMS for defining the topography and calculating vital hydrologic statistics. TINs are created by inputting digitized data, from digital topographic maps, existing digital terrain databases (e.g., digital elevation models), or from manually digitized data, and triangulating the points. After the topographic surface is modeled, WMS automatically defines the dominant streams and flow paths on the user's screen. The WMS then calculates the contributing drainage area to each of the user-defined stream junctions. The WMS employs NEXRAD radar information in assessing precipitation intensities, durations, and runoff distributions. The WMS then writes out the data into one of several formats for the watershed models (e.g., HEC-1, TR-20, or CASC2D) supported by the system. The system provides for the same GIS/CADD, grid generation, animation, and conceptual model development capabilities as those listed above for the GMS and WMS. The WMS is in use in support of resource management concerns at Picatinny Arsenal, NJ, Ft. Hood, TX, Ft. Carson, CO, and Aberdeen Proving Grounds, MD. The system also continues its nearly two-year use as the primary system for hydrologic predictions for Sava River stages in support of U.S. Forces in Bosnia. This latter study is being conducted in collaboration with the U.S. Army Corps of

Engineers Cold Regions Research and Engineering Lab and the Topographic Engineering Center.

Surface Water Modeling System (SMS)

This modeling system, developed primarily through Army Civil Works research and development, provides access to multiple incompressible Navier-Stokes (NS) solvers in both two and three spatial dimensions. The SMS has been developed for several multi-dimensional hydrodynamic and water quality models (including CH3D-WES, RMA10-WES, CEQUAL-ICM, HIVEL2D, FESWMS, ADCIRC, and the two-dimensional hydrodynamic/sediment transport system, TABS-MD) to address the need for efficient model setup, execution, and analysis. The system is mouse driven with pull-down menus and requires a minimum of manual data entry. The interface was designed to allow easy application of each of the models in the WES multi-dimensional hydrodynamic modeling family. The interface provides access to several state-of-the-art visualization and animation capabilities. Hundreds of estuarine, riverine, and wetland sites have been modeled via SMS. The SMS software runs on personal computers running Windows, as well as multiple UNIX workstations. The system provides hydrodynamic, salinity, water quality (in progress), and sediment transport M&S capabilities in support of ecosystem management in estuaries, rivers, and wetlands. The graphical user environment for the SMS is directly analogous to its GMS counterpart in look/feel and functionality. This M&S system has been used extensively in support of natural resources management strategy development for Galveston, Chesapeake, and San Francisco Bays, the Mississippi River, South Florida Ecosystem Restoration, and numerous wetland environments. Details regarding the models within the SMS are given in Holland et al (1996), Johnson et al (1994), and Cerco and Cole (1995).

Groundwater Modeling System (GMS)

The DoD GMS is among the state-of-the-art groundwater modeling systems in the subsurface modeling community. The GMS development, which is led by WES for DoD, is a highly integrated, partnered activity with collaborators from five DOE laboratories, two EPA labs, multiple DoD organizations. In addition, 20 academic institutions are partnering or have partnered in the effort. Version 2.0 of the GMS has been fielded and is in use by nearly 500 federal research groups within DOE, EPA, and DoD. The system provides access to six three-dimensional groundwater models presently (MODFLOW, MT3D, MODPATH, FEMWATER, LEWASTE, AND SEEP); this number will increase to 12 by the end of FY98. The models being incorporated into the GMS include PARFLOW, NUFT3D and RT3D (all through partnerships with DOE), ADH (developed by WES), UTCHEM (developed by the University of Texas), and SWGW (a surface water – groundwater interaction code developed by WES).

The desktop of the GMS (similar in all respects to the SMS and WMS desktops), with a contaminant plume from Aberdeen Proving Grounds, MD, visualized, is shown in Figure 1.

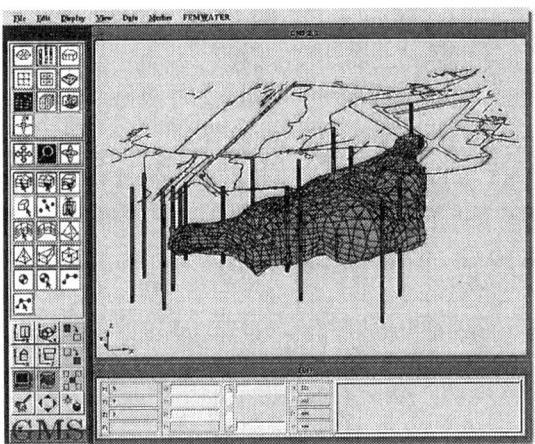

Figure 1. Desktop of DoD GMS with Visualization of a Contaminant Plume and Associated Wellfield

The GMS has state-of-the-art visualization, conceptualization, and parameterization capabilities on-board. Key components of these productivity-enhancing tools include GIS/CADD links, connectivity to several field data collection systems, structured and unstructured grid generators, a full geo-statistical library, and AVI video file animation. The GMS runs on UNIX workstations, personal computers running Windows, and supercomputers with the same look and feel. A key component of the system, the Map Module, increases the productivity of subsurface conceptualization, flow/transport modeling, and remedial design simulation by over a factor of 10. Ongoing research into the impacts of subsurface heterogeneity on subsurface flow/transport, development of remedial design/optimization modules for 10-15 cleanup technologies, and development of scalable, parallel implementations of the GMS's models will continue through at least 1999. Holland (1996) discusses these GMS activities in more detail.

Recent Advances

The DoD has recently begun a program called the Common High Performance Computing Software Support Initiative (CHSSI) which focuses on the development of scaleable, parallel implementation of codes for computing a variety of computationally intensive simulations, including water resource management issues.  Three projects will have direct impact within the numerical modeling systems addressed earlier.  These three projects will enhance the development of high-speed parallel codes for 1) Computing three-dimensional, incompressible, free surface flows; 2) Computing the interactions between surface and groundwater; and 3) A scalable version of the GMS.

To date, these three CHSSI projects have shown remarkable execution speed up using parallel computer architecture ranging from 2-5 times as fast as the single CPU usage up to 12-16 times as fast.  Obviously these codes will be most effective for very large, complex simulations, such as the modeling of an entire watershed, groundwater, and estuarine system such as the South Florida Bay-Everglades region.  Final delivery of these products is scheduled for the fall of 1999.

Hydroenvironmental Modeling and Simulation Future

Problems of the next decade would seem to demand the development of integrated modeling systems capable of simulating surface water, groundwater, and the interactions there between in a holistic framework for a number of highly-diverse DoD requirements.  To this end, the following hydrodynamic and hydrologic modeling development activities are planned at WES over the next two to five years. First, the coupling of surface water, atmospheric, and groundwater modeling tools both dynamically and passively, in two and three dimensions in support of the cleanup and potential vulnerability of environmental resources for both battlefield and civilian areas.  This coupling will require development of methods for numerically handling differing time and spatial scales associated with surface water hydrodynamics, watershed runoff, infiltration, atmospheric transport, subsurface flow and (often) highly-sophisticated multi-component geo-chemical transport. Highly-efficient numerical methods for scalable parallel computing architectures will be required to facilitate calculations on the large computational meshes (100,000's of nodes routinely; 1,000,000's often) envisioned here. Second, incorporation of integrated atmospheric / surface water /groundwater modeling tools within a comprehensive modular modeling system.  This system will be built upon the foundation created by the GMS, SMS, and WMS, but will also be designed to support other aspects of readiness and training within the DoD user community.  Interoperability with other DoD systems must be stressed and achieved as part of this development.  And third, incorporation of uncertainty and risk into all hydroenvironmental and hydrologic modeling systems to support risk-based design and assessment for both environmental quality and civil engineering concerns.

Given the nature of the studies envisioned over the next five years, new computational systems (and/or their outputs) must reach and appeal to multidisciplinary teams and decision-makers. These teams will require access to multiple conceptualization, parameterization, simulation, optimization, and visualization tools from a consistent, efficient, and geographically distributed computational environment. Developed computational systems must stress portability (across multiple computing platforms), modularity (to allow frequent updates), connectivity (to tools within and external to the given system), and consistency (to facilitate user comprehension and transition). To this end, multiple forms of computing, from personal computers to UNIX workstations to scalable parallel computing, will undoubtedly be used, necessitating dedicated distributed, heterogeneous system and software support.

Acknowledgement

This paper was prepared from research and development conducted by the U.S. Army Engineer Waterways Experiment Station, Vicksburg, MS. Permission was granted by the Chief of Engineers to publish this information.

References

Cerco, C. F., and Cole, T. M., 1995. "User's Guide to the CE-QUAL-ICM Three-Dimensional Eutrophication Model, Release Version 1.0," Technical Report EL-95-15, US Army Engineer Waterways Experiment Station, Vicksburg, MS.

Holland, J. P., 1996. "Department of Defense Groundwater Modeling Program: an Overview," Subsurface Fluid-Flow (Ground-Water and Vadose Zone) Modeling, ASTM STP 1288, J.D. Ritchey and J.O. Rumbaugh, Eds., American Society for Testing and Materials, Philadelphia, PA.

Holland, J.P., Berger, R.C., and Schmidt, J.H., 1996. "Finite Element Analyses in Surface Water and Groundwater: an Overview of Investigations at the U.S. Army Engineer Waterways Experiment Station," Third U.S.-Japan Symposium on Finite Element Modeling and Large-Scale Computing, Minneapolis, MN.

Johnson, B.H., Kim, K.W., Heath, R.E., and Butler, H.L. 1994. "Development and Application of A Three-Dimensional Hydrodynamic Model," Chapter 9, Computer Modeling of Free-Surface and Pressurized Flows, Kluwer Academic Publishers, Boston, MA.

## Continuumization of Spatial Network Data

John F. Peters[1] and Stacy E. Howington[2], Member

*Abstract*

As part of an on-going groundwater modeling research program, a numerical model based on discrete networks was developed to simulate fluid flow and contaminant transport through porous media. A fundamental problem in applying the discrete model is relationship between discrete quantities and their conceptualization in a continuum setting. This paper describes pre- and post-processing procedures for interaction between the continuum idealization of the Groundwater Modeling System and the discrete network model.

*Background*

A principal tool for design of groundwater remediation schemes is the numerical model that simulates fluid flow and contaminant transport through porous media. Experience shows that models, based on traditional continuum mechanics concepts, display scale effects in their parameters. Parameters that ostensibly represent material properties are found to vary with scale of the problem domain. It is now generally accepted that heterogeneity produces multiple scales within a medium, giving rise to this scale dependence. Thus, the apparent failure in the continuum idealization results from attempting to homogenize the real medium, with its multiple scales of heterogeneity, into an equivalent medium that is free of internal scales. At present, the only practical numerical technique to attack the problem is through very high-resolution computations in which material properties are distributed to reproduce the statistical nature of the medium being simulated. The medium's multi-scale

---

[1] Research Civil Engineer, Research Hydraulics Engineer[2], Waterways Experiment Station, 3909 Halls Ferry Road, Vicksburg, MS 39180-6199

character is captured by introducing spatial correlation into the statistical description. More formal approaches for describing heterogeneous media in continuum terms ultimately lead to so-called non-local descriptions in which the governing equations are described as integrals over space and time. Application of non-local models to real problems has been limited.

To simulate multi-scale media, with less reliance on very-large scale computations, a flow and transport model was developed in which the porous medium is described as a dense network of flow paths (called throats), along which contaminants are transported by one-dimensional advection. Dispersion occurs as a result of mixing at network connections and different travel times between connections. The general validity of networks has been demonstrated with computational experiments (see Howington et. al., 1997). In particular, the network model captures the apparent scale dependence in flow and transport parameters caused by heterogeneity. The network has provided a valuable laboratory to understand the homogenization problem at a fundamental level. Practical application of the network code depends on its compatibility with the Department of Defense Groundwater Modeling System (GMS) which is designed for problems specified in the traditional, discrete continuum (for example, finite element) format.

*Continuumization of Multi-Scale Media*

A continuum is a mathematical construct. A domain is said to be a continuum if, every point within its boundaries is occupied by a material point. All media are, at some level, composed of particles and therefore none satisfies the mathematical definition of a continuum. The applicability of the continuum concept is justified by the notion that the physical size of discrete components is small compared to the aperture size for observing the medium. Therefore, the practical definition of the continuum involves some averaging scheme in which a property averaged over some suitable, finite representative elemental volume (REV). This REV value is then assigned to each infinitesimal differential element in the continuum. For this procedure to be valid, the medium is either sufficiently fine grained to allow a single property to be defined or there is such a large separation between scales that the medium can be broken into facies, each having its own property. In each case, there exists a scale separation between the scale of averaging and the scales of material variability. The REV approach assumes that the properties are distributed such that an aperture size can be found for which the spatial average is constant. That is, the medium is statistically asymptotic. Two difficulties arise from application of the REV. First, most natural media are not statistically asymptotic. Features may exist over a range of scales such that no single aperture size is suitable to define a REV. Second, the physical and chemical phenomena of interest have natural scales that interact with those of the medium. For chemical transport, models that are artificially diffusive exaggerate the ability to support key chemical reactions.

A variation on the REV concept is to define a property as a weighted spatial average such that a discrete property such as mass can be expressed as a continuous density via a convolution integral of the domain.

$$\rho(x_i) = \int_\Omega \phi(x_i - x_i') m(x_i') dx_i'. \qquad (1)$$

This definition is advantageous from two viewpoints. From an operational standpoint, gradients of the quantity can be defined in a straight-forward fashion.

$$\frac{\partial \rho}{\partial x_i} = \int_\Omega \frac{\partial \phi}{\partial x_i} m(x_i') dx_i' \qquad (2)$$

From a conceptual viewpoint, the effect of the average is to filter information such that the continuum idealization is a coarse-grained version of the original medium. Each scale is still represented, but the contribution of finer scales is reduced. In recent work by Horner (1997), a discrete element model of granular media was approximated as a coarse-grained particulate using this smoothing concept. The method amounted to approximating the difference equations of the discrete element model with a weighted residual procedure, whereby the kernel function of Equation 1 becomes a test function.

*Continuumization of the Network*

The network is an analogy to a porous medium. To relate this discrete network to the continuous representation of the porous medium, a filter must be employed. The effect of a filter is to create a continuous space onto which quantities of the network solution and the porous medium can be mapped (Figure 1). The fine-grained detail in the network and the porous medium differs greatly but, because larger-scale structures are the same for both representations, they appear similar when viewed through the filter. This fact is most apparent when the pressure solution is considered. The network and porous media are both discrete linear systems. It can be shown formally that, when the degrees of freedom of linear networks are reduced by averaging, an infinity of these networks contain identical averaged solutions. It is this equivalence property that allows us to make conductivity measurements on two specimens that are profoundly different at the micro-scale, yet obtain the same macro-scale conductivity for both. The case for non-linear behavior is not as clear-cut. Note that, properties of a medium could not be prescribed on the basis of a finite number of specimens without some equivalence property.

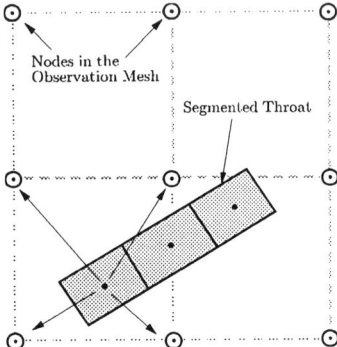

Figure 1. A schematic showing a simple, spatial averaging filter applied to an individual throat in a discrete network.

## The Inverse Problem

Specification of properties for individual throats in the network requires an inversion of the filtering procedure. The equivalence property that justifies filtering also implies that the inverse problem is ill-posed, because the fine-scale network that produces a particular filtered solution is not unique. Therefore, the goal of inversion is to find a statistical distribution of network properties that gives rise to the desired averaged medium property. To simulate flow and transport, we must create a network that displays the desired, macroscopic conductivity and apparent dispersivity.

*Conductivity.* The conductivity of a network is an equivalence continuum (average) property measured away from boundaries. The equivalence requires two levels of averaging. First, the discrete throat conductivities must be averaged to a local conductivity value at the patch scale (Figure 2). Second, these local values must be appropriately averaged to an equivalent medium property at the domain scale (Figure 2). The second of these steps is more straight-forward since the following results of Gelhar (1983) can be applied directly to relate the effective hydraulic conductivity ($K^e$) to the geometric mean ($K^g$), the problem dimension (n), and the patch-scale conductivity variance ($\sigma_f^2$):

$$K^e_{ij} = K^g_{ij}[1+(\frac{1}{2}-\frac{1}{n})\sigma_f^2] \qquad (3)$$

Patch-scale conductivities are computed as an average based on throat conductivities. This patch-average conductivity would be observed if the patch were removed from the network and placed in a constant, unit-gradient total head field.

$$K_{ij} = \frac{1}{2}\sum_{t=1}^{C_N} n_i \frac{K t}{L_t}(H_i^c - n_j L_t) \qquad (4)$$

where $H^c_j$ is computed to enforce conservation of mass at the center node.

$$H_j^c = \sum_{t=1}^{C_N} K_t n_j \bigg/ \sum_{t=1}^{C_N} \frac{K_t}{L_t}. \qquad (5)$$

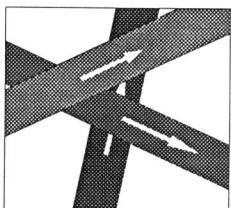

Throat Scale

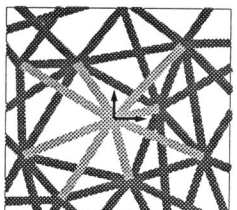

Patch or Darcy Scale

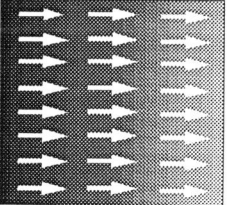

Macroscopic Scale

Figure 2. The three, distinct scales associated with the discrete network.

***Dispersion.*** Dispersion is governed by the variance in the velocity from the mean value. This variance arises from two sources. There is an essential variance caused by the various flow path directions imposed by the network topology. Even a "homogeneous" network in which all throats have the same conductive capacity will produce variations in flow-path direction. There is also variance caused by variability in throat properties. The growth in dispersivity depends on the spatial correlation among velocities (Peters and Howington, 1997) which is controlled by throat length. It was found that for all networks, the dispersivity grows from a base value, determined by the fineness of the network, to an asymptotic value that depends on the global velocity variance. The desired asymptotic values of lateral and longitudinal dispersivity dictate the variance in throat conductivity and the distribution of throat lengths in the network.

*Concluding Remarks*

The notion of spatially averaging discrete properties lies at the heart of continuum physics and the process of averaging properties is deeper than simply producing visual effects. Networks must be created to honor the site conceptualization as described by the macroscopic, observed properties. After simulation, the network model results must be spatially averaged to a format consistent with the observation techniques for comparison.

*Acknowledgements*

This work has been supported by the Army Environmental Center. Permission was granted by the Chief of Engineers to publish this information.

*References*

Gelhar, L. W., *Stochastic Subsurface Hydrology*. Prentice-Hall, Englewood Cliffs, NJ, 1993.

Horner, D. A. "Application of DEM to Micro-Mechanical Theory for Large Deformation of Granular Media," Ph.D. Dissertation, University of Michigan, 1997.

Howington, S. E., Peters, J. F., and Illangasekare, T. H., "Discrete Network Modeling for Field-Scale Flow and Transport Through Porous Media", Technical Report CHL-97-21, September 1997.

Peters, J. F. and Howington, S. E., "Pre-Asymptotic Transport Through Porous Media". On G. Delic and M. F. Wheeler, editors, *Next Generation Environmental Models: Computational Methods*, pages 271-280. SIAM, Philadelphia, 1997.

# Visualization for Analysis of Environmental Data

Peter H. Townsend[1]

## Abstract

A critical factor in an environmental investigation is the adequate analysis of site-specific data. For many projects, the challenge with understanding environmental data involves organization and visualization. This process is often neglected due to budget or schedule restrictions, resulting in insufficient interpretation and an inadequate site characterization. This paper discusses simple and cost-effective techniques for effectively evaluating environmental data.

## Introduction

The challenge with understanding environmental data involves the organization and visualization of data. This challenge can be met by utilizing appropriate graphical and statistical techniques. In many cases, applying these techniques results in the following benefits:

- Efficient organization of data
- Clear and convincing presentation of data
- Demonstration of compliance
- Cost savings.

Two general statistical methods are typically applied for the analysis of environmental data: classic statistics and geostatistics. Classic statistical methods essentially ignore locational information attached to samples, assuming that any trends or correlation will have minimal effect on the results. Classic statistics generally present a cost-effective and feasible approach for many applications. Geostatistics are used to investigate the possibility of spatial structure in environmental data. Geostatistics, when applied properly, can result in substantial cost savings during site characterization through remedial action.

The following sections present case histories which apply classic statistics and geostatistics techniques. The benefits of data visualization for presenting the results of the statistical techniques are emphasized.

---

[1]Dames & Moore, 633 Seventeenth Street, Suite 2500, Denver, CO 80202-3625; (303) 299-7868; denpht@dames.com

## Classic Statistics

Classic statistical methods focus on the characterization of the mean, variability, and "tails" of the data distributions. Applying classic statistical methods, however, require visualization and organization. Well-chosen statistical graphics are extraordinarily useful for accomplishing this task.

### Classic Statistics Case History

A recent evaluation of a groundwater and surface water monitoring program at a landfill in Colorado demonstrates the utility of well-chosen graphical techniques for visualizing environmental data. A consent decree for the site contained a clause that allowed for the evaluation of monitoring data after two years and for the scope of the monitoring program to be modified with EPA approval. Thus, the objective of the data analysis was to remove inefficiencies in the monitoring program. Removing inefficiencies from a groundwater monitoring programs involves a few obvious approaches:

- Reducing the analyte list
- Reducing sampling frequency
- Eliminating ineffective sampling locations.

Reducing the analyte list can be justified by several means, including the following:

- Demonstration of non-detects (Figure 1, where triangles represent one-half detection limit and dash represents detection limit)

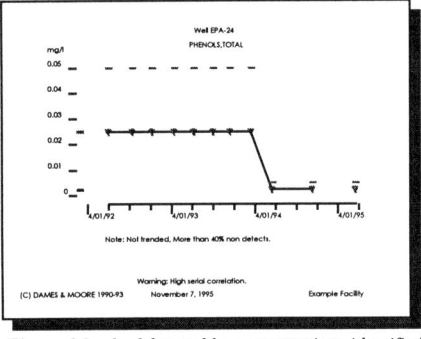

*Figure 1 Lack of detectable concentrations identified helped reduce analyte list.*

- Identification of indicator parameters (Figure 2).

Reduction in sampling frequency can be justified by the evaluation of:

- Decreasing trends below groundwater standards (Figure 3).
- Serial correlation of existing time-series data.

Evaluation of the serial correlation of time-series data provides insight to the

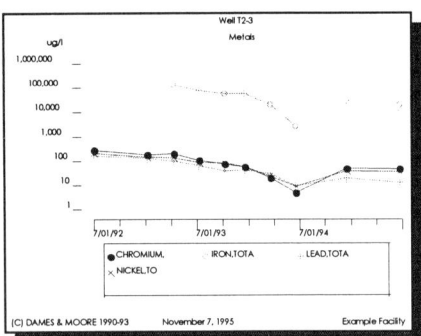

*Figure 2 Parameter overlay plot identifying potentially suitable indicator parameters*

self-similarity of the groundwater concentrations (or sequence). "High" serial correlation may indicate that the sampling interval is too frequent. In simple terms, less data may be able to characterize the same variability in the sequence. Figure 4 illustrates groundwater concentrations for chloride where a high serial correlation was identified. With the exception of the identified outlier, the data generally exhibit a variability that could be characterized with a less frequent sampling interval, such as semi-annually.

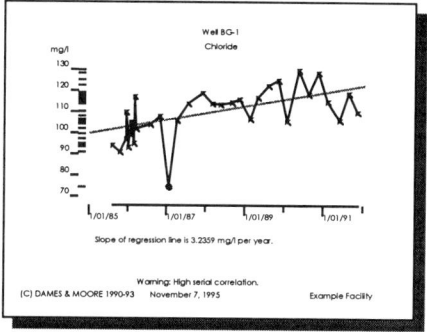

*Figure 3* Chloride concentrations in groundwater where high serial correlation was identified

Reduction in the number of monitoring points may entail the use of several techniques, including the evaluation of the spatial distribution of analyte concentrations. The number of wells and the placement of the wells can be optimized with the use of geostatistics.

The application of geostatistics for characterizing environmental data is presented in the Geostatistics Case History section.

## Geostatistics

Geostatistics are designed to manage spatially-related data that is often common to environmental data sets. When properly applied, geostatistics often

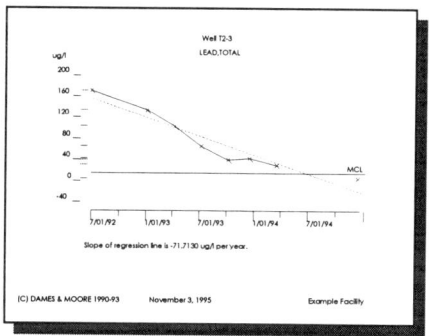

*Figure 4* Statistically significant trend of lead concentrations in groundwater helped justify reduction in sampling frequency

result in cost savings to a site owner through better definition of contaminated areas, reductions in sampling, optimal design of remedial programs, and more effective negotiations with regulators.

Many environmental sampling programs are designed to measure the nature and extent of suspected contamination. The effects of varying distance and conditions between samples causes uncertainty in estimates of the geometry and properties of the suspected plume. Regulators often expect exhaustive sampling efforts to reduce this uncertainty. The resulting data are often concentrated in contaminated areas, and form the basis for risk assessment calculations. Therefore, spatially-biased sampling can result in an over-estimate of risk. Risk calculations are a major factor in the final cost of remedial programs.

Geostatistical techniques promote unbiased sampling design and efficiency in sampling. Efficiency is achieved by reducing redundant sampling. The resulting data tend to better represent the entire site, and are less biased toward contaminated areas. The results of a properly designed geostatistical program are scientifically defensible, and are based on techniques supported by the EPA. The agency's support of geostatistical techniques, therefore, results in significant cost savings to a site owner.

**Geostatistics Case History**

Geostatistical methods were utilized in the evaluation of a proposed soil sampling program. The proposed sampling program neglected to evaluate the spatial structure of historical soil concentrations data. As a result, the proposed sampling plan was overly conservative. Figure 5 illustrates the preexisting samples and the additional samples proposed in the sampling plan.

Utilizing geostatistics, a spatial structure of the contamination was identified. Geostatistics was used optimize the number and location of samples required to characterize the contaminant distribution at the site (Figure 6). The result of the study was a large reduction in the resulting field program and lab costs.

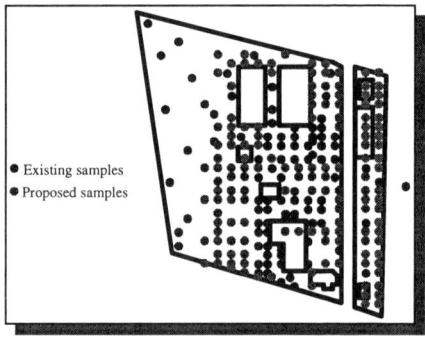

*Figure 5* *A proposed sampling plan, which neglected the spatial structure of the data was reevaluated using geostatistics.*

## Conclusion

As a result of the data analysis presented in both the Classic Statistics Case History and Geostatistics Case History, several significant characteristics of the data were quickly identified and presented in a clear and concise manner. These studies resulted in reductions in sampling requirements and cost-savings to the client.

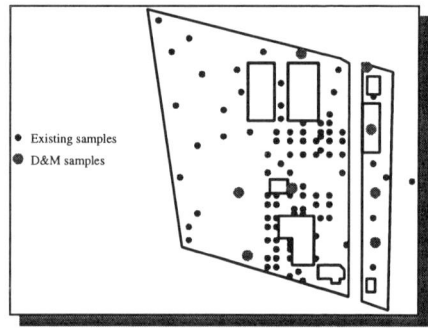

*Figure 6* Geostatistical analysis resulted in less samples and significant cost reductions.

Both case histories illustrate the importance of accomplishing an adequate analysis of site-specific data at environmental investigations. Each study utilized simple statistical techniques coupled with well-chosen data visualization tools, resulting in substantial cost-savings to the client.

Problems Applying MODFLOW-Style Models to Three-Dimensional Flow

Peter O. Sinton[1]

Finite-difference models are the most widely applied simulators for groundwater flow problems. The popular MODFLOW model is a good example. Some finite-difference models can be used to simulate groundwater flow through aquifers that have variable elevation or variable thickness. Examples are (1) a steeply-dipping sandstone layer between shale layers that thicken basinward, and (2) an alluvial valley-fill aquifer that slopes down valley coincident with a river and is thinner at its margins than its center. This work deals with the application of such models to simulate three-dimensional groundwater flow in systems with variable-elevation and variable thickness layers. Two types of approaches can be used to simulate such systems with MODFLOW and similar models: (1) with layers that change elevation or thickness coincident with natural layering, and (2) with layers of constant thickness and elevation. In the groundwater practice it is common for professionals to use the first approach, however there has been little work regarding the best way to apply the models. The two approaches are compared with each other to gain a better understanding of how models like MODFLOW should be applied to simulate three-dimensional groundwater flow systems.

The two approaches for simulating groundwater flow are called "three-dimensional" and "quasi three-dimensional." A three-dimensional (3D) model as applied in this work refers to a finite difference model with rectangular cells. The model is constructed so that there is no vertical space between the top of one cell and the bottom of the cell above. In the horizontal direction, there is no vertical offset between adjacent cells. A quasi three-dimensional (Q3D) model has layers that can change elevation and thickness.

The way in which flow is simulated in the two types of models is different. Each model must calculate flows from one cell to the next. To do this, the cross-sectional area between cells and distance between cells are used in the Darcy equation to compute the flow from one cell to the next. The distance between cells is used to

---

[1] Senior Hydrogeologist, Dames & Moore Group, Inc., 633 17$^{th}$ Denver, Colorado, 80202.

calculate the hydraulic gradient between cells. In a 3D model, each cell is aligned horizontally and vertically with all adjoining cells. Therefore, the cross-sectional area and distance are directly obtained from the geometry of the model grid.

In a Q3D model, horizontally adjacent cells can be vertically offset. If two cells are offset, the cross-sectional area between the cells is less than the area of the face of either of the cells. Similarly, the distance between cells is greater than the horizontal distance because due to vertical offset. However, in models like MODFLOW, the cross-sectional area and distance between horizontally adjacent cells is calculated using the same equations regardless of whether the model is 3D or Q3D. Hence, cross-sectional areas and hydraulic gradients are incorrectly calculated in Q3D models. The cross-sectional area and gradient are consistently too large, and flow simulated by a Q3D model will be larger than flow simulated in a similar 3D model.

To compare the two types of models (3D and Q3D), a simple two-dimensional, steady-state problem is simulated using MODFLOW. A confined aquifer 304.9 meters (1,000 feet) long is simulated with a change in hydraulic head of 0.3048 meters (1 foot). Constant-head boundaries are specified at the upgradient and downgradient ends of the model. The aquifer is about 15.2 meters (50 feet) thick, with a hydraulic conductivity of 61 meters/day (200 ft/day) and porosity of 20%. The aquifer is simulated in cross-section, so that vertical changes in aquifer geometry can be simulated. A well injecting 0.0283 meter$^3$/day (1 ft$^3$/day) is simulated at about 30.48 meters (100 feet) from the upgradient edge of the model. Particle tracking is used to assess groundwater velocities.

Two sets of geometric configurations are simulated. The first set of geometric configurations simulates aquifers with constant thickness and variable elevation. The second set simulates aquifers with variable elevation and variable thickness. For the constant-thickness aquifers, the dip of the top and bottom of the aquifer is about 0°, 1°, 2.5° and 10°. In all cases, the vertical distance from the bottom to the top of the aquifer is maintained, but within each case the saturated thickness measured perpendicular to flow direction changes (greater dip equates to smaller saturated thickness). For the variable-thickness aquifers, the dip of bottom of the aquifer is about 0°, 1°, 2.5° and 10°, and thickness varies uniformly so that the aquifer is about 20% thicker at the downgradient end of the model. Vertical distances are handled similar to the constant-thickness cases. A total of 15 cases were simulated, one 3D model for the 0° dip/constant-thickness aquifer, and two models (one 3D and one Q3D) for the other geometric configurations.

All models used the same horizontal grid spacing (1.52 meters or 5 feet). The vertical spacing between the 3D and Q3D models were matched as well as possible. Each model was simulated using the same numerical solver with the same set of closure criteria. Using this approach, the differences between the model arising from descretization or the solver are minimized.

Two measures are used to compare 3D and Q3D models. The first measure is based on the volumetric budget. The volumetric budget is a listing of the groundwater flows into and out of the model. Q3D models, which over-estimate flow, simulate the same flow no matter the geometry as dip increases (Figures 1 and 2). Flow is constant between the Q3D models because the simulated saturated thickness does not change as dip increases. To fix this error, the analyst would have to specify the saturated thickness, which is measured perpendicular to the flow direction. To do this, the analyst would have to know in advance of the simulation the flow patterns in the aquifer. However, this is generally unknown prior to modeling and so counter-productive. Indeed, it is usually the objective of a model to simulate flow patterns because they are not known. The flow overestimated by the Q3D model is consistent with the overestimated cross-sectional area and hydraulic gradient.

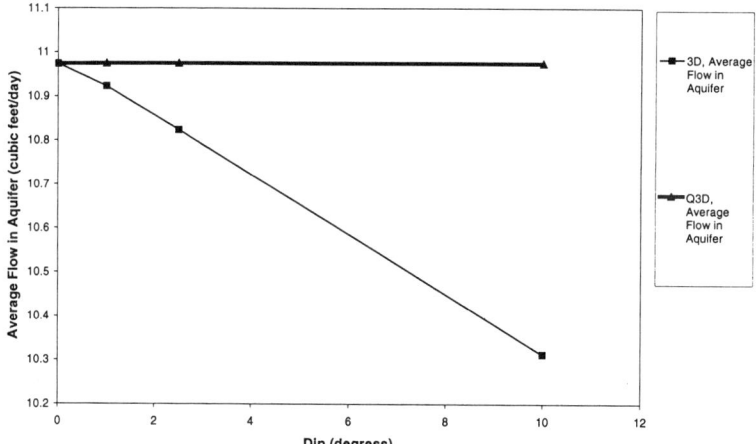

Figure 1. Effect on Volumetric Budget, Constant-Thickness Aquifer

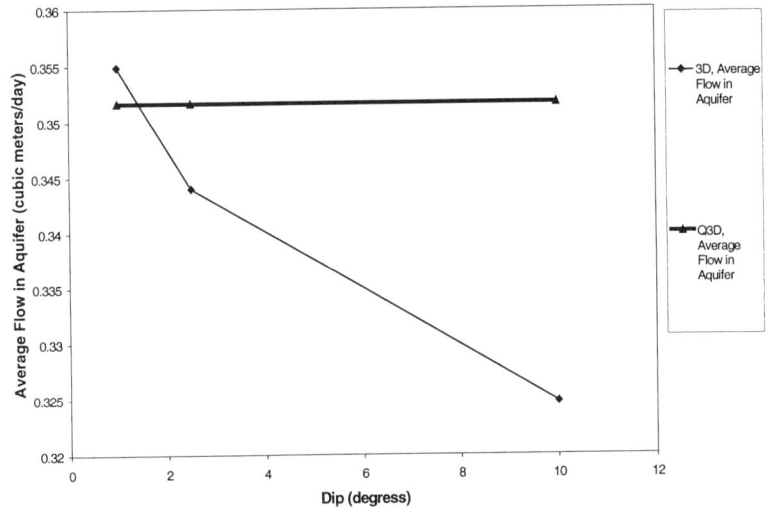

Figure 2. Effect on Volumetric Budget, Variable-Thickness Aquifer

More realistic estimates of actual flows in the aquifer are obtained using the 3D approach (Figures 1 and 2). The 3D models correctly simulate the flow pattern and the saturated thickness is correctly used in the model. The flows estimated by the 3D models decrease as dip increases because saturated thickness decreases.

The second measure is based on particle tracking. Particle tracking can be used to illustrate groundwater flow rates and patterns, which is useful in understanding contaminant transport. The time it takes particles released from the injection well to reach the downgradient edge of the model (travel time) is used as the basis for characterizing differences between 3D and Q3D models. Travel times for constant- and variable-thickness models are constant regardless of dip and are underestimated (Figures 3 and 4). This is consistent with the overestimated flows in Q3D models. Therefore, Q3D models incorrectly simulate particle travel time, which infers that contaminant transport models based on the Q3D approach will yield erroneous results.

WATER RESOURCES AND THE URBAN ENVIRONMENT 659

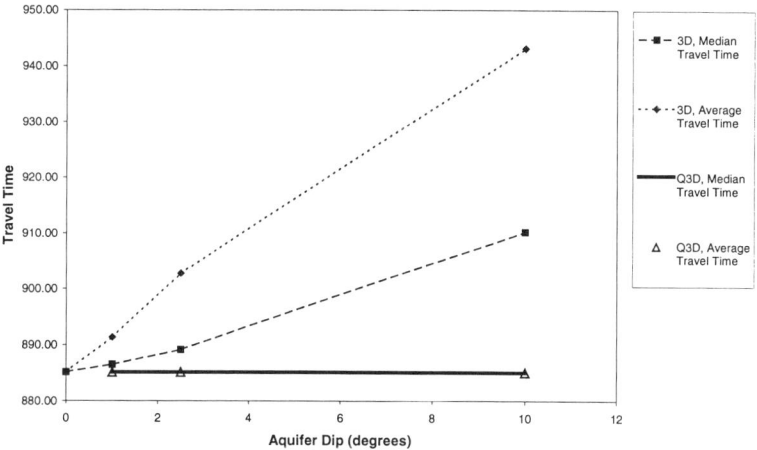

Figure 3. Travel Time Comparison, Constant Thickness Aquifer

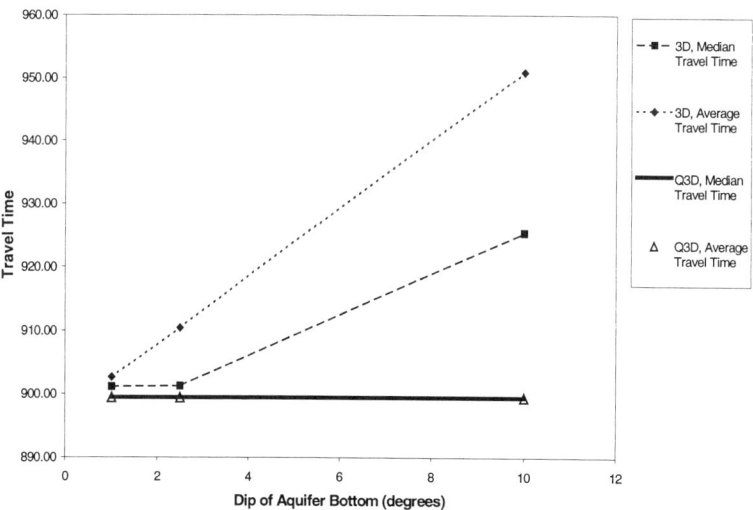

Figure 4. Travel Time Comparison, Variable Thickness Aquifer

The travel-time results for the 3D models show how descretization can affect particle-tracking calculations. The particle-tracking analyses were done with a uniformly distributed column of particles located along the injection well. In the 3D models, the top and bottom of the aquifer is approximated by a stair-step type configuration of cells. This causes particles near the top and bottom of the aquifer to travel longer paths than do particles located in the central part of the aquifer. The longer flow paths result when particles must flow around the edges of cells at a stair step. This phenomenon is shown in Figures 3 and 4 where the average and median travel times differ. The median travel time approximates the true travel time in the aquifer. The average is skewed toward longer times because of the stair-step travel paths. Therefore, particle-tracking calculations can be strongly affected by grid geometry.

Based on the results of this work, general guidelines can be developed for deciding what method to use when preparing a numerical model of an aquifer. Aquifers having dips in excess of about 2.5° or thickness changes of more than 20% probably should be modeled using the 3D approach when using a finite-differences numerical model. In applying these guidelines, the analyst should consider model-wide geometry as well as the geometry in areas of localized interest. Further work would be useful in characterizing the effects of using the Q3D approach to solve real-world problems.

## The Effect of Mass Transfer Limitations on the Optimization of Groundwater Remediation Strategies

Aysegul Aksoy[1] and Teresa B. Culver[2],A.M. ASCE

Abstract

The effect of mass transfer limitations on the optimization of groundwater remediation strategies was investigated by linking a ground water simulation model into a Genetic Algorithm (GA) optimization technique. Runs were performed to minimize pumping and treatment costs associated with pumping rates and air stripping tower performance. Remediation designs for a hypothetical aquifer, contaminated with trichloroethylene (TCE), assuming either linear equilibrium or two-site kinetic sorption, are compared. Results indicate that the equilibrium sorption assumption for groundwater remediation optimization may result in over-estimation of the clean up rate and in sub-optimal designs.

Introduction

Sorption to natural solids is an important process affecting the transport of organic compounds in ground water. Site managers typically assume instantaneous equilibrium sorption of the contaminants for modeling purposes since it is easier to obtain equilibrium sorption data experimentally. However, sorption may require weeks to many months to reach equilibrium due to slow kinetic mass transfer between the sorbed and aqueous phases (Weber and Miller, 1988; Brusseau and Rao, 1989). Especially, for sites contaminated for a long period of time, slow kinetic sorption may prevail with low mass transfer rate coefficients.

Aquifer cleanup efforts to remediate contaminated sites typically involve operation of a system of extraction wells followed by a treatment system. Although widely used, the performance of a pump-and-treat system is limited in part by the mass transfer of the contaminant between the aquifer soil material and the aqueous phase (Goltz and Oxley, 1991).

This work explores the effect of mass transfer limitations on the optimal design and costs of ground water remediation. For this purpose results obtained with linear

---

[1]Graduate Research Assistant, Department of Civil Engineering and Applied Mechanics, University of Virginia, Charlottesville, VA 22903-2442
[2]Assistant Professor, Department of Civil Engineering and Applied Mechanics, University of Virginia, Charlottesville, VA 22903-2442

equilibrium sorption model and two-site kinetic non-equilibrium sorption model are compared.

Approach

A GA was used as the optimization technique. GAs are robust optimization algorithms when applied to ground water remediation and management problems (McKinney and Lin, 1994; McKinney and Lin ,1996; Ritzel et al., 1994; Cieniawski et al., 1995). The basic idea in GAs is the representation of a population of potential solutions as strings, in a chromosome-type encoding. These encoded possible solutions are then manipulated through reproduction, crossover and mutation operators. The strings are evaluated based on an objective or fitness function. Better strings are placed in a mating pool. Strings are mated and then genetic information is exchanged between the mated strings at the crossover site with a certain probability. A mutation operator is used keep the population diverse. The generated new population is evaluated again, and this cycle is repeated until a stopping criterion is met (Goldberg, 1989).

For optimization, a GA was linked with a two-dimensional finite element simulation model, BIO2D-KE (Earles, 1996; Earles, et al., 1996). BIO2D-KE includes linear equilibrium and two-site kinetic sorption models. In the two-site model, sorption sites are assumed to be divided into two fractions. Adsorption on one fraction, F, is instantaneous, while adsorption to the other fraction, (1-F), is time dependent. The head and concentration values, at the end of the management period were simulated by BIO2D-KE.

The optimization approach was applied to a homogeneous, isotropic and confined hypothetical aquifer similar to that of McKinney and Lin (1994, 1996). The 335 m by 244 m aquifer was discritized into 352 nodes with 391 grid points. There existed no-flow conditions on the north and south boundaries of the aquifer. A steady flow in easterly direction was created by constant head boundaries of 35 m and 30.8 m on the west and east sides of the aquifer, respectively. The constant concentration boundaries on the east and west sides of the aquifer were set to 0 mg/l. The longitudinal and transverse dispersivities were set to 21.3 and 2.13 m, respectively.

The decision variables for the optimization runs were the injection and extraction rates through 5 pre-located potential wells and the concentration of the air stripping tower effluent, which was injected back into the system. Each parameter was represented by a 6-bit binary code resulting in a 36 bit long individual string composed of 6 decision variables. The population size was chosen to be 100. Iterations were carried for 100 generations. Fourteen observation wells were used in order to penalize the alternative policies that had violated the constraints at the end of the management period of 5 years.

The objective function was formulated to minimize the costs associated with pumping (injection and extraction) rates and air stripping tower performance based on the effluent concentration disposed from the treatment plant. The cost of the remediation was increased by multiplying the cost with a penalty function for policies resulting in constraint violations. Constraints for the system included;
(1)  Upper and lower bounds for pumping rates of 0.0 and 11.3 l/s, respectively,
(2)  Injection of the extracted water back into the system after treatment,
(3)  Upper and lower bounds for head values (0 and 60.96 m, respectively),
[4]  Regulatory concentration levels within the aquifer at the end of the management

period based on "EPA Drinking Water Standards" (EPA, 1996).

Example Results

The initial contaminant plume was generated by placing a point source of 1 mg/l of TCE at node 77. The transport of TCE was simulated for 20 years with both linear equilibrium and two-site kinetic sorption models. The mass transfer parameters used for the simulations were 0.18 cm$^3$/g for the distribution coefficient (Kd), and the kinetic mass transfer rate coefficient ($\alpha$) was varied. For the base case $\alpha$ was 0.0035 d$^{-1}$. The fraction of equilibrium sites (F) was 0.75 for the two-site model and 1.0 for the equilibrium model, respectively. TCE plumes obtained at the end of the contamination period using both sorption models were similar with slight variations in concentrations at far field. For optimization runs initial TCE plume was created by taking the average of the TCE concentrations obtained with linear equilibrium and two-site kinetic sorption models. The initial plume and locations of the potential and observation wells are shown in Figure 1, where the total concentration of TCE, which includes sorbed TCE, was given in terms of mg TCE per liter of aquifer. The concentration goal at the end of the management period was set to 0.005 mg/l.

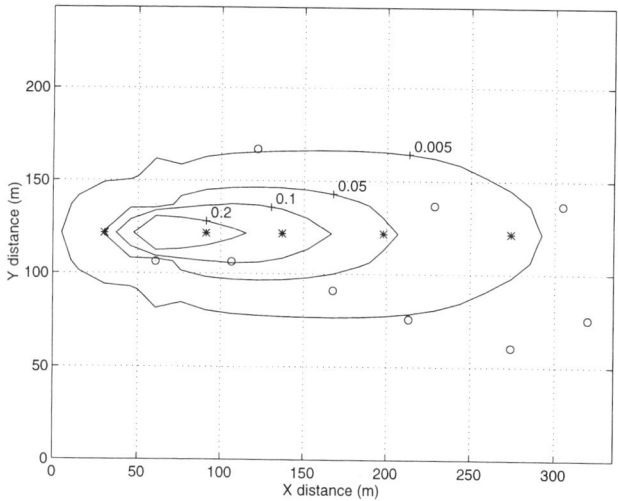

Figure 1. Initial total TCE concentration contours after contamination for 20 years (o = observation wells, * = potential wells)

Optimum policies for the base case, consisting of pumping rates (Q) for 5 potential wells numbered from left to right (- for extraction) and air stripping tower effluent concentration (CE), are stated in Table 1. The best policy obtained with linear equilibrium sorption resulted in a cost of $84782 with no violation of the constraints at the observation wells. On the other hand, optimization using two-site kinetic model ended up in a different policy of pumping rates and treatment effluent concentration. The cost was $85577. The design volume for the air stripping tower for equilibrium sorption case (2.25 m$^3$) was less than of the kinetic sorption case (2.81 m$^3$). This resulted in tower effluent concentration of almost 50% less for two-site case than for the equilibrium case. The optimum policy for two-site model resulted in a higher treatment demand in order to avoid constraint violations at the observation wells. Resulting contaminant plumes obtained using the optimum policies for equilibrium and two-site kinetic sorption assumptions are given in Figures 2 and 3, respectively. Comparison of these figures shows that at the end of the management period more contaminant mass is retained within the system for the two-site sorption model.

Table 1. Best policies obtained by the remediation system optimization for different sorption models

| sorption model | Q(1) $m^3/d$ | Q(2) $m^3/d$ | Q(3) $m^3/d$ | Q(4) $m^3/d$ | Q(5) $m^3/d$ | CE mg/l |
|---|---|---|---|---|---|---|
| linear equilibrium | 0.0 | 0.0 | -598.4 | 0.0 | 598.4 | 0.0039 |
| two-site kinetic | 0.0 | -535.4 | 0.0 | 535.4 | 0.0 | 0.0014 |

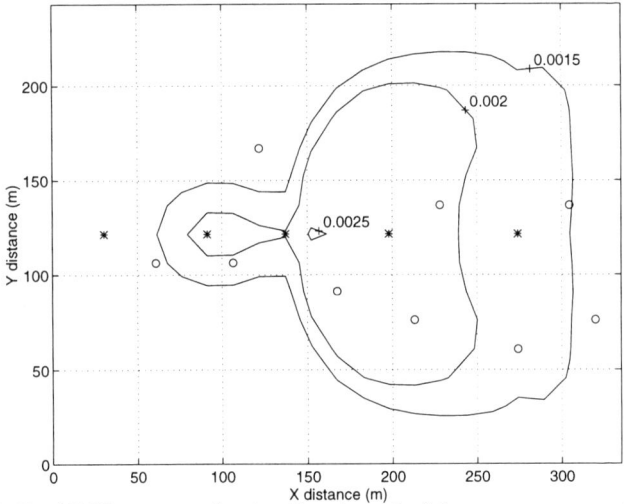

Figure 2. Total TCE concentration (mg/l) at the end of the management period for the base case; linear equilibrium sorption case

To further evaluate the significance of mass transfer limitations, the optimal policy determined using the assumption of equilibrium sorption (see Table 1) was applied to the aquifer with mass transfer limitations. Since most analysts simply assume that groundwater contaminants will have equilibrium sorption, this analysis explores the bias introduced if the assumption is inaccurate. If an "optimal" equilibrium sorption policy is applied to the system with mass transfer limited sorption, the water quality goals will not be met. For the base case example, the largest water quality violation was 37% over the water quality standard. Due to the water quality violations, the policy (with an actual cost $84782) was assigned a fitness of $284868. Furthermore, the average total concentration within the aquifer was 11% greater than that predicted using the assumption of equilibrium sorption (Figure 4).

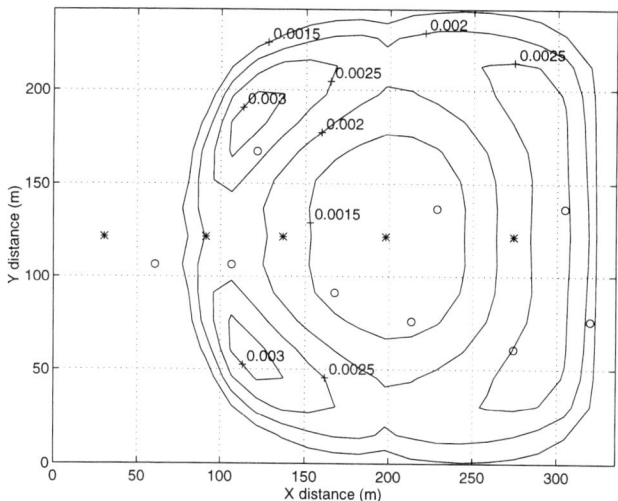

Figure 3. Total TCE concentration (mg/l) at the end of the management period for the base case; two-site kinetic sorption case

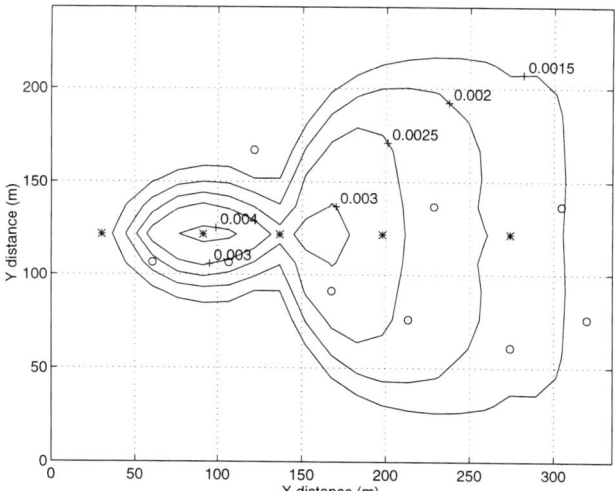

Figure 4. Total TCE concentration (mg/l) at the end of the management period for the base case; two-site kinetic sorption case with the policy obtained for the linear equilibrium sorption case

In addition to the base case, the potential impact of mass transfer limitations on the optimization performance will be presented given a range of mass transfer rates ($\alpha$).

Conclusions

The equilibrium sorption assumption for groundwater remediation optimization may result in over-estimation of the clean up rate and in sub-optimal designs. For systems where mass transfer is slow, this effect may be significant. Thus defining the sorption parameters, including sorption rates, is a crucial step for optimal, cost-effective remediation designs.

References

Brusseau, M.L., and Rao, P.S.C., Sorption Nonideality During Organic Contaminant Transport in Porous Media, *Critical Reviews in Environmental Control*, 19(1), 33-99, 1989.
Cieniawski, S.E., Eheart, J.W., and Ranjithan, S., Using Genetic Algorithms to Solve a Multiple Objective Groundwater Monitoring Problem, *Water Resources Research*, 31(2), 399-409, 1995.
Earles, T.A., Gray, J.P, and Culver, T.B., *User's Guide to BIO2D and BIO2D-KE*, U.S. Army Research Office, TCN Number: 95-066, 1996.
Earles, T.A., *Modeling Rate Limited Sorption*, MS Thesis, University of Virginia, Department of Civil Engineering, 125 pp., 1996.
EPA, http://www.epa.gov/ostwater/Tools/dwstds0.html, update date: October, 1996.
Goldberg, D.E., *Genetic Algorithms in Search, Optimization and Machine Learning*, MA: Addison-Wesley Publishing Company, Inc., 1989.
Goltz, M.N., and Oxley, M.E., Analytical Modeling of Aquifer Decontamination by Pumping When Transport is Affected by Rate-Limited Sorption, *Water Resources Research*, 27(4), 547-556, 1991.
McKinney, D.C., and Lin, M., Genetic Algorithm Solution of Groundwater Management Models, *Water Resources Research*, 30(6), 1897-1906, 1994.
McKinney, D.C., and Lin, M., Pump-and-Treat Ground-Water Remediation System Optimization, *Journal of Water Resources Planning and Management*, 122(2), 1996.
Ritzel, B.J., Eheart, J.W., and Ranjithan, S., Using Genetic Algorithms to Solve a Multiple Objective Groundwater Pollution Containment Problem, *Water Resources Research*, 30(5), 1589-1603, 1994.
Weber, W.J., and Miller, C.T., Modelling the Sorption of Hydrophobic Contaminants by Aquifer Materials-Rates and Equilibria, *Water Research*, 22(4), 457-464, 1988.

## Dynamic Ground Water Management System based on GIS

## By Werner Erhart-Schippek[1], Herbert Mascha[1], Member ASCE

## 1 Abstract

A dynamic ground water monitoring and management system, which is based on a geographical information system and a digital ground water model was developed for the extension of the protection of a drinking water well. The system was developed to reduce the risk of shortages of drinking water supply due to ground water contamination caused by a nearby industrial contamination and an old landfill. With the help of the management system daily reports of the ground water flow condition and of the ground water quality are prepared automatically. If a deviation from the specified conditions is detected, measures to avoid shortages of the drinking water supply can be taken long before the contamination reaches the well.

## 2 Introduction

The City of Kapfenberg with a population of approximately 25.000 is situated along the river Muerz in Styria, Austria. The region of the Muerz valley has a long history of industrial production on the basis of iron, which has been exploited in the region for several hundreds of years. Therefor the basic industry of the region is steel industry and associated industries.

The Public Works of the City of Kapfenberg (PWK) are operating a drinking water supply plant. The well is situated close to the river Muerz. The local ground water situation in the area is not influenced by the river Muerz. The water, which is pumped from the well, therefor is no river bank filtrate but comes from the large ground water body of the Muerz valley. In a distance of approximately 2 km upstream of the well there is a large industrial complex. In parts of this complex contamination of soil and ground water by hydrocarbons and chlorinated hydrocarbons was identified. Upstream of the drinking water well there are also a closed landfill, a local airport and a hydropower plant.

---

[1]Managing Director of Erhart-Schippek, Mascha & Partner Ressource Management, A 2340 Moedling, AUSTRIA, Wiener Strasse 9, email: office@esmp.co.at

## 3 Specification of the problem

The water supply plant started operation in 1950. At this time an area with restricted land use for the protection of the well was defined. At present there are no problems with the quality or the quantity of water from the existing well. The intensive use of the land upstream of the well and the risk of ground water contamination caused by the industrial complex, a closed landfill and several other activities are an hazard to the drinking water supply of Kapfenberg. As a consequence of higher water demand and considering soil and ground water contamination the existing protection concept had to be reviewed and revised to meet present and future demands of the supply of the population of Kapfenberg with drinking water of excellent quality.

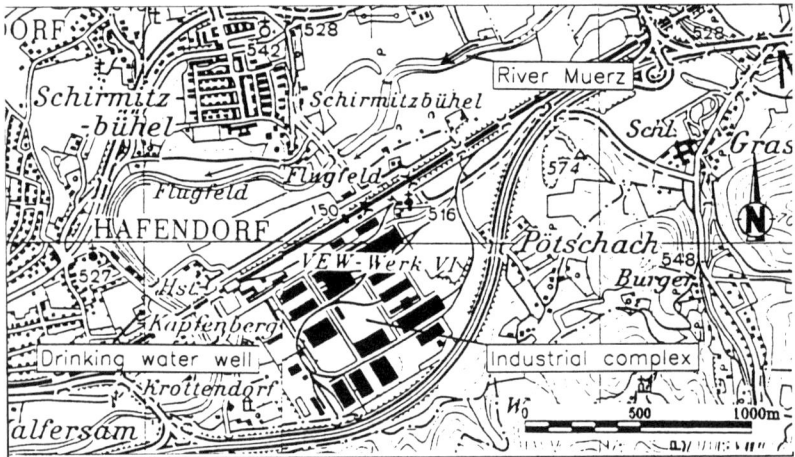

Figure 1. Area map

## 4 Project conception

For the proceeding a detailed project plan was developed. As a basis a geographical information system (GIS) was implemented. This system had to serve as a basis for the data collection and data evaluation. All geographical information (street map etc.), available information about the industrial complex and the location of the sources of the contamination related to the industrial complex, information about the landfill and other relevant information about the area upstream of the well were defined as topics and put into the GIS. The geological and hydrogeological data were first administered in a separate database (TECHBASE®). After evaluation the results were introduced to the GIS.

For the evaluation of the ground water situation a detailed ground water model was established. As a first step a 2-dimensional model was applied (WASY FEFLOW®).

The model was established on the basis of detailed hydrogeological investigations. The existing monitoring system was extended to allow accurate model calibration. The model was designed to allow a transfer of the results to the GIS for further interpretation.

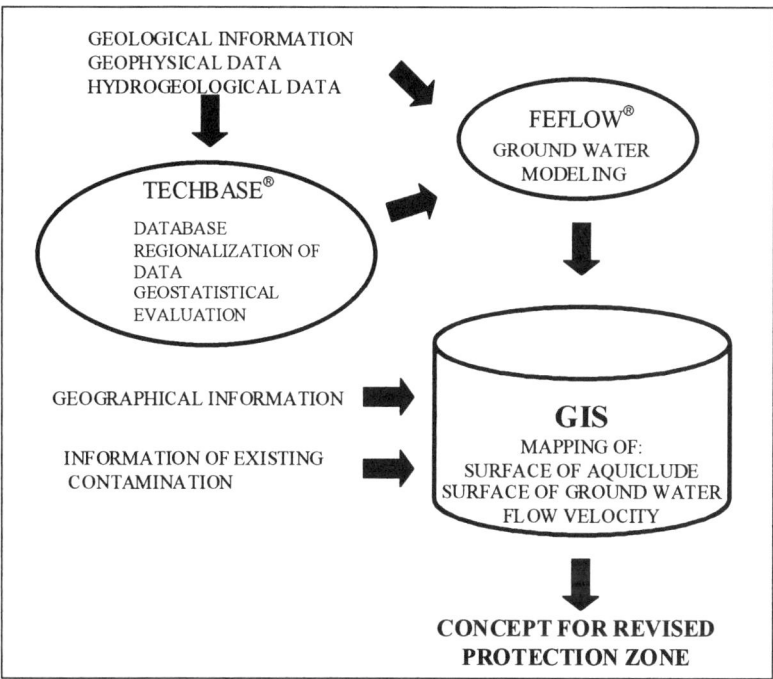

Figure 2. Concept of data administration and evaluation

## 5 Hydrogeological situation

The valley of the river Muerz is a rather narrow valley (1,0-1,5 km wide). The basic geological formations in the region of Kapfenberg are crystalline rocks, which are part of the middle and eastern alpine nappes with residuals of central alpine sediments, mostly lime stone and dolomite. The quartarian sediments are sandy gravels. They are up to 25m thick. There is only one aquifer with a thickness of approximately 10m. The general direction of ground water flow is parallel to the river Muerz. There are only specific sections of the river Muerz with a direct hydraulic contact to the ground water body. The section upstream the hydopower plant is separated from the ground water as a consequence of the intensive drawing of well water at the wells, which are used for the supply of the industrial complex. Around the river bend close to the drinking water supply plant exfiltration from the river into the ground water body takes place. In most of the other sections the river bed is situated several meters above the aquifer with a limited connection between the river and the aquifer.

## 6 Ground water situation

### 6.1 Dynamic of ground water

As a consequence of the construction of the hydropower plant and the construction of cutoff walls along the reservoir in 1988 colmation in the upstream section of the Muerz took place. This caused a reduction of the natural flow dynamics of the ground water. Downstream of the plant the ground water table was lowered up to 2m in some areas. A seasonal change of the ground water table still takes place, but it is reduced to the range of a few decimeters. The highest water table is reached in April and June, the lowest in December and January.

### 6.2 Ground water quality

Up to date there were no ground water quality data recorded, which gave evidence of a ground water quality, which would be a limitation to the use of the ground water for drinking water supply. The known sources of contamination have caused only local contamination of the ground water without negatively influencing the quality of the water in the well so far. One reason for this is the operation of the wells for industrial water supply (see also 6.3).

### 6.3 Influence of the industrial complex

The industrial complex has a history, which goes back to the end of the last century. The local industry is an important economical and social factor in the region. Due to the technical and economical developments within the last decades the steel industry in the region faces severe problems.

Within the industrial complex several sources of contamination, which are typical for the steel industry and associated branches, have been identified and investigated. The contamination of soil and ground water originate from oil spills, spills with chlorinated hydrocarbons (e.g. tetrachlorethen) as well as from seepage from sludge deposits containing PAH and heavy metals.

The detailed investigation of the ground water situation of the area and the evaluation of the results from ground water modeling showed, that the ground water flow situation is strongly influenced by the industrial water supply. There are several wells within the industrial zone, which are used for water supply. The quantity of water (150-160 l/s), which is pumped from these wells is big enough to show strong influence on the flow conditions upstream the drinking water supply well. At present the operation of these wells with the current rate of outtake assures that no contamination will be drawn from the industrial zone to the drinking water well.

The information about the contamination is still limited. Some remediation of the area has taken care only of relatively small parts of the sources of contamination and a general remediation of the area is not possible for economical and also technical reasons. Most of the remediation will have to be financed by government means as the costs of remediation would cause closure of the existing plants.

### 6.4 Land and ground water use

Apart from the industrial complex, which was discussed above, there are several other uses of land and ground water, which cause conflicts with the protection concept of the drinking water well. There are several small businesses, where products are in use, which could cause a contamination of the ground water. Although these businesses are small, their data were included, but there was no relevant input to the project. Most of these businesses get their water supply from the PWK and do not operate own wells.

There are also 2 sand and gravel pits in the area. Both pits were exploited below the ground water level. One pit has been closed down and was transferred to a lake. At the second pit the exploitation has also been finished, but there is still a gravel washing plant in operation. This pit will also be transferred into a lake.

The main traffic routes in the area (railway, Austrian highway S6) and the local airport are no relevant risk to the ground water quality. During winter time the use of salt for road conditioning may cause a local increase of salinity, but is not of relevance to the overall ground water quality.

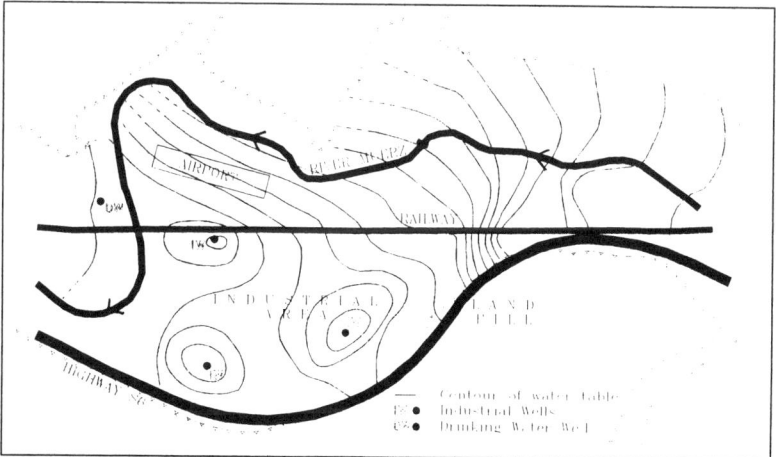

Figure 3. Ground water flow conditions

## 6.5 Contaminated sites

Besides the contaminated areas in the industrial complex there is an old landfill of greater importance to the revision of the protection concept. The landfill, which was used by a chemical company, is approximately 2 km upstream of the drinking water well. A general investigation of the landfill, which was carried out by the government in Styria in 1989, did not indicate any relevant gas emissions besides a low level of chlorinated hydrocarbons, which was below levels, which require further investigation or remediation. A contamination of the ground water from the landfill is likely, although the available quality data down stream did not indicate any contamination so far.

## 7 Revision of the area of protection using GIS

After the collection of the relevant information, the evaluation of the geological and hydrogeological information was carried out with a separate database and the results were introduced to the GIS. At this stage the GIS was mainly used for the administration of the numerous data. For the interpretation of the ground water situation several simulations with different assumptions of the operational conditions of the wells within the industrial complex were calculated. These results were also introduced into the GIS. Overlaying this information maps were produced, which showed how the operation of the wells within the industrial complex influences the risk of contamination of the drinking water well. At the present rate of operation not only the ground water flow from the industrial area, but also from other areas, where a certain risk for ground water contamination was identified, is drawn away from the drinking water well. A significant reduction of the pumping rate or the termination of the operation would cause a drastic change to the flow conditions of the ground water. As a consequence of these changes contaminants could be transported to the drinking water well and cause quality problems.

On the basis of this results the concept of the protection area for the drinking water well was revised. Besides the existing area the area of the industrial complex was proposed to be included.

### 7.1 Legal and economical constraints

The proposal of the concept caused strong opposition by the local industry. The argument of the industry was, that the consequences of the realization of the protection concept would cause additional costs from remediation of contaminated areas as well as specification of pumping rates and operational conditions for their wells and limitation of land use for future projects. These costs and restrictions would force the industry to move production to other locations and to close the plants in Kapfenberg.

The existing permissions for the operation of the wells for the industrial water supply do not include regulations, which specify the operation of the wells under all conditions or specify a minimum pumping rate. The Austrian Water Law includes regulations, which allow the change of existing permissions under certain conditions. One of these conditions is the immediate risk for ground water contamination. In the present case the risk of ground water contamination is a potential risk, because in spite of detailed ground water modeling and evaluation of the existing data the risk assessment gave as result no immediate hazard to the drinking water well. Therefor the change of the existing permissions has only a week legal position and even if it was passed in first place, a revision by a superior court is likely. As a consequences of the legal and economical constraints the concept for the protection of the drinking water well cannot be implemented as initially planned.

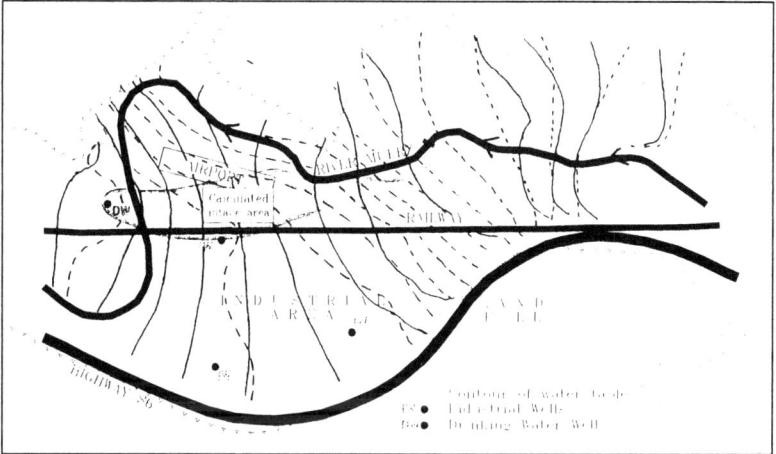

Figure 4. Results of ground water modeling

## 8 Extension of protection by implementing a dynamic ground water monitoring and management system

As the concept of the extension of the protection area upstream of the drinking water well cannot be realized, an alternative concept was developed. The already existing GIS and the ground water model will be used as a basis for the new concept. As a direct influence on the operation of the wells in the industrial complex is not possible, the protection of the drinking water well cannot rely on the protected flow conditions upstream. The ground water flow conditions are very well known as a result of the detailed modeling and it is possible to specify exactly the flow conditions, which make sure, that no ground water from areas, where contamination is already identified or suspected reaches the intake area of the well. With an extension of the monitoring system over a wide area upstream it is possible to monitor changes of the flow conditions very early. The extension of the monitoring system is planned on the basis of GIS and overlaying the result of ground water modeling. For the PWK it is therefor possible to react several weeks before a contaminated plume will reach the drinking water well. This time can be used for detailed investigation of the contamination and for the planning and evaluation of the optimum measures for protection.

The ground water flow conditions are evaluated on a daily basis. The monitoring data are collected digitally and transferred to the head quarter of the PWK, where the ground water model and the GIS are installed. The data are read into the ground water model automatically. On the basis of the latest monitoring data the present flow situation is evaluated and compared with what was specified as the desired flow conditions. This procedure is also automated. The ground water model is revised to make the automated operation possible. If there are relevant changes of the land use upstream the well or within the industrial complex the information can be read into the GIS. The specification of the ground water flow conditions and the ground water quality can be revised as needed.

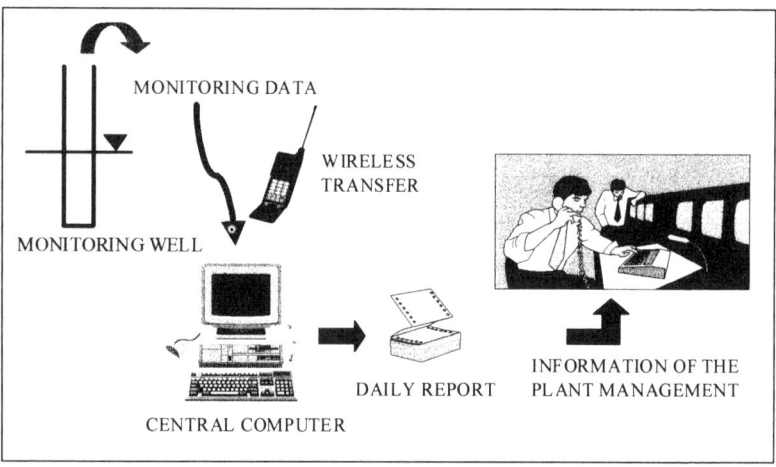

Figure 5. Schema of the ground water monitoring and management system

Additionally, ground water quality data such as pH, electric conductivity, oxygen or temperature are recorded and transferred to the central computer and put into the GIS. A report of the actual situation is generated and compared with the specified conditions automatically. If there are relevant discrepancies between the actual situation and the specified flow conditions or the desired quality situation the management of the PWK will be informed immediately.

As the collection and the transfer of all data is done automatically and the daily evaluation is also automated, there is no demand for additional personal. The data are transferred by wireless transmission. The devices for recording the data are powered by solar energy or batteries. The system is designed to have as little requirement for service as possible to keep the running costs as low as possible.

An action plan for the case of severe deviation is part of the management system. The action plan includes as measures the specification of operation parameters for the drinking water well and the wells at the industrial complex. They are calculated with the help of the ground water model. As the system is designed to have several weeks before a contamination would reach the well, there is enough time to carry out and control measures for securing the drinking water supply.

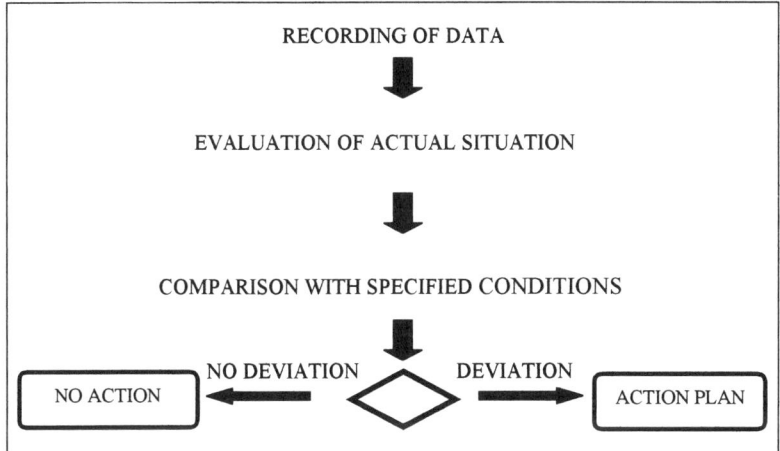

Figure 6. Flow chart of decision process

At present the start of the project is discussed by the management of the PWK. Once the financial means are designated the project will be started immediately.

## 9 Conclusions

Although due to legal, economical and political decisions the extension of the area of protection of the drinking water well of the drinking water plant of Kapfenberg, Austria, could not be realized, it is possible on the basis of a geographic information system and a detailed ground water model to establish a control system, which increases the security of the well. The automated recording, transfer and evaluation of monitoring data does not require additional personnel. The ground water model and the GIS are revised in such way, that the present personnel of the drinking water plant is able to operate the system. The daily comparison of the actual situation and the specified situation gives a high standard of security for the operation of the water supply plant. Changes of land use or other activities relevant to the ground water condition can be considered almost immediately by reading the data into the GIS and updating the basis for the evaluation.

# GROUND WATER RECHARGE PROJECT
# PRESCOTT VALLEY, ARIZONA

Reed J. Petersen[1], P.E., ASCE Member

Abstract

The Town of Prescott Valley, Arizona, recently built a Ground Water Recharge Project utilizing small lakes or ponds located in the center of town where permeable soils allow ground water recharge of excess treated wastewater effluent to gain future ground water rights. The project recharges treated effluent from a new wastewater collection and treatment system. The project will provide future water resources for development in an arid Arizona community that depends on ground water resources. This paper describes some of the aspects considered for the development of this groundwater recharge project.

Introduction

Prescott Valley, a 27 year old community, is in central Arizona near the largest stand of ponderosa pine in the world. It is 90 miles north of Phoenix at a mile high elevation. Nearby Prescott was the first Territorial Capital of Arizona. It was incorporated in 1978 and is one of the fastest growing communities in the second fastest growing state in the nation. Pronghorn antelope still live near and travel throughout the town in spite of its tremendous growth.

| YEAR | POPULATION | GROWTH RATE |
|---|---|---|
| 1980 | 2,284 | - |
| 1990 | 8,858 | 25%/year |
| 1993 | 11,440 | 15%/year |
| 1997 | 18,500 | 15%/year |

---

[1] Civil Engineer, USDA-Rural Development, Arizona State Office, 3003 North Central Avenue, Suite 900, Phoenix, AZ 85012, (602)280-8748.

The Author's views expressed in this article do not necessarily represent the views of the Agency or the United States Government.

## GROUND WATER RECHARGE PROJECT

Developments outside the town limits increase the area population to more than 25,000. Recent commercial development has supported growth. ACE Home Distribution Center just finished a 15 acre warehouse in Prescott Valley and many other commercial and light industrial facilities have been attracted to the community. Prescott Valley has a large population of retired people because of the attractive four season climate without the intense heat of the lower desert or the extreme cold and snow of higher elevations.

The Town lies above two groundwater basins, the Little Chino Valley groundwater basin and the Upper Aqua Fria groundwater basin. Both basins are in the Lonesome Valley sub-area, which receives recharge primarily from Lynx Creek. One third of the recharge flows into the Little Chino Valley basin and two thirds into the Upper Agua Fria basin. The depth to groundwater in the Town varies from 100 feet to 600 feet.

Prescott Valley recently completed a state of the art sewage collection system and 2.5 million gallons per day extended aeration wastewater treatment facility. A conventional gravity flow collection system delivers wastewater to the treatment plant. The Town was not impacted by the construction of the wastewater treatment plant due to the isolated location. The Town was temporarily inconvenienced by the construction activities to install the new collection lines and decommission septic tanks, but the overall benefit of having adequate wastewater collection and treatment systems has more than compensated for the temporary inconvenience.

A major road and street resurfacing project was completed immediately after the collection lines were installed, reducing the dust in much of the Town which previously had unpaved streets.

Treated effluent was planned to be used for irrigating golf courses and parks or other green areas. Effluent not used for irrigation was planned to be discharged into percolation beds within the Agua Fria River bed immediately downstream from the treatment plant.

The Aqua Fria River channel is approximately 30 feet deep in the vicinity of the treatment plant, with the plant site well above the 100 year flood level. The Agua Fria flood plain is largely undeveloped and has Great Basin Grassland Biotic community which is widespread through Northern Arizona, Nevada and Utah. Vegetation is dominated by short grasses with scattered forbs, shrubs and cacti. The natural landscape has been altered by grazing, but many of the native grassland species remain.

The Arizona Department of Water Resources (ADWR) determined in December 1994, near the completion of the wastewater treatment facility, that the "... proposed site (near the treatment facility in the Agua Fria River bed) is unsuitable for operation of a recharge project ...". Prescott Valley is located within the Prescott Active Management Area (AMA) established by the Arizona 1980 Groundwater Code. At some future date the ADWR will likely determine that the safe yield of the area's groundwater resources has been reached. At that time no new groundwater resource use will be approved, and any development would be dependent on water sources from outside the AMA. In arid Arizona, water importation can be very expensive, and a deterrent to development. The ADWR water budget projection for the Prescott AMA assumes full utilization of effluent in the future.

Artificial recharge of high quality effluent could increase the AMA's annual water budget significantly. Maximum use of effluent resources is essential if sufficient yield is to be achieved without the implementation of more strict water conservation measures in later management periods. A suitable site was needed because the Town had already discharged approximately 365 million gallons of treated effluent which has percolated into the groundwater of the AMA but for which the Town has received no credit. Uncredited recharge continues at a rate of 1.4 million gallons per day.

The Town needed to locate a suitable site and get ADWR approval so ADWR could issue the Town an Underground Storage and Recovery (US&R) permit. Under the terms of the permit, groundwater credits accumulate without limit until such time as the water is reclaimed from groundwater for use. A rule of thumb used by ADWR is that a US&R system permit effectively triples the water resources of an area through what is basically a large scale recycling procedure. Without such a system in place in a timely manner, the Town's potential for future growth would be severely limited.

With the help of hydrogeological consultants, the Town developed an alternative which would eliminate all the ADWR technical concerns for recharge as well as allow for economical future direct reuse. The site selected at the Mountain Valley Park Lakes was far enough away from the AMA boundary to resolve the initial ADWR concerns and at a more favorable absolute elevation with respect to the elevation of the ground water table.

The site selected is located on the top of a plateau approximately 240 feet higher than the originally proposed facility which was only 30 feet above the water table of the regional aquifer. ADWR had concerns for any location located closer to the AMA boundary than the proposed location. Hydogeological testing confirmed the desirability of the proposed location even though it will not recharge the volume desired. Future recharge sites will be located and added to the recharge effort when effluent capacity exceeds the recharge capacity of the proposed site.

The proposed site is located in a natural drainage near the center of Town at Mountain Valley Park Lakes. A tree beautification project has been underway at the site for a few years. It includes a small storm water retention basin that is dry except for periodic storm flows. Trees are small enough to be relocated as needed and the Construction for the first planned recharge site began late in 1996.

A pump station was built near the wastewater treatment plant and the pipeline for the reuse effluent built from the pump station to the recharge site at Mountain Valley Park Lakes was finished during 1997. Mountain Park Lakes were deepened and inlet and outlet control structures designed to handle the reuse effluent as well as natural flood flows. The recharge facility is expected to be completed and fully functional early in 1998.

The recharge facility consists of two ponds divided by an earth fill levy that helps to maximize the recharge area. The slopes surrounding the ponds will be landscaped for recreational use. The design of the recharge ponds includes gabions around the perimeter of the ponds to stabilize the ponds from wave and water action. The gabions also provide a level area near the water line which will increase the safety for children and others enjoying recreation activities at waters edge. The community hopes to include limited contact boating and fishing activities at the new Mountain Valley Park Lakes.

Along with the protection, preservation and replenishment of our limited, local water supplies and an increased opportunity for leisure time and recreation activities, the Prescott Valley Recharge Project at Mountain Valley Park Lakes provides an aesthetically pleasing area that adds to the betterment of the quality of life in the community. This park is also an excellent example of how a project can be constructed to perform its' intended engineering functions and still be aesthetically pleasing to the environment.

Information for this article taken from the author's personal observations and information available from the Prescott Valley construction file.

Spectral and Bispectral Analysis for Nonlinear Leaky Phreatic Aquifer Systems
Subject to Time Variable Inputs

Gwo-Fong Lin[1] and Chi-Ming Chen[2]

Abstract

In this paper, spectral and bispectral response characteristics of stochastic nonlinear partial differential equations describing leaky phreatic aquifer systems subject to time variable inputs are examined. The linear and nonlinear (quadratic) frequency response functions (FRF) for the space average hydraulic head of a leaky phreatic aquifer system subject to recharge are obtained using the second-order perturbation approach. Furthermore, for a harmonic type of recharge input, the spectrum and bispectrum of hydraulic head are derived using the linear and nonlinear FRF for hydraulic head process. Variance ratio of the hydraulic head response to the recharge input is also investigated.

Introduction

In the last two decades, the one of important methods for analyzing time variations of phreatic aquifer systems is stochastic analysis. Particularly, spectral analysis is often used to analyze the linear form of the integral balance model subject to time variable inputs (Gelhar 1974; Duffy et al. 1984; Ritzi et al. 1991a, b). In previous researches on stochastic analysis of phreatic aquifer systems subject to time variable recharge, the stochastic partial differential equation is usually assumed to be linear or can be linearized by first-order perturbation method, and then the linear FRF for the hydraulic head can be investigated in frequency domain using spectral techniques. In these cases, the variance spectrum is insufficient to analyze the nonlinear characteristics of the hydraulic head for a phreatic aquifer system. Therefore, the

---

[1] Professor, Department of Civil Engineering, and Director, Hydrotech Research Institute, National Taiwan University, Taipei 10617, Taiwan.
[2] Doctoral student, Department of Civil Engineering, National Taiwan University, Taipei 10617, Taiwan.

nonlinear filter theory is a better way to characterize nonlinearity (Subba Rao and Gabr 1984; Priestley 1988). The spectral and bispectral analysis for a nonlinear integral balance model subject to time variable inputs has also been investigated using the nonlinear filter theory (Jin et al. 1994). In this paper, not only the linear and nonlinear FRF for the hydraulic head of a leaky phreatic aquifer system subject to time variable recharge are proposed, but also the influence due to leakage is presented. Finally, assuming a harmonic input, the variance, skewness, spectrum and bispectrum for the hydraulic head process are derived.

Theoretical Analysis

Consider a stream-connected leaky phreatic aquifer system with temporal fluctuations from recharge. Assume that the free surface of a leaky phreatic aquifer is equal to atmosphere pressure. The two-dimensional equation of groundwater flow in the aquifer system can be written as (Bear 1972)

$$S\frac{\partial h}{\partial t} = \nabla \cdot (T \nabla h) + N + \frac{h-\phi}{\sigma} \tag{1}$$

where $h = h(x, y, t)$ is the hydraulic head [L], $T = T(x, y)$ is the aquifer transmissivity [$L^2/T$], $N = N(t)$ is the recharge rate per unit surface area of the phreatic aquifer [L/T], $S$ is the storage coefficient, $\phi$ is the hydraulic head of layer 2 [L], and $\sigma = B/K'$ is the coefficient of leakage [T], in which $B$ and $K'$ are the thickness and the hydraulic conductivity of the semipervious layer, respectively. Integrating (1) over the flow domain, dividing both sides by the flow domain area $A$, and then applying the divergence theorem yields

$$S\frac{\partial \tilde{h}(t)}{\partial t} + q(t) + \frac{\tilde{h}(t) - \phi}{\sigma} = N(t) \tag{2}$$

where

$$q(t) = -\frac{1}{A}\int_\Omega \nabla \cdot (T \nabla h)\, dA = -\frac{1}{A}\oint_\Gamma T \nabla h \cdot \mathbf{n}\, d\Gamma$$

is the net outflow rate from the aquifer per surface area [L/T], and

$$\tilde{h}(t) = \frac{1}{A}\int_\Omega h(x, y, t)\, dA$$

is the spatial average hydraulic head in the saturated zone [L]. Fig. 1 is a schematic illustration of the flow problem. Focusing on representation of the relationship

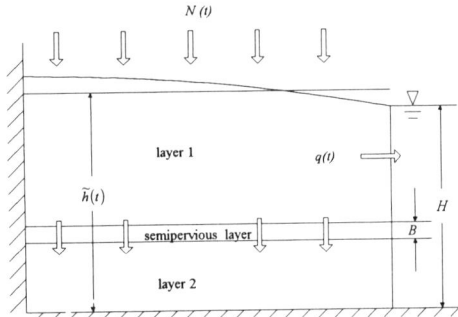

FIG. 1. Schematic Illustration of a Leaky Phreatic Aquifer System

between the recharge input and hydraulic output, we assume that the water level $H$ has no fluctuations, i.e. $H$ is a constant [L]. In addition, the outflow rate $q(t)$ can be presented as a function of the difference between the hydraulic head and the water level (Gelhar 1974), i.e. $q(t) = f(\tilde{h}(t) - H)$. The function is often nonlinear (Jin et al. 1994).

Furthermore, assume that $\tilde{h}(t)$ and $N(t)$ are stationary up to the third order, and $\tilde{h}(t)$, $N(t)$ and $q(t)$ can be expressed in terms of their temporal means plus zero-mean perturbations. Then, expand equation (3) in terms of a Taylor series around $\bar{h}(t) - H$ and neglect the third order terms. Finally, we assume that the fluctuation of recharge process $N'$ is Gaussian so that the Gaussian recharge input can produce the non-Gaussian hydraulic head output for the nonlinear governing equation. The first- and second-order perturbation equations are obtained using the second-order perturbation approach (Jin et al. 1994).

The first-order perturbation equation is written as

$$S \frac{\partial h^{(1)}}{\partial t} + q^{(1)} + \frac{h^{(1)}}{\sigma} = N^{(1)} \qquad (3)$$

and the second-order perturbation equation is expressed as

$$S \frac{\partial h^{(1)}}{\partial t} + q^{(2)} + \frac{h^{(2)}}{\sigma} = 0 \qquad (4)$$

where the exponents (1) and (2) denote the order of perturbations. Using statistical operations, the linear and quadratic FRF (LFRF and QFRF) can be respectively expressed as

$$F_{Nh}^{(1)}(\Psi) = \frac{1}{a_1(k+1)} \frac{1}{1 + \frac{\Psi}{(k+1)}} \tag{5}$$

and

$$F_{Nh}^{(2)}(\Psi, \Phi) = \frac{-a_2}{2a_1^3(k+1)^3} \frac{1}{\left(1 + i\frac{\Psi}{(k+1)}\right)\left(1 + i\frac{\Phi}{(k+1)}\right)\left(1 + i\frac{(\Psi+\Phi)}{(k+1)}\right)} \tag{6}$$

where $i = \sqrt{-1}$, $k = K'/a_1 B$, $\Psi = S\omega/a_1$ and $\Phi = S\theta/a_1$ are normalized frequencies of $\omega$ and $\theta$, respectively, and $a_1$ and $a_2$ are the first- and second-derivatives of $q(t)$ evaluated at $\bar{h}(t) - H$. The $k$ increases as the hydraulic conductivity $K'$ in a semipervious layer increases.

Spectral and bispectral analysis

To demonstrate the non-Gaussion output process produced from the nonlinear aquifer system with recharge, assume the recharge input is a Gaussian process. Consider the recharge is a single harmonic process $N(t) = N_0 \cos(\omega_0 t + \phi_0)$ where $N_0$ is the amplitude of fluctuation [L/T], $\omega_0$ is the angular frequency [1/T], and $\phi_0$ is the random phase angle in radians. The variance, skewness, spectrum, and bispectrum for the hydraulic head process can be derived as

$$\sigma_h^2 = \frac{\sigma_N^2}{a_1^2\left((1+k)^2 + \Psi_0^2\right)} \left(1 + \frac{a_2^2 \sigma_N^2}{a_1^4} \frac{\left((k+1)^2 + 2\Psi_0^2\right)}{2\left((k+1)^2 + \Psi_0^2\right)\left((k+1)^2 + 4\Psi_0^2\right)}\right) \tag{7}$$

$$v_{3,h} = -\frac{a_2 \sigma_N^4}{a_1^5} \frac{3\left((1+k)^2 + 2\Psi_0^2\right)}{\left((1+k)^2 + \Psi_0^2\right)^2 \left((1+k)^2 + 4\Psi_0^2\right)} \left(1 + \frac{a_2^2 \sigma_N^2}{a_1^4} \frac{1}{3\left((1+k)^2 + 2\Psi_0^2\right)}\right) \tag{8}$$

$$S_{hh}(\Psi) = \frac{S\sigma_N^2}{2a_1^3\left((1+k)^2 + \Psi_0^2\right)} \left[S(\Psi \pm \Psi_0) + \frac{a_0 \sigma_N^2}{4a_1^4\left((k+1)^2 + \Psi_0^2\right)}\right.$$
$$\left. \cdot \left(2\delta(\Psi) + \frac{\delta(\Psi \pm 2\Psi_0)}{1 + 4\Psi_0^2}\right)\right] \tag{9}$$

$$S_{3,h}[\Psi,\Phi] = -\frac{S^2 a_2 \sigma_N^4}{4 a_1^7 \left((k+1)^2 + \Psi_0^2\right)^2} \{\delta[\Psi]\delta[\Phi \pm \Psi_0] +$$
$$\delta[\Psi \pm \Psi_0]\delta[\Phi] + \delta[\Psi + \Psi_0]\delta[\Phi - \Psi_0] + \delta[\Psi - \Psi_0]\delta[\Phi + \Psi_0]$$
$$+ \frac{1}{1+k+2i\Psi_0} \{\delta[\Psi + \Psi_0]\delta[\Phi - 2\Psi_0] + \delta[\Psi + \Psi_0]\delta[\Phi + \Psi_0]$$
$$+ \delta[\Psi - 2\Psi_0]\delta[\Phi + \Psi_0]\} + \frac{1}{1+k-2i\Psi_0} \{\delta[\Psi - \Psi_0]$$
$$\cdot \delta[\Phi + 2\Psi_0] + \delta[\Psi - \Psi_0]\delta[\Phi - \Psi_0] + \delta[\Psi + 2\Psi_0]$$
$$\cdot \delta[\Phi - \Psi_0]\}\} - \frac{S^2 a_2^3 \sigma_N^6 \Psi_0^2}{8 a_1^{11} \left((k+1)^2 + \Psi_0^2\right)^3} \left\{ 2\delta[\Psi]\delta[\Phi] + \frac{1}{1+k+4\Psi_0^2} \right.$$
$$\cdot \{\delta[\Psi]\delta[\Phi \pm 2\Psi_0] + \delta[\Psi \pm 2\Psi_0]\delta[\Phi] + \delta[\Psi + 2\Psi_0]$$
$$\cdot \delta[\Phi - 2\Psi_0] + [\Psi - 2\Psi_0]\delta[\Phi + 2\Psi_0]\}\} \quad (10)$$

where $a_0$ is the value of $q(t)$ evaluated at $\overline{h}(t) - H$. Fig. 2 shows the relation between the variance ratio of the hydraulic head to recharge and the normalized

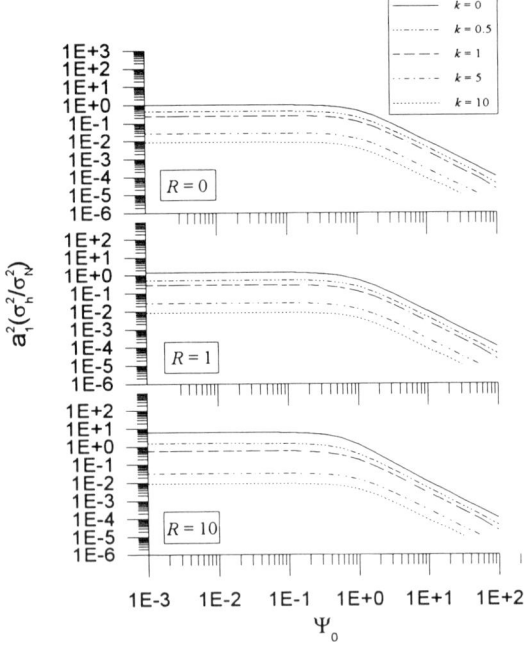

FIG. 2. Variance Ratio of Hydraulic Head to Recharge versus Normalized Frequency $\Psi_0$

frequency $\Psi_0$. In Fig. 2, $R = a_2^2 \sigma_N^2 / a_1^4$ is a dimensionless ratio. The skewness $v_{3,h}$ is equal to zero and the hydraulic head process is Gaussian as $R = 0$. For various $R$ and $k$, the low-frequency components in the recharge process contribute more variance to the hydraulic head process than the high-frequency components do. For fixed $\Psi_0$ and $k$, the variance ratio of the hydraulic head to recharge increases as $R$ increases. For fixed $\Psi_0$ and $R$, the variance ratio of the hydraulic head to recharge decreases with increasing $k$.

Conclusions

In this paper, a leaky phreatic aquifer system subject to time variable recharge is considered. For a Gaussian recharge process, the variance, skewness, spectrum, and bispectrum for the hydraulic head process have been derived using the second-order perturbation approach and statistical operations. An important factor which influences the variance ratio of the hydraulic head to recharge has been identified. Between the low and high frequency components in the recharge process, the former contributes more variance to the hydraulic head process. The methodology presented herein can be employed to analyze more complex aquifer systems.

Acknowledgments

This paper is based on work partially supported by the National Science Council of the Republic of China under Grants NSC86-2621-E-002-008 and 87-2211-E-002-052.

References

Bear, J. (1972). *Dynamics of fluids in porous media*. Elsevier, New York.
Gelhar, L. W. (1974). "Stochastic analysis of phreatic aquifers." *Water Resour. Res.*, 10(3), 539-545.
Duffy, C. J., Gelhar, L. W., and Wierenga, P. J. (1984). "Stochastic analysis of agricultural watersheds." *J. Hydrol.*, 69, 145-162.
Jin, M., and C. J. Duffy (1994). "Spectral and bispectral analysis for single- and multiple-input nonlinear phreatic aquifer systems." *Water Resour. Res.*, 30(7), 2073-2095.
Priestley, M. B. (1988). *Spectral analysis and time series*. Academic, San Diego, Calif.
Ritai, R. W., Sorooshian, S., and Gupta, V.K. (1991a). "On the estimation of parameters for frequency domain model." *Water Resour. Res.*, 27(5), 873-882.
Ritai, R. W., Sorooshian, S., and Hsieh, P. (1991b). "The estimation of fluid flow properties from the response of water levels in wells to the combined atmospheric and Earth tide forces." *Water Resour. Res.*, 27(5), 883-893.
Subba Rao, T. and Gabr, M. M. (1984). An introduction to bispectral analysis and bilinear time series models. Springer-Verlag, New York.

Construction of a New Tollway Below the Groundwater Table
With Nitrate Remediation

Scott M. Taylor[1]
Member, ASCE

Abstract

A new tollway presently under construction in Orange County, California will extend over 38 km linking three existing freeways. About 2 km of this new roadway construction will occur below the existing groundwater table. The roadway intersects a shallow unconfined aquifer comprised of a highly heterogeneous soil structure, making discharge predictions a challenge. A history of farming in the area has left the shallow aquifer with high TDS and nitrate concentrations, effectively prohibiting the discharge of the groundwater to surface waters and limiting the water's beneficial use opportunities.

Introduction

An approximately 2500 meter section of State Route 261 is depressed below the existing grade to accommodate noise and aesthetic concerns of the local community as the roadway passed through an urban area. The depth of the roadway varies from the existing grade, to about 7 meters below the existing grade. A shallow unconfined aquifer is located in the area of the roadway alignment with the free water surface about 2 meters below the ground surface (bgs). Preliminary testing indicated the aquifer contained high nitrate and TDS concentrations. Nitrate ranged from 25 to 80 mg/l as N, TDS ranged from about 2100 to 4500 mg/l. The groundwater did not contain any other harmful constituents regulated through the National Pollutant Discharge Elimination System (NPDES) permit process.

The roadway is located in the Newport Bay watershed in Orange County California. Newport Bay is listed by the Regional Water Quality Control Board (RWQCB) as a 303(d) water body. One of the impairments of this water body is excessive

---

[1] Associate, Robert Bein, William Frost and Associates, 14725 Alton Parkway, Irvine CA 92619

fertilization, caused primarily from high nitrates entering the system via San Diego Creek, the primary stream tributary to the Bay. San Diego Creek is ephemeral, though the typical base flow is from 0.3 to 0.4 m$^3$/s during dry weather. Dry weather ambient nitrate concentration in the Creek is about 16 mg/l as N.

Discharge of the groundwater to the adjacent Creek and ultimately Newport Bay was prohibited at the start of the project by the RWQCB due to the existing fertilization problem in the Bay. The existing basin plan has a nitrate objective of 13 mg/l which was subsequently recognized by the Board as "not protective". Groundwater would have to be treated by the Contractor to be discharged to surface waters.

## During Construction

State Route (SR) 261 is being constructed as a design build project. The Contractor was awarded the project in June of 1995. The tollway, consisting of portions of SR 261, SR 241 and SR 133, must be completed per the contract by December 1999. Consequently, delays in construction could result in costly liquidated damages to the Contractor.

The shallow unconfined aquifer was investigated to understand with more accuracy the hydraulic grade line and the flow rates that could be expected during construction and over the long term. Eleven continuous cores were completed as 1- and 2-inch piezometers. The results served as verification that shallow groundwater occurred consistently along this portion of the tollway. Further investigation included pumping tests to evaluate hydraulic parameters and an additional deep boring to obtain more comprehensive lithologic information.

The pumping tests indicated that the shallow aquifer was essentially a single unconfined anisotropic unit. The average horizontal hydraulic conductivity was determined to be about 6.1 m/day. The results of the field investigations were used to construct a finite element groundwater flow model to determine long-term flow rates assuming a passive subdrain system beneath the roadway, and to estimate the lateral extent of drawdown.

The roadway would be constructed an average of about 7 meters below the existing grade, groundwater was generally located about 3 meters below the ground surface (bgs). Construction dewatering would be achieved through a combination of passive and active dewatering systems. Initial discharge estimates for an active-only system were about 11,000 liters per minute, and as low as 5,600 liters per minute for a passive dewatering system.

Several alternatives for disposal of the groundwater during construction were investigated. On-site treatment was considered, but skid-mounted units would be costly to purchase or lease, take a significant lead-time to ship to the site, and would not be highly portable. It was determined that disposal to the sewer system during construction was the only alternative that would not impact the project schedule.

An agreement was reached with the local water district as well as the Sanitation District that operated the treatment works. Connections to the sewer system involve a capital facilities charge and a service fee. Charges for temporary connections can run as high as $0.4 per cubic meter. The Contractor entered into an agreement with the local water district, which in turn entered into an agreement with the Sanitation District for service for the project. Since the flow did not contain BOD or suspended solids, a 'flow only' charge base was developed and capital facilities were leased on a per year basis. The Contractor currently pays about $0.19 per cubic meter to dispose of the groundwater to the sewer system.

The regional water wholesale District in the area also levies a Basin Equity Assessment (BEA) and Replenishment Assessment (RA) for groundwater extracted within the District boundaries. The District Act allows for exceptions to the assessments provided the applicant can demonstrate that the water is not suitable for domestic use or irrigation, and that it does not recharge an aquifer used for either of these purposes. The Contractor successfully demonstrated that: 1) the high nitrate and TDS content of the groundwater precluded both domestic and irrigation use of the water; and 2) the shallow unconfined aquifer was separated from the lower principal aquifer by an aquitard. As a result, the Contractor avoided charges of $0.07 per cubic meter and an additional $0.28 per cubic meter on 25% of the total volume extracted.

Post Construction

The post-construction disposal of the groundwater posed a significant problem. The sewer system was understood, by agreement to be a temporary solution. Long-term disposal of the groundwater to the sewer system was infeasible for several reasons. First, the cost was very high, much higher than onsite treatment. Second, the groundwater contained high TDS. The local water District reclaims much of the effluent in the district. The District would not allow the long-term disposal of groundwater with such high TDS in their system. Finally, both the RWQCB and the local sewer Districts expressed their desire for the Contractor to find a long-term beneficial use for the groundwater.

A significant number of alternatives were investigated for use and/or disposal of the effluent over the long term. Table 1 lists the alternatives, which will be described in reverse order:

**Table 1**
**Alternatives for Groundwater Use/Disposal**

| Alternative | Description |
|---|---|
| 1. Onsite Treatment | Ion Exchange or Biological Treatment |

| Alternative | Description |
| --- | --- |
| 2. Advanced Onsite Treatment | Reverse Osmosis, Sell as Reclaimed |
| 3. Pump to Existing Treatment Plant | Nearby RO Plant for Domestic Water |
| 4. Treatment in Wetlands | Downstream Marsh Area |
| 5. Future Desalter Plant | Regional Desalter Plan Brine Line |
| 6. Reinjection | Reinject Groundwater in another location |
| 7. Pump to Alternate Point of Discharge | Pump Groundwater Out of Watershed |
| 8. Construct Watertight Roadway | Impervious Roadbed and Retaining Walls |
| 9. Eliminate Roadway | Terminate Roadway Prior to Aquifer |

Alternative 9 involved terminating the roadway prior to the aquifer location and about 2.4 km short of the planned location. This alternative was not feasible for two main reasons. First, the implications relative to traffic circulation were serious, by terminating the roadway prior to its planned location, circulation patterns would be compromised. Second, the local communities would not allow such a drastic change to the roadway configuration.

Alternative 8 would construct an impervious watertight roadway section eliminating the need for lowering the water table. The capital cost of this alternative was extremely high; it was estimated that about 3000 piles would be required to keep the section from floating, and the invert slab would be about 0.6 m thick. Further, the long-term maintenance and performance of the structure over such a great distance (2500 meters) was suspect.

Alternative 7 was a proposal to pump the groundwater out of the watershed to a neighboring watershed where nitrate was not a problem. The construction of a pipeline about 9100 m long would be required through a highly urbanized area. The right-of-way acquisition alone would exceed cost and schedule limitations.

Alternative 6 involved the construction of a reinjection well field. In general, reinjection of flow is likely to be difficult due to the low reinjection rate that is expected in the local area (from 19 liters per minute to 38 liters per minute per well). This would require a well field of up to 70 wells and an area of over 8 hectares (ha). Right-of-way costs would be high, as would the ongoing maintenance of a large well field.

Alternative 5 would discharge the groundwater to the brine line of a regional desalter facility. The regional reverse osmosis (RO) plant is planned as a groundwater cleanup project for the nearby El Toro Marine Corps Air Station. This alternative would be a good solution as the brine line for the plant was planned to pass near the Tollway. However, the construction schedule of the RO plant is sufficiently uncertain as to eliminate this alternative from further consideration.

Alternative 4 was a proposal to discharge the groundwater to the local creek, and intercept the flow further downstream through a diversion structure. The discharge would be diverted to a wetland area where denitrification would occur in a managed system. Preliminary studies of this system indicated that removal in the pond system would not be adequate due to limited land availability.

Alternative 3 would pump the groundwater to an existing RO plant used to produce domestic water. Concerns relative to the safety of the water supply were the primary reason this alternative was not pursued. The groundwater would be withdrawn through the roadway subdrain system with numerous locations for maintenance and cleanout. Such a system was not consistent with the normal wellhead development protocols. In addition, about 15,000 feet of pipeline would need to be constructed along with associated pumping facilities.

Alternative 2 would construct advanced treatment on site, in the form of a reverse osmosis plant to create either a potable water source or a water source suitable for irrigation. Income from selling the resulting water supply could offset the plant operating costs. It was concluded that the groundwater could be treated using an RO process with about an 80 percent recovery. Cost to produce the product water was estimated at about $0.4 per cubic meter, exclusive of BEA and RA assessments that would also be levied. Such costs would push the finish water cost to about $0.53 per cubic meter. Wholesale cost of irrigation water in the area is about $0.20 per cubic meter, making the alternative unattractive. Domestic water in the area wholesales for about $0.36 per cubic meter.

Alternative 1 was determined to be the best solution. Initially, ion exchange was planned but the process creates salt brine that must be disposed of to the sewer. The local sewer agency would not agree to this process due to the problem with increasing TDS in the system. Consequently, a biological process was selected. The biological process would incorporate denitrification using bacteria and a methanol feed as a carbon source. Filter backwash would consist of ambient TDS and process introduced BOD only, nitrates in the groundwater would be converted to $N_2$ gas.

An NPDES permit was secured from the RWQCB to operate the plant. Final effluent limitations included a 13 mg/l total nitrogen and a 20/20 standard for BOD and TSS. This allowed the filter backwash to be discharged to the creek rather than the sewer system saving significant O&M cost. Not only would normal capital costs for sewer service be required, but reclaimed water would be necessary for filter backwash to keep TDS discharge to the sewer system within acceptable levels.

The estimated operation and maintenance cost for the plant is about $100,000 per year, including capital recovery. This is equivalent to about $0.14 per cubic meter of water treated.

Conclusions

There was exceptional interest in finding a beneficial use for the groundwater encountered during the construction of this tollway. About 550 acre-feet per year will be produced by the dewatering system. Domestic and agricultural uses were precluded due to the high TDS (2500 to 3500 mg/l) in the groundwater. Reverse osmosis was determined to be one of the only technologies available for removing TDS to produce a product water suitable for retail use. The cost of producing water with a TDS of about 700 mg/l was determined to be about $0.38 per cubic meter, exceeding the local cost of both reclaimed and domestic water at the wholesale level.

Water is a finite resource in California but its efficient management and use remains an elusive goal in many instances as this case suggests. Contaminated groundwater must be viewed as a resource rather than a waste product if sustainable development is to be achieved.

Evaluation of the Technical Equivalency of Engineered Phyto-Cover Systems to RCRA Landfill Caps

Scott T. Potter[1,]
Suthan S. Suthersan[2,] Jeffery A. Smith[3,]
Timothy A. Bent[4]

Introduction

The engineered phyto-cover system is an alternative landfill cover designed to focus on the protection of groundwater quality. The phyto-cover system consists of densely-planted, deep-rooted trees and grass understory in a water-holding soil that creates a "sponge and pump" -type water removal system. The water holding capacity of the soil provides the "sponge" and is critical in storing infiltration during the dormant season. The evapotranspiration capacity of the engineered vegetative cover provides the solar-powered "pump" that depletes the soil moisture storage during the growing season. The engineered phyto-cover system covers all areas of the waste disposal portion of the site.

In the areas where the phyto cover is to be installed, the existing vegetation is removed and/or tilled into the existing soil cover as determined by site-specific soil cover requirements. Supplemental soil and non-soil amendments and adjustments in slopes are made as necessary to obtain water holding capacity requirements and maximize precipitation runoff. Supplemental non-soil amendments are incorporated into the existing site soils to provide sufficient nutrients for the trees and grasses as needed. These amendments may consist of compost and/or lime-stabilized sludge. The trees are planted into this amended cover soil layer.

---

[1]Principal Engineer, ARCADIS Geraghy & Miller, 1131 Benfield Boulevard, Suite A, Millersville, Maryland 21108
[2,3]Vice President and Senior Hydrogeologist, ARCADIS Geraghty & Miller, 3000 Cabot Boulevard West, Suite 3004, Langhorne, Pennsylvania, 19047
[4]Senior Environmental Project Manager, Bridgestone/Firestone, Inc., 50 Century Boulevard, Nashville Tennessee, 37214.

Hybrid poplar trees grow vigorously over landfills and routinely develop roots deeper than 8 feet below the surface (Licht 1997; Gatliff 1997). When the trees are stressed for water, roots continue to explore deeper and laterally such that root-growth has been measured to depths of up to 20 feet. In landfill applications where a sufficient soil/amendment cover is installed, such as proposed by this alternative, the tree roots can grow throughout the entire cover depth and deeper into the underlying landfill matrix. Densely-planted hybrid poplar trees in temperate climates have been "engineered" to maximize their growth rate and vitality, and are capable of transpiring all of the natural precipitation that infiltrates.

Alternative Landfill Cover Technology

Since 1990, scientists and engineers have been working on phyto-covers: natural, more resilient approaches to landfill cover technology that eliminate many of the maintenance and environmental problems associated with traditional membrane caps, while ensuring adequate performance through soil moisture monitoring. The goal of these efforts is to develop a self-sustaining, natural plant system (and associated wildlife habitat) designed to transpire infiltrating precipitation. An added benefit of this technology is that the phyto-covers facilitate natural processes that stabilize landfills, rendering them environmentally benign over time. Phyto-cover systems consisting of deep-rooted poplar trees with a grass understory are operating successfully at many sites across the country. According to USEPA's Office of Research and Development, installation of phyto-covers is being actively considered at still other landfills across the country (Rock, 1997).

Benefits of Phyto-Covers over Traditional RCRA Caps

In addition to satisfying the critical anti-leaching requirement, phyto-covers provide a number of significant pollution control, ecological, and economic benefits when compared to traditional RCRA caps. First, a phyto-cover actually enhances the natural biodegradation processes, instead of interfering with them, as a RCRA cap could. Second, a gas-permeable phyto-cover allows for passive venting of gaseous byproducts of biodegradation and allows oxygen to diffuse into the fill to facilitate additional biodegradation. Third, a phyto-cover would provide a forest ecosystem and an attractive alternative to a RCRA cap. Finally, a phyto-cover can be installed with less cost and less risk to public safety than a RCRA cap and once the cover is established the system has a natural stability that minimizes long-term maintenance requirements.

Hydrologic Water Balance

Moisture flow and moisture content in a landfill are extremely important to the dynamic processes of decomposition and potential leachate generation. The fundamental means to assess the moisture conditions is through an evaluation of various processes that comprise a water mass balance. A water mass balance analysis is an "accounting procedure" for tracking the moisture inputs to storage and the moisture outputs that influence the potential flux of water through the cover into the waste. The primary elements of a water mass balance include precipitation, surface runoff (R/O), potential ET (PET), infiltration (I), soil moisture storage (ST), actual ET (AET), and flux (or percolation) of water through the system. The water-shedding efficiency of a cap is then derived by calculating the percentage of flux relative to total precipitation. The phyto-cover system design concept involves maximizing efficiency by optimizing ET, runoff, and soil moisture storage to minimize infiltration, flux, and potential leachate generation. The measure of performance for the designed phyto-cover is compared to the water-shedding efficiency of traditional barrier cover systems. Presented below is a discussion of each of these steps and the basis for the general engineered phyto-cover system design.

Precipitation

Long-term precipitation data needs to be assembled from the closest weather station to evaluate local hydrologic conditions. There are no established regulatory procedures or protocols to evaluate the hydrologic performance of a phyto-cover design. Therefore, the long-term data is needed in order to characterize the long-term precipitation trends and extremes. Typically, precipitation can vary widely from site to site for a given year, season, or month. The application of these data to evaluate the phyto-cover design assumes that the daily precipitation totals are the result of individual storm events.

Runoff

Runoff from the designed phyto-cover is computed using the USDA Soil Conservation Service (SCS) Curve Number Model. The model computes direct runoff from an individual storm event as a portion of total precipitation. The Curve Number Model is widely used and is incorporated into the HELP model and other agronomic models to compute rainfall runoff and other elements that comprise a water balance. The major deficiency in the Curve Number Model is that it underestimates runoff from small precipitation events. This discrepancy in the Curve Number Model results in overestimates of infiltration and the amount of water that must be managed by the cover system (McBean et. al, 1995). Consequently, the resultant engineered phyto-cover is over-designed and conservative; the engineered phyto-cover has the ability to control more infiltration than it is designed to manage.

Potential Evapotranspiration – Measured Data

PET is a measure of the maximum rate at which ET can occur when adequate soil moisture is available for utilization by the vegetation. These data are measured in the field utilizing lysimeters planted with single species covers (usually perennial grasses).

Soil moisture levels are maintained at optimum levels and ET is measured by weighting the lysimeter. The monthly potential ET rates measured for grasses are adjusted, to best represent the supplemental ET available from the trees by incorporating a consumptive-use coefficient ($K_c$) that is applicable to the trees utilized in the phyto-cover design.

Effective Evapotranspiration

The actual or effective ET is calculated by adjusting the PET value to account for the reduction in the ET rates as the soil moisture is depleted. This adjustment is performed using a standard model of ET as a function of soil moisture (Feddes et al., 1978; Maidment, 1993). The effective ET rate occurs at the PET rate until soil moisture content is at a percentage of field capacity. The effective ET rate declines linearly at soil moisture levels drier than this value until it is approximately zero at the wilting point of the plants.

Water Balance Model

These principles were used to develop a water balance model of an engineered phyto-cover system. The water balance for a phyto-cover system begins with precipitation. It is assumed for this analysis that all precipitation is in the form of rain. This assumption causes all design considerations related to cover thickness and total water storage requirements to be conservative. The soil surface separates precipitation into runoff and infiltration. Runoff is estimated using the Curve Number model discussed previously. The Curve Number procedure was selected because it consistently under-predicts runoff volumes. This procedure adds an additional degree of conservatism in that infiltration is over estimated. The runoff analysis for the water balance assumes that daily precipitation totals correspond to individual storms. After precipitation infiltrates into the soil, water is either stored, removed through ET, or, if moisture content is in excess of field capacity, percolates through the root zone and into the waste.

Example Application

An example application was developed for a hypothetical site in central Illinois. Potential ET rates are known for this area of the United States from data collected during a 19-year lysimeter study in Coshocton, Ohio. Actual ET rates are computed based upon potential ET rates and soil moisture levels. For this example, the actual and potential ET rates are assumed to be equal until moisture contents fall to less than 50 percent of field capacity, below which ET is assumed to decline linearly to zero at the wilting point. A summary of the daily water balance calculations and results are presented in Figure 1. These curves illustrate the expected daily performance of an engineered phyto-cover design in central Illinois.

## Comparison with Conventional (ROD) Cap

The results of the hydrologic water balance calculations were used to compare the water-shedding efficiencies of the designed phyto-cover system and a RCRA cap. The efficiency of the RCRA design was computed using the HELP model, the standard tool for evaluating landfill cover performance. The HELP model was run based upon climatic conditions in Illinois for a 37-year period to assess performance. The results of the HELP model predict that a RCRA cap is 96 percent efficient in shedding precipitation.

The table presented below shows a designed water holding capacity of the phyto-cover using existing soils (1 foot), the municipal solid waste (MSW) in-place (6 feet), and 1 foot of supplemental soil and soil amendments. The existing soil cover has been assumed to have an average water holding capacity of 3.0 inches per foot for this water balance analysis (sandy to silty loams have a WHC of 2.5 to 3.5 in/ft). The water holding capacity of waste in a mature landfill is between 2.0 and 4.85 inches of water per foot depending on the percentage of municipal solid waste (Rovers, F.A. and Farquhar, 1973). For this design, the MSW is assumed to have a water holding capacity of 2.0 in/ft. Poplar trees routinely develop roots deeper than 8 feet below the soil surface. Accordingly, the engineered phyto-cover system is designed to root into the waste to capture additional water holding capacity

| MATERIAL | THICKNESS (feet) | WHC (in./foot) | TOTAL WHC (inches) |
|---|---|---|---|
| Additional Imported Soil Cover | 1.0 | 3.0 | 3.0 |
| Existing Landfill Soil Cover | 1.0 | 3.0 | 3.0 |
| Root Growth in Landfill Matrix | 6.0 | 2.0 | 10.0 |
| | | TOTAL WHC | 16.00 |

This table shows that the total water holding capacity of a design phyto-cover can easily be 16.0 inches.

Figure 2 summarizes the efficiency of a designed phyto-cover as function of total water holding capacity. This response curve is unique to the climatic data evaluated and can not be used to evaluate phyto-covers at other sites. Accordingly, the performance efficiency of the designed phyto-cover is greater than 97 percent; which exceeds the hydraulic performance efficiency of both a compacted clay or a flexible membrane RCRA cap. In summary, the above analysis demonstrates that the engineered phyto-cover system presented has the ability to achieve the objectives of a RCRA cap; to protect human health and the environment, by protecting groundwater quality through minimizing the generation of landfill leachate.

## References

Feddes, R.A., P.J.Kowalrik, and H.Zaradyn 1978. Simulation of Field Water Use and Crop Yield. Cen. For Ag. Pub. & Doc, Wageningen, Netherlands.

Licht, L.A. 1997. Ecolotree, Inc. Personal communication with Tina Stack, Geraghty & Miller, Inc., October 29, 1997.

Gatliff, E. 1997. Applied Natural sciences, Inc. Personal communication with Tina Stack, Geraghty & Miller, Inc., October 29, 1997.

Maidment, D.R. ed. 1993. Handbook of Hydrology. McGraw-Hill, Inc., New York.

McBean, E.A. , F.A.Rovers, and G.J.Farguhar 1995. Solid Waste Landfill Engineering and Design. Prentice Hall.

Rock, S. 1997. Office of Research and Development, USEPA, Personal communication with S.Suthersan, Geraghty & Miller, Inc., July 8, 1997.

Rovers, F.A. and G.J.Farquhar 1973. Infiltration and landfill behavior. J.Env.Eng.Div., ASCE 99(EE-5), pp. 671-690.

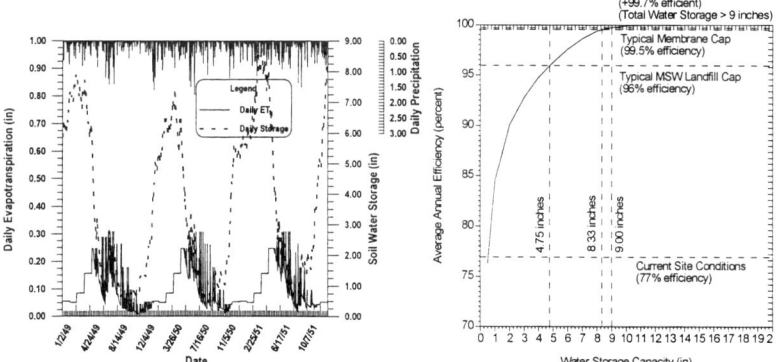

Figure 1. Example Water Balance Results.    Figure 2. Comparison of Phyto-Cover with RCRA Cover Systems.

Comparison of Steady-State and Transient Simulation of the HFBR Tritium Plume
At Brookhaven National Laboratory, Upton, New York

Michael P. Kladias[1,]
Douglas A. Smolensky[2,] Arthur J. Zahradnik Jr.[3,]
Michael G. Hauptmann[4]

Abstract

A modeling effort was undertaken at Brookhaven National Laboratory (BNL) to evaluate the observed distribution of a tritium plume located in the vicinity of the High Flux Beam Reactor (HFBR). Steady-state and long term transient model simulations were performed to explain the evolution of the current tritium distribution. The model analyses suggest that the plume has had a complex history due to on-site pumping and recharging. While transport directions in the past were highly variable, the elongate southerly trending plume is generally the result of hydraulic conditions over the past ten years. Particle based transport schemes, while computationally burdensome, may be necessary to simulate these effects.

Introduction

A comprehensive modeling effort was performed to evaluate historical, current, and future migration of the tritium plume located in the vicinity of the High Flux Beam Reactor (HFBR) at BNL (Geraghty & Miller, 1997). Both steady-state and long term transient simulations were performed to understand the current tritium distribution observed at the site. BNL is located in Upton, Suffolk County, New York, near the geographic center of Long Island.

---

[1]Principal Scientist, ARCADIS Geraghy & Miller, 1131 Benfield Boulevard, Suite A, Millersville, Maryland 21108
[2]Associate, ARCADIS Geraghty & Miller, 88 Duryea Road, Melvile, New York 11747
[3]Project Engineer, ARCADIS Geraghty & Miller, 88 Duryea Road, Melvile, New York 11747
[4]Project Manager, Brookhaven National Laboratory, Upton, New York.

## Background

Investigations at BNL detected tritium contamination in the groundwater in the center of the BNL site (just south of one of the two on-site research reactors). The tritium was found in two monitoring wells that had recently been installed approximately 100 feet south of the HFBR. Because the levels of tritium exceeded the U.S. Environmental Protection Agency (USEPA) drinking water standard, BNL immediately began an investigation of the contamination.

The suspected source of the contamination was the HFBR's 68,000-gallon spent fuel storage pool. The pool, which currently contains tritium at a concentration of 130 million picocuries per liter (pCi/L) is estimated to be leaking at a rate of 9 gallons per day. More than 1,500 groundwater samples indicate that a plume of tritium is moving southward from the reactor, in a narrow strip at depths of up to 150 feet. The plume has been characterized using temporary wells that were sampled at various depths to create a "vertical profile" of groundwater quality. Concentrations within the plume range from 1,000,000 pCi/L adjacent to the HFBR to approximately 5,000 pCi/L at the location of the downgradient leading edge of the plume.

## Hydrogeologic Framework

The hydrogeology under the BNL site consists of approximately 1,500 ft of unconsolidated sediment overlying bedrock. The unconsolidated deposits are subdivided from youngest to oldest as follows:

- Upper Pleistocene deposits (Upper Glacial aquifer).
- Gardiners Clay.
- Matawan Group-Magothy Formation (Magothy aquifer).
- Raritan Formation (Raritan confining unit and Lloyd aquifer).

This sequence dips to the southeast below eastern Long Island.

## Summary of Site-Wide Groundwater Flow Model

A groundwater flow model was constructed and calibrated for the purpose of evaluating groundwater flow at the sitewide and operable units scales (Geraghty & Miller, 1996). This model provides a quantitative tool for predicting the distribution of hydraulic heads (water levels) and groundwater velocities at various areas of interest under various hydrologic conditions. The model also provides regional-scale estimates of the volume and direction of groundwater flow within the various hydrogeologic units underlying BNL.

The modular finite-difference groundwater flow code, also known as MODFLOW, developed by the USGS (McDonald and Harbaugh, 1988), was selected for simulation of the groundwater flow field. The three-dimensional solute-transport model code MT3D (Zheng, 1992) was utilized for solute transport modeling.

## Development of the HFBR Submodel

In order to perform detailed tritium transport simulations in the vicinity of the HFBR with an acceptable degree of precision, a finer horizontal and vertical mesh spacing was needed. The refined mesh was implemented using a procedure known as Telescopic Mesh Refinement (TMR) or grid refinement to develop a site-scale HFBR submodel. The process of grid refinement involves the inscription of a subarea within the domain of a larger model to define a submodel that takes its boundary conditions from the larger model (Duffield et al., 1987).

The finite-difference grid for the HFBR local model covers an area of approximately 4 square miles. The finite-difference grid for the BNL local model consists of over one million grid cells. The grid cells in the vicinity of the HFBR measure 10 feet on each side. In order to gain accuracy in the simulation of the vertical migration of the plume, the submodel was refined vertically. The layers representing the Upper Glacial aquifer were refined since they contain the tritium plume and enable the model to accurately simulated the trajectory of the tritium plume. The lower layers of the regional model were not explicitly simulated to conserve memory allowing for greater horizontal and vertical resolution in the direct vicinity of the plume. To preserve the characteristics of the regional groundwater flow system, the local model places constant head boundaries along its external boundaries and the bottom layer of the model based on hydraulic heads simulated by the regional model. For the historical transient simulation, general head boundaries were used in place of constant heads to allow for boundary conditions that could change each stress period.

## HFBR Tritium Transport Simulations

The submodel was used to simulate the HFBR tritium plume to both steady-state and transient flow conditions. Since the distribution of the tritium plume is influenced by on-site pumping and the effects of on-site recharge basins, the transient simulation was expected to yield a better fit to the observed data than using the steady-state flow field. However, the preliminary transport analyses were begun with the steady-state flow model configured to represent long term average historical conditions. It was decided that the simpler steady-state model would be effective to estimate dispersivity coeficients for use in the transient model simulation.

## Tritium Simulation with a Steady-State Flow Field

Dispersivity was varied to improve the match between the simulated and observed tritium plume. The steady-state simulation was designed using a modified long-term average flow condition that produces groundwater flow paths similar to the axis of the observed tritium plume. A constant source at the reactor pool was simulated using three injection wells totaling 9 gallons per day at a concentration of 40 million pCi/L (the historical concentration of the tritiated water in the spent fuel pool).

Four of the simulations used the Method of Characteristics (MOC) scheme for advection while the remaining simulation used the upstream weighted finite-difference scheme. The MOC scheme is essentially free of numerical dispersion, while the finite-difference scheme contains numerical dispersion that is dependent on the model grid cell size. Simulations were run for 10 years to evaluate the following: a continuous source using the finite-difference method (upstream weighting) with no added dispersivity, the MOC scheme without dispersivity and with a longitudinal dispersivity of 3, 5, and 10 feet. Each of the simulated plumes was compared qualitatively to a composite tritium plume developed from vertical profile sampling.

The finite-difference simulation yielded the lowest concentrations of the simulations and developed a plume that was somewhat wider than has been observed in the field. The MOC scheme without dispersion developed a plume that overpredicted concentrations observed in the field and also developed an extremely narrow undispersed plume. The closest match was achieved using a dispersivity of 5 feet (Figure 1).

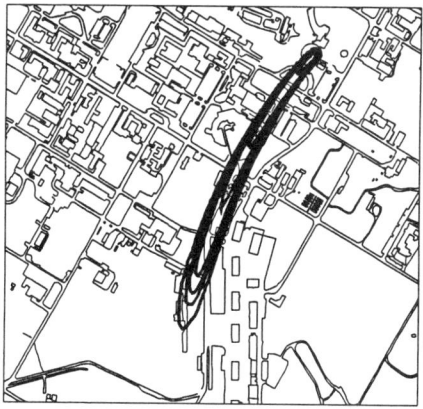

Figure 1. Simulated tritium concentration distribution after 10 years using the MOC scheme and dispersivity of 5 feet.

It is important to note that these comparisons assume that the plume could be represented by a steady-state flow field. In reality, the flow field history is rather complex due to on-site pumping and the influence of basins in the vicinity of the HFBR. The MOC schemes while producing the closest match with the observed plume are computationally burdensome. Because of the lengthy simulation time, additional resolution was built into the model to enable the use of the finite-difference scheme with reduced numerical dispersion.

Historical Transient Tritium Simulation

For the historical transient analysis, actual pumping data from on-site and off-site wells was time-discretized for input into the model. The historical flow and tritium transport simulation analyzed conditions from 1964 through 1996. Pumping data, basin flow data, and precipitation records were averaged and input to the model on a quarterly basis resulting in 132 model stress periods.

The source, assumed to begin in 1964, was developed assuming a leak rate of 9 gallons per day at a tritium concentration 40 million pCi/L, the historical pool concentration. During the last year of the simulation (1996) the source concentration was increased to130 million pCi/L. It was also assumed that no pre-existing tritium concentrations existed in groundwater.

The transient analysis revealed a complex history to tritium plume. On-site pumping and basin flows exert significant influence on the direction of transport and concentration distribution. The magnitude of the simulated concentrations in the vicinity of the HFBR are similar to those observed. The general shape and migration path of the simulated and observed plumes are comparable (Figure 2).

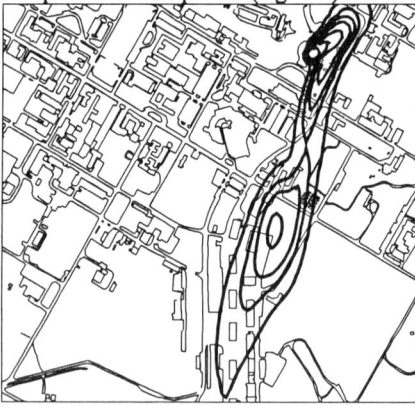

Figure 2. Simulated tritium concentration distribution after 33 years using the FD scheme and the transient flow simulation.

Results also indicate that the observed plume is due primarily to relatively recent flow pathways over the last ten years.

Summary and Conclusions

A comprehensive modeling effort was completed to evaluate historical migration of the HFBR tritium plume. This modeling study relied upon the existing conceptual and regional numerical groundwater model previously developed for the BNL site. Considerable effort was applied to develop a site-specific transport model that could adequately represent the movement of the tritium plume. Transient and steady-state model calibrations were performed to evaluate if the model was able to represent the current observed tritium plume using a constant long-term source at the HFBR spent fuel rod pool. The model results suggest that the plume has had a complex history due to on-site pumping and recharging. While transport directions in the past were highly variable, the elongate southerly trending plume is generally the result of hydraulic conditions over the past ten years.

References

Duffield, G.M., D.R. Buss, D.E. Stephenson, and J.W. Mercer. 1987. A Grid Refinement Approach to Flow and Transport Analysis of a Proposed Ground-Water Corrective Action at the Savannah River Plant. Proceedings of the Solving Groundwater Problems with Models Conference and Exhibition, National Water Well Association, Denver, Colorado.

Geraghty & Miller, Inc. 1996. Regional Groundwater Model Brookhaven National Laboratory, Upton, New York. Prepared for Brookhaven National Laboratory, November 1996.

Geraghty & Miller, Inc. 1997. Draft Groundwater Flow and Tritium Transport Modeling, Brookhaven National Laboratory, Upton, New York.

McDonald, M. G., and A. W. Harbaugh, 1988. A Modular Three-Dimensional Finite-Difference Ground-Water Flow Model, Techniques of Water-Resources Investigations, Book 6, Chapter A1, U. S. Geological Survey, Reston, Virginia.

Zheng, C., 1992. MT3D: A Modular Three-Dimensional Transport Model for Simulation of Advection, Dispersion, and Chemical Reactions of Contaminants in Groundwater Systems, prepared for the U. S. Environmental Protection Agency, Robert S. Kerr Environmental Research Laboratory, Ada, Oklahoma, Developed by S.S. Papadopulos & Associates, Inc., Rockville, Maryland.

# Sedimentation Study of Loyalhanna Lake Using a DTM

by:

Roger T. Kilgore[2], P.E., M. ASCE, Daniel C. Lucey, A.M. ASCE[2], Werner Loehlein[1], M. ASCE, P.E., Nick Fusco[2], RLS, and Daniel E. Medina[2], P.E., M. ASCE

Greenhorne & O'Mara, Inc. has completed a sedimentation survey and study of Loyalhanna Lake for the Pittsburgh District of the Corps of Engineers. Loyalhanna Lake was completed in 1942 and is used today primarily for flood control and recreation. The study was conducted in accordance with Engineer Manual (EM) 1787 and documented sedimentation occurring since the latest comprehensive survey in 1963. Since that time, increased sedimentation has resulted in the appearance of sand bars at lower pool elevations and interference with the dam's outlet works.

Site Characteristics

The Loyalhanna lake basin is located in the Commonwealth of Pennsylvania, Westmoreland County, and is shared by the Loyalhanna, Salem, Derry, and Unity Townships. Loyalhanna Creek meets with the Conemaugh River to form the Kiskiminetas River, which is tributary to the Allegheny River. Figure 1 shows the lake area and its general location.

---

[1] U.S. Army Corps of Engineers, Pittsburgh District

[2] Greenhorne and O'Mara, Inc., 9001 Edmonston, Greenbelt, MD 20770

301-882-2887; 301-220-2595 (fax); RKILGORE@G-AND-O.COM

The topography of the basin is generally rugged, particularly at the higher elevations outside the lake area. The lake itself is located in a long and relatively narrow valley reach, which resembles the sinuosity of the original meandering creek. A few minor tributaries enter the lake and have steep grades. The valley sides approaching the lake are very steep. Elevations in the basin vary from a minimum of about 265 meters (869 feet) above sea level at the dam site to a maximum of about 890 meters (2920 feet) above sea level on the southeastern divide. The basin is mostly wooded, especially in the upper areas.

Gently rolling land suitable for agricultural pursuits is located in the entire creek valley, mainly within the upper portion of the lake area. A large part of this land has been in use under agricultural lease. Extensive coal mining operations in the basin contribute wastes to the creek. Sewage from domestic developments and industrial plants in the area also pollute waters tributary to the lake.

The Loyalhanna lake basin above the dam has a gross drainage area of 751 square kilometers (290 square miles). The mean annual precipitation in the Loyalhanna lake basin for the entire period of lake operation has been about 1.1 meters (44 inches). Discharge records were obtained and assembled by the U.S. Army Corps of Engineers (USACE) for a gaging station at the Loyalhanna dam site. This station has been in operation since March 1940 and was included in the U.S. Geological Survey (USGS) streamflow data reports up to 1991. In that year the station was moved to accommodate bridge rehabilitation activities and the record is no longer reported by the USGS. Nevertheless, the USACE continues to collect streamflow data. The mean annual discharge for water years 1943 to 1996 has been 14 cubic meters (496 cubic feet) per second. A maximum flow of approximately 878 cubic meters (31,000 cubic feet) per second occurred at the dam site, prior to the time of initial lake operation, on March 18, 1936.

Surveys

The original topographic survey of the Loyalhanna lake area was developed by plane table in 1938. The topographic survey was based on the Pennsylvania Lambert grid system with reference monuments set and defined. Original topographic maps were made at a scale of 1 inch equal to 100 feet with 5 foot contours. Although the original survey was made prior to the time of initial lake operation, the topographic maps of the lake area are referred to as being in June 1942, since the topography would remain unchanged until initial storage began.

The latest survey was conducted between September 1996 and January 1997. To the average survey date of November 1996, the time elapsed is 34.58 years from the last survey and 54.41 years after the lake became operational. The existing 30 ranges were used. GPS was employed to determine the alignment of the ranges and conventional methods were used to survey each range.

The topography produced by a 1962 survey was scanned and converted into a Digital Terrain Model (DTM). The field-run topography for the ranges was used to validate the accuracy of the DTM. The bathymetric survey consisted of approximately 200 ranges to determine the profile under water. A DTM was created from this information. Thirty of these underwater ranges coincide with the original land ranges and were used to obtain complete cross sections. The 1996 DTM of the lake was obtained by combining the 1962 land surface (verified by 1996 survey) with the 1996 bathymetric survey.

Sediment Accumulation and Distribution

The two DTMs were compared to determine the distribution of sediment throughout the lake. Development of the two DTMs allowed effective analysis of sediment distribution patterns quickly using Site Works. The 1962 DTM and the 1996 DTM were merged (overlain) using Site Works to identify areas of increased sedimentation since the 1962 bathymetric study. Sediment volumes could then be calculated between range cross-sections using CADD software. The Corps, as a reference for future sedimentation studies, also will use the 1997 DTM.

Time series analyses of the sedimentation showed that the annual rate of sedimentation since 1963 is significantly lower than occurred between 1942 and 1963. This is probably a result of the reduction of mining activities in the watershed as well as improved construction and silvaculture techniques. Table 1 documents the volume of original storage capacity lost by sedimentation for all of the sedimentation surveys

Table 1. Volume loss due to sedimentation in Loyalhanna lake.

(The loss is expressed as a fraction of the original capacity of 117.7 cubic meters (95,400 acre-ft.))

| Survey Date | Time Elapsed (yrs) | Total Volume Loss (%) | Incremental Loss (%) | Average Annual Loss (%/yr) |
|---|---|---|---|---|
| June 1942 | 0.00 | 0.00 | | |
| October 1948 | 6.33 | 0.68 | 0.68 | 0.107 |
| September 1953 | 11.25 | 1.16 | 0.48 | 0.103 |
| April 1962 | 19.83 | 1.89 | 0.73 | 0.095 |
| November 1996 | 54.41 | 4.30 | 2.41 | 0.079 |

The cumulative effect on storage capacities in the reservoir is summarized in Table 2. Overall, $5.2 \times 10^6$ cubic meters (4,200 ac-ft) of storage has been lost. Most importantly, though, is that 75 percent of the original sedimentation storage has been lost and is impairing function of the outlet structure. One of the problems is that the increase in suspended sediment being released through the gates is accelerating wear on the gates.

Table 2. Lake storage capacity allocation.

| Storage Region | Original Capacity (cubic meters) | Present Capacity (cubic meters) |
|---|---|---|
| Minimum pool condition between the original streambed elevation of 869 and elevation 910 | $2.47 \times 10^6$ | $617 \times 10^3$ |
| Flood control storage, between elevations 910 and 975 | $115.2 \times 10^6$ | $12.0 \times 10^6$ |
| Total storage capacity below lake-full elevation 975 | $117.7 \times 10^6$ | $112.6 \times 10^6$ |

Conclusions and Recommendations

The overall rate of sedimentation in Loyalhanna lake is not excessive except at minimum pool. In fact the annual rate of sedimentation has decreased from $126 \times 10^3$ cubic meters (102 acre-feet) per year to $83 \times 10^3$ (67 acre-feet) per year in the period of 1962 to 1996. However, it is possible that as sedimentation has increased, more sediment is escaping through the outlet works. If this is significant, the sediment accumulation would give the appearance that the watershed contribution is decreasing when in fact more is simply being released downstream.

For older single purpose projects such as Loyalhanna Lake, the loss of 75 percent of the sediment storage pool over a 50-year lifetime is not surprising; in fact, this is the behavior for which the project was designed. However, the larger issue for the Corps is that now that the design lifetime of this, and many other projects, is reached, a plan must be developed to address the next 50 years.

If the plan is to dredge, this sedimentation survey may be used to evaluate the volume of material to be removed. Prior to implementing a dredging operation, it is recommended that additional sediment sampling be conducted focusing on the target areas. (The densities reported in this survey were significantly slower than earlier measurements.) But from a broader perspective, whatever strategy is adopted at Loyalhanna should be considered within the context that many other similar projects may require similar action. Such broader scale efforts, should be planned now.

Another important observation on this study is that the efforts to automate the computation process through the creation of DTMs reduced the amount of time required to perform laborious computations of sediment volumes. Subsequent sedimentation surveys and analyses also will be streamlined since DTMs created from the new surveys can be efficiently compared with the DTM created for the present time period.

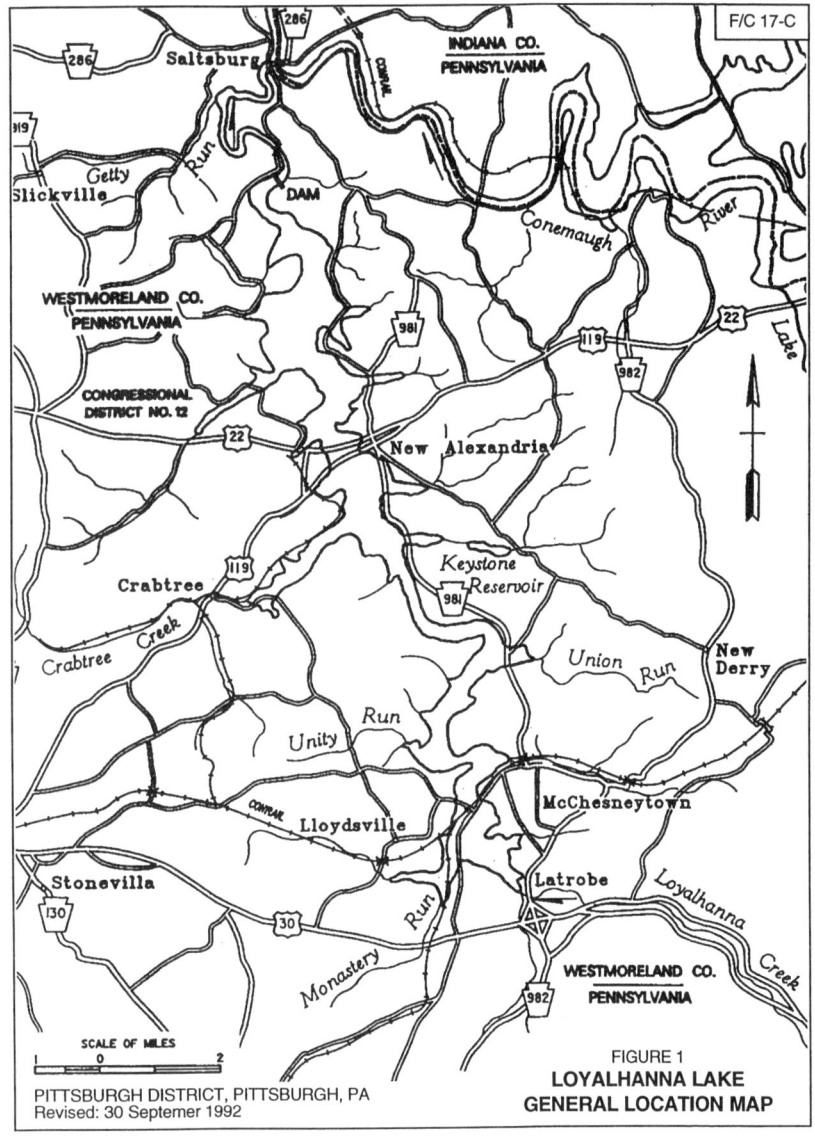

FIGURE 1
LOYALHANNA LAKE
GENERAL LOCATION MAP

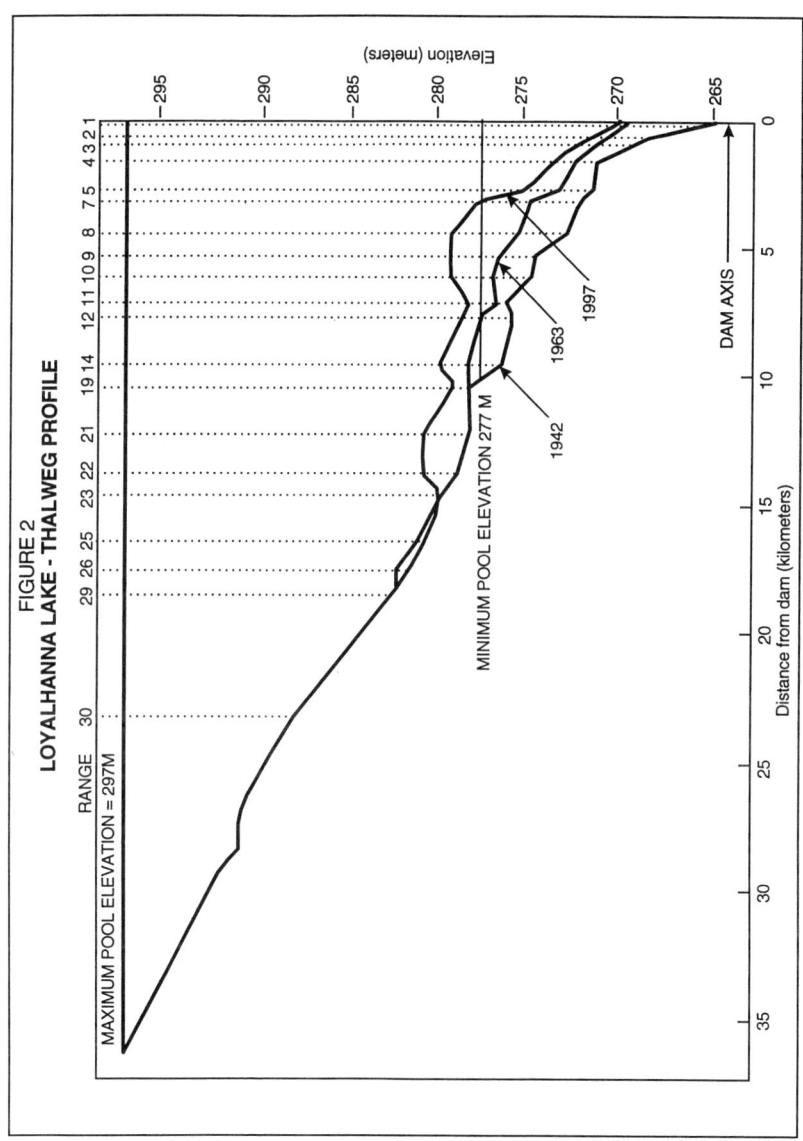

Reservoir Operations during Sedimentation Removal
at Conemaugh River Lake

Werner C. Loehlein, P.E., M. ASCE [1]

Abstract

Conemaugh River Lake was authorized by the 1936 Flood Control Act for the sole purpose of reducing flood stages on the Kiskiminetas, lower Allegheny, and upper Ohio Rivers and placed in full operation in December 1952. However, over the years sedimentation problems, seriously aggravated by erosion from extensive surface mining operations in its tributary drainage basin, has forced the Pittsburgh District to gradually increase the project's minimum pool elevation. Much of the accumulated sediment in the lake is located immediately upstream of the dam. The sediment, deposited since construction of the dam, exceeds 30 feet in thickness in the lower three miles of the reservoir and is adversely impacting the operation of the dam.

Background

The dam is located in southwestern Pennsylvania on the Conemaugh River, 7.5 miles upstream of Saltsburg, and 47 miles downstream of Johnstown. At Saltsburg, the Conemaugh River joins Loyalhanna Creek to form the Kiskiminetas River. The Kiskiminetas River then flows 27 miles to its confluence with the lower Allegheny River, 30.2 miles upstream of Pittsburgh, Pennsylvania. The drainage area above the dam is 1,351 square miles.

For many years Conemaugh operated for flood control only. The water quality of the lake was so grossly polluted by acid mine drainage that no fish or other aquatic organisms could survive in the lake or downstream in the Conemaugh and Kiskiminetas Rivers. However, project conditions, regional needs and public

---
[1]Supv. Hydraulic Engineer, U.S. Army Corps of Engineers 1000 Liberty Avenue, Pittsburgh, PA 15222-4186

expectations have changed and so has the project's water management criteria.

Conemaugh River Lake is a part of a system of nine Corps of Engineers' reservoirs in the Allegheny River basin. Since 1967, Conemaugh River Lake, in conjunction with other Corps projects in the basin, has been utilized to control adverse acid mine drainage impacts on the lower Allegheny River. This is accomplished principally by preventing the flow contribution of acid mine drainage degraded Kiskiminetas River water to the Allegheny River from becoming excessive. Past experience has shown that during low flow periods, when the flow contribution of the Kiskiminetas River was roughly 15 percent or less of the total Allegheny River flow, the Allegheny River could assimilate the Kiskiminetas acid loading by dilution and neutralization without any significant depression in downstream pH. Acidity concentrations are normally inversely related to flow and tend to be lower during the cooler months of the year. Therefore, during the higher flow periods that generally occur during the winter and early spring, a greater percentage of Kiskiminetas flow can be tolerated without an adverse effect on Allegheny River water quality.

Prior to the completion of Kinzua Dam (Allegheny Reservoir) in 1967, acute episodes of water quality degradation, including fish kills, occurred regularly in the lower Allegheny River when the contribution of Kiskiminetas flows became excessive. Allegheny Reservoir supplies most of the high quality augmentation water for dilution and neutralization of Kiskiminetas flows. Allegheny Reservoir is located 168 miles upstream of the mouth of the Kiskiminetas River. Storage in this project at summer pool is 573,000 acre-feet, 549,000 of which is available for augmentation. Since 1967, there have not been any serious episodes of water quality degradation nor reported fish kills attributable to acid mine drainage in the lower Allegheny River.

When first constructed, Conemaugh River Lake was maintained at a normal minimum pool elevation of 880 feet National Geodetic Vertical Datum (NGVD). At this elevation, the dam impounded 4,000 acre-feet of water, with a surface area of 300 acres, along a narrow 6.6 mile long slackwater reach. At its full pool elevation of 975 feet NGVD, the project could store 274,000 acre-feet of water in a 20.9 mile long lake with a surface area of 6,820 acres.

Sedimentation Surveys

Thirty-six ranges with 62 permanent monuments above

elevation 975 NGVD were established across the reservoir at more or less regular intervals, normal to the main stream valley and the principal tributaries, for use in determining sedimentation in the reservoir. Two field checks of reservoir sedimentation have been made, one in 1966 and one in 1982. The 1982 survey showed that a total sedimentation deposit of 11,340 acre-feet, representing an average annual sediment accumulation of 380 acre-feet, or 0.283 acre-feet per square mile of drainage area, per year. This amounts to a total depletion of 4.14 percent of the original storage capacity of the reservoir, or 0.139 percent per year.

Early laboratory analysis of the sedimentation samples showed that the sediment is deposited generally in a uniform pattern of coarse and fine material. The gradation analysis of the samples showed that about half of the deposited material is silt and clay.

It was concluded that the volume of sediment deposition has been considerable and that in the future, sedimentation may impose a severe condition on either the utility or operation of the gates of the reservoir.

Impact to Project Operations

Observed loss of minimum pool storage has forced the District to gradually increase the minimum pool elevation from 880 to 890 to 900 NGVD. Based on the 1982 sedimentation survey data, the lake has a surface area of 800 acres and a capacity of 5,500 acre-feet at its current minimum pool elevation of 900 feet NGVD. With the exception of a relatively narrow channel, virtually the entire reservoir from the dam upstream for approximately nine miles, had shoaled to elevation 890 or higher. Sedimentation within the lake typically results in a clarified discharge and improved downstream water quality. This occurs because the lake acts like a giant stilling basin allowing the sediment to drop out before being discharged. This stilling effect has allowed the large amount of sediment to accumulate near the dam. There have been several occasions when high discharges at relatively low pool levels have been necessary. This has resulted in large volumes of the accumulated sediment near the dam breaking loose and creating downstream mud waves. The mud waves have adversely affected the Salzburg municipal water supply by filling their sedimentation basins and clogging their intake screens.

When high levels of sediment are discharged, the sediment acts like abrasive sandpaper which has accelerated wear on the exposed surfaces of the sluice gates. In addition, as sediment is released, it gets in and around the gates making normal opening and closing operations difficult. Should a gate failure occur during a storage release when large amounts of sediment and debris break loose, the emergency gate closure could be affected causing the lake to drain.

Other Impacts

In March 1989, commercial hydropower was added to the project. The hydropower station, whose intakes are located approximately 2.4 miles upstream of the dam and powerhouse is located 1500 feet downstream of the dam, is operated as a run-of-river project, with a minimum flow of 25 cubic feet per second (cfs) maintained through the gates of the dam. Under normal operations, all discharges in excess of this minimum 25 cfs release, within the operating range of the hydropower unit, are passed through the turbines. Flows greater than the turbine capacity are passed through the dam. Conversely, if flows are less than the minimum required for proper turbine operation, the hydropower project ceases generation and all flows are passed through the dam. Hydropower discharges are subservient to the District's flood control and water quality system operations.

During periods of low runoff, the amount of water discharged from the dam and power plant normally approximates that of the inflow, with only slight changes in reservoir storage or in downstream conditions. However, during flood storage periods, when large volumes of water are detained and the rate of release varies greatly in time and magnitude from the inflow, downstream water quality and associated in-stream needs are considered in determining reservoir discharges.

In addition, the water quality has dramatically improved in recent years to the extent that the lake and tailwater area now support a limited fishery. While still degraded by acid mine drainage, the water quality of the Conemaugh River has been improving for at least the past 25 years, and aquatic life has now begun to return to its waters. With rapidly declining acidity and increasing pH, the diversity of the fish samples increased nearly five times between 1986 and 1990. Nearly 40 times as many fish were captured per unit effort of sampling time, and compared to 1986, the weight of fish collected per unit effort was 80 times higher in 1990. The dramatic recent improvement in the fishery is also apparent downstream of the project.

The lake fishery, however, was almost exclusively confined to two tributary embayments that were inundated after the reservoir pool elevation was increased to 900 feet NGVD in 1989. These two tributaries are Aultmans Run and Spruce Run. It appears that there is sufficient water quality and forage production in the bays for sport fish maintenance and growth, whereas in the main lake water quality is still only barely sufficient to provide sportfish maintenance and survival.

Cultural resources of Conemaugh River Lake include: (1) intact segments of the 1832 Pennsylvania Canal towpath and channel, canal bridge crossing piers, and a canal tunnel through Bow Ridge; (2) numerous 1852 and 1907 railroad artifacts including four intact stone arch bridges and two tunnels; and (3) an old water mill powered tannic acid production site. These historic artifacts are highly pertinent to any attempt to manage for improved lake recreation at the project not only because of the District's responsibility to protect historic structures, but also because along with numerous exposed and submerged tree stumps and an unusually intense debris problem, the old bridge piers and bridges are serious hazards to power boating. Passage under one of the stone arch bridges becomes impossible for even small boats at pool elevations higher than 905 feet NGVD.

The increased pool to elevation 900 NGVD flooded at least several hundred acres of land to depths of less than two feet, which very quickly became vegetated with wetland species of plants. The expansive new wetlands at the project have attracted abundant wildlife including resident and migratory wading birds.

## The Study

In August 1994, the Corps of Engineers Waterways Experiment Station (WES) prepared a report for the District that dealt with the sediment problem. Because sediment threatens the safe operation of Conemaugh Dam's sluice gates, WES was consulted for technical advice on how best to deal with the sediment that has accumulated in the lake. They concluded that with continued project operation, the sediment deposit in Conemaugh Lake can be expected to approach the operating level of elevation 900 NGVD. Continued raising of the pool will cause continual rises in the level of sediment deposition.

## Sediment Removal

In an attempt to temporarily remedy this problem, the District initiated plans to remove some of the accumulated sediment by dredging the area near the face of the dam extending upstream about 400 feet.

The major sediment removal work was initiated in the

spring of 1997 with the construction of a containment dike. This dike was designed to hold approximately 150,000 cubic yards of silt from the vicinity of the gates of Conemaugh Dam. The dredging operations was initiated in early September and was conducted with a hydraulic dredge. There were some concerns about the dredge's impact on the Saltsburg water supply intake. Although, the operation produced no measurable turbidity above background levels, a high flow release made after termination of dredging was unusually turbid.

The pool of Conemaugh River Lake was held very near elevation 900 NGVD for a large part of the summer of 1997. District staff monitored the project's extensive wetlands during this period. As of July, with relatively warm and dry weather in early summer, exposed flats were observed to be desiccated and barren of vegetation. However, with some wetter weather in August, volunteer smartweeds, which have high wildlife food value, began to germinate on these flats. Significant numbers of sandbar willows, a wetland shrub with high value as wildlife escape cover, also took root during this period. The net and long-term impacts of minimal pool fluctuations were probably beneficial to project wetlands and wildlife.

In 1997, prior to the dredging operations, whenever control of the discharge was switched from the dam to the hydropower plant, a large volume of sediment passed through the hydropower units, plugging its screens within minutes. This occurred for the first time, at elevations as high as elevation 902 NGVD.

In late November, dredging operations were suspended until the spring of 1998. The total volume removed to date was approximately 85,000 cubic yards. The remaining 65,000 cubic yards will be removed during the summer of 1998.

Conclusions

The long term impacts of sedimentation on the operation of the gates of the dam is a serious continuing problem. The District agrees with the WES' conclusions that sediment would continue to accumulate to the operating level of the pool. Any further increase of the level of the pool would cause an increase in the depth of accumulated sediment. Furthermore, the August 1994 WES study concluded that because the diversion of flows through the hydropower facility has reduced flow in the main channel, sediment deposition increased in the reservoir between the hydropower inlet and the dam. The District plans to pursue a permanent sedimentation control methodology at Conemaugh.

# SEDIMENT TRANSPORT CAPACITY AS AN OBJECTIVE OF RESERVOIR OPERATIONS

Robert T. Milhous[1]
Member, ASCE

## INTRODUCTION

The construction and operation of reservoirs changes the streamflow patterns in rivers. Most often the new pattern will have reduced the ability of the river to transport sediment and the supply of sediment to the river. In a gravel-bed river, fines and sand may accumulate in the gravel during periods of low flow and the pools may fill with gravel, sand and fines. If the supply of gravel is reduced or eliminated the substrate may change significantly. One objective of an instream flow requirement should be to manage the transport of sediment in the river in a matter consistent with the needs of the aquatic ecosystem. As part of a program to develop methods of flushing flow analysis, a sediment transport capacity index has been developed. This paper presents the index and how the index may be used to develop reservoir operation strategies that consider the movement of sediment (or the prevention of excess movement) as one of the reservoir management goals.

The formulation of the sediment transport capacity index is presented in the section below followed by the application of the index to an aquatic habitat concern related to the Gunnison River in western Colorado.

## THE SEDIMENT TRANSPORT CAPACITY INDEX

The sediment transport capacity index (STCI) is based on the relation between the sediment concentration in a gravel and cobble-bed river and the discharge:

$$C = a1 \; Q^{b1} + a2 \; (Q-Q_{crt})^{b2} \quad (Q > Q_{crt})$$
$$C = a1 \; Q^{b1} \quad\quad\quad\quad\quad\quad\quad\quad (Q \leq Q_{crt})$$

where C is the sediment concentration, Q is the discharge, and $Q_{crt}$ is the critical discharge for disturbance of sediment in the stream bed; a1, a2, b1 and b2 are empirically derived coefficients.

---

[1] Midcontinent Ecological Science Center. Biological Resources Division. U.S. Geological Survey. 4512 McMurry Avenue. Fort Collins, CO 80525-3400.

Sediment load is then:

$$L = k\ Q\ \{a1\ Q^{b1} + a2\ (Q-Qcrt)^{b2}\} \quad (Q > Qcrt)$$
$$L = k\ a1\ Q^{b1+1} \quad (Q \leq Qcrt)$$

where L is the sediment load, and k converts units. Sometimes the critical discharge is nearly zero and some times (often in gravel and cobble bed rivers) the left hand part of the equation is very small relative to the right. The logic used in developing a sediment transport capacity index was to divide a simplified form (either the right or left part of the top equation eliminated) of the load equation by the load at a reference discharge; hence eliminating a1 or a2 from the equation. Experience using the index has resulted in the elimination of a1 or a2 and dividing by a reference discharge - the difference is a constant and does not change the conceptual basis for applying the equation. The purpose of the reference discharge is to make the numbers manageable and to make the equation dimensionless; the value of the reference discharge is arbitrary.

The sediment transport capacity index (stci) for a day is then:

$$stci = \frac{(Qd - Qcrt)*(Qd^{b-1})}{Qref^b}$$

where Qd is the daily discharge, Qcrt is a critical discharge (sometimes zero) for some sediment transport related process in the river, Qref is an arbitrary reference discharge, and b is a power term (selected by the person doing the analysis). The stci is 0 if Qcrt is greater than Qd. The index for a period is the sum of the daily indices, the time period can be of any length but is most often a water year. There is a shift in logic from "sediment load" to a "sediment transport related process" - the importance of this shift will be shown by example in the presentation on the Gunnison River that follows. The annual index is abbreviated 'STCI'.

## APPLICATION TO THE GUNNISON RIVER

One use of the sediment transport capacity index is in the determination of an instream flow for the maintenance the substrate below a reservoir in a condition needed by a desirable ecosystem. The second possible use is in the process of investigation the impacts of reservoir on the river channel downstream of the reservoir. This paper is about the application of the index to substrate maintenance flow analysis.

A technique for calculating the magnitude of the streamflows need to maintain the substrate of gravel and cobble bed rivers in a condition needed by desirable aquatic systems has been developed. This technique was used to investigate the instream flows needed to maintain the substrate of the Gunnison River free of undesirable sediment. The method, and the application to the Gunnison River, are described in Milhous 1997a and 1997b. (An application to the Trinity River in northern California is given in Milhous, 1996). The technique uses a function relating the maximum size of the wash load, suspended load, and bed load to the discharge in the river along with the size of sediment that should not be present to determine instream flow requirements needed to maintain a desirable substrate. The results for the

Gunnison River and for an ecosystem that supports Colorado squawfish are given in Table 1. The streamflows required for different objectives given in Table 1 are the starting point of the results presented in this paper.

Table 1. The instream flows needed to maintain habitat for Colorado squawfish spawning in the Dominguez Flats reach of the Gunnison River, Colorado (from Milhous, 1997a).

| Objective | Target Size (mm) | ß | Transport Load Mode | Required Discharge (cms) |
|---|---|---|---|---|
| Flush Riffles | 4.74 | | Suspended | 355 |
| Flush River | 2.0 | | Wash | 354 |
| Maintain Riffles | 0.50 | | Wash | 27 |
| Clean Pools of gravel | | 0.021/0.016 | Bed | 484 |
| Scour Side-channels | 1.0 | | Wash | 210 |

Frequency and duration of the instream flows are also important. A summary of the results given in Milhous (1997a) for the frequency and duration, expressed either in terms of required number of days or sediment transport capacity index for the objective, is given in Table 2. The frequency and duration of the instream flows for flush the riffles, clean the pools of gravel, and to scour the side-channels was determined by a historical analysis of the annual values of the STCI. For river flushing, the frequency was based on an analysis of the historic values of the STCI but the duration of the flushing flows was determined by an analysis of the time of travel of the sediment in the stream (Milhous, 1997b).

Table 2. The frequency and duration of the instream flows needed to maintain habitat for Colorado squawfish Spawning in the Dominguez Flats reach of the Gunnison River, Colorado (from Milhous, 1997a). The power term, b, used in the calculation of the STCI was 2.0; and the reference discharge was 150 cms.

| Objective | Critical Discharge (cms) | Duration (days) | (STCI) | Frequency Years |
|---|---|---|---|---|
| Flush Riffles | 355 | | 16 | 1 in 3 |
| Flush River | 354 | 4 | | 1 in 2 |
| Clean Pools of gravel | 484 | | 6 | 1 in 3 |
| Scour Side-channels | 210 | | 20 | 2 in 3 |

The determination of the frequency and the magnitude of the STCI required considerable judgment. The starting point was the STCI variation through the years. The variation of the value calculated using the critical discharge for the flushing of riffles is given in Figure 1. There was little reservoir development prior to 1937, the first sizable reservoir was completed in 1937. Reservoirs constructed prior to 1965 provided some control of the streamflows. Blue Mesa Reservoir was completed in 1965 and provides substantial control of the streamflows. The changes in the annual streamflows, annual maximum daily flows, and STCI caused by the construction of the reservoirs are presented in Milhous (1995).

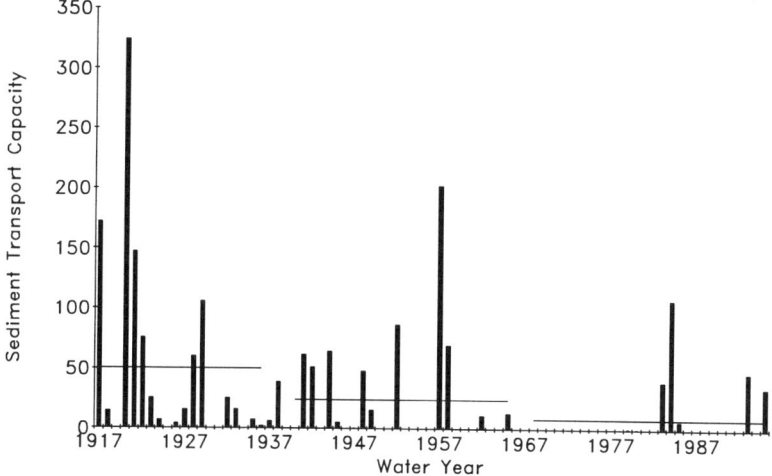

Figure 1. The sediment transport capacity index (STCI) calculated using the critical discharge required to remove 4.74 mm sediment from the riffles of the Gunnison River.

The duration diagram for the periods prior to completion of Blue Mesa Dam are given in Figure 2. In the 1993 and in 1995 the river did have the capacity to remove sediment from the river and did remove sediment. In the spring of 1993, prior to the runoff in May and June 1993, the river had much stored sediment. The 1993 event removed much of this sediment and the 1995 event completed the task. If there had been adequate flushing in the seven years prior to 1993, the 1993 event may have cleaned the river adequately. The assumption is that an STCI of at least 16 every three years would likely be adequate to keep the riffles free of sediment. The frequency of the STCI needed to clean riffles was based on a review of the frequency of the years in the pre-Blue Mesa record that the STCI was at least 16, this was equalled or exceeded 44% of the years prior to 1937 and 29% between 1940 and 1965.

The specification of a STCI along with a frequency the STCI must be exceeded allows a reservoir operator to meet the flushing flow requirements with a variety of flows all of which must exceed the critical discharge. The streamflows can be variable; never-the-less, it is useful to know how many days a constant streamflow greater than the critical discharge would have to occur in order to meet the requirements in Table 2. The required days are illustrated in Figure 3.

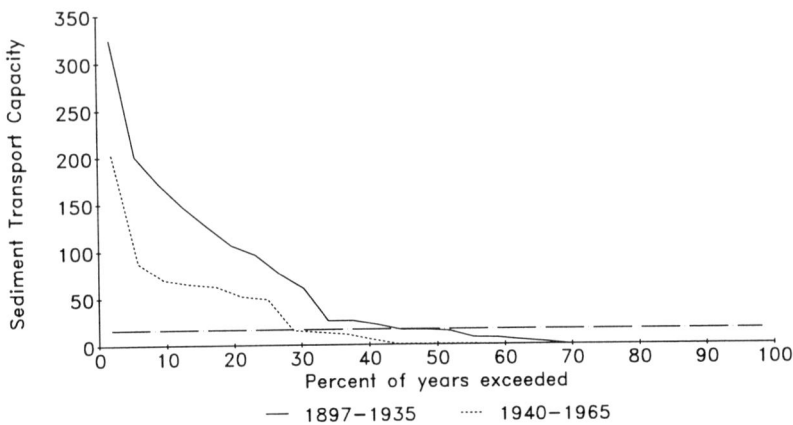

Figure 2. Relation between the percent of years the Sediment Transport Capacity Index (STCI) exceeds the STCI and the value of the STCI. There are missing years in the 1897 thru 1936 period. The horizontal line is at an STCI of 16.

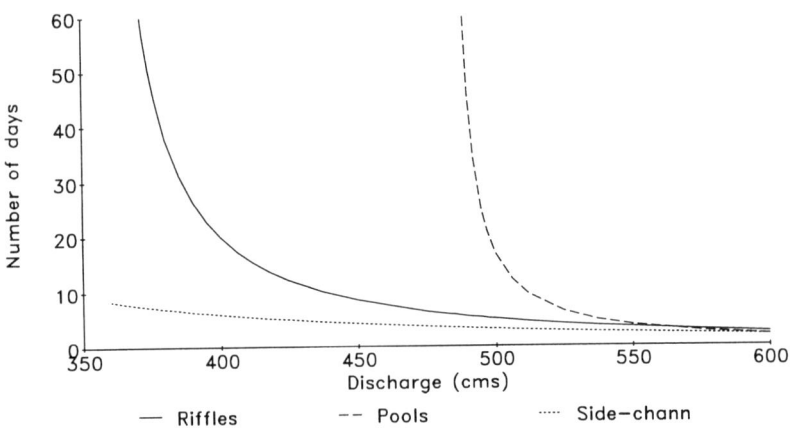

Figure 3. The number of days at a constant streamflow required to satisfy the Sediment Transport Capacity Index specified in Table 2.

The application of theses curves to determine the increases in flows that would be needed to meet the needs for substrate maintenance flow are illustrated by Table 3. The table express the volume of streamflow that would have had to be added to the peak streamflows during the water year to satisfy the STCI given in Table 2 as a percent of the total flow in the lower Gunnison River. This assumes the same annual discharge but with the flow rearranged to meet the flushing flow criteria. Table 3 shows that in most years the rearrangement of streamflows to meet a four day minimum flow would be 6 percent or less but in low flow years the rearrangement of flows would have a major impact. The frequency of the flushing events is 'on-the-average' which allows a water manager to skip low flow years so long as the deficiency is made up soon after.

Table 3. The percentage of the streamflows at the USGS hydrometric station on the Gunnison River near Grand Junction that would have to be rearranged to give the required sediment transport capacity to flush pools of sand and gravel. The number of days is the length of the period of constant streamflows.

| Water year | % of years exceeded | percent of annual streamflow rearranged | |
|---|---|---|---|
| | | 4 days | 10 days |
| 1990 | 97 | 17.5 | 43.0 |
| 1982 | 56 | 5.5 | 14.8 |
| 1987 | 19 | 3.4 | 8.5 |

## DISCUSSION

The advantage of the use of the sediment transport index is that it allows a reservoir operator the flexibility of meeting the streamflow needs with a mix of streamflows. The use of STCI also better represents the physical processes in the stream than a presentation of streamflow alone. At this time the frequency and magnitude of the STCI is based on a historic analysis of the pre-project annual values of the STCI. Criteria need to be developed to guide the section of both the frequency and the magnitude of the STCI.

## REFERENCES

Milhous, R.T. 1995. Changes in Sediment Transport Capacity in the Lower Gunnison River, Colorado, USA. in Geoffrey Petts ed. Man's Influence on Freshwater Ecosystems and Water Use. IAHS Publication No 230. International Association of Hydrological Sciences. Wallingford, Oxfordshire, UK. pp275-280.

Milhous, R.T. 1996. Streamflows for the Management of Sediment: The Trinity River, California. in Morel-Seytoux, Hubert J., ed. Sixteenth Annual Hydrology Days. Hydrology Days Publications, Atherton, CA. pp305-316.

Milhous, R.T. 1997a. Numerical Modeling of Flushing Flow Needs in Gravel-Bed Rivers. in P.C. Klingeman, R. Beschta, P. Komar, and J. Bradley eds. Gravel-bed Rivers in the Environment. Water Resources Publications. Littleton, CO. (in press)

Milhous, R.T. 1997b. Modeling of Instream Flow Needs: The Link Between Sediment and Aquatic Habitat. Regulated Rivers: Research and Management. (accepted).

# Temporary Erosion Control for Highway Construction in Kansas

Bruce McEnroe[1], M. ASCE, and Brian J. Treff[2], S. M. ASCE

## Abstract

Field observations of temporary erosion-control measures on highway construction sites revealed some common problems. Most failures result from improper location, installation, or maintenance of the erosion-control measures. A training program for field personnel could alleviate most of the misunderstandings that lead to failures. Current specifications for temporary erosion control are satisfactory, but better compliance is needed. The timely use of temporary seeding would greatly reduce soil loss.

## Introduction

Temporary erosion control is becoming an increasingly large part of highway design and construction. The Kansas Department of Transportation (KDOT) must comply with current federal and state erosion-control requirements and prepare to meet more stringent requirements in the future. Researchers at the University of Kansas recently completed a comprehensive assessment of KDOT's temporary erosion-control practices, which are based primarily on AASHTO guidelines (AASHTO, 1992).

Over a two-year period, we monitored the performance of KDOT's standard measures for temporary erosion control at several highway construction sites. We also field-tested some new non-standard measures. The field performance of erosion-control measures depends on a multitude of variables that are difficult to control and measure. Therefore, our assessments of field performance are based on qualitative observations rather than quantitative measurements. KDOT field personnel provided observations on their experiences with temporary erosion control through questionnaires and personal interviews. We also made a thorough review of

---

[1] Professor, Dept. of Civil & Environmental Engineering, University of Kansas, Lawrence, KS 66045
[2] Engineer, George Butler & Associates, 8207 Melrose Drive, Lenexa, KS 66214

the erosion-control practices of other state and federal highway agencies. The three principal products of our research are (1) recommendations for certain changes to KDOT's current practices, (2) a *KDOT Temporary Erosion-Control Manual*, and (3) a map of the relative soil-erosion potential for construction sites in Kansas.

**Findings and Recommendations**

**1. Training.** Design and construction personnel need more training on the design, installation, and maintenance of temporary erosion-control measures. Most failures of erosion-control measures result from improper location, installation, or maintenance. These mistakes indicate an inadequate understanding of how these measures work. A training program could alleviate most of the misunderstandings that lead to failures. The *KDOT Temporary Erosion-Control Manual*, developed as part of this project, provides essential instructions for successful erosion control.

**2. Temporary seeding.** Establishing vegetation is by far the most effective way to control erosion. Temporary seeding saves the contractor time and money in the long run by preventing the loss of topsoil and costly regrading. The KDOT specification requires temporary seeding within 14 days of the temporary or permanent cessation of construction on a portion of the project site, unless activities are expected to resume within 21 days. This specification is satisfactory but compliance is rare. The timely use of temporary seeding would greatly reduce soil loss.

**3. Maintenance and sediment removal.** Temporary erosion-control measures require maintenance. Water, time and construction activity cause damage and degradation. Temporary erosion-control measures must be inspected regularly for integrity. Sediment removal is another important aspect of maintenance. Once a device is nearly filled with sediment, it is ineffective. The sediment-filled ditch check is Figure 1, for example, is non-functional. KDOT specifications require inspection every 7 days and within 24 hours of a rainfall of more than 10 millimeters. Any damage must be repaired promptly. Sediment should be removed when it reaches half the exposed height of the erosion-control structure. As with temporary seeding, the specification is satisfactory but better compliance is needed.

**4. Silt-fence posts**. The spacing of silt-fence posts should not exceed be 1.2 meters. KDOT's current maximum post spacing of 1.6 meters allows for too much sag in the fence, which decreases its storage capacity and overall effectiveness. The excessive sag increases the likelihood of overtopping, which usually leads to total failure. Figure 2 shows two views of an overtopped silt fence with severe erosion below the point of failure. The new AASHTO M288-96 Silt Fence specification calls for a 1.2-meter post spacing for silt fences. Both the AASHTO and KDOT specifications require 10-cm x 10-cm hardwood posts for silt fence. This specification should be enforced. We observed numerous failures of prefabricated silt fence with 10-cm x 5-cm posts.

Figure 1. Bale ditch check full of sediment

Figure 2. Upstream and downstream views of a silt-fence failure

**5. Slope-barrier placement.** Silt-fence slope barriers should be placed along elevation contours. Otherwise, the barrier is likely to fail in one of two ways. The concentration of water at a low point decreases the barrier's storage capacity and increases the likelihood of failure by overtoppping, as in Figure 2. If no low point exists, the flow of water along the upstream base can erode a trench, as in Figure 3, and undermine the toe of the barrier.

**6. Alignment of bale ditch checks.** The original KDOT standard drawings called for the ends of hay-bale ditch checks to be angled upstream. We found that these "wings" impede sediment removal, which is often performed by a bulldozer. We recommend a linear alignment for bale ditch checks.

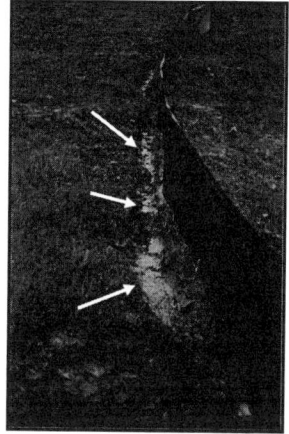 

Figure 3. Erosion along base of silt fence

Figure 4. Failure of a bale ditch check due to scour

**7. Scour apron for bale ditch checks.** We observed that bale ditch checks are often undermined by scour directly downstream of the center bale. The center bale eventually falls into the scour hole, and much of the sediment trapped behind the check is released, as in Figure 4. To prevent this common failure, we recommend the use of an erosion-control blanket on the downstream side of the check. The 1.8-m-wide "scour mat" is rolled out parallel to the check and anchored in the same trench as the bales.

**8. Rock ditch checks.** Rock ditch checks are recommended for ditches that will be rock-lined upon completion. Rock used for checks can be spread out in the ditch after final grading is complete. Rock ditch checks can withstand greater forces than bale and silt-fence structures. They require minimal maintenance beyond sediment removal.

**9. Triangular Silt Dike.** The Triangular Silt Dike (TSD) is a manufactured alternative to bales and silt fence. It consists of a triangular polyurethane foam core with a geotextile fabric cover. This product was field-tested in several installations on a highway construction project. We found that it performs particularly well as a drop-inlet barrier. This lightweight product can be installed in less than half the time of a standard bale or silt-fence structure with no machinery. It is secured with 150-millimeter landscape pins, which work well on thin soil over rock. Adjacent sections connect with overlapping sleeves. The attached fabric aprons prevent undermining. The TSD has a higher initial cost than a bale structure, but it does not degrade. Figure 5 shows a TSD barrier around a large drop-inlet before and after a rainstorm.

This TSD barrier trapped a large volume of sediment, and withstood many rainfall events over several months with no deterioration.

Figure 5. A TSD drop-inlet barrier before and after a rainstorm

**Kansas Erosion Potential Map**

The map in Figure 6 shows the relative potential for soil loss from construction sites in Kansas. The mapped quantity is the product of the soil erodibility factor and the rainfall intensity factor in the universal soil-loss equation. This product is an excellent index of the erosion potential of disturbed soils. The map was created with GIS technology. We obtained the soil erodibility factor from the STATSGO database of the NRCS, and computed the rainfall intensity factor from 2-year, 6-hour rainfall depth. The darker areas of the map have a higher potential for soil loss. In the full-size color map, a legend relates the color scale to a numerical index of erosion potential. The map is intended to serve as an index of general trends across the state and should be used only for large-scale planning. The potential for erosion is highest in the southeast and lowest in the west, mainly due to differences in rainfall.

**Conclusions**

Most failures of temporary erosion-control measures result from improper location, installation, or maintenance. A training program for field personnel could alleviate most of the misunderstandings that lead to failures. The *KDOT Temporary Erosion-Control Manual*, developed as part of this project, provides essential instructions for successful erosion control. Current specifications for temporary erosion control are satisfactory, but better compliance is needed. The timely use of temporary seeding would greatly reduce soil loss. The Triangular Silt Dike (TSD) is an attractive alternative to bales and silt fence. In Kansas, the potential for erosion is highest in the southeast and lowest in the west, mainly due to differences in rainfall.

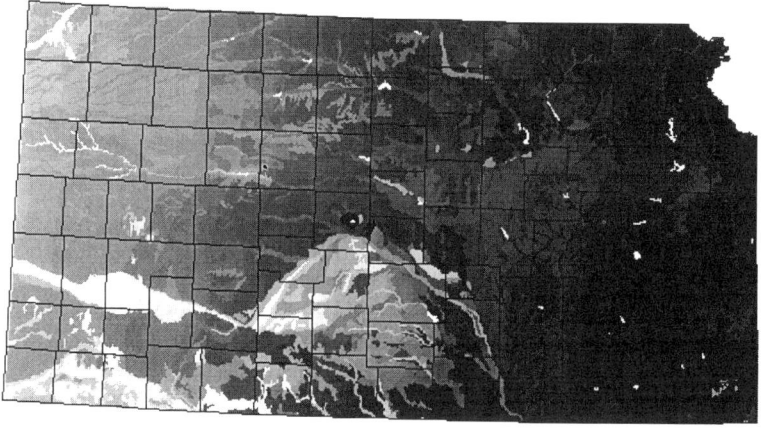

Figure 6. Kansas Erosion Potential Map

## References

American Association of State Highway and Transportation Officials (1992). *AASHTO Guidelines for Erosion and Sediment Control in Highway Construction.*

Treff, B. J. and McEnroe, B. M., *Performance of KDOT Temporary Erosion-Control Measures*, Report No. K-TRAN KU-97-2, Kansas Department of Transportation.

## Development of a Comprehensive Management Plan for Tyler Creek, Elgin, Illinois

G. C. Schaefer, P.E.[1], T. M. Denning P.E.[2], J. W. Hood P.E.[3], J. A. Wickenkamp[4]

### Abstract

A study was undertaken to develop a management plan for Tyler Creek, a tributary of the Fox River. The comprehensive management plan for the creek includes the definition of current flood risks, prevention of future flood damages, and the protection and enhancement of water quality and wildlife habitat within the Tyler Creek riparian zone within the Elgin Facilities Planning Area (FPA). The foundation for the study is a traditional hydrologic and hydraulic study for existing conditions on Tyler Creek. A future conditions modeling effort also was completed based on future land use projections and existing stormwater management regulations throughout the watershed. Assessments of flood damage, environmental resources and stormwater management regulations for Tyler Creek were prepared. The result of the study was the development of a comprehensive management plan that addresses flooding, protection of environmental resources, and stormwater management regulations for the study reach within the context of existing and future development conditions throughout the watershed.

### Introduction

The City of Elgin's concern regarding the management of Tyler Creek has spanned three decades. It includes both government and citizen initiatives. The City's major concern has been the development of a plan to provide regional stormwater storage upstream of Randall Road, located on the western side of the City. The administrative portion of this plan was implemented allowing approximately 100-acre-feet of storage to be waived for new development in exchange for payments to construct a regional reservoir at Randall Road. Due to

---

[1] President, Hey and Assoc., Inc., Libertyville, Illinois 60048
[2] Civil Engineer, City of Elgin, 150 Dexter Court, Elgin, Illinois 60120
[3] Senior Water Resources Engineer, Hey and Assoc., Inc.
[4] Water Resources Engineer, Hey and Assoc., Inc.

concerns regarding impact to the high quality ecological resources of Tyler Creek upstream of Randall Road, the reservoir was never constructed. This prompted the City to initiate the development of the Tyler Creek Management Plan to review the current state of Tyler Creek and develop a plan for its management. The Tyler Creek Management Plan begins with of a set of goals, approved and endorsed by the City of Elgin. These goals guided the development and evaluation of alternative management plan components to address problems identified in the plan. The alternatives which best addressed the identified problems were selected as plan recommendations. The purpose of this paper is to summarize the approach that was taken to develop a comprehensive management plan for Tyler Creek.

## Management Plan

The Tyler Creek Management Plan for the City of Elgin included hydrologic and hydraulic modeling, a flood damage assessment, an environmental assessment, a stormwater management assessment, a problem assessment, and the development of alternatives. All of these components were used in the development of recommendations in the plan.

**Hydrologic and Hydraulic Modeling** A floodplain mapping report was prepared to update the existing regulatory hydrologic and hydraulic models for Tyler Creek (see Figure 1) and to provide a tool for evaluating stormwater management strategies. An existing conditions hydrologic model was developed that used 61 subwatersheds. In addition, precipitation data was updated to current risk assessment standards using Bulletin 70. The Flood Insurance Study (FIS) hydraulic model was improved by the addition of fifty new surveyed sections as well as seven previously unmodeled bridges. Close to eight miles of stream were modeled. A storage floodway was defined for the study reach on Tyler Creek. The existing conditions 100-year floodplain and profiles for the 2-, 10-, 50- and 100-year flood events were developed. In general, the existing conditions floodplain was found to be one to two feet higher than the published FEMA elevations.

Future conditions hydrologic and hydraulic models also were prepared. These models represent the future conditions shown on various land use and zoning maps for the watershed. Detention storage based on current standards was modeled in the future conditions hydrologic model. The future conditions floodplain elevations in Elgin were lower than existing conditions as a result of on-site stormwater detention policies in the upstream watershed.

**Flood Damage Assessment** Potential flood damages along Tyler Creek were assessed using previous studies, historical observations and the results of the hydraulic models prepared for Tyler Creek. The 100-year floodplain was delineated onto the City of Elgin two-foot topographic maps. Structures within and adjacent to the floodplain were digitized onto the topographic maps.

**Environmental Assessment** The purpose of the environmental assessment was to identify the existing and potential environmental resources of the Tyler Creek watershed within Elgin's jurisdiction. Field surveys and historical reports were used to assess the aquatic ecology, water quality, wildlife habitat, plant communities and streambank erosion along the stream. An assessment of mitigation banking and restoration opportunities for the watershed also was performed.

According to the Illinois Biological Stream Characterization system, Tyler Creek is classified as a Highly Valued Aquatic Resource (or "B Stream") upstream of Randall Road, and as a Moderate Aquatic Resource ("C Stream") downstream of Randall Road. The corridor contains many different plant communities including true aquatic habitat, several wetland community types (marsh, sedge meadow, fen, wet prairie, wet-mesic prairie, scrub-shrub, and wooded floodplain forest), and dry-mesic woodland. The wetland vegetation, for the most part, is relegated to a thin band of vegetation along the creek channel. The riparian corridor offers significant habitat for wildlife.

The field survey completed as part of this study as well as several previous studies provided documentation on the condition of streambanks along Tyler Creek. Analysis of historic aerial photographs indicates that obvious changes in stream morphology have occurred within the stream channel over time. The 23 most significant erosion sites were mapped and representative photographs of each site were obtained. These records will be valuable as future work continues on the stream.

As part of the environmental assessment, a continuous simulation analysis was performed for the Tyler Creek watershed to determine the long-term impact from the waiver of detention for four developments within the City along Tyler Creek. For this analysis, two conditions were analyzed that included the watershed with the detention waived and the watershed prior to construction of the four developments. The specific methodology that was used in the continuous simulation analysis involved the development and execution of an XP-SWMM computer model using approximately 40 years of historical precipitation. Data were extracted from the models at six critical erosion sites on Tyler Creek. The data that were computed for each erosion site over the period of record included the number storms in which bankfull or higher flooding occurs, and the total number of hours of bankfull or higher flooding. The results of the analysis indicated that the waiver of detention for the four developments has had a minimal impact on bankfull or higher flooding at the six erosion sites. Although the developments appear to not have significantly altered stream hydrology, it does not mean that the waiver of detention had no impact to Tyler Creek. Almost every storm sewer that is greater than 24 inches leading from these developments was found to have moderate to significant erosion problems at the outfall. The energy created by discharge from these outfalls along Tyler Creek, led to increased erosion at and just downstream of the point of discharge.

**Stormwater Management Assessment** The stormwater and floodplain management regulations of neighboring communities within the Tyler Creek watershed were reviewed and compared to Elgin's current ordinances. The purpose of this comparison was to identify issues and opportunities that needed to be addressed as part of the Tyler Creek Management Plan. In general, the existing stormwater ordinances in the Tyler Creek watershed outside the City of Elgin were found to be state-of-the-art in terms of detention release rates and storage, protection of depression storage, definition of drainage systems, incorporation of water quality measures and requirements for buffers. The floodplain ordinances of the communities in the Tyler Creek watershed were examined along with ordinances from neighboring counties. The floodplain ordinances were evaluated for floodplain definition, freeboard, flood fringe compensatory storage, protection of riparian zones, water quality, and on-stream detention. As a result of the stormwater management assessment, recommendations for changes to improve the City of Elgin ordinances were made. Typical recommendations included 1.5:1.0 compensatory storage, 2- and 100-year detention release rates of 0.04 and 0.15 cfs per acre, and maintenance of a fifty-foot riparian drainage easement from the edge of the floodplain.

**Problem Assessment** Problems in the Tyler watershed were identified through the flood damage assessment, environmental assessment and stormwater management assessment. The problem assessment was developed as the framework to formulate the management plan alternatives. The project focus was on the Tyler Creek Study Area (Elgin FPA). The study area was divided into six reaches for the problem assessment. Problems such as flood damages, untreated storm sewer outfalls, streambank erosion, degraded aquatic habitat and degraded riparian corridor were identified for each reach. More general problems also were identified for the watershed area outside the Tyler Creek Study Area. These included the need to protect key existing depressional storage areas and habitat degradation.

**Development of Alternatives** Many alternatives were considered in the development of alternatives for the Tyler Creek Management Plan. Alternatives such as flood storage reservoirs required detailed hydrologic and hydraulic modeling. Other alternatives that were considered but did not require modeling included floodproofing, streambank stabilization, water quality and habitat restoration.

Three storage alternatives were developed for detailed testing using the hydrologic models prepared for the Tyler Creek. The storage alternatives included the development of flood storage facilities within the Elgin FPA, the development of flood storage upstream of the Elgin FPA, and the evaluation of eight detention options that involved combinations of four distinct detention requirements with both existing and future land use conditions.

An assessment of the benefits resulting from each storage alternative was made using five functions or values that included flooding reduction, streambank stability, water column quality, ecology, and open space. All of the storage alternatives resulted in an increase in detention or flood control storage in the Tyler Creek watershed. Additional storage is almost always beneficial to flood control. However, the results of the evaluation indicated that the storage alternatives had limited benefits to the structures in the floodplain along Tyler Creek. The storage alternatives impact water quality and streambank stability in Tyler Creek in varying ways. On-line reservoirs will do little for either water quality or streambank erosion. Off-line reservoirs will provide water quality treatment for the volume of water that enters the facilities, but this will have little or no impact to the water quality of Tyler Creek. Detention facilities and tributary storage that provide treatment closer to the point of runoff will be the most effective method to provide water quality benefits in Tyler Creek.

Additional alternatives were developed that did not require modeling to evaluate their potential benefit to Tyler Creek. Each alternative addresses a problem on Tyler Creek at the location of the problem. To address potential flood risk, the evaluation of floodproofing was recommended for all structures mapped within the floodplain of Tyler Creek. Benefits resulting from the additional alternatives are greatest at this point of implementation, however, some alternatives such as streambank stabilization and detention retrofits may have additional positive effects on downstream areas.

A variety of techniques were recommended to prevent and repair streambank erosion. Recommended streambank stabilization practices concentrated on bioengineering applications that use a combination of engineering and biological practices to control erosion and stabilize streambanks. However, traditional structural solutions such as gabions, riprap, or concrete revetments were not precluded from consideration. Recommendations to improve the water quality performance included the retrofit of existing ponds and development of detention ponds at storm sewer outfalls and tributaries to Tyler Creek. Existing ponds may be modified to change the discharge characteristics to increase storage, or they may be retrofitted to incorporate wetland vegetation. Wetland vegetation will increase the treatment efficiency of the pond. Ecological restoration of riparian areas would be accomplished through active management to remove the woody material and prevent reinvasion. To ensure quality aquatic habitats and water quality, basin-wide controls need to be designed and implemented to prevent direct inputs of polluted waters.

**Recommendations** Recommendations, organized geographically, were prepared that addressed the problems identified for Tyler Creek. Throughout the report, the watershed was divided into six reaches that comprised the Tyler Creek Study Area and the watershed outside the study area.

For the study area, general recommendations were made that included the changes to the existing Elgin ordinances. Specific recommendations were made by reach for proposed actions and projects such as streambank stabilization, detention retrofits and the acquisition of parcels for greenway enhancement. Outside of the study area recommendations for ordinance modification also were made. Key flood storage areas were identified that should be protected and acquisition parcels were identified.

## Conclusions

Detailed technical assessments led to the identification of problems and opportunities within the Tyler Creek watershed. Alternatives were devised and tested for their ability to address the identified problems. Ultimately, recommendations were developed to complete the Management Plan for Tyler Creek. The approach taken resulted in a comprehensive management plan that will serve as a blueprint for the future of Tyler Creek.

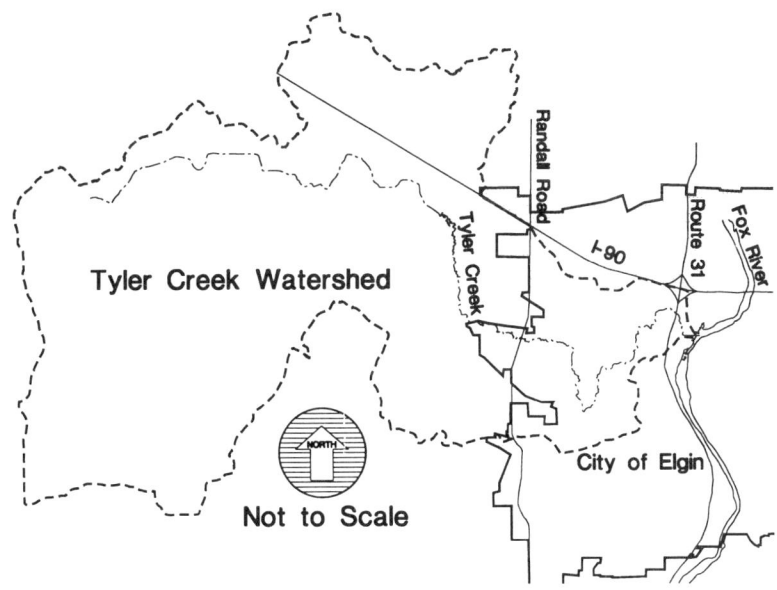

**Figure 1.** Tyler Creek watershed.

# Erosion-Controlled Radioactive Transport in Agricultural Watersheds

Reza Khanbilvardi [1]   Vyacheslav M. Shestopalov [2]   Siamak Esfandiary [3]

Abstract:

    Scientific planning for soil and water conservation requires knowledge of the relations between those factors that cause loss of soil and water and those that help to reduce such losses. Soil erosion with its elaborate complexity and vast variability of hydrological interrelationships is a major scientific and environmental problem. The parameters involved in the erosion process are not yet thoroughly tangible. Especially when the soil particles are contaminated with radionuclides or agrochemical pollutants. The Chernobyl disaster back in April 1986 was a calamitous part in the history of nuclear power usage. This contaminated site created a good opportunity for studying the transport of radioactive contaminants, attached to soil particles, in an agricultural watershed.

---

( 1 ) Director of Center of Water Resources and Environmental Research,
    Professor of Civil Eng., The City University of New York, Convent Ave. & 140th, NY, 10031

( 2 ) Director, NASU Center for Radiohydroecological Field Studies,
    Deputy Director NASU Research Institute of Geology, D.Sc. ( Geology ),
    Professor

( 3 ) Ph.D. Student, Research Assistant, Department of Civil Eng.,
    The City University of New York, Convent Ave. & 140th, NY, 10031

Introduction:

The greatest possible technological disaster of 20$^{th}$ century which occurred on April 26,1986 at the Chernobyl Nuclear Power Plant some 100 km to the north of Kiev in Ukraine not only caused global fallout of radioactive substances but also created a unique radiological polygon where elements of geological environment contain powerful radioactive indicators allowing to study various aspects of influences of such accidents on the environment.Natural and anthropogenic processes are known to take place on areas polluted with radionuclides, which lead to the cleanup and rehabilitation of these areas.
Following the Chernobyl incident, for several months the fallout of radioactive aerosols was observed around the site. $^{90}$Sr and $^{137}$Cs were among the most dangerous components of those aerosols. The large area in the vicinity of Chernobyl is being used to study the erosion- controlled transport of radioactive material within agricultural watersheds. The initial observations of the radionuclides concentrations has shown that the bulk of them is being retained by the vegetation and upper ( 5 cm - 10 cm ) top soil, while the rest goes into the aeration zone and deeper into aquifer.

Methodology:

A ( 20m X 20m ) runoff plot in the area was fixed and devegetated. It was broken into ( 1m X 1m ) elementary squares. The exposed dose intensity was measured at the central points of each elementary square. With the aim of initial soil sampling at the runoff plot a number of soil properties, as well as characteristics of pollutants, were obtained. Tensiometers were installed to measure soil suction and pore pressure in the soil. A methodology to perform artificial rainfall was developed and evaluated. The intensity and the amount of the artificial rainfall were computed based on the precipitation pattern in the area and taking into account mainly heavy showers able to induce erosion. Observations made after and during the testing clearly indicated the effect of rainfall and erosion on transport of radionuclides vertically as well as horizontally.

Analysis:

The distribution of $^{137}$Cs in the upper 5 cm topsoil on the runoff plot has been studied by analyzing results of γ-spectrometric measurements for 25 samples taken from the nodes of 4m X 4m net before and after both the entire annual cycle and each rainfall simulation. The $^{137}$Cs concentration are highly variable, which may be explained by both spotty character of fallout and nonhomogenuity of depths reached by the radionuclides within the plot.

Artificial rains caused a significant redistribution of radionuclides in the uppermost soil layer. In some places their concentration became lower, while in others it became higher.

To study the redistribution of radionuclides related to the runoff plot's microtopography, across the erosional depression extending from the northwest to the southeast corner, a profile was set and along it, 14 soil samples were taken. The concentrations of radionuclides in the samples were determined. Besides leveling survey of the profile was conducted. The measurement data are presented on Figure 1 and Table 1. As can be seen from the figure accumulation of all the radonuclides is observed on the lowest part of the profile, which can be probably explained by erosional washout of the uppermost, the most polluted by radionuclides, layer of soil from microtopographical highs to microtopographical lows. This is particularly remarkable for radioactive cesium. So, in five examples out of 14, located at or near the thalweg of the depression, there occurred 62% of the total amount of $^{137}$Cs, 65% of the total amount of $^{134}$Cs, and 50% of the total amount of $^{90}$Sr. So, microtopographic depressions are areas of accumulation of the radionuclide washed-outs of the microslopes in the process of erosion on the runoff plot.

A clear reverse correlation between the elevations of the soil surface along the profile and specific radioactivity of all three nuclides in the corresponding sampling points confirms the importance of local redistribution of contamination within the plot as compared to the insignificant removal of its initial amount.

Conclusion:

During the one-year cycle of research at the runoff plot located within the zone of intense radioactive fallout following the 1986 Chernobyl disaster, experimental data have been obtained regarding the intensity of erosional washout of radionuclides, particularly $^{137}$Cs.

It has been established that within the Chernobyl closer zone, characterized by fuel and mixed fuel-condensation types of radioactive fallout, the most important factor of self-purification of the territory is natural radioactive decay. For agricultural type of land use, the second important factor is removal of soil lumps on the roots of vegetation during weeding and harvesting.

Even with intense rain simulation causing considerable soil erosion ( 1.2 mm on the average for the runoff plot ) the annual removal of radioactive Cesium from the plot amounted to just 1% of its initial amount, that is twice as low as its losses by natural radioactive decay.

The annual amount of radioactive Cs removed by natural solid runoff has been 0.07% of its initial amount in the soil. So, in these conditions soil erosion appears to be only third important factor of self-purification of the territory from radioactive contamination.

Combining detailed leveling with continuous sampling of some soil profiles could reveal an effect of local redistribution of radionuclides within the plot with its significant accumulation in microdepressions along with the soil material washed out during rain events. In other words, the material contaminated with radioactive, which are transferred in the course of soil erosion, are migrating along the runoff plane rather poorly, and $^{137}$Cs concentration in the solid runoff are quite compatible with the concentration in the soil, although it seems that fine fractions should prevail in it, which sorbs Cs intensively.

Preliminary study of distribution of radionuclides with depths by means of sampling in trail pits on the periphery of the runoff plot implies that rainfall simulation might have intensified their vertical migration.

Acknowledgment:

This study was funded by National Science Foundation.

**Table 1**- Soil elevation and concentration of radionuclides along a 1.3 m soil profile

| Distance (cm) | $^{137}$Cs(kBq/kg) | $^{134}$Cs(Bq/k) | $^{90}$Sr (kBq/kg) | Elevation(mm) |
|---|---|---|---|---|
| 0 | 1.89 | 45.9 | 3.84 | 98304 |
| 10 | 1.48 | 44.8 | 2.67 | 98302 |
| 20 | 1.56 | 36.0 | 2.01 | 98278 |
| 30 | 2.62 | 65.6 | 1.49 | 98257 |
| 40 | 1.05 | 50.9 | 1.02 | 98231 |
| 50 | 0.94 | 23.6 | 1.04 | 98203 |
| 60 | 2.72 | 62.6 | 3.25 | 98178 |
| 70 | 7.15 | 134.0 | 3.98 | 98157 |
| 80 | 6.69 | 244.0 | 2.67 | 98154 |
| 90 | 5.07 | 121.0 | 2.90 | 98155 |
| 100 | 1.18 | 154.0 | 1.95 | 98162 |
| 110 | 1.98 | 80.4 | 0.87 | 98176 |
| 120 | 1.29 | 20.1 | 0.97 | 98185 |
| 130 | 1.44 | 29.4 | 0.73 | 98188 |

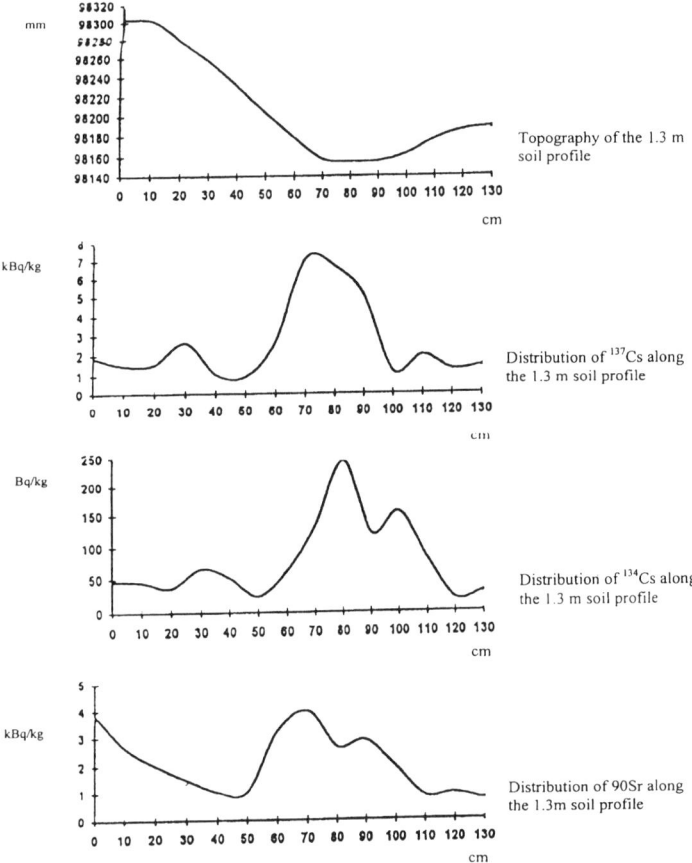

**Figure 1** - Topography and concentration of radionuclides along a 1.3 m soil profile

Engineer's Guide to the Components of GIS
and Mapping using ArcView

By Charles H. Call, Jr.[1] M. ASCE

Abstract

Geographic Information Systems (GIS) are now being used by many agencies to do their work. There are GIS computer tools on the market that can and should be used by Engineers to do mapping analysis. One of these tools is ArcView© (ESRI, 1992). This paper presents and overview of GIS concepts and definitions. Examples of possible uses of these tools in the water resources field are presented.

Components of GIS

Several different conceptual models have been proposed to describe GIS. One of the key characteristics is the ability to digitally transform spatial data into an output that is useful. In concept, GIS is the marriage of mapping cartography and database management systems (DBMS):

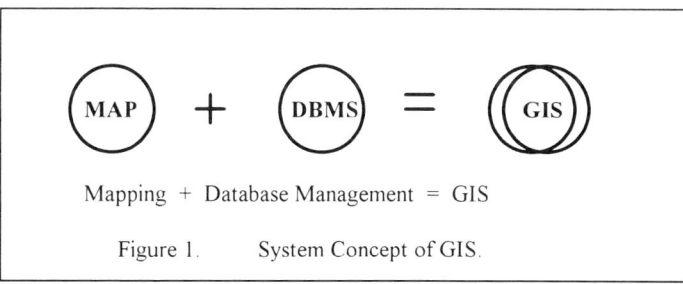

Mapping + Database Management = GIS

Figure 1.  System Concept of GIS.

---

[1]Engineering Administrator, Salt Lake City Public Utilities, 1530 S. West Temple, Salt Lake City, Utah 84115

GIS can also be viewed as a transformation of spatial data into useful output. This is presented on Figure 2 and further described on Table 1.

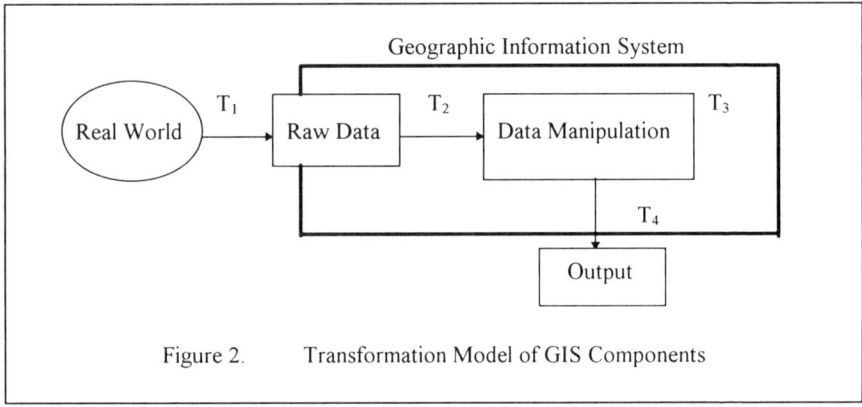

Figure 2.   Transformation Model of GIS Components

A fundamental GIS concept is topology. It is defined as the spatial relationship between connecting or adjacent spatial objects or features. In GIS, topology is defined by the arcs (Worboys, 1995). It is evaluated in terms of (1) adjacency (polygon ID #), (2) connectivity (direction of from-node and to-node) and (3) distance (left and right polygon value or end nodes with unique (x, y) coordinate value).

Table 1.   GIS Components and Transformations

| GIS Component | Transformations |
| --- | --- |
| $T_1$ -- Collection, input and correction | Clean-up and digitizing |
| $T_2$ -- Storage and retrieval | Organize and store |
| $T_3$ -- Manipulation and analysis | Quantitative analysis, map overlay and topology |
| | *Operations:* Reclassification, overlay, measurement, and characterization |
| | *Transformations:* Geometric, attribute changes and object class changes (points to areas, etc.) |
| | *Modeling:* Topological, geometric, map, table, etc. |
| $T_4$ -- Output and reporting | Maps, charts and tables |

## Definitions

*Vector GIS:* Vector based GIS consists of coordinate data and a relational database management system. Operations are digitizing, vectorizing raster images, storing, editing, analysis and transforming. Vector elements classified by length dimensions are:

1. Points -- zero dimensional
2. Lines -- one dimensional
3. Areas -- two dimensional
4. Surfaces -- two dimensional
5. Volumes -- three dimensional

*Vector Data Structures:* Vector structures predominate in Computer Aided Design (CAD) systems and GIS with strong cartographic capabilities. Topological data features are:

- Point -- Discrete locations or linked to form lines
- Vertices -- Points on a line
- Line -- A sequence of vertices
- Start Node -- Vertex at the beginning of a line
- End Node -- Vertex at the end of a line
- Chain -- A line which is part of one or more polygons
- Arc -- A chain or line
- Node -- The intersection of lines or a termination point
- Polygon -- One outer ring and zero or more inner rings
- Holes/islands -- Inner ring within a polygon
- Ring -- Consists of one or more chains

*Raster GIS:* Raster based GIS consists of pixel column and row data and a relational database management system. Operations are scanning, remote sensing, image processing, storing, editing, analysis and transforming. Raster elements are pixels.

*Raster Data Structures:* Raster structures are used in image processing systems and in Raster GIS. They predominate in analysis of spatially continuous images. Raster structures most common organization is a rectangular array of pixel values, in which row and column coordinates define a particular location. There are 7-bands or data entry values for each pixel image location. These are read in different ways. The read sequence will be one of the following:

- BIL -- Band interleaved by line -- Reads all bands on each line before moving on to next line.
- BIP -- Band interleaved by pixel -- Reads all bands on each pixel before moving on to next pixel.
- BSQ -- Band sequential -- Reads each band for complete image then moves on to next band.

Because of the large storage requirements of some of these images it is important to compress the data and conserve the available computer storage space. The data compression techniques used for raster data are as follows:

- Run-length encoding -- Data is stored by row. Adjacent pixels having the same value are combined together as a run, represented as a pair of numbers and thus compressing the data line. When more adjacent pixels have the same value, the greater the compression that is achieved.

- Quadtree structures -- Raster data structure based on recursive subdivision of a square image into quadrants using a Morton matrix (addressing or indexing) for an 8 by 8 image until each cell has a homogeneous value. This combines the two coordinate values into a single number. It provides a method of storing spatial objects based on groupings of pixels. The blocks for each level are indexed by a base 4 number, with the four quadrants of a block being 0 (SW), 1 (SE), 2 (NW) and 3 (NE). The numbering sequence is biggest quarter, second level and third level. A single spatial coordinate is used instead of two coordinates, and that point address can be used to find the attribute value at that location in any of the quadtree map layers.

- Lines and Points in Raster -- Lines can be approximated in raster by chains of connected pixels. These can be described by Freeman Code (or direction) which requires only one number per move instead of a coordinate pair. The number moves are 0 for up and go clockwise through 7 for the last position before you return to 0.

Capabilities of GIS

*GIS modeling:* Modeling is part of the analytical process of discovering, describing and predicting spatial phenomena (Bonham-Carter, 1994). The spatial operations used on spatial neighborhoods during modeling are:

- Generate new image
- Smoothing
- Remove high frequencies
- Remove disturbances
- Enhance edges
- Enhance directional features such as slope and aspect
- Derive shape information such as slope
- Enhance detail
- Perform filtering operations such as trend surface analysis
- Raster data modeling such as operations on neighboring pixels

Topological modeling is the process of changing the map characteristics to show containment, adjacency and connectivity. Geometrical modeling is the process of redrawing objects to show area, perimeter and shape. Map modeling is examining spatial patterns caused by interactions of one map to another and combining input according to a set of rules. Map modeling can produce map and/or table output. Basic output of map and table modeling are:

- Map output -- Map overlay, map modeling -- neighborhood analysis, topological modeling using adjacency, containment and connectivity for modeling.
- Table output -- Map overlay, table modeling, area analysis, correlation -- spatial cross-correlation.

Basic operations of map modeling are:

- Impose -- Turn a binary map (0,1) into multi-class map.
- Stamp -- Map A is stamped on Map B.
- Join -- In the operation Map B takes precedence over Map A defined by a set of rules or statements.
- Compare -- Where the two maps disagree the output Map is set to 0, where they have the same value this value is unchanged.
- Compare -- Using matrix overlay allows complete flexibility of assigning output classes in terms of class combination.

Applications in Water Resources

Measurements and estimations can be easily done using GIS data. There are many state and federal sources of data currently being made available. Availability of data is expanding almost exponentially. With this data the possible uses of GIS in Water Resources are limitless. Some of the uses are:

- Evaluation of average spatial values (i.e., infiltration, precipitation, snow depth, etc.)
- Measurement of drainage basin characteristics (i.e., slope, area, centroid, drainage length, etc.)
- Facility mapping (i.e., location of utilities.)
- Visually display relative location of project features for operational analysis.
- Estimation of regional values of a parameter (i.e., groundwater transmissivity, etc.)
- Determine regulatory compliance (i.e., determining if a property is in a floodplain).
- Visually display of information (i.e., weather maps)

Another application of GIS in engineering is to display maps. General map making principles can be learned from cartographers. Some of the basic principles are as follows (MacEachren, 1994 and Imhoff, 1975):

- Strive for harmony and clarity in the composition.
- Place map into geographic context by using an inset map. This inset map should be highly generalized.
- The map title should set the theme and feeling for the map.
- There is a visual hierarchy for the information shown on the map. The most important elements should be exaggerated or more prominent.

- The eye is drawn to the center of gravity of the elements shown on the map.
- The visual center should be placed about 5% of the map height below the geometric center.
- Always check the scale to make sure it is registering properly and it matches a common engineering scale value.
- Add any credits for information used in preparing the map.

Example GIS Project Set-up

An example of GIS project set-up is shown on Figure 3. The various GIS layers can be obtained from different sources and analyzed as noted.

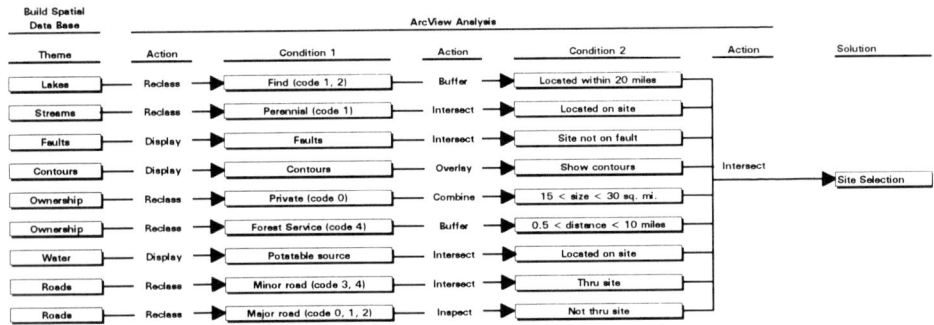

Figure 3 -- Example of ArcView Analysis for a site selection using various GIS layers

Conclusions

An overview of definitions used and capabilities of GIS were presented in this paper. A listing of applications of GIS in water resources was presented for both vector and raster GIS. No specific conclusions were presented.

References

Bonham-Carter, G.F., 1994. Geographic Information Systems for Geoscientists: Modeling with GIS. Pergamon publishers.

ESRI, 1992. ArcView$^{(C)}$ User's Guide. Copyrighted by Environmental Systems Research Institute, Inc.

Imhoff, 1975. "Positioning Names on Maps", American Cartographer, Vol. 2, pp. 128-144.

MacEachren, 1994. SOME Truth with Maps: A Primer on Symbolization and Design. Washington, D.C., Association of American Geographers Resource Publications in Geography.

Worboys, M.F., 1995. GIS: A Computing Perspective. Taylor & Francis, Inc.

# Development of a GIS-Based Watershed Assessment Model: Application to the Kentucky River Basin

## Lindell Ormsbee[1], Lee Colten[2] and Ted Stumbur[3]

Abstract

A GIS-based watershed assessment model is developed for use in the Kentucky Watershed Management Framework (KDOW, 1997). The proposed model utilizes multiple GIS coverages in prioritizing basin management units for subsequent management plan development. As part of the first year in a continuing 5-year rotating management cycle, the model is applied to the Kentucky River Basin.

Background

As part of a national EPA initiative, the Kentucky Division of Water (KDOW) has embarked on the development and coordination of a comprehensive watershed framework for use in managing and preserving the water resources of Kentucky. The proposed framework includes five basic components: 1) basin management units, 2) a basin management cycle, 3) a statewide basin management schedule, 4) a partner network and public participation, and 5) basin and watershed management plans.

Basin Management Units

In order to facilitate the application of the watershed management approach across the Commonwealth, the state of Kentucky has been subdivided into 5 basin management units. The basin management units are based on 6-digit hydrologic unit codes (HUCs), within which are nested 8, 11, and 14-digit HUCs (watersheds). HUCs were developed by the U.S. Geological Survey (USGS), the U.S. Department of Agriculture's Natural Resources Conservation Service, and others, to standardize hydrologic unit delineations for geographic description and data storage purposes. A map of the five basin management units is provided in Figure 1.

---

[1]Professor of Civil Engineering, University of Kentucky, [2]Watershed Coordinator, Kentucky Division of Water, [3]GIS Specialist, Kentucky Office of GIS

## Basin Management Schedule

In applying a watershed management approach across the state, each basin management unit will be processed through a five part basin management cycle. In order to provide for the strategic utilization of program resources, the basin management cycle for each basin management unit will be staggered by one year, and sequenced over a five year period.

## Basin Management Cycle

Kentucky's basin management cycle has five activity phases that are sequenced and repeated for each basin management unit at fixed 5-year intervals to ensure that management goals, priorities, and implementation strategies are routinely updated and progressively implemented. The five phases include: 1) scoping and data gathering, 2) assessment, 3) prioritization and targeting, 4) plan development, and 5) implementation.

## Prioritization and Targeting

As part of the prioritization and targeting phase, framework partners and other interested stakeholders work together to select priority 11-digit watersheds for subsequent management plan development. The 11-digit HUCs are selected on the basis of a two step process: 1) prioritization and 2) targeting. Priorities are be determined based primarily on technical factors related to resource impairment (i.e. severity of impacts, spatial scale or extent of impact) and threat to watershed resources (considering scale and immediacy of threats, and special protection status of certain resources). Targeted watersheds are identified based on public support, manageability, data availability, program specific funding, goals, and program constraints.

## Watershed Priority Formula

In order to prioritize the 11-digit HUCs for subsequent management plan development, a priority watershed formula was developed. The watershed priority formula is intended to summarize technical information and serve as a basis for deciding how to allocate resources to address two separate goals: 1) protection and 2) restoration. In order to accomplish these goals, the formula has been developed with two separate components: a protection (or vulnerability) component, and a restoration (or severity) component. In generating a score for each 11-digit HUC, individual 14-digit HUCs that make up the 11-digit HUCs are evaluated using the priority formula. Once these scores have been obtained they are averaged using a weighted-area approach to yield a score for the associated 11-digit HUC. Mathematically, the composite score for a particular 14-digit HUC will be the product of the two component scores. This may be expressed as follows:

14-digit HUC Priority Score = (Protection Score)*(Restoration Score)

Protection Score

The protection score is used to identify those watersheds which contain areas or streams with special designation resulting in elevated protection status above the minimum standards. These areas are identified by various programs and mandates for extra protection. The protection score for each watershed is computed based on a weighted average of the protection scores for each category. Protection categories include: 1) Wetlands, 2) Surface Drinking Water Protection Areas, 3) Well-head Protection Areas, 4) Groundwater Sensitivity Zones, 5) Fish/Wildlife Management Areas, 6) Nature Preserves Management Areas, 7) Nature Conservancy Area, 8) Reference Reach Watersheds, 9) Outstanding Resource Watersheds, 10) Recognized Resources, and 11) River Assessment Categories. Mathematically, this relationship is expressed as:

Protection Score (PS) = $(a*C1 + b*C2 + c*C3 + \ldots\ldots + k*C11)$

where $C1 \ldots C11$ are the protection scores for each category (i.e. $C1$ = wetlands impact score) and $a, b, \ldots k$ are coefficients whose sum is equal to one.

The protection scores for the Recognized Resources and River Assessment categories are themselves based on a weighted average of additional sub-categories. The sub-categories for the Recognized Resources include: 1) Rare Species, 2) National Natural Landmarks, 3) National Parks, 4) Federal Conservation Areas, and 5) University Natural Areas. The sub-categories for the River Assessment category include: 1) Agricultural Lands, 2) Botanical Resources, 3) Corridor Character, 4) Cultural Resources, 5) Fish Resources, 6) Geologic and Scenic Features, 7) Recreational Boating, 8) Water Quality, 9) Water Resources, and 10) Wildlife Resources.

The individual protection scores for each protection category are generated using a linear relationship between the category protection score and an associated category indicator score. The category protection score will range from 1 to 2 while the category indicator score limits depend upon the range of the associated category indicator (for example, see Figure 2).

Restoration Score

The restoration score is used to identify those watersheds where data indicate the system is impaired. Due to the number of 14-digit HUCs and the lack of comprehensive monitoring data in each of the HUCs, the restoration score for each HUC will be based on the maximum of either a potential impacts score (PIS) or an observed impacts score (OIS). Mathematically, this relationship could be expressed as:

$$\text{Restoration Score (RS)} = \text{MAX \{PIS, OIS\}}$$

Potential Impacts Score

The potential impacts score for a particular watershed is computed as the weighted sum of the predicted impact scores for each individual impact category. The predicted impact categories include: 1) Flooding, 2) Supply Vulnerability 3) Drought Vunerability, 4) Potential Contamination Sites (brine wells, straight pipes, landfills, dumps, underground storage tanks, hazardous waste sites, etc.), 5) Potential Pesticide Loading, 6) Potential Fertilizer Loading, 7) Agricultural Erosion Potential, 8) Livestock Operations, 9) Discharge Violations, 10) DOW Citizen Complaints, 11) Toxic Release Inventory Risk, 12) Population Projections, and 13) Unsewered Population, 14) Mining, and 15) Runoff Potential. Mathematically, the predicted impacts score can be expressed as:

Potential Impacts Score (PIS) = $(a*C1 + b*C2 +...+ o*C15)$

where a, b, ... o are coefficients whose sum is equal to one.

The predicted impacts scores for each impact category is determined using a linear relationship between the category impact score and an associated category indicator score. The category impact score ranges from 1 to 3 while the category indicator score limits are dependent upon the range of the associated category indicator (for example, see Figure 3).

Observed Impact Score

The observed impacts score for a particular watershed is computed as the sum of the observed impact scores for each individual impact indicator. The individual impact category scores are based on the maximum score of the associated sub-categories. The observed impact categories (with sub-categories in parenthesis) include: 1) Ecological Health (Aquatic life (AL), Contamination Sites (EHCS)), and 2) Human Health (Flooding (F), Supply Inadequacy (SI), Surface Drinking Water (SD), Groundwater (GD), Contamination Sites (HHCS), Tissue Consumption (TC), and Primary Contact (PC)). Mathematically, the observed impact score is expressed as:

Observed Impacts Score (OIS) =

Eco.Health(MAX{AL,EHCS})+Human Health(MAX{F,SI,SD,GD,HHCS,TC,PC})

The observed impact scores for each designated use sub-category will be generated using the following equation:

Sub-category Impact Score (i.e. AL,SD,GD,TC,PC)    =

1.0*(% of sub-category meeting designated use) +
2.0*(% of sub-category partially meeting designated use) +
3.0*(% of sub-category not meeting designated use).

The observed impact scores for the remaining sub-categories (i.e. EHCS,F,SI,HHCS) are computed using the same approach as with the potential impact scores.

GIS Utilization

Both the watershed prioritization phase and the subsequent targeting phase can be greatly facilitated through the use and application of geographic information system (GIS) technology. GIS software can be used to evaluate and integrate the various geo-referenced data coverages that make up the independent variables associated with both the protection and restoration scores. By imbedding the various model parameters directly within the GIS, the GIS software can be used to provide a visual sensitivity analysis of prioritization weights.    Such a capability provides a mechanism with which to finalize the associated model parameter values.   In addition, by developing separate coverages for the various targeting criteria, GIS can also be used to integrate and visually evaluate various targeting strategies.

Watershed Targeting

After the watersheds have been prioritized using the watershed priority formula, the next step is to determine how to allocate resources to address the associated protection or restoration goals.    Within each basin management unit, programs are expected to begin at the top of the watershed priority list and evaluate where to direct their resources based on the following types of criteria: 1) public support, 2) manageability, 3) data availability, 4) program specific funding, 5) goals, and 6) program constraints.

Classification Matrix

By cross-referencing the prioritization score and the targeting score for all watersheds, a classification matrix may be constructed which can serve to provide guidance for the type of management activity appropriate for each individual watershed (see Figure 4).

Conclusions

A watershed prioritization formula has been developed for use in selecting watersheds for subsequent targeting and plan development. The formula is evaluated using a GIS which provides an environment for efficient data manipulation and visualization .

References

Kentucky Division of Water, *Kentucky Watershed Management Framework*, (1997) Prepared by Tetra Tech, Inc., for the Kentucky Division of Water.

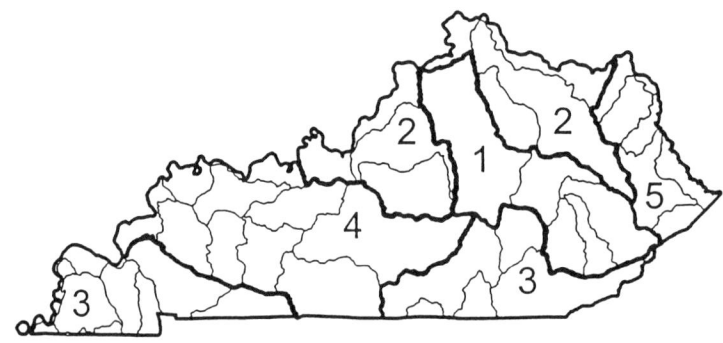

Figure 1. Basin Management Units and Major Drainage Areas

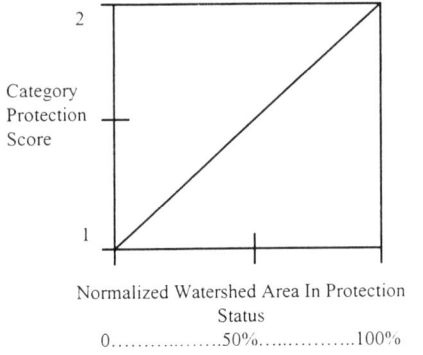

Figure 2. Protection Score Function

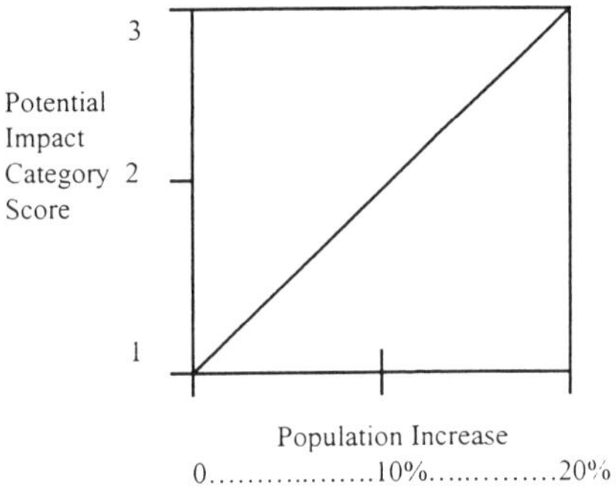

Figure 3. Potential Impact Function

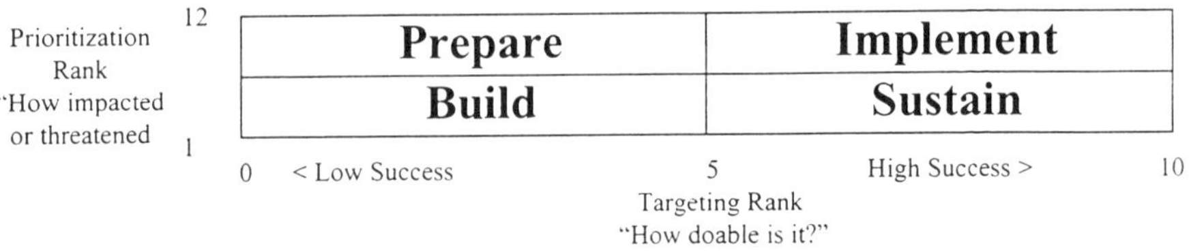

Figure 4. Watershed Classification Matrix

STORMWATER MANAGEMENT OF URBAN WATERSHED

Wing K. Tam[1], P.E., Blake N. Murillo[2], P.E.

Abstract

The City of Los Angeles contains two large and complex urban watersheds. In order to effectively manage stormwater and to facilitate the city's NPDES Stormwater permit within these watersheds, the city has enlisted the use of a Geographic Information System (GIS) to help prepare, analyze and present data for water quality evaluation. This GIS will predict water quality throughout the city's extensive storm drain network. The system will also track polluters and violations that have been identified. Ultimately, both maintenance personnel and drainage system inspectors will have a direct link back to the GIS for use in tracing and recording pollution problems in the field. Associated databases are being used to store monitoring data, to facilitate periodic NPDES permit compliance reporting and to develop information for public education and outreach programs. Some of these applications are also being provided over the Internet along with interactive public involvement.

Introduction

For most sizable municipal public works departments and flood control agencies, the past several years of effort to respond to NPDES permitting requirements have been challenging and illuminating; but also frustrating. Most applicants learned things about their systems that will help them be better managers in the future. Most of these applicants now also face the need to effectively conceptualize and manage large, complex urban watersheds. Managers must be able to collect and analyze large amounts of diverse information on a continuing basis. The City of Los Angeles has found a way to address this problem.

---

[1]Engineering Manager, Public Works Department, City of Los Angeles, 650 S. Spring Street, Suite 700, Los Angeles, CA 90014.
[2]Vice President, Director of Technical Services, Psomas and Associates, 600 S. Spring Street, Suite 1608, Los Angeles, CA 90014.

Los Angeles is home to almost four million people living within an highly urbanized 480 square miles. The city contains two large and complex urban watersheds (Figure 1). The Santa Monica Bay watershed is a highly urbanized area draining to a popular bay that supports many uses, including recreation, shellfish harvesting and fishing. The Los Angeles River watershed is also highly urbanized and additionally is subject to discharges from several treatment plants.

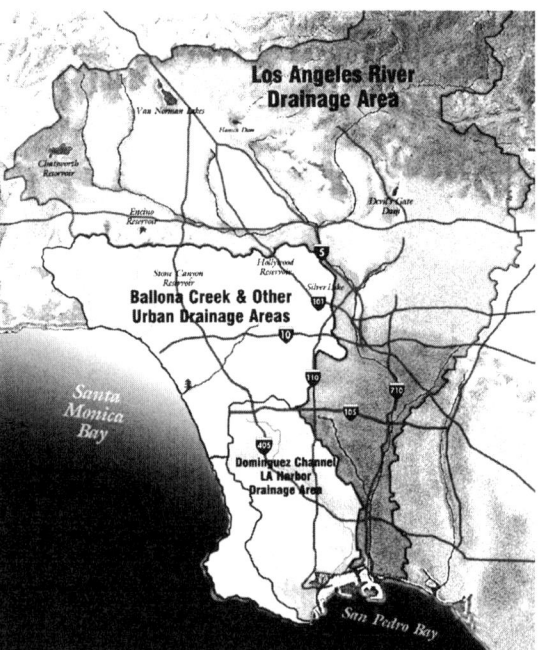

Figure 1. Watershed map

In order to effectively manage the ongoing collection of data, the city has chosen to develop a Stormwater Management Information System that is built around a Geographic Information System (GIS). Not only is the system used to track and report on current NPDES permit activities, it helps predict water quality throughout the city's extensive storm drain network. Ultimately, both maintenance personnel and drainage system inspectors will have a direct link back to the Information System for use in tracing and recording pollution problems in the field. The system will also track polluters and violations that have been identified. This and other drainage information is available via the Internet, along with programs allowing interactive public involvement.

Stormwater Management Information System

The City of Los Angeles' Stormwater Management Division (SMD) is organized into six sections: permit compliance, financial, engineering, public education, inspection and monitoring. The mission of this division is to implement and manage programs that ensure city compliance with federal, state and local flood control and stormwater pollution abatement laws and regulations — to develop programs that ensure that storm drain discharges do not interfere with the beneficial uses of the city's receiving waters. The Stormwater Information Management System is designed to respond to both the stormwater pollution regulations and the city's flood control responsibilities.

The design of the information management system stemmed from numerous workshops and meetings with each section. These interactions resulted in a conceptual database design and 12 targeted applications. Top management assigned priorities to each of the 12 applications and common data sets were identified. The common data sets identified were: (1) street network, (2) parcels, (3) storm drain conveyance system, (4) land use, (5) hydrologic boundaries (watershed and sub-areas) and (6) jurisdictional boundaries. The 12 prioritized applications were: (1) water quality model, (2) enhanced catch basin cleaning tracking, (3) catch basin stenciling tracking and reporting, (4) flood zone inquiry, (5) FEMA flood plain map revision tracking, (6) repetitive flood loss tracking, (7) school education program reporting, (8) BMP reporting by council district, (9) corporate sponsor tracking, (10) stormwater pollution abatement charge tracking, (11) industrial site investigation tracking and reporting and (12) pollution investigation trace. Users interact with these applications through a Visual Basic application (Figure 2). All of the 12 applications are accessed via this entry point.

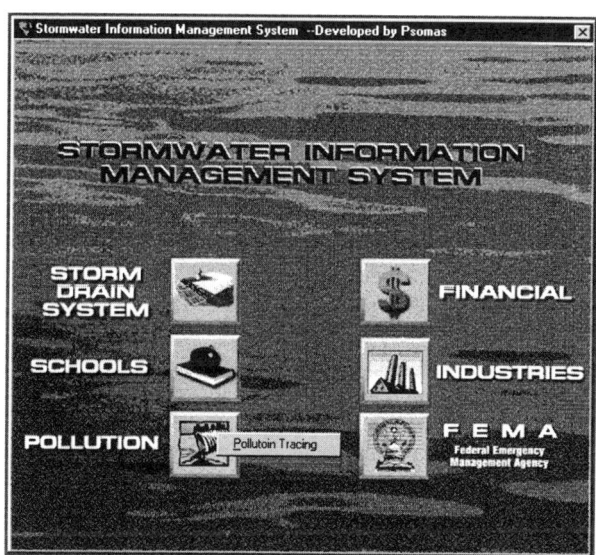

Figure 2. Application Interface

Specific Applications

The **Flood Zone Inquiry** allows SMD staff to determine the flooding potential of any parcel in the city. Previously, any request would require staff to gather both city and FEMA maps, manually overlay the two, make a determination, produce a composite map and send a letter to the requesting party. In this application, the user enters the requesting party's name, the subject parcel address and the city staff person's name. The GIS automatically generates a letter-sized map of the property and a properly addressed and formatted letter. The manual process required approximately two hours of effort and usually occurred over two days. This application reduces this process to less than five minutes. With the possible affects of "El Nino" looming over the West Coast this application was used extensively to deal with up to 200 calls a day from concerned homeowners.

A **Water Quality Model** was built to predict pollutant loads at over 2,000 points throughout the city. The model takes available rainfall data for a given storm event (from Weather Service records or local rain gauge readings), applies a basin-specific rainfall/runoff computation to estimate the runoff volume, and multiplies this volume by a seasonally adjusted pollutant concentration coefficient to compute the pollutant load from the given basin. This process is repeated using different concentration coefficients for each pollutant. The total load of a given pollutant from a given event is calculated by summing up the basin-by-basin loads for all basins that are tributary to (up-gradient) from the given location. The annual loads for a specific pollutant are calculated by summing up the loads for all events in the year. A seasonal variation in the pollutant coefficient helps to account for the typical summertime buildup of pollutants. The water quality model operates completely within the GIS environment. It uses the processing power of the GIS to retrieve, format and prepare the necessary rainfall, land use, soils and other information for the drainage area under study. The model allows for four preset rainfall events plus two user-specified events and the selection of up to 27 different pollutants to model. "What if" analyses can be done by changing land uses, pollutant coefficients and time of year.

The **Pollutant Trace** application allows the user to investigate possible pollution sources upstream from a particular location or trace the path of flow downstream from the same location. The pollutant trace application utilizes many of the capabilities of the water quality model. A user selects the outfall or other storm drain location in the city and selects the pollutants of concern. The application determines the extent of the drainage area tributary to the selected location and then compares the selected pollutants to the Standard Industrial Classifications (SIC) of each of the parcels within the drainage area. The relationship of pollutants and SIC's developed by the County of Los Angeles for the Santa Monica Bay watershed is used to select target industries (see Table 1). The GIS creates a map showing the parcels with matching pollutant source. A report listing the business address, contact name and other site specific information about each parcel also is generated.

The **Field Inspection** application is currently underway with development of the field application using pen-based computers. The application automates many field activities, including the inspection of industrial properties. The pen-based computer will access citywide maps of the storm drain system, inspection forms and historical site inspection data. The same system will assist storm drain maintenance crews in updating the inventory of drainage facilities.

The Internet is allowing SMD to carry out the City of Los Angeles' mission to better serve its citizens and Mayor Richard Riordan's intention to capitalize on technology. SMD has committed to an intense use of the Internet through the use of the World Wide Web. This use is twofold, first to provide information about the stormwater program and what the citizens can do, and secondly, to provide a vehicle for communication with the public at large. Current web applications provide an array of information about the Stormwater Management Program and where and how to get additional information. Interactive maps allow access to FEMA flood zone information and the ability to report problems with any of the city's more than 55,000 catch basins.

## FUTURE DEVELOPMENTS

While the efforts to date have provided increased analytical capabilities (water quality model) and responsiveness (pollutant trace and flood zone inquiry), many additional applications remain to be done. Some of these future applications are:

1) Access to Maintenance Data - stormwater engineers need to understand the maintenance efforts and results of maintenance operations. The information gathered by other departments will be made available on a system element basis (pipe, catch basin, debris basin). This long-term history of maintenance efforts will lead to better decisions about future system improvements as well as better management practices.

2) Ultimate Storm Drain System - while local and watershed-specific storm drain planning has been done periodically over the last 50 years, no citywide studies have been undertaken. The recent investment in GIS and associated data sets has provided a base for an engineering study of the city's extensive storm drain network. An integrated process is underway involving the use of GIS to represent, analyze and prepare data for hydrologic and hydraulic evaluation. It has proven to be an efficient and effective method for systematically evaluating the storm drain systems throughout the City. Deficiencies of the composite system are identified and replacement or upgraded systems are computed. A cost/benefit index is established for each system.

3) Advanced water quality model features - additional and better monitoring data from the Los Angeles area will provide the opportunity to add advanced features to the model. These features include preliminary BMP selection and potential benefits, dry-weather flow predictions and adjustments to the seasonality and event mean coefficients. These enhancements will allow stormwater engineers to better understand the workings of each watershed, and they will allow management to make more informed decisions.

| SIC | Industrial Description | Pb | Cr | Cd | Cu | Ni | Zn | Ag | PAH | P04 | NO3 | NH4 | TSS | BOD | COD | O&G |
|---|---|---|---|---|---|---|---|---|---|---|---|---|---|---|---|---|
| 2 | Agricultural Production (Livestock) |  |  |  |  |  |  |  |  |  | • | • | • | • |  |  |
| 13 | Oil & Gas Extraction | • | • |  |  |  | • |  |  | • |  |  | • | • | • | • |
| 15 | Building Construction |  |  |  |  |  |  |  |  |  |  |  | • |  |  |  |
| 16 | Heavy Construction |  |  |  |  |  |  |  |  |  |  |  | • |  |  |  |
| 17 | Construction (Spec. Trade) |  |  |  |  |  |  |  |  |  |  |  | • |  |  |  |
| 20 | Food & Kindred Products |  |  | • |  | • |  |  |  |  |  |  | • | • | • | • |
| 22 | Textile Mill Products |  | • | • | • | • | • | • |  |  |  |  | • | • | • | • |
| 24 | Lumber & Wood Products | • | • | • | • |  | • |  |  |  |  |  | • | • | • | • |
| 26 | Paper And Allied Products |  |  |  |  |  | • |  |  |  |  | • | • | • | • | • |
| 27 | Printing, Publishing & Allied Ind. | • | • |  |  |  |  |  |  |  | • |  | • | • | • | • |
| 28 | Chemical & Allied Prod. | • | • | • | • | • | • | • | • | • | • | • | • | • | • | • |
| 29 | Petroleum Refining & Related Ind. | • | • |  |  |  | • |  | • | • |  | • | • | • | • | • |
| 30 | Rubber & Misc. Plastics Prod. | • | • | • | • |  | • | • |  | • |  |  | • | • | • | • |
| 31 | Leather & Leather Prod. | • | • |  | • | • | • |  |  |  | • |  | • | • | • | • |
| 32 | Stone, Clay, Glass & Concrete Prod. | • | • |  | • |  | • |  |  |  |  |  | • | • | • | • |
| 33 | Primary Metal Industries | • | • | • | • | • | • | • | • |  | • | • | • |  | • | • |
| 34 | Fabricated Metal Products | • | • | • | • | • | • | • |  | • | • |  | • | • | • | • |
| 36 | Electronic & Other Electrical Equip. | • | • | • | • | • | • | • |  |  |  |  | • | • | • | • |
| 37 | Transportation Equipment | • | • |  | • |  |  |  |  |  |  |  | • | • | • | • |
| 39 | Misc. Manufacturing Indust. |  | • | • | • | • |  | • | • |  |  |  | • |  |  |  |
| 40 | Railroad Transportation |  |  |  |  |  |  |  |  |  |  |  | • | • | • | • |
| 41 | Local And Suburban Transit | • | • | • | • | • | • |  | • |  |  |  | • | • | • | • |
| 42 | Motor Freight Transportation | • |  | • | • | • | • |  |  |  |  |  | • | • |  |  |
| 43 | U.S. Postal Service | • | • | • | • | • | • |  | • |  |  |  |  |  |  | • |
| 44 | Water Transportation | • |  |  | • | • | • |  | • |  |  |  |  |  |  | • |
| 45 | Transportation By Air | • | • |  | • | • | • | • | • |  |  |  | • | • | • |  |
| 49 | Electric, Gas And Sanitary Services |  | • |  | • | • | • |  |  |  |  |  | • | • | • | • |
| 50 | Wholesale Trade | • | • | • | • | • | • |  | • |  |  |  | • |  | • | • |
| 55 | Autom. Dlrs. & Gasoline Station | • | • | • | • | • | • | • | • |  |  |  | • | • | • | • |
| 75 | Automotive Repair, Svcs. & Parking | • | • | • | • | • | • | • | • |  |  |  | • | • | • | • |

Table 1 - Pollutants vs. SIC

1) Data provided by Los Angeles County, Dept. of Public Works for Santa Monica Bay Watershed
2) Table from Santa Monica Bay Restoration Plan, Chapter 2 page 2-30

Watershed Modeling Using Remote Sensing

Nabil M. Hourani, P.E.,[1] Member ASCE

Abstract

Many watershed analysis projects are limited by the information available for modeling parameters. Topographic information is often outdated or costly to obtain. This paper will explore the benefits of combining remote sensing imagery and Geographic Information System (GIS) databases for watershed modeling. This effort makes use of a Digital Elevation Model (DEM), which is ideal for providing detailed and accurate topographic information - the key element for successful watershed modeling.

In order to illustrate the value of remote sensing applications, a case study where GIS and satellite imagery have been used to model and analyze a large ungaged watershed area will be discussed. Multispectral satellite imagery has been used to determine the land use classifications and panchromatic (black and white) satellite image stereopairs have been used to extract the DEM. Ground control points were collected using Global Positioning System (GPS) technology to rectify these images to a common coordinate space. All topographic parameters obtained from satellite image processing were then stored in a GIS database and used for hydrologic analysis. Study results have been compared with conventional methods using topographic maps and aerial photography. Remote sensing linked with GIS databases will optimize management and planning efforts in any location by enabling accurate and cost effective modeling of the study area.

---

[1] Water Resources Manager, McCormick, Taylor & Associates, Inc., 51 Haddonfield Road, Suite 200, Cherry Hill, NJ 08001

## Overview

The goal of this project was to demonstrate the effectiveness of using satellite image data (remote sensing) to perform hydrologic modeling of the Manumuskin River watershed located within Maurice River Township, Cumberland County, New Jersey. Satellite imagery and image processing software were used to generate a three-dimensional surface model, to delineate land use/land cover classifications, and to combine the land use with hydrologic soil groups. The results of this project confirm the practicality of using remote sensing to derive watershed parameters required for computing runoff hydrographs using two standard hydrologic modeling techniques: Snyder's Unit Hydrograph with percent impervious cover and the SCS Runoff Curve Number method (USACE 1990).

## Data Processing

The primary data requirements for this aspect of the project were a three dimensional surface model, land use classification, and weighted runoff curve number (CN) value calculations based on the combination of land use and soils data (USDA 1986).

## Three Dimensional Surface Model

The panchromatic stereopairs contain two scenes recorded on different dates which view the same area from different angles, thus allowing stereoscopic viewing to generate a DEM. Variations in the elevation of objects on the ground will cause a degree of distortion within an image. Relief displacement is where objects appear to tilt away from the center of the image. Image parallax refers to the apparent change in relative position of objects when viewed from different angles. By rectifying the images to a common coordinate space, the distortions can be compared and the true location and elevation of objects on the ground can be determined (Lillesand 1990).

This information can then be used to generate the surface model (Bremen 1995). The panchromatic data was projected to the New Jersey State Plane Coordinate System (North American Datum, 1983). The locations of the ground control points were chosen to fully cover the watershed area and to be easily identified on the images. The X, Y, and Z coordinates of the control points were added to the database. These points were used in conjunction with coordinate information stored with the satellite image to create a mathematical model. This model was used to register the images to the desired coordinate system.

Feedback from the image processing programs allows for manual and semi-automated refinement of the rectification process. Multiple iterations were performed to achieve the desired level of accuracy. Once properly registered, a DEM was created from the orthorectified stereopair images. The DEM was then edited to remove erroneous results, interpolate incorrect or missing elevation values, and smooth the overall surface.

The DEM was then converted to a Triangulated Irregular Network (TIN) surface model. The TIN was used for terrain modeling and automated basin delineation (BYU 1996). Once the watershed was created, geometric parameters such as slope, drainage areas, stream length, length to centroid, and overland flow distances were computed and combined with hydrologic analyses. To check the effectiveness of the DEM generated from satellite images, it was compared with a DEM created from USGS contours.

## Land Use Classification

Preprocessing - Before using the satellite imagery for classifying land use, the images had to be pre-processed to improve image quality and correct geometric distortions. Missing or inaccurate information in an image can be caused by sensor malfunctions. The errors were removed and the atmospheric effects were compensated for by a series of automated procedures. Finally, the image was geocoded by incorporating ground control points obtained from the ground survey to rectify the image to the desired coordinate system. This required multiple iterations to achieve the desired level of accuracy.

Image Enhancements - The objective of image enhancement is to modify the image to increase the amount of information that can be interpreted from the image. This included image filtering and merging the multispectral image (30 meter resolution) and the stereopair image (10 meter resolution). The merged image maintained the multiple spectral bands from the multispectral image with the increased clarity and resolution of the panchromatic stereopair imagery.

Classification - The goal of the classification stage of processing was to identify unique categories of land use based on carefully analyzed spectral signatures within the image. These signatures can be defined by two different methods, generally referred to as supervised and unsupervised classification. Supervised classification is an interactive process where the image analyst identifies areas within the image that clearly represent a particular category of land use. These areas are then used by the computer in a "training session" to generate spectral signatures based on the interpretation of these discrete locations.

Unsupervised classification relies on a number of algorithms that the image processing software uses to determine the best categories of signatures based on its interpretation of the entire image. Using the unsupervised method, the analyst would then assign land use categories to the spectral signatures generated by the program. For this project, comparison of the two methods indicated that the supervised classification method was the better choice to define land use categories.

## Weighted Runoff Curve Numbers (CN) and Percent Impervious

By making use of the image processing software's ability to combine values from multiple data layers, it was possible to compute composite runoff curve numbers for each subbasin. The land use data and hydrologic soils data were overlaid to delineate unique combinations of the eight land use categories and the four hydrologic soil groups. These combinations provided the basis for calculating a weighted CN value for each subbasin.

The hydrologic soil groups were digitized from the Soil Survey of Cumberland County published by the SCS. The digitized areas were imported into the image processing software and assigned a unique numeric attribute value for each group. Likewise, a unique attribute value was assigned to each land use category, at intervals of 4. This was necessary to create unique values for each combination of hydrologic group and land use when combined by the program. The two layers of data were then combined, which created a new data layer. This layer delineated the combinations of land use and soil groups for each subbasin, with each combination containing a unique CN value. These values were assigned based on criteria established by the SCS. Ultimately a weighted CN value was computed.

The percentage of impervious cover for different land uses and the weighted average was computed for each basin. The physical parameters - area, overland flow distance, slope, percent impervious cover, and runoff curve numbers - were stored in the GIS database (Smith 1994). These parameters were used within the HEC-1 watershed modeling program for both the Snyder's and SCS methods.

Results

Manual estimation of watershed physical parameters is time consuming and subjective, especially for large watershed areas (Shamsi 1996). The DEM generated from satellite imagery is used to describe the drainage basin topography and allows automated calculation of these parameters. These automated results compare very favorably with manual calculations. The total drainage area to Cumberland Pond Dam varies by approximately one percent between manual and automated techniques.

The traditional method of CN value calculations requires tedious calculation of the areas which make up the watershed. This step is time consuming since the land use and soil group must be overlaid. The supervised land use classification combined with a soil group database allow automated calculation of weighted CN values for each watershed, effectively reducing the amount of time required to obtain this information which is vital for the watershed model. The land use classification is very accurate when using satellite imagery since the image provides current information for the watershed.

A comparison of the CN values shows a close correlation between the manual and automated values that were obtained. However, we found that the automated values were generally higher than the manual values. A review of the breakdown of land use percentages for each subbasin showed that there was about five percent more water in many subbasins using the automated results. The satellite image revealed standing water in many locations which were not identified in the field and were not included in the manual computations. The automated results are believed to be more accurate because the water should be included in the CN calculations. Therefore, the automated method is preferred not only for time savings but also for increased accuracy.

## Conclusion

The primary reason for examining the land use classification and the generation of a DEM was for its potential application to hydrologic modeling (USACE 1980). Land use maps can be updated easily and cost effectively by reclassifying the land use from the latest satellite image to reflect new development (Ross 1993). The subbasin data file can then be updated using the GIS to include any changes within the hydrologic model. Remote sensing, combined with GPS and GIS databases, will provide accurate and cost effective watershed modeling which benefits planning, feasibility studies, environmental management, and development efforts in any location.

## References

1. Brigham Young University - Engineering Computer Graphics Laboratory, (1996). *Watershed Modeling System Reference Manual*, Provo, UT.
2. European Space Report, (1995). "Remote Sensing for Oil Exploration and Environment". *Program of Space Congress*, Bremen, Germany.
3. Lillesand, T.M. and Kiefer, R.W. (1994). *Remote Sensing and Image Interpretation*, Third Edition, John Wiley & Sons Inc., New York, NY.
4. Ross, M.A., and Tara, P.D., (1993). "Integrated Hydrologic Modeling with Geographic Information Systems." *J. Water Resour. Plng and Mgmt*, ASCE, 119(2),129-140.
5. Shamsi, U.M., (1996). "Storm-Water Management Implementation through Modeling and GIS". *J. Water Res. Plng. and Mgmt.*, ASCE 122(2), 114-127.
6. Smith, M.B. and Vidman, A., (1994). "Data Set Derivation for GIS - Based Urban Hydrological Modeling" *Photogrammetric Engrg. and Remote Sensing*, 60(1) 67-76.
7. "Determination of Land Use from Satellite Imagery for Input to Hydrologic Models," (1980). Tech. Paper No. 71, Hydrologic Engineering Center, US Army Corps of Engineers, Davis, CA.
8. "HEC-1 Flood Hydrograph Package" (1990) *User's Manual*. Hydrologic Engineering Center, US Army Corps of Engineers, Davis, CA.
9. "Urban Hydrology for Small Watersheds" (1986). *Tech. Release 55*, US Dept. of Agriculture, Soil Conservation Service, Washington, D.C. 2.5-2.8.

## Current Status and Best Management Practices for American Catchment Systems

Dennis J. Lye[1]

### Abstract

The concept of rainwater collection as a viable, alternative source of both potable and nonpotable water is gaining acceptance in several areas of the United States. A critical element for the adoption of this concept by more American households is the proper design and construction of rainwater catchment systems. Current recommendations from the states of Hawaii and Ohio are reviewed because existing state recommendations may evolve into mandatory regulations. The American Rainwater Catchment System Association (ARSCA) is addressing issues involved in the possible regulation of these systems. This paper presents initial information concerning (A) Sanitation issues, (B) Operation and Maintenance, and (C) Non-sanitation Design Standards.

### Introduction

Many areas in the United States receive adequate rainfall for viable rainwater catchment systems. The most recent attempt to determine the popularity of residential rainwater catchment systems in the United States was a 1990 survey of American State Health Departments reporting over 250,000 rooftop catchment cistern systems serving as a primary source of drinking water (Lye, 1992a). Because of the popularity of concepts such as conservation and recycling of resources, the numbers of individual, residential rainwater catchment systems will most certainly increase in the near future. The renewed interest in this ancient, simple collection device is evident in the newly organized American Rainwater Catchment System Association (ARCSA) which is particularly active in the Southwestern United States.

Among the attractions of collecting rainwater as a source of water in the United States are the following:

(A) other potable water sources are unavailable or undesirable,

---

[1]Dept. of Biological Sciences, Northern Kentucky University, Highland Heights, KY 41099

(B) the independence from centralized distribution systems which are beginning to restrict volumes of water usage to individual users,
(C) the low mineral content and high-quality of rainwater prior to collection,
(D) lower energy input with decreases in local erosion and flooding,
(E) the economics of rising water costs.

In 1996, Laurence Doxsey reported that rainwater harvesting was receiving renewed interest in the United States especially in the following three areas:
1. Rural residential - the initial cost of a rainwater harvesting system is similar to drilling a well but there are less maintenance costs with the rainwater system,
2. Urban residential - these users are attracted to the fact that low-mineral rainwater is preferred over treated distribution water for irrigation of plants, landscaping, etc.,
3. Small to medium commercial - plant nurseries, aquatic farming, architects, etc.

In the past, cistern systems have been exempt from the mandates of most drinking water legislation at both the state and federal levels in the United States. However, if cistern systems gain in popularity and become more numerous throughout all fifty states, federal agencies such as the U.S.E.P.A. may be forced to classify this type of water. Current sentiment is to classify collected rainwater as a type of raw surface water. A classification of this type would result in the application of numerous pre-existing regulations already in use throughout the United States but not entirely suitable for cistern collection systems (Ruskin, 1990).

The states of Hawaii, Ohio, and Texas currently have suggestions for the design, construction, and maintenance of rainwater collection systems. The Hawaii and Ohio guideline reports are reviewed by the State Department of Health while the Texas report is a product of the Texas Water Development Board. ARCSA currently has a committee reviewing some of these same issues and possible future regulations.

The Hawaii Department of Health Guidelines for the use and maintenance of rain-water catchment systems (Anonymous, 1993) summarizes the complexity of regulating this type of water source:

"The Department of Health does not have any vested regulatory authority over individual rain-catchment systems. There are too many variables in the construction, materials, features, environmental conditions, operation, maintenance, treatment, and uses of these types of systems to make recommendations which would assure that all systems provide water fit for drinking."

The state of Texas also recognizes that rainwater is not regulated.

"No agency authorizes or inspects private rainwater collection systems. The Texas Plumbing Code does not allow double trenching of wastewater and potable water lines and an airgap wider than the municipal line must exist between any public water line and rainwater line."

The state of Ohio "Private Water System Rules" (Anonymous, 1993) provides approximately seven pages of rules pertaining specifically to cistern water systems.

"Each private water system shall be properly maintained and operated."

## System Design

Because of the ready availability of treated community water in most areas of the United States, one issue unique to American systems is the intended use of the collected rainwater. The Texas Guide to Rainwater Harvesting (Anonymous, 1996), is the most recent American publication for educating users about this water source and recognizes that some households may construct rainwater collection systems solely for use outside the household. The sanitation requirements for non-potable uses of rainwater collection systems will be less than those for potable uses. For this report, only systems designed as a potable water supply will be addressed.
There are three major influences on the design of rainwater collection systems. In order of importance they are (1) disinfection, (2) size, and (3) maintenance.

## Disinfection Guidelines

The state of Hawaii **recommends** that users of rain-water catchment systems not use the water for drinking or cooking and **suggests** that water for all other uses be properly disinfected. The state of Ohio rules that water obtained from cisterns, ponds, or springs **shall be** continuously disinfected. The state of Texas **suggests** that appropriate filtration and disinfection practices **should be** imployed and the water **should be** tested for coliform bacteria by a certified laboratory.

Because of recent research concerning the microbial communities present in cistern systems, rainwater quality and the effects of disinfection treatments on rainwater can be effectively monitored in different ways. Possible disinfection guidelines for rainwater systems could read as follows:

(A) Water quality of rainwater catchment systems can be assured by continuous disinfection with any of the treatments given in Table One. Each system requires a **mandatory annual** test for the presence of coliform bacteria.

Table One. Continuous Treatment Techniques[A]

| Method | Location | Result |
|---|---|---|
| Boiling/Distilling | point of use | kills microorganisms |
| Chemicals | within tank or | destroys bacteria and viruses at pump |
| Ultraviolet light | after other filters | kills microorganisms but before point of use |
| Ozonation | in tank or before | kills bacteria at point of use |

A = taken from Texas report with some modification (1996)

(B) If continuous disinfection is not used, **mandatory monthly** testing must be performed for monitoring any of the following:(1) coliform bacteria (Lye, 1987), (2) Escherichia coli (Lye, 1987), (3) total bacteria (Kromoredjo and Fujioka, 1991), or (4) protozoa (Lye, 1992b)

Table Two lists devices that may be incorporated into a rainwater catchment design but are **not** to be considered as continuous disinfection treatments.

Table Two. Treatment Techniques that must be augmented with disinfection[A].

| Method | Location | Result |
|---|---|---|
| Screening devices | gutters and leaders | prevent debris from entering storage |
| Settling devices | within storage tank | sediments particulate matter |
| Filtering devices: | | |
|   In-Line, Multi-Cartridge | after pump | sieves sediment |
|   Activated Charcoal | point of use | removes chemical disinfectants |
|   Reverse Osmosis | point of use | removes contaminants |
| Mixed Media | separate tank | traps particulate matter |
| Slow Sand | separate tank | traps particulate matter |

A = taken from Texas report with some modification (1996)

### System Size Guidelines

Only the state of Ohio recommends that American rainwater catchment systems be a minimum of 20,000 liters (5,000 gallons) in total size (one or more tanks). The state of Hawaii recommends rainwater storage tanks be large enough to support the demand of the system including fire flow requirements. Hawaii, Ohio and Texas all have requirements that cistern tanks be sited at least 15 meters (50 feet) away from sources of pollution (animal stables, latrines, septic fields, etc.).

### Operation/Maintenance Guidelines

Guidelines for the effective design, operation, and maintenance of rainwater collection systems will require many specific details covering a wide range of possible systems. The materials and practices currently put forth by Hawaii, Ohio, and Texas cover the following three broad areas common to all types of rainwater catchment systems; Catchment materials, Gutter materials, and Storage materials.

### Catchment/Guttering materials

The state of Hawaii informs that most materials currently approved for use with potable water were never intended for use as an outside roofing or guttering material. Therefore, there are no known approved or certified roof or gutter materials for the collection of rain-water for drinking purposes. However, the three states require that roof material, flashing, and guttering used for potable water sources should not contain any lead content. The state of Ohio makes no specific mention of catchment materials but does require an above ground roof washer or filtering device (one for every 140 square meters [1500 sq. ft.] of roof catchment area). The state of Texas suggests that composite asphalt, asbestos, chemically treated wood shingles and some painted roofs could leach toxic materials into the rainwater and are recommended only for non-potable water uses. Roof washers are recommended by the state of Texas and should divert at least 38 liters (10 gallons) for every 90 square meters (1,000 square feet) of collection area.

## Storage materials

The state of Hawaii recognizes that all storage materials and piping should be acceptable for potable water use and refers users to the National Sanitation Foundation for information on approved or certified materials. The state of Ohio lists numerous specifics about the design and construction of not only cistern storage tanks but also for hauled water storage tanks. The state of Texas recommends that cistern tanks should be labeled as FDA-approved as well as any sealants or paints used inside the tanks. Table Three lists some of the more commonly used materials for storage of rainwater in American systems.

Table Three. Storage Tank Materials Common to American Cistern Systems[A].

| Material | Feature | Caution |
| --- | --- | --- |
| Plastic Cans | commercially available inexpensive | use only new cans |
| Fiberglass | commercially available alterable and moveable | degradable require interior coating |
| Polyethylene/ Polypropylene | commercially available alterable and moveable | degradable require exterior coating |
| Steel Drums | commercially available alterable and moveable | small capacity inspect for corrosion, toxics |
| Galvanized Steel Tanks | commercially available alterable and moveable | inspect for corrosion, rust |
| Ferrocement | durable, immoveable | potential to crack and leak |
| Stone, Concrete Block | durable, immoveable | difficult to maintain |
| Poured Cement | durable immoveable | potential to crack and leak |
| Wood | attractive, durable | expensive (redwood, Fir, Cypress) |

A = taken from Texas report with some modification (1996)

## Maintenance Guidelines

In the United States, each rainwater catchment system is unique in some manner even when identical local codes have been followed. If design regulations do eventually become implemented, they will be useful for the physical parameters of a system but may not address the actual performance of each system. As with any water supply, the performance (or water quality) of a rainwater catchment system is directly related to the quality of maintenance (Lye, 1996).

The state of Hawaii states that maintenance is extremely important to rainwater catchment systems and the frequency of maintenance activities will depend upon individual conditions. The state of Ohio endows the Department of Health with the power to order work deemed necessary to protect the public health associated with rainwater systems. The state of Texas suggests that users perform a "total coliform" test to indicate whether or not a problem exists in the system.

## Reference

Anonymous, 1993, Guidelines for the Use and Maintenance of Rain-Water Catchment Systems, State of Hawaii Department of Health, Honolulu, Hawaii

Anonymous, 1993, Private Water System Rules, The Ohio Department of Health, Columbus, Ohio.

Anonymous, 1996, Texas Guide to Rainwater Harvesting, Texas Water Development Board, Austin, Texas.

Doxsey, W. L., 1996, Residential Rainwater Catchment Systems, Publication No. 1, American Rainwater Catchment Systems Association, Austin, Texas.

Kromoredjo, P. and R. S. Fujioka, 1991, Evaluating Three Simple Methods to Assess the Microbial Quality of Drinking Water in Indonesia, Environmental Toxicology and Water Quality: An International Journal $\underline{6}$:259-270.

Lye, D., 1987, Bacterial Levels in Cistern Water Systems of Northern Kentucky, Water Resources Bulletin $\underline{23}$:1063-1068.

Lye, D., 1992a, Microbiology of Rainwater Cistern Systems: A Review, J. Environmental Science and Health $\underline{27}$:2123-2166.

Lye, D., 1992b, Legionella and Amoeba found in Cistern Systems, Proceedings of the Regional Conference on Rain Water Catchment Systems Association, Kyoto, Japan.

Lye, D., 1996, Water Quality of American Cistern Systems, Publication No. 2, American Rainwater Catchment Systems Association, Austin, Texas.

Ruskin, R.H., Lye, D., and J. H. Krishna, 1990, The Need for Separate Water Quality Standards for Cistern Water Systems: A Review, American Society for Microbiology, Annual Meeting.

Status of RWCS Development and Progress of IRCSA

Yu-Si Fok[1]
Fellow, ASCE

Abstract

The development of rainwater catchment systems (RWCS) has been gaining great momentum worldwide in recent decades. This is partly due to the United Nations General Assembly's proclamation of 1981–1990 as the International Drinking Water Supply and Sanitation Decade and partly due to the series of international conferences on rainwater catchment systems that began in June 1982. To date, eight conferences have been held at about two-year intervals in different cities of the world to promote RWCS development.

During the United Nations' water decade, RWCS was promoted as an appropriate technology to help developing countries provide their populations with a safe and adequate drinking water supply. The Thailand Jar Project (1985–1990) is an example of a national project that aimed to provide such a drinking water supply for the country's 18 million rural residents. A brief description of the Thailand Jar Project with its outstanding features is included in this paper.

Although the idea of having an international association had been conceived at the first RWCS conference, the International Rainwater Catchment Systems Association (IRCSA) was not formed until the fourth conference was held on 2–4 August 1989 in Manila, Philippines. The IRCSA constitution and by-laws were ratified by the General Assembly when the fifth conference was held on 4–10 August 1991 in Keelung, Taiwan, R.O.C. Highlights of the progress made by IRCSA in RWCS development are included in this paper.

The objective of this paper is to introduce RWCS and related IRCSA activities to help readers gain an appreciation of the benefits that can be realized from using RWCS, as discussed in other papers presented in this session.

---

[1]Professor, Department of Civil Engineering, and Researcher, Water Resources Research Center, University of Hawaii at Manoa, 2540 Dole Street, Honolulu, HI, 96822-2333

## Introduction

In recent decades rainwater catchment systems have been gaining great momentum as a viable alternative means of providing drinking water in urban and rural areas, due to the fact that water is a limiting resource for development. In the 1970s, the World Health Organization reported that more than 3,000 children die each day from sickness related to unclean water and that millions of adults are unable to work after having drunk contaminated water. It pointed out that an inadequate drinking water supply and improper sanitation facilities are the key factors hindering the economic and social advancement of developing countries. As a result, the General Assembly of the United Nations proclaimed 1981–1990 as the International Drinking Water Supply and Sanitation Decade, with the objective of providing safe drinking water and sanitary facilities for everyone in the world. During that decade most developing countries did not have enough funds to spend on developing water supply systems and waste disposal facilities, so the use of appropriate low-cost technology was promoted by the United Nations. Specifically, RWCS had been identified and promoted.

During the 1977–78 drought period in Honolulu, Hawaii, USA, a news article reported that residents in the Tantalus–Round Top district of Honolulu performed an Indian dance to pray for rainfall to refill their empty rainwater catchment tanks. This news item inspired the author of this paper to conduct research on RWCS. When an abstract of one of his papers appeared in the American Geophysical Union's EOS special issue on the 1979 fall meeting, it attracted the attention of Dr. S.J. Perrens of the University of New England, Armidale, Australia. Perrens wrote a letter to the author to tell about his work on RWCS. In his reply, the author suggested that an international conference for the exchange of RWCS information and the promotion of RWCS application would be timely. The first conference, held on 15–17 June 1982 in Honolulu, Hawaii, was the beginning of the series of international conferences on RWCS. At the first conference, it was decided that conferences should be held at two-year intervals and that an international association to promote RWCS development should be organized. The IRCSA was formed in 1989, and to date, eight conferences have been held.

## Current Status of RWCS in the World

RWCS were widely used in the world before the industrial revolution. In urban areas they were replaced by centralized water supply pipe systems. However, in recent decades, many public water supply systems have not been able to expand fast enough to keep up with the ever-increasing demand for water. As a result, water shortages have become a problem, especially during droughts. A solution is to use RWCS, which have been regaining ground as a low-cost method that is easy to develop by users in a very short time. Since in most cases RWCS catchment areas (the rooftops of water users' residences) are already available, the only need is to provide a rainwater storage tank and other small fixtures (Fok et al., 1980; Fok, 1992). Users can select from the various types of storage tanks being sold, the unit cost of which depends upon the construction material used (Fok and Leung, 1982). RWCS water tank manufacturers have increased their production all over the world, thus the unit cost of RWCS water tanks are expected to decline in the near future. This is a very favorable situation for more RWCS development.

In developed countries, RWCS nicely fits into the policy of sustainable resources development. Many metropolises have developed RWCS in their public buildings as a showcase of the public sector's effort to collect rainwater for use in washrooms, for garden irrigation, and as a water source for emergencies (Fok, 1994). In developing countries, RWCS have a wide range of development. The Thailand Jar Project is an outstanding example of good government policy on how to provide safe and adequate drinking water for 18 million rural inhabitants and on how users can use teamwork to build and finance their own RWCS (Fok and Chu, 1991). The gains in health and in the financial and social well-being of the country have greatly contributed to Thailand's economic advancement in recent years.

Thailand Jar Project

The Thailand jar is a standard reinforced thin-walled concrete container in the shape of a jar that has a volume of 2 cubic meters. Each jar can be built by two people in one day at a cost of $20 for the construction materials. The Thai government provided $7.4 million to the project for the training of two technicians per village in jar construction at one of the regional universities. The idea was that the trained technicians would, in turn, train others in their respective villages. Funds from the same source also paid for the printing of the jar construction manual and the molds for jar construction. In addition, the Thai government set up a revolving fund of $13 million so that villagers could borrow money to buy jar construction materials and then repay the loan within a set period. No loan default was reported because villagers exercised self-policing for the repayment. This project resulted in more than 9 million jars being built during the period from 1985 to 1990. Thailand was recognized for providing a safe drinking water supply for its people on time (i.e., during the United Nations' water decade). Eighteen million rural inhabitants benefited. The Thailand government was successful in inducing a $180 million investment from its private sector using only $7.4 million of its public-sector funds.

Progress of IRCSA

As mentioned earlier, IRCSA was formed during the fourth RWCS international conference. The objectives of IRCSA are to provide a communication center for interested members to work together on subjects related to RWCS and to exchange information on new RWCS developments without waiting for the next international conference. Also, IRCSA strives to provide a platform through which RWCS developments can best be communicated with members of other water-related professional organizations. The nonprofit, nonpolitical IRCSA has been gaining recognition gradually. Members of IRCSA contribute many noncompensated hours to promote RWCS, hence the grassroots approach of IRCSA has been fruitful. IRCSA has held eight international conferences so far, with the last held in 1997 in Tehran, Iran. Also, eight conference proceedings have been published, not counting the proceedings of regional conferences, national meetings, and local workshops. In due time, IRCSA membership will expand to cover every continent in the world, with chapters in every nation.

Acknowledgments

The author wishes to acknowledge the financial support received from the U.S. Geological Survey under the 1997 Regional Competitive Grant Program through the University of Guam to the University of Hawaii (account no. 53-Q-740468-R-53233). The support has made the presentation of this paper possible. This is contributed paper CP-98-02 of the Water Resources Research Center, University of Hawaii at Manoa, Honolulu.

References

Fok, Y.-S., Fong, R.H.L., Hung, J., Murabayashi, E.T., and Lo, A., "Bayes-Markov Analysis for Rain-Catchment Cisterns," Technical Report No. 133, Water Resources Research Center, University of Hawaii at Manoa, Honolulu, Hawaii, Mar., 1980.

Fok, Y.-S., and Leung, P., "Cost Analysis of Rain Water Cistern Systems," Proceedings of the [1st] International Conference on Rain Water Cistern Systems, Water Resources Research Center, University of Hawaii at Manoa, Honolulu, Hawaii, June, 1982, pp. 210-219.

Fok. Y.-S., and Chu, S.-C., "Progress of the U.N. Water Decade and Rainwater Catchment Systems," Proceedings of the 5th International Rain-Water Catchment Systems, Department of Harbor & River Engineering, National Taiwan Ocean University, Keelung, Taiwan, R.O.C. Aug., 1991, pp. 14–22.

Fok, Y.-S., "Rooftops: The Under-Utilized Resource," Proceedings of the 1992 Regional Conference, IRCSA, 2F Science Building, Kyoto Research Park, Kyoto 600, Japan, Oct. 4–10, 1992, pp. 164–174.

Fok, Y.-S., "The Present Situation of Rainwater Utilization Techniques," Proceedings of Tokyo International Rainwater Utilization Conference, Sumida City Hall, Tokyo, Japan, 1-6 Aug. 1994, pp. 43–46.

Rainwater Catchment Systems Development Guidelines

Yu-Si Fok[1]
Fellow, ASCE

*Abstract*

Most rainwater catchment systems (RWCS) are developed and managed by private users. From the public safety standpoint, RWCS are a concern because the construction of the water supply systems is unregulated since no building code or a building permit is required at present. From the public health standpoint, RWCS are often a big concern because the water quality is not monitored for potable uses. According to Wilken (1995), the number of people living in urban areas jumped from 737 million in 1950 to 2.6 billion in 1995. The urban share of the total population increased from 29 to 45 percent in just 45 years. More than 50 percent of the developed countries' populations were already urban by 1950; therefore, the increase in urban populations in recent decades largely took place in developing countries. The rapidly increasing population rate has a great impact on the public utilities. A great demand in such a short time causes frequent water shortages. This becomes a problem because traditional water supply development requires sufficient time to plan to acquire water rights and right-of-ways, and to obtain permits and funding. Most immigrants to cities are poor and are trying to earn a living in a city. The water shortage problems are just part of their daily struggles. RWCS development is a suitable solution for them. This paper presents a set of RWCS development guidelines based on the affordable principle in order to provide a common ground for the public and private sectors to solve the water supply problem.

---

[1]Professor, Department of Civil Engineering and Researcher, Water Resources Research Center, University of Hawaii at Manoa, 2540 Dole Street, Honolulu, HI 96822-2333

## Introduction

Most rainwater catchment systems (RWCS) are developed and managed by private users. From the public safety standpoint, RWCS are a concern because the construction of the water supply systems is unregulated since no building code or a building permit is required at present. From the public health standpoint, RWCS are often a big concern because the water quality is not monitored for potable uses. According to Wilken (1995), the number of people living in urban areas jumped from 737 million in 1950 to 2.6 billion in 1995. The urban share of the total population increased from 29 to 45 percent during this period. The population of developed countries was already more than 50 percent urban by 1950, so much of the urban population increase in recent decades must have occurred in developing countries. The United Nations estimates that by 2025, 60 percent of the world's population will live in urban areas. Water shortages are common problems in the urban areas. Aside from the rapid increase in population, other influential factors include frequent droughts, increased per-capita water use, degraded water quality, leakage in water systems and lack of developable water sources. The development of traditional water supply systems requires a long time to plan, to acquire water rights and right-of-ways, and to obtain permits and funding. Most of the rural-to-urban immigrants are poor. They move to the city to try to earn a living there. The water shortage problems are part of their struggles. RWCS development is a suitable solution for them. They cannot wait for the public sector to build a water supply system to meet the needs of the expanding population.

This paper presents a set of RWCS development guidelines based on the affordable principle in order to provide a common ground for the public and private sectors to solve the water supply problem. The affordable principle takes into account the user's per-capita annual income to set suitable requirements and regulations for the RWCS development guidelines. No attempt is made to provide a set of RWCS development guides for a location or a country. The idea is to have the private water users to participate in their own water supply planning, development, and management with the public sector's partnership to arrive at a set of sustainable RWCS' development guidelines to solve this water supply problem.

## Background

Regulations or guidelines for RWCS development have been reported in only a few documents. Fok (1982) reported on the Ohio Department of Public Health's rules for private water systems, which includes RWCS. Ohio's institutional policy seems only protective of the general health of users of the 63,000 privately owned water systems, most of which are farm ponds that collect and store rainwater for private domestic uses. Waller (1982) reported that Bermuda's rainwater catchment areas and storage tanks are governed by the Public Health Act of 1949 and regulations made under it. The water storage regulations (1951) require a specific storage capacity for each person. Waller and Inman (1982) reported that the Nova Scotia Department of Health recognizes roof water as a domestic source. This information is in the guidelines provided by their Division of Public Health Engineering in 1981. In the U.S. Virgin Islands, water supplies are mainly from rainwater catchments. They have the RWCS as part of the requirements in the building permit, which specifies the size of the rainwater storage tank per person. None of the RWCS regulations or guidelines mentioned here consider water quality maintenance.

In Hawaii, the island of Hawaii has more than 20,000 persons in more than 8,000 households that use RWCS for their water supply. The Hawaii Department of Health in 1993 distributed a set of draft guidelines for the use and maintenance of RWCS. This set of draft guidelines recommends that users not use RWCS for drinking or cooking and suggests that RWCS water for all other uses be properly disinfected. It also states that there are no known approved or certified roof or gutter materials for the collection of rainwater for drinking. In 1994 the Hawaii House of Representatives passed the House Concurrent Resolution No. 214, which requests that the author of this paper develop guidelines which may be adopted by County Planning and Building Departments to control the construction of private water catchment facilities. The resolution requests that the guidelines (1) Consider suitable roof catchment area and water storage capacity, (2) recommend contamination-safe construction materials, (3) recommend filter systems, (4) require that users let proper authorities use their water in the case of fire or other natural disaster, (5) include a method to guarantee that a specific volume of water be available for fire fighting, and (6) recommend a maintenance program for RWCS users to follow. After undertaking some studies, the writer produced a report indicating that the requested items for the development guidelines are too expensive for RWCS users to carry out, especially items 4 and 5. Thus, the idea of developing guidelines based on the affordable principle and of having users participate in formulating the RWCS guidelines was conceived and is applied in this paper.

*Analyses*

The background study gives four important factors for consideration of RWCS development: (1) public health is of great concern regarding the domestic use of RWCS, (2) adequate storage capacity is required, (3) contamination-safe construction materials are needed, and (4) a maintenance program is needed.

Clearly, RWCS impact on public health and the use of approved construction materials are concerns of the public sector, whereas RWCS storage capacity and maintenance are concerns of the users. The common ground for the public and private sectors to have a consensus on a set of guidelines for RWCS development and maintenance hinges on the cost for RWCS construction and maintenance. The system should be affordable. If the guidelines developed make RWCS development unaffordable for the users, they will not be followed. This is a reality that the public sector has experienced in the past.

For the application of the affordable principle in the RWCS guidelines, the participation of RWCS users with the public sector is an ideal arrangement. Public hearings of the public sector's proposed rules or regulations can produce many good results from the private sector. The spirit of cooperation and teamwork can carry the proposed guidelines for a very long time.

*Affordable RWCS Development Guidelines*

As stated previously, the affordable RWCS development guidelines are based on the user's per-capita annual income to set the suitable requirements and regulations in the paper presented by Lee et al. (1991). The type of RWCS, the per capita income groups, and the RWCS unit cost are listed in the following table.

Attributes of Rainwater Catchment Systems

| Type | Per-Capita Income Required ($) | Unit Cost ($/cu. m.) |
|---|---|---|
| Ponds | << 50 | < 10 |
| Used Container | 10-150 | < 10 |
| Cement Jar | 150-300 | 10 |
| Ferrocement Jar or Tank | 150-300 | 15+ |
| Brick Work | 250-300 | 25+ |
| Sheet Metal | 500-1000 | 100 |
| Reinforced Concrete Tank | 500-1000 | 150 |
| Fiberglass | 1000+ | 160 |
| Redwood | 1500+ | 250+ |
| Public Water System | 500+ | 200+ |

Source: World Health Organization, 1981.

The above table shows that if the RWCS user's per-capita annual income is below $50, the user has no choice but to use ponds as the water storage tank or to fetch water from other water sources. No RWCS development guidelines may be imposed on users in this income category. If the RWCS user's per-capita annual income is $150 but less than $500, the user could afford any of the first five types of water storage tanks listed in the table. The user's RWCS development guidelines can include the suitable construction materials, a proper maintenance program, and the need to boil or buy needed drinking water.

If the RWCS user's per-capita annual income is between $500 and $1000, the user can afford to have sheet metal and reinforced concrete water tanks and one drinking water purifier. The user's RWCS development guidelines can include installing a drinking water purifier or requiring the user to buy bottled drinking water. If the RWCS user's per-capita annual income is more than $1000, the user can afford to have several kinds of water filters to purify the drinking water or to have bottled drinking water delivery service. The user's RWCS development guidelines can include items developed for lower income groups. (Readers should note that the 1981 dollar value is used as the base for the above table and the classifications for all of the per-capita annual income groups. Perhaps we can use one 1981 dollar = five to ten dollars for current cost analysis. In addition, the information in the table implies the RWCS user can allocate about 5 percent of his/her annual income to pay for the RWCS storage tank.

*Acknowledgment*

The author wishes to acknowledge the financial support received from the U.S. Geological Survey under the 1997 Regional Competitive Grant Program through the University of Guam to the University of Hawaii (account no. 53-Q-740468-R-53233). The support has made the presentation of this paper possible. This is contributed paper CP-98-03 of the Water Resources Research Center, University of Hawaii at Manoa, Honolulu.

*References*

Fok, Y.-S., "Rain Water Cistern System Impact on Institutional Policy," Proceedings of the International Conference on Rain Water Cistern Systems, Water Resources Research Center, University of Hawaii at Manoa, Honolulu, Hawaii, June, 1982, pp. 227-232.

Hawaii Department of Health, "Guidelines for the Use and Maintenance of Rainwater Catchment Systems," Department of Health, Honolulu, Hawaii, 1993, Draft.

Hawaii House of Representatives, "House Concurrent Resolution No. 214," Honolulu, Hawaii, 17th Legislature, 1994, HCR LRB 94-1879.

Lee, D.L., Leung, S., Fok, Y.-S., and Chu, S.C., "Opportunities for Rainwater Cistern Systems in Rural Economic Development," Proceedings of the 5th International Conference on Rain Water Cistern Systems, Department of Harbor &

River Engineering, National Taiwan Ocean University, Keelung, Taiwan, R.O.C., 4-10 Aug., 1991, pp. 317-323.

Waller, D.H., "Rain Water as a Water Supply Source in Bermuda," Proceedings of the International Conference on Rain Water Cistern Systems, Water Resources Research Center, University of Hawaii at Manoa, Honolulu, Hawaii, June, 1982, pp. 184-193.

Waller, D.H., and Inman, D.V., "Rain Water as an Alternative Source in Nova Scotia," Proceedings of the International Conference on Rain Water Cistern Systems, Water Resources Research Center, University of Hawaii at Manoa, June, 1982, pp. 202-209.

Wilken, E., "Urbanization Spreading," VITAL SIGNS 1995, World Watch Institute, pp. 100-101.

World Health Organization, "The International Drinking Water Supply and Sanitation Decade Directory," Thomas Telford, Ltd., 1981, London, UK.

Rainwater Catchment Response Surface Ascension

Richard J. Heggen, Member, ASCE[1]

Abstract

A response surface enhances the visualization of system relationships. Rainwater catchment systems (RWCS) response surfaces can illustrate beneficial water yield as a function of catchment area and cistern volume, necessary cistern volume as a function of antecedent rainfall and drought duration, or water utilization as a function of rationing and penalty function. The nature of the response surface provides guidance in system design. A response surface technique is presented in which a loosely bounded RWCS problem can be locally optimized by an iterative gradient-based ascension. RWCS examples from various nations illustrate the concept.

Optimality

Analysis of rainwater catchment systems (RWCS) technically parallels analyses employed in other fields of engineering: time series, data synthesis, resource allocation, simulation, stochasticity, nonstationarity, operational decision-making, demand prediction, and hierarchical control. Fully optimizing even a simple RWCS would require a myriad of meteorological, engineering, economic, social and environmental health models, and then a model of the interrelationships.

Most RWCS solutions are constrained by data. The "optimization" task is a RWCS reasonably efficient within a set of basic constraints.

Response Surfaces

A response surface is the locus of points satisfying,

$$y = y(x_1, \ldots x_n) \tag{1}$$

While the totality of the complete system may never be fully documented, enough interactions can be determined from a response

---

[1]Professor, Civil Engineering, University of New Mexico, Albuquerque, NM 87131

surface to move toward (albeit never actually achieving) a "best" outcome. A response surface of two dimensions can be envisioned as a contour plot. If the objective is that of minimizing cost or maximizing performance, descending or ascending the response surface (*e.g.*, climbing a mountain on a contour map) improves the result. Restarting at several points generally resolves local vs. global optimization issues.

The ascension approach tests a small number of designs and moves toward a more productive region. If the response surface is mathematically defined over a local domain, the next move is determined by derivative,

$$\frac{dx_1}{dx_2} = \frac{dy/dx_2}{dy/dx_1} \tag{2}$$

The move distance requires testing. Sequential moves approximately 0.8 the length of the last move often converge well.

<u>Simplified Response Surfaces</u>

Three simplified response surfaces are illustrated, amenable to spreadsheet computations. More complex algorithms, *e.g.*, weighted schemes common to contour plotting, may be unnecessary.

A simple grid-based surface consists of planar triangles. Each is bounded by two-quadrant corners and a third point centered within the quadrant and assigned the mean value of that quadrant's corners. The surface at any point is,

$$z = a + bx + cy \tag{3}$$

where the coefficients stem from the triangle corner points. Fig. 1 illustrates a 16 triangle surface over four adjacent quadrants.

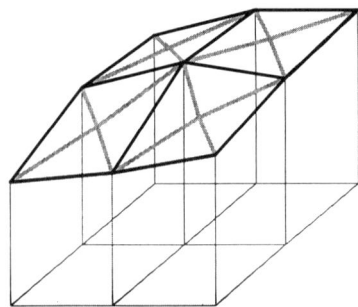

Figure 1. Triangular Surface

Adding an interactive term, the surface warps over a quadrant. The orthogonal gradient within any quadrant is linear at any point.

$$z = a + bx + cy + dxy \qquad (4)$$

where the coefficients stem from the four quadrant corners. Fig. 2 illustrates the surface for the same nine points as in Fig. 1.

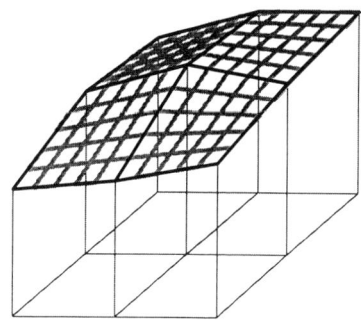

Figure 2. Warped Surface

Fitting orthogonal quadratics between a quadrant center and the centers of the four adjacent neighbors, the surface curves.

$$z = a + bx + cx^2 + dy + ey^2 \qquad (5)$$

Fig. 3 illustrates the surface over a single quadrant, again using the same nine point data set.

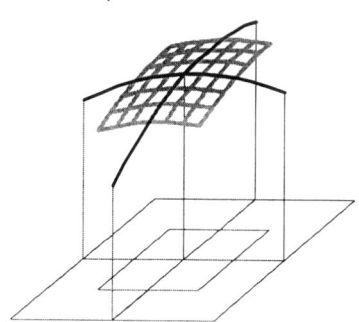

Figure 3. Orthogonal Quadratic Surface

## RWCS Application

An obvious application in RWCS design is that of allocation between catchment area $A$ and cistern storage $V$. Let $z$ be total system cost, appropriately amortized, $P$ be deficit penalty, and $a$, $b$, $c$ and $d$ be economy of scale coefficients.

Minimize $z = aA^b + cV^d + P$ (6)

$P$ is related to system deficit $D$ by one of several alternatives.

$$P = a\sum_{i=1}^{n} D_i^b, \quad D_i = \min \begin{cases} T_i - C_i \\ 0 \end{cases}$$ (7)

where $T$ is target, $C$ is consumption and $n$ is the number of periods. If a penalty is assigned for being in deficit only, not as a function of the deficit,

$$P = a\sum_{i=1}^{n} k_i, \quad k_i = \begin{cases} 1, C_i < T_i \\ 0, C_i = T_i \end{cases}$$ (8)

Alternatively, penalty can be determined by the persistence of unsatisfied demand,

$$P = a[\max(\sum n)]^b, \quad \begin{matrix} n = \text{consecutive days } C_i < T_i \\ \text{alternative } n = \text{consecutive days } C_i = 0 \end{matrix}$$ (9)

Fig. 4 illustrates a $(V, A)$ response surface ascension, by simulating a 20-year daily history. The initial point is (30, 3000); the final point is approximately (43, 3800) The surface is approximated by the warped surface (Eq. 3); the penalty function is that of Eq. 7.

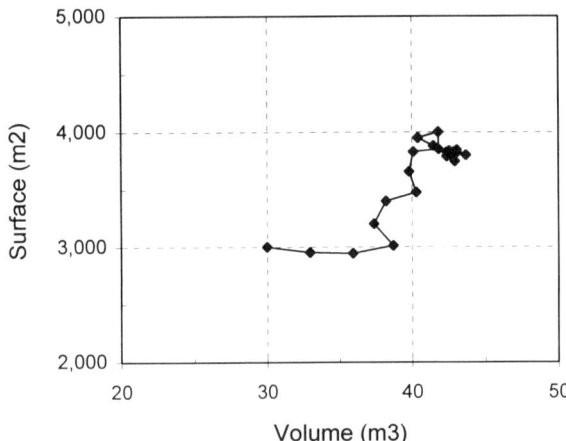

Figure 4. Ascension Path

Fig. 5 shows RWCS cost, the dependent variable, as the iterations progress. Note that the solution reasonably minimizes after eight steps.

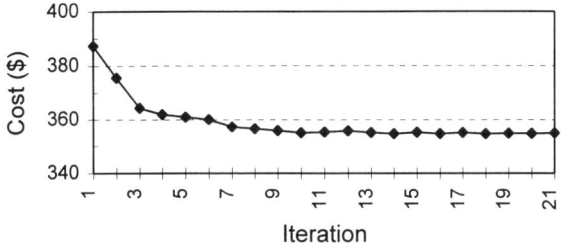

Figure 5. Iteration Result

RWCS Applications

While ascension methods are not common in RWCS analysis, Table 1 shows several response surface implementations to date.

Table 1. Response Surfaces in RWCS Studies

| Independent Variables | | Dependent Variable | Reference |
|---|---|---|---|
| Storage and Precipitation | Target | Optimal Decision | Heggen, 1982 |
| Catchment Area | Reservoir Capacity | Deficit | Heggen, 1983 |
| Normalized Volume | Normalized Discharge | Coefficient of Variation | Dwornik, 1984 |
| Storage Fraction | Demand Fraction | Reliability | Latham, 1987 |
| Catchment Area | Reservoir Capacity | Efficiency | Fewkes, 1994 |
| Catchment Area | Reservoir Capacity | Efficiency | Wang, 1995 |
| Groundwater Storage Capacity | Surface Storage Capacity | Runoff Error | Sharifi, 1997 |

Table 1 includes RWCS studies in Micronesia, the United States, Tanzania, Great Britain, China and Iran. In terms of usage, response surface analysis is well distributed. Most of the response surfaces, however, are left as simple visuals.

Ascension is not specified. Optimization is not pursued. Given the work to develop the response surfaces, this absence is unfortunate. An ascension step would move the study from one of general description to one of completed analysis.

## Conclusion

RWCS analysis is an appropriate venue for response surface analyses. Initial efforts have shown the value of visually representing complex dependencies. Future work should further incorporate surface ascension techniques.

## Appendix I. References

Dwornik, D.S., Heaney, J.P., Koopman, B. and Saliwanchik, D.R. (1984), "Stormwater Collection and Wastewater Reuse and Cooling Water Source at the Kennedy Space Center, Florida", *Second International Conference on Rainwater Cistern Systems*, H.H. Smith, ed., Caribbean Research Inst., St. Thomas, D2-1-14.

Fewkes, A. and Frampton, D.L. (1994), "Optimizing the capacity of Rainwater Storage Cisterns", *Sixth International Conference of Rainwater Catchment Systems*, G.K. Bambrah, F.O. Otieno, D.B. Thomas, eds., IRWCSA, 225-235.

Heggen, R.J. (1982), "Optimal Catchment Design by Marginal Analysis", *International Conference on Rain Water Cistern Systems*, F.N. Fujimura, ed., Water Resources Research Center, Univ. of Hawaii, Manoa, 83-91.

Heggen, R.J. (1991), "Dynamic Programming in Rain Water Catchment", *Fifth International Conference on Rain Water Catchment Systems*, S.C. Chu, ed., National Taiwan Ocean Univ., Taiwan, 251-257.

Latham, B. and Schiller, E. (1987), "Use of Rainwater Catchment in Urban Areas", *Third International Conference on Rain Water Cistern Systems*, C. Vadhanavikkit, ed., Khon Kaen Univ., Thailand, D3-1-13

Sharifi, F. (1997), "An Investigation into Rainfall Runoff Processes Aiming at Estimating Runoff from Ungaged Catchments", *Eighth International Conference on Rainwater Catchment Systems*, IRWCSA, B. Aminipoura and J. Ghoddousi, eds., Jihad-e-Sazandegi, Tehran, 500-516.

Wang, W., Yang, L. and Zhao, L. (1995), "Rainfall Collection to Develop Courtyard Economy", *Rainwater Utilization for the World's People,* Vol. I, *Seventh International Conference on Rainwater Cistern Systems*, GSC, Beijing, 5-15-18.

Integrated Dual-Mode Roofwater Collection System for
Non-Potable Uses in the NTU Complex

Adhityan Appan[1]

Abstract

The Republic of Singapore has an annual average rainfall of 2250 mm and a limited land area of about 630 km$^2$. Almost 60% of the water is imported and hence there is a continuous search for new sources. One such source is the collection of roofwater from high-rise buildings which account for almost 84% of the population. The main objectives of this paper are to study the feasibility of tapping the roofwater from the north spine of the Nanyang Technological University (NTU) Campus and utilizing the water for non-potable uses. A simple input-output model has been used which has a dual-mode system (DMS) of operation that ensures that non-potable requirements are primarily met with by the collected roofwater. In the case study involving a roof area of 38,714 m$^2$, an integrated collection and supply system has been worked out resulting in delivery of non-potable water for flushing of toilets, laboratory uses etc., The design parameters were optimized leading to a rainwater storage tank that is 41.0m x 15.5m x 4m depth with cut-in and cut-outs placed at depths of 1m, 1.5m and 3.5m from the floor of the storage tank. Incorporation of such a DMS leads to a monthly saving of S$18,400.00 which amounts to about 12.4% of the average monthly expenditure for water.

Introduction

The availability of water is largely dependent on the quanta of rainfall, area of catchment and demand. In Singapore, there is an abundant average rainfall of about 2250 mm/annum but, due to industrial development and increasing population, demands are constantly increasing. Singapore land area is 630 km$^2$ with only half the area being available as catchment as there are competing demands for land use. Besides, almost 60% of the water requirements are being imported. The search is continuously on for augmenting water resources and investigating new areas. As almost 84% of the urban population in Singapore lives in high-rise buildings, the roofs of such structures have a good potential for collection of rainwater. The main objectives of this study are:
- to review some computer simulation models with respect to Rainwater Catchment Systems (RCS) with emphasis on roofwater collection,
- to select a simple model and use existing data to determine the design parameters so that the collected rainwater can be used for non-potable uses,

---

[1] Dr Adhityan Appan, Associate Professor, School of Civil and Structural Engineering, Nanyang Technological University, Nanyang Avenue, Singapore 639798

- to integrate the proposed system with an existing system by incorporating a dual-mode facility so that potable supply can be substituted when there is insufficient stored rainwater and
- to draw conclusions and make some recommendations on the proposed system..

Review of Some Computer Models

*Computer Simulation models used in RCS:* Basically, models developed have been using rainfall, catchment area and yield to determine storage requirements. Such models have varied from simple deterministic types (Hoey & West 1982) to probabilistic (Fok et al, 1982) and stochastic models (Leung and Fok, 1982).

A program was also developed in the United Kingdom (Fewkes and Ferris, 1982) where the rainwater was only used for flushing. In this program, rainwater was simulated and using the Monte Carlo method, the percentage water saved per annum for a range of tank capacities, roof areas and family sizes were generated.

A review of computerized methods in optimizing storage volumes was done by Schiller & Latham (1982). The volume could be obtained by different methods like the conventional mass curve analysis (Rippl, 1893), yield after storage model (Jenkins et al, 1978) or the statistical method(Ree et al, 1971).

In another simulation model (Morris et al, 1984), the storage required to deliver a constant flow rate at each rainfall station was determined. A special feature of this model was the dual-mode of withdrawal with and without rationing.

*RCS models developed in NTU:* Primarily a simple input/output simulation model was used (Appan, 1982) with a minor modification:

$$Q_i = Ar_i - \{(E_i + b_i) + D_i\} \quad \text{................ Equation 1}$$

where during any duration i, $Q_i$ is the available quantity, $r_i$ the rainfall, $E_i$ the evaporation, $b_i$ the absorption, $D_i$ the draw-off and A the catchment which is independent of i. Equation 1 is applied to a known storage volume and the cistern will be empty or subject to spillage according as $Ar_i$ is less or greater than $\{(E_i + b_i) + D_i\}$.

This basic model has been developed over a period of time to suit varying field conditions. Initially a study was conducted on the feasibility of using rooftops of high-rise buildings in Singapore (Appan et al 1987). A system was proposed to utilize such water for flushing of toilets and a dual-mode of supply (DMS) was incorporated wherein potable water could be substituted when the supply of collected rain water had run out.

In another study on the collection of rainwater from a bus park cum interchange (Appan et al, 1988), a simple computer program (NTURWCS.MK1) was written to calculate the optimum tank size required to meet the daily needs. In this program there were facilities to incorporate runoff coefficients as runoff was being tapped from roofs and paved areas.

Another program (NTURWCS.MK2) was developed (Chan & Heng 1992) which was sufficiently versatile to accommodate varying discretised time intervals of rainfall. This model was further upgraded (NTURWCS.MK3) to determine both the efficiency of an existing DMS in the Singapore Changi Airport (Appan, 1993) and optimum reservoir size required to meet hourly demand for fire-fighting and toilet flushing.

## Utilization of NTU's Roof Area for a Dual-Mode System of Water Supply

*Existing Reticulation System:* The NTU campus (200 ha) lies in the south-western part of Singapore. It consists of 6 schools, administration blocks, halls of residence, canteens etc., which lie around the north and south spines. In the water supply system potable water, which is readily available, is pumped to two high-level tanks which cater for the whole campus. Water from one of these tanks (see Figure 1) is gravitated to the north spine where it is stored in a 31.0m x 15.5m x 4m deep Distribution Tank that has two inter-connected compartments. The cut-in/cut-out electrodes at bottom water level (BWL) and top water level (TWL) are at 1m and 3.5m respectively from the floor level. The water is then distributed by seven different pumps to constant head tanks located at the roofs of various buildings in the north spine.

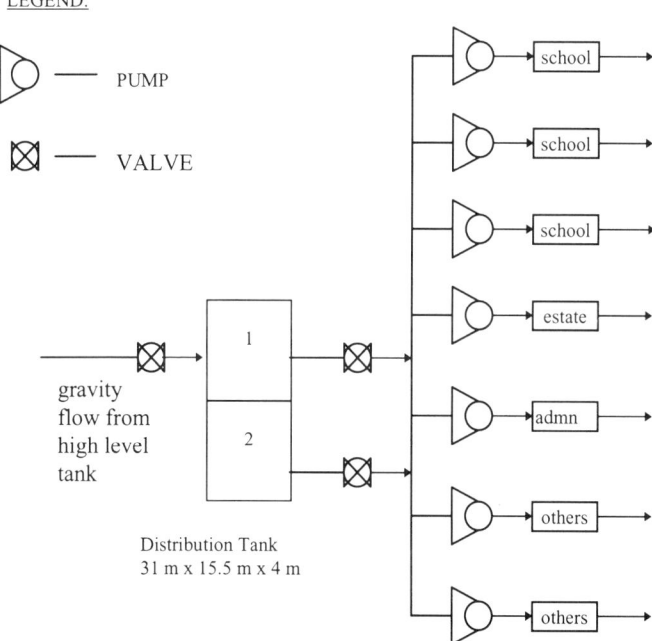

**Figure 1: Existing Reticulation System in the North Spine**

*Daily demand:* Water, which gravitates from the constant head tanks, is used by three schools, the administration building and a canteen for drinking, flushing of toilets, laboratories etc., At the outlet in the distribution Tank, eight sets of hourly readings were taken and the hourly consumption was obtained (see Table 1). The average used was 540 $m^3$/day with hourly demand varying from an off-peak of 5 $m^3$ at 2100 h to 0500 h to a maximum of 65 $m^3$ at 1200 h.

Table 1: Diurnal hourly variations of demand (m$^3$)

| Hour | Demand | Hour | Demand | Hour | Demand | Hour | Demand |
|------|--------|------|--------|------|--------|------|--------|
| 0100 | 5  | 0700 | 20 | 1300 | 65 | 1900 | 15 |
| 0200 | 5  | 0800 | 20 | 1400 | 50 | 2000 | 10 |
| 0300 | 5  | 0900 | 25 | 1500 | 35 | 2100 | 5  |
| 0400 | 5  | 1000 | 50 | 1600 | 30 | 2200 | 5  |
| 0500 | 5  | 1100 | 60 | 1700 | 20 | 2300 | 5  |
| 0600 | 10 | 1200 | 65 | 1800 | 20 | 2400 | 5  |

*Rainfall measurement and water sampling quality:* Within the NTU campus there is a rainfall station where continuous data is recorded in a tipping bucket type system. Rainfall data from September 1989 to August 1990 was used in an earlier program developed (Chan & Heng, 1992) and the time interval was discretised to one hour which is in line with the time interval of the demand data. A total of 48 rainwater samples was collected. Besides, samples were also collected periodically from storage tanks in two laboratories where it is proposed to use the collected roofwater.

*Extension/modification of existing system:* This study is confined to the collection of rainwater from the roof areas of the north spine (38,714 m$^2$). The approach in this study is to channel all roofwater into one of compartments of the Distribution Tank which may have to be enlarged. The other compartment will be connected only to those systems where potable water is warranted. The two water tanks will be separate units but they will be inter-connected by a simple DMS such that when collected roofwater is short in supply, potable water will be pumped to a pre-determined level. Hence, there will be the use of potable water for non-potable uses and in the potable water tank there will be no intrusion of rainwater.

*Choice of an appropriate model & relevant design parameters:* The model chosen was NTURWCS.MK1 which was written in Quickbasic 4.5. In the program (Appan, 1988), the basic inputs are rainfall data, size of storage tank, evaporation, absorption and the hourly demand and the output will give the rainwater collected and both quantities that overflows and the potable water that will be needed when the stored rainwater runs low. The program was run for different storage volumes following which the cut-in/cut-out levels were also varied.

Results and Discussion

A series of runs was carried out with a view to arrive at the most suitable design parameters. Some of the salient parameters arrived at are as follows:

*Optimum Tank size:* Initially a series of runs was carried out varying the known volume of the tank. The breadth and depth of the Distribution Tank were kept constant but the length was increased from 2m to 4m resulting in larger volumes. When the rain overflow was reduced considerably, the size of 40.7m was selected by trial and error to arrive at zero overflow. The results obtained are shown in Table 2. For practical purposes the length of the rainwater tank can be considered to be 41.0m which means the proposed roofwater storage tank will have to be extended by 24.5m. If site configuration is such that there is no space for the extension lengthwise, only the total width of the rainwater tank can be extended to 41.0m.

Table 2: Optimum tank size

| Tank dimensions (m) | Rain overflow (m³) | Rain water utilized (m³) |
|---|---|---|
| 33.0 x 15.5 x 4 | 9754 | 10080 |
| 33.8 x 15.5 x 4 | 4800 | 12100 |
| 37.5 x 15.5 x 4 | 1120 | 13070 |
| 40.6 x 15.5 x 4 | 40 | 15400 |
| 40.7 x 15.5 x 4 | 0 | 16360 |

*Fixing of cut-in/cut-out levels:* Though rainfall overflow has been minimized, there is the need to determine the quanta of potable water that has to be pumped into the roofwater tank whenever it is empty so that the supply of water for non-potable uses will be perpetuated. With this in mind, a series of runs was carried out to determine the relevant volumes to be used (see Table 3). Thus the most suitable level for minimizing the use of potable water will be 1.5m. Thus there will have to be at least three electrodes to cater for the BWL, TWL and the intermediate level (IML) which will be 1.5m from the floor level of the roofwater collection tank.

Table 3: Cut-in/Cut-out levels

| Cut-in/cut-out | Rainwater overflow | Potable supply (m³) | Cut-ins (No.) |
|---|---|---|---|
| 1m/3.5m | 34.4% | 10811 | 9 |
| 1m/2.5m | 19.5% | 10091 | 9 |
| 1m/1.5m | 14.6% | 9850 | 13 |

*Water quality*: Physico-chemical and bacteriological analyses were carried out for the 48 samples using standard methods and the results are shown in Table 4. The collected rain water has a generally high quality in comparison with potable water supply and the laboratory storage tanks except for pH and the total coliform counts. Besides, the rainwater also has a poorer quality than the laboratory samples in terms of colony counts. Both these parameters need be adjusted

Table 4: Rain, potable & laboratory water quality

| Parameter | Potable water* | Rainwater | Lab B3 | Lab B5 |
|---|---|---|---|---|
| pH | 7.0 - 7.5 | 4.1 (0.4) | 7.4 (0.1) | 6.3 (2.2) |
| Colour | < 5 | 8.7 (9.9) | 26.3(32.1) | 9.7(13.2) |
| Turbidity(NTU) | < 5 | 4.6 (5.7) | 41.5(16.3) | 7.9 (8.6) |
| TSS (mg/L) | 240-400 | 9.1 (8.9) | 8.5 (6.0) | 1.2 (1.1) |
| TDS (mg/L) | n.s. | 19.5(12.5) | 427(119) | 350(38) |
| Hardness as $CaCO_3$ (mg/L) | 20-40 | 0.1 (0.3) | 32.5(2.1) | 27.1(36.2) |
| $PO_4$ as P (mg/L) | n.s. | 0.1 (0.6) | 0.1 (0.1) | 0.5 (0.5) |
| Total Coliform[⊗] | 0 | 92.0(97.1) | 1.5 (2.1) | 4.0 (5.7) |
| Faecal Coliform[⊗] | n.s. | 6.7 (8.9) | 1.1 (0.1) | 8.8(10.8) |
| Total colonies[✧] | n.s. | 1766(2436) | 871(960) | 185 (56) |

Note: n.s. indicates "not specified"
Values are means and those in brackets are standard deviation
* WONG, 1986; [⊗] MPN/100 mL ; [✧] CFU/mL

by some treatment, particularly as it has been experienced in the past that such water has been the breeding ground for mosquitoes leading to dengue fever.

Conclusions and Recommendations

1. By using a simple input/output model, it is possible to arrive at basic design parameters like size of rainwater tank, cut-in and cut-out levels etc.
2. Rainwater, in the case study, has an acceptable quality for non-potable uses and it is recommended that provision be made to raise the pH and disinfect the water.
3. By extending one of the compartments in the existing Distribution Tank and operating the DMS, rainwater does not mix with the potable water supply besides which there will be continuous supply for non-potable uses.
4. The utilization of roofwater will realize savings of about S$18,400/month which is about 12.4% of the current monthly expenditure for water used in the north spine.

References

Appan, A. (1982). Some aspects of roofwater collection, Proc Conf on Rain Water Cistern Systems (RWCS), Univ. of Hawaii, Honolulu, USA, pp 220-226, June.

Appan, A., Lim, K.L. and Loh, S.K., (1987). The utilization of high-rise building rooftops for development of a dual-mode water supply in Singapore, Proc. 3rd Inter Conf. on RWCS, Khon Kaen, Thailand, pp C10-1 to 14, 14 to 16 Jan.

Appan, A., Alsogoff, F. and Tan, K.L.,(1988). A feasibility study on the utilization of surface runoff from a small paved catchment as a supplementary source for non-potable use. Proc. 6$^{th}$ IWRA World Congress on Water Resources, Vol 4, pp 260-271, Ottawa, Canada.

Chan, S.L. and Heng, H.L.C., Investigation and design of a rainwater cistern system in Nanyang Technological Univ. (unpublished) academic report, School of Civil & Structural Engrg, Nanyang Technological Univ., Singapore.

Fewkes,A and Ferris,S.A.,(1982). Rain and waste reuse for toilet flushing: A simulation model, Proc. Inter Conf. on RWCS, Honolulu, USA, pp 377-389, June.

Kok, Y.S., Fong, R.H.L., Murabayashi, E.T. and Lo, A., (1982). Deterministic and probabilistic processes of weekly rainfall, Proc. Inter Conf. on RWCS, Honolulu, USA pp 83-91, June.

Hoey, P.J. and West, S.F. (1982). Recent initiation in rain water supply systems for South Australia, Proc Inter Conf on RWCS, Honolulu, USA, pp 284-293.

Leung, P. and Fok Y.S., Determining desirable storage volume of rain catchment cistern system: A stochastic assessment, Proc Inter Conf on RWCS, Honolulu, USA, pp 128-134.

Morris, G.L., Acevido-Pimentel, R. and Ayala, G., 1984. Yield and cost of water supplies from rained cistern: Puerto Rico, 2$^{nd}$ Inter Conf on RWCS, Virgin Islands, USA, June 1984, pp FI-1 to 13.

Ree, W.O., Wimberly, F.L., Guinn, W.R. and Lauritzen, C.W., 1971. Rainwater harvesting system. ARS41-184, Agri Res Service, U.S. Dept of Agri, pp12.

Rippl, W., 1883. The capacity of storage reservoirs for water supply. Proc Inst of Civil Engrs, London, Vol 71, pp 270-278.

Schiller, E.J. and Latham, B., (1982). Computerized methods in optimizing rain water catchment systems, Proc Inter Conf on RWCS, Honolulu, USA, pp 92-101, June.

Wong, K.W., (1986). Use of ozone in the treatment of water for potable purposes, Journal Water Sc & Tech, Vol 18, No 3.

Advancement of Rainwater Uses in Taiwan

K.F. Andrew Lo[1]

Abstract

With the sustained economic growth in Taiwan, water demand has been increasing at an alarming rate. On top of the added natural environment restrictions and public awareness of environmental protection, the traditional water resource development approach had encountered many problems that resulted in water supply shortage in many regions. Exploration of alternative water sources to alleviate deficient water resources has become the top research priority in this country.

Taiwan is a rainfall-abundant country. Using rainwater catchment systems should be a logical and feasible solution to supplement public water supply. In many countries, it has been widely used and promoted because the water right controversy does not exist with the easily collected rainwater. This paper describes the many innovative techniques of utilizing rainwater in rural areas, agriculture, industries and recreation in Taiwan. Recently, even the Government of Taiwan has begun the promotion of rainwater catchment in towns and cities. The utimate goal is to be able to integrate efforts from both the public and private sectors in order to efficiently and economically utilize rainwater resource to alleviate the ever-increasing threat of water shortage problems in Taiwan.

Introduction

The mean annual rainfall in Taiwan is about 2,500 mm. It is a water-rich country, yet there has been annual water shortage in the central and southern agricultural districts, and in populated cities from the north to down south. At the same time, flooding often occurs after heavy storms and typhoons, especially for cities built on flood plains and cropland developed on temporarily dry river beds. This problem is probably due to high

---
[1] Professor, Dept. of Natural Resources, Chinese Culture University, Hwa Kang, Yang Ming Shan, Taipei, Taiwan.

population density and uneven spatial and temporal rainfall distribution that creates extreme flood and drought conditions. In addition to distributional problems, the pollution of rivers, lakes and groundwater poses yet another great threat to the water supply. Rice fields in the southern and northern regions are irrigated with water contaminated with industrial effluents and heavy metal. Aquaculture has suffered repeatly from polluted water. Reservoirs supplying drinking water to cities often have some eutrophication problems. In order to meet the greater water needs, proper management of watersheds, water pollution control, and development of new water supply are critical issues needed to be addressed in the future.

Water demand comes in the form of irrigation (including fish farming), industrial needs, and domestic uses. By the end of 1996, the population of Taiwan has reached over 20 million. The extent of industrial development, the people's living standard, and their economic abilities are all higher than those in the 1980s. Every one of these factors represent a much higher demand for water. To combat this difficiency, the Taiwan government has begun the promotion of rainwater catchment in town and cities. The utimate goal is to be able to integrate efforts from both the public and private sectors in order to efficiently and economically utilize the rainwater resource to alleviate the ever-increasing drought problem in Taiwan. This paper describes some of the many effective uses of rainwater in Taiwan.

Agricultural Rainwater Catchment Systems

Early rainwater catchment systems used for crop irrigation were mosly privately built. They were used to satisfy irrigation during drought condition and for chemical spraying. Usually they cannot fulfill large water demand. The government later recognizes its usefulness after the water problem becomes more serious and starts to construct more rainwater catchment systems. Rainwater catchment systems use for crop irrigation are commonly found in two major regions:

1. The northern part of Taiwan

Large areas of undulating tablelands are commonly found in Taoyuan and Hsinchu Counties. Elevation ranges from sea level to about 240 m with uneven rainfall distribution throughout the year. The average slope steepness is about 1 percent. The prevalent soil is the clayey soil with poor drainage. Therefore, this area is quite suitable for rainwater catchment systems development. In the old days, thousands of farm ponds were built and used by local farmers. Many ponds were small in size and privately owned. Farmers often have difficulty fulfilling their irrigation needs with rainwater alone. Water shortage problem often occurs. However, several remedial steps have been installed recently to improve the situation. The storage capacity has been increased to between 21,200 $m^3$ and 643,000 $m^3$. They are all managed by the local Irrigation Associations and operated like a series of reservoirs (Taoyuan IA, 1984). The Irrigation Associations use the linear programming theory to estimate the optimal water release and all the farm ponds have to maximize water storage for drought period use. These steps have proven successful in solving the water shortage problem in this area.

2. The southwestern part of Taiwan

The hilly area of southwestern Taiwan is dominated by mudstone soils. Rivers and streams are short and fast. They are completely dry during dry seasons. Frequent water shortage has prompted serious contraints on farmers who rely solely on rainwater for their crop production. The main seasons why this area lacks sufficient water supply are (SWCB, 1990):

(a) The annual average rainfall is about 2,100 mm. Though it is considered plentiful, it is not well distributed in time and space. More than 90 percent of the rain is concentrated during summer and typhoon seasons.

(b) The soil erosion problem is very serious. The average soil depth is less than 50 cm. The soil water retention and holding capacity is very low. Therefore, available water supply is rather limited.

(c) In most mudstone slopeland, gully formation is quite common and steep. Rainwater cannot percolate and recharge groundwater supply.

Results from field survey indicate that farm pond is an important and a special feature of rainwater catchment systems in the mudstone areas of the southwestern part of Taiwan. Besides storing rainwater during rainy seasons for supplying dry seasons need, farm ponds can also reduce surface soil removal, prevent landslides, and stablize slopeland. The silted up farm ponds can be later put to more effective use such as growing crops. The farm ponds, aside from providing irrigation water supply, can be diversely managed for animal husbandry, aquaculture, and touristic development. With the year round adequate supply of water, high cash-valued crops can be grown which will greatly enhance agriculture development and the living standard of the farmers in this area.

Rural Rainwater Catchment Systems

The Penghu Archipelago is located in the Taiwan Straits. There are altogether 64 islands, including 21 inhabited and the rest uninhabited. The total area is about 127 $km^2$. The highest elevation is only about 50 m above sea level. The mean annual rainfall on the islands is 1,024 mm. About 80% of which is concentrated between the months of April and September.

The water supply on the Penghu Archipelago is in serious trouble, that needs immediate attention (TPWCB, 1982). The Penghu islands are overlain by porous basalt rock and surrounded by elevated coral reefs. A major part of the rainfall either permeated underground or evaporates into the atmosphere due to high wind speed. Groundwater retains temporary underground and flows onwards into the ocean. Excessive groundwater pumpage always lead to sea water instrusion. Therefore, the water supply becomes a crucial problem during the dry seasons. At the moment, the surface water collected by reservoirs and groundwater pumped from deep wells are the major sources of water supply. Based on estimations, the water demand on the Penghu islands will increase from 13 million tons a year to about 20 million tons by the year 2006.

Recently, a project was initiated to construct several underground reservoirs to alleviate this water supply problem. The construction involves building a storage dam underground and a cut-off wall into an impermeable layer in order to prevent the groundwater outflow. Underground reservoir has many advantages over the conventional ones:
(a) low construction cost,
(b) short construction time,
(c) no submerged land and very low land use disturbance,
(d) negligible evaporation loss,
(e) good water quality, and
(f) simple operation and low maintenance cost.

There are three underground reservoirs being proposed at Chihkan, Houliao and Hushi. The Chikan reservoir is now under construction. The proposed annual water yield will be 700,200 $m^3$. If it proves to be successful, the next two reservoir projects will be implemented. With all three operating together, the future water demand on the Penghu islands will be easily met.

Industrial Rainwater Catchment Systems

The recent economic prosperity in Taiwan has increased the domestic and industrial demand for water. The island's natural environment contraints have almost exhausted all water resource development efforts. New water sources are extremely difficult to find. With the higher water demand, the only last resort is to develop new water sources.

Yu-chia is a new industrial park to be developed off-shore on the western coast of Taiwan. A feasibility study has been initiated in 1993 to assess the use of rainwater catchment systems to help solve the water supply problem encountered during the industrial development process. Preliminary results indicate that the most suitable method is to collect rainwater by roof-top and ground water storage. Rainfall in this area is concentrated between the months of May to September. Therefore, rainfall catchment systems should be very appropriate for alleviating the problem of uneven rainfall distribution and the best substitute for supplying water for this development project.

Recreational Rainwater Catchment Systems

The demand for residential, industrial, and agricultural water use may be marginally provided by both surface and groundwater supply. Water use for recreational purposes should turn to rainwater for satisfying their many needs. At present, rainwater catchment systems are used to provide water for game-fishing, a camping ground, and a scenic park.

1. Fishing pond

Besides water storage, a farm pond can be used for other purposes. It has been developed into a touristic attraction in addition to its normal agriculture and irrigation functions. Rainwater collected in farm ponds can be used for fish culture. Tourists are attracted to the farm pond for its aesthetic setting. Each visitor is charged an admission fee. They may enjoy boating or game fishing in the pond. Of course, they have to pay for their fish catches when they leave. The tourism business, therefore, yields additional income to the farmers.

2. Camping ground

Most camping grounds are usually located far away from populated cities. Many are situated in isolated hills. Water supply for such recreational development project has to be delivered from long distance. Some project cannot even obtain reliable source of water and has to close down. Another special water supply feature of camping ground is its high water demand during holiday seasons. Large water storage facility is necessary to satisfy the demand at peak seasons. On normal days, however, the water demand is very little.

A rainwater catchment system is designed to alleviate water supply problem in Ching Shan Camping Ground, since this site has no other alternative water sources. It is situated on a hilly area about 20 km northeast of Taipei City. To be able to store enough water supply for this development project, water catchment requires roof-top collection and underground storage. Underground water storage tanks are constructed at the foundation base of the wooden cottages and the public bath houses, and having the same dimension as the roof-top. The depth of the water tanks is about 60 cm. In addition to eliminating the water shortage problem, the rainwater catchment system may provide a considerable profit for the owner of the camping ground, with over 150 customers staying more than 2 days a week throughout the year at the site. Rainwater is collected right next to the water faucet. Only a slight cost is needed for installation of water pipes. The rainwater catchment system may also become a selling point for the owner to attract more eco-sensitive customers. Besides, the actual experience of using rainwater catchment system may impose some positive social impact on users to exercise water conservation when they return to their own household.

3. Scenic park

The Keelung City Government (1991) intended to develop Keelung Isle, which is situated only a few kilometers off the coast of Keelung City, into a scenic and recreational attraction for the local and foreign visitors. The main feature of this development plan is to design a half-day scenic and recreational activities on the Keelung Isle. Water use in this area consists of drinking, sanitation, and irrigation. Because of the long distance, public water supply is not feasible. The ground water supply is also lacking, further aggravated this water situation. Therefore, water has to be transported by water boats to satisfy the drinking water need.

It is estimated that the water boats will be able to deliver 100 tons/day.

This will be stored in underground water tanks and mainly use for drinking purposes. Rainwater is also explored as an alternative water source to supplement the inadequate water supply. This will be stored on roof-top water tanks and used for sanitary and irrigation purposes. Due to limited roof-top areas, it is necessary to seek other means to catch rainwater. The only feasbile and convenient collection device available is the cement-covered hiking trail. The entire length extends a total of 3,000 m and one meter wide. The surface of the hiking trails can be redesigned to having a slightly round shape with two drainage ditches on both sides. Rainwater is then diverted to roof-top storage tanks and provides for irrigation and toilet flushing. This not only reliefs water shortage problems, but also provides an effective means to conserve soil and control runoff.

Conclusions

Rainwater harvesting is about to come of age. It has an appropriate image about it that meshes well with the gentler ideas of the late 20th century. Because the technique makes use of an untapped resource - precipitation that would otherwise be evaporated before it had a chance to play an useful role in feeding the human population - it looks like getting something for nothing. Making use of such a resource has a certain poetry to it, particularly in a field where the resource itself can never be increased or decreased; unlike food, water cannot be grown to order, even given the right soil and the right fertilizer. But, like food, water can be harvested more efficiently. Doing so is a major priority for the twenty-first century.

References

Keelung City Government. 1991. A development plan of Keelung Isle. Y.H. cheng Architects & Engineers, Keelung, 82 pp. (in Chinese).

Soil and Water Conservation Bureau (SWCB). 1990. Mountain agricultural resources development statistical abstract. Taiwan Provincial Government, Taiwan, 112 pp. (in Chinese).

Taiwan Provincial Water Conservancy Bureau (TPWCB). 1982. A report of water resources planning and investigation in the Penghu area. Taiwan Provincial Government, Taiwan, 208 pp. (in Chinese).

Taoyuan Irrigation Association (Taoyuan IA). 1984. A record of Taoyuan Irrigation Association. Taoyuan, Taiwan. 57 pp. (in Chinese).

# Subject Index

Page number refers to the first page of paper

Acid mine water, 710
Agricultural watersheds, 392, 734
Alaska, 368
Algorithms, 633
Alums, 580
Aquifers, 597, 680, 686, 698
Archaeology, 205
Arizona, 676
ASCE Committees, 363

Baffles, 482, 511
Benefit cost analysis, 407, 574
Bids, 374
Biological treatment, 633
Boundaries, 61, 363
Bridges, 182, 188, 217

Calibration, 171, 247
California, 319
Canada, 61
Canals, 176
Caribbean, 55
Case reports, 66, 72, 111, 141, 176, 182, 194, 205, 223, 241, 368, 574, 609, 615, 627, 785
Catchments, 29, 763, 769, 773, 779, 785, 791
Channelization, 505
Chemical spills, 609
Chesapeake Bay, 488
Chlorine, 253
Cisterns, 763
Clay soils, 105, 401
Climatic changes, 123, 129
Coagulation, 580
Coastal plains, 223
Coastal structures, 182

Colorado River, 407
Combined sewers, 331, 529, 535, 547, 559
Community development, 313, 505
Community planning, 553
Comparative studies, 99, 655
Competition, 374
Computer applications, 380
Computer models, 159, 541
Concentration time, 1, 9, 16, 171, 285
Construction sites, 105
Contaminants, 419, 586
Contamination, 272
Continuum hypothesis, 645
Cost effectiveness, 633, 651
Cost minimization, 431
County government, 500
Cross sections, 159
Curricula, 380

Dams, 704
Data analysis, 229, 651
Decision making, 517
Decision support systems, 413, 471
Demographic projections, 553
Design criteria, 523
Detention basins, 86, 92, 99, 211, 482, 505, 568
Detention period, 92
Developing countries, 49, 55, 313
Digital mapping, 159, 757
Digital systems, 667
Disinfectants, 291
Disinfection, 266, 279, 285, 297
Dissolved oxygen, 465
Diversion structures, 147

Drainage, 29
Drainage basins, 453
Drainage systems, 72, 141, 194, 247
Dredge spoil, 78, 392
Droughts, 153, 591

Economic analysis, 574
Economic development, 36
Economic factors, 49
Economic growth, 55
Economic impact, 153
Ecuador, 49
Education, 380
Effluent reuse, 676
Egypt, 42
Elevation, 757
Emergency services, 303
Energy development, 36
Energy losses, 217
Environmental factors, 129
Environmental impacts, 407
Environmental planning, 42, 49, 55
Environmental Protection Agency, 541
Environmental quality, 639
Environmental research, 36
Environmental surveys, 651
Erosion control, 105, 722
Evaporation, 135
Evapotranspiration, 692
Expert systems, 443

Federal agencies, 200
Feedback control, 297
Field investigations, 609, 722
Finite difference method, 655
Finite element method, 182
Flood control, 147, 194, 211, 247, 453, 476, 505, 704, 710, 728
Flood damage, 66, 111, 194, 200, 728
Flood forecasting, 425

Flood frequency, 425
Flood level, 425
Flood Management, 66, 72, 505
Flood plain insurance, 200
Flood plain planning, 111, 194
Flood plain studies, 159, 165, 223
Flood plains, 176, 205, 217
Flood routing, 211
Flood stages, 211
Flooding, 339
Floodways, 500
Florida, 453, 580
Flow control, 325, 331
Flow distribution, 188, 217
Flow patterns, 279
Flow rates, 260
Flow simulation, 153, 627, 655
Frequency response, 680

Geographic information systems, 165, 380, 443, 667, 739, 745, 751, 757
Global warming, 129
Great Lakes, 117, 135, 339
Ground water, 92, 205
Ground-water flow, 135, 591, 603, 627, 645, 655, 667, 698
Ground-water management, 603, 667
Ground-water pollution, 609, 615, 621, 627, 633, 661, 667, 686
Ground-water quality, 667, 692
Ground-water recharge, 676, 680
Ground-water supply, 597

Highway construction, 686, 722
Highway planning, 494
Highways, 562
Hospitals, 307
Hydraulic models, 141, 159, 188, 241, 279
Hydraulic performance, 279
Hydraulic properties, 217

Hydraulic structures, 147
Hydrodynamics, 413
Hydroelectric power generation, 153
Hydroelectric powerplants, 407
Hydrogeology, 603
Hydrographs, 86, 141, 568
Hydrologic aspects, 568
Hydrologic models, 117, 147, 437, 639
Hydrologic properties, 1, 9, 16, 339, 476
Hyetographs, 235

Illinois, 351, 500
Industrial wastes, 42
Infiltration, 1, 692
Infiltration rate, 171
Input, 229
Institutional constraints, 368
Instream flow, 345, 351, 357, 716
Insurance, 200
International commissions, 61, 769
International compacts, 36
International development, 313
International waters, 363
Interstate commissions, 488
Irrigation systems, 431

Kansas, 722
Kazakstan, 459
Kentucky, 745
Knowledge-based systems, 42
Korea, 471, 476

Laboratory tests, 609
Lakes, 704
Land development, 99
Land usage planning, 488
Landfills, 78, 401, 443, 692
Latin America, 55
Leaching, 401
Legislation, 351

Levees, 211
Local governments, 368, 374, 500
Low flow, 153

Maintenance costs, 482
Malaysia, 22, 29
Manholes, 511
Mapping, 200, 739
Marinas, 78
Mass transfer, 661
Master plans, 111, 241, 453
Mathematical programming, 357
Mixing, 279, 285, 413
Monitoring, 272, 291, 562, 586
Monte Carlo method, 419
Multimedia, 380, 443
Municipal wastes, 78, 443
Municipal water, 272, 368, 374, 431

Nebraska, 505
Neural networks, 229
New Hampshire, 345
Nitrates, 686
Nonaqueous phase liquids, 609, 615, 621
Nonpoint pollution, 392, 395, 586
Numerical models, 645

Optimal design, 779
Optimization, 529, 661
Optimization models, 266, 297, 591, 597
Ordinances, 99
Overflow, 217, 529, 535, 547, 559

Paleohydrology, 425
Parameters, 247, 253
Peak runoff, 86
Performance evaluation, 92, 123, 547, 559
Performance standards, 303
Permits, 386

Pipe networks, 253, 431
Plumes, 698
Pollutants, 517, 586
Pollution control, 16, 42, 49, 55, 523
Population growth, 313, 773
Porous media flow, 645
Potable water, 260, 266, 291, 297, 449, 667, 763, 769
Potomac River, 437
Precipitation, 135
Private sector, 773
Privatization, 368, 374
Probabilistic methods, 419
Probability, 153
Probability distribution, 260, 425
Project evaluation, 200
Public health, 773
Public information programs, 16, 494
Public participation, 111, 751
Pumping, 597

Radioactive materials, 734
Rain water, 763, 769, 773, 779, 785, 791
Rainfall, 165
Rainfall-runoff relationships, 147, 171, 235, 331
Receiving waters, 331, 559, 574
Reclamation, 78
Recreational facilities, 78, 205
Regional analysis, 129
Regulation, 763
Rehabilitation, 307
Remedial action, 603, 609, 615, 621, 627, 633, 661, 686
Remote sensing, 757
Research and development, 541, 639
Reservoir management, 716
Reservoir operation, 407, 437, 465, 471, 476, 710, 716
Reservoir performance, 285, 465
Reservoir storage, 459

Reservoir system regulation, 471
Reservoir systems, 123
Reservoirs, 147
Residential location, 66
Resource allocation, 401
Retrofitting, 72, 482, 511
Return flow, 459
Riparian land, 728
Risk analysis, 419
Risk management, 437
River basins, 171, 471, 745
River systems, 159, 407
Roofs, 785
Runoff, 1, 9, 16, 494, 562
Runoff coefficient, 235

Salt water, 580
Salt water intrusion, 591
San Francisco, 449
Sandbars, 704
Sanitary sewers, 313, 535
Scale models, 285
Scour, 182, 188
Seasonal variations, 117, 586
Seattle, 61
Sediment control, 105, 716
Sediment deposits, 710
Sediment load, 413
Sediment transport, 716
Sedimentation, 392, 535, 704, 710
Service life, 307
Sewage disposal, 553
Sewer design, 553
Sewer maintenance, 553
Simulation models, 523, 591, 597, 627, 639, 661, 698
Singapore, 785
Site evaluation, 1, 9, 205, 443, 615
Socioeconomic data, 129
Soil analysis, 621
Soil erosion, 734
Soil loss, 722

Soil pollution, 734
Solid wastes, 401
Solutes, 603
Sorption, 661
Spatial data, 645
Spectral analysis, 680
State action, 488
State government, 500
Statistical analysis, 651
Steady state, 253, 698
Stochastic models, 117
Stochastic processes, 680
Storm drains, 751
Storm runoff, 86, 99, 223, 247, 325, 511
Storm sewers, 86, 395, 580
Stormwater, 535, 562, 568, 574
Stormwater management, 1, 9, 16, 22, 29, 66, 92, 99, 147, 325, 386, 395, 453, 482, 488, 500, 505, 511, 517, 523, 547, 580, 728, 751
Streamflow, 357, 459
Streamflow forecasting, 229, 247
Streamflow records, 345
Subsurface drainage, 92
Subsystems, 303
Surface runoff, 135
Surface waters, 123, 591
Suspended sediments, 413
Suspended solids, 580
System analysis, 779

Taiwan, 597, 791
Teaching methods, 380
Technology transfer, 36
Tennessee Valley Authority, 465
Thailand, 36
Three-dimensional flow, 655
Three-dimensional models, 603
Tidal waters, 176, 223
Time series analysis, 117, 165
Topographic mapping, 757

Topographic maps, 159
Transient flow, 698
Transport phenomena, 734
Travel time, 260
Tributaries, 22
Tunnel construction, 401
Two-dimensional models, 176, 182, 188

Unit hydrographs, 235
Unsteady flow, 141, 211
Urban areas, 111, 223, 313, 505, 541, 568, 751
Urban development, 1, 325, 331
Urban renewal, 22, 29
Urban roads, 176
Urban runoff, 22, 29, 86, 141, 247, 325, 386, 395, 517, 535, 568
Urbanization, 597, 773
Utilities, 368

Vadose zone, 621
Vegetation, 692, 722
Virginia, 494
Visual perception, 651

Wastewater treatment, 307, 529, 559, 676
Wastewater use, 676
Water allocation policy, 357
Water conservation, 319, 476
Water consumption, 345, 357
Water demand, 123
Water distribution, 253, 260, 266, 285, 291, 297, 303, 431
Water law, 351
Water level fluctuations, 217
Water levels, 135
Water management, 345, 459
Water policy, 319, 363
Water pollution control, 272, 392, 395, 482

Water pollution sources, 419, 535
Water pressure, 241, 303
Water quality, 61, 253, 260, 303, 339, 751, 763
Water quality control, 72, 105, 291, 386, 449, 453, 465, 471, 488, 494, 511, 541, 559, 562, 574, 728
Water quality standards, 363, 517
Water resources, 129, 363, 380
Water resources management, 61, 117, 229, 319, 351, 357, 437, 449, 471, 639, 676, 739, 745
Water reuse, 459
Water rights, 351
Water services, 241
Water shortage, 791
Water storage, 211, 529, 779
Water supply, 135, 763, 769, 791
Water supply systems, 123, 241, 253, 260, 266, 272, 307, 313, 374, 437, 773
Water surface, 217
Water table, 686
Water treatment, 279, 307, 449, 511
Water use, 319, 785, 791
Water yield, 779
Watershed management, 72, 325, 331, 453, 500, 517, 586, 745
Watersheds, 1, 9, 16, 22, 111, 135, 165, 223, 235, 339, 488, 494, 535, 541, 745, 751, 757
Wetlands, 205, 339, 523, 568
Wildlife habitats, 728
Wisconsin, 386, 395

# Author Index

Page number refers to the first page of paper

Adeel, Zafar, 603
Ahn, Taejin, 431
Akanbi, Abiola A., 211
Aksoy, Aysegul, 661
Alavian, Vahid, 42
Al-Omari, A. S., 253
Ames, P. L., 459
Anderson, David J., P.E., 553
Appan, Adhityan, 785
Ashe, K. W., 92
Athow, Robert F., Jr., 639

Bachhuber, James A., 395
Baker, V. R., 425
Barrett, Kenneth M., 374
Bartolini, Paolo, 171
Belk, Ellen Hoffman, 401
Bent, Timothy A., 692
Berdanier, Bruce W., P.E., 307
Bishop, Richard C., 407
Boccelli, Dominic L., 266
Boehm, Greg, P.E., 500
Bogardus, Ellen R., 374
Borst, Michael, 541
Bottomley, Glenn, P.E., 494
Boulos, Paul F., 279
Brock, W. Gary, 465
Buchberger, Steven G., 260, 291
Burke, Thomas W., 627
Burn, Donald H., 229
Burton, Michele Good, 111, 159

Cabezas, L. Moris, P.E., 453
Call, Charles H., Jr., 739
Carter, Jason T., 260, 291
Casanave, Eric A., 99, 241
Cazenas, Pat, 562

Chang, C. C., 615
Chang, H. Sherrie, 141
Charlton, T. J., 147
Chaudhry, M. H., 253
Chen, Chi-Ming, 680
Chen, Ching-Ho, 86
Chen, Mow-Soung, 9
Cheng, Mow-Soung, 1
Chien, Calvin, 609
Chin, K. K., 22, 29
Choo, A., 22, 29
Chung, C. H., 117
Clar, Michael, 1, 9, 16
Clark, Gary R., 351
Coffman, Larry, 1, 9, 16
Colten, Lee, 745
Cooper, Paula, 141
Corley, R. Wayne, 182
Corssmit, C. (Kees) W., 368
Culver, Teresa B., 633, 661

D'Antuono, James R., 386
Davis, D. R., 425
Dee, David D., Jr., P.E., 488, 494
DeGroot, William G., P.E., 194
Deininger, Rolf A., 285
Denning, T. M., P.E., 728
DePue, P. Michael, 182
DePue, P. Michael, II, 176
Doneker, Robert L., 413
Dorsch, William R., 627
Duckstein, L., 425

England, Gordon, P.E., 482, 511
Erhart-Schippek, Werner, 667
Escobar, Isabel C., 580
Esfandiary, Siamak, 734

803

Fan, Chi-Yuan, 331, 517, 535
Faust, Charles, 603
Fennessey, Neil M., 345
Field, Richard, 331, 517, 523, 529, 535
Fok, Yu-Si, 769, 773
Fontane, Darrell G., 471
Frevert, D. K., 117
Fusco, Nick, 704

Glenn, Mark V., P.E., 241
Goggin, Jim, 42
Gontasz, Karen T., 547
Gordon, Bill, 368
Gowin, Dirk L., P.E., 217
Graziano, Frank R., 562

Hafs, William C., 392
Hagemeyer, Todd, 603
Halmi, Steven, 176
Halmi, Steven R., 182
Hannoun, Imad A., 279
Hasni, M., 22, 29
Hauptmann, Michael G., 698
Headrick, M. R., 459
Heany, James P., 325
Heggen, Richard J., 779
Hey, D. L., 339
Higgins, John M., 465
Hilton, Amy B. Chan, 633
Hoeke, Mark P., 547
Holland, Jeffrey P., 639
Hood, J. W., P.E., 728
Hourani, Nabil M., P.E., 757
Howington, Stacy E., 645
Hsu, Nien-Sheng, 597
Hsu, Shiang-Kueen, 597
Hulbert, William H., 182

Ishii, A. L., 147
Israel, Morris S., 55

Jackson, Gilbert S., 55
Jacobs, Jennifer M., 357
Jacobson, John A., 105
Jirka, Gerhard H., 413
Johnson, Karen E., 449

Kao, Jehng-Jung, 443
Kessler, Avner, 272
Keyes, Conrad G., Jr., P.E., 363
Khanbilvardi, Reza, 734
Kilgore, Roger T., P.E., 704
Killgore, Mark W., P.E., 61
Kim, Kyung Tak, 476
Kirshen, Paul, 123, 129
Kladias, Michael P., 627, 698
Klink, John C., 135
Ko, Ick Hwan, 471
Kuch, Anthony W., 247
Kudrna, Frank L., Jr., 78
Kwon, Oh Ig, 476

Labadie, John W., 471
Lane, Melissa, 123, 129
Lane, W. L., 117
Lantzy, Ron, 603
Larson, Douglas M., 574
Lee, S. W., 22, 29
Lee, YeongHo, 260, 291
Lester, Barry, 603
Lew, Dan, 574
Liao, Shih-Long, 517, 523
Liaw, Shu-Liang, 86
Lily, Sehayek, 609
Lin, Gwo-Fong, 680
Lloyd, David W., P.E., 194
Lo, K. F. Andrew, 791
Loehlein, Werner, P.E., 704
Loehlein, Werner C., P.E., 710
Loganathan, G. V., 431
Loucks, Eric D., 135
Loucks, Orie L., 135
Lucey, Daniel C., 704

Lund, Jay R., 574
Lye, Dennis J., 763

Male, James W., 401, 419
Manbeck, D. M., 459
Marshall, Frank E., III, P.E., 580
Martin, Peter, 591
Mascha, Herbert, 667
McCollum, William, 205
McEnroe, Bruce, 722
McEnroe, Bruce M., 165
McGahey, Christopher, 313
Medina, Daniel E., P.E., 704
Milhous, Robert T., 716
Mohsen, Mohammad F. N., 621
Montereggio, Federica, 171
Morgan, Veronica J. B., P.E., 111, 159
Morris, Charles, P.E., 217
Murillo, Blake N., P.E., 751

Ng, K. Y., 22, 29
Nishikawa, Tracy, 591
Nnadi, F. N., 92
Nzewi, Emmanuel U., 380

O'Connor, Thomas P., 529, 535
Olenik, John, 494
Ormsbee, Lindell, 745
Ortel, T. W., 147
Ostfeld, Avi, 272, 303
Ovbiebo, Tai, 247

Page, Roger H., 621
Peng, C., 425
Peters, John F., 645
Petersen, Reed J., P.E., 676
Pew, Kenneth A., 559
Pistilli, James P., P.E., 553
Pitt, Robert, 331
Polycarpou, Marios M., 297
Potter, Scott T., 692

Quinn, Rebecca J., 165

Randall, Andrew A., P.E., 580
Rangarajan, Srinivasan, P.E., 153
Rao, A. Ramachandra, 99, 235
Rice, Eugene W., 291
Richard, Gilbert, 42
Riebau, Mark A., P.E., 200
Rossman, Lewis A., 266, 291
Royal, Fred, 66

Salas, J. D., 117
Salazar, Mario, 49
Sample, David, 325
Santini, Andrew D., 285
Schaefer, G. C., P.E., 728
Schuller, Daniel J., 99, 235
Schwartz, Stuart S., 437, 586
Seger, Wayne, P.E., 188
Sela, Erez, P.E., 223
Shafer, Kevin, 505
Shamir, Uri, 303
Sharek, R. C., 92
Sherman, Robert A., 488
Shestopalov, Vyacheslav M., 734
Shim, Myung Pil, 476
Sieber, Edith, 621
Simonovic, Slobodan P., 229
Singh, Krishan P., 211
Sinton, Peter O., 655
Sleep, Brent, 609
Smith, Jeffery A., 692
Smolensky, Douglas A., 627, 698
Soehren, Rick, 319
Stein, Stuart M., 562
Stewart, Edward H., 449
Stinson, Mary K., 535
Streiner, Carol F., 368
Stumbur, Ted, 745
Subramaniam, Prathiba, 297
Sullivan, Daniel, 36, 517
Sullivan, Michael P., 547

Surminski, Harold M., P.E., 153
Sutherson, Suthan S., 692
Suttle, Rick K., 205

Tafuri, Anthony N., 36
Tam, Wing K., P.E., 751
Taylor, Scott M., 686
Townsend, Peter H., 651
Traver, Robert G., 568
Treff, Brian J., 722
Tryby, Michael E., 266

Uber, James G., 266, 297

Valdés, Juan B., 171
Vandell, Terry, 609
Vogel, Richard M., 123, 129, 357
Vonnahme, C. C., 147

Wahba, Salwa, 42
Watkins, Edwin W., P.E., 188
Watkins, Michael L., 621

Weinstein, Neil, 1, 9, 16
Welsh, Michael P., 407
Werner, Patricia S., 72
Wickenkamp, J. A., 339, 728
Wilchfort-Kalman, Orit, 574
Wolf, Bob, 505
Wright, Leonard, 325
Wu, Ray-Shyan, 86

Yang, Shu-li, 597
Yeh, William W-G., 597
Yezzi, James J., Jr., 36
Young, C. Bryan, 165
Young, G. Kenneth, 562
Young, Greg, P.E., 319
Yu, Chun, 86
Yu, Shaw L., 523

Zahradnik, Arthur J., 627
Zahradnik, Arthur J., Jr., 698
Zealand, Cameron M., 229